Radiative
Corrections
Results and
Perspectives

NATO ASI Series

Advanced Science Institutes Series

A series presenting the results of activities sponsored by the NATO Science Committee, which aims at the dissemination of advanced scientific and technological knowledge, with a view to strengthening links between scientific communities.

The series is published by an international board of publishers in conjunction with the NATO Scientific Affairs Division

A	**Life Sciences**	Plenum Publishing Corporation
B	**Physics**	New York and London
C	**Mathematical and Physical Sciences**	Kluwer Academic Publishers Dordrecht, Boston, and London
D	**Behavioral and Social Sciences**	
E	**Applied Sciences**	
F	**Computer and Systems Sciences**	Springer-Verlag
G	**Ecological Sciences**	Berlin, Heidelberg, New York, London,
H	**Cell Biology**	Paris, and Tokyo

Series B: Physics

Radiative Corrections

Results and Perspectives

Edited by

N. Dombey and F. Boudjema

University of Sussex
Brighton, United Kingdom

Plenum Press
New York and London
Published in cooperation with NATO Scientific Affairs Division

Proceedings of a NATO Advanced Research Workshop
on Radiative Corrections: Results and Perspectives,
held July 10–14, 1989,
in Brighton, United Kingdom

Library of Congress Cataloging-in-Publication Data

NATO Advanced Research Workshop on Radiative Corrections: Results and
   Perspectives (1989 : Brighton, England)
     Radiative corrections : results and perspectives / edited by N.
Dombey and F. Boudjema.
        p.   cm. -- (NATO ASI series. Series B. Physics ; v. 233)
     "Proceedings of a NATO Advanced Research Workshop on Radiative
Corrections: Results and Perspectives, held July 10-14, 1989, in
Brighton, United Kingdom"--T.p. verso.
     Includes bibliographical references and index.

     ISBN 978-1-4684-9056-5   ISBN 978-1-4684-9054-1 (eBook)
     DOI 10.1007/978-1-4684-9054-1

     1. Radiative corrections--Congresses.   I. Dombey, N.
II. Boudjema, F.   III. Title.   IV. Series.
QC794.8.W4N38 1989
539.7'544--dc20   90-7921
                                                            CIP

© 1990 Plenum Press, New York
Softcover reprint of the hardcover 1st edition 1990

A Division of Plenum Publishing Corporation
233 Spring Street, New York, N.Y. 10013

PREFACE

The Workshop on Radiative Corrections: Results and Perspectives was held at the University of Sussex in fine weather between July 9 and 14 1989. The Workshop was well timed: the day after its concluding session the first beam at LEP was circulated.

The Original aims of the Workshop were twofold: first to review the existing theoretical work on electroweak radiative corrections in the light of the initial experiments at SLC and LEP, and to attempt to obtain a consensus on the best means of carrying out the calculations of the various processes. This aim became Working Group A on *Renormalisation Schemes for Electroweak Radiative Corrections*.

The second aim was to review the experimental implementation of radiative corrections and this became Working Group B. Here the problem was to obtain a consensus on the use of Monte Carlo event generators.

At the time (March 1987) when Friedrich Dydak wrote to one of us (ND) to suggest a Workshop on the subject of electroweak radiative corrections to take place just before experiments at LEP were to begin, the main theoretical problem was that there was no agreement among theorists on the use of a specific renormalization scheme. Similarly, it was already becoming clear that it was going to be very difficult to compare the experimental results of different groups because they would use different event generators and experimental cuts of their data. Again it was important to obtain a consensus on how to present the data and how to assess different event generators. So these were the primary problems which Working Groups A and B were expected to tackle.

In our opinion Working Groups A and B carried out their tasks admirably. In Working Group A there was general (if not complete) agreement that the appropriate renormalization scheme for electroweak radiative corrections should be performed on-shell in a fully gauge-invariant manner taking as parameters the electric charge e, the mass of the Z M_Z, the various fermion masses m_f and the muon decay constant G_μ. There was also general agreement that the ρ-parameter should not be assumed to be one (as it is, of course, in the minimal standard model) but that a more general formulation should be used where the current-mixing angle and the vector meson mass-mixing angle were not necessarily identical. Veltman, however, in his stimulating introductory talk warned us against using renormalization schemes and advocated instead that we stick to the use of observables derived from the bare Lagrangian. He also asked us to think the unthinkable: what do we do if the top quark is not found below 200 GeV?

In Working Group B several different Monte Carlo event generators were presented, but for once, their proponents were willing to admit that all had merits as well as shortcomings. Although the various approaches to the implementation of the radiative corrections of the various physical observables at the Z peak were in very good agreement when the same parameters in the event generators were assumed, and the theoretical results recovered when cuts were not applied, some important conceptual differences emerged regarding the analysis and presentation of the data. These problems refer mainly to the non-QED ("weak correction") part which is dependent on the top mass m_t (and to a lesser extent on the Higgs mass m_H) which at the time of writing we do not know. Some generators for instance are specific to the *minimal* standard model (in spite of the warning from Working Group A) with a *fixed* value for the top mass while others take a more general structure for the weak part which is then "dressed" by QED which depends only on the geometry of the particular detector. In this respect some participants reiterated the proposal of presenting the data using common agreed "canonical cuts" at different experiments, while Kleiss suggested presenting the corrected data alongside the "raw" data together with storing the version of the programme used for the analysis and presentation of the data in the CERN library, so that it was available to all. (It would then presumably have to be written complete with explanatory notes!).

Another problem of the event generators was that the QED part of those programmes is based essentially on soft photon theorems (via the Yennie–Frautschi–Suura approach for example). Yet in practice, photons need not be soft (the threshold k_0 problem) and in theory, there are QED corrections which cannot in principle be simulated by soft photons alone (Bhabha scattering at large angles and $ee{-}{-}{>}\gamma\gamma$, γZ for example). So it seems to us that there are problems still to be solved.

During the preparation of the Workshop, the Organising Committee realised that QCD as well as electroweak radiative corrections would play an important role in the processes to be studied at SLC and LEP, and QCD radiative corrections would be dominant at HERA. We therefore added Working Group C on *The Interplay of QCD and Electroweak Radiative Corrections*. The speakers in this Working Group showed the importance of a well-defined value of the QCD expansion parameter $\alpha_S(q^2)$ especially when applying radiative corrections to ep collisions. For Z-physics the QCD corrections are relatively small and not particularly dependent on the experimental cuts. Nevertheless it is necessary to systematically include the effects of fragmentation and hadronisation. The interplay between QCD and electroweak can be very subtle, for instance even when the process does not contain initial or final quarks, QCD corrections can be present. One difficulty, however, is that QCD corrections are always calculated off-shell, whereas the consensus for electroweak radiative corrections is to use an on-shell scheme. Perhaps, some thought should be given to using a unified picture. The structure function approach for QED of Trentadue and his collaborators may well be a step in this direction.

The problem associated with large coefficients in perturbation theory, whether QED or QCD, was addressed by West who suggested a systematic way of deciding when to truncate the perturbation series. This is clearly an important issue which deserves more attention.

There were several informative and timely talks outside the Working Group structure, in particular one session devoted to the Z peak. Finally, Alberto Sirlin in his Summary talk admired the progress made in radiative corrections over the past few years, especially by the contributions of the many young physicists who had been attracted to work in this once obscure subject.

Both of us greatly enjoyed the Workshop and would like to thank all the participants and speakers who helped to make it such an enjoyable and successful meeting and in particular Frits Berends, Friedrich Dydak and Claudio Verzegnassi who planned the Workshop with us in the Organising Committee. We also thank NATO Scientific Affairs Division, the Science and Engineering Research Council and CERN who provided the necessary funds which allowed the Workshop to take place.

FAWZI BOUDJEMA

NORMAN DOMBEY

CONTENTS

CHAPTER 4
WORKING GROUP B: IMPLEMENTATION OF RADIATIVE CORRECTIONS

THE IMPORTANCE OF RADIATIVE CORRECTIONS

M. Veltman

University of Michigan
Ann Arbor
Michigan, USA

I did not invent the title of this talk. It was invented by the Organising Committee. I thought it was a bit funny because everyone who is here must know why they are working on radiative corrections. It seems the oddest place to have to explain why radiative corrections are important. Also most of you have probably spent more time than I have recently on the subject. So what do I do? I am sure you do not want me to talk on very technical matters.

Thinking about what I should say, I felt a tendency to be nostalgic: to dwell on the past. This of course happens when you are getting old and have spent a long time in a subject. But I think it may not be such a bad idea to try to remember what the motives of people working in the subject were in the old days. Twenty years ago my motives were certainly different than they are now; our objectives today are somewhat different. Whenever you do work of this type which is often tedious you need to have some sort of goal in mind which motivates that work. So I will try to clarify why we used to do radiative corrections and then give my opinion of why we do them now.

Well when we go back 20 years to 1969 the known particles were:

$$u \quad u \quad u \qquad\qquad s \quad s \quad s$$

$$d \quad d \quad d$$

$$\nu_e \qquad\qquad\qquad \nu_\mu$$

$$e \qquad\qquad\qquad\quad \mu$$

We did not know then that quarks came in three colours but that is the way I have written them anyway. We also had a theory then called SU(3) which united the up, down and strange quarks into one triplet; you see how much things have changed as we don't think in that way any more. Anyway, we were on the verge of having the renormalisabilty of the standard model being proven. Then came one of the spectacular successes of the renormalisability idea: the GIM mechanism. Then charm was discovered in 1974. You have to understand that that was a spectacular success of the idea of renormalisability.

Radiative Corrections, Edited by N. Dombey and
F. Boudjema, Plenum Press, New York, 1990

When we talk about radiative corrections we have to keep two things in mind: first we can compute to a high order of accuracy and thus verify the theory that we are working with. To a large extent doing radiative corrections means verifying the renormalisability of the theory. Second, in the case of a gauge theory renormalisability means that coupling constants of different particles are the same; that you have complete multiplets, and all kinds of other relations which follow from the gauge invariance of the theory. So, for example, in the standard model not only should the W's be where you expect them to be but there should also be the one thing that we have not yet discovered which is the four—point vector boson interaction.

So when you speak about radiative corrections being a verification of the renormalisability of a theory you are losing information, for renormalisbility of a gauge theory implies much more than just the finiteness of the perturbative calculations. Now we can see that history is repeating itself for in 1969 renormalisability predicted the charmed quark, while in 1989 renormalisability predicts the top quark. Except that now no one talks of SU(5) as unifying the quarks: that is a very different point of view. Few people now actually doubt that the top quark exists. This shows the difference between now and twenty years ago because although the argument is precisely the same as it was in 1969 for the prediction of the charmed quark, most people then did not believe it.

Round about 1974 or so, the main prediction arising from gauge theories apart from the existence of charm was the existence of neutral currents in weak interactions. At that time I well remember that the theory we worked with then called the Weinberg Model—now called the Standard Model— predicted not only neutral currents, but a neutral current with a very well-defined strength when compared to the charged current. There were many neutral current experiments of different types then under way but the experiments were difficult and sometimes there was too much neutral current and sometimes too little. So I started to think about it and thought that this was crazy: there must be sufficient flexibility in the theory to encompass different strengths of neutral current. You needed to use the Higgs mechanism in any event to generate the vector boson masses, and a sufficiently complicated symmetry breaking would generate any neutral vector boson mass you like. So it must be possible to work with the neutral vector boson mass as a free parameter. Douglas Ross and I worked on that[1] and we tried to understand the general structure of the Higgs sector which would allow this. That was the aim of the enterprise at that time: we wanted to establish that if there was a neutral current with a strength not in agreement with the standard model, then we could still have a renormalisable gauge theory.

That was much against the conventional wisdom of the time. I well remember an episode which illustrates this. There was a neutrino conference in Paris around Christmas 1975 and I had twenty minutes to talk. I decided to use one transparency, and on that transparency I had just three lines in a box. These said (I can't recall the exact words):

IN A GAUGE THEORY OF ELECTROWEAK INTERACTIONS
THE Z MASS IS A FREE PARAMETER
THEREFORE NEUTRAL CURRENTS CAN BE MADE IN ANY STRENGTH

For twenty minutes I just repeated that over and over again. There is not a single person in the world today who remembers that talk! Everyone must have been thinking about food, or sex, or just sleeping.

I am also telling this story to illustrate our motivation. Why did Douglas Ross and I do this work? Well, we were thinking about the cosmological constant and how to keep it zero. That led to the study of a general type of spontaneous symmetry breaking through the Higgs mechanism, and that led in turn to the freeing of the Z mass from its Weinberg model value and therefore to the introduction of a new parameter β which we now call the rho-parameter (following Sakurai and Hung[2] who later came to a similar conclusion from a non-gauge theory point of view). So it was then that I began to recognise that the Higgs sector is a very special part of the standard model.

Well in the 1970s charm was discovered, then the tau, and then the bottom quark. So we tried at Utrecht to establish a limit on the number of generations that we can have. Now if you look at the picture that was building up you can see that while the generations go on and the quarks get correspondingly heavier, to each generation there is a massless neutrino. Thus rather than looking for all the heavy quarks, you can look for all the light neutrinos. There was then a paper by a Russian[3] that you can count neutrinos using cosmological arguments, so at least it was realised that the number of generations was finite. Then at some point in trying to do radiative corrections in this theory which was so different to QED I discovered this extraordinary radiative correction which is sensitive to high mass particles, even if this mass is large compared to the energies you are working with. This is, of course, the radiative correction to the rho-parameter which depends on the top mass squared.

This is where we are now and it is this result in the minimal standard model which gives us the upper bound on the top mass. So let us go back to the original question of why we need radiative corrections. Of course we need to check the theory, and of course we need to test that the theory is renormalisable, but as I have already said renormalisability predicts a lot of other things and you check it in many more ways than by just calculating radiative corrections. Now we also have to keep in mind that not all radiative corrections are equally important and that usually they are not sensitive to a high energy scale. For example, the radiative corrections to g − 2 for an electron or muon, even if calculated to many places of decimals, are not very sensitive to strong interactions. In practice, nothing with a mass of over 200 MeV or so will affect g − 2. So g − 2 is a poor indicator of high energy processes and we will not find out anything about the mass of vector bosons or the top quark from it.

Much better, and much better than many people think, is the radiative correction to β-decay which has been calculated by Professor Blin-Stoyle[4] and Professor Sirlin[5], both of whom are here. That calculation and its verification are much more of a test of the standard model than is commonly understood. It extends the energy scale in which we have confidence in the standard model to about 60 GeV. These calculations also show that the vector bosons are not composite: I do not have time to elaborate on this but if the W were composite, there would be an additional diagram to calculate in the radiative correction to beta-decay which would spoil the present excellent agreement between theory and experiment.

Now more information is available to us from the calculations of the radiative corrections δM to the vector boson mass, but as has been emphasised by many authors those results follow predominantly from the running of the fine structure constant from QED between 0 and 80 GeV. It does not result from any special details of the standard model. On the other hand the radiative corrections to the rho-parameter are sensitive to such details. The sensitivity is there because if you start with a gauge model there will always be cancellations between different terms of the theory. For example if the top does not exist the correction diverges: moreover it is a quadratic divergence and therefore the radiative correction depends on the square of the top mass. Similarly, if there is no Higgs in the theory it blows up. But now this is only a logarithmic divergence and hence the Higgs mass enters only as a logarithm. That is why it is so difficult to obtain information on the Higgs: we have a limit on the top mass but no limit on the Higgs mass.

The rho-parameter is important but can only give you one number. You require further information from, for example, e^+e^- to muon pairs. That will of course be studied at LEP as time goes on. At LEP2 one of the first results which will come out in full clarity will be the mass of the W. This is the region where you will find the strongest influence of the Higgs. But unless we are able to go up to an energy of 250 or 300 GeV as I have always advocated very strongly, we will only be able to examine the small range shown. Maybe LEP2 sometime will be considered to be an enterprise as mistaken as NINA or Frascati. Only the future will say.

What are the things that we do not know at the present time? They are the existence of the top; the existence of other neutrino families, and the possibility of other heavy multiplets, not just the ones we are used to. And then there is the Higgs. Now in my view there are only two things to learn about the top quark and that is the date when it will be discovered and its mass. The number of neutrino families will be determined very soon from the total width of the Z. I have no idea how you find out about other multiplets.

So if you take this view, the future seems meagre and bleak. But I don't think that is right. You have to have some vision of what you are after, otherwise you die of boredom. My vision of the situation is this: we are looking at the standard model which has many characteristics, but we do not understand any of it. No one has explained to us why there are only three generations or why mixing is important or why we have $SU(2)_L \times U(1)$. We have absolutely no clue. Then there is another fact which has not been advertised very much: you can introduce any mass simply by hand, or you can get a mass by spontaneous symmetry breakdown. Now in the standard model every particle gets a mass through the Higgs mechanism. Why is this? For some reason or other, Nature has constructed the world in a way where all masses come via the Higgs mechanism. So the Higgs mechanism is a shorthand for how particles get their mass. To me it is a door in the wall: when we learn how to open it we will discover how particles acquire mass. I do not think that the way we describe the Higgs mechanism today is right. We are missing something. There is a threshold energy where the theory changes and we will go through the door and enter another world.

Where will it be? At the moment we are seeing a mass spectrum of particles which stops somewhere around 100 GeV. So it seems that there is a dividing line somewhere like 200 or 300 GeV with the property that all particle masses are below that line. It will be the limiting mass scale in our world.

We must take this into account when we consider radiative corrections. This is a necessary constraint on their size. The top quark seems to have a mass no smaller than 100 GeV, but from radiative corrections we have an upper limit which is something like 190 GeV. So maybe the mass is 150 GeV.

There is an amusing question which I ask myself from time to time: what if the top quark was not found below 200 GeV. This is called thinking the unthinkable. There may, of course, be other contributions to the rho-parameter. Just almost all contributions that you can think of: scalars, fermions, a large class of multiplets, all give a positive contribution[6] to the rho-parameter. So the limit of 190 GeV is not something that can be easily disposed of. Hence if the top is not found at the limit given by radiative corrections you have a serious conflict. There seem to me to be just two ways out of this. The trivial way out is that the Higgs sector is not what you think it is and there is an admixture of another Higgs sector. Then the rho-parameter simply becomes an arbitrary number as the ratio of vector boson masses becomes a free parameter and there is nothing to explain. That is the end of the story. You will just have to explain why the ratio of vector boson masses is so close to the value given by the standard model.

There is another way out: perhaps the way we do our calculations is not correct. Politzer and Woltram and Hung some time ago[7] demonstrated that if the Higgs mass is heavier than 100 GeV and the top mass heavier than 100 GeV then the theory possesses certain non-perturbative effects. You look at the Higgs sector at the minimum of the potential where you obtain the vacuum but you get radiative corrections to Higgs-Higgs scattering via a fermion loop. That is something you can treat according to the ideas of Coleman and Weinberg[8]. You can ask what contribution this scattering makes to the Higgs potential. If the Higgs is in some mass range and if the top is sufficiently heavy then the minimum of the Higgs potential simply disappears. So you no longer have spontaneous symmetry breaking.

This paper has not been taken sufficiently seriously in the past because nobody believed that the top mass could be so heavy. Everyone expected the top to be just around the corner. But these ideas become important now. I would expect that if something like this is true, then it must be associated with some new and interesting physics. So now we have something to look forward to in the development of LEP.

I want to discuss another matter now and in doing so to take issue with Alberto Sirlin. That is this subject of renormalisation schemes which is advertised to be a hot topic at this conference: all kinds of people are set to talk about these schemes. But this is a subject which always makes me mad. So I will try to explain why I feel about it as I do so that maybe you will get mad too. This is the situation: the standard model Lagrangian has a number of parameters in it; you fit the data including radiative corrections to this Lagrangian and you get the value of all the parameters which appear in the Lagrangian: g, sin θ_W, M_Z, M_W, etc. As far as I am concerned that is the whole affair.

The only thing more that I can think of in this connection is to relate this to how renormalisation was carried out in the old days. Then there really was some physics connected with the idea of the subtraction of infinities. The idea was that every infinity was literally related to some physical quantity. There was a one-to-one correspondence. There was a bare electron mass and then came the infinity and this resulted in the dressed mass that we could observe. That defined the physical electron mass. There was real physics behind this interpretation: there was the bare particle surrounded by a cloud of pairs from radiative corrections. So that attitude led to a correspondence between infinities and measurable physical parameters like the electron mass or the coupling constant in QED which was unrenormalised at zero momentum transfer so here nature was even kinder.

But now we know this is too narrow a view. In a non-Abelian gauge theory many relations hold in the Lagrangian but to which radiative corrections must be computed before the theory is checked against experimental data. The most notorious example is sin θ_W. This occurs all over the place in the Lagrangian: then you can do experiments, for example a neutral current experiment, and you find that it occurs in the ratio of the axial to vector current strengths g_A/g_V; but of course it also comes in through the ratio of vector boson masses: $\cos\theta_W = M_W/M_Z$ in the minimal standard model. So you can define sin θ_W in either way: from the ratio of neutrino–electron to antineutrino–electron scattering where the vector boson mass cancels out, or from measurements of the vector boson masses. This is clearly a domain of confusion. At least the ratio of scattering cross sections is not sensitive to the Higgs sector. So which is the real sin θ_W?

I can play the same game with the coupling constant which also appears in many places in the Lagrangian. It appears in the coupling of a W to an electron–neutrino pair (for example in muon–decay). It also appears in the trilinear interaction of vector bosons and the interaction of the Higgs with the vector bosons. Now all these terms give rise to different radiative corrections. So you may say that there are three separate coupling constants, but the requirements of the gauge theory are that they should be equal when you remove the radiative corrections. So you can define the ratios of these coupling constants as different types of rho–parameter and study those radiative corrections. So I can confuse you as much as you like. But you see that we have lost any semblance of being able to say that a given infinity is related directly to a given physical quantity. This is now seen to be too narrow minded a view.

Also we do not have to say "this is the coupling constant" of the theory, or "this is the sin θ_W." That is the quarrel I have with Sirlin. He makes a big point of having a renormalisation scheme in which sin θ_W is defined as the ratio of vector boson masses. If he takes that to be the definition it is fine by me. I don't need a definition in the first place as I hope I have shown. The second reason I have against this particular definition is that he has precisely obscured the one place where I hope to get information about the theory, because the ratio of vector boson masses is the parameter which tells us about the Higgs sector.

Of course, if the rho-parameter is defined to be one as it is in the Sirlin definition, the problem arises somewhere else. Furthermore, the definition of sin θ_W in terms of the vector boson masses assumes that these masses are precisely defined. Yet these particles are unstable and so do not have a uniquely-defined mass. I am not saying that you cannot work with unstable particles but it is a matter of principle. The width of the Z is about 3% of its mass. We are now entering a new era; the era of precision measurements, 0.1% measurements. I therefore ask you how you measure the weak mixing angle to 0.1% from a measurement of the W and Z masses, neither of which are well-defined to the level of 1%. You can of course locate a peak to a precision of 0.1% but there are many ways of parametrising it and the mass will depend on the parametrisation.

We don't need to do any of this. When we start computing our radiative corrections we compute to the greatest precision in the region of the whole peak: then you can fit to the parameters of the Lagrangian. But maybe I have misunderstood the problem. That is always possible. So to conclude this section I would advocate the simple procedure of expressing everything in terms of the "bare" parameters in a specific gauge (the 't Hooft-Feynman gauge). Then throw away $1/(n - 4)$ (MS-scheme) or $1/(n - 4)$ + ln c ($\overline{MS}$- scheme).

The important issue is how to confront theory and experiment. The precision measurements that will be available in the near future are the position of the Z peak and the asymmetry measurement using polarised beams. As long as the top mass is unknown it may be used as a free parameter which may be predicted as a result of these measurements.

Finally, I would like to ask a different question: how sure can we be that the Higgs system of the real world is the simplest one assumed in the standard model. This I want to discuss from a somewhat different point of view. Traditionally, for example see the account by our host Professor Dombey[8], fermion-W coupling is described by a left-handed doublet L and a right-handed singlet R with

$$L = i \bar{L} \gamma^\mu \left[\frac{1}{2} g \underline{\tau} . \underline{W}_\mu - \frac{1}{2} g' B_\mu \right] L - ir \bar{R} \gamma^\mu R B$$

showing the SU(2) × U(1) structure of the model where g' and r are free parameters. Now introduce mixing

$$Z = W^3 \cos \theta_W - B \sin \theta_W$$

$$A = W^3 \sin \theta_W + B \cos \theta_W$$

(for the e, ν_e case)

(i)   the neutrino has no charge, so

$$g^1 = g \tan \theta_W$$

(ii)   electromagnetism preserves parity, so

$$r = g^1$$

Thus conservation of parity is an accident.

Let us now assume that

(a) all masses derive through the Higgs mechanism,

and

(b) the Higgs mechanism comes about through the linear σ–model:

$$L_H = -\frac{1}{2}\,(\partial_\mu \varphi_\alpha)^2 - \frac{1}{2}\,\mu^2 \varphi_\alpha^{\,2} - \frac{1}{8}\,\lambda\,\varphi_\alpha^{\,2}$$

$\alpha = 0,1,2,3,$ where $\sigma = \varphi_0$

This is obviously invariant under 0(4) which is a six parameter group.

Rewrite L_H :

$$L_H = -\frac{1}{4}\,T_r\left[\partial_\mu \Phi^+ \partial_\mu \Phi\right] - \frac{1}{2}\,\mu\,Tr\left[\frac{1}{2}\,\Phi^+\Phi\right] - \frac{1}{8}\,\lambda\,Tr\left[\frac{1}{2}\Phi^+\Phi\right]^2$$

$\Phi = \sigma\,\tau^0 + i\,\varphi_\alpha\,\tau^\alpha$

$$= \begin{bmatrix} \sigma + i\varphi_3 & i\varphi_1\,\varphi_2 \\ i\varphi_1 - \varphi_2 & \sigma - i\varphi_3 \end{bmatrix}$$

$Tr\left[\Phi^+\Phi\right]$ is easily shown to be invariant under

$$\Phi \rightarrow G\,\Phi\,H^+$$

where G and H are $SU(2)_L$ and $SU(2)_R$. This is also six–parameter, and is of course equivalent to the O(4) description.

We can gauge $SU(2)_L$ with three parameters, while rotations about the 3–axis of $SU(2)_R$ gives one additional parameter.

If $SU(2)_R$ is not gauged we would have a strict global $SU(2)_L \otimes SU(2)_R$ symmetry. This is ordinary isospin. The vector bosons are an isospin triplet. It follows that

$$M(W^+) = M(W^-) = M(W^0)$$

Mixing modifies this to

(A) $\qquad\qquad\qquad \rho = M_W^{\,2}\,/\,(M_Z^{\,2}\,\cos^2\theta_W) = 1$

Now fermion masses. They arise from a term of the form

$$\bar{\psi}_L\,\psi_R\,\Phi + h.c.$$

Since Φ transforms as a doublet under $SU(2)_L$ it follows that ψ_L cannot be in the same representation of $SU(2)_L$.

(B) Parity is broken in weak interactions:

The relevant two coupling constants are g from $SU(2)_L$, and g' from $SU(2)_R$ restricted to rotations about the 3-axis. Define θ_W by

$$g' = g \tan \theta_W$$

Take

$$\langle \Phi \rangle_0 = \begin{bmatrix} f_0 & 0 \\ 0 & f_0 \end{bmatrix}$$

$$L_H = -\frac{1}{4} f_0^2 g^2 W_\mu^+ W_\mu^- - \frac{1}{8} f_0^2 g^2 \left[W_\mu^3 \cos \theta_W - B_\mu^0 \sin \theta_W \right]^2 / \cos^2 \theta_W$$

The quantity in the brackets above is just

$$Z_\mu = W_\mu^3 \cos \theta_W - B_\mu \sin \theta_W$$

and clearly parity is not conserved in charged weak interactions, nor in neutral weak interactions for general θ_W. Furthermore the Z has a mass.

(C) There is one zero mass vector boson:

The Higgs system has only 4 degrees of freedom of which at least one Higgs particle must remain physical. So at most three vector bosons can acquire a mass. So

$$A_\mu = W_\mu^3 \sin \theta_W + B_\mu \cos \theta_W$$

which is orthogonal to Z must remain massless.

It is now easy to check that the photon A_μ couples identically to left-handed and right-handed fermions, so that

(D) Parity is conserved in electromagnetic interactions.

References
1. D.A. Rose and M. Veltman, Nucl. Phys. B59 135 (1975).
2. P.Q. Hung and J.J. Sakurai, Nucl. Phys. B143 81 (1978).
3. V.F. Shvartsman, Sov. Phys. JETP Lett. 9 315 (1969). The first paper on this subject was F. Hoyle and R. Tayler. Nature 203 1108 (1964). For a recent treatment which gives a maximum of three neutrinos see J.D. Barrow and J. Morgan, Mon. Not. R. Ast. Soc. 203, 393 (1983).
4. R.J. Blin-Stoyle and J. Freeman, Nucl. Phys. 150A 369 (1970).
5. A. Sirlin, Rev. Mod.Phys. 50 573 (1978) is a general review of the author's work on the subject.
6. M.B. Einhorn, D.R.T. Jones and M. Veltman, Nucl. Phys. B191 146 (1981).
7. H.D. Politzer and S. Wolfram, Phys. Lett. 82B 242; [Erratum 83B 421 (1979)]; P.Q. Hung, Phys. Rev. Lett. 9 42, 873 (1979).
8. N. Dombey, Surveys in High Energy Physics 3 1 (1982).

THE Z LINE SHAPE

F.A. Berends

Instituut-Lorentz, University of Leiden
P.O. Box 9506, 2300 EA Leiden, The Netherlands

1. INTRODUCTION

With the advent of the first data from SLC and the prospect of LEP data in the near future the need for an accurate description of the Z line shape has become obvious. In the last few years considerable progress has been made towards an evaluation of the line shape which is accurate in size at the 0.3% level and in shape around the 10 MeV level (i.e. the positions of the maximum and half maxima).

The initial progress was on exact evaluations both of the weak part and the QED part. Some time ago the two evaluations were combined to get a realistic answer for the line shape[1]. Since then the efforts have been towards detailed comparisons between exact evaluations of the electroweak part as performed by different groups. This clarified several differences in approach and input. Also the calculations which initially focussed mostly on mupair production directed their attention to other fermion pair production channels. Moreover, gradually the interest increased in having a qualitative understanding of the line shape through some analytic for-mulae, which approximate the exact calculations reasonably well. The development is reflected in the stepwise progress in a number of reviews refs. 2,3,4. The last reference is the most complete account of the present status. The purpose of this paper is in the first place to summarize the main points of the established results. In the second place the attention is focussed on the effect of cuts. Up to now all calculations deal with the ideal case of no experimental cuts on the measurement of the line shape. This was done in order to be able to compare different calculations in the simplest case thus avoiding the additional differences in event generators. Once there is a standard for the ideal no cut case event generators should reproduce this standard result. Subsequently one can impose cuts on the generated events, and study the effects on the line shape. Here we include some results of such studies.

The calculations in the standard model are performed in the on-shell (OS) renormalization scheme. The results depend on the following parameters: M_Z, m_t, M_H and α_s. The lepton masses are known experimentally. The quark masses represent a suitable parametrization of the dispersion relation results for the hadronic vacuum polarization. The values used are (in GeV)

Radiative Corrections, Edited by N. Dombey and
F. Boudjema, Plenum Press, New York, 1990

$$m_u = m_d = 0.041, \quad m_c = 1.5, \quad m_s = 0.15, \quad m_b = 4.5 \ . \tag{1.1}$$

In the calculations we also require M_W, which is obtained from the muon decay constant and the quantity Δr as will be indicated below. In section 2.3 the input for α_s will be given. The free parameters M_Z, m_t and M_H will usually be restricted as follows. In most results M_Z will be 92 GeV, but sometimes 91 GeV. The masses m_t and M_H will be in the ranges 60–230 GeV and 10–1000 GeV.

We separate the corrections in non-photonic and photonic ones. They can also be referred to as electroweak and QED corrections.

The actual outline of the paper is as follows. In section 2 both exact and approximate analytical results for non-photonic corrections will be discussed. Section 3 deals with the exact photonic corrections and analytic approximations. The last section gives results when cuts are applied.

2. TOTAL CROSS SECTIONS WITH NON-PHOTONIC CORRECTIONS

2.1. The Outline of the Section

Firstly, lowest order formulae for widths and total cross sections will be given. The latter can be rewritten in terms of partial and total widths. Such expressions are useful for the discussion of qualitative features and for the derivation of approximate analytical formulae. Secondly, the first order non-photonic corrections to the widths and total cross sections are discussed and numerical results are presented. Approximate analytical expressions for the electroweak corrected cross sections are derived and used for a qualitative description of the resonance shape without QED corrections.

2.2. Lowest Order Widths and Cross Sections

As mentioned in the introduction we shall need M_W or $\sin^2\theta_W$ as input parameter besides M_Z, m_t and M_H. The muon decay constant G_μ or equivalently

$$A = \frac{\pi\alpha}{\sqrt{2}\, G_\mu} = (37.281 \text{ GeV})^2 \tag{2.1}$$

is used to determine M_W. In lowest order the required relation reads

$$M_W^2 \sin^2_W = A \ , \tag{2.2}$$

but the inclusion of electroweak corrections in muon decay modifies this into [5]

$$M_W^2 \sin^2\theta_W = \frac{A}{1-\Delta r} \ . \tag{2.3}$$

The actual value of Δr and its dependence on the parameters of the theory can be found for instance in ref. 6. In the OS scheme one uses as definition of the weak mixing angle

$$s_W^2 \equiv \sin^2\theta_W = 1 - \frac{M_W^2}{M_Z^2} \ . \tag{2.4}$$

This relation combined with (2.3) gives the values for M_W and s_W^2.

10

In lowest order the Z decays only into fermion pairs. The Born term expression for the partial width is

$$\Gamma^{(0)}_{Z \to \bar{f}f} = \frac{\alpha}{6} N_c M_Z \sqrt{1 - \frac{4m_f^2}{M_Z^2}} \left((g_f^-)^2 + (g_f^+)^2 + \frac{m_f^2}{M_Z^2} (6g_f^- g_f^+ - (g_f^-)^2 - (g_f^+)^2) \right).$$

(2.5)

with

$$g_f^- = (I_f^3 - Q_f s_W^2)/(s_W c_W)$$

(2.6)

$$g_f^+ = -Q_f s_W^2/(s_W c_W) \ ,$$

(2.7)

where $c_W = \cos\theta_W$, N_c represents the number of colours whereas Q_f and I_f^3 refer to the charge and weak isospin of the fermion.

Making use of the tree level relation (2.2) one may rewrite eq. (2.5) in terms of G_μ

$$\bar{\Gamma}^{(0)}_{Z \to \bar{f}f} = N_c \frac{G_\mu M_Z^3}{24\pi\sqrt{2}} \sqrt{1 - \frac{4m_f^2}{M_Z^2}} \left(1 - \frac{4m_f^2}{M_Z^2} + (2I_f^3 - 4Q_f s_W^2)^2(1 + \frac{2m_f^2}{M_Z^2})\right).$$

(2.8)

For fixed M_Z, M_H and m_t one can determine from eqs. (2.3) and (2.4) M_W and s_W^2. Inserting the latter into eqs. (2.5) and (2.8) gives different values for the widths. It will turn out that for $m_t \lesssim 120$ GeV expression (2.8) gives a good description of the corrected partial width (cf. section 2.3). They would be the same when the lowest order relation (2.2) is used to determine s_W^2.

The total lowest order width is denoted by

$$\Gamma_Z^{(0)} = \sum_f \Gamma^{(0)}_{Z \to \bar{f}f} \ .$$

(2.9)

For massless fermions (in practice all but the b quark) the partial width simplifies to

$$\Gamma^{(0)}_{Z \to \bar{f}f} = \Gamma_-(f) + \Gamma_+(f) = \frac{\alpha}{6} N_c M_Z \left(|g_f^-|^2 + |g_f^+|^2 \right) \ ,$$

(2.10)

where the − and + signs refer to the helicity of f, the helicity of $\bar{f}$ being opposite. For massive quarks the helicity of $\bar{f}$ is not necessarily opposite and one has four helicity amplitudes.

The total cross section in lowest order for

$$e^+e^- \to \bar{f}f$$

(2.11)

with massless fermions reads

$$\sigma_0(s) = \frac{4\pi\alpha^2}{3s} \frac{N_c s^2}{4} \left\{ \left| \frac{g_e^- g_f^-}{s - M_Z^2 + iM_Z\Gamma_Z} + \frac{Q_e Q_f}{s} \right|^2 \right.$$

11

$$+ \left| \frac{g_e^- g_f^+}{s - M_Z^2 + iM_Z\Gamma_Z} + \frac{Q_eQ_f}{s} \right|^2 + \left| \frac{g_e^+ g_f^-}{s - M_Z^2 + iM_Z\Gamma_Z} + \frac{Q_eQ_f}{s} \right|^2$$

$$+ \left| \frac{g_e^+ g_f^+}{s - M_Z^2 + iM_Z\Gamma_Z} + \frac{Q_eQ_f}{s} \right|^2 \Biggr\}. \tag{2.12}$$

The four terms in the cross section correspond to the following helicity combinations for e^+e^-, $\bar{f}$ and f: $(+-\ +-)$, $(+-\ -+)$, $(-+\ +-)$ and $(-+\ -+)$. It is clear that in the neutrino case only the first and third terms contribute. For Γ_Z the lowest order total width (2.9) is used. For massive fermions the expression is somewhat more involved and can be found in the literature (see e.g. ref. 7).

Another representation for the total lowest order cross section will turn out to be useful as basis for an approximate expression to the corrected total cross section. It uses the partial (helicity) widths and follows from eqs. (2.10) and (2.12). Firstly we rewrite eq. (2.10) in the form

$$\sigma_0(s) = \frac{N_c s}{48\pi} \sum_{\lambda_e \lambda_f} \left| \frac{R_{\lambda_e \lambda_f}}{s - M_Z^2 + iM_Z\Gamma_Z} + \frac{Q_eQ_f e^2}{s} \right|^2, \tag{2.13}$$

with

$$R_{\lambda_e \lambda_f} = \pm \lambda_e \lambda_f \frac{24\pi\ \Gamma_{\lambda_e}^{\frac{1}{2}} \Gamma_{\lambda_f}^{\frac{1}{2}}}{N_c^{\frac{1}{2}} M_Z}, \tag{2.14}$$

where the $\pm$ sign corresponds to $I_f^3 = \mp \frac{1}{2}$ of the final state and λ_e, λ_f refer to the helicities of the e^- and f. Carrying out the helicity summations one finds

$$\sigma_0(s) = \frac{sN_c}{(s-M_Z^2)^2 + M_Z^2\Gamma_Z^2} \left[\frac{12\pi\ \Gamma_e\Gamma_f}{M_Z^2 N_c} + I\ \frac{s-M_Z^2}{s} \right] + \frac{4\pi Q_f^2 \alpha^2 N_c}{3s} \tag{2.15}$$

where

$$I = \pm \frac{4\pi Q_eQ_f \alpha}{N_c^{\frac{1}{2}} M_Z} (\Gamma_+^{\frac{1}{2}}(e) - \Gamma_-^{\frac{1}{2}}(e))(\Gamma_+^{\frac{1}{2}}(f) - \Gamma_-^{\frac{1}{2}}(f)). \tag{2.16}$$

The first term in eq. (2.15) is the Breit-Wigner form for a spin 1 resonance, the last term is the pure QED cross section. The latter term still is of relevance for mupair production. The interference term is positive for realistic s_W^2 values and is less relevant for mupair production than for quark pair production.

Although we have not indicated this explicitly in the expressions (2.10)-(2.16) all partial and total widths are here lowest order expressions.

At this point some qualitative features of the lowest order total cross section (2.15) may be noticed. For neutrino pair production only the pure Breit-Wigner formula contributes

$$\sigma_0(s) = \frac{12\pi}{M_Z^2} \frac{\Gamma_e \Gamma_f}{(s-M_Z^2)^2 + M_Z^2 \Gamma_Z^2} \cdot$$ (2.17)

The peak value of

$$\sigma_{max} = \frac{12\pi}{M_Z^2 \Gamma_Z^2} \Gamma_e \Gamma_f (1 + \tfrac{1}{4}\gamma^2) \; ,$$ (2.18)

where

$$\gamma = \Gamma_Z/M_Z$$ (2.19)

is obtained for

$$(\sqrt{s})_{max} = M_Z(1+\gamma^2)^{\frac{1}{4}} \; .$$ (2.20)

The right and left half maxima positions are at

$$(\sqrt{s})_{\pm} = M_Z(1 + \tfrac{3}{8}\gamma^2) \pm \frac{\Gamma_Z}{2}(1 - \tfrac{1}{8}\gamma^2) \; ,$$ (2.21)

when we expand up to γ^2. Thus the peak position is about 17 MeV above M_Z, whereas the average value of the half maxima positions is slightly greater than (2.20). The distance between the half maxima positions is slightly less than the width Γ_Z.

When one photon exchange is present the shift in the peak position remains the same but the distance between the half maxima becomes larger. This is only noticeable for mupair production, since the pure QED term is most important for this channel. It contributes about 0.5% to the peak value and forms a slowly varying background to the Breit-Wigner resonance.

These qualitative effects will be somewhat modified after the inclusion of electroweak corrections. This will be illustrated by actual numbers and formulae in subsections 2.4 and 2.5.

2.3. The Corrected Partial and Total Widths

As a first step to the corrections to the total cross section we discuss corrections to the partial width $\Gamma_{Z \to f\bar{f}}^{(0)}$. The corrections can be divided into four classes:

1. Non-photonic loop corrections to the decay into fermion pairs.
2. Photonic loop corrections and radiative decay.
3. QCD corrections.
4. Decay into three or more particles.

In this list the second class may look out of place since photonic corrections are dealt with in section 3. It is nevertheless also included here since it contributes to the total width, which through the Z-propagator affects the total cross section. Moreover it is similar to the QCD corrections.

The photonic loop correction and radiative decay gives in first order the correction

$$\delta_{QED} = \frac{3\alpha}{4\pi} \, Q_f^2 = 0.0017 \, Q_f^2 \, , \qquad (2.22)$$

which is a tiny effect. The QCD correction is larger due to the size of α_s. From electron-positron experiments at 34 GeV the QCD correction is known and can be extrapolated to M_Z through the running coupling constant. One then obtains for massless quarks

$$\delta_{QCD} = 0.040 \pm 0.007 \qquad (2.23)$$

and for b-quarks, using the expressions of ref. 9

$$\delta_{QCD} = 0.045 \pm 0.007 \, . \qquad (2.24)$$

The errors will eventually lead to an error of about 10 MeV on the total width, which is the main source of uncertainty in the shape of the Z.

The decay into three or more particles has been considered in the literature [10,11,12]. The most important decay is

$$Z \to H\bar{f}f, \qquad (2.25)$$

but it is only relevant for $M_H \lesssim 10$ GeV. Using the formulae of ref. 13 and adding all fermion combinations the partial width for the decay (2.25) is of the order of 5 MeV. The sum of all other decays e.g.

$$Z \to \bar{f}f \, \bar{f}f \qquad (2.26)$$

where at least two fermions are leptons, is expected to have a width of at most a few MeV [11]. It may be that smaller phase space decays like

$$Z \to \pi^0 \, \bar{\ell}\ell \qquad (2.27)$$

which belong to the class (2.26) could give slight deviations from the expectations. On the other hand the decay $Z \to \pi^0\gamma$ is already included in the above mentioned QED corrections, just like the decays into four quarks is a QCD correction.

In view of the required accuracy we neglect the above rare decays, keeping in mind however that for small M_H the decay (2.25) should be taken into account.

We now turn to the non-photonic loop corrections, which have been discussed extensively in the literature [14,15,16]. These corrections come from vertex corrections and wave function renormalizations. In ref. 14 the correction to eq. (2.5) is sizeable, mainly due to the wave function renormalization of the Z. In the scheme of ref. 15 the correction to eq. (2.8) is considered which is small, at least for $m_t \lesssim 120$ GeV. The corrected partial and total widths are however the same and are given in table 1. These values include the effects of the first three classes of corrections. The agreement is within 0.1%. For all massless fermion channels also the results of ref. 16 are in agreement with the table. The difference between the partial widths for decays into d- and b-quarks is due to three effects: phase space, the QCD corrections and the contribution of a massive top quark to the vertex correction. Table 2 gives the results of ref. 14 for $M_Z = 91$ GeV.

Table 1. Partial and total Z widths for $M_Z = 92$ GeV and δ_{QCD} from eqs. (2.23) and (2.24). The first line lists the results of ref. 14, the second of ref. 15. In the second line only the digits differing from line 1 are printed. In the cases where $M_H = 10$ the decay (2.25) has a width of 5.3 and 5.8 ($m_t = 230$), which is not included in the listed value for Γ_Z.

m_t (GeV)	M_H (GeV)	$\sin^2\theta_W$	Γ_Z (MeV)	$\Gamma_{Z\to\nu\bar{\nu}}$ (MeV)	$\Gamma_{Z\to e^+e^-}$ (MeV)	$\Gamma_{Z\to u\bar{u}}$ (MeV)	$\Gamma_{Z\to d\bar{d}}$ (MeV)	$\Gamma_{Z\to b\bar{b}}$ (Mev)
60	100	0.2296	2562	170.5	85.8	306.3	394.4	392.1
			1			.1	.2	.0
90	10	0.2242	2564	170.4	85.8	307.1	395.0	391.9
						6.9	4.8	.8
90	100	0.2258	2567	170.8	86.0	307.2	395.4	392.3
		7		.7	5.9	.0	.2	.2
90	1000	0.2288	2559	170.5	85.7	305.9	394.0	390.9
		7		.3		.6	3.6	.6
150	100	0.2186	2581	171.6	86.4	309.6	398.2	391.6
			80			.4	7.9	
200	100	0.2109	2596	172.5	86.9	312.5	401.5	390.5
				.4		.3	.3	90.8
230	10	0.2035	2604	172.8	87.1	314.3	403.5	389.1
			5				.4	.8
230	100	0.2052	2607	173.2	87.2	314.5	403.9	389.6
			8				.8	90.2
230	1000	0.2086	2599	172.9	87.1	313.3	402.6	388.5
			8	.7	.0	.0	.2	.7

Table 2. Partial and total Z widths for $M_Z = 91$ GeV and δ_{QCD} from eqs. (2.23) and (2.24). The results are those of ref. 14.

m_t (GeV)	M_H (GeV)	$\sin^2\theta_W$	Γ_Z (MeV)	$\Gamma_{Z\to\nu\bar{\nu}}$ (MeV)	$\Gamma_{Z\to e^+e^-}$ (MeV)	$\Gamma_{Z\to u\bar{u}}$ (MeV)	$\Gamma_{Z\to d\bar{d}}$ (MeV)	$\Gamma_{Z\to b\bar{b}}$ (MeV)
60	100	0.2366	2461	165.0	82.7	292.7	378.3	376.0
90	10	0.2312	2463	164.9	82.7	293.4	379.0	375.8
90	100	0.2328	2466	165.3	82.9	293.6	379.3	376.2
90	1000	0.2360	2458	165.0	82.7	292.3	377.9	374.8
150	100	0.2257	2479	166.0	83.3	295.9	382.0	375.5
200	100	0.2181	2493	166.9	83.7	298.6	385.2	374.5
230	10	0.2106	2501	167.2	83.8	300.4	387.2	373.2
230	100	0.2123	2504	167.6	84.0	300.6	387.6	373.7
230	1000	0.2158	2497	167.3	83.8	299.4	386.2	372.5

2.4. Total Cross Section with Electroweak Corrections

The lowest order total cross section for massless fermions is given in eq. (2.12). The cross section is of order α^2 except at the resonance position where it is of order α^0. As stressed in ref. 17 one should therefore consider the $O(\alpha)$ corrections to Γ_Z. Since Γ_Z is related to the imaginary part of the self-energy it means that the one-loop corrections to the propagator are not sufficient but that two-loop corrections should be taken into account in the resonance region.

Besides the modifications of coupling constants by vertex corrections, the introduction of very small box diagrams, the electroweak corrections amount to the following replacements. We use here the results of ref. 6. Several related earlier discussions exist in the literature e.g. refs. 18, 19.

$$\frac{1}{s} \rightarrow \frac{1}{s + \Sigma_\gamma(s)} \quad , \tag{2.28}$$

$$\frac{1}{s - M_Z^2 + iM_Z\Gamma_Z} \rightarrow \frac{1}{s - M_Z^2 + \Sigma_Z(s)} \quad , \tag{2.29}$$

where

$$\Sigma_\gamma(s) = \Sigma_{\gamma\gamma}(s) - \frac{\Sigma_{\gamma Z}^2(s)}{s - M_Z^2 + \Sigma_{ZZ}(s)} \tag{2.30}$$

$$\Sigma_Z(s) = \Sigma_{ZZ}(s) - \frac{\Sigma_{\gamma Z}^2(s)}{s + \Sigma_{\gamma\gamma}(s)} \quad . \tag{2.31}$$

These expressions are obtained from a Dyson series summation involving the renormalized one particle irreducible self-energies $\Sigma_{\gamma\gamma}$, Σ_{ZZ} and $\Sigma_{\gamma Z}$. Besides the above propagators one has also to include a γ-Z mixing propagator, which takes the form

$$D_{\gamma Z}(s) = \frac{-\Sigma_{\gamma Z}(s)}{s[s - M_Z^2 + \Sigma_Z(s)]} \quad . \tag{2.32}$$

In these expressions the real parts of Σ_γ and Σ_Z are taken in first order. The imaginary part of Σ_Z is considered up to second order. That is, also the imaginary part of Σ_{ZZ} should be evaluated in second order. This is done by the following approximation

$$\text{Im } \Sigma_{ZZ}^{(2)}(s) = \frac{s}{M_Z^2} \text{ Im } \Sigma_{ZZ}^{(2)}(M_Z^2) \quad , \tag{2.33}$$

where the latter expression is related to the first order corrections to the width. All corrections to the width contribute to eq. (2.33) except for the wave function renormalization of the Z and γZ mixing contributions. This can be seen by expanding (2.29) in the resonance region

$$\frac{1}{s - M_Z^2 + \Sigma_Z(s)} = \frac{1}{1 + \Pi_Z(M_Z^2)} \frac{1}{s - M_Z^2 + \dfrac{i \text{ Im } \Sigma_Z(s)}{1 + \Pi_Z(M_Z^2)}} \quad , \tag{2.34}$$

Table 3. The total cross section including electroweak corrections for $e^+e^- \rightarrow \mu^+\mu^-$ for $M_Z = 92$ GeV

m_t (GeV)	M_H (GeV)	Γ_Z (MeV)	σ_{max} (nb)	$\sqrt{s_{max}}$ (GeV)	$\sqrt{s_-}$ (GeV)	$\sqrt{s_+}$ (GeV)
60	100	2562	1.951	91.983	90.704	93.280
90	10	2564	1.949	91.983	90.703	93.281
90	100	2567	1.951	91.983	90.701	93.283
90	1000	2559	1.953	91.983	90.706	93.278
150	100	2581	1.954	91.983	90.695	93.290
200	100	2596	1.958	91.982	90.687	93.279
230	10	2604	1.963	91.982	90.683	93.301
230	100	2607	1.963	91.982	90.681	93.303
230	1000	2599	1.963	91.982	90.685	93.299

Table 4. The total cross section including electroweak corrections for $e^+e^- \rightarrow \mu^+\mu^-$ for $M_Z = 91$ GeV

m_t (GeV)	M_H (GeV)	Γ_Z (MeV)	σ_{max} (nb)	$\sqrt{s_{max}}$ (GeV)	$\sqrt{s_-}$ (GeV)	$\sqrt{s_+}$ (GeV)
60	100	2460	2.008	90.983	89.755	92.229
90	10	2463	2.005	90.983	89.753	92.230
90	100	2466	2.007	90.983	89.752	92.232
90	1000	2458	2.011	90.983	89.756	92.227
150	100	2479	2.009	90.983	89.746	92.238
200	100	2493	2.012	90.983	89.738	92.245
230	10	2501	2.014	90.983	89.734	92.249
230	100	2504	2.015	90.983	89.733	92.250
230	1000	2497	2.017	90.983	89.736	92.247

Table 5. The total cross section including electroweak
corrections for $e^+e^- \to$ hadrons for $M_Z = 92$ GeV

m_t (GeV)	M_H (GeV)	Γ_Z (MeV)	σ_{max} (nb)	$\sqrt{s}_{max}$ (GeV)	$\sqrt{s}_-$ (GeV)	$\sqrt{s}_+$ (GeV)
60	100	2562	40.627	91.984	90.712	93.275
90	10	2564	40.634	91.984	90.711	93.277
90	100	2567	40.633	91.984	90.710	93.278
90	1000	2559	40.634	91.984	90.714	93.274
150	100	2581	40.678	91.983	90.703	93.286
200	100	2596	40.752	91.983	90.696	93.294
230	10	2604	40.833	91.983	90.692	93.297
230	100	2607	40.821	91.983	90.690	93.299
230	1000	2599	40.797	91.983	90.694	93.295

Table 6. The total cross section including electorweak
corrections for $e^+e^- \to$ hadrons for $M_Z = 91$ GeV

m_t (GeV)	M_H (GeV)	Γ_Z (MeV)	σ_{max} (nb)	$\sqrt{s}_{max}$ (GeV)	$\sqrt{s}_-$ (GeV)	$\sqrt{s}_+$ (GeV)
60	100	2460	41.560	90.984	89.762	92.225
90	10	2463	41.556	90.984	89.761	92.226
90	100	2466	41.562	90.984	89.760	92.227
90	1000	2458	41.572	90.984	89.764	92.223
150	100	2479	41.594	90.984	89.754	92.234
200	100	2493	41.653	90.984	89.747	92.241
230	10	2501	41.718	90.984	89.743	92.245
230	100	2504	41.711	90.984	89.741	92.247
230	1000	2497	41.696	90.984	89.745	92.243

where

$$\Pi_Z(M_Z^2) = \frac{\partial \, \text{Re} \, \Sigma_Z}{\partial s} (M_Z^2) \tag{2.35}$$

and

$$M_Z \Gamma_Z = \frac{\text{Im} \, \Sigma_Z(M_Z^2)}{1 + \Pi_Z(M_Z)} \, . \tag{2.36}$$

The denominator in eq. (2.36) represents the wave function renormalization of the Z. It gives a first order correction to Γ_Z. When one also considers the real part of the second term in eq. (2.31) one effectively takes a part of the second order correction of the real part of Σ_Z into account. The effect of this has been discussed in ref. 6. For high top masses ($m_t \geq 120$ GeV) slight deviations from the results presented here occur. It should be stressed that for a conclusive discussion of this effect the full second order calculation of Re Σ_Z should be performed.

In the replacement (2.29) it is not only of importance that corrections to Γ_Z or equivalently to $\Sigma_Z(s)$ are taken into account. Also the energy dependence of Im $\Sigma_Z(s)$ is crucial. It shifts the peak position with −35 MeV with respect to a constant width formula[1,20]. In a qualitative discussion below we come back to this point.

Besides the propagator effects vertex corrections replace the couplings g^-_f and g^+_f by s dependent form factors. The size of the line shape is affected by these corrections. The box diagrams turn out to be very small and can be neglected in the total cross section. It should be noted that the QCD correction (2.23) is included, but that the QED correction (2.22) is omitted as correction on the final state. It is however a tiny effect which can easily be included.

Results for the total cross sections for mupair and hadron production are given in tables 3-6. They have been obtained from the program ZSHAPE[21], which uses results from refs. 1, 2, 6, 14 and 22. This program can be considered as a rewritten and improved ZBATCH[2] program. It should be noted that the numbers agree within 0.1% with those obtained from ref. 23. The main features of the results are that the peak position is about 17 MeV lower than M_Z. Moreover for mupair production the distance between the half maxima positions exceeds Γ_Z by about 15 MeV. This can be understood from the qualitative discussion in the next section.

2.5. Approximate Expressions

As mentioned in the previous section the modification of the total cross section (2.12) due to electroweak corrections is in essence due to an introduction of s-dependent form factors which replace the coupling constants $g^\pm$, and the changes (2.28), (2.29) and (2.32) of the propagators. Thus in several places in the original formula s-dependent quantities replace the original constant ones. Also the values for s = M are different from the lowest order quantities.

In the region of the resonance the s-dependence of $\Sigma_Z(s)$ in eq. (2.29) is crucial, the other s-dependences have a small influence. The s-dependence of Im $\Sigma_Z(s)$ near the resonance can be well approximated by the replacement

$$\frac{1}{s - M_Z^2 + iM_Z\Gamma_Z} \rightarrow \frac{1}{s - M_Z^2 + is\Gamma_Z/M_Z} \, . \tag{2.37}$$

The other s-dependent corrections can be taken at $s = M_Z^2$. Effectively we get new coupling constants $g^{\pm}(M_Z^2)$ and

$$e^2(M^2) = \frac{e^2}{1 + \Pi_{\gamma}(M_Z^2)} \, , \tag{2.38}$$

where

$$\Pi_{\gamma}(s) = \frac{\text{Re } \Sigma_{\gamma\gamma}(s)}{s} \, . \tag{2.39}$$

In the constants $g^{\pm}(M_Z^2)$ also form factor effects are incorporated. This is not done for the coupling to the photons. The imaginary parts of the form factors and Σ_{γ} are neglected at this point.

We now end up with the following approximation for the electroweak corrected total cross section for massless fermions.

$$\sigma(s) = \left\{ \frac{s}{(s-M_Z^2)^2 + s^2\Gamma_Z^2/M_Z^2} \left[\frac{12\pi\Gamma_e\Gamma_f}{M_Z^2} + \frac{IN_c(s-M_Z^2)}{s} \right] \right.$$

$$\left. + \frac{4}{3}\pi Q_f^2 \frac{\alpha^2(M_Z^2)N_c}{s} \right\} (1 + \delta_{QCD}) \, , \tag{2.40}$$

where Γ_e, Γ_f are the electroweak corrected partial widths without QED and QCD corrections (i.e. the values of tables 1 or 2 with the factors $1 + \delta_{QED}$ and $1 + \delta_{QCD}$ removed). The quantity I is given by eq. (2.16), where α is repladed by $\alpha(M_Z^2)$ and the helicity partial widths contain electroweak corrections. For b-quarks the pure QED part now becomes the massive QED part, but in I we use the electroweak corrected $\Gamma_{\pm}$ for massless b-quarks .

The approximation (2.40) describes the exact electroweak corrected cross section within 0.2% in the range $(M_Z - \Gamma_Z, M_Z + \Gamma_Z)$. The values for maxima and half maxima obtained from (2.40) are within 1 MeV from the values in the tables.

Although the approximation (2.40) describes the standard model prediction sufficiently well one can also look for a representation which uses more parameters and which can accommodate deviations from the standard model. This has been done in ref. 24. The differences with (2.40) are firstly that in the propagator (2.38) one allows for a more general s-dependence

$$\text{is } \frac{\Gamma_Z}{M_Z} \to iM_Z\Gamma_Z \left[\frac{s}{M_Z^2} + \varepsilon \frac{s - M_Z^2}{M_Z^2} \right] \, . \tag{2.41}$$

Secondly, besides the Breit-Wigner term and the interference between real parts a third term $\sim \Gamma_Z J/M_Z s$ is added in the square bracket in (2.40) arising from the interference of the imaginary part of the propagator with other imaginary parts (e.g. Im Σ_{γ}). The parameters ε and J can be neglected in the standard model. Instead of calculating the values of all parameters in this extended parametrization one can try to determine them from the data thus allowing for a more general model than the standard model.

From the approximate formula some qualitative features can be derived. The pure Breit-Wigner part in eq. (2.41) originates from (modulo a factor)[20].

$$\chi(s) = \frac{1}{s - M_Z^2 + is\gamma} = \frac{1}{1 + i\gamma} \frac{1}{s - \tilde{M}_Z^2 + i\tilde{M}_Z\tilde{\Gamma}_Z} \tag{2.42}$$

with

$$\tilde{M}_Z = \frac{M_Z}{(1+\gamma^2)^{\frac{1}{2}}} \tag{2.43}$$

$$\tilde{\Gamma}_Z = \frac{\Gamma_Z}{(1+\gamma^2)^{\frac{1}{2}}} . \tag{2.44}$$

The qualitative features of our approximation are the same as the ones of eq. (2.15) when we make the replacement

$$M_Z \rightarrow \tilde{M}_Z, \quad \Gamma_Z \rightarrow \tilde{\Gamma}_Z . \tag{2.45}$$

Therefore the peak position of σ is given by a modified eq. (2.20)

$$\sqrt{s}_{max} = \frac{M_Z}{(1+\gamma^2)^{\frac{1}{4}}} . \tag{2.46}$$

Wheres eq. (2.15) gives a peak position about 17 MeV above M_Z the form with an energy dependent width (2.40) gives a peak position of 17 MeV below M_Z.

One can see from the numbers of the tables that the exact calculation has these quantitative features.

The simple Breit-Wigner form does not give the complete description. For muon pairs the relative importance of the pure QED term is largest. The latter makes the distance between the half width positions in muon pair production greater than the Breit-Wigner expectation (cf. (2.21)). For hadrons this effect is tiny.

3. THE PHOTONIC OR QED CORRECTIONS

3.1. The Outline of the Section

In the first place it is argued that for the total cross section without cuts the initial state photonic corrections are the most important ones. The expressions for these effects are given. Exact and approximate results for the line shape including both photonic and non-photonic corrections are presented.

3.2. Initial state photonic corrections

When one considers the full first order QED correction to the line shape, one finds contributions from initial state radiation, final state radiation and the interference. When no cuts on the outgoing fermions are imposed the final state radiative correction is just eq. (2.22) and is therefore small. Also the interference term is negligible, refs. 26, 27, 28. Thus the initial state radiative corrections remain and they are sizeable due to the occurrence of large logarithms of the type

$$L = \ln \frac{s}{m_e^2} \tag{3.1}$$

It should be noted that in general such large logarithms could occur in the final state corrections with m_e replaced by e.g. m_μ but that for the total phase space they cancel due to the KLN theorem [25]. We shall see this explicitly in section 4.2.

In first order in α only single photon emission has to be considered, in second order not only double photon emission occurs but also the emission of an additional fermion pair. For instance as correction to mupair production one has

$$e^+e^- \rightarrow \mu^+\mu^-e^+e^- \ . \tag{3.2}$$

In ref. 29 a discussion of the effects of photon emission and of (3.2) on $d\sigma/ds'$ is given, where s' is the square of the invariant mass of the mupair. The conclusion is that the effect of (3.2) can be neglected with respect to photon emission, except for $s'/s < 0.3$, where s is the square of the laboratory energy. In the region of low invariant mass the so-called two photon production of a mupair, which is one of the mechanisms in reaction (3.2), becomes more important than the bremsstrahlung for $d\sigma/ds'$. In the following it is assumed that the specific characteristics of the two photon mechanism allow for the removal of that type of events from the data. What is then left in $d\sigma/ds'$ comes essentially from photon emission. In case a cut on s' is used for the removal of the events one should in principle take into account a modification of the final state correction (2.22). We ignore this point at this moment and consider here as QED correction only multiple photon emission from the incoming e^+ and e^- and the corresponding vertex corrections.

From explicit calculations up to order α^2 and the resummation of soft photons [29], which takes a part of all higher order corrections into account the following formula for the initial state photonic corrections can be derived

$$\sigma_T(s) = \int_{z_0}^1 dz \ \sigma(sz)G(z) \ , \tag{3.3}$$

with $s'=sz$ and $4m_f^2/s \leq z_0$. In eq. (3.3) the cross section $\sigma(s)$ is the electroweak corrected one from section 2.4. The function $G(z)$ takes the form [29]

$$G(z) = \beta(1-z)^{\beta-1}\delta^{V+S} + \delta^H \ , \tag{3.4}$$

where

$$\beta = \frac{2\alpha}{\pi} (L-1) \ , \tag{3.5}$$

$$\delta^{V+S} = 1 + \delta_1^{V+S} + \delta_2^{V+S} \ , \tag{3.6}$$

$$\delta^H = \delta_1^H + \delta_2^H \ , \tag{3.7}$$

$$\delta_1^{V+S} = \frac{\alpha}{\pi} \left(\frac{3}{2} L + 2\zeta(2) - 2 \right) \ , \tag{3.8}$$

$$\delta_2^{V+S} = (\frac{\alpha}{\pi})^2 \left[(\frac{9}{8} - 2\zeta(2))L^2 + (- \frac{45}{16} + \frac{11}{2} \zeta(2) + 3\zeta(3))L \right.$$

$$\left. - \frac{6}{5} \zeta(2)^2 - \frac{9}{2} \zeta(3) - 6\zeta(2) \ln 2 + \frac{3}{8} \zeta(2) + \frac{19}{4} \right] , \qquad (3.9)$$

$$\delta_1^H = - \frac{\alpha}{\pi} (1+z)(L-1) , \qquad (3.10)$$

$$\delta_2^H = (\frac{\alpha}{\pi})^2 \left\{ \chi - (1+z)[2\ln(1-z)(L-1)^2 + (L-1)(\frac{3}{2}L + 2\zeta(2) - 2)] \right\}, \qquad (3.11)$$

$$\chi = \left(- \frac{1+z^2}{1-z} \ln z + (1+z) \frac{1}{2} \ln z + z - 1 \right)L^2$$

$$+ \left[\frac{1+z^2}{1-z} \left(\text{Li}_2(1-z) + \ln z \ln(1-z) + \frac{7}{2} \ln z - \frac{1}{2} \ln^2 z \right) \right.$$

$$\left. + (1+z) \frac{1}{4} \ln^2 z - \ln z + \frac{7}{2} - 3z \right] L$$

$$+ \frac{1+z^2}{1-z} \left(- \frac{1}{6} \ln^3 z + \frac{1}{2} \ln z \, \text{Li}_2(1-z) + \frac{1}{2} \ln^2 z \ln(1-z) \right.$$

$$\left. - \frac{3}{2} \text{Li}_2(1-z) - \frac{3}{2} \ln z \ln(1-z) + \zeta(2)\ln z - \frac{17}{6} \ln z - \ln^2 z \right)$$

$$+ (1+z) \left(\frac{3}{2} \text{Li}_3(1-z) - 2S_{1,2}(1-z) - \ln(1-z)\text{Li}_2(1-z) - \frac{1}{2} \right)$$

$$- \frac{1}{4} (1-5z)\ln^2(1-z) + \frac{1}{2} (1-7z)\ln z \ln(1-z) - \frac{25}{6} z \, \text{Li}_2(1-z)$$

$$+ (-1 + \frac{13}{3} z)\zeta(2) + (\frac{3}{2} - z)\ln(1-z) + \frac{1}{6} (11+10z)\ln z$$

$$+ \frac{2}{(1-z)^2} \ln^2 z - \frac{25}{11} z \ln^2 z - \frac{2}{3} \frac{z}{1-z} \left(1 + \frac{2}{1-z} \ln z + \frac{1}{(1-z)^2} \ln^2 z \right).$$
$$(3.12)$$

In these definitions the polylogarithms [30,31] $\text{Li}_n(x)$ and $S_{n,p}(x)$ have been introduced and the Riemann zeta function $\zeta(2) = \pi^2/6$ and $\zeta(3) = 1.202$. the terms δ_1^{V+S} and δ_1^H originate respectively from virtual, soft photon, and hard photon corrections of order α^1. Due to the resummation of soft photons the exponentiated 1-z term arises. It should be noted that the form (3.3) was first introduced in the structure function approach to this problem [32-35]. However in these papers only the leading log terms $(\alpha/\pi)^n L^n$ are completely taken into account but not the subleading log terms. The result (3.4) goes beyond this. It includes up to order α^2 also the subleading log and non-log terms. In ref. 29 the leading and subleading log terms were calculated in two independent ways, one by the structure function approach and the other by an explicit evaluation of Feynman diagrams.

The numerical results [21] following from the convolution (3.3) are given in tables 7-10. The peak height decreases by a factor 0.74 and the position is shifted by about 112 MeV. Furthermore the shape is distorted, the half maxima shift by typically 60 and 430 MeV, the asymmetry being caused by the radiative tail of the resonance. The convolution decreases the Breit-Wigner peak but increases the pure QED cross section. The first is a soft photon effect the second a hard photon effect. This is the cause of the widening of the mupair line shape with respect to the hadronic one. In fig. 1 the line shape for mupairs is shown for the minimum $\sqrt{s'}$ value of 1 GeV in the convolution. Besides the result of a convolution with the full expression (3.4) the results for two other cases are given i.e. $\delta_2^{V+S} = \delta_2^H = 0$ (first order exponentiated) and the pure second order result. The three cases are the solid, dashed and fine dashed lines.

Table 7. The total cross section including all corrections for $e^+e^- \to \mu^+\mu^-$ for $M_Z = 92$ GeV and a minimum for $\sqrt{s'}$ of 0.2 GeV

m_t (GeV)	M_H (GeV)	Γ_Z (MeV)	σ_{max} (nb)	$\sqrt{s_{max}}$ (GeV)	$\sqrt{s_-}$ (GeV)	$\sqrt{s_+}$ (GeV)
60	100	2562	1.453	92.094	90.754	93.724
90	10	2564	1.452	92.094	90.752	93.726
90	100	2567	1.453	92.094	90.751	93.728
90	1000	2559	1.454	92.094	90.755	93.722
150	100	2581	1.456	92.095	90.745	93.737
200	100	2596	1.460	92.095	90.737	93.747
230	10	2604	1.464	92.095	90.733	93.752
230	100	2607	1.464	92.095	90.732	93.755
230	1000	2599	1.464	92.095	90.736	93.749

Table 8. The same as table 7. However $M_Z = 91$ GeV

m_t (GeV)	M_H (GeV)	Γ_Z (MeV)	σ_{max} (nb)	$\sqrt{s_{max}}$ (GeV)	$\sqrt{s_-}$ (GeV)	$\sqrt{s_+}$ (GeV)
60	100	2460	1.492	91.090	89.802	92.656
90	10	2463	1.490	91.091	89.801	92.657
90	100	2466	1.492	91.091	89.800	92.659
90	1000	2458	1.493	91.090	89.804	92.654
150	100	2479	1.493	91.091	89.794	92.668
200	100	2493	1.497	91.091	89.786	92.678
230	10	2501	1.499	91.092	89.782	92.682
230	100	2504	1.500	91.092	89.781	92.685
230	1000	2497	1.500	91.091	89.785	92.680

Table 9. The total cross section including all corrections for $e^+e^- \to$ hadrons and a minimum for $\sqrt{s'}$ of 10 GeV

m_t (GeV)	M_H (GeV)	Γ_Z (MeV)	σ_{max} (nb)	$\sqrt{s_{max}}$ (GeV)	$\sqrt{s_-}$ (GeV)	$\sqrt{s_+}$ (GeV)
60	100	2562	30.043	92.095	90.772	93.705
90	10	2564	30.051	92.095	90.771	93.707
90	100	2567	30.054	92.095	90.770	93.709
90	1000	2559	30.054	92.095	90.773	93.703
150	100	2581	30.102	92.096	90.763	93.718
200	100	2596	30.174	92.097	90.756	93.729
230	10	2604	30.244	92.097	90.753	93.734
230	100	2607	30.238	92.097	90.751	93.736
230	1000	2599	30.211	92.097	90.755	93.731

Table 10. The total cross section including all corrections for
e^+e^- hadrons for M_Z=91 GeV and a minimum for $\sqrt{s'}$ of 10 GeV

m_t (GeV)	M_H (Gev)	Γ_Z (MeV)	σ_{max} (nb)	$\sqrt{s}_{max}$ (GeV)	$\sqrt{s}_-$ (GeV)	$\sqrt{s}_+$ (GeV)
60	100	2460	30.652	91.091	89.820	92.637
90	10	2463	30.653	91.092	89.819	92.639
90	100	2466	30.660	91.092	89.817	92.641
90	1000	2458	30.658	91.091	89.821	92.635
150	100	2479	30.699	91.092	89.811	92.649
200	100	2493	30.761	91.093	89.805	92.659
230	10	2501	30.818	91.093	89.801	92.665
230	100	2504	30.817	91.093	89.800	92.667
230	1000	2497	30.796	91.093	89.803	92.662

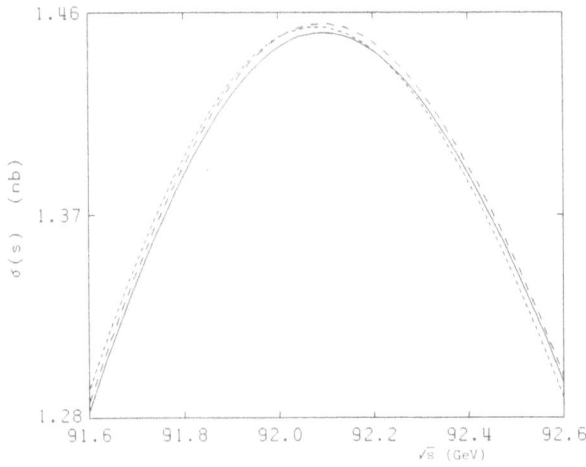

Fig. 1

3.3. Analytical Results for the Line Shape

In order to obtain approximate analytical results for the line shape
we successively make the following approximations. Related approaches to
the Z line shape are given in refs. 36, 24 and 37.

Firstly, in the integrand of eq. (3.3) we take for $\sigma(s')$ the
approximation of subsection 2.4 and introduce $\tilde{M}$ and $\tilde{\Gamma}$. One finds with
$s' = s(1-x)$

$$\sigma(x) = \left\{ \frac{s(1-x)(C_R+C_I) - (\tilde{M}_Z^2+\tilde{\Gamma}_Z^2)C_I}{s^2[(1-x-\tilde{M}_Z^2/s)^2 + \tilde{M}_Z^2\tilde{\Gamma}_Z^2/s^2]} + \frac{C_Q}{s(1-x)} \right\}(1 + \delta_{QCD}), \quad (3.13)$$

where

$$C_R = \frac{12\pi \; \Gamma^e_\Gamma{}^f}{M_Z^2(1+\gamma^2)} \quad , \quad C_I = \frac{IN_c}{1+\gamma^2} \qquad (3.14)$$

$$C_Q = 4\pi \; Q_f^2\alpha^2(M_Z^2)N_c/3 \; . \qquad (3.15)$$

The partial widths (also in I) contain electroweak corrections, but not the QED and QCD corrections. The latter is included as overall factor in eq. (3.13).

Secondly, the function G in eq. (3.4) is taken in the first order exponentiated form, which means that δ_2^{V+s} and δ_2^H are omitted (cf. fig. 1). For the convolution of the pure QED term even the exponentiation can be omitted and the first order result[38] is used.

Thirdly, the integration region for the exponential part of G is extended to $(0,\infty)$, for the other part of G $(0,1)$ is used and for the QED part the original integration region is kept. Introducing partly the same procedure as ref. 36 which is inspired by the J/ψ line shape[39] one has the integrals

$$J_\beta = \beta \int_0^\infty dx \; \frac{x^{\beta-1}}{x^2-2\eta x \; \cos\zeta + \eta^2} = \eta^{\beta-2} \; \phi(\cos\zeta,\beta) \; , \qquad (3.16)$$

where

$$\phi(\cos\zeta,\beta) = \frac{\pi\beta \; \sin[(1-\beta)\zeta]}{\sin\pi\beta \; \sin\zeta} \; , \qquad (3.17)$$

with

$$\eta = \sqrt{a^2+b^2}, \quad \cos\zeta = a/\eta \; , \qquad (3.18)$$

$$a = \tilde{M}_Z^2/s-1 \; , \quad b = \tilde{M}_Z\tilde{\Gamma}_Z/s \; , \qquad (3.19)$$

and from the δ_1^H part

$$J_1 = \int_0^1 dx \; \frac{2-x}{(x+a)^2 + b^2} = \frac{2+a}{b} A - \frac{B}{2} \; , \qquad (3.20)$$

$$J_2 = \int_0^1 dx \; \frac{x(2-x)}{(x+a)^2 + b^2} = -\frac{a^2+2a-b^2}{b} A + (1+a)B = 1 \; , \qquad (3.21)$$

with

$$A = \arctan\frac{a+1}{b} - \arctan\frac{a}{b} \; , \qquad (3.22)$$

$$B = \ln\frac{2a+1+a^2+b^2}{a^2+b^2} \; . \qquad (3.23)$$

The analytic approximate cross section is

$$\sigma(s) = \left\{ \left(\frac{C_R + C_I}{s} - \frac{\tilde{M}_Z^2 + \tilde{\Gamma}_Z^2}{s^2} C_I \right) \left(\eta^{\beta-2} \phi(\cos\zeta, \beta)(1 + \delta_1^{v+s}) - \frac{\beta}{2} J_1 \right) \right.$$

$$- \frac{C_R + C_I}{s} \left(\frac{\beta}{\beta+1} \eta^{\beta-1} \phi(\cos\zeta, \beta+1)(1 + \delta_1^{v+s}) - \frac{\beta}{2} J_2 \right)$$

$$+ \frac{C_Q}{s} \left[1 + \frac{\alpha}{\pi} L \left(2\ln(1 - \frac{s_m}{s}) - \ln \frac{s_m}{s} + \frac{1}{2} + \frac{s_m}{s} \right) \right.$$

$$\left. \left. + \frac{\alpha}{\pi} \left(\frac{\pi^2}{3} - 1 - 2\ln(1 - \frac{s_m}{s}) + \ln \frac{s_m}{s} - \frac{s_m}{s} \right) \right] \right\} (1 + \delta_{QCD}) , \quad (3.24)$$

where $s_m > 4m_f^2$.

The expression (3.23) is within 0.4% a good approximation to the line shape in the region $(M_Z - 3\Gamma_Z, M_Z + \Gamma_Z)$. It applies to the lepton and quark channels. For the lepton channels we take $\sqrt{s_m}$ to be 200 MeV, for the hadrons 10 GeV.

Essential features of the line shape after the convolution (3.3) are the reduction ρ of the peak height and the shift $\Delta\sqrt{s_{max}}$ of the peak position. From the approximate analytical expression (3.24) one derives

$$\rho = \left(\frac{\Gamma_Z}{M_Z} \right)^{\beta} (1 + \delta_1^{v+s}) \quad (3.25)$$

and

$$\Delta\sqrt{s_{max}} = \frac{\beta\pi}{8} \Gamma_Z \quad (3.26)$$

or in terms of the peak position after the convolution

$$\sqrt{s_{max}} = M_Z + \frac{\beta\pi}{8} \Gamma_Z - \frac{1}{4} \gamma^2 M_Z . \quad (3.27)$$

4. THE INFLUENCE OF CUTS

4.1. Outline of the Section

The question which is pursued in this section is whether cuts applied to the produced fermions mainly affect the size of the total cross section and not so much the shape. If this were the case then the experimental line shape could be fitted with either the exact expression (3.3) or the approximate formula (3.24) leaving an overall scale factor adjustable.

A cut on s' can be dealt with rather easily even if we take into account final state radiation, since a semi-analytical approach is possible. This is done in section 4.2. Other cuts like on the acollinearity angle or the production angles of the muons are not easily studied in a semi-analytic fashion. One has to rely on an event generator which is done in section 4.3. All results in this section are obtained for $M_Z=92$, $M_H=100$ and $m_t=60$ GeV.

4.2. A Cut on the Invariant Mass of the Fermion Pair

Suppose a cut on the invariant mass $\sqrt{s'}$ of the muon pair is applied e.g.

$$s' > s_{min} . \tag{4.1}$$

For the initial state radiation it means that z_0 in eq. (3.3) now reads

$$z_0 = s_{min}/s . \tag{4.2}$$

For the final state radiation we do not have anymore the expression (2.22), but a result derived from integrating the final state radiation spectrum [38]

$$\frac{d\sigma}{ds'} = \frac{\sigma(s)}{s} \left\{ \delta(1-z) + \frac{\alpha}{\pi} \left(\delta(1-z)[2(L_\mu-1)\ln\epsilon + \frac{3}{2} L_\mu + 2\zeta(2) - 2] \right. \right.$$

$$\left. \left. + \theta(1-z-\epsilon)[2(L_\mu-1+\ln z) \cdot (\frac{1}{1-z} - \frac{1+z}{2})] \right) \right\} , \tag{4.3}$$

where

$$L_\mu = \ln \frac{s'}{m_\mu^2} . \tag{4.4}$$

The $\frac{\alpha}{\pi} \delta(1-z)$ term represents the effect of virtual and soft photon corrections, the $\theta(1-z-\epsilon)$ originates from the hard bremsstrahlung. The invariant mass of the fermion pair would be $\sqrt{s}$ without final state radiation but becomes $\sqrt{s'}$ when a photon is radiated from the final state. Upon integration over the available s' range one finds

$$C_1(s_{min},s) = 1 + \delta_{QED}(s_{min},s) = \frac{1}{\sigma_0(s)} \int_{s_{min}}^{s} ds' \frac{d\sigma}{ds'} =$$

$$= 1 + \delta_1^{V+S}(m_\mu) + \beta_f \ln(1-z_0) + Y , \tag{4.5}$$

where

$$\beta_f = \frac{2\alpha}{\pi} (L_\mu-1) , \tag{4.6}$$

$$Y = - \frac{\beta_f}{2} (\frac{3}{2} - z_0 - \frac{1}{2} z_0^2) + \frac{2\alpha}{\pi} \left[-Li_2(1-z_0) \right.$$

$$\left. + \frac{5}{8} + \frac{z_0}{2} (1 + \frac{z_0}{2}) \ln z_0 - \frac{z_0}{2} (1 + \frac{z_0}{4}) \right] , \tag{4.7}$$

and $\delta_1^{V+S}(m_\mu)$ is given by (3.8) with m_e replaced by m_μ. One sees that for $s_{min} = 4m_\mu^2$ the L_μ terms cancel and one recovers eq. (2.22) since $m_\mu^2/s \simeq 0$.

For a very stringent cut the first order result (4.5) is not realistic since it becomes large and negative. One should then use a first order exponentiated spectrum

$$\frac{d\sigma}{ds'} = \frac{\sigma(s)}{s} [\beta_f(1-z)^{\beta_f-1} (1 + \delta_1^{V+S}(m_\mu)) + \delta_1^H] \tag{4.8}$$

with

Table 11. Effect of invariant mass cut on mupair production. For every s_{min} three cases are given, the first line gives a cut without final state radiation, the second line with first order final state radiation and the third line with first order exponentiated final state radiation. The masses are $M_Z=92$, $M_H=100$ and $M_t=60$ GeV. Units of $\sqrt{s}$ in GeV, σ_{max} in nb.

$\sqrt{s}_{min}$	σ_{max}	$\sqrt{s}_{max}$	$\sqrt{s}_-$	$\sqrt{s}_+$	$\sqrt{s}_+ - \sqrt{s}_-$
1	1.451	92.094	90.756	93.722	2.966
1	1.453	92.094	90.756	93.722	2.966
1	1.453	92.094	90.756	93.722	2.966
45	1.446	92.094	90.761	93.715	2.954
45	1.438	92.094	90.761	93.716	2.955
45	1.438	92.094	90.761	93.716	2.955
82	1.443	92.094	90.764	93.712	2.948
82	1.364	92.098	90.772	93.720	2.948
82	1.363	92.098	90.771	93.720	2.949

Table 12. Effect of invariant mass cut on hadron production. Same conventions as in table 11.

$\sqrt{s}_{min}$	σ_{max}	$\sqrt{s}_{max}$	$\sqrt{s}_-$	$\sqrt{s}_+$	$\sqrt{s}_+ - \sqrt{s}_-$
10	30.043	92.095	90.772	93.705	2.933
10	30.053	92.095	90.772	93.705	2.933
10	30.053	92.095	90.772	93.705	2.933
45	30.034	92.095	90.772	93.705	2.933
45	29.997	92.095	90.773	93.705	2.932
45	29.995	92.095	90.773	93.705	2.932
82	30.010	92.095	90.774	93.704	2.930
82	29.612	92.096	90.775	93.705	2.930
82	29.611	92.096	90.775	93.705	2.930

$$\delta_1^H = \frac{\alpha}{\pi} \left\{ -(L_\mu +1)(1+z) + 2\ln z \left[\frac{1}{1-z} - \frac{1+z}{2} \right] \right\} . \qquad (4.9)$$

This leads to a correction factor

$$C(s_{min},s) = (1-z_o)^{\beta_f}(1 + \delta_1^{V+S}(m_\mu)) + Y . \qquad (4.10)$$

When the final state photonic correction should be taken into account one makes the following replacement in the convolution (3.3)

$$\sigma(sz) = \sigma(s') + \sigma(s')C(s_{min},s') \qquad\qquad (4.11)$$

In this way one still has a semi-analytical result since there is only one integration which has to be carried out.

The effects of a cut on s' are displayed in tables 11 and 12. The size of the resonance shape is mainly affected and not so much the shape. The apparent increase of σ_{max} for a cut of 1 GeV in the mupair case is an artifact due to the absence of any δ_{QED} of the final state in the first line and to the presence of it in the second line (i.e. a correction of about 0.17%). For the mupair case the line shape is somewhat displaced to the right for a cut of 82 GeV and it becomes narrower. It is questionable whether this effect can be measured. For hadrons the results are given for completeness. Since they depend on the quark masses (actually the values (1.1) are used), there is a rather crucial theoretical uncertainty in these results. When taken at face-value the hadronic line shape seems to be even less affected by these cuts than the muon line shape.

4.3. Cuts on generated events

In this section cuts on the production angle θ of the muons (i.e. angle with respect to the beam) and on the acollinearity angle ζ between the muons are studied. Since these cuts cannot easily be incorporated in a semi-analytical approach an event generator is used. The event generator[40] in the version used here contains first and second order initial state corrections.

The weak corrections are treated in exactly the same way as in the program ZSHAPE[21]. When we switch off the exponentiation i.e. use G(z) in the convolution (3.3) without soft photon resummation the two programs should give the same line shape, also when a cut on s' is applied. This is shown

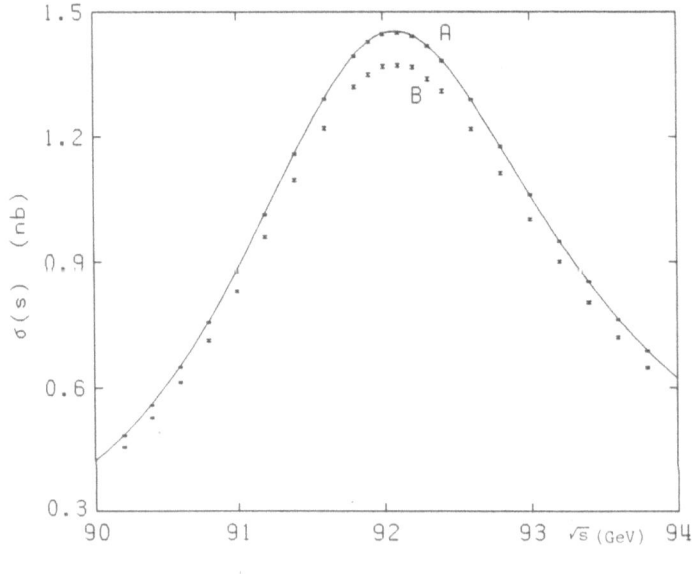

Fig. 2

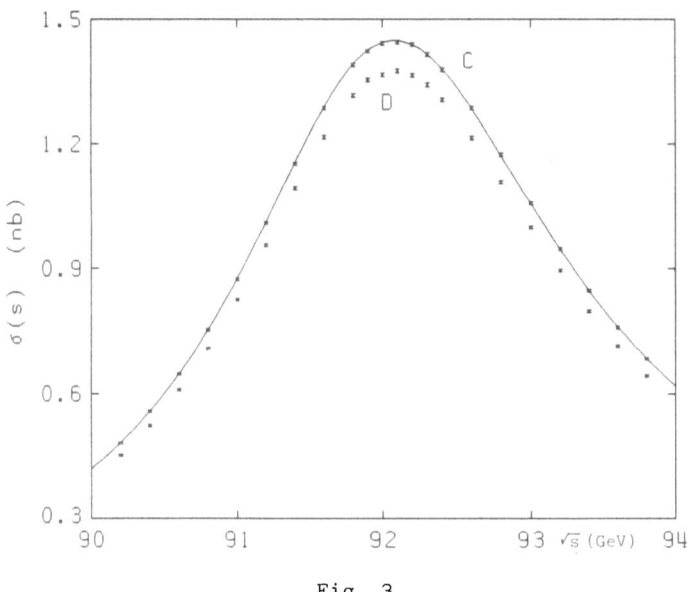

Fig. 3

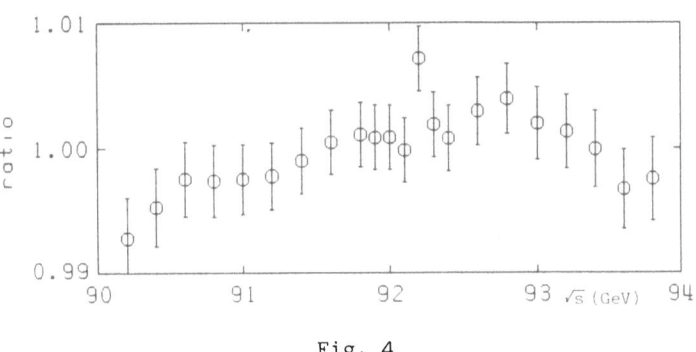

Fig. 4

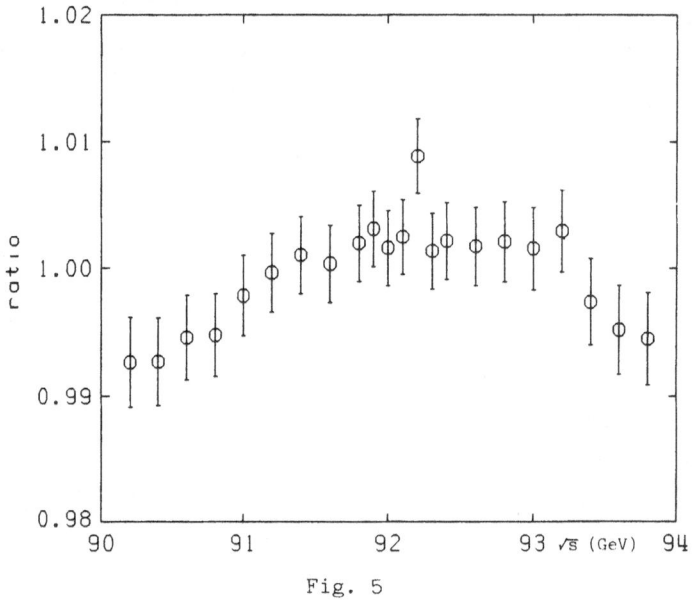

Fig. 5

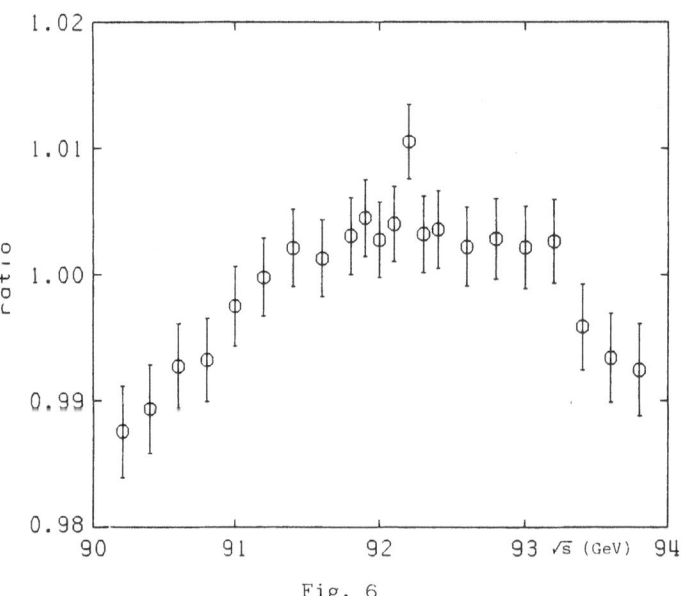

Fig. 6

in figures 2 and 3. In fig. 2 the line A is the line shape from the semi-analytic calculation and $\sqrt{s_{min}}$ = 1 GeV. The event generator points lie along this curve when also $\sqrt{s_{min}}$ = 1 GeV is imposed. When on these events the additional requirement is imposed that θ is in the interval $(15^0, 165^0)$ one obtains set B. In fig. 3 the line C shows the results of the semi-analytical calculation and event generator when $\sqrt{s_{min}}$ = 45 GeV. The set D consists of events which satisfy $\sqrt{s_{min}}$ = 1 GeV, $15^0<\theta<165^0$ and $\zeta<10^0$. The cuts imposed on sets B and D decrease the size of the cross section. In order to assess a possible change of the shape the cross section points from the sets C, B and D are divided by scaled down ZSHAPE results for a total cross section with almost no cut i.e. $\sqrt{s_{min}}$ = 1 GeV. The ratios are displayed in figs. 4, 5 and 6. The case of fig. 4 is similar to the case of table 11 with a cut of $\sqrt{s_{min}}$ = 45 GeV. Since it corresponds to a line shape which is 12 MeV narrower we know what type of effect fig. 4 corresponds to. In fig. 5 the muon angles obey $15^0<\theta<165^0$ (set B) and in fig. 6 the muons (set D) also satisfy the additional requirement of being back to back within 10^0. In both cases the line shape becomes narrower in particular when both the θ and ζ restrictions are imposed. On the other hand, moderate cuts on the invariant mass of the fermion pair leave the line shape unaffected.

ACKNOWLEDGEMENT

I gratefully acknowledge the fruitful collaboration with W. Beenakker, G. Burgers, W. Hollik, S. van der Marck and W.L. van Neerven. Moreover, the efforts of D. Bardin, T. Riemann and M. Sachwitz contributed greatly to obtain a consensus on the numerical results for the electroweak corrections. The results from the event generator were kindly provided by S. van der Marck.

REFERENCES

1. F.A. Berends, G. Burgers, W. Hollik and W.L. van Neerven, Phys. Lett. B203:177 (1988).
2. G. Burgers in: Polarisation at LEP, CERN 88-06, p. 121, ed. by G. Alexander, G. Altarelli, A. Blondel, G. Coignet, E. Keil, D.E. Plane and D. Treille.
3. F.A. Berends in: Radiative Corrections in e^+e^- Collisions, ed. J.H. Kühn, Springer, Berlin, 1989.
   T. Riemann, D. Bardin, M. Bilenky, M. Sachwitz, ibidem.
4. D. Bardin, W. Beenakker, F.A. Berends, M. Bilenky, G. Burgers, W.Hollik, T. Riemann, M. Sachwitz, S. van der Marck, W.L. van Neerven, contribution to the LEP workshop 1989, to be published as CERN report.
5. A. Sirlin, Phys. Rev. D22:2695 (1980).
6. W. Hollik, DESY 88-188 (1988), to be published in Fortschritte Physik.
7. F.A. Berends and A. Böhm in: High Energy Electron-Positron Physics, ed. A. Ali and P. Söding, World Scientific, Singapore (1988).
8. W. de Boer, 10th Warsaw Symposium on Elementary Particles, Kazimierz (1987).
9. T.H. Chang, K.J.F. Gaemers and W.L. van Neerven, Nucl. Phys. B202:407 (1982).
10. P. Kalyniak, J.N. Ng and P. Zakarauskas, Phys. Rev. D29:502 (1984).
11. P. Kalyniak, J.N. Ng and P. Zakarauskas, Phys. Rev. D30:123 (1984).
12. E. Franco in Physics at LEP, CERN 86-02, p. 187, ed. by J. Ellis and R. Peccei.
13. F.A. Berends and R. Kleiss, Nucl. Phys. B260:32 (1985).
14. W. Beenakker and W. Hollik, Z. Phys. C40:141 (1988).
15. A.A. Akhundov, D.Y. Bardin and T. Riemann, Nucl. Phys. B276:1 (1986).
16. D.C. Kennedy, B.W. Lynn, C.J.-C Im and R.G. Stuart, SLAC-PUB 4128 (1988).
17. W. Wetzel, Nucl. Phys. B227:1 (1983); CERN 86-02:40 (1986).

18. L. Baulieu and R. Coquereaux, An.. Phys. 140:163 (1982).
19. R.G. Stuart, Rutherford Appleton Lab report RAL T008 (1985).
20. D.Y. Bardin, A. Leike, T. Riemann and M. Sachwitz, Phys. Lett. B206:539 (1988).
21. ZSHAPE, authors W. Beenakker, F.A.Berends and S. van der Marck.
22. W. Beenakker and W. Hollik, ECFA Workshop on LEP 200, CERN 87-08:185, ed. A. Böhm and W. Hoogland.
23. D.Y. Bardin, M.S. Bilenky, P. Christova, T. Riemann, M. Sachwitz and H. Vogt, Berlin-Zeuthen preprint PHE 89-09, submitted to Comp. Phys. Comm.
24. A. Borelli, M. Consoli, L. Maiani and R. Sisto, preprint Univ. of Rome.
25. T. Kinoshita, J. Math. Phys. 3:650 (1962);
    T.O. Lee and M. Nauenberg, Phys. Rev. 133 :B1549 (1964).
26. F.A. Berends, R. Kleiss and S. Jadach, Nucl. Phys. B202:63 (1982).
27. D.Y. Bardin, O.M. Fedorenko, T. Riemann, Dubna preprint E2-87-663;
    D.Y. Bardin, M.S. Bilenky, O.M. Fedorenko and T. Riemann, Dubna preprint E2-88-324.
28. S. Jadach, J.H. Kühn, R.G. Stuart and Z. Was, Z. Phys. C38:609 (1988).
29. F.A. Berends, G. Burgers and W.L. van Neerven, Nucl. Phys. B297:429 1988); E B304:921 (1988).
30. R. Barbieri, J.A. Mignaco and E. Remiddi, Nuovo Cim. 11A:824,865 (1972).
31. A. Devoto and D.W. Duke, Riv. Nuovo Cim. 7, No. 6 (1984).
32. E.A. Kuraev and V.S. Fadin, Sov. J. Nucl. Phys. 41:466 (1985).
33. G. Altarelli and G. Martinelli, Yellow Report CERN 86-02:47 (1986).
34. O. Nicrosini and L. Trentadue, Phys. Lett. B196:551 (1987).
35. V.S. Fadin and V.S. Khoze, Novosibirsk preprint 87-157.
36. R.N. Cahn, Phys. Rev. D36:2666 (1987).
37. F. Aversa and M. Greco, LNF-89/025(PT).
38. F.A. Berends and R. Kleiss, Nucl. Phys. B177:237 (1981).
39. D.R. Yennie, Phys. Rev. Lett. 34:239 (1975);
    J.D. Jackson and D.L. Scharre, Nucl. Instr. 128:13 (1975);
    M. Greco, G. Pancheri-Srivastava and Y. Srivastava, Nucl. Phys. B101: 234 (1975);
    F.A. Berends and G.J. Komen, Nucl. Phys. B115:114 (1976).
40. R. Kleiss and S. van der Marck, to be published;
    S. van der Marck, these Proceedings.

IMPLICATIONS ON THE ELECTROWEAK PARAMETERS OF A PRECISE MEASUREMENT OF THE Z MASS

A. Blondel

CERN, Geneva, Switzerland

Abstract

The implications of the presently most precise low energy neutral current measurements on the W mass and Z peak observables are investigated in the framework of the $SU(2)_L \times U(1)$ Standard Model, where the free parameters are the top quark mass m_t, the Higgs boson mass m_H, and the ρ parameter. It is found that, once m_Z is precisely measured,

i) the W mass is predicted to ± 315 Mev;

ii) the coupling constants of light fermions $(e, \mu, \tau, u, d, s, c)$ to the Z are predicted with a precision equivalent to knowing the effective weak mixing angle $sin^2 \vartheta_w^* \equiv \frac{e^2}{g^2}(m_Z^2)$ to ± 0.002.

iii) the decay width of the Z into $b\bar{b}$, and thereby the total width, do not exhibit the degeneracy between the effects of large values of m_t and departure of ρ_0 from 1 which is the rule for other observables. A precise measurement of these quantities could be used to set limits on m_t and ρ_0 independently.

Numerical predictions as a function of the Z mass are given.

1 Introduction: $sin^2 \vartheta_w$ and $sin^2 \vartheta_w^*$

The measurement of the Z mass with a precision of $\Delta m_Z = \pm 50$Mev or better will be one of the first outcomes of LEP/SLC. Combined with the QED coupling constant α and the muon decay constant G_μ, it can be related to the electroweak mixing angle in its most common definition, $sin^2 \vartheta_w \equiv 1 - \frac{m_W^2}{m_Z^2}$, by the relation:

$$m_Z^2 = \frac{\pi \alpha}{\sqrt{2} G_\mu sin^2 \vartheta_w (1 - sin^2 \vartheta_w)(1 - \Delta r)} \qquad (1)$$

where Δr is the usual 'radiative correction parameter' [1]. In the Minimal Standard Model (MSM) with only one doublet of Higgs bosons $(\rho = 1)$, three families of quarks and leptons, and if the masses of the top quark and of the Higgs boson were known,

Δr can be calculated with a small error, $\Delta(\Delta r) = \pm 0.0009$ [2]; this turns out to dominate the error in extracting $sin^2\vartheta_w$ from equation (1), $\Delta sin^2\vartheta_w = \pm 0.0003$. However we do not know m_t and m_H, and have no experimental evidence for the Higgs mechanism, so that $sin^2\vartheta_w$ is simply a free parameter.

In the following the symbol $\rho_0 \equiv 1 + \Delta\rho$ will be used to describe not only the tree level effect, due for instance to triplets of Higgs bosons, but also radiative effects that could occur from any new $SU(2)$-breaking set of particles: further families of fermions or supersymmetric particles with large isotopic splitting would produce similar effects [3]. It is worth noting that the tree level effect can result in $\Delta\rho \leq 0$, while radiative effects lead to $\Delta\rho \geq 0$. The resulting contribution to Δr is $-cotg^2\vartheta_w\Delta\rho$.

It has been shown by Lynn et al. [3] that there is not a one-to-one relation between the Neutral Current (NC) coupling constants and $sin^2\vartheta_w$ when varying m_t and m_H or ρ_0. It has been long noticed [4], and recently emphasized [5] that NC couplings at the Z can be extracted from improved Born approximation formulae, the so-called effective lagrangian, including a redefinition of a running $sin^2\vartheta_w$, denoted $sin^2\vartheta_w^*$

$$sin^2\vartheta_w^* \equiv \frac{e_*^2}{g_*^2}(-m_Z^2) \tag{2}$$

where e_*^2 and g_*^2 are the running electron charge and weak coupling respectively, as defined by Lynn et al. [5].

A large class of radiative effects to Z peak observables, i.e. the renormalization of couplings and masses, (also called *oblique corrections*), can be absorbed in this parameter. In the case of four-fermion processes involving light quarks and leptons, e, μ, τ, u, d, s, c, which belong to complete doublets of $SU(2)_L \times U(1)$, these oblique corrections contain most of the dependence on the unknown parameters of the model. For these light fermions, photonic or gluonic effects as well as purely weak boxes and vertices are large, but in principle well calculable and independent on unknown parameters. This is not the case for the b quark, which couples directly to the top quark. The couplings of the b have a specific m_t dependence [6] which could lead to surprising effects, discussed in section 2.4.

It seems that $sin^2\vartheta_w^*$ is the relevant parameter for the interpretation of the Z partial widths, interference effects and asymmetries [8] involving light fermions. Measurements of these observables can be related to $sin^2\vartheta_w^*$ without assumptions on unknown parameters, but only assuming the algebraic structure of $SU(2) \times U(1)$ to be correct [7]. Once m_Z is measured, the variation of $sin^2\vartheta_w^*$ with m_t, m_H and ρ_0 differs from that of $sin^2\vartheta_w$. It has been investigated numerically using the program EXPOSTAR [5] [8].

The results are shown in table 1 and drawn in figure 1. It is remarkable that the dependence of $sin^2\vartheta_w^*$ on unknown parameters is smaller than that of $sin^2\vartheta_w$ in all cases, by a factor of 2 for the Higgs mass, and by a factor of 3.3 for heavy top masses and ρ_0. In fact, if one defines a 'radiative correction parameter' for Z observables, Δr_*, in a way similar to Δr,

$$m_Z^2 = \frac{\pi\alpha}{\sqrt{2}G_\mu sin^2\vartheta_w^*(1 - sin^2\vartheta_w^*)(1 - \Delta r_*)} \tag{3}$$

the contribution of $\Delta\rho$ to Δr_* is only $-\Delta\rho$, smaller than the contribution to Δr by a factor $cotg^2\vartheta_w = 3.3$.

Table 1. Values of $sin^2\vartheta_w$ and $sin^2\vartheta_w^*$ for various values of m_Z, m_H and ρ_0, calculated with EXPOSTAR [8]

m_Z	ρ_0	m_H	m_t	$sin^2\vartheta_w \equiv 1 - \frac{m_W^2}{m_Z^2}$	$sin^2\vartheta_w^*(-m_Z^2)$
91	1	10	30	0.23497	0.23438
			60	0.23533	0.23406
			90	0.23148	0.23343
			120	0.22821	0.23266
			150	0.22467	0.23173
			200	0.21765	0.22980
91	1	100	30	0.23657	0.23529
			60	0.23692	0.23497
			90	0.23307	0.23434
			120	0.22979	0.23357
			150	0.22625	0.23263
			200	0.21980	0.23069
91	1	1000	30	0.23980	0.23659
			60	0.24014	0.23626
			90	0.23627	0.23562
			120	0.23298	0.23483
			150	0.22942	0.23388
			200	0.22235	0.23191
93	1	100	60	0.22318	0.22110
93	0.98	100	60	0.24507	0.22751
93	0.98	100	230	0.22327	0.22220

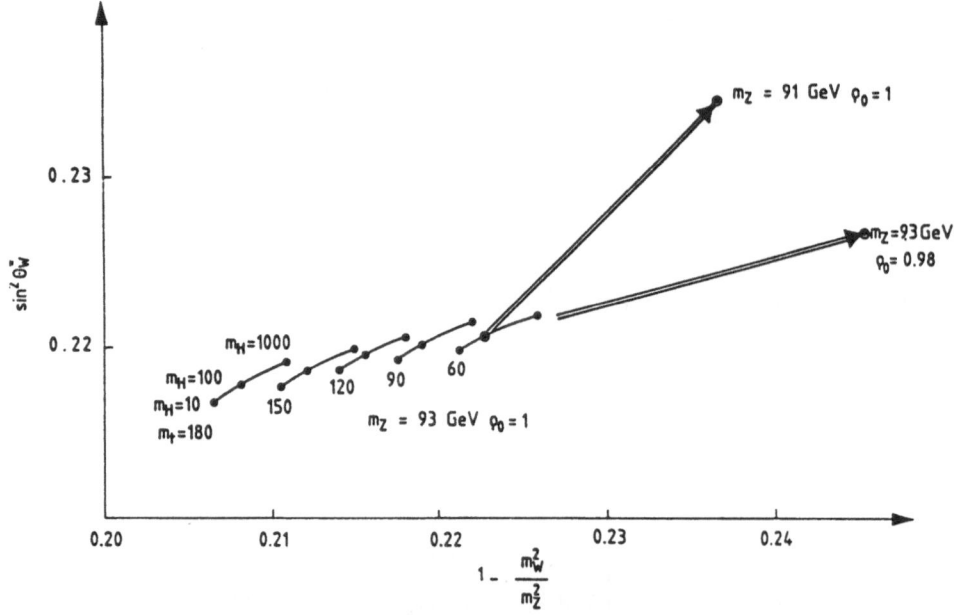

Figure 1. The relation between $sin^2\vartheta_w$ and $sin^2\vartheta_w^*$ when varying m_Z, m_t, m_H and ρ_0. The variations upon the various parameters are practically independent of each other and have been represented only once.

We show here that the most precise measurements of $sin^2\vartheta_w$ presently available project very differently on $sin^2\vartheta_w$ and $sin^2\vartheta_w^*$ once m_Z is known, leading to predictions for the asymmetries at the Z peak that are much more precise than naively expected from the present error on $sin^2\vartheta_w$ $\Delta sin^2\vartheta_w = \pm 0.006$, which will not change once m_Z is known.

2 Implications of lower energy experiments

2.1 Electroweak parameters from neutrino nucleon scattering

Neutrino-nucleon scattering experiments provide at present the most precise measurement of $sin^2\vartheta_w$. Only the two most precise experiments, CDHS [9] and CHARM [10], will be taken into account here, but it has been shown [11] that the result is not significantly changed when using all available measurements.

It has been shown by Stuart [12] that the ratio R_ν of neutrino Neutral Current (NC) to Charged Current (CC) cross-sections on isoscalar nuclei is a measurement of $sin^2\vartheta_w$ with very little dependence on the top and Higgs masses. The same argument holds for ρ_0.

The NC and CC cross-sections written at tree level can be interpreted in the presence of radiative effects by using the effective lagrangian notations. In order to do this properly, one has to keep the relation between vector boson masses and the weak couplings free, allowing for unknown radiative effects. At tree level $sin^2\vartheta_w$ is defined as a ratio of couplings, and in order to keep this in mind it will be written here $sin^2\vartheta_w^0$. One has:

$$\sigma^{\nu}_{NC} \propto \left(\frac{e^2}{sin^2\vartheta^0_w cos^2\vartheta^0_w} \right)^2 \left(\frac{1}{m^4_Z} \right) \left(\mathcal{L}g^2_L + \mathcal{R}g^2_R \right)$$

$$\sigma^{\nu}_{CC} \propto \left(\frac{e^2}{sin^2\vartheta^0_w} \right)^2 \left(\frac{1}{m^4_W} \right) (\mathcal{L})$$

$$\sigma^{\bar{\nu}}_{NC} \propto \left(\frac{e^2}{sin^2\vartheta^0_w cos^2\vartheta^0_w} \right)^2 \left(\frac{1}{m^4_Z} \right) \left(\mathcal{L}g^2_R + \mathcal{R}g^2_L \right)$$

$$\sigma^{\bar{\nu}}_{CC} \propto \left(\frac{e^2}{sin^2\vartheta^0_w} \right)^2 \left(\frac{1}{m^4_W} \right) (\mathcal{R}), \tag{4}$$

where

$$\begin{aligned}
g^2_{L,R} &= g^2_{uL,R} + g^2_{dL,R} \\
g_{uL} &= \frac{1}{2} - \frac{2}{3}sin^2\vartheta^0_w \\
g_{dL} &= -\frac{1}{2} + \frac{1}{3}sin^2\vartheta^0_w \\
g_{uR} &= -\frac{2}{3}sin^2\vartheta^0_w \\
g_{dR} &= \frac{1}{3}sin^2\vartheta^0_w
\end{aligned} \tag{5}$$

In the quark-parton model, the quantities $\mathcal{L}$ and $\mathcal{R}$ are given by integrals over the kinematic variables of the quark distributions:

$$\begin{aligned}
\mathcal{L} &= (U + D) + (\overline{U} + \overline{D})(1 - y)^2 \\
\mathcal{R} &= (U + D)(1 - y)^2 + (\overline{U} + \overline{D}).
\end{aligned}$$

However, the validity of the equations (4) does not depend on the structure of the target provided it is isoscalar [13]. The dependence of these cross-sections on m_W and m_Z comes from the W and Z propagators. From this one can derive:

$$\begin{aligned}
R_\nu &= \frac{\sigma^\nu_{NC}}{\sigma^\nu_{CC}} = \frac{m^4_W}{m^4_Z cos^4\vartheta^0_w} \left(g^2_L + rg^2_R \right) \\
&= \frac{m^4_W}{m^4_Z cos^4\vartheta^0_w} \left(1/2 - sin^2\vartheta^0_w + 5/9(1 + r)sin^4\vartheta^0_w \right)
\end{aligned} \tag{6}$$

$$\begin{aligned}
R_{\bar{\nu}} &= \frac{\sigma^{\bar{\nu}}_{NC}}{\sigma^{\bar{\nu}}_{CC}} = \frac{m^4_W}{m^4_Z cos^4\vartheta^0_w} \left(g^2_L + \bar{r}g^2_R \right) \\
&= \frac{m^4_W}{m^4_Z cos^4\vartheta^0_w} \left(1/2 - sin^2\vartheta^0_w + 5/9(1 + \bar{r})sin^4\vartheta^0_w \right)
\end{aligned} \tag{7}$$

with

$$\begin{aligned}
r &= \frac{\mathcal{R}}{\mathcal{L}} = \frac{\sigma^{\bar{\nu}}_{CC}}{\sigma^\nu_{CC}} \\
\bar{r} &= 1/r
\end{aligned} \tag{8}$$

It is possible to re-write the equations (6) and (7) either as functions of ρ and $sin^2\vartheta^0_w$ or as function of $\frac{m^2_W}{m^2_Z}$ and $sin^2\vartheta^0_w$:

$$R_\nu = \rho^2 \left(1/2 - sin^2\vartheta^0_w + 5/9(1 + r)sin^4\vartheta^0_w \right) \tag{9}$$

Table 2. Measured values of R_ν, $R_{\bar\nu}$ and r and values of R_ν^* and $R_{\bar\nu}^*$ extracted by applying corrections for non-isoscalarity in the quark-parton model, photonic effects, weak boxes and vertices, but no oblique corrections. The errors on R_ν^*, $R_{\bar\nu}^*$ and r^* include the additional errors due to these corrections, and in particular that due to the charm-quark mass. The resulting correlations between these variables were taken into account when extracting the weak parameters.

	CDHS	CHARM
R_ν	0.3072 ± 0.0031	0.3093 ± 0.0031
R_ν^*	0.3104 ± 0.0045	0.3045 ± 0.0045
$R_{\bar\nu}$	0.375 ± 0.011	0.390 ± 0.014
$R_{\bar\nu}^*$	0.372 ± 0.012	0.388 ± 0.016
r	0.393 ± 0.016	0.456 ± 0.011
r^*	0.376 ± 0.016	0.433 ± 0.011

$$R_{\bar\nu} = \rho^2 \left(1/2 - sin^2\vartheta_w^0 + 5/9(1+\bar r)sin^4\vartheta_w^0\right) \tag{10}$$

$$R_\nu = \frac{1}{2}\frac{m_W^4}{m_Z^4}\frac{1 - 2sin^2\vartheta_w^0 + 10/9(1+r)sin^4\vartheta_w^0}{1 - 2sin^2\vartheta_w^0 + sin^4\vartheta_w^0} \tag{11}$$

$$R_{\bar\nu} = \frac{1}{2}\frac{m_W^4}{m_Z^4}\frac{1 - 2sin^2\vartheta_w^0 + 10/9(1+\bar r)sin^4\vartheta_w^0}{1 - 2sin^2\vartheta_w^0 + sin^4\vartheta_w^0} \tag{12}$$

In all these equations, $\rho = \frac{m_W^2}{m_Z^2 cos^2\vartheta_w^0}$.

It can be seen from the above equations and from the numerical value of $r \approx 0.4$ that:

- R_ν is directly related to $\frac{m_W^2}{m_Z^2}$ with very little dependence on $sin^2\vartheta_w^0$, equation (11);

- $R_{\bar\nu}$ is directly related to ρ with little dependence on $sin^2\vartheta_w^0$, equation (10).

It follows that one can extract from R_ν and $R_{\bar\nu}$ the W-Z mass ratio, or $sin^2\vartheta_w = 1 - \frac{m_W^2}{m_Z^2}$, and ρ.

Radiative effects include three classes that are calculated in one block in available radiative correction programs i) QED photonic corrections, ii) Weak boxes and vertices, iii) Renormalization of the photon, Z and W propagators and of the coupling constants (oblique corrections). The effects i) and ii) can be calculated without assumptions on unknown particles and masses, at least for R_ν and $R_{\bar\nu}$, since the direct coupling of u and d quarks to the top or heavier families and to Higgs particles is very small. The oblique radiative effects can be taken into account either by assuming a given value for $m_t, m_H, \rho_0 = 1$, which is usually done, or by expressing the result as a function of effective lagrangian quantities at the relevant $Q^2, Q_\nu^2 \approx 20 Gev^2$:

$$R_\nu^* = \rho_*^2 \left(1/2 - sin^2\vartheta_w^* + 5/9(1+r)sin^4\vartheta_w^*\right) \tag{13}$$

$$R_{\bar\nu}^* = \rho_*^2 \left(1/2 - sin^2\vartheta_w^* + 5/9(1+\bar r)sin^4\vartheta_w^*\right) \tag{14}$$

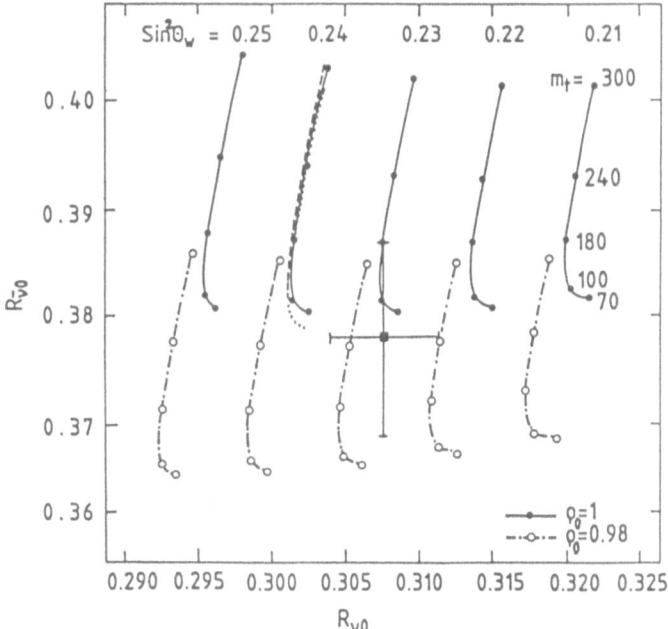

Figure 2. Relation between R_ν and $R_{\bar\nu}$ for various values of m_Z, m_t, m_H and ρ_0. For display purposes the quantity R_ν^0 has been formed: R_ν^0 is the average NC/CC ratio extracted from CDHS and CHARM data, corrected for non-isoscalarity in the quark-parton model, photonic effects and weak boxes and vertices, reduced to the kinematical conditions of the CDHS experiment. Masses in GeV. Unless otherwise specified, $m_H = 100$. The effect of changing m_H to 10 (dashed line) or 1000 (dotted line) is also shown for $sin^2\vartheta_w = 0.24$ $\rho_0 = 1$.

where R_ν^* and $R_{\bar\nu}^*$ are the experimentally measured quantities corrected only for usual experimental and quark-parton model effects, charm mass etc.. and for QED, boxes and vertices. The raw experimental results and the respective values of R_ν^* and $R_{\bar\nu}^*$ are given in table 2. The relation between R_ν and $R_{\bar\nu}$ is shown in figure 2 and compared with Standard Model predictions for fixed $sin^2\vartheta_w$. They can also be expressed as an ellipse in the $\rho_* - sin^2\vartheta_w^*$ plane as shown in figure 3. The lines of constant m_W, m_Z or $sin^2\vartheta_w$ are also drawn in this plane for various values of m_t, m_H, ρ_0. The basic degeneracy between the effects of ρ_0 and m_t is clearly apparent, whereas the effect of the Higgs mass is to thicken these lines.

2.2 Implication on the W mass

From this, one can derive first a value of the mass ratio or of $sin^2\vartheta_w \equiv 1 - \frac{m_W^2}{m_Z^2}$ which is very much independent on any assumption on m_t, m_H, ρ_0:

$$
\begin{aligned}
sin^2\vartheta_w &= 0.232 \pm 0.006 \\
\frac{m_W}{m_Z} &= 0.876 \pm 0.0035 \\
m_W &= 0.876 \cdot m_Z \pm 0.315 Mev.
\end{aligned}
\tag{15}
$$

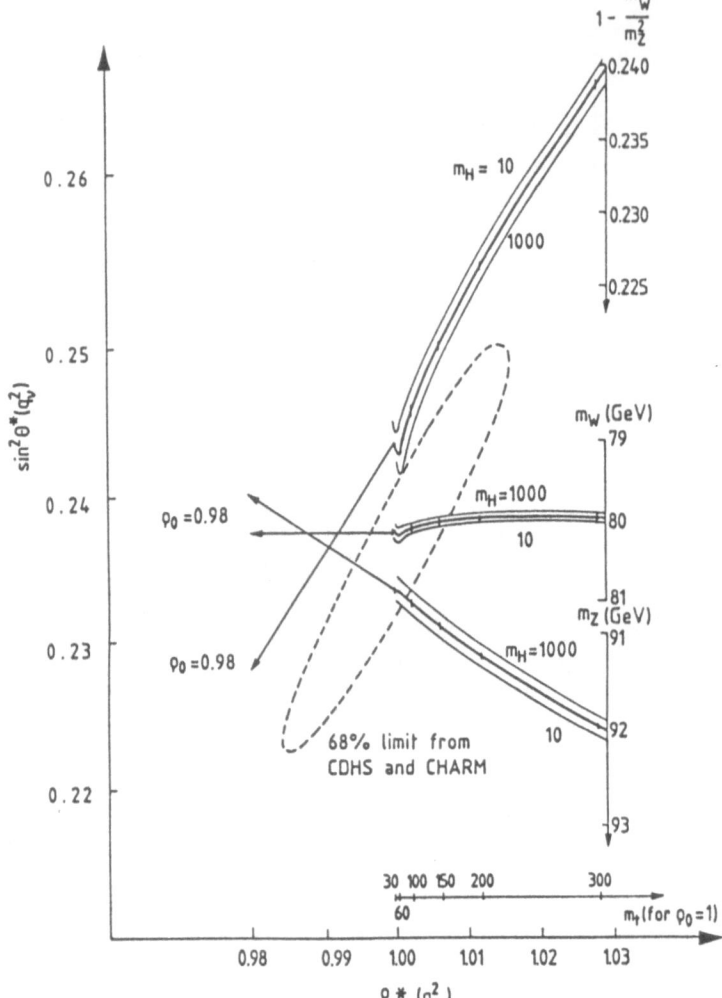

Figure 3. Result of the fit to the neutrino NC/CC ratios in the $sin^2\vartheta_w^*(Q_\nu^2) - \rho^*(Q_\nu^2)$ plane. The lines of constant m_W, m_Z, $sin^2\vartheta_w$, are indicated. Also shown is the dependence of ρ^* on m_t. Full lines correspond to $m_H = 100$ Gev.

The relation of R_ν with m_W is shown on figure 4, and the relation with m_W/m_Z on figure 5. The comparison with the direct measurements of the W and Z masses and of their ratio by the UA1 and UA2 collaborations [14] is also given. The upper limit on m_t, $m_t \leq 200$ Gev, which can be derived from figure 4, is valid only for $\rho_0 = 1$; the same is true for the limit extracted from global fits to NC data [11].

2.3   Implication on Z peak observables

One can also derive the constraint that these measurements would give on the effective couplings at the Z peak, once m_Z is known. In order to do this, one has to relate $sin^2\vartheta_w^*(Q_\nu^2)$ to $sin^2\vartheta_w^*(-m_Z^2)$. This relation turns out to be independent on

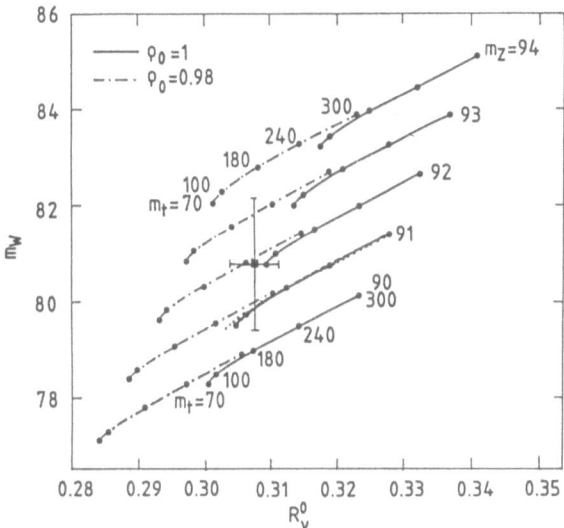

Figure 4. Relation between R_ν and m_W for various values of m_Z, m_t, m_H and ρ_0. Masses in Gev. Unless otherwise specified, $m_H = 100$. The effect of changing m_H to 10 (dashed line) or 1000 (dotted line), is also shown for $m_Z = 91$, $\rho_0 = 1$.

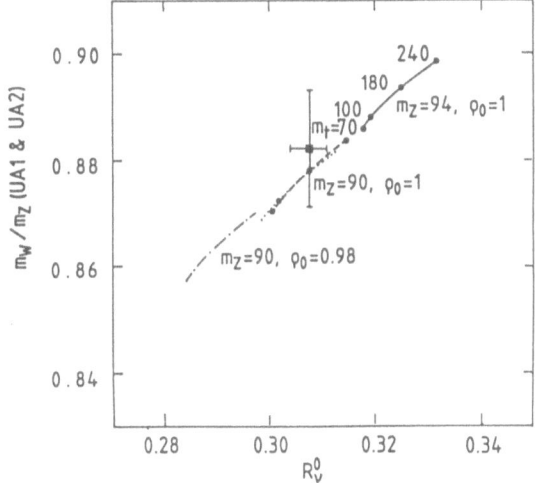

Figure 5. Relation between R_ν^0 and the mass ratio m_W/m_Z, showing the equivalence of these two quantities in the $SU(2)_L \times U(1)$ gauge structure. Masses in Gev. Unless otherwise specified, $m_H = 100$. The effect of changing m_H to 10 (dashed line) or 1000 (dotted line), is also shown for $m_Z = 90$, $\rho_0 = 1$.

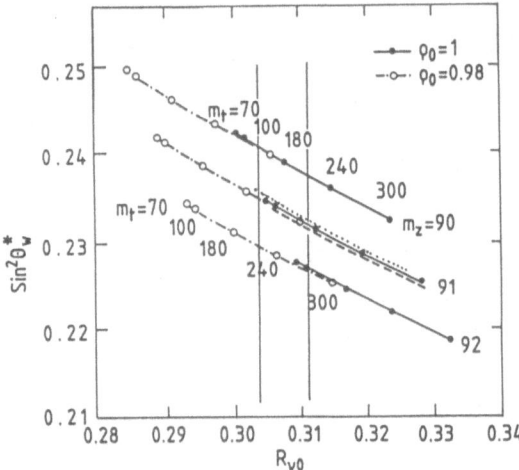

Figure 6. Relation between R_ν and $sin^2\vartheta_w^*$ for various values of m_Z, m_t, m_H and ρ_0. Masses in Gev. Unless otherwise specified, $m_H = 100$. The effect of changing m_H to 10 (dashed line) or 1000 (dotted line), is also shown for $m_Z = 91$, $\rho_0 = 1$.

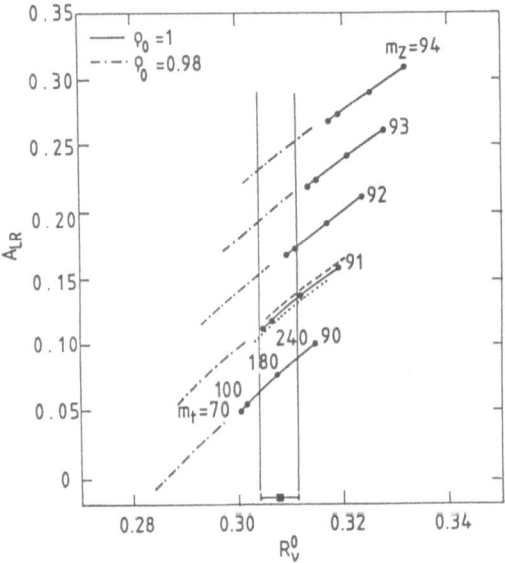

Figure 7. Relation between R_ν and A_{LR} for various values of m_Z, m_t, m_H and ρ_0. Masses in Gev. Unless otherwise specified, $m_H = 100$. The effect of changing m_H to 10 (dashed line) or 1000 (dotted line), is also shown for $m_Z = 91$, $\rho_0 = 1$.

Table 3. Modification to the Z width due to the $b\bar{b}$ vertex correction. Couples of values for m_t and ρ_0 have been chosen such that the neutral current phenomenology remains mostly unchanged.

ρ_0	$m_t(\text{Gev})$	$\Delta\Gamma_Z(\text{Mev})$
1.	120.	0.
0.98	300.	-18.
0.90	600.	-80.
0.84	780.	-140.
0.80	970.	-170.

m_t, m_H, ρ_0: $sin^2\vartheta_w^*(Q_\nu^2) = sin^2\vartheta_w^*(-m_Z^2) + 0.0050 \pm 0.0003$. The ellipse of figure 3 projects on the lines of constant m_Z to give:

$$sin^2\vartheta_w^*(-m_Z^2) = 0.2282 - 0.0053(m_Z - 92) \pm 0.0016(exp) \pm 0.001(m_H). \qquad (16)$$

The first error reflects the experimental errors, the second one the uncertainty due to the Higgs mass in the range $m_H = 10\text{-}1000$ Gev. This result can be seen in a slightly different approach in figure rnustar, where the relation between $sin^2\vartheta_w^*$ and R_ν is shown.

The asymmetries at the Z peak – corrected for experimental acceptance and QED photonic effects – can be predicted as well:

$$
\begin{aligned}
A_{LR} &= A_\tau = & 0.168 \pm 0.015 + 0.042(m_Z - 92) \\
A_{FB}^{(\mu)} &= & 0.021 \pm 0.004 + 0.011(m_Z - 92) \\
A_{FB}^{(d)} &= & 0.118 \pm 0.011 + 0.030(m_Z - 92) \\
A_{FB}^{(u)} &= & 0.085 \pm 0.008 + 0.023(m_Z - 92)
\end{aligned}
\qquad (17)
$$

Figure 7 shows the behavior of the left-right asymmetry.

These errors are similar to what can be obtained in LEP, at least until longitudinally polarized beams are available [15].

Similar results can be obtained for the Z partial widths: figure 8 shows the behavior of Γ_{ee}. The effects of varying m_t and ρ_0 are again identical. The same property holds for the other partial widths of the Z into light fermions. This result can be summarized as:

$$
\begin{aligned}
\Gamma_{ee} &= 0.0854 \pm 0.0007 + 0.0025(m_Z - 92) \\
\Gamma_{\nu\bar{\nu}} &= 0.170 \pm 0.0015 + 0.0045(m_Z - 92) \\
\Gamma_{u\bar{u}} &= 0.305 \pm 0.0025 + 0.011(m_Z - 92) \\
\Gamma_{d\bar{d}} &= 0.393 \pm 0.0030 + 0.013(m_Z - 92)
\end{aligned}
\qquad (18)
$$

in Gev. The QED and QCD – when appropriate – final state radiation effects have been included. A value of the strong coupling constant $\alpha_s(m_Z^2) = 0.12 \pm 0.02$ has been used.

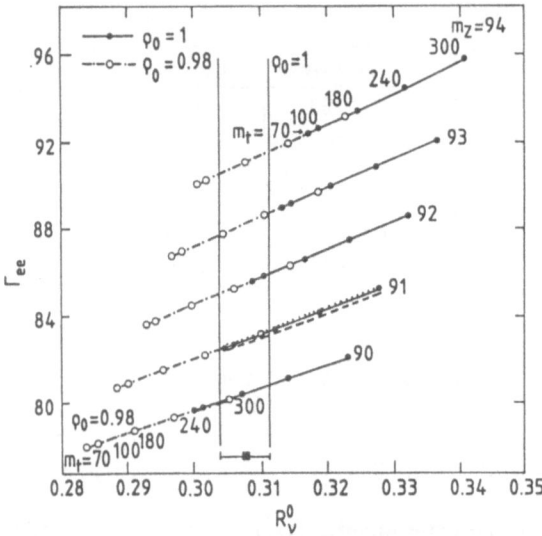

Figure 8. Relation between R_ν and Γ_{ee} for various values of m_Z, m_t, m_H and ρ_0. Masses in Gev. Unless otherwise specified, $m_H = 100$. The effect of changing m_H to 10 (dashed line) or 1000 (dotted line), is also shown for $m_Z = 91$, $\rho_0 = 1$.

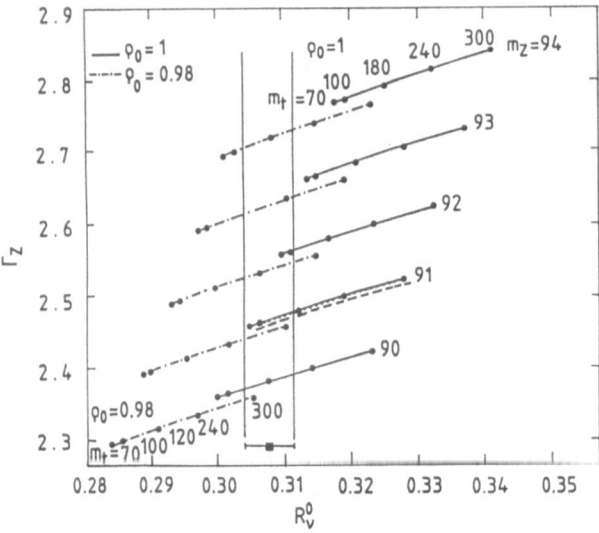

Figure 9. Relation between R_ν and Γ_Z for various values of m_Z, m_t, m_H and ρ_0. Masses in Gev. Unless otherwise specified, $m_H = 100$. The effect of changing m_H to 10 (dashed line) or 1000 (dotted line), is also shown for $m_Z = 91$, $\rho_0 = 1$.

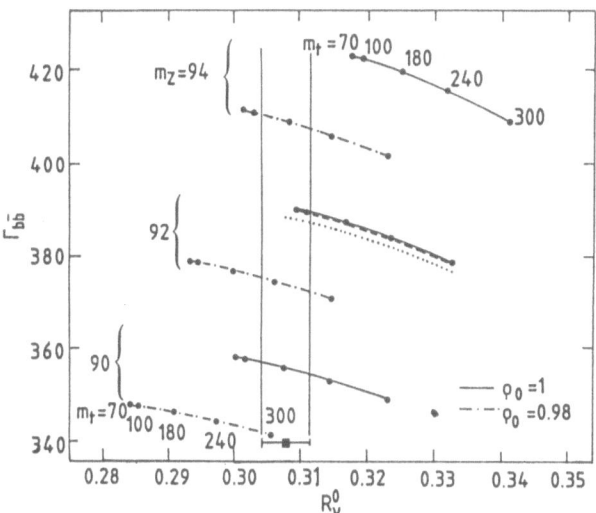

Figure 10. Relation between R_ν and $\Gamma_{b\bar{b}}$ for various values of m_Z, m_t, m_H and ρ_0. Masses in Gev. Unless otherwise specified, $m_H = 100$. The effect of changing m_H to 10 (dashed line) or 1000 (dotted line), is also shown for $m_Z = 91$, $\rho_0 = 1$.

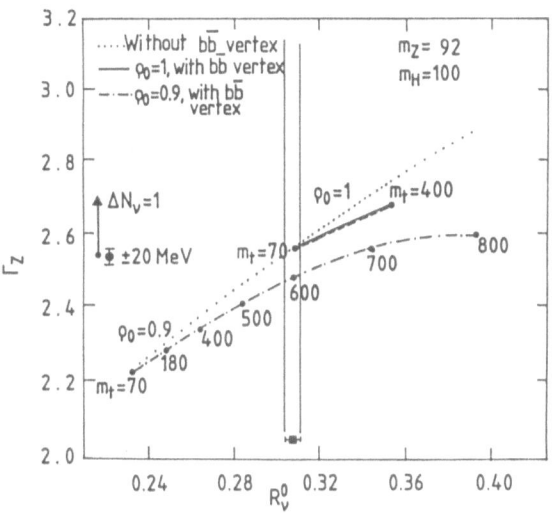

Figure 11. The relation between R_ν and Γ_Z for $\rho_0 = 1$ and $\rho_0 = 0.9$ for various values of m_t, showing the effect of the $b\bar{b}$ vertex correction.

2.4 The particular case of the Z→ $b\bar{b}$ partial width

When comparing R_ν with the total width of the Z, figure 9, the degeneracy between m_t and ρ_0 effects is not observed. The origin of this different behavior is found in the $b\bar{b}$ partial width, figure 10, which is affected by vertex corrections sensitive to the top quark mass. This *direct* effect, calculated by Beenakker and Hollik [6], singles out the $b\bar{b}$ partial width.

This interesting property allows to disentangle the top quark mass effect from the ρ_0 effect. A scenario where a large cancellation between these two effects would take place cannot be ruled out by existing data. It would produce a considerable deficit in the total width with respect to the MSM expectation.

Deficits in the Z width for various values of ρ_0 are shown in table 3. The example of $\rho_0 = 0.9$ is shown in figure 11. One has to be aware, however, that these results have been obtained from first order calculations; given the size of the effect on the $b\bar{b}$ partial width, it is likely that higher orders would not be negligible.

This effect can potentially obscure neutrino counting from the total width measurement: for $\rho_0 = 0.8$, neutral current phenomenology would be respected for $m_t \simeq$ 970 Gev, but the Z width would be then reduced by 170 Mev – a full neutrino family! On the other hand, the method proposed by Feldman [16] is immune to this uncertainty: it measures the invisible width $\Gamma_{inv} = N_\nu \Gamma_{\nu\bar{\nu}}$ in units of the charged lepton width, Γ_{ee}, and the ρ_0-m_t cancellation is valid for both.

3 Discussion

Neutrino experiments project on the LEP physics a set of very precise predictions. It is important to summarize the assumptions under which these have been obtained:

- The only intermediate vector bosons are the photon, the W and the Z.

- Quarks and leptons belong to the standard multiplets with coupling constants given by equations (5).

- The correct running of coupling constants from Q_ν^2 to $-m_Z^2$ is essential.

- It was not assumed that $\rho_0 = 1$. These predictions do not depend on the value of the top quark mass, or on all effects that are comparable to it, such as split multiplets in new families of quarks and leptons or their supersymmetric partners.

- The predictions for the asymmetries at the Z peak assume that there is only one Higgs particle with a mass in the range 10-1000 Gev. Unsplit multiplets of fermions or further degenerate Higgs multiplets would produce effects similar to that of the standard Higgs and typically of the same order of magnitude or smaller. More than one family or extra Higgs would be necessary to obtain a significant deviation.

- The prediction for the mass ratio $\frac{m_W}{m_Z}$ does not depend on the Higgs mass or on the absence of unsplit doublets, but only on the proper algebraic structure of the model.

One could conceivably blame a departure of the Z peak measurements from these predictions on radiative effects, but this would involve the existence of new particles beyond those already known. A departure of the mass ratio to be measured at the $p\bar{p}$ colliders [17] would imply a real breakdown of the structure of the theory. A comparison of the Z peak measurements amongst themselves could be used as a probe of the correctness of the structure of the theory as well, since all observables should correspond to the same value of $sin^2\vartheta_w^*$. Such scenario has been developed in ref. [7] and [18].

One should not ignore that these measurements are performed on nuclear targets. Intrinsic doubts could exist regarding their validity, but it has been shown by Llewellyn-Smith [13] that possible uncertainties should not exceed the quoted experimental error, especially since the experiments impose rather high energy cuts on the events. The measurement of the NC/CC ratio on purely leptonic target will be performed by the CHARM-II collaboration [19] and at a later stage by the water Cerenkov experiment at Los Alamos [20]. Reaching a precision of $\pm 1\%$ will, however, be rather difficult.

4 Conclusion

The measurements of neutrino NC/CC ratios on nuclear targets have been shown to constrain very severely the measurements of asymmetries at the Z peak and the mass ratio $\frac{m_W}{m_Z}$ to be performed at LEP/SLC or at the $p\bar{p}$ colliders respectively. As soon as the Z mass is measured, $sin^2\vartheta_w \equiv 1 - \frac{m_W^2}{m_Z^2}$, obviously redundant with the knowledge of the W mass, has a completely different meaning than $sin^2\vartheta_w^* \equiv \frac{e_*^2}{g_*^2}(m_Z^2)$, which determines the asymmetries at the Z pole. Their values differ, but also their error bars and their sensitivity to physical effects.

5 Acknowledgements

It is a pleasure to thank B.W. Lynn and P. Langacker [21] for many clarifying discussions, in particular on the ρ parameter. The numerical calculations were made with help of computer programs provided by D.Yu. Bardin [22], the CDHS collaboration [9], and D.C. Kennedy [8].Special thanks to G. Burgers for his strong encouragement, and help in the implementation of the $b\bar{b}$ vertex correction. The time spent by G. Altarelli on several lively discussions is gratefully acknowledged.

References

[1] A. Sirlin, Phys. Rev. D22 (1980), p. 971.
W.J. Marciano and A. Sirlin, Phys. Rev. D22 (1980), p. 2695; Phys. Rev. D29 (1984) p. 75. and p. 945.

[2] H. Burkhardt, F. Jegerlehner, G. Penso and C. Verzegnassi, in 'Polarization at LEP', G. Alexander et al. (ed.), CERN 88-06, Geneva (1988), vol.I, p. 145.

[3] B.W. Lynn, M.E. Peskin and R.G. Stuart, SLAC-PUB 3725 (1985) and in 'Physics at LEP' CERN 86-02, Geneva (1986), p. 90.

[4] M. Böhm and W. Hollik, Nucl. Phys. B204 (1982), p. 45.

[5] D.C. Kennedy and B.W. Lynn, SLAC-Pub 4039 (1986, rev. 1988), Nucl. Phys. B322 (1989) p. 1.
See also B. W. Lynn, SU-ITP-867 (1989) for a discussion of the gauge invariance of the effective Lagrangian quantities.

[6] W. Beenakker and W. Hollik, DESY-88-007 (1988).

[7] M. Cvetic and B.W. Lynn, Phys. Rev. D35 (1987), p. 51.

[8] D.C. Kennedy, J.M. Im, B.W. Lynn and R.G. Stuart, Nucl. Phys. B321 (1989) p. 83. Computer programm EXPOSTAR 1.0, courtesy of D.C. Kennedy.

[9] CDHS Collaboration, H. Abramowicz et al., Phys. Rev. Lett. 57 (1986), p. 298. A. Blondel et al., CERN/EP 89-101 (1989), submitted to Z. Phys. C.

[10] CHARM Collaboration, J.V. Allaby et al., Phys. Lett. 177B (1986), p. 446. J.V. Allaby et al., Z. Phys. C36 (1987), p. 611.

[11] U. Amaldi et al. Phys.Rev. D36 (1987) p. 1385.
A. Blondel, in 'XXII Rencontres de Moriond', J. Tran Thahn Van (ed.), Editions Frontieres (1987), p. 3; CERN-EP/87-174.
G. Costa et al., Nucl. Phys. B297 (1988), p. 244.

[12] R.G. Stuart, Z. Phys. C34 (1987), p. 445.

[13] C.H. Llewellyn-Smith, Nucl. Phys. B228 (1983), p. 205.

[14] L. DiLella, summary talk at the 7th $p\bar{p}$ workshop, Fermilab, 1988; CERN-EP/89-12.

[15] A. Blondel, in 'Polarization at LEP', G. Alexander et al. (ed.), CERN 88-06, Geneva (1988), vol.I, p. 1.

[16] G. J. Feldman, SLAC-Report-315 (1987), p. 169.

[17] See for instance: D. Froidevaux and P. Jenny, CERN-EP/88-111 (1988).

[18] C. Verzegnassi, in 'Polarization at LEP', G. Alexander et al. (ed.), CERN 88-06, Geneva (1988), vol.I, p. 204;
A. Djouadi, ibid, p. 215;
F. Boudjema and F.M. Renard, ibid, p. 223;
F.M. Renard and C. Verzegnassi, Phys. Lett. B217 (1989), p. 199.

[19] CHARM-II collaboration, C. Busi et al., CERN/SPSC/83-24, SPSC P 186 (1983).

[20] D.H. White (Spokesman), Large Water Cerenkov Proposal, Los Alamos proposal (1986).

[21] P. Langacker, W.J. Marciano and A. Sirlin, Phys. Rev. D36 (1987), p. 2191.

[22] D.Yu. Bardin and O.M. Dokuchaeva, Preprint JINR-E2-86-260 (1986).
D.Yu Bardin and O.M. Federenko, Yad. Phys. 30 (1070), p. 811, Sov. J. Nucl. Phys. 30 (1979), p. 418.
D.Yu Bardin and O.M. Dokuchaeva, Yad. Phys. 36 (1982), p. 482, Sov. J. Nucl. Phys. 36 (1982), p. 282.

RESULTS ON R FROM PEP, PETRA, TRISTAN, AND SLC

Wim de Boer *

Max–Planck–Institut für Physik und Astrophysik
Werner-Heisenberg-Institut für Physik
D 8000 Munich 40

Abstract

We discuss the determination of the Standard Model parameters α_s, M_Z, and $\sin^2\theta_W$ from the total hadronic cross section in e^+e^- annihilation at centre of mass energies between 7 and 93 GeV. At the highest TRISTAN energies, the tail of the Z^0 resonance does increase R by 50%. Such a sizeable increase allows a direct measurement of M_Z. However, this measurement turns out to be somewhat below the direct value measured at SLC at energies around the Z^0-resonance. This has led to speculation about possible new physics, although statistically the effect is not too significant (less than 2 standard deviations). We study in detail if other effects, like normalization errors or higher order radiative corrections could reduce the difference. The radiative corrections at TRISTAN energies can be calculated more precisely with the new knowledge of the Z^0- mass and the correspondingly improved value for the top mass ($120\pm35\pm20$ GeV/c^2), resulting from the combination with the value of $\sin^2\theta_W$ from low energy measurements.

If third order QCD contributions are taken into account, one finds $\Lambda_{\overline{MS}}=0.26^{+0.16}_{-0.13}$ GeV, which corresponds to $\alpha_s(34$ GeV$) = 0.144 \pm 0.015$ and $\alpha_s(91$ GeV$) = 0.122 \pm 0.011$. From the combined R measurements one finds that near the Z^0-resonance the infinite QCD series $1 + \alpha_s/\pi + ..$ is 1.046 ± 0.006. We study the effect of changing the renormalization scale of α_s and find that the third order QCD corrections, which are larger than the second order ones for the usual scale $Q = \sqrt{s}$, are smaller than the second order contributions at other scales.

1 Introduction

The most precise measurement of the Z^0 mass (91.11 ± 0.23 GeV/c^2) has been obtained from the recent observation of the Z^0 s-channel resonance in e^+e^- annihilation by the MARK-II Collaboration[1].The data has been obtained at the Stanford Linear Collider (SLC). A comparison with other values is shown in Fig. 1 and Table 1. The M_Z values from SLC and the $p\bar{p}$ collider[2] experiments are in good agreement. However, the M_Z value from R data up to 60.8 GeV is about 2 standard deviations below the one from SLC data, implying a somewhat higher electroweak contribution to the hadronic cross section than expected from the Standard Model. On the other hand, the combined leptonic and hadronic data from TRISTAN yields M_Z values much closer to the SLC value (see Table 1), implying that the leptonic sector has a smaller electroweak contribution than the expectation from the Standard Model, as has been

---

*Talk at the Brighton Conference on Electroweak Radiative Corrections, July 9-14, 1989.
Mailing address: DESY FH1K, Notkestraße 85, D 2000 Hamburg 52.
Bitnet address: user F36WDB at node DHHDESY3

Table 1. M_Z **values from various experiments.**

Data	Ref	$M_Z(\text{GeV/c}^2)$
$MARK\ II$	[1]	91.11 ± 0.23
$UA1$	[2]	$93.1 \pm 1.0 \pm 3.1$
$UA2$	[2]	$91.5 \pm 1.2 \pm 1.7$
CDF	[2]	$90.9 \pm 0.3 \pm 0.2$
$R + leptons$	[13]	$90.4^{+1.7}_{-1.9}$
R	[thiswork]	88.9 ± 1.2

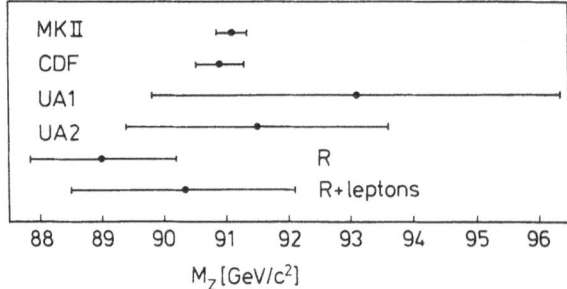

Figure 1. **Comparison of** M_Z **values.**

published[3,4] and discussed at various Conferences[5]-[13]. This has led to speculations about a possible Z' heavy boson, which through its interference with the Z^0 could lead to a decrease in the leptonic cross section, but simultaneously an increase in the hadronic cross section[14]. In this paper we try to see if more mundane effects could have contributed to the difference between 'low' (i.e. below $\sqrt{s}=61$ GeV) and 'high' energy data. In particular, we consider

- New constraints on the top mass from the new values of M_Z combined with the low energy measurements from $\sin^2 \theta_W$ (θ_W is the electroweak mixing angle or Weinberg angle). Knowledge of M_Z and M_{top} is important for a precise knowledge of the radiative corrections at 'low' energies.

- Possible effects from higher order radiative corrections, which have not been applied neither to the published low energy data nor to the SLC data.

- Possible effects from common normalization errors.

The report has been organized as follows: we first summarize the Standard Model (SM) formulae and radiative corrections, then discuss the results on M_Z from 'low' and 'high' energy data. We conclude with the determination of α_s and the scale dependence of α_s.

2 Standard Model Formulae

The normalized cross section R is defined as the ratio

$$R \equiv \frac{\sigma[e^+e^- \rightarrow \gamma, Z^0 \rightarrow hadrons]}{\sigma[e^+e^- \rightarrow \gamma \rightarrow \mu^+\mu^-]}$$

The $\mu^+\mu^-$ cross section is the lowest order pointlike QED cross section of massless spin $\frac{1}{2}$ particles, and is equal to $4\pi\alpha^2/3s$, where s is the square of the centre of mass energy.

Table 2. Summary of the couplings between the Z^0 and various leptons for $\sin^2\theta_W = 0.23$.

Fermion	I_3^L	I_3^R	a	v	e_f
neutrino	$1/2$	0	1	1	0
e, μ, τ lepton	$-1/2$	0	-1	$-1 + 4\sin^2\theta_W = -0.08$	-1
u, c, t quarks	$1/2$	0	1	$+1 - \frac{8}{3}\sin^2\theta_W = 0.39$	$+2/3$
d, s, b quarks	$-1/2$	0	-1	$-1 + \frac{4}{3}\sin^2\theta_W = -0.69$	$-1/3$

The total hadronic cross section at the Born level is given by the sum of the following contributions:

$$\sigma_{had}^{\gamma} = \frac{4\pi\alpha^2}{3s} r_{QCD} \sum_{q=1}^{5} \epsilon_\epsilon^2 \epsilon_q^2 \tag{1}$$

$$\sigma_{had}^{\gamma Z} = 8\pi\alpha r_{QCD} \frac{K(s - M_Z^2)}{(s - M_Z^2)^2 + s^2\Gamma_{tot}^2/M_Z^2} \sum_{q=1}^{5} \epsilon_\epsilon \epsilon_q v_\epsilon v_q \tag{2}$$

$$\sigma_{had}^{Z} = 12\pi r_{QCD} \frac{K^2 s}{(s - M_Z^2)^2 + s^2\Gamma_{tot}^2/M_Z^2} (v_\epsilon^2 + a_\epsilon^2) \sum_{q=1}^{5} (v_q^2 + a_q^2) \tag{3}$$

The superscripts indicate the contribution from photon exchange, Z^0 exchange and their interference and the sum is taken over five quark flavours. thus assuming the top quark is too heavy; v and a represent the vector and axial vector couplings of the quarks (subscript q) and electrons (subscript ϵ) and Γ_{tot} is the total width of the Z^0. For simplicity we have neglected small mass effects in the formulae above, but they have been taken into account in the analysis, using the formulae in Ref.[15]. The factor r_{QCD} represents the effect from gluon radiation and is given in the $\overline{MS}$ scheme by[16,17]:

$$r_{QCD} = 3[1 + \frac{\alpha_s}{\pi} + (1.986 - 0.115 n_f)\left(\frac{\alpha_s}{\pi}\right)^2 +$$
$$\left(70.985 - 1.2 n_f - 0.005 n_f^2 - 1.679\frac{(\sum \epsilon_q)^2}{3\sum \epsilon_q^2}\right)\left(\frac{\alpha_s}{\pi}\right)^3]$$

The factor 3 on the righthand side accounts for the colour of the quarks. The total Z^0 width is determined by:

$$\Gamma_{tot} = K M_Z N_g \left[a_\nu^2 + v_\nu^2 + a_e^2 + v_e^2\right] + K M_Z r_{QCD} \sum_{q=1}^{5} (v_q^2 + a_q^2) \tag{4}$$

while the partial width into a fermion f is:

$$\Gamma_f = K M_Z r_{QCD}(v_f^2 + a_f^2) \tag{5}$$

N_g is the number of generations and the sum in the total width has been taken over 5 quarks, again assuming the top quark is too heavy to contribute.

The constant K can be either defined as:

$$K_1 = \frac{\sqrt{2} G_F M_Z^2 \kappa_1}{48\pi} \tag{6}$$

or

$$K_2 = \frac{\alpha \kappa_2}{48 \sin^2\theta_W \cos^2\theta_W} \tag{7}$$

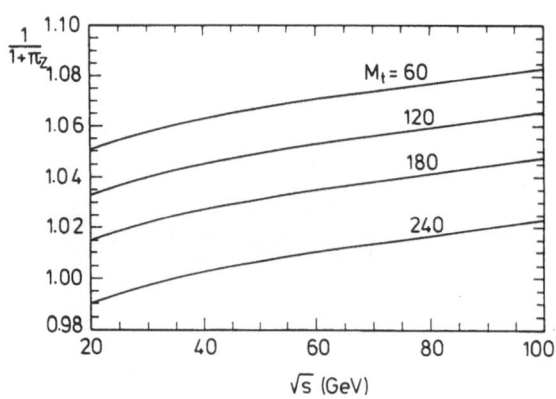

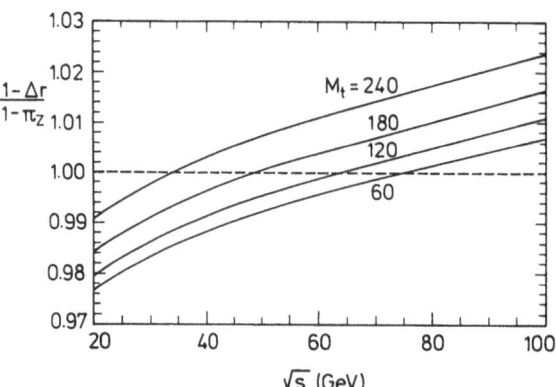

Figure 2. **Electroweak correction factors** (κ_1 and κ_2 **in text) for the different parametrizations of the Born cross section.**

Here G_F is the Fermi constant, which is well known from muon decay; $\sin^2\theta_W$ defines the electroweak mixing angle, which can be used to define the coupling constants between a pair of fermions and the Z^0 gauge boson (see Table 2):

$$v_f = 2(I_3^L - I_3^R) - 4\,e_f \sin^2\theta_W \qquad (8)$$

$$a_f = 2(I_3^L - I_3^R) \qquad (9)$$

$I_3^{L(R)}$ is the 3th component of the weak isospin. In the definitions of K we have explicitly included the factor κ which represents the loop corrections to the Z^0 propagator. For example, practically all data from the PEP and PETRA experiments have been corrected with the LUND Monte Carlo program[18], which uses the radiative corrections from Berends et al.[19]. These corrections include the loop corrections for the photon propagator, but not the loop corrections for the Z^0 propagator; the latter can be taken into account by defining[20,21]:

$$\kappa_1 = \frac{1 - \Delta r}{1 + \Pi_Z(s)} \qquad (10)$$

or

$$\kappa_2 = \frac{1}{1 + \Pi_Z(s)} \qquad (11)$$

where

$$1 - \Delta r = \frac{\alpha(0)}{\alpha(M_W)} + \delta_r(M_t, M_H) \qquad (12)$$

and

$$1 + \Pi_Z(s) = \frac{\alpha(0)}{\alpha(M_Z)} + \delta_\Pi(M_t, M_H, s) \qquad (13)$$

Here Δr represents the electroweak corrections to the charged gauge boson exchange in muon decay and $\Pi_Z(s)$ represents the electroweak loop corrections to the neutral gauge boson exchange. One sees that the first term in both cases is given by the running of the QED coupling constant coming from the light fermion loops in the photon propagator (hence the indication of the scale in α). For top quark masses below the gauge boson masses this term is dominant in both expressions. E.g. for a top mass of 70 GeV Δr is about 7 % and 6% is coming from the first term alone. However, for a top mass of 230 GeV the latter term $\delta_r(M_{top}, M_H)$ is as large as the first term, but of opposite sign, so the total correction Δr is about zero. $\Pi_Z(s)$ shows a similar behaviour, so that the ratio in κ_1 is much less dependent on the top mass and furthermore close to 1 (see Fig. 2b). This is the advantage of the parametrization with K_1: one can neglect the electroweak corrections to a large extent and the results are insensitive to the unknown top mass. This was the reason why in previous fits to data on R this parametrization has been used, e.g. to determine the strong coupling constant [15]. What was considered further as an advantage compared to the K_2 parametrization was the insensitivity to the Z^0-mass at PETRA energies: the dominant term in both the numerator and denominator in Eq. 3 is proportional to M_Z^6, thus largely canceling the uncertainty in M_Z. However, at TRISTAN-energies one observes the tail of the Z^0-resonance and it becomes possible to make a direct measurement of the Z^0-mass. In this case one obtains much more sensitivity with K_2, since one can measure the pure propagator effect without the compensation from the M_Z^6 factor in the numerator. However, with the K_2 parametrization the electroweak corrections cannot be neglected anymore (κ_2 in this case, see Fig. 2a), since the correction to the total hadronic cross section is of the order of 3% at 60 GeV.

From the definitions of K_1 and K_2 one can deduce the following well known relation between $\sin^2\theta_W$ and M_Z, first derived by Sirlin[22]:

$$\sin^2\theta_W = \frac{1}{2}\left[1 - \sqrt{1 - \frac{4\pi\alpha}{\sqrt{2}G_F M_Z^2(1 - \Delta r)}}\right] \qquad (14)$$

3 Definition of the strong coupling constant

The strong coupling constant α_s is in first order given by

$$\alpha_s(Q^2) = \frac{12\pi}{(33 - 2n_f)\ln(Q^2/\Lambda^2)}. \tag{15}$$

The number of flavours n_f is 5 for the energies considered, and so the Λ value determined is $\Lambda^{(5)}$, while e.g. deep inelastic scattering studies refer to $\Lambda^{(4)}$, the energies there being below the bottom threshold. They are related to each other by $\Lambda^{(5)} \approx 0.7\, \Lambda^{(4)}$[23].

In higher order one not only has to specify the renormalization scheme, but in addition several approximations of α_S have been in use. Within a given scheme (we will use the widely used $\overline{MS}$ scheme) one can define the higher order contributions in terms of the first order coupling constant, i.e. an expansion in $1/ln(Q^2/\Lambda^2)$:

$$\alpha_S(Q^2) = \frac{12\pi}{(33 - 2n_f)ln(Q^2/\Lambda^2_{\overline{MS}})} \left[1 - 6\frac{153 - 19n_f}{(33 - 2n_f)^2} \frac{ln(ln(Q^2/\Lambda^2_{\overline{MS}}))}{ln(Q^2/\Lambda^2_{\overline{MS}})} \right] \tag{16}$$

or one can sum the leading logarithms in this expansion, i.e. terms proportional to $\alpha_S^n\, ln^n(Q^2/\Lambda^2)$, which yields:

$$\alpha_S(Q^2) = \frac{12\pi}{(33 - 2n_f)ln(Q^2/\Lambda^2_{\overline{MS}}) + 6\frac{153-19n_f}{(33-2n_f)}ln[ln(Q^2/\Lambda^2_{\overline{MS}})]} \tag{17}$$

The two definitions are numerically somewhat different. For example for a given value of $\alpha_S = 0.15$ and $n_f = 5$ the Λ values from the two definitions differ by 14%. Both definitions have been widely used: the first one in the Particle Data Book[24], the second one e.g. in the LUND Monte Carlo[18]. The difference is of some importance in practice. For example, if one would use the definition from the Particle Data Book in the LUND Monte Carlo, one should simultaneously increase the default Λ value from 500 to 570 MeV in order to get the same value of α_S. If one includes the next higher order term, one finds[23]:

$$\alpha_s(Q^2) = \frac{4\pi}{\beta_0 L} \left[1 - \frac{\beta_1}{\beta_0^2}\frac{lnL}{L} + (\frac{\beta_1}{\beta_0^2})^2 \frac{1}{L^2} \left\{ (lnL - \frac{1}{2})^2 + \frac{\beta_2\beta_0}{\beta_1^2} - \frac{5}{4} \right\} \right] \tag{18}$$

with

$$
\begin{aligned}
L &= ln(Q^2/\Lambda_{\overline{MS}}^{\,2}) \\
\beta_0 &= 11 - \frac{2}{3}n_f \\
\beta_1 &= 2(51 - \frac{19}{3}n_f) \\
\beta_2 &= \frac{2857}{2} - \frac{5033}{18}n_f + \frac{325}{54}n_f^2 \, .
\end{aligned}
$$

Instead of using a series expansion in α_s, one can solve the renormalization group equation exactly[25]. However, from such a solution one can obtain α_s only numerically for a given $\Lambda_{\overline{MS}}$, which is rather inconvenient and slow in fitting programs. Instead, we have checked that the third order expression Eq. 18 agrees extremely well with the exact solution (better than 0.1% for the energies of interest in this paper). Therefore, we will use this third order expression everywhere, even if we compare with observables calculated only up to second order, since in principle one always should use the correct energy dependence of α_s for all observables, independent of their unknown higher order corrections.

4 Radiative Corrections

4.1 Introduction

Radiative corrections can be subdivided into the following classes:

- 'QED corrections', which consist of those diagrams with an extra photon added to the BORN diagrams either as real bremsstrahlung or as a virtual photon in vertex - or box diagrams (see Fig. 3a).

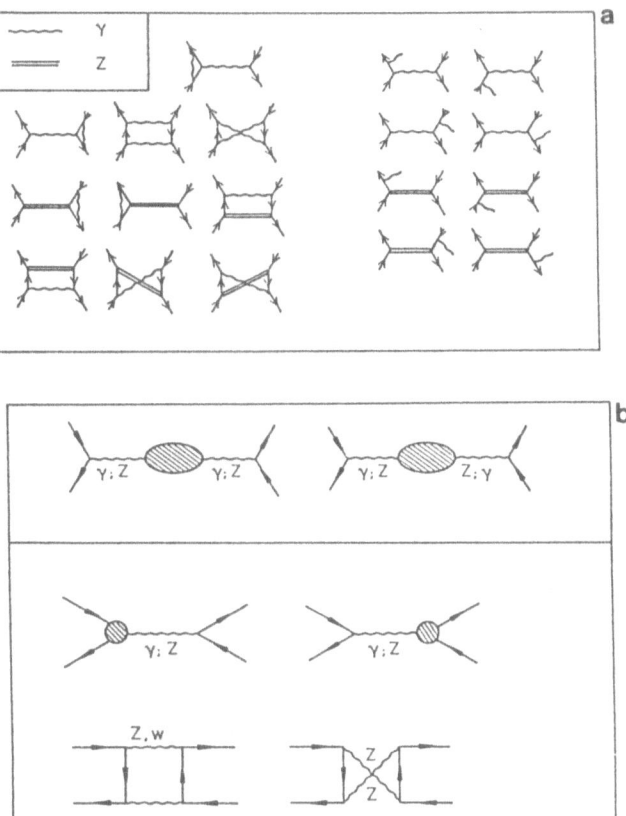

Figure 3. a) Feynman diagrams for the 'QED corrections' b) Feynman diagrams for the 'electroweak corrections': loop corrections to the propagators and box- and vertex corrections.

- 'Electroweak corrections', which collect all other diagrams up to 1 loop: the loop corrections to the propagators, the vertex corrections (excluding the ones with virtual photons), and the box diagrams with two massive gauge bosons (see Fig. 3b).

These two subsets are both gauge invariant[20] and it has sometimes been advertised that experiments should only make the 'QED corrections', which depend on the experimental cuts. Furthermore, the 'electroweak corrections' are unknown as long as the top - and Higgs mass are unknown, but they do not depend on experimental cuts, so in principle one could have 'QED corrected' data, which could be compared with theory for any given top or Higgs mass. This sounds nice. However, in practice the 'QED corrections' depend on the loop corrections, simply through the fact that the cross section after initial state radiation has a reduced centre of mass energy, but this cross section depends on the loop corrections. Therefore, corrected data is only useful, if the original data with applied cuts and correction factors have all been completely specified, including the values used for M_Z, M_{top}, and M_H. Usually the detector acceptances do not change too much, if the top or Higgs mass are changed, so one can than easily recalculate the correction for different values, if the original values have been specified.

The 'electroweak corrections' are very well approximated by the factors κ in Eq. 6 or 7, since the loop corrections are the dominant ones. Higher order contributions to the loop corrections can be estimated by summing the leading logarithms to a geometric series. The change in cross section by the 'electroweak corrections' depends on energy: at low energy only the loop corrections to the photon (vacuum polarization) are important; they change the cross section typically by 10 %; near the Z^0 resonance peak the effect is small, since on resonance the K^2-factor in both the numerator and denominator cancel each other (see Eqs. 3 and 4). Off-resonance, the loop corrections depend on the choice of K-factor and the top mass as shown in Fig. 2: at $\sqrt{s}$=80 GeV the effect is less than 0.5% with κ_1, but 12% with κ_2 for M_{top}=120 GeV/c^2.

From the 'QED corrections' the initial state radiative corrections are dominant. Final state electromagnetic radiation from the quarks is small, since the Kinoshito-Lee-Nauenberg theorem[26] assures that the procedure of summing over all $q\bar{q}$ final states with an arbitrary number of photons, as is done in the detection of multihadronic events, will cancel all leading logarithms and the remaining radiative correction is of order $O\left(\frac{\alpha}{\pi}\right) \approx 0.2\%$.

For initial state radiation an exact second order calculation has been made by Berends, Burgers and van Neerven[27], which allows a comparison with approximate procedures, based on exponentiation. Such a comparison is important, since it allows to choose an 'optimum' exponentiation procedure, i.e. the one which is closest to the exact second order calculation. Such an 'optimum' procedure can than be applied to e.g. Bhabha scattering, for which no exact second order calculation exists. We describe such an 'optimum' choice in the next section.

4.2 Higher order initial state radiative corrections

In first order the total cross section can be written as:

$$\sigma_1(s) = \sigma_0(s) \left(1 + \delta_1^v + \beta \ln k_0\right) + \int_{k_0}^{k_{max}} \beta \left(\frac{1 + (1 - k)^2}{2 k}\right) \sigma_0(s')\, dk \tag{19}$$

while in second order this is modified to [27]:

$$
\begin{aligned}
\sigma_2(s) \;=\; & \sigma_0(s) \left(1 + \delta_1^v + \delta_2^v + \beta \ln k_0 + \delta_1 \beta \ln k_0 + \frac{1}{2}\, \beta^2 \ln^2 k_0\right) \\
& + \int_{k_0}^{k_{max}} \sigma_0(s') \left[\beta \left(\frac{1 + (1 - k)^2}{2 k}\right)\left(1 + \delta_1^v + \beta \ln k\right) \right. \\
& + \left. \left(\frac{\alpha}{\pi}\right)^2 \left\{\left(\frac{1 + (1 - k)^2}{k}\right) A(k) + (2 - k)\, B(k) + (1 - k)\, C(k)\right\}\right] dk
\end{aligned}
\tag{20}
$$

Here $s(s')$ is the centre of mass energy before (after) energy loss from initial state radiation and

$$\beta = 2\alpha/\pi \left(\ln s/m_e^2 - 1\right) \tag{21}$$

$$k = 1 - s'/s \tag{22}$$

$$\delta_1^v = \frac{\alpha}{\pi} \left[\frac{3}{2} \ln\left(s/m_e^2\right) - 2 + \pi^2/3\right] \tag{23}$$

The expressions for $A(k), B(k), C(k)$ and δ_2^v have been given in Ref. [27].

Note that in first order the energy loss parameter k is just the fractional photon energy E_γ/E_{beam}. In higher orders, when the energy loss is distributed over more photons, these latter have an invariant mass $(M_{\gamma\gamma})$ and the relation between ΣE_γ and k becomes: $k \equiv 1 - s'/s = \Sigma E_\gamma/E_{beam} - M_{\gamma\gamma}^2/s$. Since $M_{\gamma\gamma}^2/s$ is usually small, the difference between k and the total radiated fractional energy $\Sigma E_\gamma/E_{beam}$ becomes small too. This implies that most of the additional photons are either soft or collinear. Therefore, 'simple' Monte Carlo simulations, in which the total radiated energy is given to a single 'effective' photon, do not differ significantly from more elaborate simulations, in which the radiated energy is distributed over multiple photons. In particular, for efficiency determinations of hadronic cross sections they yield similar results, as has been checked explicitly for the generators compared in Ref. [28].

The cross sections in Eqs. 19 and 20 have been split into 2 parts. The first part is the part with only soft photons, so $\sigma(s) = \sigma(s')$; the second part includes hard photon radiation, in

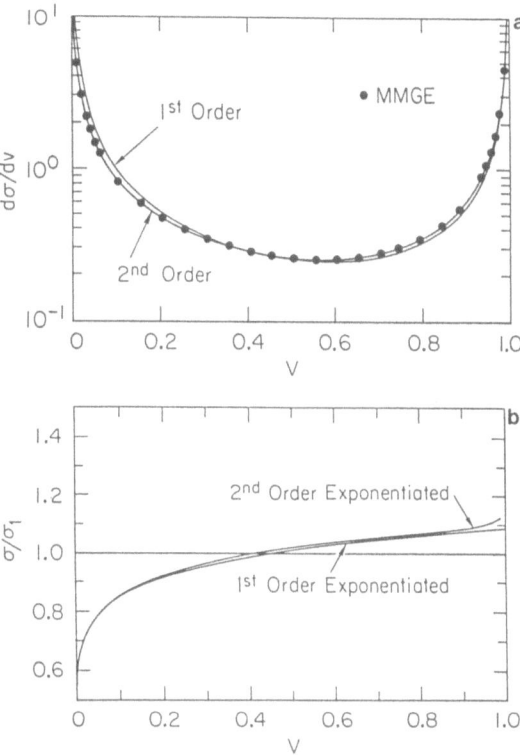

Figure 4. a) The first and second order cross sections as function of $v = 1 - s'/s$, where s and s' are the invariant masses squared of the initial and final state, respectively. In first order, v is the photon energy normalized to the beam energy. b) The first and second order exponentiated cross sections normalized to the first order non-exponentiated cross section; the latter corresponds to the horizontal line. The first order exponentiated cross section is already close to the second order calculation; exponentiating the second order calculation hardly changes it. One sees, that the higher order calculations are about 30 % below the first order calculation for small v-values, but 10 % above for large v-values.

which case $\sigma(s') \neq \sigma(s)$, so one has to convolute the energy loss spectrum with $\sigma(s')$, as is done by the expesssions below the integrals. The separation between the soft and hard parts of the spectrum is defined by k_0. For $k_0 \to 0$ the soft part goes to $-\infty$, while the hard part goes to $+\infty$. As long as k_0 is small, the sum of the two parts is independent of the choice of k_0.

It can be shown [29] that the leading terms of the real photon emission always lead to terms $\frac{1}{n!} \beta^n \ln^n k$, so summing the leading logs to all orders implies " exponentiating" the cross section, which yields:

$$\sigma_{soft}^{exp}(s) = \sigma_0(s) \left(1 + \delta_1^v + \delta_2^v + ... \right) k^\beta \tag{24}$$

since

$$k^\beta = e^{\beta \ln k} = 1 + \beta \ln k + \frac{1}{2!} \beta^2 \ln^2 k + ... \tag{25}$$

This expression clearly reproduces the soft parts of the first and second order cross sections (Eqs. 19 and 20), if the higher order terms are dropped in the expansion.

For the virtual corrections no such simple formulae exist. Therefore they have to be calculated, a difficult task already in second order. Fortunately, numerically the second order vertex correction is already small, so one may hope that the higher orders are small too. For example at $\sqrt{s} = 29$ GeV , $\delta_1^v = 0.08$ and $\delta_2^v = -0.005$.

If one neglects the small terms proportional to $A(k), B(k)$ and $C(k)$ in Eq. 20, one sees that the hard part in second order contains the factor $(1 + \delta_1^v + \beta \ln k)$ which can be replaced again by the exponentiated version $(1 + \delta_1^v + ...) k^\beta$ (see Eq. 25), so one finds:

$$\sigma_1^{exp}(s) = \int_0^{k_{max}} \beta \left(\frac{1}{k} - 1 + \frac{k}{2} \right) (1 + \delta_1^v) k^\beta \sigma_0 (s') dk \tag{26}$$

Note that the factor k^β regularizes the infinities, if $k \to 0$, so the integral for $k = 0$ is well behaved and one does not have to split the cross section in a hard- and soft part anymore, but integrates from 0 to the kinematical limit. One can regulate the divergencies in the second order cross section too by exponentiating the soft part:

$$\sigma_2^{exp}(s) = \int_0^{k_{max}} \sigma_0(s') \left[\beta \left(\frac{1}{k} \right) (1 + \delta_1^v + \delta_2^v) k^\beta \right] dk + \sigma_{2H} \tag{27}$$

where σ_{2H} contains the finite part of the cross section (everything except the $1/k$ pole).

Note that in the first order exponentiated cross section (Eq. 26) we have exponentiated the finite part ($-1 + k/2$ term) too, while in second order the finite part is treated exactly. Exponentiating the finite part of the first order cross section is usually not done [30,31], but its justification stems from the fact that in second order this finite part is multiplied by $1 + \delta_1^v + \beta \ln k$ (see Eq. 20), and secondly, that σ_1^{exp} is numerically now very close to σ_2^{exp}, as can be seen from Table 3. Thus a very simple procedure for exponentiating a first order Monte Carlo is: weight the hard part of the cross section with the factor $(1 + \delta_1^v) k^\beta$. This works perfect in the case of μ-pairs and is probably the best guess in case of Bhabha scattering, for which no exact second order calculation exists. However, in that case the β factor has to be modified to include final state radiation and the interference, as will be discussed hereafter.

The effect of the higher order contributions on the energy loss is shown in Fig. 4a (taken from Ref. [28]). As expected, one observes a shift of events towards higher energy losses (a decrease at the left hand side and increase at the right hand side). On a logarithmic plot the effect seems small, but if one plots the ratio between the higher order and first order curves, the difference is -30% at small v and 10% at large v (see Fig. 4b). These rather drastic effects on the radiated energy change the cross section considerably near a resonance and exponentiating the second order cross section still changes the cross section a few % on the high side of the Z^0 resonance (see Table 3). However, at all energies the first order exponentiated cross section is very close to the second order exponentiated one, thus exponentiating the first order cross section in the way described before is enough.

60

Table 3. The total μ-pair cross sections after initial state radiation and electroweak radiative corrections, expressed in units of the Born cross section σ_B, defined as the cross section without any loop- or vertex corrections. The maximum energy loss is limited to $0.99 E_{beam}$. Various levels of initial state radiation are compared (see text). Note the close agreement between σ_1^{exp} and σ_2^{exp}, indicating that simply exponentiating the first order cross section is enough. Note too the difference between first and second order, which is appreciable at all energies, not only near the resonance. The input parameters correspond to $M_Z = 91$ GeV/c^2, $M_H = M_t = 100$ GeV/c^2, and $\alpha_s = 0.12$.

$\sqrt{s}$	σ_B (nb)	$\dfrac{\sigma_1}{\sigma_B}$	$\dfrac{\sigma_1^{exp}}{\sigma_B}$	$\dfrac{\sigma_2}{\sigma_B}$	$\dfrac{\sigma_2^{exp}}{\sigma_B}$	$\dfrac{\sigma_1^{exp}}{\sigma_2}$	$\dfrac{\sigma_2^{exp}}{\sigma_2}$
20	0.2172	1.341	1.367	1.369	1.367	0.998	0.999
30	0.0966	1.360	1.390	1.393	1.390	0.998	0.998
40	0.0546	1.372	1.406	1.408	1.406	0.998	0.998
50	0.0355	1.378	1.413	1.416	1.413	0.998	0.998
60	0.0258	1.370	1.405	1.408	1.405	0.998	0.998
70	0.0222	1.319	1.351	1.352	1.350	0.999	0.998
80	0.0326	1.135	1.152	1.152	1.150	1.000	0.998
87	0.1598	0.882	0.897	0.898	0.894	0.999	0.995
88	0.2636	0.836	0.856	0.857	0.852	0.998	0.994
89	0.5074	0.778	0.808	0.811	0.805	0.996	0.992
90	1.1593	0.701	0.751	0.755	0.748	0.995	0.990
91	2.0002	0.694	0.743	0.742	0.740	1.001	0.997
92	1.1416	0.980	0.943	0.930	0.940	1.014	1.010
93	0.5126	1.311	1.215	1.195	1.211	1.017	1.013
94	0.2732	1.597	1.472	1.451	1.468	1.015	1.012
95	0.1688	1.841	1.706	1.687	1.703	1.011	1.010
96	0.1155	2.050	1.915	1.899	1.913	1.008	1.007
97	0.0849	2.228	2.099	2.088	2.098	1.005	1.005

4.3 Higher order initial state radiative corrections for Bhabha scattering

Unfortunately there exists no higher order calculation for Bhabha scattering, i.e. at most one photon is allowed from initial or final state radiation. Therefore, the effect of higher order contributions can only be estimated by exponentiation. Since exponentiation procedures are not unique, the exponentiation procedure which 'works best' for μ-pairs has been choosen, i.e. the procedure which gives results closest to the exact second order calculation. As mentioned in the description of the exponentiation of the μ-pair generator, this corresponds to weighting each radiative event (both with soft AND hard photons, not only the soft part) with a factor $(1 + \delta_1^v) k^\beta$ and exponentiating the soft part as described before. For Bhabha scattering final state radiation is important, in which case β should be defined is $\beta_i + \beta_f + 2\beta_{int}$, where $\beta_i = \beta_f$ equals β as defined by Eq. 21, and [32]

$$\beta_{int} = \frac{4\alpha}{\pi} \ln \tan \frac{\theta}{2}, \qquad (28)$$

where θ is the polar scattering angle.

In the neighbourhood of a resonance the exponentiation procedure becomes somewhat more complicated[32], but this does not play a major role in the luminosity measurements, since even at SLC and LEP the resonance contribution is very small in the angular range considered for luminosity measurements.

4.4 Effect of higher order radiative corrections on R

R is calculated from the number of multihadron events - N_{MH} - and the number of Bhabha events - N_{BB} - in the following way:

$$R = \frac{\sigma_{BB}}{\sigma_{\mu\mu}} \frac{N_{MH}}{N_{BB}} \frac{\epsilon_{BB}(1+\delta)_{BB}(1+\delta_{VP})_{BB}}{\epsilon_{MH}(1+\delta)_{MH}(1+\delta_{VP})_{MH}}$$

Here ϵ is the detection efficiency, δ_{VP} is the vacuum polarization correction, δ is the correction for initial and final state radiation and vertex graphs, and σ_{BB} and $\sigma_{\mu\mu}$ are the Born cross sections for Bhabha scattering and μ-pair production.

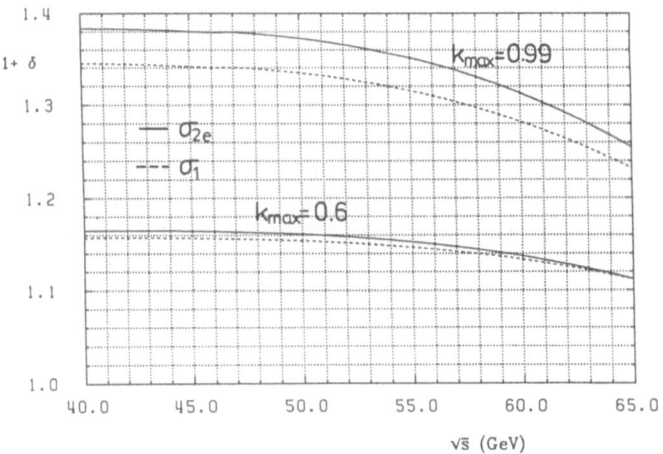

Figure 5. Radiative correction factors calculated in first and exponentiated second order for k_{max} = 0.99 and 0.6. The M_Z mass is 90 GeV/c² and the top mass 60 GeV/c².

We have factorized the effects of loop corrections and other radiative corrections; this yields a higher order contribution $\delta\delta_{VP}$, which should be neglected in first order, but can be large for multihadrons, since typically $\delta = 0.2$ and $\delta_{VP} = 0.1$. Such a contribution, representing the radiative corrections to graphs including a fermion pair in the propagator, yields the main difference between the first and second order corrections, shown by the upper two curves in Fig. 5. Experimental cuts require typically a visible energy of $0.4\sqrt{s}$, which corresponds to $k_{max} \approx 0.6$. For such a cutoff the difference between first- and second order becomes much smaller as shown by the two lowest curves in Fig. 5. The reason is simple: initial state radiative corrections are rather small in that case and therefore the product $\delta\delta_{VP}$ is small too.

In order to estimate more precisely the effect of higher order radiative corrections, one should consider possible changes in ϵ, since the energy loss distribution is changed considerably by the higher order corrections, as has been shown in Fig. 4b.

We have calculated the change in efficiency for a wide range of experimental cuts and center of mass energies using the LUND fragmentation program[18] after implementing the higher order radiative corrections in that program[33]. The result is that indeed the product of efficiency and $1 + \delta$ hardly changes by the higher orders, as expected from the small differences in the lower curves of Fig. 5.

For Bhabha scattering the complete second order calculation has not been done. We have estimated the higher order radiative corrections by applying the exponentiation procedure described before to the standard Bhabha generator from Berends and Kleiss[34]. Different experiments usually make quite different cuts on acolinearity and visible energy to select the Bhabha sample. It turns out that the higher order corrections are somewhat sensitive to the various cuts, since one integrates over quite different regions of phase space and different parts are affected differently as shown by the curves in Fig. 4. It was found that the product $\epsilon_{BB}(1+\delta)_{BB}$ drops between 0 and 1%, if one considers the various cuts. The experiment dependence makes it difficult to correct the R-values. Instead we use these values to make an estimate. If we vary all experimental points between the limits, the refitted value of α_s drops between 0 and 11%. Estimating the error to be half this range or less, we see that the uncertainty from higher order radiative corrections are smaller than the quoted experimental error of 11 %.

The effect of higher order corrections on M_Z is appreciably smaller, since the energy dependence is rather smooth, so the shape of R versus energy is not changed significantly.

5 Analysis method

Combining the data from different experiments is always a delicate procedure. It requires that:

- all data are have been corrected to the same level and their errors have a similar meaning;

- correlations between the data points within the same experiment and eventually between different experiments, must be considered.

Correlated errors between measurements can be taken into account by defining the χ^2 via an error correlation matrix[15]:

$$\chi^2 = \Delta^T V^{-1} \Delta \tag{29}$$

Here Δ is a column vector containing the residuals between the measurements R_i and its estimators R_i^*; V is the NxN error correlation matrix between N measurements. The elements of V can be estimated as follows: Assume the true R values deviate from the measured values by a common normalization factor f, which causes a correlation between the measurements and will make the off-diagonal elements of V nonzero. In this case the best estimator R_i^* from the fit including the correlations will deviate from the best estimator excluding the correlations - called r_i^*- by the same factor f, so $R_i^* = f\, r_i^*$. If the estimator is efficient, r_i^* will just be the averaged R value. The variance of f around 1 is called σ_n^2, where σ_n is the relative normalization error, so $< (f-1)^2 >= \sigma_n^2$. Then

$$
\begin{aligned}
V_{ii} &= \; < (R_i - R_i^*)^2 > \\
&= \; < (R_i - r_i^* + r_i^* - R_i^*)^2 > \\
&= \; < (R_i - r_i^*)^2 > \; + \; < (r_i^* - R_i^*)^2 > \\
&= \; < (R_i - r_i^*)^2 > \; + \; < (r_i^* - f r_i^*)^2 > \\
&= \; \sigma_i^2 \; + \; < (1-f)^2 > \; (r_i^*)^2 \\
&= \; \sigma_i^2 \; + \; \sigma_n^2 \, (r_i^*)^2 \\
V_{ij} &= \; < (R_i - R_i^*)(R_j - R_j^*) > \\
&= \; < (R_i - r_i^* + r_i^* - R_i^*)(R_j - r_j^* + r_j^* - R_j^*) > \\
&= \; < (R_i - r_i^* + (1-f)r_i^*)(R_j - r_j^* + (1-f)r_j^*) > \\
&= \; < (1-f)^2 > \; r_i^* \, r_j^* \\
&= \; \sigma_n^2 \, r_i^* \, r_j^*
\end{aligned}
$$

All terms containing $< (R_i - r_i^*) >$ have not been written, since they do not contribute and $\sigma_i^2 =< (R_i - r_i^*)^2) >$ contains the uncorrelated part of the error, which is the sum of both the statistical error - and point to point systematic error squared, but excludes the overall

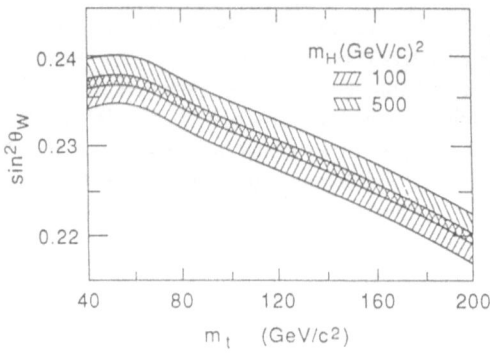

Figure 6. **Display of Sirlin relation between** $\sin^2 \theta_W$ **and** M_{top}

normalization error. Note that σ_n is the error on f, so it is the relative normalization error, which has to be multiplied by r_i^*, while σ_i is the absolute error on R_i.

It should be noted that the procedure of taking correlated errors into account via an error correlation matrix has distinct advantages over a likelihood method, in which the correlations are taken into account by fitting a renormalization constant to the data, as is done e.g. in Ref. [35]: in the first case one has only the physical parameters as free parameters, in the latter case one would have to fit 16 additional parameters in our problem. But what is more important, with a correlation matrix one can define a correlation between every pair of experimental points, thus taking correctly the energy dependence of the correlations into account, which exists for some experiments Furthermore one can study the effects of possible correlations between experiments, e.g. correlations from uncertainties in Monte Carlo programs or radiative corrections. This can be done by setting matrixelements connecting different experiments nonzero. It was found that there is practically no change in the fit results if one includes an overall correlation at the percent level[15]. We have not included the uncertainty from higher order QED radiative corrections in the covariance matrix, since we believe that treating it in a probabilistic way is uncorrect. We prefer to quote the variation of the final results for a given assumption on this correction.

For some experiments the separation into point-to-point and common error has not been given explicitly. In these cases it was checked that the numerical values of the fitted parameters were very stable against even large variations of these splittings.

6 Estimating the top mass in the minimal SM

The Sirlin relation (Eq. 14) gives a connection between M_Z, $\sin^2 \theta_W$ and Δr. The latter depends on the top and Higgs mass, so if M_Z and $\sin^2 \theta_W$ are known, one gets a handle on the unknown values of M_{top} and M_H, of course under the assumptions of the minimal SM. The relation between $\sin^2 \theta_W$ and M_{top} is shown in Fig. 6 for the best value of M_Z from MARK-II (Fig. from Ref. [1]). The measured values of $\sin^2 \theta_W$ usually depend on M_{top} too via the radiative corrections[36].

However, the most precise value of $\sin^2 \theta_W$ stems from the ratio of neutral and charged current cross sections in deep inelastic neutrino quark scattering. This ratio has been measured to a 1% accuracy and the extracted value of $\sin^2 \theta_W$ from such a ratio is insensitive to M_{top}[36,37,38]. For large top masses and assuming $\rho = 1$ one finds: $\sin^2 \theta_W = 0.2293 \pm 0.0030 \pm 0.0028$[38]. The

Table 4. M_{top} values for various M_H values from the Sirlin relation with $M_Z=91.11\pm0.23$ GeV/c^2 and $\sin^2\theta_W =0.2293\pm0.003\pm0.0028$. The errors on M_{top} stem from the uncertainty in $\sin^2\theta_W$. The uncertainty from M_Z introduces an additional error of 15 GeV/c^2.

M_H (GeV/c^2)	M_{top} (GeV/c^2)
10	100 ± 35
100	115 ± 35
500	134 ± 33
1000	143 ± 32

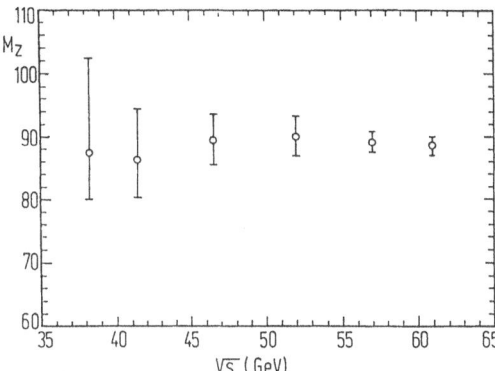

Figure 7. Fitted M_Z value as function of the maximum fitted centre of mass energy.

corresponding values of M_{top} are shown in Table 4 for various Higgs masses. The values range between 100 and 140 GeV, so taking this range into account in the errors, one estimates:

$$\mathbf{M_{top} = 120 \pm 35 \pm 20 \ GeV/c^2}.$$

The latter error covers the present uncertainty in M_Z as well.

7 Determination of M_Z from PEP/PETRA/TRISTAN data

At the highest TRISTAN energies Z^0 exchange does increase R already by 50 %, thus allowing a direct measurement of M_Z. The sensitivity to this direct propagator effect is demonstrated in Fig. 7 (from Ref.[4]): it shows the error on the fitted M_Z as function of the maximum available centre of mass energy for fixed values of the couplings, i.e. $\sin^2\theta_W$, so the only sensitivity is through the propagator.

The results on M_Z are rather sensitive to radiative corrections, which depend on M_Z and M_{top}, as shown in Fig. 8. The curves represent the ratio of the cross section including radiative

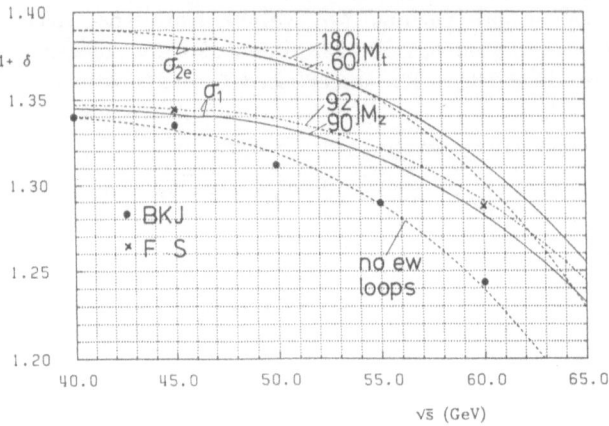

Figure 8. **Radiative correction factors calculated in first and second order for $k_{max} = 0.99$. The M_Z mass is 90 GeV/c^2 and the top mass 60 GeV/c^2, unless indicated differently. The lowest curve excludes the Z^0 selfenergy and is comparable with the BKJ radiative corrections as implemented in the Lund Monte Carlo (black dots). The first order full electroweak corrections from Fujimoto et al. (FS) (crosses) are in good agreement with the results from Burgers and Hollik (curves).**

corrections and the Born cross section, defined by Eqs. 1-3 with $K = K_2$ and $\kappa_2 = 1$. The second order exponentiated curve has been calculated with the program ZHADRO from Burgers and Hollik[21] and the other ones with suitably modified versions. The differences originate from the following physics:

- the lower curve corresponds to initial state and vertex corrections for both γ and Z^0 exchange, but only loop corrections for γ exchange, thus excluding the self energy of the Z^0 propagator. The Z^0-mass was taken to be 90 GeV, the Higgs mass 100 GeV and the top mass 60 GeV. Second order QCD corrections have been included using $\alpha_s = 0.12$). Only first order QED graphs have been taken into account ($O(\alpha^3)$. The maximum photon energy allowed corresponds to $k_{max} = E_\gamma/E_{beam} = 0.99$, except for the b-quark, where k_{max} corresponds to the kinematical limit assuming $m_b = 4.7$ GeV. For the other light quarks $u, d, s,$ and c we assumed masses of 0.04, 0.065, 0.3, and 1.5 GeV, respectively. These quark masses have been used to calculate the vacuum polarization. Since the vertex corrections for the Z^0-propagator are small in the on-shell renormalization scheme of Böhm et al.[39], this curve is very close to the results from the well known Berends, Kleis, Jadach (BKJ) results[19] as implemented in the Lund Monte Carlo. These BKJ results, indicated in the figure as solid dots, ignore the Z^0 self energy too.

- The two middle curves for two Z^0 masses (90 and 92 GeV/c^2) include in addition the loop corrections from the Z^0-propagator. The sensitivity comes mainly from the graphs with initial state radiation, which depend on the cross section at the energy after radiation and are therefore sensitive to the shape of the cross section, i.e. to the Z^0 mass. The results agree with the calculations from Fujimoto et al.[40], which have been indicated as crosses in the figure. These crosses were calculated for $M_Z = 91.9$ and $M_{top} = 45$ GeV/c^2 and

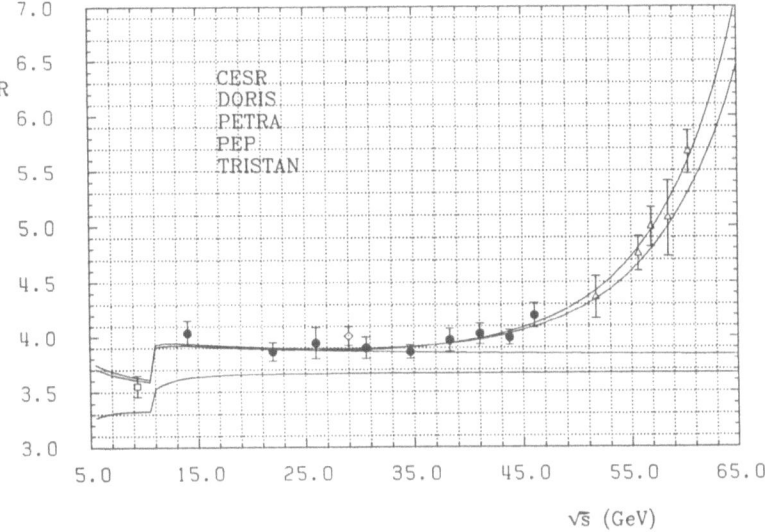

Figure 9. **Experimental data and the Standard Model expectations for M$_Z$ = 88.9 (highest curve) and M$_Z$ = 91.11 GeV/c^2, both using $\sin^2 \theta_W$ = 0.2293 and Γ_{tot}=2.5 GeV/c^2.**

include final state radiation. It should be noted that in these programs final state radiative corrections are large and cannot be neglected, as is the case for the calculations of Hollik et al.[20]. The reason is the different wave function renormalization, which shifts part of the electroweak loop corrections into the vertex corrections; however, the final corrections in both calculations are very close.

As long as M_Z is poorly known, the radiative corrections cannot be calculated accurately. One only can use the available data to estimate M_Z, apply the corresponding radiative corrections, and iterate this procedure untill a stable value of M_Z has been found. Since the radiative corrections depend on M_{top} too, this procedure has to be repeated for every top mass. The M_Z values found for M_{top} between 60 and 180 GeV/c^2 range between 88.1 and 89.3 GeV/c^2[4, 12]. Since the error on each value of M_Z is at least 1 GeV/c^2, there is no strong discrepancy (less than 3 s.d.) with the M_Z value of 91.11 GeV/c^2 from SLC. Therefore, we will use this latter value to calculate the radiative corrections more precisely and use for M_{top} the value given in the previous section ($M_{top} = 120 \pm 35 \pm 20$ GeV/c^2).

The result from a 3 parameter fit to the data on R between 7 and 60.8 GeV is:

$$
\begin{aligned}
M_Z &= 89.3 \pm 1.4 \text{ GeV/c}^2 \\
\alpha_s(34 \ GeV) &= 0.142 \pm 0.018 \\
\sin^2 \theta_W &= 0.217^{+0.024}_{-0.019}
\end{aligned}
$$

The value of $\sin^2 \theta_W$ agrees with the more precise value from deep inelastic neutrino scattering[38]: $\sin^2 \theta_W = 0.2293 \pm 0.003 \pm 0.0028$. If we use this value for $\sin^2 \theta_W$, we find:

$$
\begin{aligned}
M_Z &= 88.9 \pm 1.2 \text{ GeV/c}^2 \\
\alpha_s(34 \ GeV) &= 0.137 \pm 0.017
\end{aligned}
$$

The uncertainty from the top mass entering into the radiative corrections is less than $0.1\,\text{GeV}/c^2$, if we use $M_{top} = 120 \pm 35 \pm 20\ \text{GeV}/c^2$ (see previous Section).

Here we included the data as presented at the Topical Conference at KEK[41]. These data, which are of a preliminary nature and the results should be considered accordingly, included full electroweak corrections using $M_{top}=45\ \text{GeV}/c^2$ and $M_Z=91.8\ \text{GeV}/c^2$. However, the results are in excellent agreement with older data, as presented at the Munich Conference[5] and after taking the different radiative correction factors into account [4].

The correlation coefficients between the 3 parameters are (independent of the top mass): $\rho(\sin^2\theta_W, M_Z) = -0.57$, $\rho(\Lambda^{(5)}_{\overline{MS}}, \sin^2\theta_W) = -0.43$ and $\rho(\Lambda^{(5)}_{\overline{MS}}, M_Z) = 0.41$. The χ^2/DF is $70/101$; this excellent χ^2 comes mainly from the fact that the common normalization errors might have been overestimated. If we calculate the χ^2 only from the diagonal elements of the matrix V^{-1}, thus ignoring the correlations, but including the complete errors, we find χ^2_D/DF is $94/101$.

Fig. 9 shows the fit results together with the expectations from the SM with $M_Z=88.9$ and $91.11\ \text{GeV}/c^2$, $\Gamma_{tot}=2.5\ \text{GeV}/c^2$, and $\sin^2\theta_W =0.2293$. For clarity we have averaged the data points within certain energy bins in the following way: we have fitted a constant value to the data points within a certain energy bin using the complete error correlation matrix. This procedure is equivalent to calculating a weighted average by taking correctly into account independent and correlated errors. So the error bars represent the total errors including the correlation and the data have not been renormalized.

One clearly sees that the TRISTAN data is above the fit expected for $M_Z=91.11\ \text{GeV}/c^2$. The discrepancy is about 2 standard deviations. There are several possibilities to explain these differences:

- The difference is just a statistical fluctuation: 2 s.d. effects occur in 5% of all comparisons of independent measurements.

- The TRISTAN data is systematically too high for some unknown reason. In the fit we have assumed a 5% normalization error for each experiment and this error is included in the error on M_Z. We have checked that all experiments are consistent with the common fit, so it is not a single experiment, which causes M_Z to come out low. It is hard to explain the difference by a common normalization problem for all TRISTAN experiments for the simple reason that M_Z is determined mainly by the increase in the cross section at the highest energies, so it is the shape, which is important, not the overall normalization. If one lowers all TRISTAN data simultaneously by as much as 6%, one gets $M_Z=91\ \text{GeV}/c^2$. However, the χ^2 gets considerably worse in that case, since the lowest TRISTAN data start to disagree with the highest PETRA data.

- The most interesting explanation would be the contribution of new physics. One possible suggestion is the existence of a new heavy neutral gauge boson (Z')[14], which would be able to explain simultaneously a too high hadronic cross section and a too low leptonic cross section; The μ-pair cross section from the combined TRISTAN data is indeed on the low side[13]. If one uses both hadronic and leptonic data from TRISTAN in the fit for M_Z, the value becomes $90.4\ \text{GeV}[13]$.

8 Comparison between PEP/PETRA/TRISTAN and SLC data on R

Fig. 10 shows all data on R for centre of mass energies between 7 and 93 GeV. In order to compare the 'low' energy data with the published SLC data, we subtracted the small leptonic 'background' contribution in the latter and applied the second order exponentiated radiative

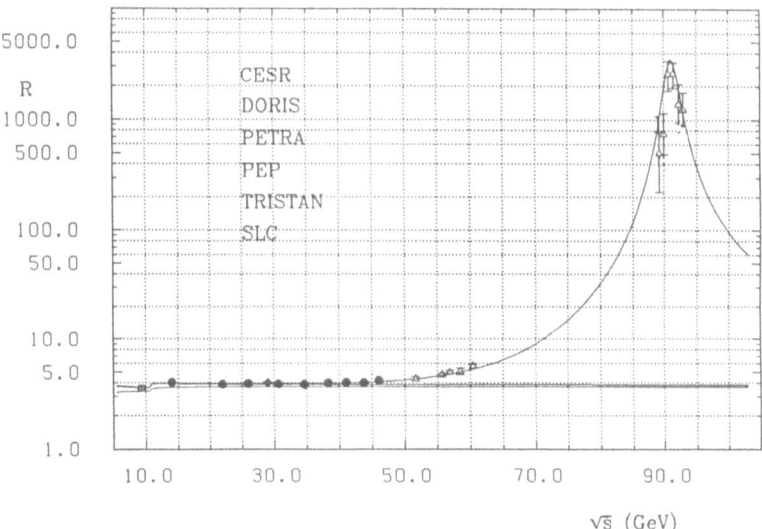

Figure 10. R values between 7 and 93 GeV. The curves indicate the contributions from the quark parton model (QPM), the QCD contribution, and the electroweak contribution for $\sin^2 \theta_W = 0.2293$, $M_Z = 91.11$, **and** $\Gamma_{tot} = 2.5$ **GeV/c^2.**

corrections using the program ZHADRO from G. Burgers and W. Hollik[21]. We used M_{top}=115 GeV/c^2 and M_H=100 GeV/c^2 and assumed that the detection efficiency, calculated for first order radiative corrections, does not change in second order, which is certainly a good approximation with the present statistical errors.

The curve in Fig. 10 corresponds to $M_Z = 91.11$ GeV/c^2 and $\Gamma_{tot} = 2.5$ GeV/c^2. Here we have used the SM width (Γ_{tot}=2.5 GeV/c^2), which is close to the width calculated from the fitted number of neutrino species: $\Gamma_{tot} = 2.63 \pm 0.23$ GeV/c^2 corresponding to $N_\nu = 3.8 \pm 1.4$[1]. The directly fitted width ($1.61^{+0.60}_{-0.43}$ GeV/c^2[1]) is slightly lower, because of the different assumptions involved: in the first case one assumed all couplings known (Table 2), while in the fit of Γ_{tot} one only assumed the lepton couplings to be known. In the latter case the couplings to hadrons are determined by: $\Gamma_f = \Gamma_h + f(\Gamma_\mu + \Gamma_\tau) = \Gamma_{tot} - (1 - f)(\Gamma_\mu + \Gamma_\tau) - \Gamma_e - N_\nu \Gamma_\nu$ with f=0.556. Γ_{tot}=1.61 leads to $\Gamma_h \approx 0.8$ GeV/c^2, which has to be compared with 1.6 GeV expected for 5 quarks and 3 colours. However, neither value of Γ_{tot} (1.6 or 2.6) deviates by more than 1.5 s.d. from the expected SM value of 2.5 GeV/c^2, so there is no disagreement. One just has to wait for more statistics, especially to get better limits on the number of neutrino species.

9 Determination of α_s

For the determination of α_s it is convenient to use the G_F parametrization (i.e. combining Eqs. 1-3 and 6,10), which makes the results insensitive to M_Z and the unkown top mass owing to the compensation in κ_1. In Table 5a we give the fit results of α_s with the G_F parametrization. We fixed $\sin^2 \theta_W$ to the world average, since all fits for $\sin^2 \theta_W$ are consistent with that value (0.2293[38]). We used M_Z=91.11 GeV/c^2 and M_{top}=100 GeV/c^2, but as mentioned before, the results are insensitive to the latter values. The values of α_s obtained from the different energy regimes are consistent. From the fitted scale parameter $\Lambda_{\overline{MS}} = 0.260^{+0.160}_{-0.130}$ GeV one finds in third order:

$$\alpha_s(34 \ GeV) = 0.144 \pm 0.015$$

Table 5. $\alpha_s(34\ GeV)$ and $\Lambda_{\overline{MS}}^{(5)}$ fitted with the G_F parametrization.

Data	Energy range	$O(\alpha_s^2)$	$O(\alpha_s^3)$
PEP, PETRA	$14 - 47\ GeV$	$\alpha_s = 0.167 \pm 0.025$	$\alpha_s = 0.151 \pm 0.020$
		$\Lambda_{\overline{MS}}^{(5)} = 580^{+470}_{-330}\ MeV$	$\Lambda_{\overline{MS}}^{(5)} = 330^{+240}_{-180}\ MeV$
PEP, PETRA, TRISTAN	$14 - 61\ GeV$	$\alpha_s = 0.172 \pm 0.024$	$\alpha_s = 0.155 \pm 0.018$
		$\Lambda_{\overline{MS}}^{(5)} = 650^{+450}_{-340}\ MeV$	$\Lambda_{\overline{MS}}^{(5)} = 370^{+230}_{-190}\ MeV$
CESR, DORIS, PEP, PETRA, TRISTAN	$7 - 61\ GeV$	$\alpha_s = 0.159 \pm 0.019$	$\alpha_s = 0.144 \pm 0.015$
		$\Lambda_{\overline{MS}}^{(5)} = 460^{+290}_{-230}\ MeV$	$\Lambda_{\overline{MS}}^{(5)} = 260^{+160}_{-130}\ MeV$

$$\alpha_s(91\ GeV) = 0.122 \pm 0.011$$

Here we have taken the third order QCD corrections into account, which lower the α_s values about 10% (see Table 5) with respect to the second order one.

It is interesting to determine the QCD contribution independently from the definition of α_s. If one assumes a linear energy dependence within the energy range, one finds this contribution to be[4]: $f_{QCD}(\sqrt{s} = 34\ GeV) = 1.057 \pm 0.008$. An extrapolation of this value to the LEP/SLC energy range yields[4] $f_{QCD}(\sqrt{s} = 91\ GeV) = 1.046 \pm 0.006$, which is an experimental number for the infinite series $(1 + \alpha_s/\pi + ...)$.

10 Scale dependence of α_s

In the definition of α_s we have made the usual choice $Q^2 = \sqrt{s}$, since this is the only large scale in the process. However, one may argue that gluon radiation occurs at a much smaller scale and there is no reason to use such a large scale. A large amount of literature exists with arguments for the choice of scale[25,42]. The main arguments are based on the size of the higher order corrections and/or the sensitivity to the renormalization scheme. It is interesting to study the scale dependence for R, since this is the first quantity for which both the second and third order contributions have been calculated, so one can study the scale dependence in second order and check if there are scales for which the third order corrections are small. We have studied these contributions as a function of the scale in contrast to previous results for a few specific scales[43]. This can be done easily as follows. Suppose a variable is given in a certain renormalization scheme and for a given Q^2 scale to be:

$$R = r_1\alpha_s + r_2\alpha_s^2 + r_3\alpha_s^3 + O(\alpha_s^4) \tag{30}$$

If we choose the coupling at a different scale, we get:

$$R' = r_1'\alpha_s' + r_2'\alpha_s'^2 + r_3\alpha_s'^3 + O(\alpha_s'^4) \tag{31}$$

If one neglects the terms of order $O(\alpha_s^4)$, then

$$R' - R = dR = r_1 d\alpha_s + \alpha_s dr_1 + \alpha_s^2 dr_2 + 2\alpha_s r_2 d\alpha_s + \alpha_s^3 dr_3 = 0 \tag{32}$$

This can only be zero, if the coefficient for each power of α_s equals zero, which yields 3 equations for the 3 unknowns r_1', r_2', r_3'. After calculating $d\alpha_s$ from the renormalization group equation:

$$\mu \frac{\partial \alpha_s(\mu)}{\partial \mu} = -\beta_0\alpha_s^2(\mu) - \beta_1\alpha_s^3(\mu) - \beta_2\alpha_s^4(\mu) + O(\alpha_s^5), \tag{33}$$

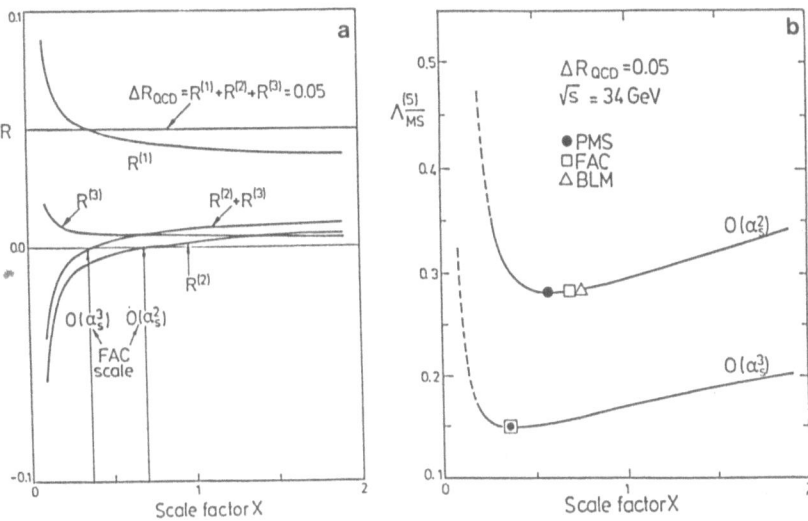

Figure 11. **QCD contributions to R in first, second, and third order ($R^{(1)}$, $R^{(2)}$, $R^{(3)}$, respectively) (a) and Λ determinations in second - and third order as a function of the renormalization scale factor x for $R_{QCD}=0.05$ (b).**

we find for the coefficients at a different scale $Q' = xQ$:

$$
\begin{aligned}
r_1' &= r_1 \\
r_2' &= r_2 + r_1\beta_0 ln\ x \\
r_3' &= r_3 + r_1\beta_1 ln\ x + 2r_2\beta_0 ln\ x
\end{aligned}
\tag{34}
$$

The β-factors are given by[44]:

$$
\begin{aligned}
\beta_0 &= \frac{1}{6\pi}\left[33 - 2n_f\right] \\
\beta_1 &= \frac{1}{12\pi^2}\left[153 - 19n_f\right] \\
\beta_2 &= \frac{1}{64\pi^3}\left[2857 - 5033n_f/9 + 325n_f^2/27\right]
\end{aligned}
$$

The first two β-factors are independent of the renormalization scale. The scale dependence of the last one can be calculated by means of Stevenson's invariants[25]. The QCD contributions to R in first, second, and third order have been displayed in Fig. 11a as function of the renormalization factor x, assuming the total contribution to be constant ($\Delta R_{QCD} = 0.05$). One observes that at $x = 1$ the third order contribution is indeed larger than the second order contribution, but at small and large x the absolute value of the second order contribution is larger. However, at all scales the first order contribution is dominant. We have indicated the scales at which the second order or the second plus third order contributions become zero. These are the so-called FAC scales (Fastest Apparent Convergence)[45]. After recalculating the coefficients at a new scale, one can redetermine the corresponding α_s from the measured R-value and recalculate the corresponding Λ value. The result is shown in Fig. 11b. The minimum in this curve is the PMS scale corresponding to the point of minimal sensitivity, a concept introduced by Stevenson[25]. We have indicated too the scale advocated by Brodsky, Lepage, and Mackenzie (BLM)[46]. One observes that at all scales the Λ values in third order are roughly a factor two below the Λ values in second order, indicating that for this reaction there is nothing like an optimum scale, where the higher orders are not important.

This clearly indicates that all the heated discussions about choosing a certain scale mean nothing more than betting on the future: you only can say something seriously about higher order contributions by calculating them, not by fiddling with renormalization scales or schemes. Nevertheless, studying the scale dependence in a given order is still very useful, since a strong scale dependence indicates large higher order contributions. On the other hand, a small scale dependence does not guarantee that the higher orders are negligible, as can be seen from the above example of R.

11   Summary

We have determined the Standard Model parameters α_s, M_Z, and $\sin^2 \theta_W$ from a 3 parameter fit to the data on R between 7 and 60.8 GeV. The result is:

$$
\begin{aligned}
M_Z &= 89.3 \pm 1.4 \text{ GeV}/c^2 \\
\alpha_s(34 \ GeV) &= 0.142 \pm 0.018 \\
\sin^2 \theta_W &= 0.217^{+0.024}_{-0.019}
\end{aligned}
$$

The value of $\sin^2 \theta_W$ is in good agreement with the more precise value from deep inelastic neutrino scattering[38]: $\sin^2 \theta_W = 0.2293 \pm 0.003 \pm 0.0028$. If we use this value for $\sin^2 \theta_W$, we find:

$$
\begin{aligned}
M_Z &= 88.9 \pm 1.2 \text{ GeV}/c^2 \\
\alpha_s(34 \ GeV) &= 0.137 \pm 0.017
\end{aligned}
$$

The uncertainty from the top mass entering into the radiative corrections is negligible, if we use

$$
M_{top} = 120 \pm 35 \pm 20 \text{ GeV}/c^2,
$$

as obtained from a fit to M_Z and $\sin^2 \theta_W$ from deep inelastic scattering using the Sirlin relation Eq. 14.

The SM parametrization of the cross section with the Fermi constant is insensitive to M_Z and M_{top}. This can be exploited to determine the strong coupling constant almost independently of M_Z and M_{top}. From the fitted scale parameter $\Lambda_{\overline{MS}} = 0.260^{+0.160}_{-0.130}$ GeV one finds:

$$
\begin{aligned}
\alpha_s(34 \ GeV) &= 0.144 \pm 0.015 \\
\alpha_s(91 \ GeV) &= 0.122 \pm 0.011
\end{aligned}
$$

Here we have taken the third order QCD corrections into account, which lower the α_s values about 10% (see Table 5) with respect to the second order one.

The QCD series $1 + \alpha_s/\pi + ...$ has been determined from a direct fit to the data in a model independent way. Extrapolating to the Z^0 region, one finds this factor to be 1.046 ± 0.006[4].

The M_Z value from 'low' ' energy data is about 2 standard deviations below the one from 'on-resonance' data. if one only takes hadronic data. On the other hand, if one includes both hadronic and leptonic data from TRISTAN in the fit for M_Z, the value becomes 90.4 GeV/c^2[13]. This indicates that the hadronic data is below the expectation of the SM, while the leptonic data is below, which is the possible signature for a second neutral heavy gauge boson[14]. A second boson will have little effect on the resonance, but could show up below and should show up in $p\bar{p}$ collisions. Clearly, the whole effect is statistically not very compelling, but it will be interesting to watch future data from TRISTAN at somewhat higher energies.

References

[1] MARK-II Coll., G.S. Abrams et al., Phys. Rev. Lett. **63**(1989) 724

[2] UA1 Coll., C. Albajar et al., CERN-EP/88-168, November 1988, to be published in Z. Physik C,
UA2 Coll., R. Ansari et al., Phys. Lett. **186B** (1987) 440.
CDF Coll., F. Abe et al., Phys. Rev. Lett.**63** (1989) 720

[3] AMY Coll., T. Mori et al.. Phys. Lett. **218B** (1989) 499

[4] G. d'Agostini, W. de Boer, and G. Grindhammer, DESY Report 89-057, to be published in Phys. Lett. B

[5] T. Kamae, UT-HE-Preprint-88-05, published in Proc. of 24th Int. Conf. on High Energy Physics, Munich (1988), Eds. R. Kotthaus and J.H. Kühn, p. 156

[6] T. Tauchi, KEK-Preprint-88-39, to be published in Proc. of Multiparticle Dynamics, Arles, France (1988)

[7] T. Nozaki, presented at Renc. de Moriond, Les Arcs, France (1989)

[8] G. d'Agostini, W. de Boer, and G. Grindhammer, Contr. to 24th Int. Conf. on High Energy Physics, Munich (1988)

[9] G. d'Agostini, Proc. Renc. de Moriond (1987), to be published

[10] W. de Boer, SLAC-Pub 4482, Proc. of the Xth WARSAW Symposium on Elementary Particle Physics, Kazimierz, Ed. Z. Ajduk, Poland, (1987), p. 503

[11] W. de Boer, Proc. of 24th Int. Conf. on High Energy Physics, Munich (1988), Eds. R. Kotthaus and J.H. Kühn, p. 905

[12] W. de Boer, DESY 89-067. to be published in Proc. of the Ringberg Workshop on Radiative Corr., Munich (1989), Ed. H. Kühn

[13] S. Iwata, TRISTAN Results presented at the IXth Int. Conf. on Phys. in Collision, Jerusalem, June, 1989

[14] A.A. Pankov and C. Verzegnassi, CERN-TH.5373/89

[15] CELLO Coll., H.J. Behrend et al., Phys. Lett. **183B** (1987) 400

[16] M. Dine, J. Sapirstein, Phys. Rev. Lett. **43** (1979) 668
K.G. Chetyrkin et al. , Phys. Lett. **85B** (1979) 277
W. Celmaster, R.J. Gonsalves, Phys. Rev. Lett. **44** (1980) 560

[17] S.G. Gorishny, A.L. Kataev, and S.A. Larin, Hadron Structure '87, Proc., Smolenice, Czechoslovakia, Physics and Applications. Vol. **14** (1988) 180, and preprint JINR, E2-88-254

[18] T. Sjöstrand, Comp. Phys. Comm. **27** (1982) 243, ibid. **28** (1983) 229
T. Sjöstrand and M. Bengtsson, Comp. Phys. Commun. **43**(1987)367

[19] F.A. Berends, R. Kleiss, S. Jadach, Comp. Phys. Commun. **29** (1983) 185

[20] W. Hollik, DESY 88-188, December 1988, and references there in

[21] G. Burgers, CERN-TH/5119/88, G. Burgers and W. Hollik, CERN-TH5131/88, both published in the Yellow Book on Polarization at LEP (CERN- 88-06, Vol. 2)

[22] A. Sirlin, Phys. Rev. **D22**(1980) 971

[23] W.J. Marciano, Phys. Rev. **D29** (1984) 580

[24] Particle Data Group, M. Aguilar-Benitez et al.. Phys. Lett. **B204** (1988) 1

[25] P.M. Stevenson, Phys. Rev. **D23** (1981) 2916

[26] T. Kinoshita, J. Math. Phys. **3** (1962) 650
T.D. Lee and M. Nauenberg Phys. Rev. **D133** (1964) 1549

[27] F. A. Berends, G. J. H. Burgers, and W. L. van Neerven, Phys. Lett. **185B** (1987) 395; Nucl. Phys. **B297** (1988) 429

[28] W. de Boer, SLAC-Pub 4682, Nucl. Instr. Meth. A278(1989)687

[29] D.R. Yennie, S.C. Frautschi and H. Suura, Annals of Phys.**13** (1961) 379

[30] F. A. Berends, G. J. H. Burgers, and W. L. van Neerven, Nucl. Phys. **B297** (1988) 429

[31] J. P. Alexander, private communication and J. P. Alexander, G. Bonvicini, P. S. Drell and R. Frey, Phys. Rev. **D37** (1988) 56

[32] M. Greco, Riv. Nuovo Cim. **11**(1988) 5:1 and Phys. Lett. **177B**(1986) 97
See also Berends and Komen, Nucl. Phys. **B115**(1976) 114

[33] The higher order calculations have been implemented in the standard μ-pair generator of F.A. Berends, R. Kleiss, and S. Jadach (Comp. Phys. Commun. **29** (1983) 185) by J.P. Alexander (private communication) and adapted for quarks by W. de Boer, Mark-II Note 210

[34] F. A. Berends and R. Kleiss, Nucl. Phys. B228(1983) 537. This Bhabha generator was first exponentiated by M. Levi (private communication) according to a prescription from Tsai, SLAC Pub 3129 (1983). We modified it in order to have the same exponentiation for hadrons and Bhabhas; we know that this exponentiation procedure agrees with the exact second order calculation in case of μ-pair production

[35] R. Marshall, RAL-Preprint-88-049, submitted to Z. Phys. **C**

[36] J. Ellis and G.L. Fogli , Phys. Lett **B213** (1988) 526

[37] G.L. Fogli and D. Haidt, Z. Physik **C40** (1988) 379

[38] G.L. Fogli. Bari Preprint BA-TH/40-89 (1989), to be published

[39] M. Boehm, W. Hollik, H. Spiesberger, Fort. der Physik **34**(1986)687

[40] J. Fujimoto, Y. Shimizu, Mod. Phys. Lett. **A3** (1088) 581

[41] AMY Coll., H. Sagawa et al., Phys. Rev. Lett. **60** (1988) 93, Phys. Lett. **218B** (1989) 499, and G. Kim, Topical Conference, KEK (1989);
TOPAZ Coll., I. Adachi et al., Phys. Rev. Lett. **60** (1988) 97
and S. Suzuki, Topical Conference, KEK (1989);
VENUS Coll., H. Yoshida et al., Phys. Lett. **198B** (1987) 570
and K. Ogawa, Topical Conference, KEK (1989)

[42] D.W. Duke, R.G. Roberts, Phys. Reports **120** (1985) 275

[43] C.J. Maxwell and J.A. Nicholls, Phys. Lett. **B213** (1988) 217
A.P. Contogouris and N. Mebarki, Phys. Rev. **D39** (1989) 1464

[44] D. Gross and F.Wilczek, Phys.Rev. Lett. **30**(1973)1323; Phys. Rev. **D8** (1973) 3633
H.D. Politzer, Phys. Rev. Lett. **30** (1973) 1346
O. Tarasov, A. Vladimirov, and A. Zharkov, Phys. Lett. **93B**(1980)429

[45] G. Grunberg, Phys. Lett. **B95** (1980) 70

[46] S. Brodsky, G.P. Lepage and P. Mackenzie, Phys. Rev. **D28** (1983) 493

THE SLD DETECTOR AND THE POLARIZATION PROJECT AT SLC

Traudl Hansl—Kozanecka

Massachusetts Institute of Technology
Cambridge, Massachusetts 02139

Abstract

The SLD is the second generation detector for the SLAC Linear Collider (SLC). A short review of the detector and its status is given. The availability of longitudinally polarized beams will be a unique opportunity at SLC. The polarization project and its status is described.

1. Introduction

Since March 1989 the SLAC Linear Collider has started to deliver events. The Mark II detector at SLC has for the first time performed an energy scan of the Z° resonance. This has opened a new area of high precision measurements of the parameters of the Standard model.

SLD is the second generation detector at SLC and is well optimized for Z° physics and exploration of new physics in the Z° energy region. We give a short review of the detector and its status (section 2) and describe the most important parameters of the SLC at the interaction point (section 3). The polarization project and its status are described in section 4.

2. The SLD detector

The SLD is optimized for the study of e^+e^- interactions at Z^0 energies. The detector is being built by an international collaboration of about 130 physicists and 30 engineers from 32 institutions[1]. The important design criteria are

- uniform coverage over the full solid angle;
- hadronic and electromagnetic calorimetry, with good energy resolution and fine segmentation;
- robust particle identification over a wide range of momenta;
- good vertex detection;

The detector is shown schematically in Fig. 1 . The SLD will have a CCD vertex detector, high resolution drift chambers, a Čerenkov Ring Imaging Detector for particle identification, and calorimetry based on lead/liquid argon followed by an iron/Iarocci tube system. The 0.6 Tesla magnetic field is produced by a conventional aluminum solenoid. We will describe the major subsystems from the interaction point outward, with special emphasis on those components that distinguish SLD from other large new generation detectors.

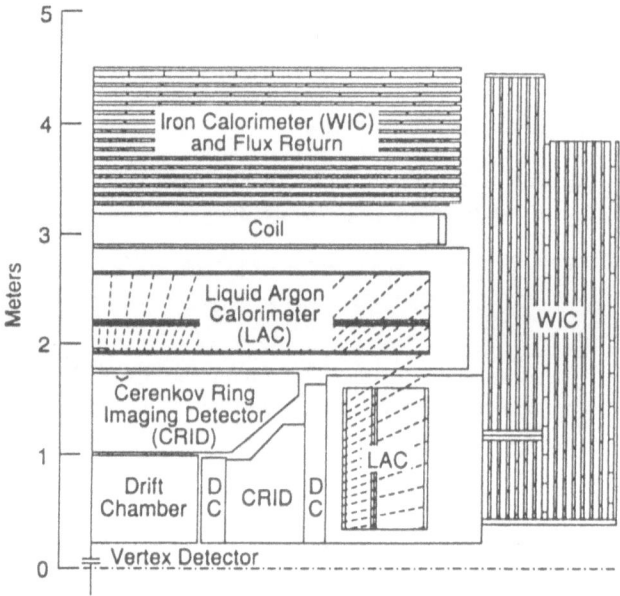

Figure 1.   Vertical section in the plane including the $e^\pm$ beams of one quadrant of the SLD.

2.1. THE VERTEX DETECTOR

The small size of the SLC beam pipe allows the vertex detector to be placed very close to the interaction point; the collision frequency of not more than 180 Hz permits long readout times. Therefore high resolution Charge−Coupled Devices (CCD's) can be used[2]. They offer spatial resolution of $\approx 6\mu m$ as unambiguous space points with high efficiency ($\geq 98\%$). The vertex detector consists of four cylindrical layers of ceramic ladders on which the precisely located CCD chips are mounted.

Two versions of the vertex detector have been designed: a first version to be used for initial data taking with a steel beam pipe of 25 mm (Fig. 2), and a final version that will be used for final data taking with a beryllium beam pipe of 16 mm. The initial version will use 60 ladders and a total of 480 CCD's or 10^8 pixels. It will achieve a resolution of the impact parameter of $38\mu m \oplus \frac{45\mu m}{p}$ for low momenta and $11\mu m \oplus \frac{87\mu m}{p}$ for high momenta ($\geq 2\ GeV$). The average number of pixels per track is 2.2 for $|\cos\theta| \leq 0.7$. The coverage extends up to $|\cos\theta| \leq 0.8$.

Given the experience of Mark II with beam related backgrounds it seemed to be prudent to plan for two different beam pipe radii. However, when SLC achieves stable operations with clean beams, the original 16 mm design will be installed.

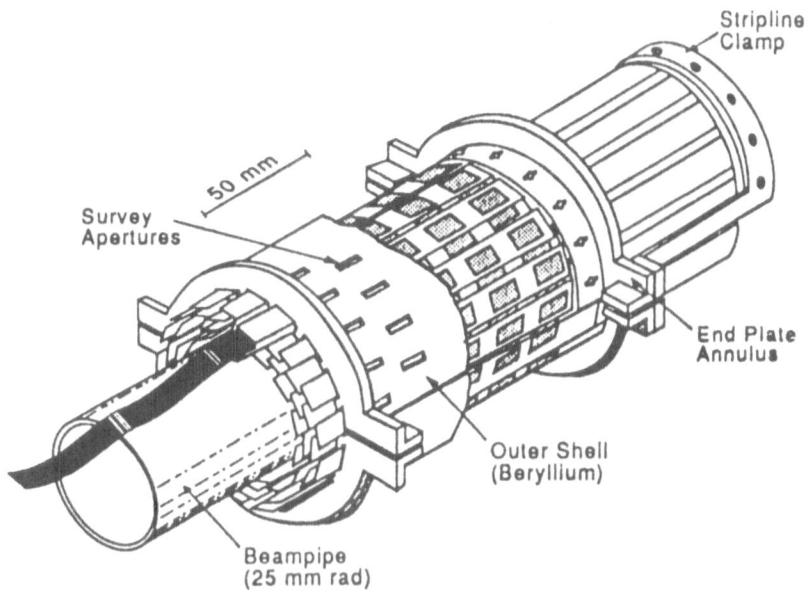

Figure 2. Layout of the initial version of the vertex detector, designed for a 25 mm beam pipe. 8 CCD's are mounted per ladder. The ladders are arranged in cylinders and held in precise position by slotted annulii. The final version of the vertex detector will use a 16 mm beam pipe.

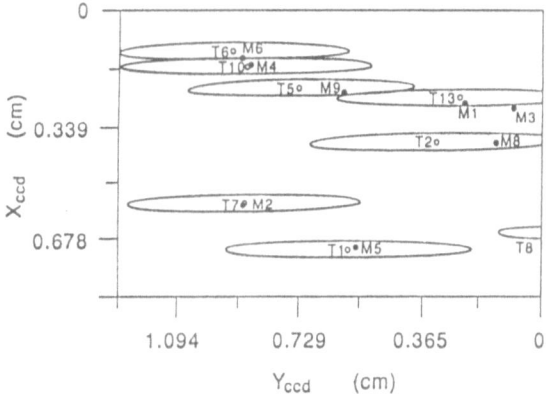

Figure 3. Error ellipse (4 sigma) of drift chamber tracks projected onto the plane of an outer CCD chip of the vertex detector (courtesy of J. Richman).

2.2. TRACKING – THE CENTRAL AND ENDCAP DRIFT CHAMBERS

The Central Drift Chamber extends from 20 cm radius to 1 m radius. The sense wires are arranged in 80 layers which are subdivided into 10 superlayers of vectored cells. A slow gas (75% CO_2, 21% Ar, 4% $Isobutan$, 0.4% H_2O) is used to achieve high spatial resolution. Prototype tests[3] have found $55\mu m$ spatial resolution and $0.3\% \times L$ resolution for charge division. Digitization of the wave form on each end of the sense wire results in a two−track resolution of 1 mm. Charge division and four layers of 40 mrad stereo provide 2×10^{-3} in the tangent of the dip angle. In the full−sized drift chamber it is expected that the resolution will be degraded by systematic uncertainties, and an overall resolution of $70\mu m$ is expected. This translates into a momentum resolution of

$\frac{\sigma(p_T)}{p_T} \approx 1\% \oplus 0.2\,p_T$ over 85% of the solid angle, dominated by multiple scattering in the drift gas.

A pair of planar drift chambers for each endcap extends the tracking coverage to 97% of 4π. Each of these endcap chambers consists of three 60° stereo chambers with spatial resolution of $100\mu m$.

To link the drift chamber information to the hits found in the vertex detector the tracks reconstructed in the drift chamber are extrapolated to the outer CCD planes of the vertex detector. The extrapolated track errors generate an error ellipse on the CCD−plane within which the search for candidate hits is performed. Projection of the accepted candidate hit onto the next inner vertex layer reduces the search area to about 70 pixels at low momentum and to a few pixels at high momentum. Fig. 3 shows the four−sigma error contours from the initial extrapolation to an outer layer.

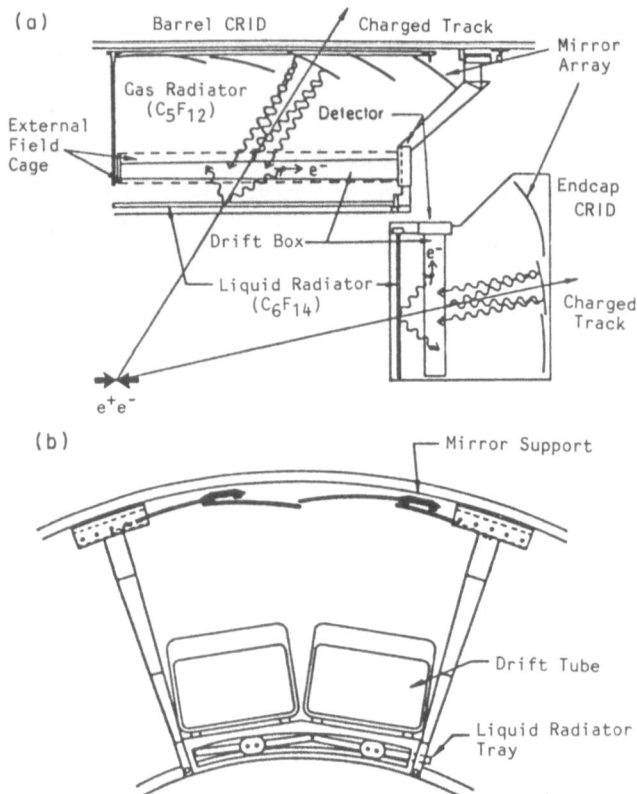

Figure 4. The SLD Čerenkov Ring Imaging detector system a) Principal components of the barrel and endcap systems. b) Engineering detail of the barrel CRID vessel and internal components: azimuthal arrangement at one end.

2.3. PARTICLE IDENTIFICATION: THE ČERENKOV RING IMAGING DETECTOR

Particle identification is crucial for tagging and selecting special decay channels. The Čerenkov Ring Imaging detector (CRID) will provide almost complete particle identification over $\geq 90\%$ of the solid angle. By making use of both liquid and gaseous radiators, $\pi/K/p$ separation will be possible up to about 30 GeV/c, and e/π separation up to about 5 GeV. Above this momentum the SLD calorimetry becomes effective at e/π separation. Čerenkov photons, generated by the passage of charged tracks enter the drift boxes through quartz windows (Fig. 4). In case of photons from the gas radiator, the Čerenkov light is reflected by quartz mirrors onto the drift boxes. The photons will ionize a drift gas mixture with TMAE as gaseous photo cathode. The resulting photoelectrons, which drift up to 1.25 m in a uniform electric field, are detected by a proportional wire plane of $7\mu m$ carbon sense fibres. Charge division along the sense wires provides the conversion depth of the photon, while the drift time and wire number give its longitudinal and transverse position. The ring images can be sharpened up by exploiting this space point information. This technique has been demonstrated by the SLD [4] and DELPHI [5] groups.

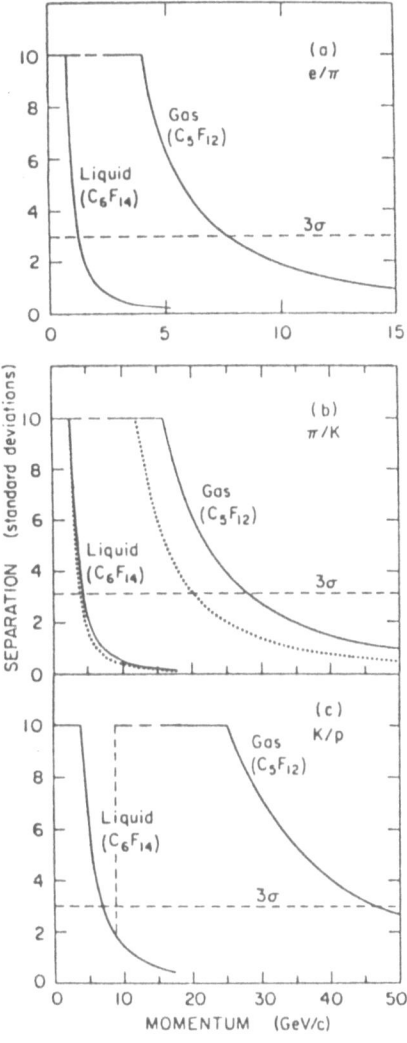

Figure 5. Particle separation capability of the CRID system (at middle of the half barrel) as function of momentum. The performance of the liquid (C_6F_{14}) and gas (C_5F_{12}) radiators are shown separately. The dashed line indicates the region where the heavier particle is below radiator threshold. The dotted line in (b) is for doubled measurement errors.

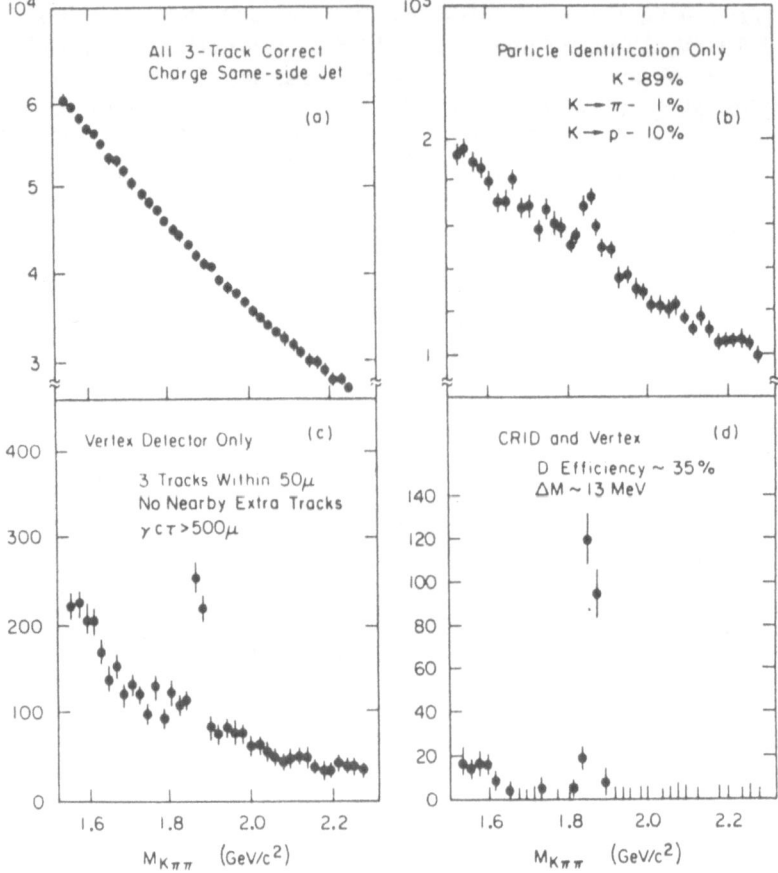

Figure 6. The number of events versus $K\pi\pi$ mass for three particles from same side jets and total charge $= \pm 1$. a) All three—track combinations; b) cut based on the CRID without use of the vertex detector; c) cut based on the vertex detector without use of the CRID; d) cut based on the CRID and vertex detector.

To separate between particle types a combined loglikelihood is formed using the extrapolated driftchamber tracks in search for the Čerenkov rings and the expected number of photoelectrons. Fig. 5 shows the separation that can be achieved.

Both, particle identification and the precise detection of the vertex are needed to find efficiently exclusive final states, especially from heavy quarks. For example, a secondary vertex and an identified $K\pi\pi$ combination can give an essentially background—free D sample with 35% efficiency (Fig. 6).

2.4. CALORIMETRY: LIQUID ARGON CALORIMETER, WARM IRON CALORIMETER AND LUMINOSITY MONITORS

Liquid argon sampling was chosen, since this technique is unaffected by radiation and magnetic field, and channel—to—channel variations can be calibrated very accurately. As high—Z absorber lead proved to be in performance only slightly inferior to depleted uranium, which is more expensive and difficult to handle [6].

Table 1.
Energy resolution and readout geometry of the SLD calorimeters

calorimeter (absorber)	$\sigma(E)\sqrt{E}$ (%)	depth x_{rad} or Λ_{abs}	tower size	$\sigma(pos)$ (elm only)	coverage (mrad)
LAC elm (2 mm lead)	10	$(6+15)x_{rad}$	$2^o \times 2^o$	$1.4\ cm/\sqrt{E}$	down to 150
LAC had (6 mm lead)	55	$(1+1)\Lambda_{abs}$	$4^o \times 4^o$	$--$	down to 150
WIC had (5 cm iron)	80	$(2+2)\Lambda_{abs}$	$4^o \times 4^o$	$--$	down to 150
LMSAT elm (3.5 mm tungsten)	18	$(6+17)x_{rad}$	$1 \times 1\ cm^2$ $\Delta\theta \approx 6\ mrad$	$0.6\ mrad/\sqrt{\frac{46.5}{E}}$	28 to 65
MASC elm (7 mm tungsten)	18	$(6+17)x_{rad}$	$1 \times 1\ cm^2$ $\Delta\theta \approx 30\ mrad$	$2.8\ mrad/\sqrt{\frac{46.5}{E}}$	65 to 190

For good containment of high−energy electromagnetic showers, the total depth of the electromagnetic section is 22 radiation lengths. This fine−grained section is extended to about three interaction lengths, using coarser sampling (Fig. 1). Thus the Liquid Argon Calorimeter (LAC) contains on average 85% of the shower energy for most hadrons. The coil is placed out in the shower tails where the unsampled energy will be small.

The coil is then surrounded by an iron−plate calorimeter sampled with plastic streamer tubes (Warm Iron Calorimeter, WIC)[7]. With 5−cm−thick iron plates and total depth of more than four interaction lengths, this device measures the energy emerging from the LAC plus coil with adequate resolution and also serves to identify and track muons. The muon identification is fully efficient for muons of $p_T \geq 2\ GeV/c$ in the barrel and $p_{Long} \geq 1.5\ GeV/c$ in the endcaps.

Endcap liquid argon calorimeters complete the electromagnetic and hadronic coverage to 150 mrad from the beamline. For smaller angles the luminosity monitor and small angle mask extend the coverage for electromagnetic showers down to 28 mrad. These devices use tungsten as absorber and sample the energy with silicon wafers[8].

Energy resolution and angular segmentation are summarized in Table 1. All calorimeters use projective tower geometry, that is adapted in transvers segmentation to the shower type to be measured, electromagnetic or hadronic.

2.5. STATUS

The magnet iron, the coil and the liquid argon calorimeter have been assembled in the detector in the collider hall. The installation of the WIC Iarocci chambers is complete. The drift chambers are finished and passed the high voltage tests. At present the outer vessel of the CRID is being installed. For cosmic ray tests in early 1990 most of the detector will be operational, including half of the Barrel CRID and a six ladder vertex detector. The full Barrel CRID will be available in summer 1990.

The time for installation of SLD on beam line is expected to be decided end of 1989.

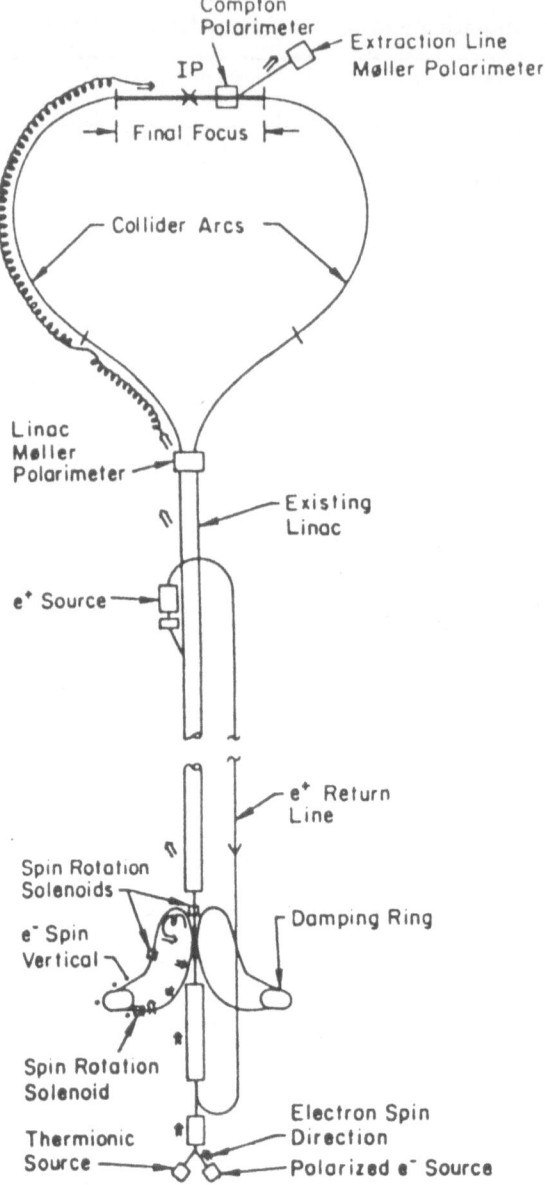

Figure 7. A layout of the SLC with emphasis on polarization. The orientation of the electron spin vector is shown as the electron is transported from the source to the interaction point.

3. The SLC beams at the interaction point

The SLC is the first operating single−pass e^+e^- collider [9]. An electron bunch and a positron bunch, each containing of order 10^{10} particles, are accelerated simultaneously in the SLAC linear accelerator (Fig. 7), magnetically separated and bent around 1 km arcs, then brought into collisions after being focused to an rms radius of about 3 μm, and finally extracted to dumps. Typical luminosities are 3×10^{27} $cm^{-2}s^{-1}$.

The SLC was proposed for two purposes: to provide a test of linear collider technology and to produce large numbers of Z^0 's to provide precise tests of the Standard Model. Although the SLC has yet to produce large number of Z^0 's, much has been learned about linear collider issues and a sufficient number of Z^0 's have been produced to enable a measurement of the Z^0 parameters more precise than previous measurements [10,11].

3.1. EXTRACTION−LINE SPECTROMETERS

The determination of weak parameters requires absolute and relative measurements of the center−of−mass energy to high precision over long periods of time. Spectrometers have been installed in both the electron and positron extraction lines. The method of measurement is to indirectly observe the deflection of charged beams via the narrow beams of synchrotron radiation they emit [12]. The systematic errors in each beam energy measurement are 20 MeV. Including an allowance for offset beams and a finite dispersion the absolute error on E_{cm} is 40 MeV, the relative error relevant for the lineshape measurement is 35 MeV [13]. The energy spread within each beam is about ±0.3%.

The polarization facility will eventually give a complementary measurement of the electron beam energy with comparable or slightly better accuracy. This provides an important systematic check.

3.2. BEAM SPOT SIZE AT THE INTERACTION POINT

The beam intensity profile is measured at the interaction point by scanning the beam in steps as small as 1 μm across fine carbon fibers (4, 7, and 30 μm diameter). The cross section of the beam is determined from the bremsstrahlung photons emitted by the beam as it passes through the wire [14]. Beamsizes of $\sigma_{x,y} \approx 3$ μm are routinely achieved.

Beams are aligned by measuring the deflection of one beam as it passes the electric field of the other beam [15]. This procedure is sufficiently accurate to routinely maintain the beams in collision with their centers aligned within 1 μm or better.

For SLD the magnets close to the interaction point will be replaced by superconducting magnets, permitting stronger gradients and smaller spot size (1.8 μm), which should increase the luminosity by a factor 1.6 to 3.2. Other improvements of the luminosity are expected from increasing the rate from 60 Hz to 120 Hz, increasing the positron current (new rotating positron target), increasing the electron current (better control of wakefield effects in the linac) and better reliability and faster tuning procedures.

4. The Polarization Project at the SLC

The SLC has the capability of accelerating polarized electrons and transporting these through the damping ring to the interaction region with small depolarization of the beam [16]. We give a short description of the project and review its status.

A detailed review of the status at the time of autumn 1988 can be found in reference 17, a pedagogical in−depth review is given in reference 18.

The layout of the polarized SLC is shown in Fig. 7. Commissioning of the SLC has allowed only partial installation of the polarization system. The schedule for completion of the installation is expected to be decided in autumn of 1989.

4.1. Polarized Electron Source

The electrons are produced longitudinally polarized by irradiating a GaAs crystal with circularly polarized light. The polarization sense is reversible from pulse to pulse by selecting between left and right circularly polarized light. Polarization of about 45% is expected, limited by the band structure of the crystal. The use of aluminum−dobed GaAs has opened the possibility to work at lower wavelengths that are accessible to dye lasers of long lifetime ($\tau \approx 210$ hours for Oxazine 720).

The polarized source is in place. The laser system will be ready for installation in early autumn 1989. Improvement of the accelerator vacuum will be needed to increase the lifetime of the GaAs photocathode.

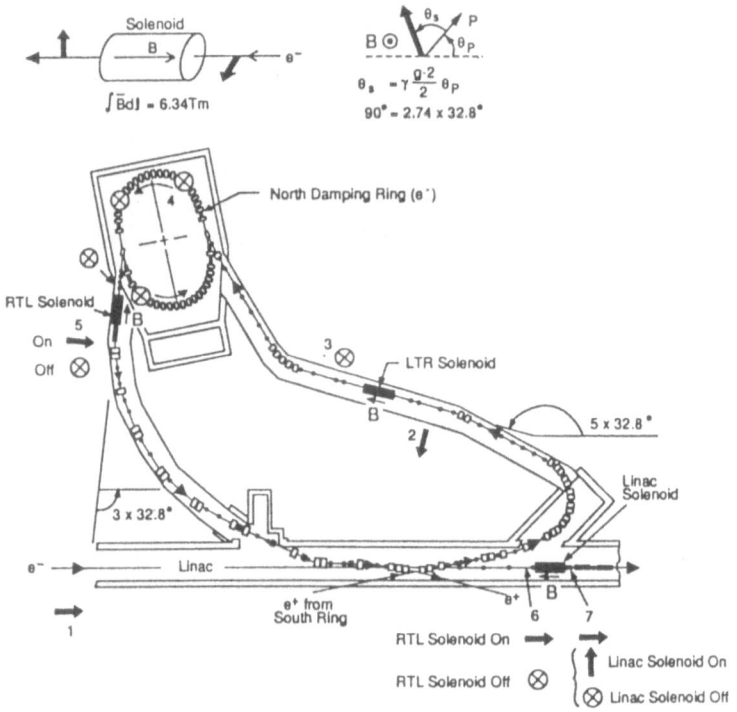

Figure 8.   The spin rotation system for the north damping ring. The orientation of the electron spin vector is shown by the full arrow.

4.2. Spin rotation

Electrons, before being delivered to the Linac, are stored in the damping ring, where their nominal energy is 1.21 GeV (1.153 GeV at present). To preserve the polarization while in the damping ring, the spin direction must be transverse to the plane of the damping ring (Fig. 8). The spin direction must be reestablished after leaving the damping ring so that it points in the desired direction for acceleration to the end of the linac and transportation around the arc to the interaction point where longitudinal polarization is required. The angular bend in the transfer line to the damping ring (LTR) is such that for electrons of nominal energy the spin precesses in−plane to a position perpendicular to the direction of flight. An axial solenoidal field is then needed to rotate the spin perpendicular to the damping ring plane. To reestablish longitudinal polarization after damping another solenoid will be installed close to the exit of the damping ring (RTL). With a third solenoid, located in the Linac arbitrary spin directions can be achieved. The three spin rotating solenoids are ready and have been operated under simulated SLC conditions with all associated instrumentation. There exists the possibility to install initially only the LTR solenoid and the Linac solenoid, to shorten the installation time from 15 to 10 weeks and minimize modifications to the transfer lines. Such a two—rotator system would be sufficient to achieve full source polarization at the interaction point at center—of—mass energies of 90 and 91.8 GeV (Fig. 9). The $Z°$ cross section at the latter energy is more than 80% of its peak value.

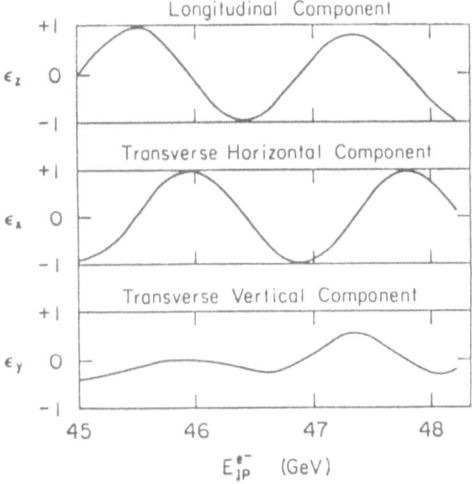

Figure 9.   Degree of longitudinal polarization at the interaction point as function of the center—of—mass energy.

4.3. POLARIMETRY

Compton and Møller polarimeters will be used to monitor and measure the polarization near the interaction point (Fig. 7).

The Compton polarimeter uses circularly polarized laser light that intercepts the electron beam downstream of the interaction point. Recoiling electrons at the maximum Compton angle emerge directly forward in the laboratory with momentum less than half that of the beam from which they are separated in the field of bend magnets. The asymmetry of the reaction rate, when either photon or electron polarization is reversed, is high ($\geq 75\%$). The counting rates are also high. The accuracy of the polarization measurement is limited by the accuracy with which the polarization of the laser light can be determined (1%). The SLC Compton polarimeter can only measure the longitudinal component of the beam polarization.

The Møller polarimeters measure all three components of the polarization. Electrons are scattered on a magnetized iron foil. The accuracy that can be reached is $\Delta P/P = \pm 5\%$, limited by the small target polarization (8%) and the uncertainty of the target polarization ($\Delta P \approx \pm 2\%$).

The Møller polarimeters that are located in the Linac (for monitoring) and in the extraction line are installed and commissioning is in progress. A prototype Compton polarimeter is in place for background studies and the laser has been installed.

The availability of a polarized electron beam represents a unique physics opportunity. It is expected that polarization will be available latest at the time when SLD is moving on beam line.

5. Outlook

Polarized beams allow measurements of the weak couplings with a precision otherwise unreachable[19]. The jewels among these measurements are the left−right asymmetry A_{LR} and the polarized forward−backward asymmetry $A_{FB}^{Pol,f}$, which permit to isolate the effective electron couplings at the production vertex of the Z° boson from the effective fermion couplings at the decay vertex. Without polarization similar information cannot be obtained. A_{LR} and A_{FB}^{Pol} show very small dependence on the center−of−mass energy and are therefore less sensitive to the exact knowledge of the beam energy and to initial state radiation effects.

These measurements will be a unique contribution to the precise tests of the Standard Model that will be performed in the coming years at the new e^+e^- machines SLC and LEP. With better measurements of the mass of the W boson at the $\bar{p}p$ colliders or possibly even the discovery of the top quark, this will be a very interesting period of time.

Acknowledgements: I want to thank the organizers for the opportunity to participate in this very stimulating workshop. I thank also the Polarization group for allowing me to report on their work. My thanks go especially to K. Moffeit and M. Swartz for update on the latest status of the polarization project.

This note has described only a small fraction of the SLD which is a large undertaking with significant contribution from many of those listed in the papers of reference 1.

REFERENCES

1. SLD design Report, SLAC−Report−273, UC-34D (1984); M. Breidenbach, Overview of the SLD, SLAC−PUB−3798 (1985) presented at Nuclear Science Symp., San Francisco, CA (1985); SLD Collaboration (M. Breidenbach et al.), A status report on the SLD Data Acquisition, SLAC−PUB−4786 (1985) presented at Nuclear Science Symp., Orlando, FL (1988).

2. C. J. S. Damerell et al., CCDs for Vertex Detection in High-energy Physics, Nucl. Instrum. Methods A253 (1987) 478.

3. Performance of the SLD Central Drift Chamber Prototype, Nucl. Instrum. Methods A252 (1986) 295, and SLAC−PUB−3910 (1986).

4. D. Aston et al., Development and construction of the SLD Čerenkov Ring Imaging Detector, to be published in Nucl. Instrum. Methods , also SLAC−PUB−4795 (1989) and references therein.

5. R. Arnold et al., Photosensitive gas detectors for the RICH technique and the DELPHI barrel RICH prototype, Nucl. Instrum. Methods A252 (1986) 188.

6. A. C. Benvenuti et al., The SLD Calorimeter System, SLAC−PUB−5028 (1989) and references therein.

7. A. C. Benvenuti et al., The Iron calorimeter and Muon identifier for SLD, Nucl. Instrum. Methods A276 (1989) 94.

8. S. C. Berridge et al., The Small Angle Electromagnetic Calorimeter at SLD: A $2\ m^2$ application of Silicon Detector Diodes, OREXP−88−1102 (1988).

9. B. Richter, SLAC Linear Collider Conceptual Design Report SLAC−Report−229 (1980); D. L. Burke, SLAC−PUB−4851 (1989).

10. Mark II Collaboration, Initial measurement of the Z Boson resonance parameters in e^+e^-annihilation, SLAC−PUB−5037 (1989), submitted to Phys. Rev. Lett. .

11. P. Rankin, The first Results from SLC, contribution to this workshop.

12. Mark II Collaboration and Final Focus Group, Extraction Line Spectrometers for SLC Energy Measurement, SLAC−SLC−PROP−2 (1986).

13. J. Kent et al., Precision measurement of the SLC beam energy, SLAC−PUB−4922 (1989).

14. R. Fulton et al., Nucl. Instrum. Methods A274 (1989) 74; G. Bowden et al., Nucl. Instrum. Methods A278 (1989) 664.

15. P. Bambade et al., SLAC−PUB−4767 (1989), submitted to Phys. Rev. Lett. .

16. D. Blockus et al., Proposal for Polarization at SLC, SLAC−PROPOSAL−SLC−01 (1986).

17. Polarization at the SLC, K. C. Moffeit, SLAC−PUB−4764 (1988).

18. M. L. Swartz, Polarization at SLC, SLAC−PUB−4689 (1988) and references therein; M. L. Swartz, Physics with Polarized electron beams, SLAC−PUB−4656 (1988).

19. The virtues of A_{LR} have been discussed in many papers, see for example A. Blondel, Physics with polarized beams, CERN 88−06 (1988) 1 and references therein.

MULTIPLE SOFT PHOTON BREMSSTRAHLUNG
ON THE Z RESONANCE

H. Spiesberger

II. Institut für Theoretische Physik
Universität Hamburg, Luruper Chaussee 149, 2000 Hamburg 50. FRG

Abstract

We discuss different prescriptions for the exponentiation of multiple photonic corrections and their application to the cross section for e^+e^- annihilation near the Z resonance. Our results are based on a modified soft photon approximation that respects the resonant character of the matrix element. Also the influence of a cutoff for the total radiated energy is investigated.

1. Introduction

The high precision that is expected to be reached at the new e^+e^- colliders LEP [1] and SLC requires the development of adequate techniques for the calculation of higher order corrections. There is a variety of methods that have been applied in order to improve the accuracy of theoretical predictions for the Z production cross section in electron positron annihilation. In addition to explicit fixed order calculations, renormalization group techniques have been used to find the $\mathcal{O}(\alpha^2)$ initial state corrections in [2]. The so-called structure function approach, which is well-known in QCD, has recently been applied in the calculation of electromagnetic corrections [3,4,5,6,8]. There IR as well as collinear logarithmic contributions can be summed to all orders. In [6] this method has been extended to calculate both initial and final state corrections to e^+e^- annihilation on the Z resonance. The complexity of the experimental setup calls also for the use of Monte Carlo programs that can simulate the process under consideration with the required accuracy. A first step towards the inclusion of multiple bremsstrahlung has been done in [7]. An event generator based on the formalism of structure functions has been described in [8]

For the particular class of infrared photonic corrections a technique for the summation of higher order contributions has been found long ago by Yennie, Frautschi and Suura [9] known under the notion of exponentiation. It has been applied for the calculation of the e^-e^- annihilation cross section in [10]. Exponentiation of IR photonic corrections appears in the limit of vanishing photon energy. The treatment with soft photons, where the energy of the photons ΔE is neglected, is a good approximation if the emitted energy is small compared to other energies that are relevant in the description of the process under consideration. In the vicinity of a resonance it is the width of the resonance Γ that sets the energy scale with which the photon energy must be compared and in this case a soft photon approximation leads to sensible results only if ΔE is small compared to Γ. This is a strong restriction in the case of the Z production in e^+e^- annihilation and the naive exponentiation prescription of [9] does not yield necessarily good results. A discussion of exponentiation in the soft photon approximation near a resonance has first been given by [11]. There the concept of coherent states which had been used in the description of non-resonant QED processes was applied also for a resonant cross section.

The aim of this note is to discuss a modification of the soft photon approximation that takes care of the resonant character of the underlying scattering process. In this modification it is not the ratio $\Delta E/\Gamma$ but $\Delta E/E_{cms}$ that characterizes the quality of the approximation.

If the cross section varies rapidly with the energy then also a discussion of the phase space becomes crucial. Imposing a cutoff for the total radiated energy will modify the corrections. In this case also the final state corrections will become important because they influence indirectly the probability for photon emission from the initial state. We will discuss the two idealized cases where the phase space integration is performed independently for each photon and where the integration is restricted by a total energy cutoff.

2. Calculation of multiple bremsstrahlung corrections

In the following we consider the electron positron annihilation into μ pairs. The reasoning may easily be transferred to other reactions. The process

$$e^+e^- \rightarrow \mu^+\mu^-$$

is described in lowest order by a cross section consisting of three parts:

$$\sigma = \sigma_{NR} + \sigma_{INT} + \sigma_{RES}, \tag{1}$$

namely one purely non-resonant QED part, the interference of γ and Z annihilation and the purely resonant Z contribution. Neglecting fermion masses the cross section to lowest order reads

$$\sigma^0 = \frac{4\pi\alpha^2}{3s}\left\{1 + 2v^2\mathrm{Re}\chi(s) + (v^2 + a^2)^2\,|\chi(s)|^2\right\} \tag{2}$$

where the vector and axial vector coupling constants are

$$v = \frac{4s_W^2 - 1}{4s_W c_W}, \quad a = -\frac{1}{4s_W c_W}, \tag{3}$$

and the reduced propagator χ is

$$\chi(s) = \frac{s}{s - M^2}. \tag{4}$$

$s = (p_+ + p_-)^2 = E_{cms}^2$ is the center of mass energy squared. The momenta of the incoming positrons (outgoing μ's), resp. electrons are denoted by $p_+(q_+)$, resp.

$p_-(q_-)$. The mass M of the Z boson contains an imaginary part determined by the width of the resonance:

$$M^2 = M_Z^2 - i M_Z \Gamma_Z. \tag{5}$$

If one wants to include self energy corrections then $M_Z \Gamma_Z$ has to be replaced by $\mathrm{Im}\Sigma^Z(s)$. In the soft photon approximation the matrix element for the 1-γ bremsstrahlung factorizes into the matrix element for the elastic process times a radiation factor

$$\mathcal{M}^{1\gamma} = \mathcal{M}^0 i \left(J_{in}^\mu(k) - J_{fin}^\mu(k) \right) \varepsilon_\mu(k) \tag{6}$$

given by the currents of the initial and final state

$$
\begin{aligned}
J_{in}^\mu &= ie \left(\frac{p_+^\mu}{kp_+} - \frac{p_-^\mu}{kp_-} \right), \\
J_{fin}^\mu &= ie \left(\frac{q_+^\mu}{kq_+} - \frac{q_-^\mu}{kq_-} \right)
\end{aligned}
\tag{7}
$$

multiplied by the polarization vector $\varepsilon(k)$ of the emitted photon with 4-momentum k.

For the emission of n photons with momenta $k_1, \cdots, k_n$ and polarization vectors $\varepsilon(k_1), \cdots, \varepsilon(k_n)$ the calculation is easily iterated and results for the non-resonant case in the factorization into independent factors for each photon:

$$
\begin{aligned}
\mathcal{M}_{NR}^{n\gamma} = \mathcal{M}_{NR}^0 \times i^n \quad &\left(J_{in}(k_1)\varepsilon(k_1) - J_{fin}(k_1)\varepsilon(k_1) \right) \times \\
&\cdots \\
&\times \left(J_{in}(k_n)\varepsilon(k_n) - J_{fin}(k_n)\varepsilon(k_n) \right).
\end{aligned}
\tag{8}
$$

These results are adequate to describe the bremsstrahlung of photons whose energies are small compared to the energy that is characteristic for the variation of the elastic cross section.

For the resonant part of the cross section this prescription is valid only if the photon energy is restricted to very small values. To do better we treat explicitly that part of the matrix element which is responsible for the strong energy dependence near the Z pole and find

$$\mathcal{M}_{RES}^{1\gamma} = \mathcal{M}_{RES}^0 i \varepsilon (R J_{in} - J_{fin}), \tag{9}$$

with

$$R(k) = \frac{(p_+ + p_-)^2 - M^2}{(p_+ + p_- - k)^2 - M^2}. \tag{10}$$

In the centre of mass system we have $k(p_+ + p_-) = k^0 \sqrt{s}$ and

$$R(k) = \left(1 - 2 \frac{\sqrt{s} k^0}{s - M^2} \right)^{-1}. \tag{11}$$

In the resonant case one finds for multiphoton emission

$$
\begin{aligned}
\mathcal{M}_{RES}^{n\gamma} = \mathcal{M}_{RES}^0 \times i^n \Big(& J_{in}(k_1)\varepsilon(k_1) \cdots J_{in}(k_n)\varepsilon(k_n) R(k_1 + \cdots + k_n) \\
& - J_{fin}(k_1)\varepsilon(k_1) J_{in}(k_2)\varepsilon(k_2) \cdots J_{in}(k_n)\varepsilon(k_n) R(k_2 + \cdots + k_n) \\
& \cdots \\
& + (-1)^n J_{fin}(k_1)\varepsilon(k_1) \cdots J_{fin}(k_n)\varepsilon(k_n) \Big).
\end{aligned}
\tag{12}
$$

It can be shown that omitting terms $\sim k_i k_j$ in R means only neglecting non-leading contributions[1]. Therefore we write

$$R(k_1 + \cdots + k_n) \simeq \left(1 - 2\frac{\sqrt{s}(k_1^0 + \cdots + k_n^0)}{s - M^2}\right)^{-1}. \tag{13}$$

Then using the representation

$$Z = \frac{1}{i}\int_0^\infty d\xi \exp(i\xi/Z), \qquad \mathrm{Im}\,Z < 0,$$

$$R(k_1^0 + \cdots + k_n^0) = \frac{1}{i}\int_0^\infty d\xi \exp(i\xi)\prod_{i=1}^n \exp\left(-i\xi\frac{2\sqrt{s}\,k_i^0}{s - M^2}\right),$$

the condition for the convergence of the integral being fulfilled because $\mathrm{Im}\,M^2 < 0$, we can factorize the matrix elements in the resonant case, too:

$$\mathcal{M}_{RES}^{n\gamma} = \mathcal{M}_{RES}^0 \times$$

$$\frac{1}{i}\int_0^\infty d\xi\, e^{i\xi} i^n \;\left(J_{in}(k_1)\varepsilon(k_1)\exp\left(-i\xi\frac{2\sqrt{s}\,k_1^0}{s - M^2}\right) - J_{fin}(k_1)\varepsilon(k_1)\right)$$

$$\cdots$$

$$\times \left(J_{in}(k_n)\varepsilon(k_n)\exp\left(-i\xi\frac{2\sqrt{s}\,k_n^0}{s - M^2}\right) - J_{fin}(k_n)\varepsilon(k_n)\right). \tag{14}$$

As a result we have obtained a generalization of the soft photon approximation which is valid for a resonant cross section.

The n-γ bremsstrahlung contribution to the inclusive cross section can now be calculated from

$$\frac{d\sigma}{d\Omega}\bigg|^{n\gamma} = \frac{1}{64\pi^2 s}\frac{1}{n!}\int \prod_{i=1}^n \frac{d^3 k_i}{(2\pi)^3 2k_i^0}\,|\mathcal{M}^{n\gamma}|^2, \tag{15}$$

using eq. (14). The infrared divergences of eq. (15) have to be extracted using, for example, a regularization with a small photon mass λ. The IR contributions will cancel in any given order with the corresponding virtual corrections. We include the virtual soft photon corrections, summed to all orders, according to the Yennie-Frautschi-Suura prescription [9] by defining:

$$\frac{d\sigma}{d\Omega}\bigg|^v = \frac{d\sigma}{d\Omega}\bigg|^0 \exp\{2\alpha\mathrm{Re}B\}, \tag{16}$$

---

[1] The neglected terms in R^{-1} are bounded from above:

$$\frac{\sum k_i k_j}{(p_+ + p_-)\sum k_i} \leq \frac{\sum k_i^0}{2E_{cms}}.$$

Therefore the relative error introduced by this approximation is of the order of $\Delta E/E_{cms}$, where ΔE is the maximally allowed radiated energy. This means that correlations due to the finite angles between emitted photons are neglected.

with $B = B_{in} + 2B_{int} + B_{fin}$, and

$$2\alpha \mathrm{Re} B_{in} = \frac{\alpha}{2\pi} \left\{ 4\left(\ln \frac{s}{m_e^2} - 1 \right) \ln \frac{\lambda}{\sqrt{s}} + \ln^2 \frac{s}{m_e^2} - \ln \frac{s}{m_e^2} + \frac{2\pi^2}{3} \right\},$$

$$2\alpha \mathrm{Re} B_{fin} = \frac{\alpha}{2\pi} \left\{ 4\left(\ln \frac{s}{m_f^2} - 1 \right) \ln \frac{\lambda}{\sqrt{s}} + \ln^2 \frac{s}{m_f^2} - \ln \frac{s}{m_f^2} + \frac{2\pi^2}{3} \right\}, \qquad (17)$$

$$2\alpha \mathrm{Re} B_{int} = \frac{\alpha}{2\pi} \left\{ 4\ln \frac{t}{u} \ln \frac{\lambda}{\sqrt{s}} + \ln \frac{t}{u} + \ln^2 \frac{-u}{s} - \ln^2 \frac{-t}{s} \right\},$$

$t = (p_+ - q_+)^2$, $u = (p_+ - q_-)^2$, $t/u = (1 - \cos\theta)/(1 + \cos\theta)$, and θ is the scattering angle of the outgoing μ^- with respect to the beam axis. $d\sigma/d\Omega^0$ contains all other IR finite virtual corrections (e.g. self energies).

To proceed we have to discuss the limits of the phase space integration in eq. (15). This, of course, cannot be done once for all but depends on the details of the experiment. The upper limits of the k^0 integration may be complicated functions of the emission angles. In this case the cross section can most easily be calculated using a Monte Carlo approach. We shall focus here on two idealized, extreme situations:

i) The experiment is able to detect single photons if their energy is bigger than a threshold $\Delta\omega$. Then all the k integrals in eq. (15) are performed up to $\Delta\omega$, independently of each other,

$$k_i^0 \le \Delta\omega.$$

ii) The detector can measure the total energy of the outgoing $\mu^+\mu^-$ pair and all events are counted if the missing energy is below a threshold ΔE. This means that the integral eq. (15) should be performed with the condition

$$\sum k_i^0 \le \Delta E.$$

In case ii) we can factorize the photon momentum integration in eq. (15) provided the θ function that expresses the condition for the limited total energy is represented by:

$$\theta \left(\Delta E - \sum_{i=1}^{n} k_i^0 \right) = \frac{1}{\pi} \int_{-\infty}^{\infty} \frac{d\tau}{\tau} \sin(\tau) \prod_{i=1}^{n} \exp(-i\tau k_i^0/\Delta E).$$

This then allows for exponentiation of the soft photon bremsstrahlung corrections as in the non-resonant case but now with the complication of additional parameter integrals.

Explicitly we find:

case i), non-resonant:

$$\left. \frac{d\sigma}{d\Omega} \right|_{NR}^{\Delta\omega} = \left. \frac{d\sigma}{d\Omega} \right|_{NR}^{v} \exp \left\{ -S_{in}^{\Delta\omega}(0) + 2S_{int}^{\Delta\omega}(0) - S_{fin}^{\Delta\omega}(0) \right\}, \qquad (18)$$

case i), resonant:

$$\left. \frac{d\sigma}{d\Omega} \right|_{RES}^{\Delta\omega} = \left. \frac{d\sigma}{d\Omega} \right|_{RES}^{v} \int_0^\infty d\xi \int_0^\infty d\eta \, e^{i(\xi-\eta)}$$
$$\exp \Big\{ -S_{in}^{\Delta\omega} \left(\frac{2\sqrt{s}\xi}{s - M^2} - \frac{2\sqrt{s}\eta}{s - M^{*2}} \right)$$
$$+ S_{int}^{\Delta\omega} \left(\frac{2\sqrt{s}\xi}{s - M^2} \right) + S_{int}^{\Delta\omega} \left(-\frac{2\sqrt{s}\eta}{s - M^{*2}} \right) - S_{fin}^{\Delta\omega}(0) \Big\},$$
$$(19)$$

case ii), non-resonant:

$$\frac{d\sigma}{d\Omega}\bigg|_{NR}^{\Delta E} = \frac{d\sigma}{d\Omega}\bigg|_{NR}^{v} \frac{1}{\pi}\int_{-\infty}^{\infty}\frac{d\tau}{\tau}\sin(\tau)\exp\left\{-S_{in}^{\infty}(\tau/\Delta E) + 2S_{int}^{\infty}(\tau/\Delta E) - S_{fin}^{\infty}(\tau/\Delta E)\right\},$$

(20)

case ii), resonant:

$$\frac{d\sigma}{d\Omega}\bigg|_{RES}^{\Delta E} = \frac{d\sigma}{d\Omega}\bigg|_{RES}^{v} \frac{1}{\pi}\int_{-\infty}^{\infty}\frac{d\tau}{\tau}\sin(\tau)\int_{0}^{\infty}d\xi\int_{0}^{\infty}d\eta\, e^{i(\xi-\eta)}$$

$$\exp\left\{ -S_{in}^{\infty}\left(\frac{2\sqrt{s}\xi}{s-M^2} - \frac{2\sqrt{s}\eta}{s-M^{*2}} + \frac{\tau}{\Delta E}\right)\right.$$

$$\left. +S_{int}^{\infty}\left(\frac{2\sqrt{s}\xi}{s-M^2} + \frac{\tau}{\Delta E}\right) + S_{int}^{\infty}\left(-\frac{2\sqrt{s}\eta}{s-M^{*2}} + \frac{\tau}{\Delta E}\right) - S_{fin}^{\infty}\left(\frac{\tau}{\Delta E}\right)\right\}.$$

(21)

We have introduced the abbreviation

$$S_i^A(x) = \int^A \frac{d^3k}{(2\pi)^3 2k^0} e^{-ik^0 x} J_a J_b^*$$

(22)

with

$$\begin{aligned}
i &= in &&\text{for} &&a = b = in,\\
i &= fin &&\text{for} &&a = b = fin,\\
i &= int &&\text{for} &&a = in, b = fin.
\end{aligned}$$

$S_i^{\Delta\omega}(0)$ are the well-known bremsstrahlung integrals of Yennie, Frautschi and Suura (denoted by $2\alpha\tilde{B}_i$ there). If the infrared divergences are regularized with a small finite photon mass λ they read:

$$S_{in,fin}^{\Delta\omega}(0) = -\beta_{in,fin}\ln\Delta\frac{\sqrt{s}}{\lambda} + S_{in,fin}^{f.p.}, \qquad S_{int}^{\Delta\omega}(0) = +\beta_{int}\ln\Delta\frac{\sqrt{s}}{\lambda} + S_{int}^{f.p.},$$

(23)

with

$$\Delta = \begin{cases} \dfrac{2\Delta\omega}{\sqrt{s}}, & \text{for case } i,\\[2ex] \dfrac{2\Delta E}{\sqrt{s}}, & \text{for case } ii, \end{cases}$$

(24)

and

$$\begin{aligned}
\beta_{in} &= \frac{2\alpha}{\pi}\left(\ln\frac{s}{m_e^2} - 1\right),\\[1ex]
\beta_{fin} &= \frac{2\alpha}{\pi}\left(\ln\frac{s}{m_f^2} - 1\right),\\[1ex]
\beta_{int} &= \frac{2\alpha}{\pi}\ln\frac{t}{u}.
\end{aligned}$$

(25)

The finite parts $S_i^{f.p.}$ are:

$$\begin{aligned}
S_{in}^{f.p.} &= -\frac{\alpha}{2\pi}\left\{-\ln^2\frac{s}{m_e^2} + 2\ln\frac{s}{m_e^2} - \frac{2\pi^2}{3}\right\},\\[1ex]
S_{fin}^{f.p.} &= -\frac{\alpha}{2\pi}\left\{-\ln^2\frac{s}{m_f^2} + 2\ln\frac{s}{m_f^2} - \frac{2\pi^2}{3}\right\},\\[1ex]
S_{int}^{f.p.} &= +\frac{\alpha}{2\pi}\left\{\ln^2\frac{-t}{s} - \ln^2\frac{-u}{s} - 2\mathrm{Li}_2\left(\frac{-t}{s}\right) + 2\mathrm{Li}_2\left(\frac{-u}{s}\right)\right\}.
\end{aligned}$$

(26)

For eqs. (19, 20, 21) we need the generalization of eq. (23) with an additional phase factor $\exp(-ik_0x)$. A straightforward calculation yields [12]:

$$S_i^{\Delta\omega}(x) = -\beta_i \left(i\frac{\pi}{2} - \gamma_E - \ln(-x\Delta\omega) - E_1(ix\Delta\omega) \right) + S_i^{\Delta\omega}(0), \qquad (27)$$

(with the exponential integral $E_1(x) = \int_1^\infty dt/t \, \exp(-xt)$) and

$$S_i^\infty(x) = -\beta_i \left(i\frac{\pi}{2} - \gamma_E - \ln(-x\Delta E) \right) + S_i^{\Delta E}(0). \qquad (28)$$

In the last equation the cutoff ΔE has been introduced artificially. This enables us to extract the standard QED corrections also in the case of resonant scattering and in the presence of more complicated phase space cuts. Using again the notation of Yennie, Frautschi, and Suura we can write the non-resonant and the resonant contributions to the cross section in the form ($I = NR, RES$):

$$\left.\frac{d\sigma}{d\Omega}\right|_I^A = \exp\left\{ 2\alpha(\text{Re}B + \tilde{B}) \right\} \delta_I^A \times \left.\frac{d\sigma}{d\Omega}\right|_I^0. \qquad (29)$$

The first factor is the well-known YFS-factor [9] and reads:

$$\begin{aligned}
2\alpha(\text{Re}B + \tilde{B}) = \ & \beta_{in} \ln\Delta + \frac{\alpha}{2\pi} \ln\frac{s}{m_e^2} \\
& + \beta_{fin} \ln\Delta + \frac{\alpha}{2\pi} \ln\frac{s}{m_f^2} \\
& + 2\beta_{int} \ln\Delta + \frac{\alpha}{\pi}\left\{ -\ln\frac{t}{u} + 2\text{Li}_2\left(\frac{-u}{s}\right) - 2\text{Li}_2\left(\frac{-t}{s}\right) \right\}.
\end{aligned} \qquad (30)$$

The second factor δ_I^A in eq. (29) describes the modifications due to the resonant character of the cross section and the phase space condition. These factors read:

$$\delta_{NR}^{\Delta\omega} = 1, \qquad (31)$$

$$\begin{aligned}
\delta_{RES}^{\Delta\omega} = \ & \int_0^\infty d\xi \int_0^\infty d\eta\, e^{i(\xi-\eta)} \\
& \exp\left\{ -\beta_{in}\left(\ln(A^*\eta - A\xi) + E_1(-iA^*\eta + iA\xi) - i\frac{\pi}{2} + \gamma_E \right) \right. \\
& \quad - \beta_{int}\left(\ln(A^*\eta) + E_1(-iA^*\eta) - i\frac{\pi}{2} + \gamma_E \right) \\
& \quad \left. - \beta_{int}\left(\ln(-A\xi) + E_1(+iA\xi) - i\frac{\pi}{2} + \gamma_E \right) - \beta_{fin}\left(-i\frac{\pi}{2} + \gamma_E \right) \right\},
\end{aligned} \qquad (32)$$

$$\delta_{NR}^{\Delta E} = \frac{1}{\pi}\int_{-\infty}^\infty \frac{d\tau}{\tau} \sin(\tau) \exp\left\{ -(\beta_{in} + 2\beta_{int} + \beta_{fin})\left(-i\frac{\pi}{2} + \gamma_E + \ln\tau \right) \right\}, \qquad (33)$$

$$\begin{aligned}
\delta_{RES}^{\Delta E} = \ & \frac{1}{\pi}\int_{-\infty}^\infty \frac{d\tau}{\tau} \sin(\tau) \int_0^\infty d\xi \int_0^\infty d\eta\, e^{i(\xi-\eta)} \\
& \exp\left\{ -\beta_{in}\left(\ln(A^*\eta - A\xi + \tau) - i\frac{\pi}{2} - \gamma_E \right) - \beta_{int}\left(\ln(A^*\eta + \tau) - i\frac{\pi}{2} + \gamma_E \right) \right. \\
& \quad \left. - \beta_{int}\left(\ln(-A\xi + \tau) - i\frac{\pi}{2} + \gamma_E \right) - \beta_{fin}\left(\ln(\tau) - i\frac{\pi}{2} + \gamma_E \right) \right\},
\end{aligned} \qquad (34)$$

with

$$A = \Delta \frac{s}{s - M^2}.$$ (35)

For the γZ interference contributions to the cross section one finds in a similar way the following correction factors:

$$\delta_{INT}^{\Delta\omega} = \frac{1}{i} \int_0^\infty d\xi\, e^{i\xi} \exp\left\{ -(\beta_{in} + \beta_{int}) \left(\ln(-A\xi) + E_1(iA\xi) - i\frac{\pi}{2} + \gamma_E \right) \right\},$$ (36)

$$\delta_{INT}^{\Delta E} = \frac{1}{i\pi} \int_{-\infty}^\infty \frac{d\tau}{\tau} \sin(\tau) \int_0^\infty d\xi\, e^{i\xi} \exp\left\{ \begin{array}{l} -(\beta_{in} + \beta_{int}) \left(\ln(-A\xi + \tau) - i\frac{\pi}{2} + \gamma_E \right) \\[2mm] -(\beta_{int} + \beta_{fin}) \left(\ln \tau - i\frac{\pi}{2} + \gamma_E \right) \end{array} \right\}.$$ (37)

In contrast to (32) - (34) the δ_{INT}^A are complex. That part which describes the resonant energy dependence and builds up the ξ integral had to be factored out of the resonant matrix element only. The δ_{INT}^A appear in the γZ contribution to the cross section in the form $2\mathrm{Re}\{\delta_{INT}^A \mathcal{M}_{RES}^* \mathcal{M}_{NR}\}$.

3. Discussion

The significance of the correction factors eqs. (32, 34) is twofold. Firstly, they describe the distortion of the cross section for single photon bremsstrahlung due to the strong energy dependence of the lowest order resonant cross section (radiative tail). The corresponding results in the lowest nontrivial order [11] are easily recovered from eq. (32) (see eq. (44)).

Secondly, the factors $\delta_{RES}^{\Delta\omega}$, $\delta_{RES}^{\Delta E}$, and $\delta_{NR}^{\Delta E}$ describe the correlation of the emission probabilities if there is more than one photon. This correction of $\mathcal{O}(\alpha^n), n \geq 2$ has its origin in two effects, namely again the energy dependence of the Born cross section and the restriction of the available phase space by a total energy cutoff. Both circumstances physically mean that photons are not emitted independently of each other. Also in the non-resonant case the condition $\sum k_i^0 \leq \Delta E$ leads to a correlation. In this case the resulting correction factor

$$\delta_{NR}^{\Delta E} = \frac{e^{-\beta\gamma_E}}{\Gamma(1 + \beta)} \simeq 1 - \frac{\pi^2}{12}\beta^2,$$ (38)

with

$$\beta = \beta_{in} + 2\beta_{int} + \beta_{fin}$$ (39)

has been known for a long time [9] and is discussed at full length in the literature [13]. The result eq. (38) is also the limiting case of eq. (34) for $\Delta E \to 0$. The corresponding limit of eq. (32) for $\Delta\omega \to 0$ is of course 1. Note that the correlation resulting from a restriction of the total energy works also between initial and final state radiation, whereas the resonance effect influences only the initial state corrections.

This is in contrast with the results of the coherent state approach [11]. In fact, the characteristic feature of the coherent states is the assumption that photons are emitted independently which is strictly true only in the limit $k_i^0 \to 0$. An inspection of the calculation of [11] reveals, that the use of coherent states corresponds to replacing

$$R\left(\sum_{i=1}^{n} k_i\right) \to \prod_{i=1}^{n} R(k_i)$$

in eq. (12) and thus can only lead to trivial exponentiation of the $\mathcal{O}(\alpha)$ results. Of course, the differences are only visible at the level of $\mathcal{O}(\alpha^2)$, but these corrections are exactly those we want to learn something about.

Finally we give analytical results for the special cases of eq. (34) for $\beta_{int} = 0$ (*i.e.* $\theta = 90^0$ scattering) and of eq. (37). These correction factors can be expressed with help of the hypergeometric function $_2F_1$:

$$\delta_{RES}^{\Delta E}(\beta_{int} = 0) = \frac{e^{-\beta\gamma_E}}{\Gamma(1+\beta)} \frac{\mathrm{Im}\left[\frac{s\Delta}{s-M^2} \, _2F_1(1, \beta_{in}; 1 + \beta_{in} + \beta_{fin}; \frac{s\Delta}{s-M^2})\right]}{\mathrm{Im}\frac{s\Delta}{s-M^2}}, \qquad (40)$$

$$\delta_{INT}^{\Delta E} = \frac{e^{-\beta\gamma_E}\Gamma(1 + \beta_{int} + \beta_{fin})}{\Gamma(1+\beta)\Gamma(1 + \beta_{int} + \beta_{in})} \, _2F_1(1, \beta_{in} + \beta_{int}; 1 + \beta; \frac{s\Delta}{s-M^2}). \qquad (41)$$

For the general case we do not know such simple formulas. In order to discuss the different prescriptions for the exponentiation of soft photon corrections we transformed eqs. (32, 34) into a form that was accessible to a numerical evaluation. For $\delta_{RES}^{\Delta E}$ this is straightforward. In the other case we found for eq. (34) the representation (see the appendix):

$$\delta_{RES}^{\Delta E} = \frac{e^{-\beta\gamma_E}}{\Gamma(1+\beta_{int})\Gamma(1+\beta_{fin})\Gamma(1 + \beta_{in} + \beta_{int})}$$

$$\times \int_0^1 dx \, (1-x)^{\beta_{fin}} x^{1+\beta_{in}+2\beta_{int}} \int_0^1 dy \, (1-y)^{\beta_{int}} y^{\beta_{int}+\beta_{in}}$$

$$\mathrm{Re}\left\{\frac{F_1}{1-xyA^*}\left[(2+\beta)(1+\beta) + 2A^*\left(\frac{3}{2}+\beta\right)\frac{2xy-1}{1-xyA^*} - 4A^{*2}\frac{xy(1-xy)}{(1-xyA^*)^2}\right]\right.$$

$$\left. + \frac{F_2}{1-xyA^*}\left[A - 2 + \frac{|A-1|^2 + 1}{1-xyA^*}\right]\right\}, \qquad (42)$$

with

$$F_1 = \frac{1}{1-xA} \, _2F_1\left(1, 1 + \beta_{int}; 1 + \beta_{in} + \beta_{int}; \frac{-xyA}{1-xA}\right), \qquad (43)$$

$$F_2 = \frac{1}{(1-xA)^2} \, _2F_1\left(2, 1 + \beta_{int}; 1 + \beta_{in} + \beta_{int}; \frac{-xyA}{1-xA}\right).$$

The remaining integrations have been performed numerically.

4. Numerical Results

We start with the discussion of some numerical results for the correction factors eqs. (32, 34). In the following we have chosen

$$M_Z = 93.0 \, GeV, \quad \Gamma_Z = 2.5 \, GeV.$$

Fig. 1 shows $\delta_{RES}^{\Delta E}$ for $\beta_{int} = 0$ (*i.e.* $\theta = 90^0$) for some choices of the energy cutoff ΔE. We scaled ΔE with the beam energy $E_{beam} = E_{cms}/2$, *i.e.* we used a fixed value for

Table 1: Position of the maximum of σ_{tot} $(e^+e^- \rightarrow \mu^+\mu^-)$ including initial state corrections only. Numbers are in GeV. $\Delta = \Delta E/E_{beam}$. Case i) refers to the condition $k_i^0/E_{beam} \leq \Delta$ for independent phase space integration and case ii) to the total energy condition $\sum k_i^0/E_{beam} \leq \Delta$.

Prescription	$\Delta = 0.02$	$\Delta = 0.04$	$\Delta = 0.10$	$\Delta = 0.20$
$O(\alpha)$, no exponentiation	93.140	93.180	93.195	93.196
Exponentiation of [11]	93.101	93.131	93.143	93.144
Exponentiation case i)	93.096	93.121	93.130	93.131
Exponentiation case ii)	93.092	93.119	93.130	93.131

Table 2. Position of the maximum of σ_{tot} $(e^+e^- \rightarrow \mu^+\mu^-)$ including initial and final state corrections. Numbers are in GeV. $\Delta = \Delta E/E_{beam}$.

Prescription	$\Delta = 0.02$	$\Delta = 0.04$	$\Delta = 0.10$	$\Delta = 0.20$
$O(\alpha)$, no exponentiation	93.210	93.251	93.246	93.229
Exponentiation of [11]	93.102	93.133	93.145	93.145
Exponentiation case i)	93.097	93.122	93.131	93.132
Exponentiation case ii)	93.091	93.117	93.125	93.135

$\Delta = \Delta E/E_{beam}$. This is the reason for the unsymmetric feature of the curves. Fig. 1 shows that the cross section below the resonance is slightly lowered but strongly increased above M_Z – this is the radiative tail effect. The effect is strongly enhanced for larger cutoffs. However, close to the resonance ($\sqrt{s} \simeq M_Z \pm \Gamma_Z$) the corrections are moderate.

In fig. 2 we compare the different prescriptions for the phase space integration for $\theta = 30^0$ and $\Delta = 0.04$ with the pure $\mathcal{O}(\alpha)$ result

$$\delta_{RES}^{\Delta\omega}\Big|_{\mathcal{O}(\alpha)} = 1 + (\beta_{in} - 2\beta_{int}) \left\{ \ln \left| \frac{M^2 - s}{M^2 - s + 2\Delta\omega\sqrt{s}} \right| \right\} + \beta_{in} \frac{s - M_Z^2}{M_Z\Gamma_Z}(\varphi - \varphi_0), \quad (44)$$

where

$$\varphi = \arctan \frac{M^2 - s + 2\Delta\omega\sqrt{s}}{M_Z\Gamma_Z},$$

$$\varphi_0 = \arctan \frac{M^2 - s}{M_Z\Gamma_Z},$$

and with Greco's prescription [11]

$$\delta_{RES}^{\Delta\omega}\Big|_{Greco} = \exp\left\{ \beta_{in} \ln \left| \frac{M^2 - s}{M^2 - s + 2\Delta\omega\sqrt{s}} \right| + 2\beta_{int} \ln \left| \frac{-s}{M^2 - s + 2\Delta\omega\sqrt{s}} \right| \right\}$$

$$\times \left(1 + \beta_{in} \frac{s - M_Z^2}{M_Z\Gamma_Z}(\varphi - \varphi_0) + 2\beta_{int} \ln \left| \frac{M^2 - s}{s} \right| \right). \tag{45}$$

Our result for an independent photon phase space cutoff is quite close to Greco's results[11]. For case ii), $\delta_{RES}^{\Delta E}$, the results are smaller because of the tighter phase space. Also shown is the result for the non-resonant correction $\delta_{NR}^{\Delta E}$ which is almost

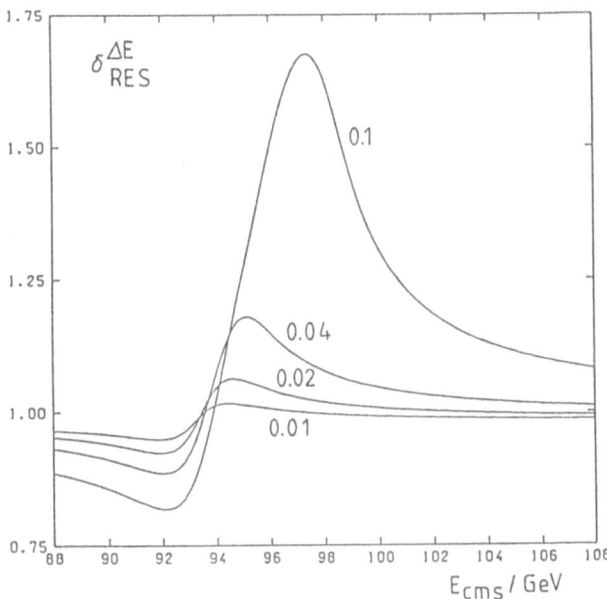

Fig. 1 $\delta_{RES}^{\Delta E}$ from eq. (42) for $\theta = 90^0$ and different cutoffs ΔE with $\Delta = \Delta E/E_{beam} = 0.01,\ 0.02,\ 0.04,\ 0.1.$ $M_Z = 93.0\ GeV,\ \Gamma_Z = 2.5\ GeV.$

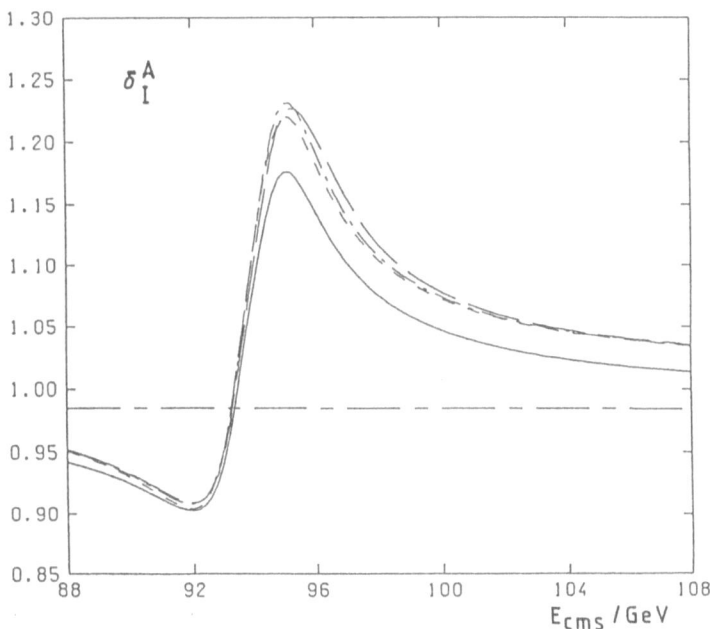

Fig. 2 Comparison of the different exponentiation prescriptions for $\Delta = 0.04,\ \theta = 30^0$ (parameters as in fig. 1). ———— $\delta_{RES}^{\Delta E}$, ————— $\delta_{RES}^{\Delta \omega}$, — · — · — results of [11], · · · · · · · · $\mathcal{O}(\alpha)$ results, ——— · ——— $\delta_{NR}^{\Delta E}$.

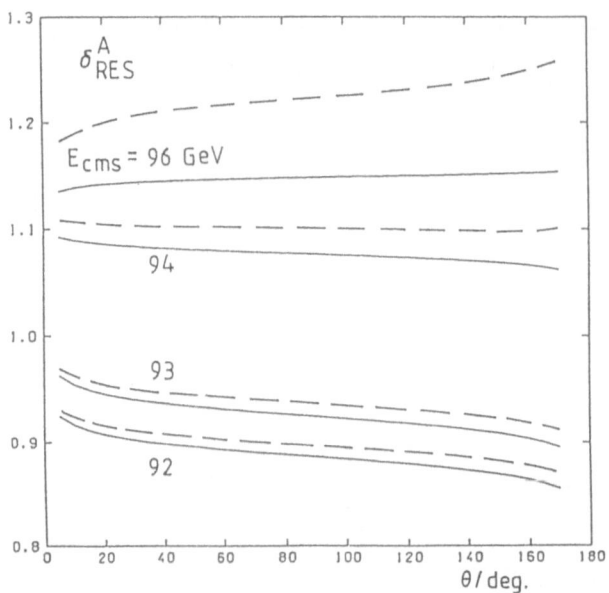

Fig. 3 θ dependence of $\delta_{RES}^{\Delta E}$ (————) and $\delta_{RES}^{\Delta \omega}$ (— — — — —) for $\Delta = 0.04$ and 4 different energies: $E_{cms} = 92, 93, 94$ and 96 GeV (same parameters as in fig. 1).

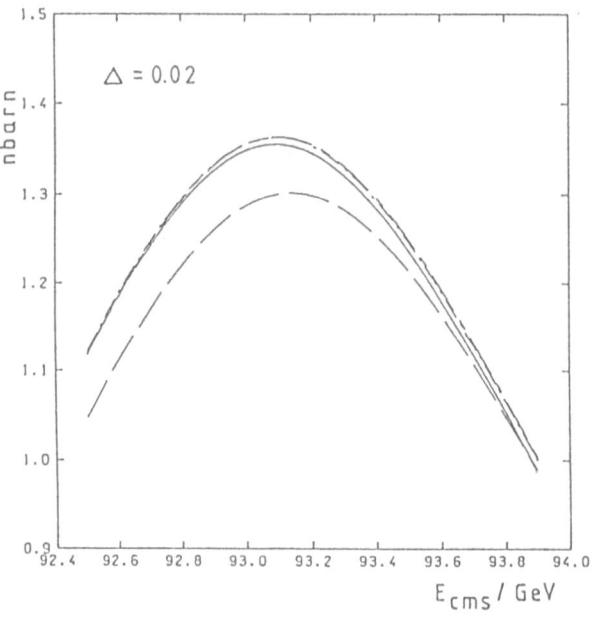

Fig. 4 Total cross section σ_{tot} for $e^+e^- \to \mu^+\mu^-$ near the resonance peak for $\Delta = 0.02$. Only initial state corrections are included. ———— case *ii*, — — — — $\mathcal{O}(\alpha)$, — · — · — case *i*, —— · —— results of [11].

Parameters are chosen as before and $s_W^2 = 0.223$.

Table 3. Lower value of $E_{cms} = \sqrt{s}$ where σ_{tot} ($e^+e^- \to \mu^+\mu^-$) reaches half of its peak value and the full width at half maximum for $M_Z = 93\ GeV, \Gamma_Z = 2.5\ GeV$. Including initial state corrections only. Numbers are in GeV.

Prescription	$\Delta = 0.02$	$\Delta = 0.04$	$\Delta = 0.10$	$\Delta = 0.20$
$O(\alpha)$, no exponentiation	91.860	91.876	91.876	91.873
	94.479	94.685	94.899	94.922
	2.619	2.809	3.023	3.049
Exponentiation of [11]	91.829	91.842	91.844	91.843
	94.430	94.609	94.773	94.780
	2.601	2.767	2.929	2.936
Exponentiation case i)	91.824	91.833	91.831	91.830
	94.433	94.591	94.726	94.741
	2.609	2.758	2.895	2.911
Exponentiation case ii)	91.822	91.833	91.832	91.830
	94.416	94.573	94.724	94.740
	2.594	2.740	2.892	2.911

constant over the range of energy considered here. It is clear from fig. 2 that we cannot reproduce $\delta_{RES}^{\Delta E}$ quantitatively by simply multiplying $\delta_{RES}^{\Delta \omega}$ by $\delta_{NR}^{\Delta E}$, as was claimed in [11].

From fig. 3 we learn that the dependence on the scattering angle θ (via β_{int}) of $\delta_{RES}^{\Delta E}$ and $\delta_{RES}^{\Delta \omega}$ is weak as is expected because $|\beta_{int}| \ll \beta_{in}$. However, in the calculation of the forward-backward asymmetry this θ dependence should be taken into account (see below).

We have also investigated the predictions for the total cross section of the e^+e^- annihilation into μ pairs. Fig. 4 shows the total cross section integrated over the μ scattering angle close to the resonance peak including initial state corrections only. These results contain the non-resonant γ annihilation and the γZ interference parts as well. The final state corrections and initial-final state interference are included in fig. 5. These results contain also the complete $\mathcal{O}(\alpha)$ results for the $\gamma\gamma$ and γZ boxes but no self energies. We find small corrections and small differences between the various exponentiation prescriptions as could have been foreseen from the above results. The differences for the value of σ_{tot} near the maximum are of the order of $1 - 2\%$ and comparable with the expected experimental accuracy. However, the position of the peak is expected to be measured with a precision of at least $\delta M \simeq \pm 50\ MeV$ [1] and it is planned to improve this value to $\mathcal{O}(10\ MeV)$ [14]. Table 1 and 2 contain the values of $\sqrt{s}$ where σ_{tot} is maximal. Again the differences can be of the order of the experimental accuracy.

In fig. 2 we have seen that some GeV above the resonance the different presriptions deviate more significantly from each other. Therefore in the determination of the width of the resonance we expect a bigger effect. In fact, the cross section above the resonance comes out quite differently (see fig. 6). This is quantitatively shown in table 3, and 4 where those values of $\sqrt{s}$ are given where σ_{tot} reaches half of σ_{max} together with the full width at half maximum. Shifts of these values reach up to $\mathcal{O}(50\ MeV)$. Of course, this is only an indication for the importance of an adequate choice of the exponentiation prescription in the determination of the width because ultimatively

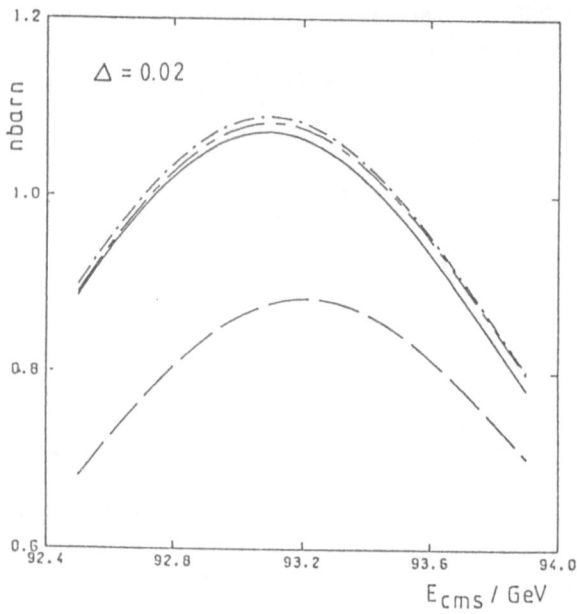

Fig. 5 Same as fig. 4 but including final state corrections and initial-final state inter-ference.

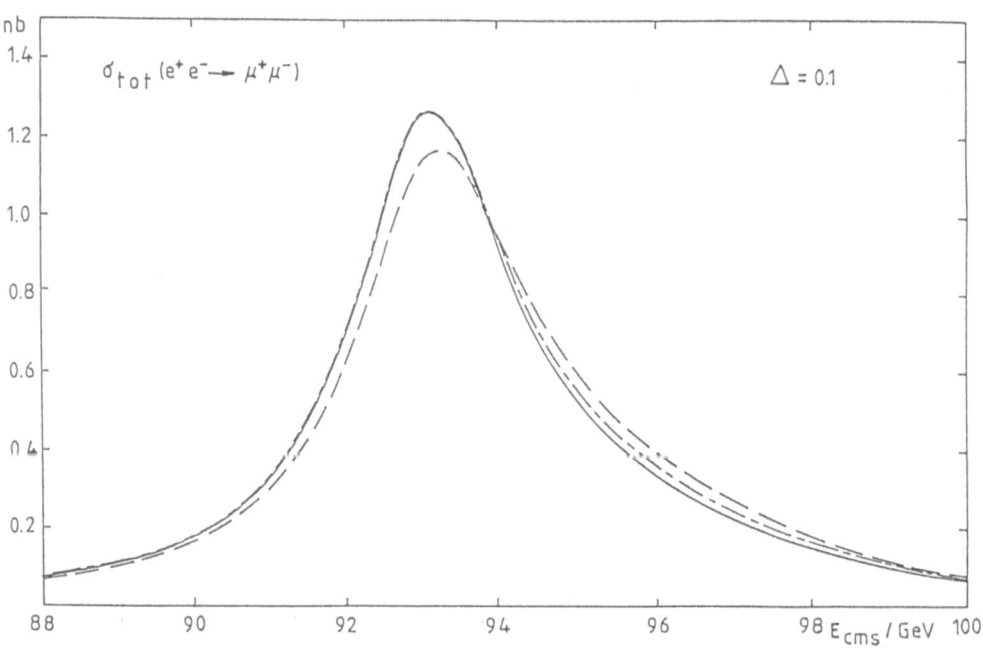

Fig. 6 Same as fig. 5 but over a wider range of energy and for $\Delta = 0.1$. In this figure the results for phase space prescription case i) have been omitted.

Table 4. Lower value of $E_{cms} = \sqrt{s}$ where σ_{tot} ($e^+e^- \to \mu^+\mu^-$) reaches half of its peak value and the full width at half maximum for $M_Z = 93\ GeV, \Gamma_Z = 2.5\ GeV$. Including initial and final state corrections. Numbers are in GeV.

Prescription	$\Delta = 0.02$	$\Delta = 0.04$	$\Delta = 0.10$	$\Delta = 0.20$
O(α), no exponentiation	91.927	91.932	91.912	91.896
	94.564	94.808	95.060	95.034
	2.637	2.876	3.148	3.148
Exponentiation of [11]	91.833	91.846	91.847	91.845
	94.434	94.612	94.775	94.781
	2.601	2.766	2.928	2.936
Exponentiation case i)	91.819	91.836	91.835	91.833
	94.436	94.592	94.727	94.741
	2.617	2.756	2.892	2.908
Exponentiation case ii)	91.826	91.837	91.835	91.833
	94.414	94.560	94.713	94.736
	2.588	2.723	2.878	2.903

this should be done by performing a combined fit to the experimental data.

Finally, in fig. 7 we present results for the forward-backward asymmetry

$$A_{FB} = \frac{\int_0^1 d\cos\theta\, d\sigma/d\Omega - \int_{-1}^0 d\cos\theta\, d\sigma/d\Omega}{\int_0^1 d\cos\theta\, d\sigma/d\Omega + \int_{-1}^0 d\cos\theta\, d\sigma/d\Omega} \tag{46}$$

for fixed $\Delta E = 1$ GeV. The various prescriptions result in values for A_{FB} which differ by about 1% at $\sqrt{s} = M_Z$. If the angular integration in eq.(46) is restricted the differences become smaller.

All these results are obtained from calculations based on the soft photon approximation. For large values of the cutoff $\Delta\omega$, resp. ΔE, the hard photon bremsstrahlung cross section has to be included. If soft photon corrections - exponentiated or not - and hard photon contributions are combined, care has to be taken that the combined results do not depend on the arbitrary cutoff that separates hard from soft photons. This is guaranteed by calculating soft photon corrections up to the experimentally allowed maximum of the photon energy and then to correct for the difference:

$$\frac{d\sigma}{d\Omega} \sim \sum_n \int \prod_{i=1}^n \frac{d^3k_i}{(2\pi)^3 2k_i^0} \left(|\mathcal{M}^{n\gamma}|^2_{exact} - |\mathcal{M}^{n\gamma}|^2_{soft} \right) + \frac{d\sigma}{d\Omega}\bigg|_{soft}.$$

In this expression the lower limit of the $|k|$ integration can be set to 0, the infrared divergent contributions having been subtracted. For initial state radiation the bremsstrahlung contribution can be written in the form

$$\frac{d\sigma}{d\Omega}\bigg|^{n\gamma}(s) = \int_{z_{min}}^{z_{max}} dz\, \frac{d\sigma}{d\Omega}\bigg|^0 (sz)f^{(n)}(z), \tag{47}$$

where to order $\mathcal{O}(\alpha)$, i.e. 1-γ-bremsstrahlung[2]

$$f^{(1)}(z) = -\frac{1}{2}\beta_{in}(1 + z) + \mathcal{O}(\alpha^2), \tag{48}$$

---

[2] An $\mathcal{O}(\alpha^2)$ calculation for the initial state corrections has been performed by [2].

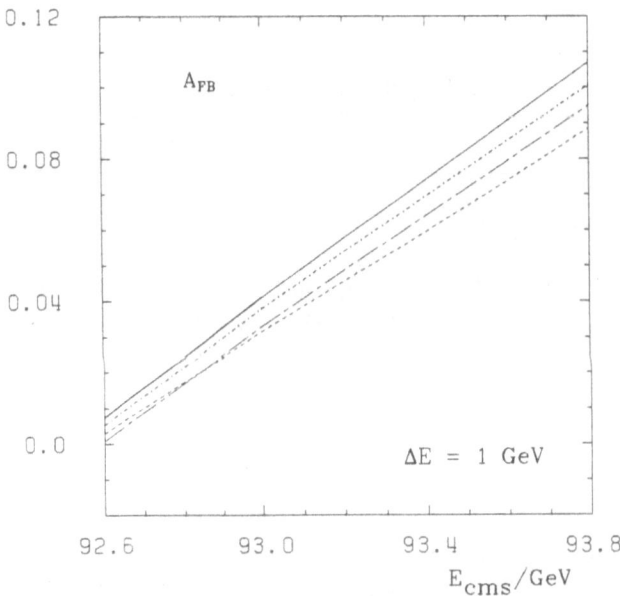

Fig. 7 Forward-backward asymmetry eq.(46) for $\Delta E = 1\,\mathrm{GeV}$. Parameters are as in fig. 4. ———— case *ii*, — · — · — case *i*, - - - - - $\mathcal{O}(\alpha)$, —— · —— results of [11].

with $z_{min} = 1 - k_{max}^0$, $z_{max} = 1 - 4m_f^2/s$, and for k_{max}^0 one can take any desired value up to the kinematical limit E_{beam}. We have also performed a calculation that includes the $\mathcal{O}(\alpha)$ hard initial state bremsstrahlung and verified that in the determination of the position of the peak this contribution can be neglected. The additional shift of the maximum due to this hard bremsstrahlung is well below 3 MeV. The effect on the width of the resonance is bigger but the differences between the various exponentiation prescriptions remain unchanged. The distortion of the resonance is obviously mainly a soft photon effect [16].

5. Conclusion

We have discussed the emission of soft photons in the presence of a resonance. We found corrections to the non-resonant case that were presented by parameter integrals. A discussion of the photon phase space integration has shown that it is important whether the energy is cut off for each photon independently or whether a total energy cutoff is imposed.

Summing up multiple soft photon emission leads to an improved 'exponentiation' prescription for the case of resonance scattering with a total energy cutoff. As expected from general physical reasons the resulting cross section cannot be interpreted as coming from coherent state radiation. Our calculations are approximative in the sense that hard photons other than in the Z propagator have been omitted. We want to point out that in any order there are still leading logarithmic contributions that cannot be found by exponentiation. They either have to be evaluated by explicit higher order calculations or by other summation techniques like renormalization group applications [15]. For the initial state corrections to e^+e^- annihilation into μ pairs this has been done in [2,3,5].

We do not claim that our prescriptions give better results than those of explicit calculations given earlier. But rather we find that our results give an answer to the question of how accurate theoretical predictions can be unless complete higher order calculations have been done. These calculations need to include final state bremsstrahlung and a detailed discussion of the phase space conditions — either through analytic calculations or via Monte Carlo studies.

References

[1] Physics at LEP. Eds. J. Ellis, R. Peccei, CERN 86-02 (1986).

[2] F. A. Berends, G. J. H. Burgers, W. L. van Neerven, Nucl. Phys. B297 (1988) 429.

[3] E. A. Kuraev, V. S. Fadin, Sov. J. Nucl. Phys. 46 (1985) 466.

[4] G. Altarelli, G. Martinelli, Physics at LEP, Eds. R. Peccei, CERN 86-02, 1986.

[5] O. Nicrosini, L. Trentadue, Phys. Lett. 196B (1987) 551.

[6] O. Nicrosini, L. Trentadue, Z. Phys. C39 (1988) 479.

[7] S. Jadach, preprint of the Max-Planck-Institut, München, MPI-PAE/PTh 6/87;
S. Jadach, B. F. L. Ward. Phys. Ref. D38 (1988) 2897.

[8] J. E. Campagne, R. Zitoun. LPNHE-88-06. LPNHE-88-08. preprint. Paris 1989. and Phys. Lett. 222 (1989) 497.

[9] D. R. Yennie, S. C. Frautschi, H. Suura, Ann. of Physics 13 (1961) 379.

[10] J. D. Jackson, D. L. Scharre, Nucl. Instr. and Meth. 128 (1975) 13.

[11] M. Greco, G. Pancheri-Srivastava, Y. Srivastava, Nucl. Phys. B101 (1975) 234; M. Greco, G. Rossi, Nuovo Cim. 50 (1967) 168.

[12] A. Vogt, Diploma thesis, Würzburg, 1988.

[13] E. Etim, G. Pancheri, B. Touschek, Nuovo Cim. 51B (1967) 276.

[14] R. Leiste et al., preprint PHE 89-02 (Berlin-Zeuthen). 276.

[15] K. E. Eriksson, Nuovo Cim. 27 (1963) 178.

[16] J. P. Alexander et al., Phys. Rev. D37 (1988) 56.

THE FORWARD BACKWARD ASYMMETRY A_{FB} IN e^+e^-- ANNIHILATION[1]

M. Bilenky* and M. Sachwitz**

* Joint Institute for Nuclear Research, Dubna
**Institut fuer Hochenergiephysik
  Berlin-Zeuthen,GDR

ABSTRACT

The forward-backward asymmetry A_{FB} in e^+e^-- annihilation in order $O(\alpha)$ QED with weak corrections and soft photon exponentiation is described. Numerical results of the dependence on the free parameters of the Standard Theory are discussed.

1. INTRODUCTION

In these days the e^+e^--collider LEP at CERN will go into operation. The cross section of the reaction

$$e^+e^- \longrightarrow (\gamma, Z) \longrightarrow \mu^+\mu^- \tag{1}$$

provides the best possibility to check the electroweak Standard Theory [1]. The mass production of the Z-boson in the region of its resonance will allow a high precision measurement of the free parameters of the Standard theory. It is expected to determine the mass M_Z (and the width Γ_Z) with an accuracy of 20-50 MeV [2] and in addition - with adequate choice of input parameters - the electroweak mixing angle $sin^2\Theta_W$ within .003 [3]. Such a precision necessitates to take into account higher order corrections. So by confronting the theoretical prediction of (1) with the experimental data one can confirm the standard theory or look for deviations as a hint of the existence of new physics.

A further consistency check is the measurement of independent observables e.g. the forward backward asymmetry A_{FB} , the left right asymmetry A_{LR} or the τ -lepton polarization A_{pol}. In this contribution we will describe the possibilities of the determination of A_{FB} especially with respect to the unknown free parameters of the Standard theory.

2. A_{FB} IN BORN APPROXIMATION

The definition of A_{FB} is:

$$A_{FB} = \frac{\sigma_F - \sigma_B}{\sigma_F + \sigma_B} \tag{2}$$

with $\sigma_F = \int_0^1 dc \frac{d\sigma}{dc}$ and $\sigma_B = \int_{-1}^0 dc \frac{d\sigma}{dc}$ and where $c = cos\theta$ is the scattering angle of the fermions

---

[1] Presented by M. Sachwitz

Radiative Corrections, Edited by N. Dombey and
F. Boudjema, Plenum Press, New York, 1990

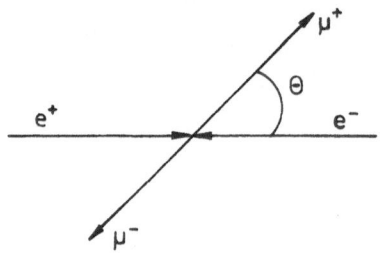

Figure 1. Scattering angle Θ in e^+e^--annihilation

In Born approximation the differential cross section is:

$$\frac{d\sigma}{dc} = \frac{\pi\alpha^2}{2s} \qquad *$$ (3)

$$
\underbrace{\left.
\begin{array}{l}
\{Q_e^2 Q_f^2 [1\,(1+c^2) \qquad +0\ 2c] \\
+\ 2Re\chi(s)Q_e Q_f [v_e v_f (1+c^2) \qquad +a_e a_f 2c] \\
+\ \chi(s)\chi^*(s)[(v_e^2+a_e^2)(v_f^2+a_f^2)(1+c^2)
\end{array}
\right.}_{C-even}
\underbrace{
\begin{array}{l}
\\
\\
+\ 4v_e a_e v_f a_f 2c]\}
\end{array}}_{C-odd}
\left.
\begin{array}{l}
\\
\\
\\
\end{array}
\right\}
\begin{array}{l}
\gamma\gamma \\
\gamma Z \\
ZZ
\end{array}
$$

The couplings are defined as:

$$v_f = 2(I_3^f - 2s_W^2 Q_f)$$ (4)

$$Q_f = 2I_3^f Q_f$$ (5)

$$a_f = 2I_3^f.$$ (6)

The s_W^2 stands for the weak mixing angle $sin^2\Theta_W$, Q_f is the final fermion charge, $Q_\mu = -1$ and $I_3^f = \pm\frac{1}{2}$ is the third component of the weak isospin. With this definition one gets automatically the right couplings when e.g. changing from up to down fermions.

The resonating factor $\chi(s)$ in (3) is defined as a product of the Breit-Wigner function with the amplitude k:

$$\chi(s) = k\frac{s}{s - M_Z^2 + iM_Z\Gamma_Z}$$ (7)

with

$$k = \frac{(1-\delta r)^{-1}}{16s_W^2(1-s_W^2)} = \frac{G_\mu M_Z^2}{8\sqrt{2}\pi\alpha}$$ (8)

Therefore the $\gamma\gamma$, γ Z and ZZ parts in (3) have a very different behaviour in the energy s. The γ part decreases with $\frac{1}{s}$. The γ Z part is proportional to

$$Re\chi(s) = k\frac{s(s - M_Z^2)}{(s - M_Z^2)^2 + \Gamma_Z^2 M_Z^2}$$ (9)

While negative below the pole $(M_Z^2 > s)$ it changes the sign next to the pole and becomes positive. The Z-part with

$$\chi(s)\chi^*(s) = k\frac{s}{(s - M_Z^2)^2 + \Gamma_Z^2 M_Z^2}$$ (10)

has the typical resonance behaviour at the pole (See fig.2).

The calculation of A_{FB} implies a integration of the forward resp. backward hemisphere. Therefore only c-odd parts with 2cos Θ of the differential cross section are showing up in the denominator $(\sigma_F - \sigma_B)$ whereas the total cross section $(\sigma_F + \sigma_B) = \sigma_T$ which is in fact an integration over

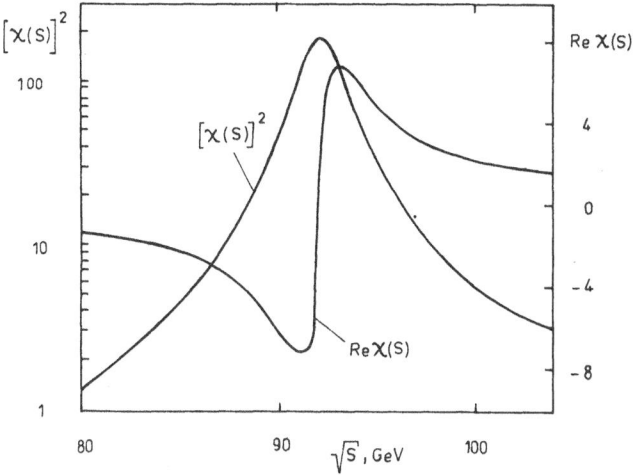

Figure 2. Re $\chi(s)$ and $\chi(s)\chi^*(s)$ for $M_Z = 92$ GeV, $\Gamma_Z = 2.5$ GeV, $\sin^2 \Theta_W = .23$

the whole phase space consists of c-even parts only. Of course this behaviour holds for experimentally determined *symmetric* angular cuts as well. The forward-backward asymmetry A_{FB} in Born approximation without angular cut is:

$$A_{FB} = \frac{\frac{3}{2} Re\chi(s) Q_e Q_f a_e a_f + 3\, \chi(s)\chi^*(s) v_e a_e v_f a_f}{Q_e^2 Q_f^2 + 2 Re\chi(s) Q_e Q_f v_e v_f + \chi(s)\chi^*(s)(v_e^2 + a_e^2)(v_f^2 + a_f^2)} \tag{11}$$

Because $v_e v_f$ is very small ($\leq 10^{-2}$) and $a_e a_f = 1$, A_{FB} is governed by the behaviour of $Re\chi(s)$ in the numerator and influenced by $\chi(s)\chi^*(s)$ only in the immediate vicinity of the Z-peak. At the pole ($M_Z^2 = s$) $Re\chi(s)$ is zero and for μ-production $Q_e^2 Q_f^2 = 1$ and neglegible with respect to $\chi(s)\chi^*(s)$. There A_{FB} in this approximation is simply

$$A_{FB} = \frac{3 v_e v_f}{(v_e^2 + 1)(v_f^2 + 1)} \tag{12}$$

This gives with (5) a dependence on the weak mixing angle $\sin^2 \Theta_W$:

$$\Delta A_{FB} = 24(1 - 4\sin^2 \Theta_W)\Delta \sin^2 \Theta_W \tag{13}$$

what is about $\Delta A_{FB} = 2\Delta \sin^2 \Theta_W$ at $\sin^2 \Theta_W = .23$.

It is well known that the width of the Z-boson is energy dependent. This can be taken into account by the very good approximation at LEP energies

$$\Gamma(s) = \frac{s}{M_Z^2} \Gamma_Z \tag{14}$$

It was shown in [4] that a parameter transformation

$$M' = \sqrt{1 + (\frac{\Gamma_Z}{M_Z})^2}\, M_Z$$

$$\Gamma' = \sqrt{1 + (\frac{\Gamma_Z}{M_Z})^2}\, \Gamma_Z$$

$$G'_\mu = [1 + i\frac{\Gamma_Z}{M_Z}]^{-1}\, G_\mu \tag{15}$$

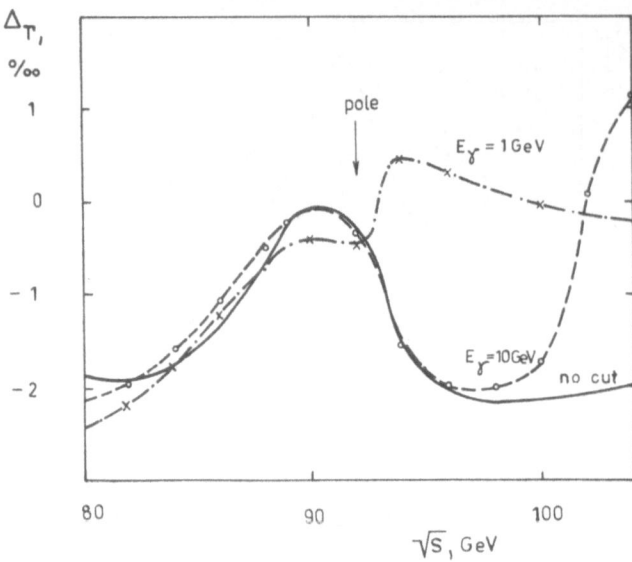

Figure 3: Difference between the exact $\Gamma(s)$ and $\Gamma =$ constant treatment of A_{FB}

allows to take into account the energy-dependence of the width in a quite simple form. In this way the centre of gravity of the resonance and in the same manner the mass of the Z-boson appears to be shifted by $-\frac{\Gamma^2}{2M_z} \sim -35 MeV$.

Figure 3 shows the difference between a $\Gamma =$ constant and the proper $\Gamma(s)$ treatment. At the pole where the expected statistical error in a 100 day run with a luminosity of $10^{31}\ cm^{-2}\ s^{-1}$ is about $\frac{1}{\sqrt{N}} \sim .002$ the influence of $\Gamma(s)$ is smaller than .1% and therefore not as important as in the case of the Z-boson line shape.

3. RADIATIVE CORRECTIONS TO A_{FB}

The radiative corrections are separated into QED and weak corrections. The QED corrections are dependent on experimental cuts (e.g. E_γ and angular cuts). As global input only M_Z, Γ_Z and $\sin^2 \Theta_W$ are needed. There exist complete analytic $O(\alpha)$ calculations [5] and leading logarithmic corrections to the initial state radiation of the next order [6]. Several Monte Carlo results have been presented recently during the 1989 LEP physics workshop [11]. The general form of the $O(\alpha)$ correction is as in equation 3, but with additional initial, final and interference terms of the photon bremsstrahlung characterized by an overall factor of $\frac{\alpha}{\pi}$. These terms have a c-even and a c-odd behaviour in such a way that three even and three odd functions give an additional component to the $\gamma\gamma-$, $\gamma Z-$ and ZZ-term of the differential cross section 3. Again after integration only the c-odd terms are involved in the numerator of A_{FB}.

Here one should make the following comment:

After a trivial replacement of the couplings in the differential cross section in order to implement a possible polarization of the beams and/or final helicity states h for the observation of final polarization ($\tau-$leptons) one gets immediately the corresponding observables [8]

- Left-Right Asymmetry

$$A_{LR} = \frac{\sigma_L - \sigma_R}{\sigma_L + \sigma_R} \tag{16}$$

where σ_{LR} are the left,right handed cross sections.

- Polarization Asymmetry

$$A_a^{pol} = \frac{\sigma_a(h_f = 1) - (h_f = -1)}{\sigma_a(h_f = 1) + (h_f = -1)} \tag{17}$$

110

where a means total, forward or backward asymmetry. The forward and backward cross sections may be calculated via

$$\sigma_{F,B} = \frac{1}{2}\sigma_T(h)[1 \pm A_{FB}(h)] \qquad (18)$$

with h= ±1.

Figure 4 shows the different sensitivities for the asymmetries described.

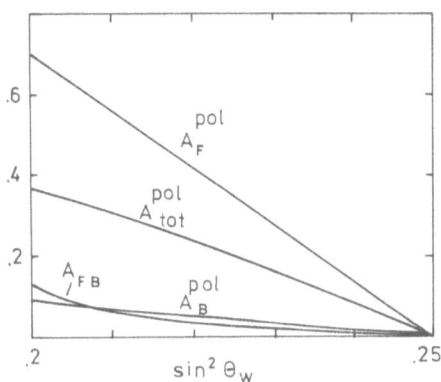

Figure 4. Sensitivity of the asymmetries to $\sin^2 \Theta_W$

4. THE CONVOLUTION INTEGRAL FOR A_{FB}

In [5] the bremsstrahlung functions are given after complete integration over the unrestricted photon momentum phase space. Going one step back to the functions still dependent on the energy of the photons emitted we can see much better underlying structures. Here [7] for the $\gamma\gamma, \gamma Z$ and ZZ case only one common c-even and one c-odd function of each initial, final and interference part exists. The bremsstrahlung correction can be seen as an energy dependent correction of the Born term. As a result, one can introduce a convolution integral for the QED corrections of A_{FB} with the corresponding c-odd functions ρ as weights. These differ from the well-known function for the total cross section [12].

$$\sigma_T(s)A_{FB}(s) = Re \int_0^1 dv \ \sigma_{F-B}^0(s, s')\rho(v), \qquad (19)$$

where σ_T is the total cross section, $v = 1 - \frac{s'}{s}$, s' the remaining energy after the emission of the photons. The difference of the forward backward Born cross section is σ_{F-B}^0 and $\rho(v)$ are the weight functions. The integrand is a sum of three terms :

- initial radiation with $\sigma_{F-B}^0(s')$ and $\rho_{initial}(v)$

- interference with $\sigma_{F-B}^0(s, s')$ and $\rho_{interference}(v)$

- final radiation with $\sigma_{F-B}^0(s)$ and $\rho_{final}(v)$.

$$\rho_{initial}(v) = Q_e^2 \frac{\alpha}{\pi} \frac{1+(1-v)^2}{v} \frac{1-v}{(1-\frac{v}{2})^2}[(L_e - 1) - \ln \frac{1-v}{(1-\frac{v}{2})^2}], \qquad (20)$$

$$\rho_{interference}(v) = \frac{2}{3v}[2(1-v)(v^2+2v-2) + (1-v)(5v^2-10v+8)\ln(1-v) + (5v^3-18v^2+24v-16)\ln(2-v)] \qquad (21)$$

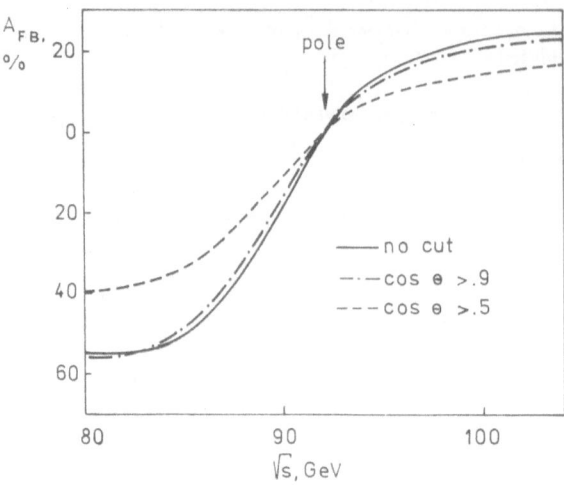

Figure 5. A_{FB} with angular cut

$$\rho_{final}(v) = \frac{2}{v}[(1-v)(L_f - 1) + \ln(1-v) + \frac{1}{2}v^2 L_f] \tag{22}$$

where

$$L_{e,f} = \ln \frac{s}{m^2_{e,f}} \tag{23}$$

If one applies a cut on the photon energy (e.g. reject photons not seen in the detector) the cross section and therefore the A_{FB} may even become unphysical (negative cross sections). The reason is that the remaining radiative corrections are negative and may become large. The *hard* bremsstrahlung weight functions contain still soft photon parts like

$$\lim_{v \to 0} \frac{1 + (1+v)^2}{v} = \frac{2}{v}. \tag{24}$$

In order to take this into account one has to rearrange the weight functions by taking away the *soft* part of the hard bremsstrahlung [14]. An integration over the resulting soft part implies all orders of correction and gives a $v^{\beta-1}$ behaviour [13] with

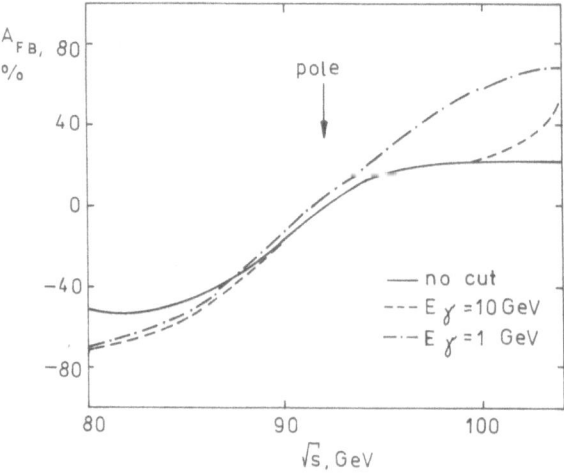

Figure 6. A_{FB} with photon energy cut E_γ

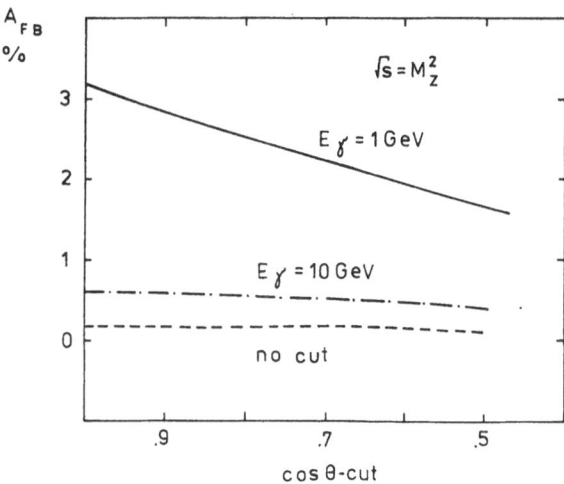

Figure 7. Dependence of the value of A_{FB} at the pole from a E_γ cut and an angular cut

$$\beta = Q^2 \frac{2\alpha}{\pi}(\ln \frac{s}{m_e^2} - 1). \tag{25}$$

It is shown in [6] that a $O(\alpha)$ correction with soft photon exponentiation is very close to the second order leading log calculations. Fig. 5 shows A_{FB} in order $O(\alpha)$ correction with soft photon exponentiation in dependence on a symmetric angular cut.

As it can be seen from figure 6 at the pole the photon energy cut E_γ has a great influence on the value of A_{FB}. This is shown in detail in figure 7. Whereas no big variation arises due to a modest E_γ cut a severe soft photon cut of 1 GeV causes great deviation.

5. WEAK CORRECTIONS

The weak corrections have a small but visible effect on the value of A_{FB}. Additional free parameters (Higgs mass M_H and top mass m_t) are needed. A classification can be done into

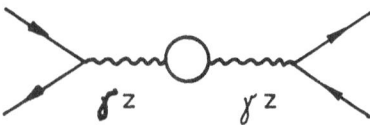

- propagator corrections (self energies, 2-point functions)

- vertex corrections (3-points functions)

113

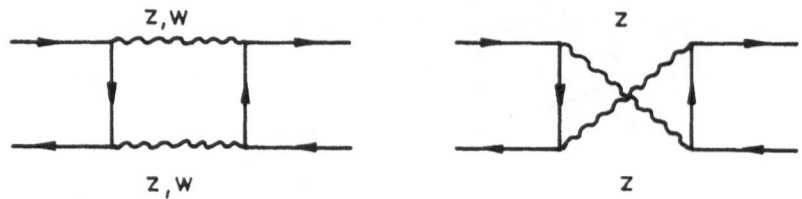

- Box corrections (4-point functions)

In analogy to the nucleon formfactors it is very convenient to describe the weak contributions in the language of weak formfactors [10]. In order to be gauge independent one has to combine the s- dependent 2- and 3-point functions with the s- and t (cos Θ) dependent box contributions. Therefore the weak corrections depend not only on internal structures but also to a numerically small amout on the experimental acceptance cut. After calculation of the complex gauge invariant weak formfactors $\rho_Z(s)$, $\kappa_{e,f}(s)$, κ_{ef} [9] one has to replace the couplings in the cross section (3) by the dressed complex couplings

$$v_a = 1 - 4\sin^2\Theta_W \ Q_a \ \kappa_a(s), \qquad a = e, f, \tag{26}$$

$$v_{ef} = v_e + v_f - 1 + 16\sin^4\Theta_W \ Q_e Q_f \ \kappa_{ef}(s), \tag{27}$$

In this way all electroweak corrections are implemented in the cross section. The complex formfactors are explicitely defined in [9]. The formfactors and the differential cross section from wich A_{FB} is derived have been calculated with the combined Fortran code DIZET [15] and MUCUTCOS [16] In a physical sense the formfactors are finite renormalizations of

- G_μ in the case of the ρ_Z and

- the weak mixing angle $\sin^2\Theta_W$ in the case of the formfactors κ

As an example for $M_Z = 92$ GeV, $M_H = 100$ GeV and $m_t = 90$ GeV one gets for ρ_Z and κ which are formally one in the Born approximation the following weak corrections :

$$\rho_Z = .999 - i\,0.013$$
$$\kappa_e = 1.009 + i\,0.013$$
$$\kappa_{e,f} = 1.018 + i\,0.027 \tag{28}$$

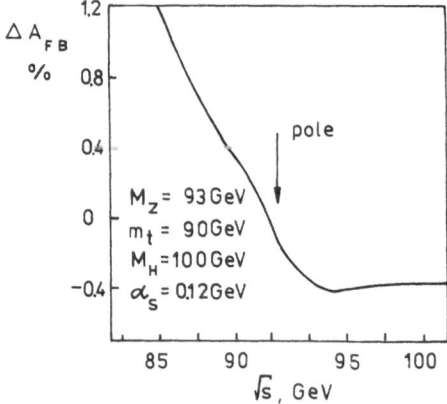

Figure 8. Δ A_{FB} due to use of Born couplings in the QED contributions instead of the dressed complex couplings

Table 1

	ini.	exp.ini.	fin.	interf.
no cut	0.3118	1.0864	1.0838	1.1165
	-.1338	0.6907	0.6891	0.7226
10 GeV	0.4145	1.1849	1.0666	1.1232
	-.0270	0.7936	0.6676	0.7264
1GeV	1.5740	2.0612	1.8301	3.1917
	1.2088	1.7280	1.4815	2.9488

Forward backward asymmetry in percent at $\sqrt{s} = M_Z$ as function of the photon energy cut E_γ for QED and QED+weak corrections. The contributions are stepwise included:

- ini. : Born plus initial state radiation

- exp.ini. : exponentiated soft photon initial state radiation

- fin. : final state radiation

- interf. : interference bremsstrahlung

The first row includes QED corrections only, the second QED+weak.

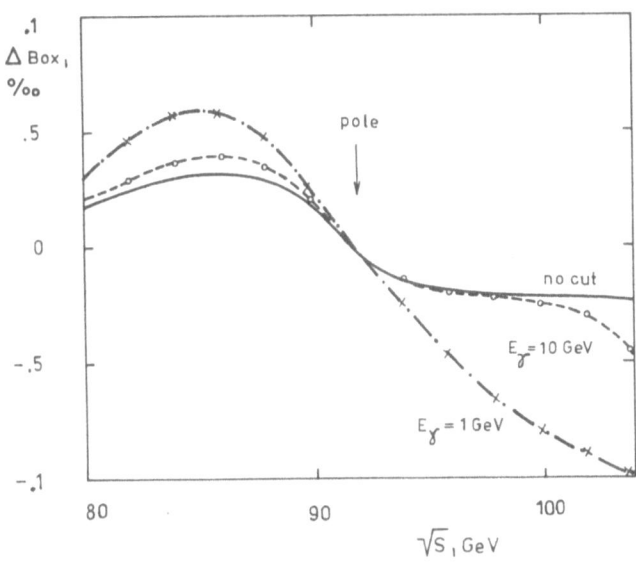

Figure 9. The contribution of box diagrams to A_{FB}

Furthermore one has to replace the fine structure constant $\alpha(0) = \frac{1}{137}$ by $\alpha(s = M_Z^2)$. This gives an additional correction factor of

$$\frac{\alpha(s)}{\alpha(0)} = 1.060 - i\, 0.016 \qquad (29)$$

Figure 8 shows the influence of the proper complex determination of the weak formfactors. Taking only the real part one would make at the pole an error greater than the expected statistical error.

The contribution of the W,Z box functions is shown in figure 9. One can see that the absolute value of the box part is very small and at the pole less than .02%.

The modification of the QED values by the weak correction for $M_Z = 92$ GeV, $m_t = 90$ GeV and $M_H = 100$ GeV is seen in table 1.

The first row shows the QED values and the second one the QED+weak corrections. The first column in the table gives the values for the initial state radiation only. Then in addition the soft photon exponentiation is applied. The next columns describe the implementation of the final and the interference term. Here one can see that a photon energy cut E_γ increases the role of the interference part. The inclusion of the interference is nearly neglegible without cut ($\Delta \sim .03\%$) but with a cut $E_\gamma = 1$ GeV the difference grows up to $\Delta \sim 1.3\%$.

The top mass m_t { $60 \leq m_t \leq 230$ GeV } and the Higgs mass M_H {$10 \leq M_H \leq 1000$ GeV'} are unknown within a large region. The influence possible variation of these free parameters on A_{FB} at the pole ($M_Z = \sqrt{s}$) for M_Z values from 88 GeV to 94 GeV can be taken from figure 10. Within the experimental errors expected the unknown parameter M_H seems to be no problem for the determination of A_{FB}. But if the value of M_Z is not at ~ 88 GeV there is a strong variation of the A_{FB} value with m_t. Therefore one has to look for other possibilities to determine with A_{FB} the free parameters of the Standard Theory for example by taking more than the value at the pole. The motivation is that the precise determination of M_Z from the line shape requires at least two points not far away (within about $\pm \frac{\Gamma_Z}{2}$) from the pole [17].

6. QCD CORRECTIONS

QCD corrections are influencial in two different ways:

• vertex and bremsstrahlung corrections

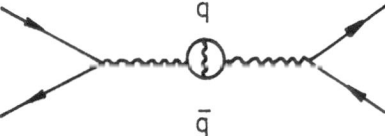

• self energies with $(\gamma, \gamma), (\gamma, Z)$, (Z,Z) and (W,W)

Additional contributions are coming from the oblique diagrams which are diagrams with gluon insertions to the polarization operators of the intermediate vector bosons. These two loop self energy contributions are of sizesible order only for heavy quark production.

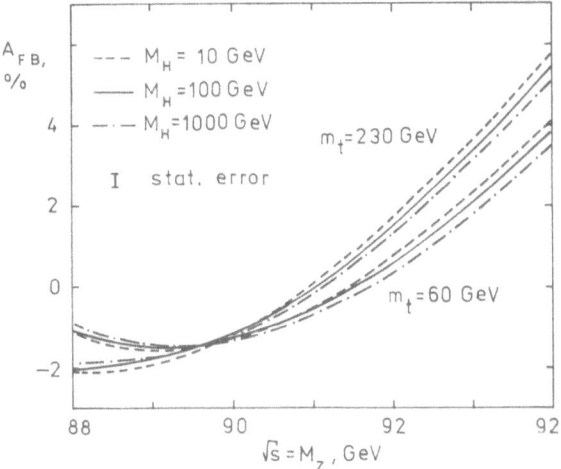

Figure 10. Variation of A_{FB} at the pole (M_Z 88 - 94 GeV) for $m_t = 60$ and 230 GeV and $M_H = 10,100,1000$ GeV.

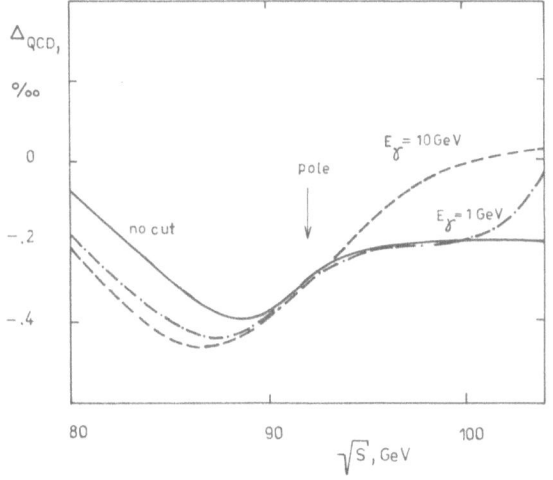

Figure 11. Variation of A_{FB} due to QCD corrections

For μ production the vertex and bremsstrahlung QCD corrections [9] have to be taken into account for the determination of the Z decay width. Here approximately

$$\Gamma_Z = \Gamma_Z^{lept} + \Gamma_Z^{hadr}(1 + \frac{\alpha_s}{\pi} + \ldots) \tag{30}$$

where $\alpha_s = 0.12 \pm 0.02$ is the strong coupling constant.

In [18][19] $O(\alpha_{em}\alpha_s)$ corrections of the strong virtual interaction are calculated. These quantities contain the polarization functions of the photon and the intermediate vector bosons with Z - γ mixing. Because here the contribution of the light quarks are very small only the b- and t-quarks are considered. The QCD corrections are calculated for the decay processes of the Z- and W-Boson, four fermion scattering and annihilation in a wide region of the transfer momenta Q^2. As a result these QCD contributions can be incorporated into the weak formfactors ρ and κ and δr as an additional correction leaving the former organization of the cross section as it is. Here $\delta r(M_Z, M_W, M_H, m_{fermions})$ is the $O(\alpha)$ non-QED correction to the μ - decay. As an example for $m_t = 90$ GeV , $M_Z = 92$ GeV ,$\sin^2\Theta_W = .23$ and $\alpha_s = .12$ one gets the following corrections in percent:

$\Delta\delta\,r^{QCD} = .17$, $\Delta\rho^{QCD} = -.03$ and $\delta\kappa^{QCD} = -.18$.

From figure 11 one can see the contribution of the described QCD corrections to A_{FB}, which turns out to be very small.

7. DETERMINATION OF $\sin^2\Theta_W$ FROM THE ZERO OF A_{FB}

The determination of $\sin^2\Theta_W$ from A_{FB} in a 100 day run at the pole gives a statistical error $\Delta\sin^2\Theta_W \sim .001$. With $M_Z = 92$ GeV the systematical error from the possible variation in m t is about double the statistical error. The difference of the efficiency $\Delta\epsilon$ of the forward-backward detector is very critical because the different efficiencies go linearly into the value of A_{FB}. A $\Delta\epsilon$ of 1% give a additional error in $\sin^2\Theta_W \sim$ of .005. So with very small values around the pole inefficiencies are very dangerous.

Whereas a fit the line shape of A_{FB} with points below and above the pole seems to give no reasonable result due to the big statistical errors one can look for a extraordinary point like $A_{FB} = 0$. In Born approximation the energy s at $A_{FB} = 0$ is

$$s|_{A=0} = M_Z^2 \frac{1}{1 + 2v^2 k} \tag{31}$$

where k is taken from (8). In figure 12 one can see the dependence of $\sqrt{s}$ at $A_{FB} = 0$ from the weak mixing angle $\sin^2\Theta_W$ for $M_Z = 93$ Gev. We assume a three point linear least square fit for the zero of A_{FB} with data points at the pole and $\sim \pm 1$ GeV each in a 10 days run. The error in $\sin^2\Theta_W$ from ΔM_Z is .001 for $\Delta M_Z = 20$ Mev and .0025 for $\Delta M_Z = 50$ MeV. The error in the fit of the zero of A_{FB} is about .003. The error in the determination of $\sin^2\Theta_W$ goes like $\frac{1}{N}$ with N = number of points or number of days per data point. The detector inefficiency $\Delta\epsilon$ is not as dangerous as at the pole, because with a fit more or less the slope of A_{FB} is determined and therefore a shift to positive or negative values is only indirect related to the zero of A_{FB}. A $\Delta\epsilon = $ 1% gives an error of $\sim .002$. If one includes the electroweak corrections one has to study the m_t and M_H dependence as well. The influence of m t and M_H can be seen from figure 13. The unknown Higgs mass causes only a small variation of A_{FB} and therefore of $\sin^2\Theta_W$. If the top mass is not known better than 50 Gev one can not expect a determination of $\sin^2\Theta_W$ better than .005.

8. CONCLUSION

The forward backward asymmetry A_{FB} with electroweak corrections is an important factor for the determination of the free parameters of the Standard Theory. There is a strong dependence on the free parameters which is a source of uncertainties but on the other side also a possibility to determine these parameters with an input from other experiments. Great attention has to be paid on the experimental circumstances as there are the angular cuts, the photon energy cut and detector inefficiencies in the forward backward hemisphere.

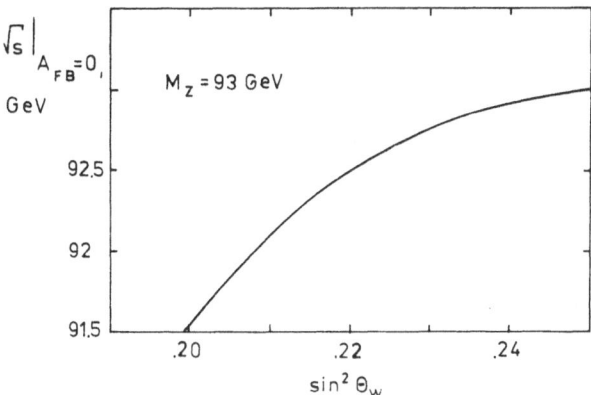

Figure 12. Dependence of the energy at $A_{FB} = 0$ from $\sin^2 \Theta_W$, $M_Z = 93$ GeV.

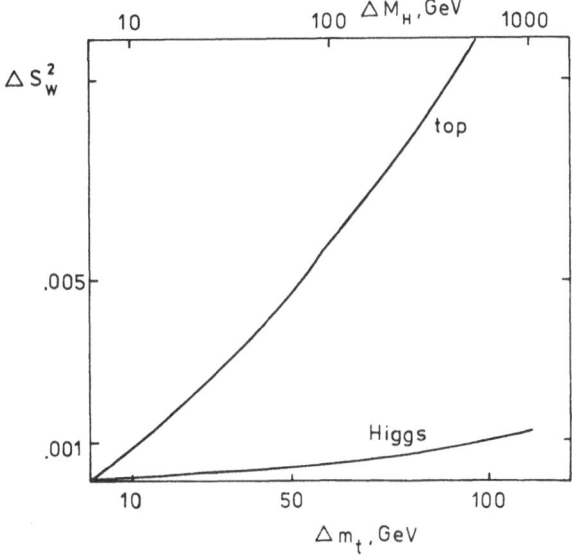

Figure 13. The influence of m_t and M_H on the determination of $\sin^2 \Theta_W$ in a three point fit.

The numerical results presented here have been obtained with the Fortran codes DIZET, MU-CUT and MUCUTCOS which are available at CERN VM.

9. ACKNOWLEDGEMENT

We would like to thank all our collegues from the Dubna-Zeuthen radiative correction group for the fruitful collaboration and A. Boehm for valuable discussions.

References

[1] S.L.Glashow, Nucl. Phys. 22: 579 (1961);

S.Weinberg, Phys. Rev. Letters 19: 1264 (1967);

A. Salam, in *Elementary Particle Theory*, p. 361 ed. N. Svartholm (Stockholm 1968).

[2] J. Ellis and R. Peccei, eds., Physics at LEP, CERN report CERN 86-02 (1986).

[3] T. Riemann and M. Sachwitz, Phys. Letters 212B: 488 (1988).

[4] D. Bardin, A. Leike, T. Riemann, M. Sachwitz, Phys. Letters 206B: 539 (1988).

[5] O. Federenko and T. Riemann, Acta Phys. Polonina B18: 761(1987); D. Bardin, M. Bilenky, O. Fedorenko, T. Riemann, JINR Dubna prepr. E2-87-663(1987), E2-88-324(1988) and unpublished work.

[6] F. Berends and W. van Neerven, Ringberg workshop April 1989 to appear in the Proceedings.

[7] D.Bardin, M. Bilenky, A. Chizov, A. Sazonov, Y. Sedykh, T. Riemann and M. Sachwitz preprint CERN-Th.5411/89 subm. to Phys. Letters B.

[8] T. Riemann and M. Sachwitz, in Proc. Seminar *Physics of $e^+ e^-$ Interactions* Dubna 1987 E2-88-363 and PHE 87-12(1987).

[9] D. Bardin, M. Bilenky, G. Mitselmakher, T. Riemann and M. Sachwitz Berlin-Zeuthen prepr. PHE 89-05(1989), to appear in Z.Physik C(1989).

[10] D. Bardin and T. Riemann,PHE 89-11(1989) and these Proceedings.

[11] G. Altarelli, ed. to be published as Yellow Report.

[12] G. Bonneau and F. Martin, Nucl. Phys. B27: 381 (1971).

[13] M. Greco, G. Pancheri-Srivastava and Y. Srivastava, Nucl. Phys. B171: 543 (1982).

[14] E. Kuraev and V. Fadin, Sov. J. Nucl. Phys. 41:466 (1985).

[15] D. Bardin, M. Bilenky, T.Riemann M. Sachwitz, H. Vogt, P. Christova Berlin-Zeuthen prep. PHE 89-09.

[16] D. Bardin et. al. in preparation.

[17] R. Leiste, T. Riemann, M. Sachwitz, H. Vogt, D. Bardin, M. Bilenky, G. Mitselmakher. , in Proc. Seminar *Physics of $e^+ e^-$ interactions* Dubna 1988 E2-89-462 and PHE 89-02(1989).

[18] A. Djouadi and C. Verzegnassi, Phys. Letters 195B: 265 (1987);

A. Djouadi, Montpellier prepr., PM/87-35, PM/87-53 (1987).

[19] D. Bardin and A. Chizhov , in Proc. Seminar *Physics of $e^+ e^-$ Interactions* Dubna 1988 E2-89-462.

HIGH PRECISION MEASUREMENT OF THE TOTAL Z° WIDTH

USING LARGE ANGLE e⁺e⁻ EVENTS

R. Battiston, M. Pauluzzi, L. Servoli, and A. Santocchia

INFN Sezione di Perugia and Universita' di Perugia
I-06100 Perugia, Italy

Presented by R. Battiston

Introduction

The operation of the two planned Z° factories, SLC and LEP, will soon allow accurate tests of the validity of the Standard Model of electroweak interactions, possibly revealing the presence of new, unexpected physical effects.

The experiments planned at these machines, in addition to the precise measurement of the Z° mass, will also perform precise measurements of other physical quantities like the Z° width and (un)polarized asymmetries that are directly related to the particle content of the theory and to $\sin^2\theta_W$. The more precise the measurements of these quantities, the deeper will be our understanding of the validity of the Standard Model: it follows that every effort should be made to reach the best possible accuracy. Thanks to the high luminosity ($L \approx 10^{31} cm^{-2} sec^{-1}$) that is expected at these machines, it is foreseen that, for the measurements of the Z° parameters, systematic errors will fairly quickly come to dominate over statistical errors, thereby setting the limit to the ultimate precision.

It is instructive for instance to look at the case of the measurement of the Z° mass and total width. It has been shown[1] how, after a few weeks of data taking close to the peak with $L \approx 10^{31} cm^{-2} sec^{-1}$, the *statistical errors* on these quantities, obtained in the fit of the line shape, will be < 10 MeV and <15 MeV respectively, while the corresponding *systematic errors*, due to uncertainties on beam energies and luminosity measurements, range from 20 to 40 MeV, thus dominating the overall error.

In this paper we show how, at or close to the peak, the study of the reaction $e^+e^- \rightarrow e^+e^-$ at large angles may give precise informations on the Z° mass and width. Various quantities may be used to perform this study. As an example we introduce the (unpolarized) forward-backward electron asymmetry, A^e_{FB}, and we show how this quantity is sensitive to Z° parameters, in particular to the width. In fact due to the presence of the Bhabha t-channel, A^e_{FB} depends not only on

Radiative Corrections, Edited by N. Dombey and
F. Boudjema, Plenum Press, New York, 1990

$\sin^2\theta_W$ (like FB asymmetries for other fermions, A^f_{FB}) but also on the $e^+e^- \to Z^\circ \to e^+e^-$ cross section, that is on the Z° total width. We also show that A^e_{FB} is remarkably insensitive to systematic errors on beam energies and luminosity measurements, that affect the direct measurement of the line shape. Moreover most detector dependent systematics show a tendence to cancel in the expression of δA^e_{FB}. Other variables having similar properties and not making use of the charge measurement are also discussed at the end of the paper. To fully exploit the intrinsic accuracy of this method, it is important to understand the size of the QED radiative correction to these variables: we show for instance how A^e_{FB} is moderately sensitive to these corrections. In conclusion, it turns out that, by performing a precise measurement of A^e_{FB} or similar quantities, very close to the peak, the Z° total width can in principle be measured with an ultimate overall accuracy of few MeV after $\approx 10^7$ Z°. The accuracy that can be obtained with this method is then roughly one order of magnitude better than what can be obtained with other methods discussed in the literature.

A nice feature of this method, is that one needs only to run at (or very close to) the peak: thus, this way of measuring the Z° width (and mass) takes advantage of the total luminosity delivered by the machine without affecting other measurements that need the maximum avaliable statistics.

An example of the method proposed: the unpolarized forward-backward electron asymmetry

It is well known that the study of the forward-backward asymmetry, $A^f_{FB}(x)$

$$A^f_{FB}(x) = \frac{{}_0\!\int^x dz \; d\sigma/dz - {}_{-x}\!\int^0 dz \; d\sigma/dz}{{}_{-x}\!\int^x dz \; d\sigma/dz} \tag{1}$$

where f is the final state fermion, f≠e, z=cosθ and x is an integration limit given by the detector acceptance, will allow accurate comparison with the SM predictions[2]. To the lowest order, this quantity depends only on axial and vector couplings of the initial and final fermions to the Z°. One obtains, at the peak:

$$A^\mu_{FB}(x) = F(x) \frac{3r_A r_V |\chi(M_Z)|^2}{1 + |\chi(M_Z)|^2} \approx F(x) \; 3r_A r_V \tag{2}$$

where:

$$|\chi(M_Z)|^2 = |(M^2_Z/16\mu^2)(v_e^2+a_e^2)s/(s-M^2_Z+i\,M_Z\Gamma_Z)|^2_{s=M^2_Z} =$$

$$(M^2_Z/16\mu^2)^2(v_e^2+a_e^2)^2\,M^2_Z/\Gamma_Z^2\;,$$

$$\mu = (\pi\alpha/\sqrt{2}\;G_F)^{1/2} = 37.281 \text{ GeV},$$

$$r_V = v_e v_f/(v_e^2+a_e^2)\;, \quad r_a = a_e a_f/(v_e^2+a_e^2)\;,$$

$v_f = -1 - 4Q_f \sin^2\theta_w$, $a_f = -1$ for   negatively charged fermions,

$v_f = 1 - 4Q_f \sin^2\theta_w$, $a_f = 1$ for   positively charged fermions,

$F(x) = 4x/(3+x^2)$.

In (2) we neglected unity with respect to the resonant term $|\chi(M_z)|^2 \approx 200$. That is the reason why, at the peak, $A^f_{FB}(x)$ does not depend explicitly on the Z^o shape and then is completely insensitive to the total Z^o width.

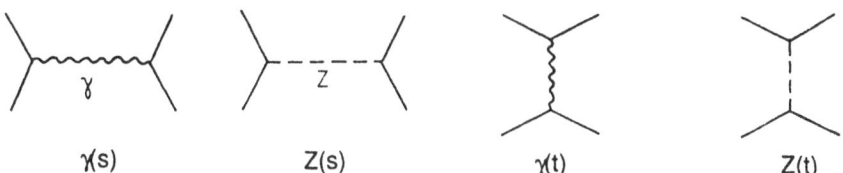

$\gamma(s)$ $Z(s)$ $\gamma(t)$ $Z(t)$

Fig. 1. Born diagrams for the Bhabha scattering. Throughout the paper we number the ten corresponding contributions to the total cross section as follows: #1 $\gamma(s)\gamma(s)$, #2 $\gamma(s)\gamma(t)$, #3 $\gamma(t)\gamma(t)$, #4 $\gamma(s)Z(t)$, #5 $\gamma(t)Z(t)$, #6 $Z(t)Z(t)$, #7 $Z(s)\gamma(s)$, #8 $Z(s)\gamma(t)$, #9 $Z(s)Z(t)$ and #10 $Z(s)Z(s)$.

    Let us analyze what happens to FB asymmetry in the process $e^+e^- \rightarrow e^+e^-$. This process has been widely studied in the context of small angle luminosity monitors (Bhabha)[3], but it has not apparently received the same attention in the context of precision tests of the SM, as for instance in the process $e^+e^- \rightarrow \mu^+\mu^-$. The differential cross section, calculated from the four Feynman diagrams shown in fig. 1, contains ten contributing amplitudes: due to the presence of the diverging t channel the total cross section depends strongly on the minimum angle, θ_{min}, chosen in the integration. Fig. 2a shows the lowest order total cross section together with the ten different contributions for the case θ_{min}=15°: in fig. 2b the corresponding F-B cross sections are shown. Unless otherwise specified, we use throughout the paper the following SM parameters; M_z=92 GeV/c^2, $\sin^2\theta_W$=0.2290, M_{top}=65 GeV/c^2, Γ_z=2.493 GeV.

    In   fig. 2   we note that at the peak many of the terms are negligible. First of all, all interference terms with the $Z(s)$ diagram (#7,8 and 9), vanish because the real part of the Z^o propagator goes to zero at the pole. One should note that these contribution to the total cross section and to the asymmetry are non-negligible when we slightly move from the pole. Of the remaining contributions, only three contribute significantly to the total cross section: these are $\gamma(t)\gamma(t)$ (#3), $Z(s)Z(s)$ (#10) and $\gamma(t)\gamma(s)$ (#2). As it is shown in fig. 3, at the pole the relative size of the first two terms depends on

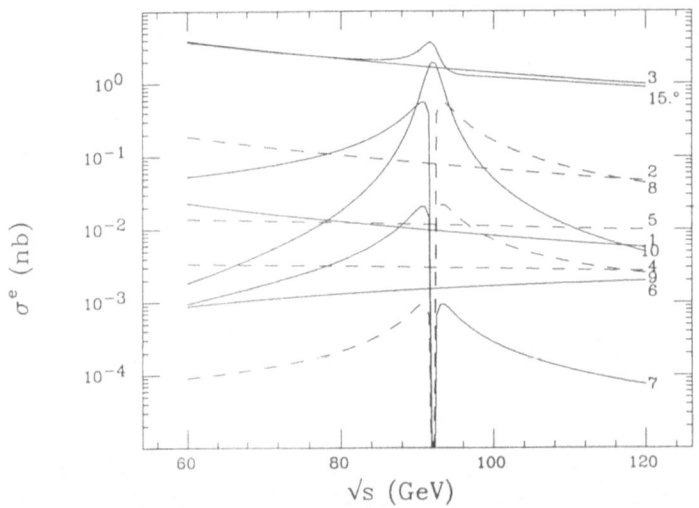

Fig. 2a. Lowest order Bhabha cross section (labelled 15°) together with the ten contributions versus the c.o.m. energy. Dashed lines correspond to negative contributions from interference terms. Minimum angle 15°.

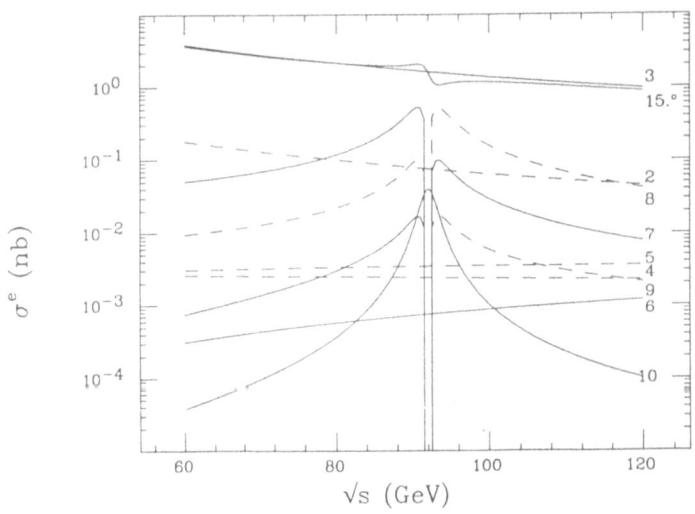

Fig. 2b. Forward - backward hemisphere Bhabha cross section together with the ten contributions. Minimum angle as in 2a.

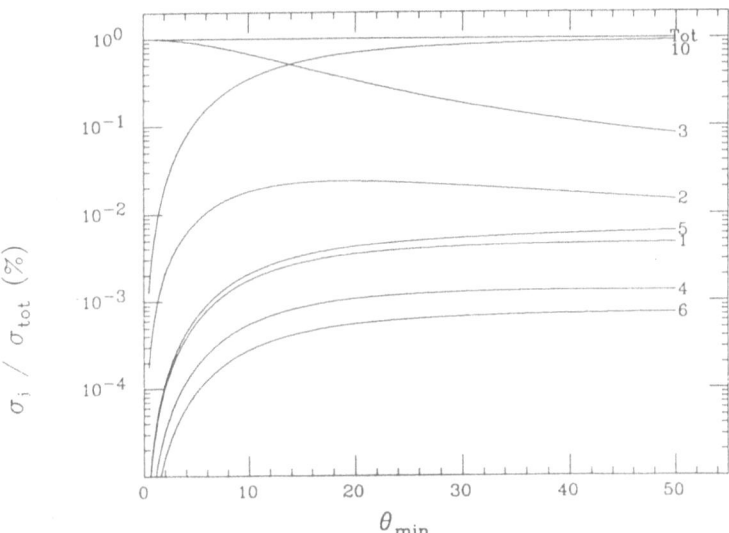

Fig. 3. Relative size of the contributing term to the total
cross section at the peak versus θ_{min}. Lowest order calculation.

θ_{min}. At lowest order, they are equal for $\theta_{min} \approx 13°$: at smaller θ_{min}, $\gamma(t)\gamma(t)$ dominates, while at larger θ_{min}, $Z(s)Z(s)$ is the most important. The $\gamma(t)\gamma(s)$ term is always much smaller.

The total cross section as a function of the energy is shown in fig. 4 for different values of θ_{min}.

Throughout the paper, we show, where relevant, the radiatively corrected distributions obtained using the analytical formulas discussed in ref [4] with the following choice of parameters: energy threshold for photon detection $\Delta E\gamma/E_{beam}=10^{-2}$ and angular resolution $\delta=1°$. These formulas are correct up to O(1%) terms, and are then useful for a first comparison with the Born results. The formulas used in this paper do not include hard (visible) photon emission to any order.

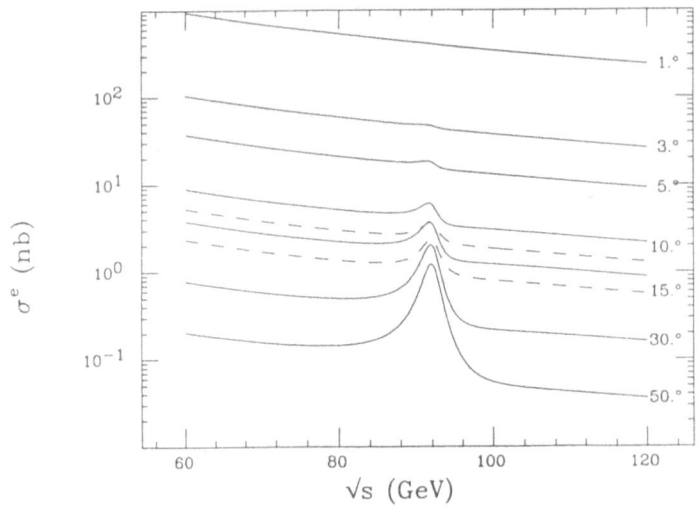

Fig. 4. Lowest order Bhabha cross section versus of the c.o.m. energy, for different values of θ_{min}. For two values of θ_{min} also the radiatively corrected result is shown (dashed lines)(ref.4).

If we now compute $A^e{}_{FB}(x)$ following definition (1), we obtain the results shown in fig. 5, for different values of θ_{min}. Qualitatively, far from the peak, $A^e{}_{FB}(x)$ is close to unity, because the Bhabha contributions dominates, while near to the peak, the total cross section at the denominator suddenly increases, while the corresponding $Z°$ contribution to the numerator remains small. As a result, $A^e{}_{FB}(x)$ shows a dip at the resonance, whose shape and size are directly related to the $Z°$ line shape and to θ_{min}.

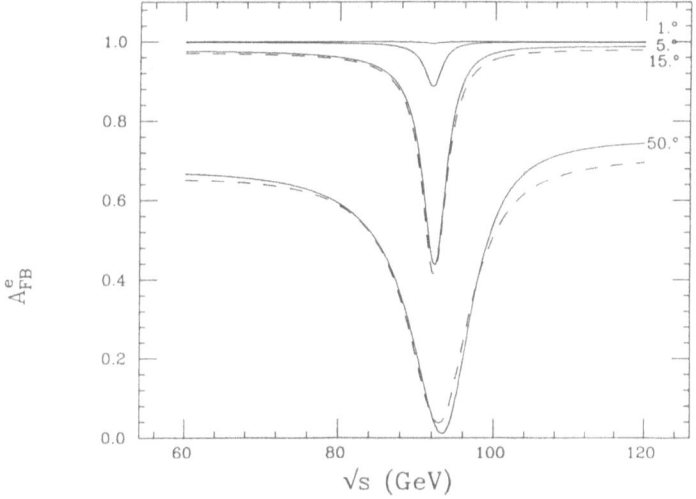

Fig. 5. $A^e_{FB}(x)$ versus the c.o.m. energy for different values of θ_{min}. For two values of θ_{min} also the radiatively corrected result is shown (dashed lines)(ref.4).

$A^e_{FB}(x)$ has then the property of mapping the Z^o line as a ratio with a non-resonant cross section. This fact is very important from an experimental point of view, because it allows a luminosity independent shape measurement. In addition, at (or very near to) the resonance, $A^e_{FB}(x)$ shows a minimum: this means that if we run close to the peak $A^e_{FB}(x)$ must be insensitive to the uncertainty on the beam energy, another kind of systematics that limits the measurement accuracy of the line shape.

It is instructive to write an approximate expression for $A^e_{FB}(x)$ at the Z^o pole by keeping only the three most important contributions to the cross section (note that this approximation fails as we move frome the pole). By defining

$$d\sigma^*/dz = d\sigma^{\gamma(t)\gamma(t)}/dz + d\sigma^{Z(s)Z(s)}/dz + d\sigma^{\gamma(t)\gamma(s)}/dz \qquad (3)$$

we obtain a fairly simple analytic expression:

$$A^e_{FB}(x) \approx \frac{{}_0\!\int^x dz\, d\sigma^*/dz - {}_{-x}\!\int^0 dz\, d\sigma^*/dz}{{}_{-x}\!\int^x dz\, d\sigma^*/dz} = \frac{B_-(x)+Z_-(x)\ r_A r_V |\chi(s)|^2}{B_+(x)+Z_+(x)\ |\chi(s)|^2} \qquad (4)$$

where we introduced four acceptance dependent functions, two relative to the Bhabha $\gamma(t)\gamma(t)$ and $\gamma(t)\gamma(s)$ contributions:

$$B_-(x)=16/(1-x^2) + 8\ln(1-x^2) -16 + x^2$$
$$B_+(x)=16x/(1-x^2) + 8\{x+\ln[(1-x)/(1+x)]\} \qquad (5)$$

and two to the $Z(s)Z(s)$ term:

$$Z_-(x) = 4x^2$$
$$Z_+(x) = x(1+x^2/3) \qquad\qquad (6)$$

It results that over most of the θ_{min} range $B_-(x) \approx B_+(x)$ and $Z_-(x) r_A r_V \approx 2.10^{-2} Z_+(x)$, as expected from the fact that $A^\mu_{FB}(x)$ is very small at the resonance. In conclusion, due to the presence in the asimmetry of the Bhabha terms, by themselves insensitive to the Z° parameters, we obtain a physical quantity explicitly dependent on the Z° shape, which, as we will show in the next paragraphs, has a number of interesting experimental properties.

Sensitivity of $A^e_{FB}(x)$ to M_Z $(\sin^2\theta_W)$ and Γ_Z

A straightforward calculation allows us to determine the sensitivity of $A^e_{FB}(x)$ to the Z° line parameters. . If we call $N^{+-}(x) = L_{tot}\sigma(x)$ the number of e^+e^- events seen by a detector covering down to $\cos\theta_{min} = x$ and after an integrated luminosity L_{tot} (pb^{-1}), then the *statistical error* on $A^e_{FB}(x)$ turns out to be:

$$\delta A^e_{FB}(x)/A^e_{FB} \approx \{A^e_{FB}\sqrt{N^{+-}(x)}\}^{-1} \sqrt{(1-A^e_{FB}{}^2)} \qquad\qquad (7).$$

As expected, for a given L_{tot}, the statistical error *decreases* if A^e_{FB} is closer to unity, that is if we accept a larger fraction of the t-channel contribution.

Correspondingly the *sensitivity* of A^e_{FB} to $|\chi(s)|^2$ also decreases when A^e_{FB} is close to unity. If one differentiates equation (4) with respect to $|\chi(s)|^2$, neglecting the Z° contribution at the numerator, one obtains at the peak, the approximate relation

$$\delta A^e_{FB}(x)/A^e_{FB} \approx -(1-A^e_{FB}(x)) \; \delta|\chi(s)|^2/|\chi(s)|^2 \qquad\qquad (8)$$

which clearly shows how the sensitivity to the Z° shape goes to zero both if the asymmetry approaches unity (i.e. when Bhabha dominates) or if the Z° dominates (i.e. when A°_{FB} goes to zero). As a matter of fact the total *accuracy* of the measurement of the Z° parameters, that is the precision of the measurement after a given integrated luminosity, keeps improving as θ_{min} is reduced as is shown in Table I for the case of Γ_Z. The data reported in the table correspond to lowest order calculation. From the table, it is also evident that the improvement in accuracy below $\theta_{min} \approx 13^\circ$, the angle where the

Z(s)Z(s) and γ(t)γ(t) Born terms are equal, is very small, as expected.

Similarly, the error induced on A^e_{FB} by uncertainties on the Bhabha cross section (for instance due to detector or radiative correction systematics) gives, in the case where $B_-(x) \approx B_+(x) = B(x)$:

$$\delta A^e_{FB}(x)/A^e_{FB} \approx (1- A^e_{FB}(x)) \, \delta B(x)/B(x) \qquad (9a).$$

From the fact that expressions (8) and (9) have opposite signs it follows that A^e_{FB} is unsensitive to quantities that affect both the Bhabha and the Z° terms in the same way, like for instance the luminosity.

It is now easy to compute the sensitivity of A^e_{FB} to deviations of Γ_Z from the SM prediction, Γ_Z^{SM}:

$$\Gamma_Z^{SM} = M^3_Z G_F/(24\pi\sqrt{2}) \sum \beta_i (N_c)_i \, (a_i^2 + v_i^2) \, (C_f)_i \qquad (9b)$$

where the sum is over all known pointlike fermions. The kinematic factor $\beta_i = \sqrt{1-4m^2_i/s}$ are close to one for the known fermions and $(N_c)_i$ is the number of colors (three for quarks, one for leptons). The final-state correction $(C_f)_i$ depend on the phase enlargement due to final-state radiation [5-6]:

Table 1. Expected statistical accuracy on the measurement of the total Z° width for different values of the minimum angle. Lowest order result.

θ_{min}	A^e_{FB}	σ_{e+e-} (nb)	N^{e+e-} (1 pb^{-1})	δA^e_{FB} (1 pb^{-1})	$\delta\Gamma/\Gamma$ (1 pb^{-1})	N^{e+e-} (100 pb^{-1})	δA^e_{FB} (100 pb^{-1})	$\delta\Gamma/\Gamma$ (100 pb^{-1})
1°	0.995	400	$4.0 \, 10^5$	$1.6 \cdot 10^{-4}$	$1.6 \, 10^{-2}$	$4.0 \, 10^7$	$1.6 \, 10^{-5}$	$1.6 \, 10^{-3}$
3°	0.968	45	$4.5 \, 10^4$	$1.2 \, 10^{-3}$	$1.6 \, 10^{-2}$	$4.5 \, 10^6$	$1.2 \, 10^{-4}$	$1.6 \, 10^{-3}$
5°	0.92	18	$1.8 \, 10^4$	$3.0 \, 10^{-3}$	$1.7 \, 10^{-2}$	$1.8 \, 10^6$	$3.0 \, 10^{-4}$	$1.7 \, 10^{-3}$
10°	0.66	6	$6.0 \, 10^3$	$9.7 \, 10^{-3}$	$2.2 \, 10^{-2}$	$6.0 \, 10^5$	$9.7 \, 10^{-4}$	$2.2 \, 10^{-3}$
15°	0.46	3.7	$3.7 \, 10^3$	$1.4 \, 10^{-2}$	$3.0 \, 10^{-2}$	$3.7 \, 10^5$	$1.4 \, 10^{-3}$	$3.0 \, 10^{-3}$
30°	0.17	2.	$2.0 \, 10^3$	$2.2 \, 10^{-2}$	$7.8 \, 10^{-2}$	$2.0 \, 10^5$	$2.2 \, 10^{-3}$	$7.8 \, 10^{-3}$
50°	0.07	1.1	$1.1 \, 10^3$	$3.0 \, 10^{-2}$	$2.2 \, 10^{-1}$	$1.1 \, 10^5$	$3.0 \, 10^{-3}$	$2.2 \, 10^{-2}$

$$(C_f)_i = (1 + 3\alpha_{QED}/4\pi \ q^2_i + O(\alpha^2_{QED}))(1 + 0.5(N_c-1)(\alpha_s/\pi +$$

$$(\alpha_s/\pi)^2(1.98 - .115 \ N_f) + O(\alpha_s^3)), \tag{10}$$

where N_f is the number of light flavour, and α_s corrections are in the modified minimal subtraction renormalization scheme[7].

Any deviation of Γ_z^{SM} is directly related to the presence of additional decay channels of the Z^0 in to real undetected particles, such as additional neutrino families, light SUSY particles and so on, or to unexpected virtual effects that modify the width. With good approximation we can write at the peak, at fixed M_z $(\sin^2\theta_w)$:

$$\delta A^e_{FB}(x)/A^e_{FB} \approx 2(1 - A^e_{FB}(x))(\Gamma_z - \Gamma_z^{SM})/\Gamma_z^{SM} \tag{11}$$

The dependence of the derivative $\partial A^e_{FB}/(\Gamma_z - \Gamma_z^{SM})$ on Γ_z^{SM}, when running at energies close to the peak, is shown in fig. 6. As one can see this dependence is very weak and rather insensitive to the effects of the radiative corrections. At the contrary it is very sensitive to the choice of θ_{min} in good agreement with the approximate expression (11).

As shown in Table I, when running at the peak with $\theta_{min}=10°$ the statistical accuracy on Γ_z^{SM} obtained after 1 pb^{-1} is about

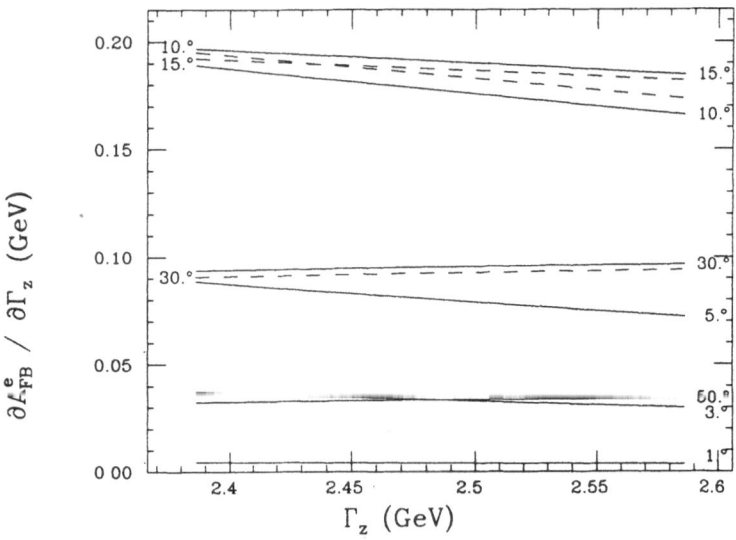

Fig. 6. A^e_{FB} sensitivity to Γ_z versus Γ_z for different values of θ_{min}. For three values of θ_{min} also the radiatively corrected result is shown (dashed lines)(ref.4).

50 MeV (lowest order), and it improves by an order of magnitude if a reasonable estimate of the total statistics expected at LEP is used (100 pb⁻¹). The gain in sensitivity when going to smaller θ_{min} is very marginal, in particular if we take into account systematic effects due to the detection efficiency, discussed in the next paragraph. In fact the exploitation of this additional sensitivity is not trivial from an experimental point of view given that the also the systematics on A^e_{FB} have to be kept well below the statistical error, i.e. at the 10^{-4} level or less for $\theta_{min}<10°$ and after 100 pb⁻¹.

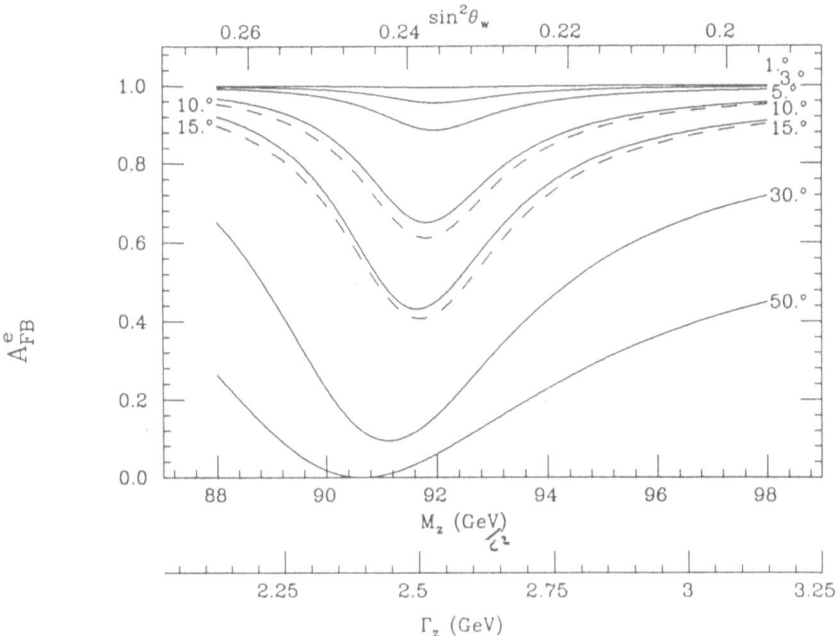

Fig. 7. Dependence of A^e_{FB} on M_Z for different values of θ_{min}.

On the other hand, the error on the width increases by one order of magnitude if θ_{min} is greater than 30°. The effect of the radiative corrections on the asymmetry is remarkably small - $\delta A^e_{FB}/A^e_{FB}\approx-3.7\%$ at the peak and for $\theta_{min}=15°$- given that they modify the differential cross section by about -30%. An experiment capable of unambiguosly detect the electron charge above $\theta_{min}\approx10° - 20°$, is then in a very good position to perform an high precision measurement of the total $Z°$ width by a careful measurement of A^e_{FB}.

Similar considerations should be made for the accuracy that could be obtained on M_Z ($sin^2\theta_w$) by a precision measurement of A^e_{FB}. As we move from the peak, we must take into account, when calculating A^e_{FB}, all the terms that contribute to the cross section: some of the terms that vanish at the peak play a non-negligible role in this case, such as term #7 $(Z(s)\gamma(s))$ and #8 $(Z(s)\gamma(t))$.

The dependence of A^e_{FB} on M_Z is shown in fig. 7; by chosing a suitable scale for $sin^2\theta_w$ and Γ^{SM} we are able to show in the figure also the dependence of A^e_{FB} with respect to these quantities.. The derivative $\partial A^e_{FB}/\partial M_Z$ in the vicinity of the peak is plotted in fig. 8. The phase space dependence of the width, $G(\sqrt{s}) = G(M_Z)\sqrt{s}/M_Z$ is included in fig. 7 and 8. We note that the sensitivity of A^e_{FB} to these parameters depends both on θ_{min} and on the distance from the peak. In particular, for a given θ_{min} ther exists a value of the c.o.m. energy where the first derivative with respect to M_Z vanishes. As we shall see in the following, this is very relevant when discussing the error induced on A^e_{FB} by the beam energy systematics. The dependence of A^e_{FB} on the beam energy and on M_Z is very similar: it follows that the precision on M_Z is limited by the exact knowledge of the c.o.m. energy, as in the case of the direct shape measurement. Running at an energy where the sensitivity to the mass is large implies that the influence of the beam systematics is also important. Nevertheless, a measurement of M_Z ($sin^2\theta_w$) performed with this technique would represent an interesting cross-check with other methods.

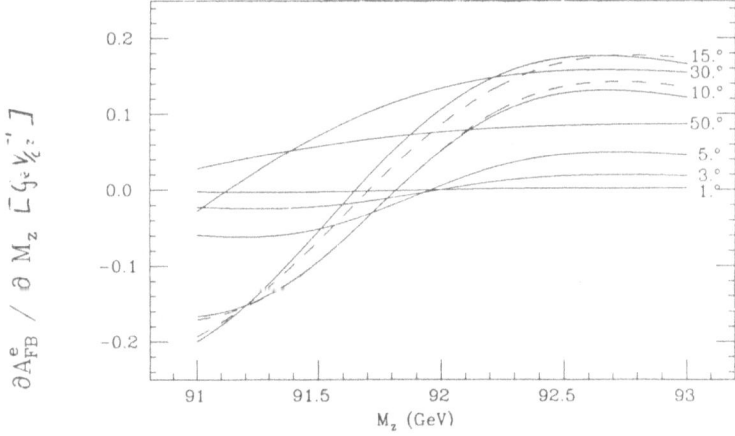

Fig.8. Solid line: derivative of A^e_{FB} with respect to M_Z for different values of θ_{min}. Dashed line, radiatively corrected result.

Systematics

The systematics that will affect the measurement of A^e_{FB} comes from three possible sources - knowledge of the machine parameters, understanding of the experiment and understanding of the theory. We recall that from eq. (7) and (11) it follows that in order to reach the 10^{-3} level on the total width, we have to keep the systematics on A^e_{FB} below $\approx 5. \, 10^{-4}$.

a) <u>Machine parameters (luminosity, beam energies)</u>

As discussed in the first paragraph, one strong point in favour of the use of A^e_{FB} is the fact that it is a luminosity-independent quantity. Calling $N^+(x)$ $(N^-(x))$ the number of electrons detected after an integrated luminosity L_{tot}, in the positive (negative) hemisphere by an experiment having an electron detection efficiency ε_e, we have:

$$A^e_{FB} = \frac{N^+(x)/e_eL_{tot} - N^-(x)/e_eL_{tot}}{N^+(x)/e_eL_{tot} + N^-(x)/e_eL_{tot}} = \frac{N^+(x) - N^-(x)}{N^+(x) + N^-(x)} \qquad (12)$$

that is independent both of the luminosity and of the detection efficiency (assumed constant over the interesting angular region, see below).

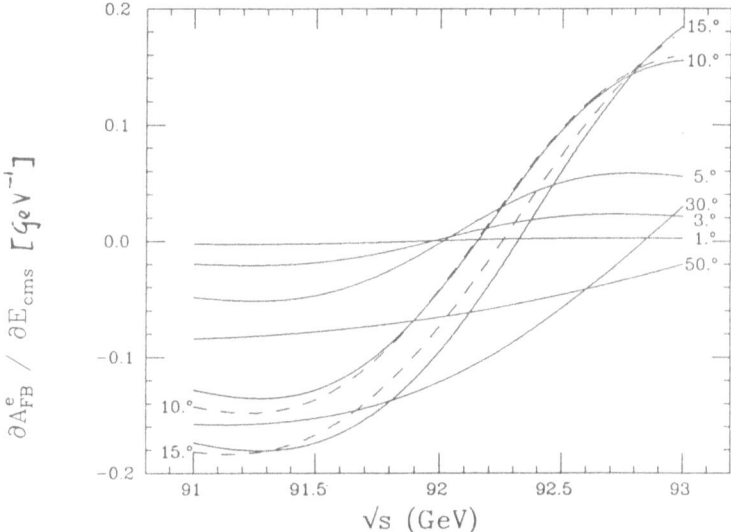

Fig. 9. Solid line: derivative of A^e_{FB} with respect to $E_{c.o.m.}$ for different values of θ_{min}. Dashed line: radiatively corrected result.

The dependence of A^e_{FB} on the c.o.m. energy is shown in fig.5, where $\sqrt{s}$ is varied around the peak while M_Z is kept constant, $M_Z=92$ GeV. The minimum in the asymmetry is very close to the pole for small θ_{min}, while it drifts to higher energy for larger θ_{min}. The quantity $\partial A^e_{FB}/\partial E_{c.o.m.}$ is shown in fig.9 : for a given θ_{min}, the first derivative of A^e_F with respect to the beam energy could be made equal to zero, by running just slighlty above the pole. Typical values are, at lowest order, $\Delta E=\sqrt{s}-M_Z\approx$ 325 MeV for $\theta_{min}=15°$, $\Delta E\approx160$ MeV for $\theta_{min}=10°$: the radiative corrections reduce the first value to 265 MeV and leave essentially unchanged the second. In that way, the systematic error on the beam energy, that would otherwise severely limit the accuracy of the width measurement, can be substantially reduced. Recalling that the radiative corrections move the peak of the $\mu^+\mu^-$ cross section ≈ 130 MeV above the Z^0 pole we note that these requirements do not affect significantly the Z^0 production rate. A simple calculation shows that, taking $\theta_{min}=15°$ and running 265 MeV above the Z^0 pole (92 GeV), the error induced on A^e_{FB} by a systematic uncertainty on $\delta\sqrt{s}=50$ MeV (25 MeV) is $\delta A^e_{FB}/A^e_{FB}\approx 1.3 \ 10^{-3} \ (3.4 \ 10^{-4})$

In conclusion, A^e_{FB} exhibits the remarkable property of being insensitive to the two main sources of error that affect the direct measurement of the total width, i.e. the systematic error on the luminosity and on the beam energies.

b) Experimental parameters (detection efficiency, charge measurement, minimum angle)

If the efficiency for detecting a high energy electron (positron) and, at the same time, for measuring the sign of its charge does not show a significant angular dependence in the angular region of interest then the systematic uncertainty on this quantity should not affect A^e_{FB} because it cancels in the ratio. The sophisticated LEP/SLC detectors will be able to collect a large sample of high momentum tracks that should allow detailed knowledge of the angular dependence of the efficiency: systematic effects can be monitored using independent channels like the μ-pairs, reversing, when possible, the magnet polarity and using equal sign di-muons and di-electrons for a precise measurement of the probability of assigning the wrong sign to the charge. For each detector a very careful study is clearly needed, but one would expect that for θ_{min} above 10°-20° and for the very simple di-electron final state, the systematics on the detection efficiency should not limit the accuracy on the measurement of the width at the level of 10^{-3}.

Another systematic uncertainty affecting the asymmetry is related to the knowledge of θ_{min}. The derivative of A^e_{FB} with respect to θ_{min} is shown in fig. 10 As one can see, in the interesting angular region $\theta_{min}\approx10-20°$ the asymmetry depends strongly on the minimum angle.

The accuracy requested on the knowledge of θ_{min} (≈ 0.2 mrad) is of the order of the typical accuracy on the track angle expected in good central detectors. When necessary one can statistically correct for the angular smearing induced by the tracking detectors.

c) Theory (QED, QCD and electroweak radiative corrections)

The radiative corrections enter at various level in the measurement of A^e_{FB}: they are needed in order

- to correct the measured value of A^e_{FB} for experimental effects, for instance the QED radiatively corrected Montecarlo needed to determine the acceptance and the efficiency of the detector through simulation;

- to extract Γ_Z from the measured value of A^e_{FB}, that is QED radiative corrections and electroweak radiative corrections to A^e_{FB};

- to take into account the effects on Γ_Z due to final state gluon (QCD) and photon (QED) radiation, as described in expression (10).

Only after all these corrections have been applied can the resulting value for Γ_Z be compared with the existing electroweak corrected prediction of the SM.

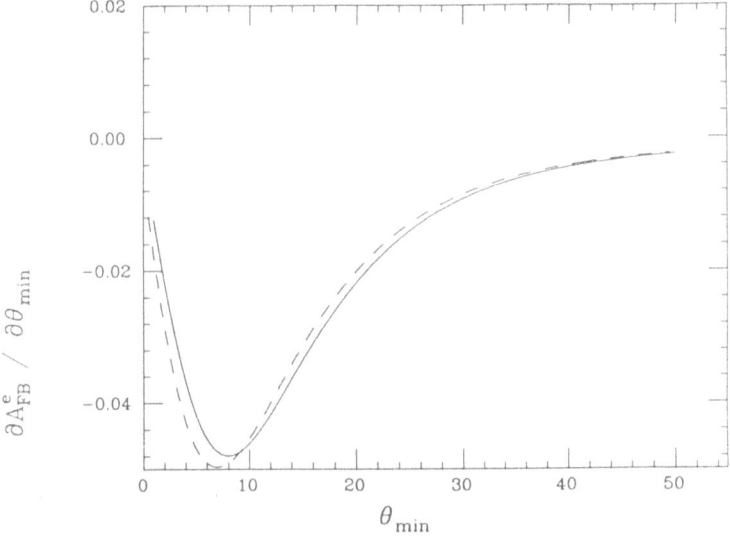

Fig. 10. Solid line: derivative of A^e_{FB} with respect to θ_{min}. Dashed line, radiatively corrected result.

In order to fully exploit the sensitivity of the proposed measurement of Γ_z, all the above corrections must be known with a good accuracy. The precise evaluation of the accuracy needed is under way, but it is possible to make the following statements:

a) only systematic uncertainties on the radiative corrections showing an angular dependence will affect A^e_{FB}. Scale factors will drop in the ratio;
b) an accurate knowledge of the Bhabha radiative corrections at intermediate angles is needed, but, extrapolating the results obtained with the O(1%) calculation ($\delta QED/QED \approx 35\%$ => $\delta A^e_{FB}/A^e_{FB} \approx$ 7%), we expect that 1% uncertainties will affect A^e_{FB} at the 2 10^{-3} level. An accuracy of few parts in 10^{-3} on the knowledge of the QED radiative corrections would probably be sufficient;
c) $O(\alpha_{QED})$ and $O(\alpha^2_s)$ corrections to the final state radiation, must be included in the expression of the total width: at this time the relatively poor precision on the measurement of α_s limits the accuracy of the QCD correction at little less than 1% level. It follows that an improved value of α_s is badly needed to reach the 10^{-3} level of accuracy.

Other quantities that can be used to perform an high accuracy measurement of the width

The study performed on A^e_{FB} illustrates how by measuring large angle e^+e^- pairs in different angular regions and by making a ratio of cross sections one can reach an excellent accuracy in the measurement of the total width. As we did anticipate, other variables can be defined with similar properties. In particular we did study the sensitivity to the width of three quantitites that do not depend on the measurement of the charge and are then even less sensitive to the systematics due to the angular dependence of the error on the charge measurement efficiency.
These quantities are ($x_1 = \cos\theta_1 = \cos 15° > x_2 = \cos\theta_2$) :

$$R_1(x_1, x_2) = \frac{\int_0^{x1} dz \, d\sigma/dz - \int_{-x2}^0 dz \, d\sigma/dz}{\int_{-x2}^{x1} dz \, d\sigma/dz} \qquad (13)$$

$$R_2(x_1, x_2) = \frac{\int_{-x2}^{x2} dz \, d\sigma/dz}{\int_{-x1}^{x1} dz \, d\sigma/dz} \qquad (14)$$

and

$$R_3(x_1, x_2) = \frac{{}_{-x_2}\!\int^{x_2} dz \; d\sigma/dz}{{}_{x_2}\!\int^{x_1} dz \; d\sigma/dz + {}_{-x_1}\!\int^{-x_2} dz \; d\sigma/dz} \qquad (15)$$

where we assumed an experimental capability of measuring electrons and positrons accurately (good energy and angular measurement without charge measurement) in the region above 15°.

Fig. 11 shows the accuracy on the measurement of the width after 1 pb^{-1} as a function of θ_2 for the three quantities defined above and $A^e_{FB}(\cos 15°)$. The values of θ_2 for best accuracy are 43.7°, 32° and 34.2° for R_1, R_2 and R_3 respectively, but the dependence on θ_2 is rather weak for a large interval of values around the minimum. The sensitivity of R_2 is slightly better than A^e_{FB} ($\Delta\Gamma_z/\Gamma_z \approx 2.6\%$ versus 3.3% after 1 pb^{-1}). Fig. 12 compares the expected statistical accuracy for the four variables as a function of the integrated luminosity. The dependence of the relative accuracy of the four quantities we have defined on the choice of the minimum angle $\theta_1 = 15°$ should be investigated in detail, but this example suggest that R_2 is likely a good variable to exploit the e^+e^- decays of the $Z°$ for measuring its width, without even measuring the charge of the electrons (positrons). In addition one should note, that being R_2 the ratio of cross section integrated over angular ranges symmetric with respect to $\theta=\pi/2$, this quantity is expected to be insensitive also to the interference terms in the cross section.

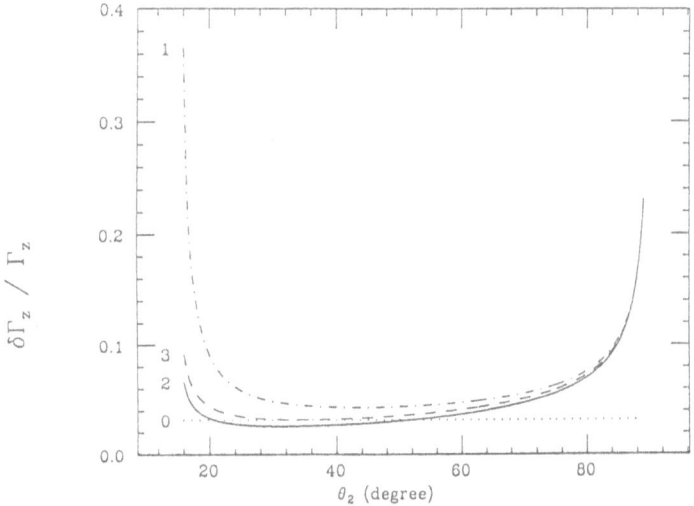

Fig. 11. Accuracy on the total width versus θ_2 for the various quantities defined in the text: (0) $A^e_{FB}(\cos 15°)$, (1) R_1, (2) R_2 and (3) R_3.

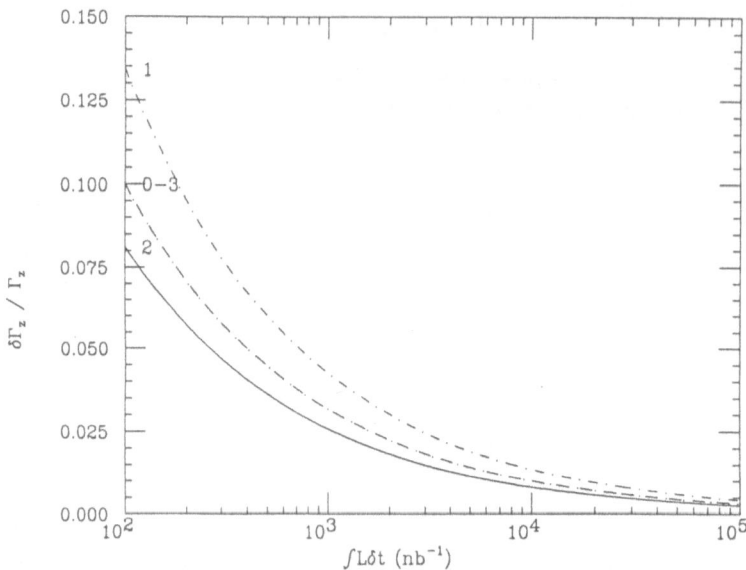

Fig. 12. Statistical accuracy on the measurement of the total width as a function of the integrated luminosity : (0) $A^e_{FB}(\cos 15°)$, (1) R_1, (2) R_2 and (3) R_3. Curves (0) and (3) overlap.

Conclusions

In this paper we have shown how it is possible to perform a precise measurement of the $Z°$ width by a study of the e^+e^- pairs produced at large angles. With the full statistics that will be delivered at the $Z°$ factories in the next few years, the ultimate sensitivity of this method, including the experimental systematic effects, is of the order of few MeV. This sensitivity is about one order of magnitude better than what one expects to obtain with other methods.

We introduced various quantities that can be used and we have shown how they are moderately sensitive to the electromagnetic radiative corrections to intermediate angles Bhabha, while being very sensitive to the width of the $Z°$. Work is in progress to study in detail the dependence on the QED radiative corrections. In addition the precise knowledge of the strong coupling constant plays an important role in the precise measurement of the total width. Due to the importance of a very accurate measurement of the width any effort to improve the knowledge of the radiative corrections to the Bhabha differential cross section below the 1% level is thus well motivated.

References

[1] See for instance the contribution of A. Blondel et al. in *Physics at LEP*, edited by J.Ellis and R.Peccei, CERN 86-02 (1986)

[2]For reviews of LEP/SLC physics see for instance, in addition to ref. [1]:*Proceedings of the 1978 LEP Summer Study,* CERN 79-01 (1979);*SLC Workshop,* SLAC Report 247 (1982)

[3]M.Consoli *Nucl Phys B,* **160**, 268 (1979); M.Consoli, S. Lo

Presti and M.Greco, *Phys. Lett. B,* **113**, 415 (1982);R.Sommer, M.Bohm and W.Hollik: Wurzburg preprint (1983);M.Greco, *Phys. Lett.B,* **177**, 97 (1986);S. Kuroda,T. Kamitami, K. Tobimatsu, S.Kawabata and Y. Shimizu, *KEK preprint,* 87-66, August 1987; F. Berends, R. Kleiss and W.Hollik, *Nucl. Phys.B,* **304**, 687 (1988) and *Nucl. Phys.B,* **304**, 712 (1988)

[4]M. Greco, *Riv. del Nuovo Cimento,* **11**, 1 (1988)

[5]F.A. Berends and R.P.Kleiss, *Nucl.Phys B,* **177**, 237,(1981)

[6]M.Dine and J.Sapirstein, *Phys. Rev. Lett.,* **43**,668(1979); W. Celmaster and R.J.Gonsalves, *Phys.Rev.D,* **21**, 3112 (1980); K.G.Ketrychin *et al., Phys. Lett.,* **85B**, 277(1979)

[7]W.A.Bardeen *et al., Phys.Rev. D,* **18**, 3998(1978)

RENORMALIZATION SCHEMES: WHERE DO WE STAND?[*]

B. F. L. Ward

Theory Division, CERN, Geneva, Switzerland; and
Stanford Linear Accelerator Center, Stanford University, Stanford, CA
94309, USA; and
Department of Physics and Astronomy, University of Tennessee, Knoxville
TN 37996–1200, USA

ABSTRACT

We consider the status of the current approaches to the application of the renormalization program to the standard $SU_{2L} \times U_1$ theory from the standpoint of the interplay of the scheme chosen for such an application and the attendant high-precision tests of the respective loop effects. We thus review the available schemes and discuss their theoretical relationships. We also show how such schemes stand in numerical relation to one another in the context of high-precision Z^0 physics, as an illustration.

1. INTRODUCTION

As it is by now well-known,[1] a primary objective of the SLC/LEP Z^0 physics program is to test the standard $SU_{2L} \times U_1$ theory at a level below 1%. When one reaches the precision of 0.1%, one is indeed at the level of the so-called electroweak quantum loop effects, so that these loop effects must be calculated in a precise way if this aspect of the SLC/LEP physics program is to be successful. We are aware also, of course, that the large so-called QED effects[1,2] in the theory must be controlled at the 0.1% level, but these effects are indeed so understood, even on an event-by-event basis[3]; thus, we do not focus on them here.

Regarding then the pure electroweak loop effects, we emphasize that several authors[4] have calculated such effects with various data sets in mind. Indeed, one can distinguish those calculations which have looked at corrections to processes involving space-like four-momentum transfers and those which have looked at corrections to processes involving only time-like four-momentum transfers. In all cases, it is necessary to choose a so-called scheme for the interpretation of the ultraviolet divergences in the loop corrections to the $SU_{2L} \times U_1$ theory via the renormalization program. If we knew how to sum up over all orders in this resultant renormalized theory, the question of scheme dependence (or independence) would be moot: if one sums to all orders, all schemes lead to the same physical observable values, so that they lead to the same uncertainty in extracting these observables from the data; assuming infinite numerical precision on the evaluation of the renormalized theory, this uncertainty is entirely determined by the experimental errors. In the $SU_{2L} \times U_1$ theory, while it is possible to sum over the infrared singularities to all orders in α, it is as yet not practical to do

---

[*]Work supported in part by Department of Energy contracts DE–AC03–76SF00515 and DE–AS05–76ER03956.

this in a complete way for the infrared finite, ultraviolet finite parts of the respective S–matrix elements; so far, one has been able only to compute these finite parts to a finite order in α. To any finite order, what one compares with experiment can greatly affect how accurate the parameters of the $SU_{2L} \times U_1$ theory are determined. But, a choice of what one compares with experiment is precisely correlated with one's scheme for implementing the renormalization at a finite order in α. Hence, it is for this reason that we may speak about the appropriateness or lack of appropriateness of a given scheme: the errors on the defining physical parameters of the theory, as determined from experiment, may be reduced if an appropriate scheme is used, for any given finite order in α calculation of the various physical processes. Hence, in the context of high-precision Z^0 physics, the issue of the scheme is indeed important for the comparison of one- (and two-) loop predictions with Z^0 production and decay phenomena; for example, in $e^+e^- \rightarrow Z^0 \rightarrow X$. Hence, in what follows, we consider the current status of the implementation of renormalization schemes in the context of one- (and two-) loop calculations of high-precision $SU_{2L} \times U_1$ physics.

Our discussion is organized as follows. In the next Section, we define precisely what actually constitutes a renormalization scheme. In Sec. 3, we then discuss the existent schemes which have been used to compute complete one (or two) loop pure electroweak corrections to $SU_{2L} \times U_1$ processes for high precision tests of the theory. In Sec. 4, we turn to the status of the use of these schemes and their possible extensions. Section 5 contains some concluding remarks.

2. DEFINITION OF A RENORMALIZATION SCHEME

In this section, we wish to define precisely what we intend by a renormalization scheme for a theory like the $SU_{2L} \times U_1$ theory. We begin by recalling the definition of renormalization itself.

Specifically, given a Lagrangian $\mathcal{L}$ dependent on a collection of fields $\{\phi_i\}$ (we suppress possible Lorentz and internal symmetry labels into the single label i), we separate $\mathcal{L}$ into a free part $\mathcal{L}_0$ and its interaction part $\mathcal{L}_{\text{int}}$:

$$\mathcal{L} = \mathcal{L}_0 + \mathcal{L}_{\text{int}} \quad . \tag{1}$$

Focusing on the case that $\{\phi_i\}$ contains a single scalar field ϕ of rest mass m, we write

$$\mathcal{L} = \frac{1}{2} \left(\partial_\mu \phi \, \partial^\mu \phi - m_0 \phi^2 \right) - \frac{\lambda_0 \phi^4}{4!} \tag{2}$$

for purposes of illustration so that

$$\mathcal{L}_0 = \frac{1}{2} \left(\partial_\mu \phi \, \partial^\mu \phi - m_0 \phi^2 \right) \tag{3}$$

and

$$\mathcal{L}_{\text{int}} = -\frac{\lambda_0 \phi^4}{4!} \quad . \tag{4}$$

The need for renormalization is then made manifest by any effort to compute quantum loop corrections in the theory in Eq. (2): the first quantum correction to the ϕ propagator is shown in Fig. 1. (We imagine a path-space approach to the Feynman rules so that, in ϕ^4 theory, the contraction of a ϕ with itself at a single point of space-time

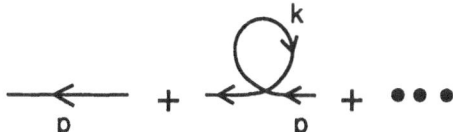

Fig. 1. Quantum corrections to the two-point connected Green function in ϕ^4 theory.

need not be trivial.) The standard methods would give for the loop term Fig. 1 [here, we use the n–dimensional method of Ref. 5 to define the divergent integral for $n \uparrow 4$]:

$$
i\Delta_F^{(1)} = \frac{i}{p^2 - m_0^2 + i\epsilon} \frac{(-i\lambda_0)}{2} \int \frac{d^4 k}{(2\pi)^4} \frac{i}{k^2 - m_0^2 + i\epsilon} \frac{i}{p^2 - m_0^2 + i\epsilon}
$$

$$
\equiv \frac{i}{p^2 - m_0^2 + i\epsilon} \frac{(\lambda_0/2)\left\{ \left[-i\pi^{n/2} \Gamma(1 - n/2)(m_0^2 - i\epsilon)^{n/2-1} \right] / \Gamma(1) \right\}}{(2\pi)^n} \frac{i}{p^2 - m_0^2 + i\epsilon}.
$$

$$(5)$$

Hence, the most elementary loop graph in the theory is formally infinite. It must be cut off, as we have illustrated (this we will refer to as regularization), and then interpreted in terms of the physically observable quantities (parameters) in the theory (this we will refer to as a renormalization program). This then is all quite well known.

Hence, a renormalization scheme is a choice of definite procedures and prescriptions for implementing the interpretation of the infinities in a theory (presumed to admit this) in terms of its physically measurable parameters. It means that a regularization procedure has been chosen, a sufficient number of gauge-specifying terms and their respective compensating ghosts terms have been added to $\mathcal{L}$ to define all free contributions of the type in Eq. (3), and that a set of normalization conditions has been chosen at which one may identify the physically observable parameters in the theory in such a way that all S–matrix elements are finite order-by-order in perturbation theory when expressed in terms of these physical parameters.

In the trivial example of Eq. (2), we can use the standard power counting to identify that the two- and four-point 1PI vertices are, in fact, the only primitively divergent graphs, as it is well known. Hence, a wavefunction renormalization constant Z, a mass counter-term $\delta m^2 \equiv m_0^2 - m^2$, and a vertex renormalization constant Z_1 are sufficient to maintain the unit residue of the two-point function at its physical mass pole at $p^2 = m^2$ (Z and δm^2 are adjusted for this purpose), and to maintain the physical value λ of the four-particle coupling at the appropriate point; which may be taken, for example, at some point where $p_i^2 = \mu^2 = m^2$—for example, with $p_i = (m, \vec{o})$ [zero momentum $\phi\phi$ scattering—Z^2/Z_1 is adjusted for this purpose]. Hence, the three constants Z, Z_1 and δm^2 are uniquely fixed by the two conditions on the renormalized propagator and the one condition on the physical four-particle scattering vertex. All infinities in the theory are then absorbed into Z, Z_1 and δm^2, order-by-order in perturbation theory in the renormalized parameter λ, in a familiar way.

From this simple example, we see one very important thing:

If we sum to all orders in the renormalized coupling, we can always write every contribution to a given S–matrix element in terms of the respective all-orders series in the bare coupling λ_0. Since this bare theory does not know anything about our renormalization program, this series must be independent of this program also. Hence, in this sense, we indeed have a predictive framework in principle. More precisely, given the complete all orders in λ_0 unrenormalized 1PI vertices $\{\Gamma_u^{(n)}(\{p\}, \lambda_0, m_0)\}$ and a complete set of all orders in λ renormalized vertices $\{\Gamma^{(n)}(\{p\}; \lambda, m, \mu)\}$ (μ may not be equal to m), we know that ($\Gamma^{(n)}$ are amputated)

$$\Gamma_u^{(n)} = \left(Z^{1/2} \right)^{-n} \Gamma^{(n)} \quad , \tag{6}$$

so that any two different realizations of the renormalization program, $\{\Gamma_1^{(n)}\}$ and $\{\Gamma_2^{(n)}\}$, will be related by

$$\left(Z(1)^{1/2} \right)^{-n} \Gamma_1^{(n)} = \left(Z(2)^{1/2} \right)^{-n} \Gamma_2^{(n)} \quad . \tag{7}$$

This expresses the equivalence of all consistent renormalization schemes when the respective theories are summed to all orders in their respective couplings, and implies that the 1PI vertices satisfy the famous renormalization group equation:

$$\left(\mu \frac{\partial}{\partial \mu} + \beta(\lambda) \frac{\partial}{\partial \lambda} - \gamma_\Theta \, m \, \frac{\partial}{\partial m} - \gamma_\Gamma \right) \Gamma^{(n)} = 0 \tag{8}$$

where, from Eqs. (6) and (7), we see that β, γ_Θ and γ_Γ are computable order-by-order in perturbation theory in λ. In fact, one can use Eq. (8) as a check of the consistency of a calculation in a given scheme by verifying that Eq. (8) is satisfied order-by-order in perturbation theory. For a computation to order λ^n, Eq. (8) should be respected to the same order in λ. Such checks of the various calculations in schemes[4] for the $SU_{2L} \times U_1$ are in progress[6] and have been done in part by others to some degree.

Equation (8) makes it clear why precision tests of a theory like the $SU_{2L} \times U_1$ theory need to pay attention to the issue of the respective renormalization scheme. One sees that $\Gamma^{(n)}$ at different μ are connected by solving Eq. (8). Thus, to pin down the theory most precisely, one should choose as parameters to be compared to experiment those values or functions of the $\{\Gamma^{(n)}\}$ which can be most accurately measured at the respective scales $\{\mu\}$. This will then pick out a scheme. Deviation from this may propagate unnecessary errors from the comparison of theory and experiment into the defining parameters of the theory.

With this definition and elucidation of what we mean by a scheme, we turn now to the existent schemes which have been used for the order α (and α^2) corrections to the $SU_{2L} \times U_1$ theory for precision tests of the pure electroweak effects therein. This we do in our next Section.

3.   RENORMALIZATION SCHEMES FOR THE ELECTROWEAK THEORY

The effort to use the comparison between theory and precise experiments to check the one-loop (or two-loop) corrections to the $SU_{2L} \times U_1$ theory in the electroweak sector has a long history. Each effort in this history has, in fact, employed a renormalization scheme by necessity, although in some cases, this scheme is only partially implemented, due to the nature of the specific calculation under study. We then wish to identify the various schemes that have been successfully used or advanced for precision tests of loop effects in the standard $SU_{2L} \times U_1$ theory. This is the purpose of our discussion in this section.

Specifically, we consider the various schemes in Refs. 4 and 7, in turn. We consider first the formal schemes employed by Lee and Zinn-Justin and 't Hooft and Veltman[7] in their arguments for the explicit renormalizability of the $SU_{2L} \times U_1$–type theory and the equivalence of its R and U gauge formulations.

More precisely, the regularization at m–loops may be either effected with gauge-invariant Pauli–Villars type (higher-derivative) regulators or with the n–dimensional methods of 't Hooft (some questions still remain about the higher-derivative regulator in higher loops). The normalization point may be the Symanzik-type point $p_i^2 = -\mu^2$, $p_i p_j = (n-1)^{-1} \mu^2$ for an n-point irreducible vertex which is primitively divergent. The vacuum expectation value of any renormalized field is then maintained at its physical value order-by-order in perturbation theory. Counterterms are then used to remove the respective divergent terms in the theory's S–matrix so that this S–matrix is finite. A gauge of either the so-called renormalizable type, in which the vector boson propagator projection $P^{\alpha\beta}$ in $iD_{F\,ab}^{\alpha\beta} = -i\delta_{ab}\, P^{\alpha\beta}(k)/(k^2 - M_V^2 + i\epsilon)$ is bounded for $|k^2| \to \infty$ (in the Euclidean region), or of the U–type, in which all unphysical fields are removed from the theory and $P^{\alpha\beta} = -g^{\alpha\beta} + k^\alpha k^\beta/M_V^2$, may be chosen. This is really a whole class of schemes. Indeed, whenever the infrared singularities of the $\{\Gamma^{(n)}\}$ allow it, one may consider that some of $\{p_i\}$ are at $p_i^2 \geq 0$ (i.e., are on-shell) at the normalization point. The arguments of Lee, Lee and Zinn–Justin, and 't Hooft and Veltman then show that any two schemes of this type are equivalent. This is a powerful result for the precision tests of the loop corrections to the standard $\mathrm{SU}_{2L} \times \mathrm{U}_1$ theory.

Indeed, the freedom created by the general theorems of Lee and Zinn-Justin has resulted in the specific choices of a convenient scheme by several authors in the practical calculations.[4] We list in Table 1 a compendium of such choices, together with the specific physical process analyzed. The key characteristics of these schemes are their normalization points (they are all on-shell, in that the basic parameters are defined by subtracting the ultraviolet divergences at the points in the respective external momenta where the particles involved are on their mass shell). The existent practical schemes either use the renormalizable R_ξ–type gauge or the unitary gauge. Most of the work in the R_ξ gauge has had $\xi = 1$, so that it is work in the 't Hooft-Feynman gauge. The regularization procedures are all based on the 't Hooft n-dimensional methods.[5] The key differences arise in the field renormalization.

Specifically, it is well known from Ref. 7 that the infinities in the spontaneously broken gauge theory are identical to those in the unbroken theory. In the symmetric theory, there is but one field renormalization constant for each multiplet of the symmetry group of the theory. Hence, all particles in the multiplet have the same wavefunction renormalization constant. In the broken theory, these particles may, in general, all get a different mass: m_a = rest mass of a. If the subtractions are made on-shell, then the symmetric wavefunction renormalization constant, for a totally broken multiplet, can only give one particle from the original multiplet the unit residue at its propagator pole: all others from the multiplet will have a nonunit residue if only the symmetric wavefunction renormalization constant is used. There is, however, no theoretical reason to restrict oneself to a single wavefunction renormalization constant: the only requirement is that the divergences in the respective constants for the members of a multiplet must be the same. The finite parts of these constants may differ, however, in such a way that the residue of each member's propagator pole is unity. In doing this, one gains in the simplicity of the particle wavefunction normalizations, but one has more Z_{2a}'s to calculate, where Z_{2a} is the wavefunction renormalization constant for particle a. As one can see from Table 1, some authors prefer fewer Z_{2a}'s, and some prefer simple external line normalizations. The physics does not depend on these preferences.

The issue of field renormalization is amplified in the A–Z^0 mixing phenomenon. Any one-loop calculation must treat this effect. If one only uses the symmetric wavefunction renormalization constants, mass counterterms for $W^\pm$ and Z^0, and the requirement that the A is massless and couples with the usual Thomson limit charge at zero four-momentum transfer squared, one in general cannot have $\widehat{\pi}_{AZ^0}(q^2 = 0) = 0$ and unit residues for the poles in the A, W, and Z^0 propagators, where a hat denotes a renormalized quantity. However, in the case that we treat each independent field as truly independent, we can, in fact, arrange that π_{AZ^0} vanishes for $q^2 = 0$ and the vector particles have unit residue. Again, to do this or not is a matter of taste; the physics cannot depend on such issues of taste.

Table 1. Existent Practical Schemes.

Authors	Gauge	Normalization Point	Regularization Method	$Z_L^{up} = Z_L^{down}$	$\pi_{AZ}|_{phys} = 0$	$\pi'_{VV}|_{phys} = 0$	Process
Ross & Taylor	Feynman-'t Hooft	On-shell	't Hooft	Yes	No	No	General
M. Igarashi et al.; Aoki et al.	Feynman-'t Hooft	On-shell	't Hooft	No	Yes	Yes	$e^+e^- \to X$; $\nu_\ell e \to \nu_\ell e(\ell\nu_e)$
Wetzel	Feynman-'t Hooft	On-shell ($\nu\bar{\nu}Z^0$)	't Hooft	No	Yes	Yes	$e^+e^- \to \mu^+\mu^-$
Paschos & Wirbel; Sakakibara; Brown et al.	Feynman-'t Hooft	On-shell ($f\bar{f}Z^0$)	't Hooft	Yes	Yes	No	$\nu N \to \nu(\ell)+X$; $e^+e^- \to \mu^+\mu^-$
Antonelli et al.; Consoli	Feynman-'t Hooft	On-shell ($f\bar{f}Z^0$)	't Hooft	No	Yes	No	General
Llewellyn-Smith & Wheater	Feynman-'t Hooft	$\overline{MS}$ (On-shell)	't Hooft	No	Yes	Yes	$\nu N \to \nu(\ell)+X$; $ed \to e+X$
Marciano & Sirlin (II)	Feynman-'t Hooft	On-shell	't Hooft	No	Yes	Yes	General
Bardin et al.	Unitary	On-shell	't Hooft	No	Yes	Yes	$ff' \to f''f'''$; $Z^0 \to f\bar{f}$, etc.
Cole	R_ξ	On-shell	't Hooft	Yes	No	No	Neutral Current Processes
Hollik & Timme	Feynman-'t Hooft	On-shell Off-shell	't Hooft	Yes	Yes	No	$e^+e^- \to \mu^+\mu^-$

Authors	Gauge	Normalization Point	Regularization Method	$Z_L^{up} = Z_L^{down}$	$\pi_{AZ}|_{phys} = 0$	$\pi'_{VV}|_{phys} = 0$	Process
Lynn et al.	Feynman–'t Hooft	On-shell s_*^2	't Hooft	No	Yes	No	$f\bar{f} \to f'\bar{f}'$
Fleischer & Jegerlehner	R_ξ	On-shell	't Hooft	No	Yes	Yes	$H \to \tau^+\tau^-$; $H \to W^+W^-$; $H \to Z^0Z^0$
Passarino & Veltman	Feynman–'t Hooft	On-shell	't Hooft	No	Yes	No	$e^+e^- \to \mu^+\mu^-$
Bohm et al.; Hollik	Feynman–'t Hooft	On-shell	't Hooft	Yes	Yes	No	$\mu \to \nu_\mu + e\bar{\nu}_e$, $\overset{(-)}{\nu_\mu} e \to \overset{(-)}{\ell_\mu}\ell_e$; $e^+e^- \to \mu^+\mu^-$ $e^+e^- \to W^+W^-$
Marciano & Sirlin (I)	Unitary	On-shell $W\bar{\nu}_\mu\mu$	't Hooft	No	Yes	Yes	$W \to \ell + \bar{\nu}_\ell$
Appelquist et al.	Unitary	On-shell $W\bar{\nu}_\mu\mu$	't Hooft	No	Yes	Yes	$W \to \bar{\mu}\nu_\mu$; $\mu \to \nu_\mu e\bar{\nu}_e$
Salomonson & Ueda	Unitary	On-shell $W\bar{\nu}_\mu\mu$	't Hooft	No	Yes	Yes	$\overset{(-)}{\nu_\mu} e \to \overset{(-)}{\nu_\mu} e$
Philippe	Feynman–'t Hooft	On-shell	't Hooft	No	Yes	Yes	$e^+e^- \to W^+W^-$

Finally, we re-emphasize that the choice of gauge in Table 1 is the famous R_ξ gauge such that the vector meson propagator is, for four-momentum k,

$$iD_{F_{\mu\nu}}(k) = \frac{-i\left[g_{\mu\nu} - (1-\xi)k_\mu k_\nu/(k^2 - \xi M^2)\right]}{k^2 - M^2 + i\epsilon} \quad, \tag{9}$$

where either ξ is left as a free parameter, or $\xi = 1$, or $\xi \to \infty$. Conspicuously absent are the axial and the background field gauges, with

$$D^\mu_{cl} V_\mu = 0 \quad \text{and} \quad n \cdot V = 0 \qquad V = A, Z^0, W^\pm \quad, \tag{10}$$

respectively, for some fixed classical covariant background derivative and for some fixed vector n^μ; also, the Landau gauge $\xi = 0$ is not very plural in Table 1. ($\xi \to \infty$ gives the unitary gauge.) Hence, the gauge choices are not complete in this sense. However, the independence of those calculations in the general R_ξ gauge of the choice of ξ is a strong proof of the gauge invariance of the respective calculations in practice. We emphasize here that we do not question that the $SU_{2L} \times U_1$ theory can be calculated in a gauge-invariant framework. Rather, we advocate the eventual check of the work on the theory in several types of gauges, such as axial versus R_ξ gauge, etc., as a practical check on what is done in efforts to effect the calculations in such a framework.

The regularization method is that due to 't Hooft.[5] The usual problematic issue of the meaning of γ_5 in n-dimensions gets circumvented here by the absence of anomalies in the $SU_{2L} \times U_1$ theory. Hence, any method of defining γ_5 in n-dimensions should suffice, such as a totally anticommuting γ_5 in n-dimensions.[8] Here, we see no reason to pursue alternatives. However, there are alternatives, such as the zeta function regularization[9] or the higher derivative-Pauli-Villars methods[7] (for one-loop), etc. In principle, these should be pursued as eventual checks on what one does in practice in the $SU_{2L} \times U_1$ radiative correction program.

Regarding the normalization point, the on-shell prescription is overwhelmingly preferred. Again, there is nothing intrinsically lacking here. Indeed, this is the physically relevant normalization point. For some purposes, however, it may be more convenient to normalize at a Euclidean point $-\mu^2$ off the mass shell and, subsequently, to make a final finite renormalization transformation to the on-shell normalized theory. When this intermediate step is used, the parameters of the theory become $g'(\mu)$, $g(\mu)$, $M_{Z^0}(\mu)$, $m_{h,R}(\mu)$, $m_{f,R}(\mu)$; and on passing to the on-shell theory, we obtain the physical parameters α, G_F and $M_{Z^0}, \ldots$, as functions of the intermediate scale parameters. A particularly useful application of this intermediate scale device (in the textbooks[10] this is referred to as intermediate renormalization) is in conjunction with the 't Hooft–Weinberg[11] renormalization group improvement of $\mathcal{O}(\alpha)$ radiative corrections. For, when the theory is viewed at the Euclidean normalization point, its partial differential equations for the 1PI vertices $\{\Gamma\}$ take the simple form of Eq. (8):

$$\left[\mu\frac{\partial}{\partial\mu} + \beta\frac{\partial}{\partial g} + \beta'\frac{\partial}{\partial g'} - \sum_A m_{A,R}\,\gamma_{\theta_A}\frac{\partial}{\partial m_{A,R}} - \gamma_\Gamma\right]\Gamma = 0 \quad. \tag{11}$$

This equation has the well-known solution

$$\Gamma\left(\{\lambda p_j\}, g, g', \{m_{A,R}\}, \mu\right) =$$

$$\Gamma\left(\{p_j\}, g(\lambda), g'(\lambda), \{m_{A,R}(\lambda)\}, \mu\right) \lambda^{D_\Gamma} \exp\left\{-\int\limits_1^\lambda \frac{d\lambda'}{\lambda'}\,\gamma_\Gamma\left[g'(\lambda'), g(\lambda')\right]\right\} \quad, \tag{12}$$

where D_Γ is the engineering dimension of Γ, $g'(\lambda)$ and $g(\lambda)$ are the famous running charges[11] for the U_1 and SU_{2_L} groups, and $\{m_{A,R}(\lambda)\}$ are the respective running mass parameters[11] of the theory. Hence, from Eq. (12), we learn immediately how to get the rigorous renormalization group improvement of an $\mathcal{O}(\alpha)$ result. We do not have to "guess" that certain parts of an $\mathcal{O}(\alpha)$ expression are a part of a geometric series and other parts are not. This device, then, can be of substantial practical use.

Finally, we remark that the processes considered in Table 1 cover the key places of experimental accessibility. The LEP/SLC physics has been amply addressed from the standpoint of the number of efforts to treat the radiative correction issues which pertain thereto. Similarly, the lepton-lepton and lepton-hadron scattering physics issues have had considerable attention from the standpoint of the attendant radiative corrections. And, indeed, Sirlin, Marciano and others[12] have shown that, with the radiative corrections taken into account, the different scattering experiments become more consistent with one another and yield the result, for example, $M_{Z^0} = 91.8 \pm 0.9$ GeV for 44 GeV $\leq m_t \leq$ 200 GeV, and 10 GeV $\leq m_H \leq$ 1 TeV. Regarding the SLC/LEP Z^0 physics, the corresponding data are only beginning to be accumulated. Therefore, it is quite fitting to address the issue of the agreement of the respective theoretical work in that area. To this we now turn in the next section.

4. RENORMALIZATION SCHEMES IN PRACTICE

Given the scenario illustrated by Table 1, the question naturally arises as to what happens in practice when different schemes are used to compute the same physical process. Hence, in this section we explore this question.

Specifically, two schemes may be said to be numerically equivalent [at some order (α^n)] for a given process if the difference between their respective predictions for that process is below one-third of the level of the respective experimental errors in the measurement of that process. Here, the reader may think of a "process" as a cross section or a decay rate, etc. Accordingly, for comparing schemes in practice, it is convenient to use the high precision Z^0 physics processes as benchmarks for the numerical equivalence of different schemes. It is known[1] that such physics has as a goal the 0.1% precision checks of the $SU_{2L} \times U_1$ theory. Thus, the $\mathcal{O}(\alpha)$ pure electroweak corrections to such physics in two different schemes are numerically equivalent if they agree to better than 0.03%, for example. Let us consider a simple quantity, such as the cross section for $e^+e^- \to \mu^+\mu^- + n(\gamma)$ near the Z^0 resonance, as our pedagogical example. We now wish to compare the results of several schemes for this process to get some idea about the practical aspects of "scheme" effects. Here, we have in mind that the number of real photons may be arbitrarily large.

We have shown elsewhere that the various approaches to the multiple photon effects on the line shape in $e^+e^- \to \mu^+\mu^- + n(\gamma)$ are in agreement below the level of 0.1%. Thus, it is indeed fitting to inquire as to the level of agreement of the various approaches to the pure electroweak corrections themselves at the one-loop level, for example.

Specifically, as one can see from Table 1, most of the efforts on this process have used the on-shell scheme in the strict sense of the word scheme. Where the efforts differ is in the amount of resummation or improvement of the respective perturbation series by the various techniques.

For example, one can use the Dyson–Schwinger equations to improve the expression for a given Green function (for example, a two-point function, at one-loop) to an all-orders sum of all one-particle reducible graphs built from the respective one-particle irreducible parts, where the one-particle irreducible part is computed to one loop. The classic example is the photon propagator, $iD'_{F\mu\nu}(q)$, which can be represented as

$$iD'_{F\mu\nu}(q) = \frac{-ig_{\mu\nu}}{[1 + e^2\widehat{\pi}^{(1)}(q^2)](q^2 + i\epsilon)} + \text{gauge terms} \quad , \tag{13}$$

where $e^2 \widehat{\pi}^{(1)}(q^2)$ is the renormalized one-loop vacuum polarization function. The important point is that in Eq. (13) we have summed an infinite number of graphs but that, since we only have computed $\widehat{\pi}^{(1)}$, we have also omitted the n–loop 1PI contributions for $n > 1$. Thus, we have also *omitted* an infinite number of graphs from Eq. (13). Of course, one can hope that the omitted higher-order graphs are small; and, indeed, for most applications, they are.

Hence, depending on the amount of resummation one does, one can clearly get different results in the same scheme with the identical 1PI parts. This is not a scheme dependence in the classic sense; it is a calculational procedural difference in the same scheme.

Particularly interesting in this connection is the so-called starred scheme of Lynn et al.[4] In the reduced notation, the squared unrenormalized charge e_0^2 which enters into the strength of the Thompson amplitude is replaced by the improved Schwinger–Dyson series

$$ e_0^2 \;\to\; \frac{e_0^2}{1 + e_0^2\,\pi^{(1)}(q^2)} \;\equiv\; e_*(q^2)^2 \tag{14} $$

to obtain the starred charge squared. Here, $\pi^{(1)}(q^2)$ is unrenormalized. Nothing prevents one from computing a formal expression for $\pi(q^2)$ in Eq. (1) to all orders in e_0^2. The key task is to interpret systematically the respective infinities in such an expression, together with those in the analog vertex and box-type graphs for the Thomson amplitudes, in such a way that the resulting perturbation series in $e_*(q^2)$ is finite order-by-order in $e_*(q^2)$. [When $\pi^{(1)}$ is used in Eq. (14), some explicit arguments for such finiteness exist.[4]] For $q^2 \to 0$, one then imposes $e_*(q^2)^2 \to 4\pi\alpha$, where α is the fine-structure constant at $q^2 = 0$. In effect, one expands the $SU_{2L} \times U_1$ theory in terms of Dyson–Schwinger series-improved parameters. Since the resulting perturbation series still is renormalizable, one may still use the renormalization group to exploit these improved series. The key point, however, is to isolate properly what combinations of amplitudes in the starred-type series correspond to the respective $\{\Gamma\}$ that enter into Eq. (11) without double counting, for example.

Hence, the application of the various "schemes" in Table 1 can yield different results for several reasons: (1) the two compared schemes are intrinsically different; (2) the two schemes are actually the same but the calculational procedures are different; (3) the schemes and calculational procedures are both the same but there is an error in the arithmetic. The reader should keep these three possibilities in mind in what follows.

Specifically, we consider the results of the Stuart et al.,[4] pure weak corrections library and the Hollik et al.,[4] pure weak corrections library in conjunction with the YFS2 multiple-photon Monte Carlo event generator,[3] as realized in the program KORALZ.[13] The result[14] for the line shape under study, for the same typical model parameters, is shown in Fig. 2, in reference to the semianalytic comparison program ZBATCH,[13] where ZBATCH corresponds to the "exponentiated exact order α^2" improved line shape calculation of Berends et al., Ref. 2. We see that the two pure weak libraries agree at the level of 0.1% in the line shape near the Z^0 pole and that, further, the Hollik et al., library is closer to the ZBATCH prediction at this pole, where the pole is located at $\sqrt{s} = M_{Z^0} = 92$ GeV. Clearly, this type of agreement, which is worse for $\sqrt{s} \sim 102$ GeV, would allow us to test the $SU_{2L} \times U_1$ theory below the level of 1%; it would have to be improved to explore the level below 0.1% in these $SU_{2L} \times U_1$ tests, in general. Such improvement is in progress.[14] Further, we should note that some controversy still persists regarding Fig. 2, and the resolution of this controversy is also in progress.[14]

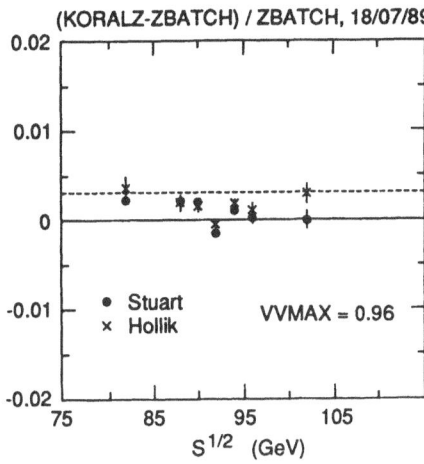

Fig. 2. Comparison of Stuart and Hollik et al., pure electroweak libraries in $e^+e^- \to \mu^+\mu^-$. Here, $M_{Z^0} = 92$ GeV, $m_t = 60$ GeV, and $m_H = 100$ GeV. In both cases, the libraries were obtained from its principal author via private communication. Note that *vvmax* is the maximum value of $1 - s'/s \equiv v$, where $s = (p_e + p_{\bar{e}})^2$ and $s' = (p_\mu + p_{\bar\mu})^2$.

Turning next to the popular asymmetry A_{FB}, we illustrate the results of three different calculations in Table 2: at 90 GeV, we show the results of Bardin et al. (row one), Hollik et al. (row two), and of Lynn et al. (row three). We see a general agreement of rows one and two for low values of m_t, and a qualitatively and quantitatively different set of predictions in row three. We and others are currently investigating this latter difference, as well as the large m_t comparison of rows one and two.[15]

We emphasize that it is indeed encouraging that the various efforts on our prototypical process are all relatively close on such a complicated calculation but we, of course, feel that the high-precision Z^0 physics requires that the current residual differences be understood.

Table 2. Comparison of A_{FB} for $e^+e^- \to \mu^+\mu^-$; $M_{Z^0} = 90$ GeV, $m_H = 100$ GeV.

Authors	m_t (GeV)	50	100	150	200	
Bardin et al.		0.0036	0.0044	0.0050	0.0054	
Hollik et al.		0.0037	0.0043	0.0054	0.0071	
Lynn et al.	*	0.0038*	0.0041*	0.0035*	0.0012*	
	m_t (GeV)	60	90	130	180	230
Lynn et al.		0.0037	0.0041	0.0040	0.0028	−.0012
*Obtained by linear interpolation from the published results at $m_t = 30, 60, 90, 130, 180$ and 230 GeV.						

5. CONCLUSIONS

We conclude that the current status of EWRC renormalization schemes is encouraging. The various efforts do tend to indicate that schemes have been identified and used to compute processes of physical interest to an accuracy such that the remaining higher-order corrections are indeed small. This is the basic requirement for high-precision tests of the $SU_{2L} \times U_1$ model at SLC and LEP near the Z^0 resonance. There remain, however, some interesting questions of the actual practice in this connection.

Specifically, we have shown that both in the Z^0 line shape in $e^+e^- \to \mu^+\mu^- + n(\gamma)$ and in the attendant A_{FB} for μ–pairs, there are significant deviations in the respective available results which need to be resolved for the high-precision Z^0 physics. Such a resolution is then the objective of various researchers in the field; we and others hope to report its attainment in the not-too-distant future.

ACKNOWLEDGMENTS

The author is indebted to the Organizing Committee for giving him the opportunity to review this very interesting subject at this very stimulating workshop. He thanks also the MarkII and SLD Collaborations for their hospitality and stimulating Physics Working Group Activities in the area of Radiative Corrections. He has benefitted from discussions with B. Lynn, R. Stuart, D. Kennedy, W. F. L. Hollik, D. Bardin, T. Riemann, F. A. Berends, R. Kleiss and, of course, from his collaborators, Prof. S. Jadach and Dr. Z. Was. Further, the author thanks the CERN EP and TH Division Directors, Profs. F. Dydak and J. Ellis, for their support while this manuscript was written and for the opportunity to have stimulating interactions with the CERN LEP Physics Working Groups on Radiative Corrections, and the author thanks Prof. G. Feldman for the kind hospitality of SLAC Group H while this manuscript was completed.

REFERENCES

1. See, for example, B. F. L. Ward, *Acta Phys. Pol.* B19:465 (1988); R. Rankin, *Proc. MkII SLC Pajaro Dunes Workshop*, J. Hu, ed., SLAC, Stanford, CA (1987); G. Altarelli, *Physics at LEP*, J. Ellis and R. Peccei eds., CERN, Geneva, Switzerland (1985); G. Altarelli, *Proc. HERA Experiments*, Genoa, Italy (1984); *Proc. La Thuile Workshop 1987, Vol. 1*, J. H. Mulvey, ed., CERN, Geneva, Switzerland (1987) and references therein.

2. F. A. Berends *et al.*, *Nucl. Phys.* B 297:249 (1988); M. Greco, *La Rivista del Nuovo Cimento* 11 n.5 (1988) and references therein.

3. See, for example, S. Jadach, preprint MPI–PAE/P Th 6/87; S. Jadach and B. F. L. Ward, *Phys. Rev.* D 38:2897 (1988); preprint UTHEP–88–0801; *Phys. Lett.* 220B:611 (1989); SLAC–PUB–4834, to be published in Comp. Phys. Commun. (1989).

4. T. W. Applequist, J. R. Primack and H. R. Quinn, *Phys. Rev.* D 7:2998 (1973); W. J. Marciano and A. Sirlin, *Phys. Rev.* D 8:3612 (1973) (I); P. Salomonson and Y. Ueda, *Phys. Rev.* D 11:2606 (1975); R. Philippe, *Phys. Rev.* D 26:1588 (1982); D. A. Ross and J. C. Taylor, *Nucl. Phys.* B 51:125 (1973); M. Veltman, *Phys. Lett.* 91B:95 (1980); G. Passarino and M. Veltman, *Nucl. Phys.* B 160:151 (1979); A. Sirlin, *Phys. Rev.* D 22:971 (1980); W. J. Marciano and A. Sirlin, *ibid.*, D 22:2695 (1980); *Nucl. Phys.* B 189:442 (1981) (II); J. Fleischer and F. Jegerlehner, *Phys. Rev.* D 23:2001 (1981); D. Yu. Bardin, P. Ch. Christova and O. M. Fedorenko, *Nucl. Phys.* B 175:435 (1980); *ibid.*, B 197:1 (1982); preprint PHE 88–15 and references therein; C. H. Llewellyn–Smith and J. F. Wheater, *Phys.*

Lett. 105B:486 (1981); J. F. Wheater and C. H. Llewellyn–Smith, *Nucl Phys. B* 208:27 (1982); S. Sakakibara, *Phys. Rev. D* 24:1149 (1981); K. Aoki, Z. Hioki, R. Kawabe, M. Konuma and T. Muta, *Prog. Theor. Phys.* 64:707 (1980); *ibid.,* 65:1001 (1981); M. Igarashi, N. Nakazawa, T. Shimada and Y. Shimizu, *Nucl. Phys. B* 263:347 (1986); F. Antonelli, G. Corbo, M. Consoli and O. Pellegrino, *Nucl. Phys. B* 183:195 (1981); E. A. Paschos and M. Wirbel, *Nucl. Phys. B* 194:189 (1982); M. Consoli, S. Lo Presti and L. Maiani, *Nucl. Phys. B* 223:474 (1983); W. Wetzel, *Nucl. Phys. B* 227:1 (1983); J. P. Cole *in* "Trieste Workshop on Radiative Corrections in SU(2)$_L$ × U(1)," B. W. Lynn and J. F. Wheater , eds., World Scientific Publ. Co., Singapore (1984); Sussex preprint (March 1983); R. W. Brown, R. Decker and E. A. Paschos, *Phys. Rev. Lett.* 52:1192 (1984); W. Hollik and H.-J. Timme, *Z. Phys. C* 33:125 (1986); B. W. Lynn *et al.,* SLAC–PUB–4128 (1988); B. W. Lynn and R. G. Stuart, *Nucl. Phys. B* 253:216 (1985); B. W. Lynn and D. C. Kennedy, SLAC–PUB–4039 (1986) and references therein; M. Bohm, H. Spiesberger and W. Hollik, *Fortschr. Phys.* 34:687 (1986); W. F. L. Hollik, DESY preprint 88–188 and references therein.

5. G. 't Hooft and M. Veltman, *Nucl. Phys. B* 44:189 (1972).

6. S. Jadach and B. F. L. Ward, to appear.

7. B. W. Lee and J. Zinn-Justin, *Phys. Rev. D* 5:3121,3137,3155 (1972); Benjamin W. Lee, *ibid.,* 9:933 (1974); G. 't Hooft and M. Veltman, *Nucl. Phys. B* 50:318 (1972).

8. See, for example, M. Chanowitz, M. Furman and I. Hinchliffe, *Nucl. Phys. B* 159:225 (1979).

9. See, for example, A. Rebhan, *Phys. Rev. D* 39:3101 (1989); A. Shiekh, IC/88/346 (1988) and references therein.

10. See, for example, J. D. Bjorken and S. D. Drell, "Relativistic Quantum Fields," McGraw–Hill Book Co., New York (1965).

11. S. Weinberg, *Phys. Rev. D* 8:3497 (1973); G. 't Hooft, *Nucl. Phys. B* 61:455 (1973).

12. See, for example, W. J. Marciano and A. Sirlin, *Phys. Rev. D* 36:2191 (1987).

13. See, for example, The CERN Electro-Weak Generators Working Group Report, R. Kleiss, ed., to appear.

14. R. Stuart, private communications (1987,1988); Z. Was, presentation at this meeting (1989).

15. Most of this large m_t difference between rows 1 and 2 is due to a difference in the degree of resummation in the two calculations (W. Hollik, private communication, 1989).

EWRC AT TRISTAN

Masataka Igarashi

Department of Physics, Tokai University
Hiratsuka, Kanagawa 259-12, Japan

Abstract

I explain the renormalization scheme in the $SU(2) \times U(1)$ standard model adopted by the TRISTAN e^+e^- working group and present some of the recent results of the Electro-Weak Radiative Correction applied to e^+e^- annihilation processes.

1. Introduction

Since the early 80's, a theory group at KEK has made efforts to calculate the one-loop correction to e^+e^- annihilation processes before the construction of TRISTAN. Our purpose at that time was to calculate the QED corrections by ourselves to understand the details of the results for theoretical and experimental application of them. Meanwhile an on-shell renormalization scheme in the standard model was formulated in Japan, which was welcomed by us because the energy of TRISTAN reaches to the region where the effects of weak interactions are fairly large. We have calculated the corrections in the full $SU(2) \times U(1)$ standard model to various e^+e^- annihilation processes in the energy range of TRISTAN to LEP II. This is a report of our group at present.

In section 2, I explain the adopted renormalization scheme and its features. In section 3, I state a recent result on the correction to heavy fermion pair production. In section 4, I report the present status of our efforts toward the full correction to the neutrino counting reaction.

Radiative Corrections, Edited by N. Dombey and
F. Boudjema, Plenum Press, New York, 1990

2. Renormalization

We take the $SU(2) \times U(1)$ standard model with three generations and one Higgs doublet as the basis of the calculation of one-loop full correction[1]. Among the particles contained in this model, the top quark and the Higgs particle are not found yet; thus these masses are free parameters at present.

For the renormalization scheme we adopt the on-shell scheme formulated by Kyoto group[2]. The divergent diagrams at loop level are the vertex, the fermion self-energy, the gauge boson self-energy and the Higgs particle self-energy ones. These divergences are cancelled with the divergences in the following counterterms.

$$
\begin{aligned}
e_0 &= Ye, \\
M_{W_0}^2 &= M_W^2 + \delta M_W^2, \\
M_{Z_0}^2 &= M_Z^2 + \delta M_Z^2, \\
m_{f_0}^2 &= m_f^2 + \delta m_f^2, \\
m_{\phi_0}^2 &= m_\phi^2 + \delta m_\phi^2,
\end{aligned}
\tag{2.1}
$$

i.e.,the electric charge e, the W-boson mass M_W, the Z-boson mass M_Z, the fermion mass m_f and the mass of Higgs particle m_ϕ as is usual in the on-shell scheme. Here the quantities with and without 0 mean the unrenormalized and renormalized ones respectively. Besides them, one must introduce the wave function renormalization constants to obtain the finite Green's functions. A feature of the Kyoto scheme is to introduce these constants after the mixing of neutral gauge bosons, which is remarkably different from other on-shell schemes[3].

$$
W_0^\pm = Z_W^{1/2} W^\pm,
$$

$$
\begin{pmatrix} Z_0 \\ A_0 \end{pmatrix} = \begin{pmatrix} Z_{ZZ}^{1/2} & Z_{ZA}^{1/2} \\ Z_{AZ}^{1/2} & Z_{AA}^{1/2} \end{pmatrix} \begin{pmatrix} Z \\ A \end{pmatrix},
\tag{2.2}
$$

$$
\phi_0 = Z_\phi^{1/2} \phi, \qquad f_{L,R}^0 = Z_{L,R}^{1/2} f_{L,R},
$$

where the letter means the corresponding field operator.

In order to see more details, let me consider the renormalized self-energy amplitudes of the neutral gauge boson sector. For the photon field, it is given by

$$
A_R^A(q^2) = A^A(q^2) + A_C^A(q^2),
\tag{2.3}
$$

i.e., the sum of the amplitude at one-loop level and the counterterm. The latter is given by

$$A_C^A(q^2) = -2q^2 Z_{AA}^{1/2}. \tag{2.4}$$

Since the one-loop amplitude $A^A(q^2)$ is proportional to q^2 due to the remaining $U(1)$ gauge invariance, the property

$$A_R^A(q^2 = 0) = 0 \tag{2.5}$$

automatically follows. The renormalization constant $Z_{AA}^{1/2}$ is determined by the residue condition

$$A^{A\prime}(0) - 2Z_{AA}^{1/2} = 0. \tag{2.6}$$

Thus we have

$$\frac{1}{q^2} + \frac{1}{q^2} A_R^A(q^2) \frac{1}{q^2} \bigg|_{q^2 \to 0} \to \frac{1}{q^2} \tag{2.7}$$

and do not need any renormalization for the external photon field just like QED.

For the Z-boson field we have similarly

$$\begin{aligned}
A_R^Z(q^2) &= A^Z(q^2) + A_C^Z(q^2), \\
A_C^Z(q^2) &= \delta M_Z^2 + 2Z_{ZZ}^{1/2}(M_Z^2 - q^2),
\end{aligned} \tag{2.8}$$

and set the following on-shell renormalization conditions

$$\begin{aligned}
A^Z(M^2) + \delta M_Z^2 &= 0, \\
A^{Z\prime}(M_Z^2) - 2Z_{ZZ}^{1/2} &= 0.
\end{aligned} \tag{2.9}$$

Thus the renormalized amplitude takes the single pole behaviour at the Z-boson pole.

For the Z-A transition we have

$$\begin{aligned}
A_R^{ZA}(q^2) &= A^{ZA}(q^2) + A_C^{ZA}(q^2), \\
A_C^{ZA}(q^2) &= 2Z_{ZA}^{1/2}(M_Z^2 - q^2) - Z_{AZ}^{1/2} q^2,
\end{aligned} \tag{2.10}$$

and set the renormalization conditions

$$A^{ZA}(M_Z^2) - Z_{ZA}^{1/2} M_Z^2 = 0,$$
$$A^{ZA}(0) + Z_{AZ}^{1/2} M_Z^2 = 0. \tag{2.11}$$

Therefore the renormalized Z-A transition amplitude

$$\frac{1}{q^2 - M_Z^2}[A^{ZA}(q^2) + A_C^{ZA}(q^2)]\frac{1}{q^2} \tag{2.12}$$

does not change the single pole properties of both the photon and Z-boson fields at their on-shell and require no wave function renormalization for the external fields. The charge renormalization constant Y is determined also in the same way as QED.

$$\bar{u}(m_e)\Gamma^\mu(k)u(m_e)\mid_{k^\mu=0} = 0. \tag{2.13}$$

As a result of these prescriptions, the renormalization is the same as in the usual treatment of QED. We do not need any extra factor for external lines for any fields. In the framework of this scheme we have calculated the one-loop correction to various electron-positron annihilation processes in the 't Hooft-Feynman gauge by treating the ultraviolet divergence in the dimensional regularization and by regularizing the infrared divergence by a fictitious photon mass λ.

In the one-loop correction there appears a large correction from the counterterm Y.

$$Y = 1 - \frac{\alpha}{4\pi}\left\{\frac{7}{2}(C_{UV} - \ln M_W^2) + \frac{1}{3} - \frac{2}{3}\sum_i Q_i^2(C_{UV} - \ln m_i^2)\right\} \tag{2.14}$$

where Q_i is the electric charge of fermion i and the color factor should be taken into account for the quark contribution. The finite part of δY is about 3.5% around the Z-boson mass. In the reaction $e^+e^- \rightarrow \mu^+\mu^-$, this large correction can be seen in the following way[4]. We have constructed the blobs of initial electron and final muon; they are composed of the vertex correction, half of the photon self-energy, half of the Z-boson self-energy and half of the γ-Z transition amplitude. In Table 1, I show the contributions of the QED and the weak interaction to the blob separately. Around the Z-boson pole, the photon channel can be neglected in comparison with the Z-boson channel and, furtheremore, the vector coupling of Z-boson is quite small compared with the axial-vector coupling.

Table 1. The numerical values of the blob in unit of 10^{-4} in the $SU(2) \times U(1)$ standard model; F and G are the vector and axial-vector blobs, respectively. The numbers in parentheses are those in QED. The numbers in the first column are those of the corresponding bare coupling constants for $M_Z = 93 GeV$ and $M_W = 81 GeV$. They are obtained by subtracting the infrared divergent terms proportional to $\ln(m_e/\lambda)$.

W(GeV)		70	93	120	150
$F^{e\gamma}$ (−3028)	Re	765 (765)	809 (802)	839 (838)	873 (870)
	Im	−221 (−220)	−223 (−222)	−229 (−227)	−234 (−232)
$G^{e\gamma}$ (0)	Re	1 (0)	−195 (0)	1 (0)	0 (0)
	Im	6 (0)	0 (0)	−13 (0)	−9 (0)
F^{eZ} (−60.75)	Re	17 (17)	17 (18)	18 (19)	20 (19)
	Im	5 (−5)	8 (−5)	11 (−5)	14 (−5)
G^{eZ} (1769)	Re	−439 (−493)	−465 (−519)	−489 (−543)	−508 (−564)
	Im	141 (143)	143 (147)	144 (150)	144 (153)

Thus the weak correction stems mainly from the axial-vector blob of Z-boson. Its magnitude can be read off from Table 1.

$$\frac{-465 + 519}{1769} = 3.05\% \qquad (2.15)$$

In the corrected cross section, a factor 2 due to the interference and another factor 2 of the electron and muon channels should be multiplied. Thus the weak correction around the Z-boson peak reaches to about 13% as expected from the magnitude of δY. It can be also seen in Figure 1. If the same correction is applied to the total decay width of the Z-boson propagator, the weak correction in the numerator is cancelled by that in the denominator as easily seen from the formula of cross section around Z-boson.

$$\sigma(M_Z^2) \sim 12\pi \frac{1}{M_Z^2} \frac{\Gamma_{ee}\Gamma_{\mu\mu}}{\Gamma_{tot}^2} \qquad (2.16)$$

3. EWRC to heavy fermion pair production[5]

Recently the correction to the heavy fermion pair production has been completed by keeping all the masses appearing in virtual corrections. Numerical examples for $t\bar{t}$ production on the Z-boson pole are shown in Table 2. The input parameters are taken as

$$\left.\begin{array}{l} M_Z = 90.7 GeV, \\ m_t = 40 GeV, \\ m_\phi = 100 GeV \end{array}\right\} \quad \begin{array}{l} (M_W = 79.126 GeV), \\ (\sin^2\theta_W = 0.239) \end{array} \qquad (3.1)$$

and a fixed value of $\Gamma_Z = 2.5 GeV$, the photon mass $\lambda = 10^{-6} GeV$ and the soft photon cut $k_c = 0.3 GeV$. Here the value of M_W is calculated by the well-known M_Z-M_W relation[6]. The hard photon contribution is obtained by integrating over the whole phase space. The Born cross section is asymmetric with respect to angular distribution. This is due to the V-A interference; thus its effect is quite small in the lepton, say, muon pair production but sizable in the quark pair production. Among the quark vertex corrections the main contribution stems from the W-W-b vertex shown in Figure 2. In our scheme its largeness is due to the counterterm. The box contributions are fairly small and give correction opposite with respect to the angular asymmetry. The total correction lowers the forward peak. This tendency is the same as in the case of muon pair production.

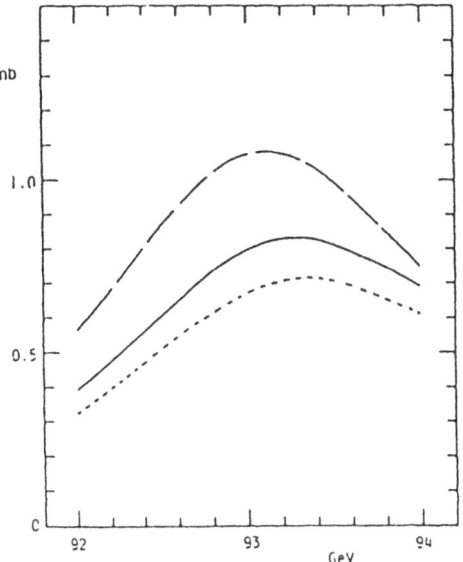

Figure 1. The shift of the Z-boson peak due to the initial photon emission. The solid curve is the angular distribution $d\sigma/d\cos\theta$ at $\theta = 30°$ and with cuts $\zeta_c = 10°$ and $E_{th} = 2GeV$. The dotted curve is the corresponding cross section with the QED correction only. The dashed one is that in the Born approximation. Input parameters are $M_Z = 93, M_W = 81, m_t = 35, m_\phi$ and $\Gamma_Z = 2.5$ in GeV unit.

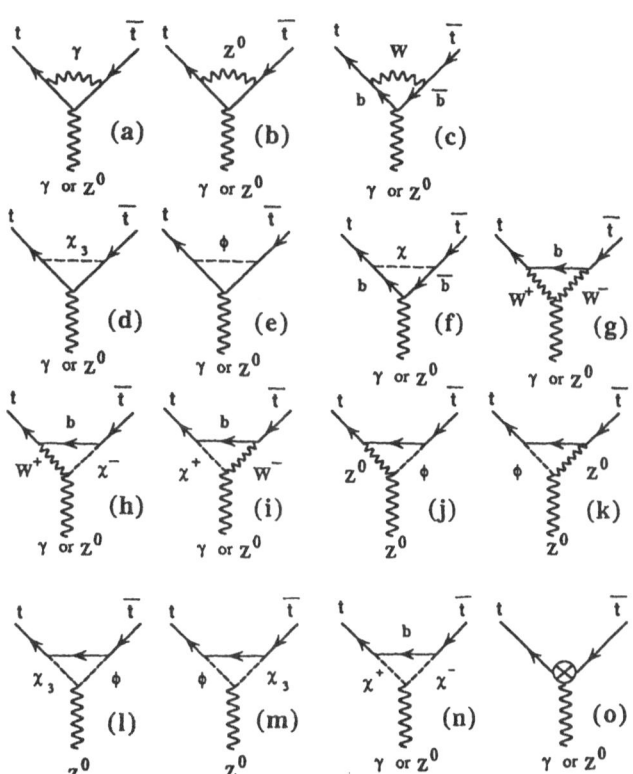

Figure 2. The Feynman diagrams giving rise to the sizable final state correction in $e^+e^- \to t\bar{t}$.

Table 2. The contents of the corrected $d\sigma/d\cos\theta$ (nb) for $e^+e^- \to t\bar{t}$ at $\sqrt{s} = 90.7$ and $100GeV$. The input parameters are given in Eq.(3.1).

$\underline{\sqrt{s} = 90.7 \text{ GeV}}$ (on-pole)

$\cos\theta = -0.9$

Born	Hard	Soft	LVTX+VP	QVTX	Box	Total
0.533	0.079	0.395	−0.636	0.0398	0.0409	0.451
	(14.8%)	(74.1%)	(−119.3%)	(7.47%)	(7.67%)	(−15.4%)

$$\text{Box}\begin{cases} \gamma\gamma & 0.338 \times 10^{-3} \\ \gamma Z & 0.406 \times 10^{-1} \\ ZZ & 0.531 \times 10^{-6} \\ WW & -0.283 \times 10^{-4} \end{cases}$$

$\cos\theta = +0.9$

0.628	0.083	0.552	−0.742	0.0464	−0.0479	0.519
	(13.2%)	(87.9%)	(−118.15%)	(7.38%)	(−7.63%)	(−17.3%)

$$\text{Box}\begin{cases} \gamma\gamma & -0.590 \times 10^{-3} \\ \gamma Z & -0.473 \times 10^{-1} \\ ZZ & -0.780 \times 10^{-6} \\ WW & -0.478 \times 10^{-4} \end{cases}$$

$\underline{\sqrt{s} = 100 \text{ GeV}}$

$\cos\theta = -0.9$

7.559	21.48	5.491	−9.314	0.719	0.875	26.81
	(284%)	(72.64%)	(−123.2%)	(9.51%)	(11.57%)	(254.7%)

$$\text{Box}\begin{cases} \gamma\gamma & -0.370 \\ \gamma Z & 1.267 \\ ZZ & -0.218 \times 10^{-2} \\ WW & -0.169 \times 10^{-1} \end{cases}$$

$\cos\theta = +0.9$

42.77	39.39	38.93	−50.54	1.557	−5.577	66.52
	(92.1%)	(91.0%)	(−118.2%)	(3.64%)	(−13.04%)	(55.5%)

$$\text{Box}\begin{cases} \gamma\gamma & -1.773 \\ \gamma Z & -3.584 \\ ZZ & -0.496 \times 10^{-2} \\ WW & -0.216 \end{cases}$$

Another example at $\sqrt{s} = 100GeV$ is also shown in Table 2. It should be noticed that the quark vertex correction shows rather strong energy and angle dependence. It can be clearly seen in Figure 3. Thus the initial state correction only is not sufficient for this case. This effect is also seen in the b-quark pair production due to the virtual t-quark contribution. For heavy quark pair production, the initial state correction is not sufficient and the final state correction and possibly the interference should be taken into account. This phenomenon is brought about in our scheme by the vertex correction to the final state.

4. QED correction to neutrino counting reaction

The process has been proposed by Ma and Okada[7] and its experimental feasibility compared with the backgrounds has been studied by Barbiellini et al.[8]. The result shows that, although the cross section is reduced by the photon tagging, it becomes easily sizable and almost proportional to the number of neutrino species (N_ν) around or a little over the Z-boson peak. Actually N_ν cannot be deduced from this reaction unless the QED correction is applied to the Z-boson because of the nature of narrow resonance of Z-boson[9,10].

Since the photon is detected, the electron mass in the propagator as well as in the numerator can be set equal to zero. The Born cross section is then given by the five Feynman diagrams in Figure 4. We have performed the exact analytic calculation of them and compared it with the cases taking the Z-boson channel only and the Z-boson channel plus the W-boson channel in the point interaction approximation. These are compared in Figure 5. If one goes to a little higher energy over the Z-boson peak to make use of the information from the whole peak, the approximation taking the Z-boson channel only is not quite enough for the deduction of N_ν.

The QED virtual correction is given by the diagrams in Figure 6. Since the QED virtual correction does not flip the electron helicity, the QED correction to the Z-boson channel can be also applied to the W-boson channel in the point interaction approximation by making use of the Fiertz transformation and a simple replacement of the corresponding factors[11]. In the QED correction the possible three point loop contribution can be neglected because of the anomaly cancellation in the standard model[12].

For the real photon emission we include the soft photon with $k \leq \Delta_c$, the collinear photon emitted around the beam axis within $\theta_c = 5°$ and the effect of

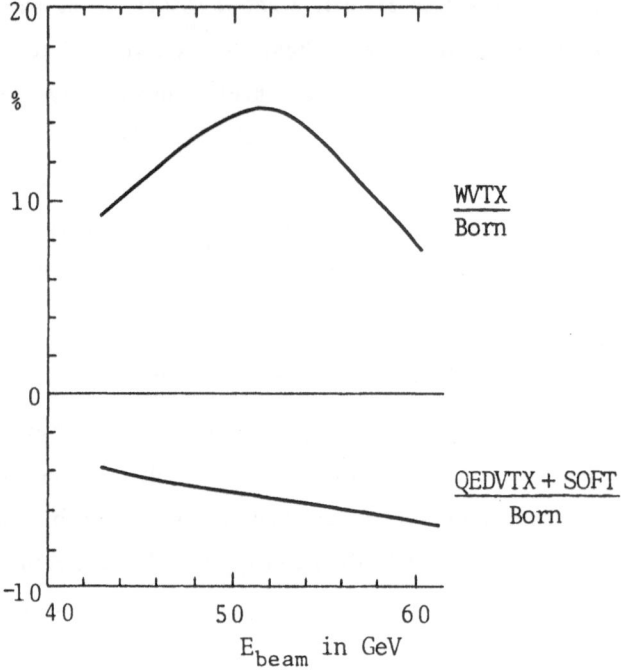

Figure 3. The energy dependence of some parts of the correction to $d\sigma/d\cos\theta$ for $e^+e^- \to t\bar{t}$ at $\cos\theta = -0.9$.

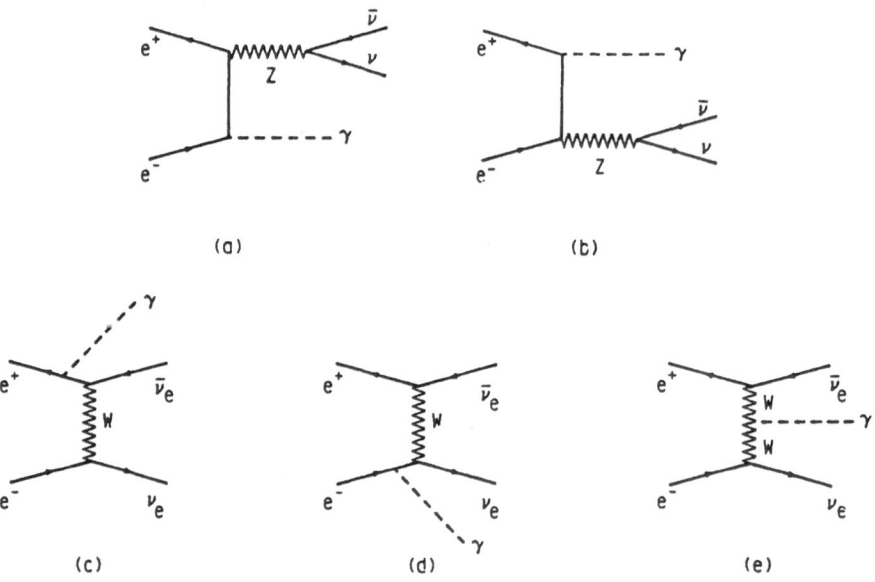

Figure 4. The diagrams contributing the neutrino counting reaction.

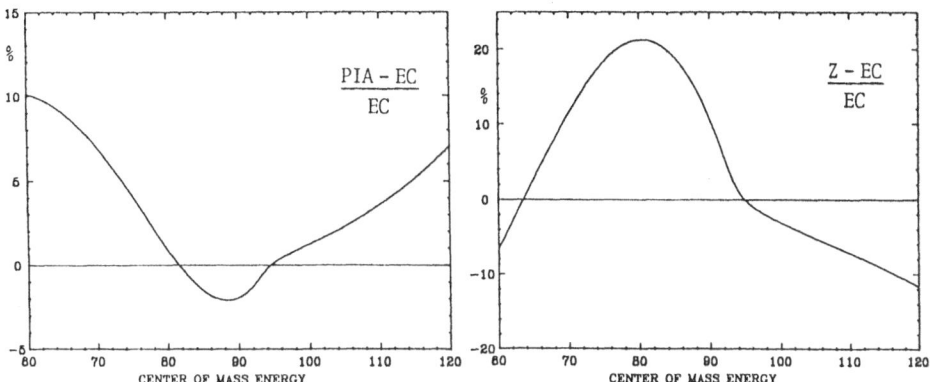

Figure 5. Comparison of various treatments of the Born cross section; EC, PIA and Z mean the exact calculation, W-channel in point interaction approximation and Z-channel only, respectively. The input parameters are $M_W = 81.0$, $M_Z = 92.4$ and $\Gamma_Z = 2.7$ in GeV unit and $N_\nu = 3$. The detected photon is integrated over $k \geq 1 GeV$ and $|\cos\theta| \leq 0.96$.

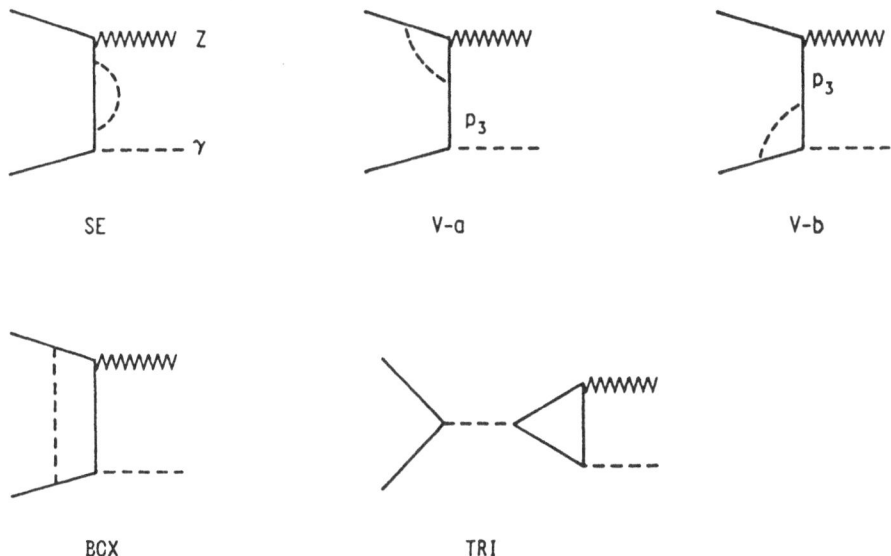

Figure 6. The QED correction to the Z-boson channel.

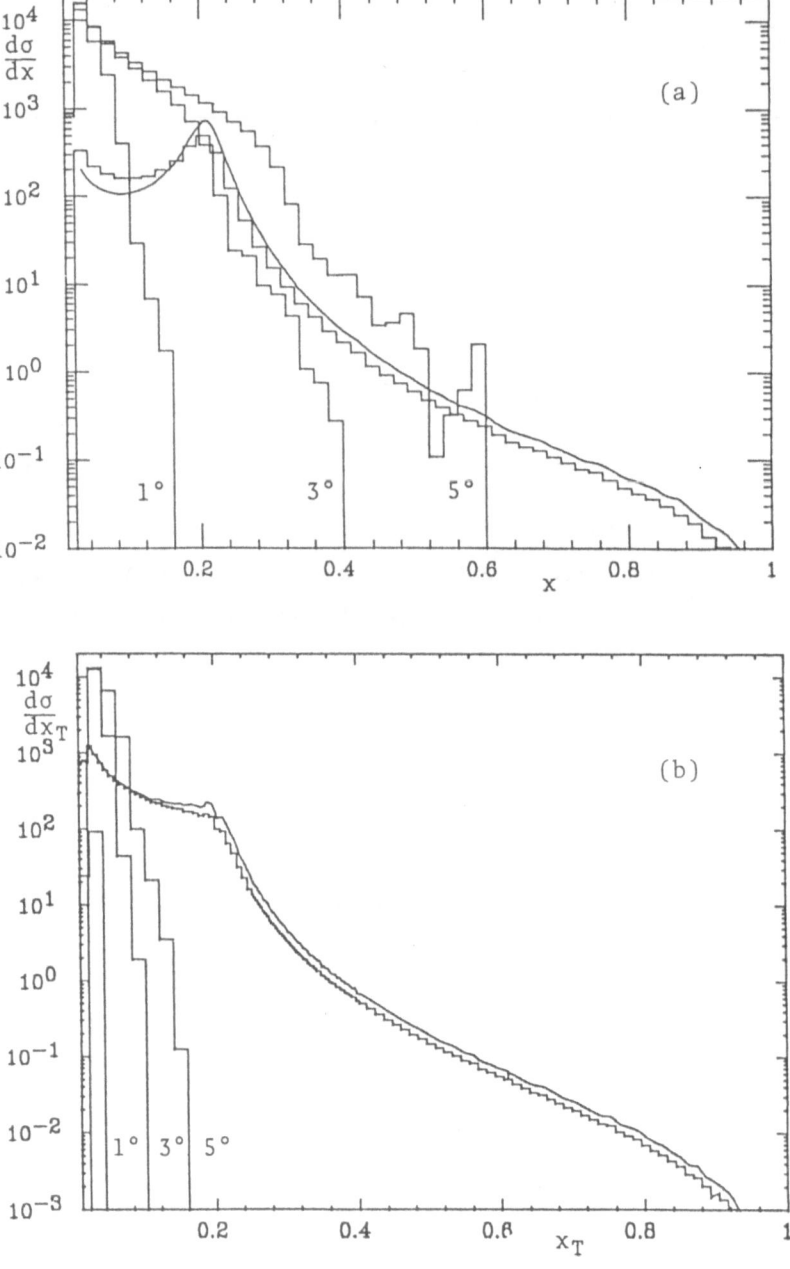

Figure 7. Typical photon spectra of the neutrino counting reaction and the radiative Bhabha scattering at $\sqrt{s} = 104 GeV$; (a) $x = k/E$ distribution, (b) $x_T = x \sin\theta$ distribution and (c) $y = \cos\theta$ distribution. The vetoing angle for $e^{\pm}$ is shown in Figure. The solid line is the Born cross section and the histgram is the cross section in PIA for W-channel with the QED correction. The input parameters are the same as in Figure 5 and the conditions on the radiative correction are given in text.

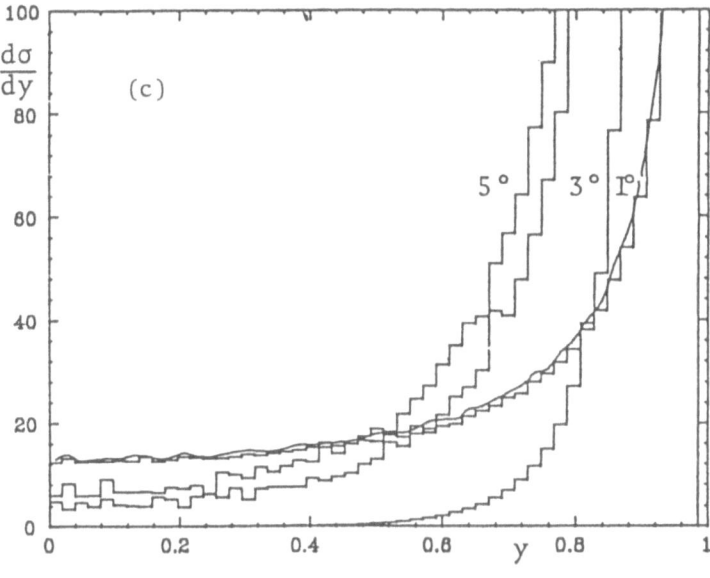

Figure 7. *continued*

the photon emitted parallel to the detected photon within 1°. The three typical photon spectra are given in Figure 7 together with the spectra due to the main background, i.e., the radiative Bhabha scattering[11]. For x distribution (a), the minimum scattering angle θ_m is 10° and $\Delta_c = 0.5 GeV$. For x_T distribution (b), $\theta_m = 10°$, $\Delta_c = 0.2 GeV$ and the photon emitted with $\Delta_c = 0.2 GeV$ to $k_T \leq 1 GeV$ is also integrated. For y distribution (c), $\Delta_c = 1 GeV$.

The calculation of the correction in $SU(2) \times U(1)$ is in progress.

5. Summary

The features of our calculations are that the adopted renormalization scheme is simple and transparent just like QED. However it gives rise to a large counterterm in vertex because of the fact that the electric charge is a low energy parameter. For the heavy quark pair production the final state correction and possibly the interference should be taken into account. For the neutrino counting reaction, it is very important to calculate the precise correction to it since it also gives information on the weakly interacting particles other than neutrino. At the first stage of the effort toward this direction we have succeeded to apply the QED correction to the W-channel in the point interaction approximation.

I sincerely thank the members of the TRISTAN e^+e^- project at KEK for and discussions. I also thank T.Kobayashi and N.Dombey for giving a chance for me to attend this meeting.

References

[1] S. L. Glashow, Nucl. Phys. 22 (1961) 579; S. Weinberg, Phys. Rev. Lett. 19 (1967) 1264; A. Salam, Elementary particle theory, ed. N. Svartholm (Almqvist and Wiksell, 1968) 367.

[2] K. Aoki, Z. Hioki, R. Kawabe, M. Konuma and T. Muta, Prog. Theor. Phys. Supplement No. 73 (1982).

[3] Radiative corrections in $SU(2) \times U(1)$ ed. B. W. Lynn and J. F. Wheater (World Scientific, 1984).

[4] M. Igarashi, N. Nakazawa, T. Shimada and Y. Shimizu, Nucl. Phys. B263 (1986) 347.

[5] J. Fujimoto and Y. Shimizu, Mod. Phys. Lett. 3 (1988) 581 and Nagoya University preprint DPNU-89-18 and KEK preprint (1989).

[6] W. J. Marciano and A. Sirlin, Nucl. Phys. B189 (1981) 442 and also Z. Hioki, Acta Phys. Pol. B17 (1986) 1037.

[7] E. Ma and J. Okada, Phys. Rev. Lett. 41 (1978) 287 and 1759(E).

[8] B. Barbiellini, B. Richter and J. L. Siegrist, Phys. Lett. 106B (1981) 414.

[9] M. Igarashi and N. Nakazawa, Nucl. Phys. B288 (1987) 301.

[10] F. A. Berends, G. Burgers, C. Mana, M. Martinez and W. L. Neerven, Nucl. Phys. B301 (1988) 583.

[11] M. Igarashi, N. Nakazawa and K. Tobimatsu, preprint TKU-HEP 89/01 and KUDP 89/01 (1989).

[12] A. Barroso, F. Boudjema, J. Cole and N. Dombey, Z. Physik C28 (1985) 149.

ELECTROWEAK RADIATIVE CORRECTIONS AT THE Z PEAK[*]

D.Yu. Bardin[1]) and T. Riemann[2])

[1]) Joint Institute for Nuclear Research, Dubna
[2]) Institute for High Energy Physics
Zeuthen/Berlin, GDR

ABSTRACT

Electroweak radiative corrections to the Z peak are determined in the unitary gauge. Peculiarities of this gauge are discussed. A gauge-invariant expression for the weak loop corrections is presented. An approximate cross section formula based on partial Z widths may be derived from the exact one loop calculations.

INTRODUCTION

Electroweak radiative corrections (EWRC) in the standard theory[1] for the reaction

$$e^+e^- \rightarrow (\gamma, Z) \rightarrow \mu^+\mu^- \qquad (1)$$

have been first published in[2]. Subsequent calculations[3-6] applied also to the region of the Z resonance production. Here, an accuracy of theoretical predictions of about 0.1 % seems necessary in order to compete with the precision of measurements which will be based on a sample of several millions of produced Z bosons at LEP[7].

In this contribution, we comment on a calculation of the EWRC for reaction (1) in the unitary gauge[6]. Renormalisation is performed in the on mass shell renormalisation scheme[8] with consequent wave function renormalisation[9]. In search for a gauge invariant formulation of the cross section, we generalised the form factor approach developed in[10] for neutrino scattering to the case of charged particles. The use of these form factors is very convenient for e.g. a combination with Monte Carlo programs for bremsstrahlung calculations but also for phenomenological analyses due to the similarity of the corrected cross section with the Born approximation. Further, the true physical degrees of freedom are explicitly exhibited.

----
[*] presented by T.Riemann

Radiative Corrections, Edited by N. Dombey and
F. Boudjema, Plenum Press, New York, 1990

Besides EWRC that are specific for a reaction like (1), there are those connected with the use of the muon decay constant G_μ instead of α for a normalisation of the weak amplitude and those permitting a calculation of M_W from α, G_μ, M_Z and the other masses of the theory. Both may be reduced to the renormalisation Δr of the muon life time [8,11]. Further, the Breit-Wigner formula for the Z boson propagator containing the Z width Γ_Z [12] is a prediction of quantum field theory which is based on a careful analysis of the common renormalisation [3,13] of both photon and Z propagator and results in an energy dependent width function which has observable consequences [5,14] and necessitates some care in case of an analytic integration of QED bremsstrahlung [14].

In the present contribution, we concentrate on the gauge invariant formulation of EWRC for (1). A recent review of most of the relevant literature may be found in [15].

LAGRANGIAN IN THE UNITARY GAUGE

The minimal standard theory [1,16] has the symmetry group SU(2) x U(1) with one triplet of gauge fields (W_μ^+, W_μ°, W_μ^-), a singlet B_μ, a complex-valued Higgs doublet $\Phi(x) =$ (Φ^+, Φ°) with vacuum expectation value $\Phi_{vac} = (0, V/\sqrt{2})$. The Higgs field can be written as

$$\Phi(x) = e^{i \frac{1}{2V} \tau_a \theta^a(x)} \cdot \left(\begin{array}{c} 0 \\ \frac{1}{\sqrt{2}} [V + H(x)] \end{array} \right), \tag{2}$$

where $H(x)$ is a physical scalar field while the three $\theta^a(x)$ correspond to "would-be" Goldstone bosons and can be completely eliminated by chosing an appropriate gauge (unitary or U-gauge). There exists a class of gauges in which vector boson propagators are ($V = W$, Z, A):

$$D^{\mu\nu} = \frac{i}{q^2 - M_V^2} \left[-g^{\mu\nu} + \frac{(1-\xi^V) q^\mu q^\nu}{q^2 - \xi^V M_V^2} \right], \tag{3}$$

where $M_A = 0$ (massless photon) and Z, A are mass eigenstates composed from the neutral fields W°, B°. Connected with gauge fixing, ghost fields are introduced with propagators

$$d^{\mu\nu} = \frac{i \sqrt{\xi^V}}{q^2 - \xi^V M_V^2} (-g^{\mu\nu}). \tag{4}$$

Often used gauges are characterised by:

$\xi^V = 1$:   t'Hooft-Feynman gauge
$\xi^V = 0$:   Landau gauge
$\xi^V = \infty$:   unitary gauge

Advantages of the unitary gauge:
- no ghost fields in one-loop diagrams
- unphysical Higgs components are completely absorbed by longitudinal W and Z components

- transparent on mass shell renormalisation based only on physical fields

Disadvantages of this gauge due to the $q^\mu q^\nu / M_V^2$ terms in (3):
- individual diagrams are highly divergent, e.g.

$$z \mathrel{\sim\!\!\!\bigcirc\!\!\!\sim} z \;\sim\; \begin{cases} \ell n\, \Lambda,\; \Lambda^2,\; \Lambda^4 & \text{- cut off language} \\ 1/(n\text{-}4),\; 1/(n\text{-}2),\; 1/n & \text{- dim. regularisation} \end{cases}$$

- loop diagrams with vector propagators become complicated
- separate diagrams have non-unitary behaviour - the Λ^2, Λ^4 terms are accompanied by terms rising with q^2 like q^2, q^4.

Further, it is not possible to take advantage from the Ward-Takahashi identities in this gauge. Due to gauge invariance of the theory, finally all gauges yield identical results so that the choice of a gauge is merely an internal theoretical, technical point.

The full lagrangian $\mathcal{L}_{ew}$ consists of a classic lagrangian part, a gauge fixing term, a ghost lagrangian and the counter terms:

$$\mathcal{L}_{ew} = \mathcal{L}_{c\ell} + \mathcal{L}_{fix} + \mathcal{L}_{gh} + \mathcal{L}_{ct}. \tag{5}$$

The lagrangian is very compact in the unitary gauge (we leave out here the counter terms; see e.g. [18]):

$$
\begin{aligned}
\mathcal{L}_{ew} = {}& -\tfrac{1}{4} F_{\mu\nu} F^{\mu\nu} + \sum_f \left[\, \bar{f}(i\partial_\mu \gamma^\mu - m_f) f - ie\, Q_f\, \bar{f} \gamma^\mu f A_\mu \,\right] \\
& -\tfrac{1}{2} W_{\mu\nu}^{+} W^{\mu\nu-} - M_W^2\, W_\mu^{*} W^\mu - \tfrac{1}{4}\, Z_{\mu\nu} Z^{\mu\nu} - \tfrac{1}{2} M_Z^2\, Z_\mu Z^\mu \\
& + i\frac{e}{\sqrt{2}\, s_w} \sum_{i,j} \left[\, \bar{f}_i^u\, \gamma^\mu\, \tfrac{1}{2}(1-\gamma_5)\, K_{ij}\, f_j^d\, W_\mu^{*} + h.c. \,\right] \\
& + i\frac{e}{2\, s_w c_w} \sum_f \bar{f}\, [\gamma^\mu(v_f - a_f \gamma_5)] f\, Z_\mu \\
& + ie\left(\frac{c_w}{s_w} Z_\nu - A_\nu\right)\left[W_\mu W^{\mu\nu*} - W^{\mu*} W_{\mu\nu} + \partial_\mu\left(W^\mu W^{\nu*} - W^\nu W^{\mu*}\right)\right] \\
& + e^2\left(g^{\mu s} g^{\nu\sigma} - g^{\mu\nu} g^{s\sigma}\right) \cdot \\
& \quad \cdot \left[\tfrac{1}{2 s_w^2} W^{\mu*} W_\nu\, W^{s*} W_\sigma + \left(A^\mu - \frac{c_w}{s_w} Z^\mu\right)\left(A_\nu - \frac{c_w}{s_w} Z_\nu\right) W^{s*} W_\sigma\right] \\
& - \tfrac{1}{2} \partial_\mu H\, \partial^\mu H - \tfrac{1}{2} M_H^2 H^2 - \frac{e}{2 M_W s_w} \sum_f m_f\, \bar{f} f H \\
& - \frac{e M_W}{s_w} W^\mu W_\mu^{*} H - \frac{e M_Z}{2 s_w c_w} Z^\mu Z_\mu H - \frac{e^2}{4 s_w^2} W^\mu W_\mu^{*} H^2 \\
& - \frac{e^2}{8 s_w^2 c_w^2} Z^\mu Z_\mu H^2 - \frac{e^2 M_H^2}{4 M_W s_w} H^3 - \frac{e^2 M_H^2}{32 M_W^2 s_w} H^4,
\end{aligned}
\tag{6}
$$

$$F_{\mu\nu} = \partial_\mu A_\nu - \partial_\nu A_\mu, \qquad W_{\mu\nu} = \partial_\mu W_\nu - \partial_\nu W_\mu, \qquad (7)$$

$$W_\mu = W_\mu^-, \qquad\qquad W_\mu^* = W_\mu^+,$$

$$s_w^2 = 1 - c_w^2 = 1 - M_w^2 / M_Z^2, \qquad (8)$$

and K_{ij} is the Cabibbo-Kobayashi-Maskawa mixing matrix.

From this lagrangian, the radiative corrections in one loop approximation for process (1) may be derived.

THE EWRC TO MUON PRODUCTION AT LEP

The matrix elements for muon production in Born approximation correspond to the diagrams

$$(9)$$

and are

$$M_0(\gamma) = e^2 Q_e Q_\mu \frac{1}{s} \gamma_\alpha \otimes \gamma_\alpha, \qquad (10)$$

$$M_0(Z) = \frac{e^2}{4 s_w^2 c_w^2} \frac{1}{s - M_Z^2 + i M_Z \Gamma_Z(s)} \gamma_\alpha (v_e - a_e \gamma_5) \otimes \qquad (11)$$
$$\gamma_\alpha (v_\mu - a_\mu \gamma_5),$$

$$v_f = I_3^f - 2 s_w^2 Q_f, \qquad a_f = I_3^f, \qquad (12)$$
$$a_e = 2 I_3^e |Q_e|, \qquad\qquad I_3^e = -\tfrac{1}{2}.$$

The QED corrections consist of diagrams with one additional virtual photon (vertex and box diagrams) or real photon (bremsstrahlung). For the neutral current process (1) they form a gauge invariant subset and are not discussed here [15].

The propagator corrections are due to self-energy insertions:

$$(13)$$

The vertex corrections are due to insertions at the electron or muon vertex:

$$\tag{14}$$

Weak box corrections are due to double massive vector boson exchange:

$$\tag{15}$$

SELF ENERGY CORRECTIONS AND COUPLING CONSTANTS RENORMALISATION

The photon and Z field renormalisation is performed with the following counter term matrix [9] :

$$\begin{pmatrix} Z_o \\ A_o \end{pmatrix} = \begin{pmatrix} \sqrt{Z_Z} & 0 \\ \sqrt{Z_M} & \sqrt{Z_A} \end{pmatrix} \cdot \begin{pmatrix} Z \\ A \end{pmatrix}. \tag{16}$$

In that way we ensure the non-renormalisation of external lines and get the following renormalised self energies:

$$\hat{\Sigma}_Z(s) = (s - M_Z^2)\,\hat{\Pi}_Z(s) + i\,\mathrm{Im}\,\hat{\Sigma}_Z(s), \tag{17}$$

$$\hat{\Sigma}_\gamma(s) = s \cdot \hat{\Pi}_\gamma(s), \tag{18}$$

$$\hat{\Sigma}_{\gamma Z}(s) = s \cdot \hat{\Pi}_{\gamma Z}(s), \tag{19}$$

where $\hat{\Pi}_Z$ is real, $\hat{\Pi}_\gamma$ and $\hat{\Pi}_{\gamma Z}$ are complex and

$$\hat{\Pi}_Z(M_Z^2) = \hat{\Pi}_\gamma(0) = \mathrm{Re}\,\hat{\Pi}_{\gamma Z}(M_Z^2) = 0. \tag{20}$$

We split self energies into a gauge-invariant fermionic and a gauge-dependent bosonic part, where only the first one contains large logarithms which requires Dyson summation for them:

$$\frac{1}{1 + \hat{\Pi}_V(s)} \;\Rightarrow\; \frac{1 - \hat{\Pi}_V^B(s)}{1 + \hat{\Pi}_V^F(s)}\;. \tag{21}$$

As a result, for the amplitudes one gets:

$$\overline{M}_\gamma(s) = Q_e Q_\mu\, e_F^2(s)\, \bar{\varsigma}_\gamma(s)\, \frac{1}{s}\, \gamma_\alpha \otimes \gamma_\alpha, \tag{22}$$

$$\bar{M}_Z(s) = \sqrt{2}\, G_\mu M_Z^2\, \bar{\varsigma}_Z(s)\, \frac{1}{s-M_Z^2+i\frac{s}{M_Z^2}\Gamma_Z}\, \gamma_\alpha\left[\bar{v}_e(s)-a_e\gamma_5\right]\otimes\gamma_\alpha\left[\bar{v}_\mu(s)-a_\mu\gamma_5\right], \quad (23)$$

$$e_F^2(s) = e^2\left\{1-\frac{e^2}{2\pi^2}\sum_f c_f Q_f^2 \int_0^1 dx\, x(1-x)\,\ell n\left[\frac{-sx(1-x)+m_f^2-i\varepsilon}{m_f^2}\right]\right\}^{-1}, \quad (24)$$

$$\bar{v}_f(s) = v_f + 2Q_f\, s_W c_W\, \hat{\bar{\Pi}}_{\gamma Z}(s), \quad (25)$$

$$\bar{\varsigma}_\gamma(s) = 1 - \hat{\bar{\Pi}}_\gamma^B(s), \quad (26)$$

$$\bar{\varsigma}_Z(s) = (1-\Delta r)(1+\Delta r')\,\frac{\left[1-\hat{\bar{\Pi}}_Z^B(s)\right]}{\left[1+\hat{\bar{\Pi}}_Z^F(s)\right]}, \quad (27)$$

where $\Delta r'$ is due to the coupling constant renormalisation,

$$\frac{e_o^2}{s_W^{o2} c_W^{o2}} = \frac{e^2}{s_W^2 c_W^2}\,(1+\Delta r'), \quad (28)$$

and in the product $(1-\Delta r)(1+\Delta r')$ the influence of light fermions is cancelled out. As a consequence, all barred quantities do not contain large logarithms.

EXPLICIT VERTEX CORRECTIONS

The most general structure of the explicit vertex corrections to the coupling of light fermions to the Z boson is:

$$Z \!\!\!\sim\!\!\!\!\bigcirc\!\!\!\!\!\!\!\begin{smallmatrix}\nearrow f\\\searrow \bar{f}\end{smallmatrix} = i\frac{e}{2s_W c_W}\left[\Gamma_{Z,f}^V(s)\cdot\gamma_\alpha - \Gamma_{Z,f}^A(s)\cdot\gamma_\alpha\gamma_5\right]. \quad (29)$$

The sum of the two insertions (see (10)) with $\bar{M}_Z(s)$ of (23) leads to the following replacements defining $\tilde{M}_Z(s)$:

$$\bar{v}_f(s) \;\rightarrow\; \tilde{v}_f(s) = \bar{v}_f(s) + \Gamma_{Z,f}^V(s) - \frac{v_f}{a_f}\Gamma_{Z,f}^A(s), \quad (30)$$

$$\bar{\varsigma}_Z(s) \;\rightarrow\; \tilde{\varsigma}_Z(s) = \bar{\varsigma}_Z(s) + \frac{\Gamma_{Z,e}^A(s)}{a_e} + \frac{\Gamma_{Z,f}^A}{a_f}. \quad (31)$$

The quantities marked with tilde are yet gauge-dependent. While the barred couplings $\bar{v}_e(s)$, $\bar{v}_f(s)$ could yet be described by only one new quantity, an effective (gauge-dependent) weak mixing angle,

$$s_w^2 \; \rightarrow \; \bar{s}_w^2(s) \; = \; s_w^2 \; - \; s_w c_w \; \hat{\Pi}_{\gamma z}(s),$$ (32)

this is no longer true for $\tilde{v}_{e,f}(s)$. Nevertheless, the Born-like structure is yet maintained.

The corresponding vertex insertions to the photonic couplings yield two corrections $\Gamma_{\gamma,f}^V(s)$, $\Gamma_{\gamma,f}^A(s)$. The matrix element $\bar{M}_\gamma(s)$ becomes then:

$$\tilde{M}_\gamma(s) = Q_e Q_\mu \; e_F^2(s) \; \frac{1}{s} \left[\gamma_\alpha \otimes \gamma_\alpha \left(\bar{s}_\gamma + \Gamma_{\gamma,e}^V + \Gamma_{\gamma,\mu}^V \right) \right.$$
$$\left. - \gamma_\alpha \otimes \gamma_\alpha \gamma_5 \; \Gamma_{\gamma,\mu}^A \; - \; \gamma_\alpha \gamma_5 \otimes \gamma_\alpha \; \Gamma_{\gamma,e}^A \right].$$ (33)

These loop insertions are due to massive gauge boson and higgs insertions. Therefore, we decided to rearrange them in order to combine with the Z boson amplitude corrections:

$$\tilde{M}_\gamma(s) = \; Q_e Q_\mu \; e_F^2(s) \; \frac{1}{s} \; \gamma_\alpha \otimes \gamma_\alpha$$
$$+ \; \frac{\sqrt{2} \, G_\mu \, M_z^2}{s - M_z^2 + i \frac{s}{M_z} \Gamma_z} \left[\gamma_\alpha \otimes \gamma_\alpha \; \tilde{\Gamma}_{\gamma,e\mu}^V - \gamma_\alpha \otimes \gamma_\alpha \gamma_5 \; \tilde{\Gamma}_{\gamma,\mu}^A - \gamma_\alpha \gamma_5 \otimes \gamma_\alpha \; \tilde{\Gamma}_{\gamma,e}^A \right].$$ (34)

From a comparison of (33) and (34) it is evident that the corrections $\tilde{\Gamma}_{\gamma,\alpha}^{V,A}(s)$ are nonresonating at the peak.

As a result of the above rearrangements, one gets:

$$\hat{M}_z(s) = \sqrt{2} \, G_\mu \, \hat{g}(s) \frac{M_z^2}{s - M_z^2 + i \frac{s}{M_z^2} \Gamma_z} \left[\gamma_\alpha \otimes \gamma_\alpha \; \hat{V}_{e\mu}(s) - \gamma_\alpha \otimes \gamma_\alpha \gamma_5 \; \hat{V}_e(s) \, a_\mu \right.$$
$$\left. - \gamma_\alpha \gamma_5 \otimes \gamma_\alpha \; \hat{V}_\mu(s) \, a_e \; + \; \gamma_\alpha \gamma_5 \otimes \gamma_\alpha \gamma_5 \, a_e a_\mu \right],$$ (35)

$$\hat{g}(s) = \tilde{g}(s),$$ (36)

$$\hat{V}_{e(\mu)}(s) = \tilde{V}_{e(\mu)}(s) \; + \; \tilde{\Gamma}_{\gamma,\mu(e)}^A(s) \; / a_{\mu(e)},$$ (37)

$$\hat{V}_{e\mu}(s) = \tilde{V}_e(s) \cdot \tilde{V}_\mu(s) \; + \; \tilde{\Gamma}_{\gamma,e\mu}^V(s).$$ (38)

In (38), the factorisation property for the renormalised

175

vector couplings is lost. Nevertheless, it is realised yet with high precision at the Z peak. The corrections are yet gauge-dependent (and even divergent in the U-gauge).

THE WW AND ZZ BOX CORRECTIONS

The general structure of the box corrections is:

$$M_{Box} \sim G_\mu \left[\gamma_\alpha \otimes \gamma_\alpha \ B_1 - \gamma_\alpha \otimes \gamma_\alpha \gamma_5 \ B_2 - \gamma_\alpha \gamma_5 \otimes \gamma_\alpha \ B_3 + \gamma_\alpha \gamma_5 \otimes \gamma_\alpha \gamma_5 \ B_4 \right], \quad (39)$$

where the B_i depend on both s and the scattering angle. The box diagrams are gauge-dependent because they contain the gauge boson propagators (3). One can decompose them into a gauge-dependent, in the U-gauge UV-divergent part and a finite, gauge-independent but angular dependent genuine box contribution:

$$B_i (s, \cos\theta) = \hat{B}_i (s, \xi) + b_i (s, \cos\theta). \quad (40)$$

Combining the $\hat{v}_e$, $\hat{v}_\mu$, $\hat{v}_{e\mu}$ with the corresponding $\hat{B}_i$ (s, ξ), one arrives finally at the gauge-invariant form factors $s_z(s)$, $v_a(s)$, a = e, μ, eμ. Their explicit expressions are given in 6.

DISCUSSION AND SOME NUMERICAL RESULTS

The above derived form factors are smooth functions of the gauge boson masses M_z , M_W , the Higgs boson mass M_H and of the fermion masses m_f . In Figs. 1-3, we show them as functions of s over a wide energy range:

$$\alpha (s) = \frac{1}{4\pi} \ e_F^2 (s), \quad (41)$$

$$v_a (s) = I_3^a \cdot \left[1 - 4|Q_a| \ \sin^2\theta_W \ æ_a (s) \right], \quad a = e, \mu, \quad (42)$$

$$v_{e\mu} (s) = I_3^e \cdot I_3^\mu \left[v_e(s) + v_\mu(s) - 1 + 16 \sin^4\theta_W |Q_e Q_\mu| æ_{e\mu}(s) \right], \quad (43)$$

$$\sin^2\theta_W = 1 - M_W^2 / M_z^2 . \quad (44)$$

The form factors s_z (s) and $s_W^2(s) = \sin^2\theta_W \cdot æ_e$ (s) exhibit at large s the interesting feature of getting infrared divergent because there M_V^2 /s approaches zero. This behaviour must be repaired then by including also massive gauge boson bremsstrahlung. Further, we indicated in the figures particle production thresholds. In Fig. 3, the

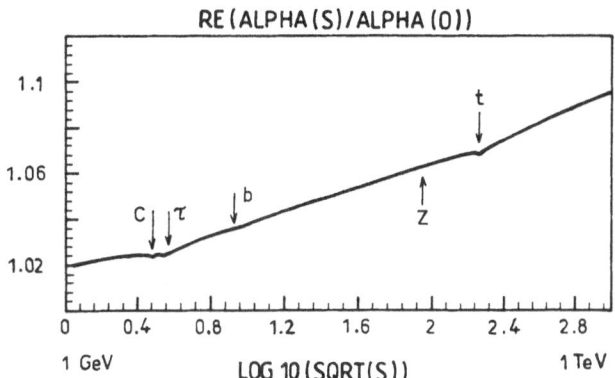

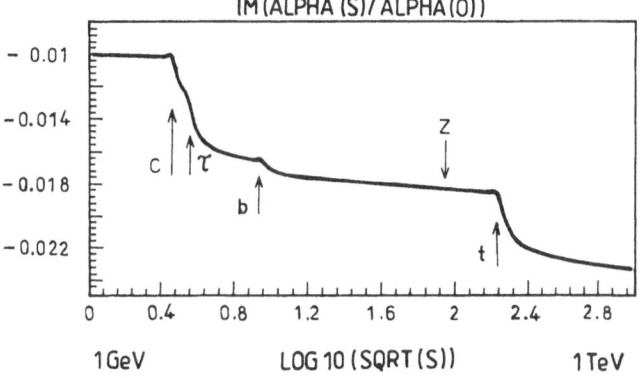

Fig. 1. Energy-dependence of the running electromagnetic coupling as defined in (41).
Fermion thresholds are indicated.

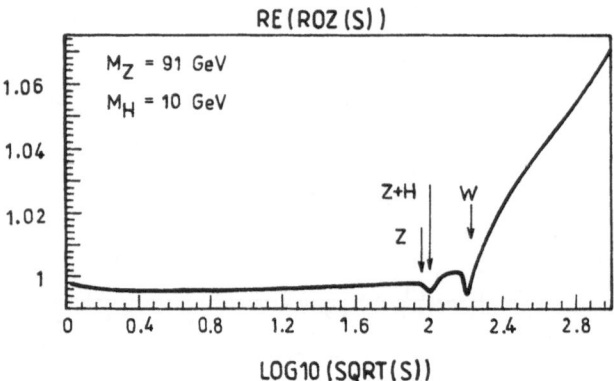

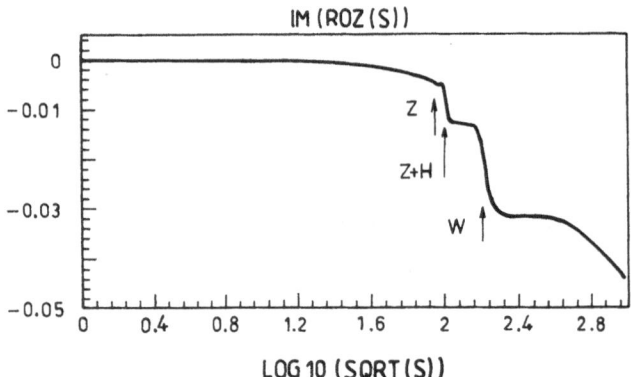

Fig. 2. Electroweak correction $\varrho_z(s)$ to the muon decay constant for (1) as defined in (36, 39) as function of energy; M_z = 91 GeV, m_t = 90 GeV, M_H = 10 GeV.

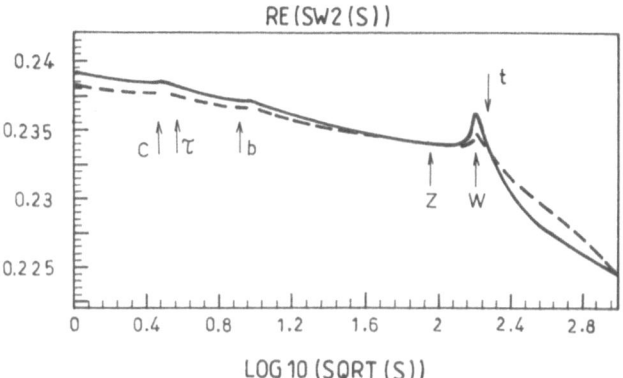

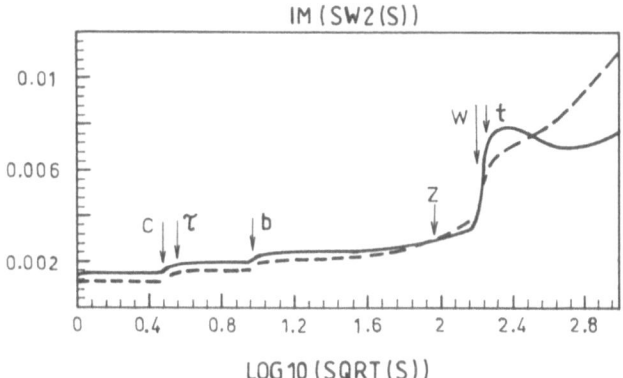

Fig. 3. Electroweak correction of the weak neutral
vector coupling $v_e(s)$ as defined in (37,
39,42) as function of energy. Full line:
$s_W^2(s) = \sin^2\Theta_W \cdot \mathcal{H}_e(s)$; broken line:
the corresponding quantity for $v_{e\mu}(s)$ as a
measure of the deviation from the factorisation property ($v_{e\mu} = v_e \cdot v_\mu$); parameters
as in Fig. 2.

deviations of $v_{e\mu}(s)$ from the factorisation property are shown (broken line). As discussed above, at resonance factorisation is realised with high precision.

The total cross section for muon production expressed by the form factors introduced above is:

$$\sigma_T = \frac{4}{3}\frac{\pi\alpha(o)^2}{s}\sum_{B=\gamma,I,z}Re\left[K(B)\cdot V(B)\right] + \sigma_{Box} + \sigma_{QED}, \tag{45}$$

$$K(B) = \left\{ |Q_eQ_\mu\frac{\alpha(s)}{\alpha(o)}|^2, \ 2Q_eQ_\mu\chi(s)\varrho_z(s)\frac{\alpha(s)^*}{\alpha(o)}, \ |\chi(s)\cdot\varrho_z(s)|^2 \right\}, \tag{46}$$

$$V(B) = \left\{ 1, \ v_{e\mu}(s), \ a_e^2 a_\mu^2 + a_e^2|v_\mu(s)|^2 + |v_e(s)|^2 a_\mu^2 + |v_{e\mu}(s)|^2 \right\}, \tag{47}$$

$$B = \gamma, I, z,$$

$$\chi(s) = \frac{G_\mu}{\sqrt{2}}\frac{M_z^2}{2\pi\alpha(o)} \ \frac{s}{s-M_z^2+iM_z\Gamma_z(s)}, \tag{48}$$

$$\Gamma_z(s) = \frac{s}{M_z^2}\Gamma_z. \tag{49}$$

The cross section is shown in Fig. 4 in comparison to the Born cross section. The QED bremsstrahlung is taken into account here to order $0(\alpha)$ with soft photon exponentiation and cut $\Delta = 2E_\gamma^{max}/\sqrt{s}$; for details and further references see [19].

In the vicitinity of the Z peak, $\sqrt{s} = M_z \pm 0(1\div 2 \ GeV)$, the electroweak form factors may be assumed to be constant (though not being equal to one) aiming at a precision of about 0.1 %. From the above derivation, it is further evident that in this region self-energy insertions yield very small contributions (being non-resonating or vanishing on the Z mass shell).

In Fig. 5, the smallness of the genuine weak box diagrams is visible. The remaining vertex corrections are closely connected to those of the Z decay widths into electrons or muon (see (14)). Evidently, the form factors introduced here are the natural generalisation - derived on a well-founded field theoretic basis - of the intuitive ansatz for the cross section in terms of partial decay widths:

$$\sigma_T \sim \frac{\Gamma_e\Gamma_\mu}{\Gamma_z^2}\left|\frac{s}{s-M_z^2+iM_z\Gamma_z(s)}\right|^2. \tag{50}$$

So, it is quite natural that (50) and its more sophisticated versions are numerically absolutely satisfactory for the analysis of LEP1 physics [20].

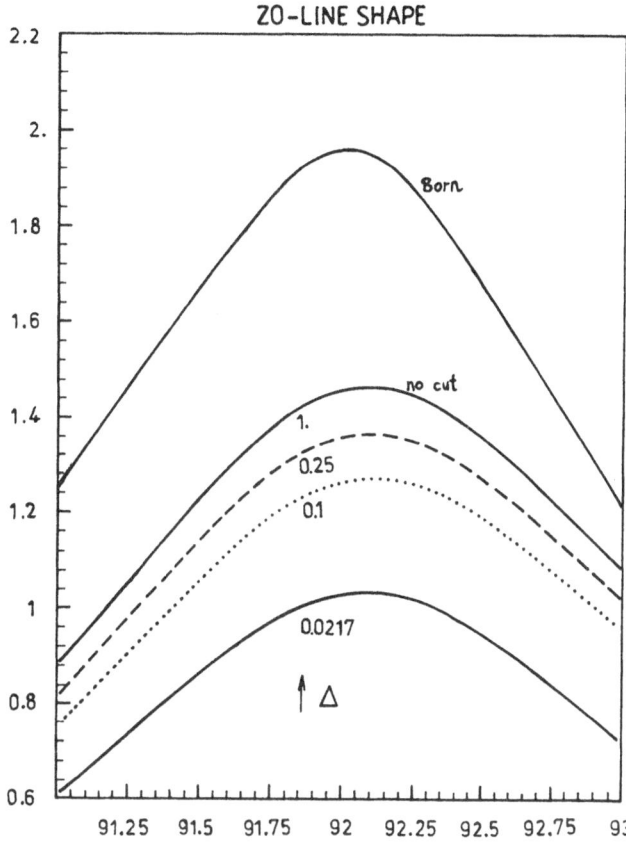

Fig. 4. The Z line shape in muon production (45) as
function of energy in the high precision
region around the peak. Input parameters
are M_Z = 92 GeV, m_t = 90 GeV, M_H = 100 GeV,
α_s = 0.12. These correspond to $\sin^2 \theta_W$ =
0.2258 and Γ_Z = 2.567 GeV; Parameter:
photon cut energy $E_\gamma = \Delta \cdot \sqrt{5} /2$.

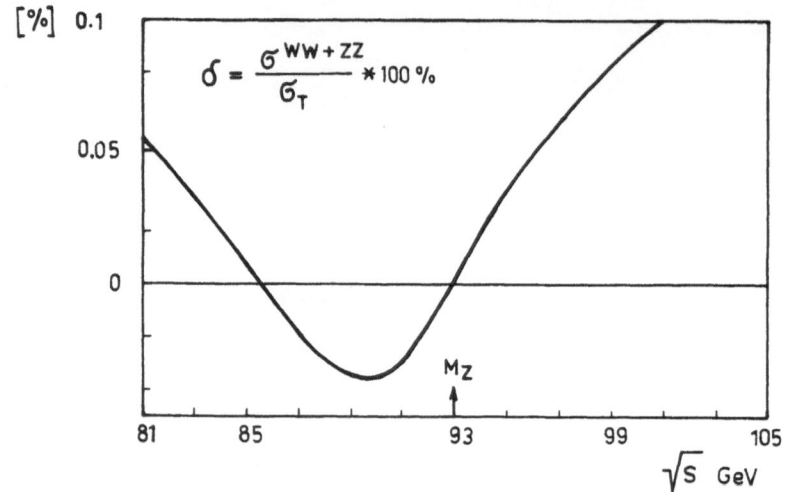

Fig. 5. Relative contribution of the WW- and ZZ-box
diagrams to the total muon production
cross section (45) as function of energy;
M_Z = 93 GeV, m_t = 90 GeV, M_H = 100 GeV.

If, e.g., the muon production cross section (45) is parametrised as

$$\sigma_T = \sigma_0 \left\{ A + Re(B \cdot \mathcal{x}) + D |\mathcal{x}|^2 \right\}, \qquad (51)$$

$$\mathcal{x} = \frac{M_Z^2 S}{S - M_Z^2 + i M_Z \Gamma_Z(S)}, \qquad (52)$$

then (see (47)):

$$A = \left| Q_e Q_\mu \, \alpha(S) / \alpha(0) \right|^2, \qquad (53)$$

$$B = 2 Q_e Q_\mu \frac{\alpha^*(S)}{\alpha(0)} \, \rho_Z(S) \, k \cdot V(1), \qquad (54)$$

$$D = \left| \rho_Z(S) \cdot k \right|^2 \cdot V(2), \qquad (55)$$

$$k = \frac{G_\mu}{\sqrt{2}} \frac{1}{2\pi\alpha(0)}. \qquad (56)$$

This then can be compared to ad-hoc ansatzes based on the partial Z widthes, e.g.[21]:

$$\overline{\sigma}_T = \sigma_0 \cdot A + Real \left\{ \frac{12\pi \Gamma_e \Gamma_\mu \left[\frac{S}{M_Z^2} + \frac{S - M_Z^2}{M_Z^2} R + \frac{S}{M_Z^2} I \right]}{(S - M_Z^2)^2 + M_Z^2 \cdot \Gamma_Z(S)^2} \right\}, \qquad (57)$$

which rests on the Z boson width formula [12] :

$$\Gamma_f = \Gamma_f^{ew} \left(1 + \frac{3}{4}\frac{\alpha}{\pi}Q_f^2\right) \cdot \left(1 + \frac{d_f}{3}\frac{\alpha_s(M_Z^2)}{\pi}\right), \tag{58}$$

$$d_f = 0 \; (3) \qquad \text{for leptons (quarks)}, \tag{59}$$

$$\Gamma_f^{ew} = C_f \frac{G_\mu}{12} \frac{M_Z^3}{6\pi} \rho_f(M_Z^2)\left[w_f^2(M_Z^2) + q_f^2\right], \tag{60}$$

where $\rho_f(s)$ renormalises G_μ and $w_f(s)$ are vector couplings for the decay $Z \longrightarrow f\bar{f}$ corrected by weak loop effects [12] which generally do fulfill $w_f(s) \neq v_f(s)$, $\rho_f(s) \neq \rho_Z(s)$ and $v_f(s)$ and $\rho_Z(s)$ are the weak loop corrected couplings (42). Evidently, there exist one-to-one relations between parameters A,B,D of the exact expression and the ad-hoc parameters A,R,I. For the leading term, e.g., one gets

$$G_0 \cdot D \cdot s \cdot M_Z^2 = 12 \pi \Gamma_e \Gamma_\mu \cdot \delta(s), \tag{61}$$

$$\delta(s) = \frac{V(z)}{[q_e^2 + w_e^2(M_Z^2)][q_\mu^2 + w^2(M_Z^2)]} \frac{|\rho_Z|^2}{\rho_e(M_Z^2) \cdot \rho_\mu(M_Z^2)} \tag{62}$$

The correction factor $\delta(s)$ is at the Z peak very close to one and depends only weakly on s. The same may be shown for the parameters R,I. An ansatz like (57) is justified and quantitatively controlled.

Much of the material of this talk is a densed version of lectures held by the authors for experimentalists at CERN, May/June 1989, on "Weak corrections in $e^+e^- \longrightarrow \mu^+\mu^-$" (D.B.) and "The Z line shape" (T.R.). Numerical results have been produced with the packages DIZET[12] based on [6] and MUCUT[23].

REFERENCES

1. S.L.Glashow, Nucl.Phys. 22: 579 (1961);
    S.Weinberg, Phys.Rev. Letters 19: 1264 (1967);
    A.Salam, in: "Elementary Particle Theory", p. 367,
    N.Svartholm, ed., Almquist and Forlag, Stockholm (1968).
2. G.Passarino and M.Veltman, Nucl.Phys. B160: 151 (1979).
3. W.Wetzel, Nucl.Phys. B227: 1 (1982).
4. B.W.Lynn and R.G.Stuart, Nucl.Phys. B253: 216 (1985).
5. W.Hollik, Phys. Letters B152: 121 (1985);
    F.A.Berends, G.Burgers, W.Hollik and W.L.van Neerven,
    Phys. Letters B203: 177 (1988);
    G.Burgers, in: "Polarization at LEP", p.121, G.Alexander
    et al., eds., CERN report CERN 88-06 (1988).

6. D.Bardin, M.Bilenky, G.Mitselmakher, T.Riemann and M.Sachwitz, Berlin-Zeuthen Prepr. PHE 89-05 (1989), to appear in Z.Physik C (1989).
7. J.Ellis and R.Peccei, eds., "Physics at LEP", CERN report CERN 86-02 (1986).
8. A.Sirlin, Phys. Rev. D22: 971 (1980).
9. D.Bardin, P.Christova and O.Fedorenko, Nucl.Phys. B197: 1 (1982).
10. S.Sarantakos, A.Sirlin and W.Marciano, Nucl.Phys. B217 : 84 (1983).
11. M.Consoli, W.Hollik and F.Jegerlehner, prepr. CERN-TH 5395/89 (1989), subm. to Phys. Letters B.
12. A.Akhundov, D.Bardin and T.Riemann, Nucl.Phys. B276: 1 (1986);
    W.Wetzel, in ref. 7, p. 40;
    F.Jegerlehner, Z.Physik C32: 425 (1986); E: C38: 519 (1988);
    W.Beenakker and W.Hollik, Z.Physik C40: 141 (1988).
13. T.Riemann, D.Bardin and M.Sachwitz, in: "New Theories in Physics", p. 238, eds. Z.Ajduk et al., World Scientific, Singapore (1989).
14. D.Bardin, A.Leike, T.Riemann and M.Sachwitz, Phys. Letters B206: 539 (1988).
15. W.Hollik, prepr. DESY 88-188 (1988), to appear in Fortschr. Phys. (1989).
16. E.S.Abers and B.W.Lee, Phys. Reports 9: 1 (1973);
    L.Maiani, in: "Proc. of the 1976 CERN School of Physics", p. 23, CERN report CERN 76-20 (1976).
17. K.Fujikawa, B.W.Lee and A.I.Sanda, Phys. Rev. D6: 2923 (1972).
18. D.Bardin, Lecture at XVIII Int. School on High Energy Physics, Dubna, Dec. 1986, Dubna report P2-88-189 (1988).
19. D.Bardin et al., Prepr. CERN-TH. 5411/89 (1989), subm. to Phys.Letters B.
20. F.A.Berends, contribution to this conference and references quoted therein.
21. G.Altarelli, private communication.
22. D.Bardin et al., prepr. PHE 89-09 (June 1989), subm. to Comp. Physics Comm.
23. D.Bardin et al., unpublished.

RENORMALIZATION SCHEME DEPENDENCE OF
ELECTROWEAK RADIATIVE CORRECTIONS

Fred Jegerlehner

Paul Scherrer Institute
CH-5232 Villigen PSI
Switzerland

INTRODUCTION

Precision tests of the electroweak Standard Model will be a major goal at LEP. The effects we want to establish are the *genuine* weak corrections, the self-interactions of gauge fields, Higgs boson and top quark interactions or similar effects from *new physics*, expected to be of the order of about 1%. Their detection requires both experimental and theoretical uncertainties to be not more than 0.1%. Typically the expected level of accuracy for various observables at LEP [1] is $\delta M_Z \lesssim 50$ MeV from the Z line-shape, $\delta \sin^2 \Theta_W \lesssim 0.0015$ from the forward-backward asymmetry A_{FB}, $\delta \sin^2 \Theta_W \lesssim 0.0004$ from left-right asymmetry A_{LR} and $\delta M_W \sim 100$ MeV from the W-mass measurement at LEP2.

In order to be able to establish clearly small effects an estimate of the theoretical uncertainties is necessary. The study of the scheme dependence (SD) of predictions is a tool to get an estimate for the quality of perturbative approximations used in standard model calculations.

In perturbative quantum field theory models, like the standard model of electroweak interactions, the starting point is a *classical* Lagrangian with some free mass and coupling parameters. The classical Lagrangian is suitable only for tree level calculations of limited accuracy. If we aim to make precise predictions, we have to take into account higher order corrections. Their calculation is possible only if we properly quantize the Lagrangian. The theory needs be regularized and a gauge fixing prescription is necessary. The validity of the Slavnov-Taylor identities infers the gauge invariance of the S-matrix elements. The parameters of the Lagrangian now become the bare parameters. The relation between the bare and the physical (renormalized) parameters takes the form

$$e_b = e + \delta e, \quad M_b = M + \delta M, \quad \cdots \tag{1}$$

where the shifts δe, δM, $\cdots$ between the bare (b) and the renormalized quantities are the *counter terms*. The latter are fixed either by some formal prescription or by defining how a particular physical parameter is measured. We will not dwell on

Radiative Corrections, Edited by N. Dombey and
F. Boudjema, Plenum Press, New York, 1990

"intermediate" renormalization schemes, like the minimal subtraction scheme familiar from QCD, which work with unphysical auxiliary quantities at intermediate steps. We are interested to discuss cases where physical quantities are calculated in terms of physical quantities.

- Before we can make predictions, a set of independent parameters must be determined from experiment.

- A specific choice of experimental data points used as an input parameter set defines a renormalization scheme (RS).

Parametrizations frequently used are the following:

1) A natural choice of "basic" parameters is the QED-like parametrization in terms of the fine structure constant α and the physical particle masses

$$\alpha, M_W, M_Z, m_f, m_H \qquad (I)$$

often referred to as the "on-shell scheme". We shall refer to it as the α-scheme. It allows for a natural separation of the QED part of the electroweak radiative corrections which is dominated often by large soft photon effects accompanying external charged particles.

2) In the Standard Model , which unifies weak and electromagnetic interactions, we can use as a coupling parameter as well the Fermi constant G_μ instead of α. We then have

$$G_\mu, M_W, M_Z, m_f, m_H \qquad (II)$$

as an independent set of parameters. This set is suitable for processes which are dominated by neutral (NC) or charged (CC) current transitions. An important property of G_μ is that it is not running from low energy up to the vector boson mass scale $M_W(M_Z)$. This G_μ-scheme thus is a genuine high energy scheme in the sense that no large logarithms show up in the calculation of vector boson processes in the LEP energy region (Z and W-pair production).

We know that the parameters of the two schemes are related by [2]

$$\sqrt{2}G_\mu = \frac{\pi\alpha}{M_W^2 \sin^2\Theta_W} \frac{1}{1-\Delta r}, \qquad (2)$$

where Δr is the non-QED correction to μ-decay calculated in the α-scheme. If not stated otherwise, we use the definition

$$\sin^2\Theta_W = s_W^2 = 1 - \frac{M_W^2}{M_Z^2} \qquad (3)$$

for the weak mixing angle.

A disadvantage of the parametrizations (I) and (II) is that they require a precise knowledge of M_W which will be measured precisely at LEP2 only. In order to keep the input parameter errors as small as possible we have to replace M_W by G_μ in (I).

3) The scheme to be used as a starting point for precise calculations of radiative corrections uses

$$\alpha, G_\mu, M_Z, m_f, m_H \qquad (III)$$

as input parameters, with M_Z measured from the Z line-shape at LEP1.

4) Another interesting possibility would be to predict quantities in terms of the low energy parameters

$$\alpha, G_\mu, \sin^2\Theta_{\nu_\mu e}, m_f, m_H \qquad (IV)$$

where $\sin^2\Theta_{\nu_\mu e}$ is determined from neutrino-electron scattering (by CHARM II for example).

Scheme-dependence can be investigated by predicting an observable in terms of different input parameter sets. Since not all the parameters are known to the same precision we proceed as follows: We first predict M_W and $\sin^2\Theta_{\nu_\mu e}$ in the scheme (III) and then take any 3 parameters which are independent at tree level to calculate quantities like the vector boson widths $\Gamma_{Zf\bar{f}}$, $\Gamma_{Wf\bar{f}'}$, or the cross-sections $\sigma(e^+e^- \to f\bar{f})$, $\sigma(e^+e^- \to W^+W^-)$ e.t.c.

Predictions of physical quantities of course should not depend on the specific choice of the input parameters and they if fact do not if we include all orders of the perturbation expansion. Actually, the reparametrization invariance is infered by renormalization group invariance. However, practical perturbative calculations are *approximations* obtained by truncation of the perturbation series. The accuracy of the finite order approximations depends on the choice of the input parameters i.e. finite order results are scheme dependent.

Let us illustrate this point by an example: Suppose we compute a matrix element M in the α-scheme (I) to one-loop order yielding a result

$$M^{(1)} = \alpha^n C[1 + b\alpha].$$

Now, suppose we calculate the same quantity in the G_μ-scheme (II) which amounts to a replacement of $\alpha \simeq 137^{-1}$ by $\alpha' = \frac{\alpha}{1-\Delta r} \simeq 127^{-1}$ i.e. to one-loop order $\alpha' = \alpha[1+a\alpha]$ and

$$M'^{(1)} = \alpha'^n C[1 + b'\alpha'].$$

Inserting α' we get

$$M'^{(1)} = M^{(1)} + \delta M$$

with $b' = b - na$ and

$$\delta M = \alpha^n C \left[(\frac{n(n-1)}{2}a^2 + (n+1)ab')\alpha^2 + \cdots + a^{n+1}b'\alpha^{n+2} \right].$$

Thus the result differs by δM. If we do not actually calculate the higher orders

$$\delta M = M'^{(1)} - M^{(1)} \qquad (4)$$

must be considered as an uncertainty due to unknown higher order effects.

For LEP experiments one-loop calculations are insufficient to get the precision of 0.1% and one has to go to resummation improved calculations by including leading higher order effects. The study of the scheme dependence of resummation improved results is a way to estimate missing higher order contributions (*educated guess*). Of course only an actual n-loop calculation can tell us what the full n-loop answer is.

SCHEME DEPENDENCE OF $\sigma(e^+e^- \to W^+W^-)$

Some important points concerning the SD of higher order predictions can be well illustrated by a calculation of $\sigma(e^+e^- \to W^+W^-)$ in the α- and G_μ-scheme respectively. For a more detailed discussion of W-pair production we refer to A. Denner's contribution [3]. At the Born, level the cross-sections differ by

In Fig. 1, the total cross-section is shown as a function of the c.m. energy $\sqrt{s}$ for both the G_μ and the α scheme [4]. At the Born, level the cross-sections differ by

$$\delta\sigma = \frac{\sigma(\alpha) - \sigma(G_\mu)}{\sigma(G_\mu)} \simeq -13.7\%$$

in the two schemes. This difference, as expected, gets substantially smaller with

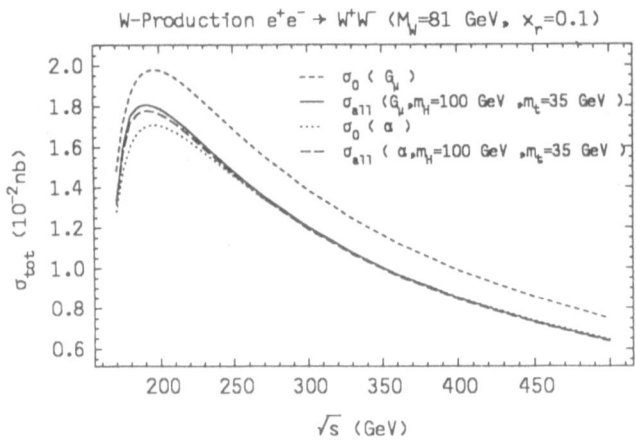

Figure 1. Total cross-section in lowest order(σ_0) and including radiative corrections (σ_{all}) in the G_μ- and α-scheme respectively

the inclusion of the one-loop corrections. Nevertheless, there remains a non-negligible difference which signals missing higher order effects. The difference is largest, $\simeq$ -1.7%, at about 180 GeV ($\simeq$ 12 GeV below the peak), -1.1% at 170 GeV and -1.2% at 200 GeV. For a required precision of better than 1% the missing leading higher order effects must be included. Since we know that the parameters of the two schemes are related by Eq. (2) we have to linear order $(O(\alpha))$ in the cross-section

$$\sigma_0(G_\mu) \simeq \sigma_0(\alpha)(1 + 2\Delta r).$$

This is what is usually included at $O(\alpha)$ in the α-scheme. To second order we get

$$\sigma_0(G_\mu) \simeq \sigma_0(\alpha)(1 + 2\Delta r + 3\Delta r^2).$$

and the additional term $3\Delta r^2$ is the missing next leading term. Typically for $\Delta r \simeq$ 7% we get $3\Delta r^2 \simeq 1.5\%$ which essentially accounts for the above mentioned difference.

This example shows *one* important aspect of the dependence of a prediction on the choice of parameters.

- in the α-scheme large leading logarithms are found which must be resummed using renormalization group arguments in order to get a prediction of acceptable accuracy, whereas

- in the G_μ-scheme no large logs show up and one gets a good approximation without any resummation of higher order effects.

- Even after summing leading logs in the α-scheme the results differ and the difference actually is energy-dependent and hence cannot be solely a matter of change of parameters. The remaining difference is merely due to the omission of higher order effects and unless we perform the next order calculation must be taken as the theoretical uncertainty of the prediction.

Perhaps an even more important drawback of a particular choice of parametrization is the different sensitivity to unknown parameters like m_t and m_H.

The sources of m_H- and m_t- dependence are:

- contributions from γ, Z and W self-energy diagrams

- contributions from ZWW and γWW form factors

Using the G_μ-scheme, with G_μ, M_Z and M_W as input parameters, the dependence of the cross-section on m_H and m_t is very weak, as can be seen from Tab. 1 [4].

Table 1. σ_{tot} (pb) for some values of m_t and m_H. Energies and masses in GeV

	(m_H, m_t)	$E_{c.m.}= 163$	170	180	200
G_μ-scheme	(100,30)	5.042	13.392	17.566	18.005
	(100,200)	4.998	13.297	17.461	17.908
	(100,300)	5.001	13.314	17.496	17.965
	variation	0.85 %	0.71 %	0.60 %	0.54 %
α-scheme	(100,30)	5.039	13.240	17.268	17.793
	(100,200)	4.489	11.908	15.613	16.041
	(100,300)	3.938	10.576	13.959	14.293
	$(E_{c.m.}, m_t)$	$m_H= 10$	100	500	1000
G_μ-scheme	(195,60)	17.861	18.000	17.988	17.986

Of particular importance is the possible top-mass dependence of the W-mass measurement [5]. A crucial fact is that the threshold region is dominated by the t-channel exchange terms, where only the renormalization effects (counter terms) depend on unknown physics, mainly showing up in the vector boson self-energies. Only the t-channel amplitude $T_1^{(-)}$, which is proportional to the ν-exchange diagram at the tree level, exhibits such terms:

$$T_1^{(-)} = -\sqrt{2}G_\mu \frac{2M_W^2}{t}\{1 + 2\Sigma_r^{\nu_e \nu_e}(t) + 2A_{1r}^{We\nu_e}(t)\} + \cdots \tag{5}$$

$$= -\sqrt{2}G_\mu \frac{2M_W^2}{t}\{1 + \Delta C^W + \cdots\} + \cdots$$

where the m_H and m_t dependence of ΔC^W has been analyzed in Ref. 6. Formally,

$$\Delta C^W = \frac{Re\Pi^{WW}(M_W^2)}{M_W^2} - \frac{\Pi^{WW}(0)}{M_W^2} - Re\frac{d\Pi^{WW}}{dq^2}(M_W^2)$$

$$= -M_W^2 \frac{d\pi^{WW}}{dq^2}(M_W^2)$$

is determined solely by the (twice subtracted) W self-energy function

$$\Pi^{WW}(q^2) = \Pi^{WW}(0) + q^2\pi^{WW}(q^2).$$

One finds for the top contribution $\Delta C^{Wt} = -2K$ for $m_t \gg M_W$ and $\Delta C^{Wt} = 2K$ for $m_t \ll M_W$ with $K = \frac{\sqrt{2}G_\mu M_W^2}{16\pi^2}$, i.e. not even a logarithmic dependence on m_t is present!

Similarly, for the Higgs dependence, we find $\Delta C^{WH} = 0$ for $m_H \gg M_W$ and $\Delta C^{WH} = 2K(\ln \frac{m_H^2}{M_W^2} + \frac{47}{12})$ for $m_H \ll M_W$. The potentially interesting infrared log for a light Higgs disappears if Higgs Bremsstrahlung off the final state W's is taken into account. Actually, one finds $\delta\Gamma_{W\to Hff'}/\Gamma_{W\to ff'} = -\Delta C^{WH}$ such that $\Delta C^{WH,virtual+soft} = 0$ for $m_H \ll M_W$. Hence there is a very weak dependence only on both m_t and m_H near threshold. The situation is quite different if we use α, M_Z and M_W as input parameters. Using Eq. (2) we get

$$T_1^{(-)} = -\frac{\pi\alpha}{M_W^2(1 - M_W^2/M_Z^2)}\frac{2M_W^2}{t}\{1 + \Delta r + \Delta C^W + \cdots\} + \cdots \tag{6}$$

instead of Eq. (5), the following quadratic m_t- and logarithmic m_H-dependence results [6]

$$\Delta r^{top} = \frac{\sqrt{2}G_\mu M_W^2}{16\pi^2}\left\{-3\frac{c_W^2}{s_W^2}\frac{m_t^2}{M_W^2} + 2\left(\frac{c_W^2}{s_W^2} - \frac{1}{3}\right)\ln\frac{m_t^2}{M_W^2} + \cdots\right\}$$

$$\Delta r^{Higgs} \simeq \frac{\sqrt{2}G_\mu M_W^2}{16\pi^2}\left\{\frac{11}{3}(\ln\frac{m_H^2}{M_W^2} - \frac{5}{6})\right\} \quad (m_H \gg M_W).$$

In Figs. 2 and 3 we illustrate the m_t-dependence of the W-production production cross-section in the threshold region (see also Ref. 5 for plots of results which include finite widths effects).

The α-scheme is therefore completely inadequate for a model independent determination of the W-mass since the cross-section for given α and M_Z depends in an essential way on *two parameters*, M_W and m_t. In contrast for given G_μ and M_Z the cross-section is a function of M_W only to high accuracy i.e. this scheme is very good for a *model independent* fit of M_W (independent on M_Z, m_t, m_H and possible new physics)!

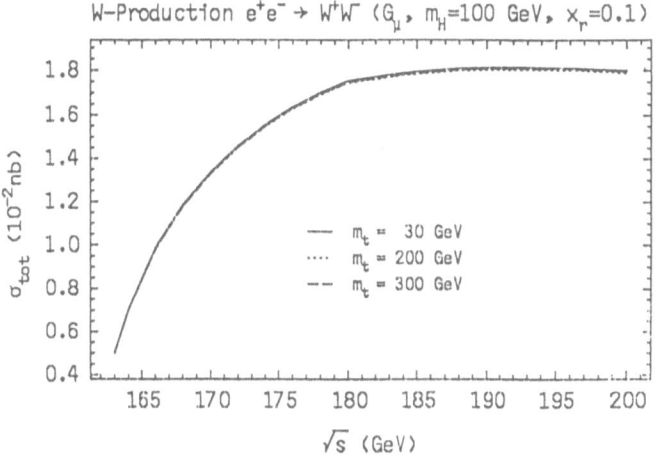

Figure 2. Total cross-section including radiative corrections in the G'_μ-scheme for various values of m_t

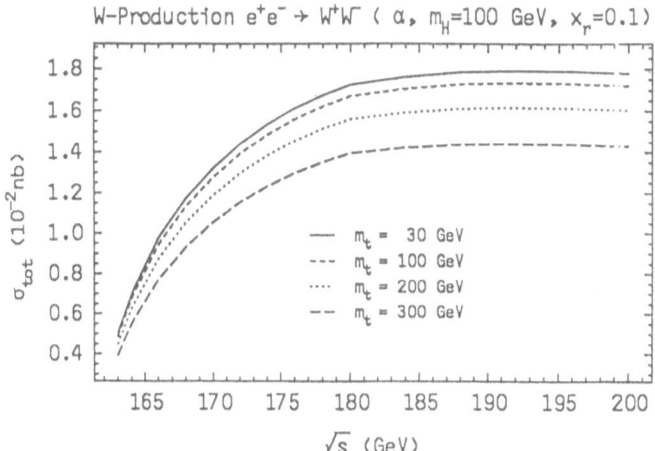

Figure 3. Total cross-section including radiative corrections in the α-scheme for various values of m_t

SCHEME DEPENDENCE OF W-MASS PREDICTION

Of particular interest is the precise relationship between different parameter sets. The relationship between the schemes (I), (II) and (III) is determined by the quantity Δr as introduced in Eq. (2). For example, once Δr is given, the W mass can be predicted by using the values of α, G_μ and M_Z from LEP1. According to Eqs. (2) and (3) we obtain

$$M_W^2 = \frac{M_Z^2}{2} \left(1 + \sqrt{1 - \frac{4A_0^2}{M_Z^2} \frac{1}{1 - \Delta r}}\right) \qquad (7)$$

with $A_0 = \left(\frac{\pi\alpha}{\sqrt{2}G_\mu}\right)^{1/2} = 37.2802(3)$ GeV . $\sin^2\Theta_W$ is given then by Eq. (3).

We have mentioned earlier that a straight forward one-loop calculation would not be sufficient to obtain the precision we need. Indeed, in Eqs. (2) and (7) we have resummed the one-loop result as prescribed by the renormalization group (RG)

$$1 + \Delta r^{(1)} \to \frac{1}{1 - \Delta r^{(1)}} = 1 + \Delta r^{(1)} + (\Delta r^{(1)})^2 + \cdots. \tag{8}$$

Since $\Delta r^{(1)} \simeq 0.07$ we have $(\Delta r^{(1)})^2 \simeq 0.005$ which is not negligible i.e. omission of such higher order terms would show up as a substantial SD. Since we know how to include leading higher order terms we can substancially reduce the SD which would result from a straight forward one-loop calculation. Before we discuss the SD of Δr we therefore have to say how large higher order terms have to be taken into account. To this end we exhibit the large and potentially large terms in Δr by writing

$$\Delta r = \Delta\alpha - \frac{c_W^2}{s_W^2}\Delta\rho + \Delta r_{rem} \tag{9}$$

with

$$\begin{aligned}
\Delta\alpha &= \Pi^{\gamma\gamma}(0) - \Pi^{\gamma\gamma}(M_Z^2) \\
\Delta\rho &= \frac{\Sigma^Z(0)}{M_Z^2} - \frac{\Sigma^W(0)}{M_W^2}.
\end{aligned} \tag{10}$$

and a small but non-negligible remainder Δr_{rem}. By Σ^i ($i = W, Z, \gamma$, or γZ) we denote the unrenormalized vector-boson self-energies, $\Pi^\gamma = \Sigma^\gamma/s$.
$\Delta\alpha$ is the photon vacuum polarization contribution, large due to the large change in scale from zero momentum (Thomson limit) to the Z-mass scale:

$$\Delta\alpha = \frac{\alpha}{3\pi}\Sigma_f N_{cf} Q_f^2 (\ln\frac{M_Z^2}{m_f^2} - \frac{5}{3}) \tag{11}$$

where the sum extends over the light fermions. N_{cf} is the color factor. In contrast to $\Delta\alpha$, $\Delta\rho$ is minuscule for light fermions but large for heavy fermions with a light iso-doublet partner [7]. A typical large term is the heavy top contribution

$$\Delta\rho^{top} = \frac{\sqrt{2}G_\mu}{16\pi^2}3m_t^2. \tag{12}$$

1. Leading log summation

The leading logarithms (11) may be resummed according to the renormalization group. The result may be cast into the form

$$\Delta\alpha^{-1} = \frac{1}{\alpha} - \frac{1}{\alpha(M_Z)} \simeq \frac{1}{3\pi}\Sigma_f N_{cf} Q_f^2 \ln\frac{M_Z^2}{m_f^2} \tag{13}$$

which shows that the leading log summation is scheme independent because the right hand side does not depend on the choice of the parametrization. Beyond the leading log approximation, assuming $\Delta\rho$ to be small,

$$\Delta\alpha^{-1} = \frac{1}{\alpha} - \frac{1}{\alpha(M_Z)} = \frac{\Delta r^{(1)}(\alpha, \cdots)}{\alpha} \tag{14}$$

is "weakly scheme dependent" because only subleading terms depend on the actual parametrization. The RG solution precisely leads to the substitution (8) and thus

$$\alpha(M_Z) = \frac{\alpha}{1 - \Delta r^{(1)}}. \tag{15}$$

This expression includes all terms of the form $\alpha^n \ln^n \frac{M_Z^2}{m_f^2}$ in a scheme independent way. The two-loop irreducible contributions are known and may be included by substituting

$$\alpha \ln(\cdot) \to \alpha(1 + \frac{3\alpha}{4\pi})\ln(\cdot)$$

in $\Delta\alpha^{(1)}$. With $\frac{3\alpha}{4\pi} \simeq 1.7 \times 10^{-3}$, this yields a negligible effect only.

2. Summation of heavy particle effects

In contrast to $\Delta\alpha$ the summation of large $\Delta\rho$ contributions does not follow from the RG. According to Ref. 8, we have to modify Eq. (2) by replacing

$$\frac{1}{1 - \Delta r} \to \frac{1}{1 - \Delta\alpha}\frac{1}{1 + \frac{c_W^2}{s_W^2}\Delta\rho_G} + \Delta r_{rem} \tag{16}$$

where $\Delta\rho_G$ represents the leading irreducible contribution to the ρ parameter defined by the ratio of neutral to charged current amplitudes at low energy, i.e.

$$\frac{G_{NC}}{G_\mu}(0) = \rho_G = \frac{1}{1 - \Delta\rho_G}. \tag{17}$$

By $\Delta\rho_G$ we specifically denote $\Delta\rho$ calculated in the G_μ-scheme. It is important to notice that, in contrast to $\Delta\alpha$, which is not significantly modified by the inclusion of two-loop irreducible contributions, ρ as defined in Eq. (17), can differ sizeably from the one-loop result. In fact as shown in Ref. 8, by including the two-loop irreducible terms calculated in Ref. 9, one finds

$$\Delta\rho_G = N_{cf}x_f[1 - (2\pi^2 - 19)x_f + \cdots] \tag{18}$$

with

$$x_f = \frac{\sqrt{2}G_\mu \Delta m_f^2}{16\pi^2} \tag{19}$$

and $N_{cf} = 1$ for leptons, $N_{cf} = 3$ for quarks. Notice that the negative sign of the two-loop contribution $(2\pi^2 - 19 \simeq 0.739)$ leads to a screening of heavy particle effects and signals a possible restoration of decoupling (see Fig. 4). This effect, if confirmed

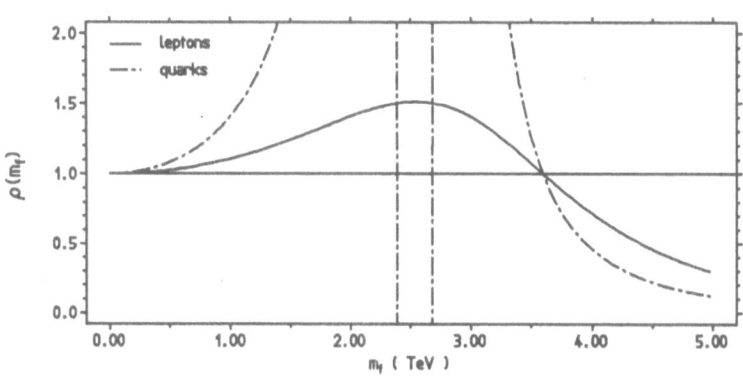

Figure 4. The ρ-parameter as a function of a heavy fermion mass at two-loop order

by higher order calculations would have serious drawbacks for our understanding of the Standard Model .

Using Eqs. (2) and (16) the W mass is determined by expression

$$M_W^2 = \frac{\rho_G M_Z^2}{2} \left(1 + \sqrt{1 - \frac{4A_0^2}{\rho_G M_Z^2} \left(\frac{1}{1-\Delta\alpha} + \Delta r_{rem}\right)}\right). \tag{20}$$

This relation has been obtained in Ref. 10 for the case of a more general Higgs structure with $\rho \neq 1$ at the tree level. The resummation makes sense for the leading fermion contributions, which form a gauge invariant subset. Since terms like the two-loop irreducible contribution proportional to $\frac{\alpha}{4\pi}\sqrt{2}G_\mu m_t^2 \ln(m_t^2/M_Z^2)$ are unknown, non leading fermionic contributions and the bosonic contributions , including the vertex and box corrections, should be added perturbatively, i.e. included in the term Δr_{rem}.

It is important to stress that Eqs. (16) and (20) properly include the higher order terms only if $\Delta\alpha = \Delta\alpha(\alpha, M_W, M_Z, m_f, m_H)$ is calculated in the α-scheme and $\Delta\rho_G = \Delta\rho(G_\mu, M_W, M_Z, m_f, m_H)$ in the G_μ-scheme!

By defining $\tilde{s}^2$ through the relation

$$\tilde{s}^2 \tilde{c}^2 = \frac{A_0^2}{M_Z^2}$$

$(\tilde{c}^2 = 1 - \tilde{s}^2)$ and using Eqs. (2) and (16) we can obtain the $M_W - M_Z$ relation to second order in the large terms

$$M_W^2 = \tilde{c}^2 M_Z^2 \left\{1 - \frac{\tilde{s}^2}{\tilde{c}^2 - \tilde{s}^2}\left(\Delta\alpha - \frac{\tilde{c}^2}{\tilde{s}^2}\Delta\rho_\alpha + \Delta r_{rem}\right) - \frac{\tilde{s}^2(1 - 3\tilde{s}^2 + 3\tilde{s}^4)}{(\tilde{c}^2 - \tilde{s}^2)^3}\Delta\alpha^2 \right. \tag{21}$$
$$\left. + \frac{2\tilde{c}^2\tilde{s}^4}{(\tilde{c}^2 - \tilde{s}^2)^3}\Delta\alpha\Delta\rho_\alpha + \frac{\tilde{c}^4(1 - 3\tilde{s}^2)}{(\tilde{c}^2 - \tilde{s}^2)^3}(\Delta\rho_\alpha)^2\right\}$$

derived in Ref. 11. Here

$$\Delta\rho_\alpha = N_{cf} y_f [1 - (2\pi^2 - 19)y_f + \cdots] \tag{22}$$

where

$$y_f = \frac{\alpha}{16\pi s_W^2 c_W^2} \frac{\Delta m_f^2}{M_Z^2} \tag{23}$$

must be parametrized in the α-scheme with an implicit m_f dependence through $\sin^2\Theta_W$. Another version of Eq. (21) has been presented recently in Ref. 12.

Since the parametrizations of $\Delta\alpha$ and $\Delta\rho_G$ are fixed in order Eqs. (16) and (20) to be valid, the remaining SD is due to Δr_{rem}. Numerical results illustrating the remaining SD are given in Tab. 2. We may summarize the result of Tab. 2 by giving the maximal deviations from the mean value

$$\Delta r = 0.07028 \pm \begin{matrix} 0.00121 \\ 0.00131 \end{matrix}.$$

The SD thus suggests a theoretical uncertainty of about [13]

$$\delta(\Delta r)^{higher-order} \simeq 0.0013(0.0009) \tag{24}$$

leading to $\delta M_W \simeq 25(17)$ MeV and $\delta\sin^2\Theta_W \simeq 0.0005(0.0003)$. In brackets we have

Table 2. Scheme dependence ($M_Z = 91.17$ GeV, $m_H = 100$ GeV, $m_t = 60$ GeV)

Input Parameters	Δr	$\sin^2 \Theta_W$	M_W
α, M_W, M_Z	0.06963	0.23490	79.747
α, G_μ, M_Z	0.07056	0.23524	79.729
α, G_μ, M_W	0.07149	0.23558	79.711
G_μ, M_W, M_Z	0.06897	0.23466	79.759
$\alpha, G_\mu, \sin^2 \Theta_{\nu_\mu e}$	0.07074	0.23530	79.725

given as a reference the hadronic uncertainty $\delta(\Delta \alpha)^{hadronic} \simeq 0.0009$ of the light quark contributions to the photon vacuum polarization [14].

SCHEME DEPENDENCE OF NC COUPLINGS AT THE Z RESONANCE

We now consider LEP1 observables directly related to the NC process $e^+e^- \to f\bar{f}$ near the Z peak. The tree level widths and cross-sections are given by

$$
\begin{aligned}
\Gamma_{Zff} &= \frac{\sqrt{2}G_\mu M_Z^3}{3\pi}(v_f^2 + a_f^2)N_{cf} \\
\sigma_{peak}^{f\bar{f}} &\simeq \frac{12\pi}{M_Z^2}\frac{\Gamma_e \Gamma_f}{\Gamma_Z^2}
\end{aligned}
\tag{25}
$$

where

$$
v_f = \frac{I_3^f}{2} - Q_f \sin^2 \Theta_W, \quad a_f = \frac{I_3^f}{2}
\tag{26}
$$

are the neutral current (NC) couplings for fermions with flavor f. All the asymmetries are functions of the NC couplings ratios

$$
A_f = \frac{2v_f a_f}{v_f^2 + a_f^2}
\tag{27}
$$

and thus provide accurate determinations of the weak mixing angle $\sin^2 \Theta_W$. At tree level the on-resonance asymmetries are given by

$$
A_{FB}^{f\bar{f}} = \frac{3}{4}A_e A_f, \quad A_{LR} = A_{pol}^\tau = A_e, A_{FB,pol}^{f\bar{f}} = \frac{3}{4}A_f.
\tag{28}
$$

Because of the factorization of the non-QED corrections at the resonance, the *weak* corrections of the $Zf\bar{f}$ vertex

$$
(\sqrt{2}G_\mu)^{1/2}2M_Z\gamma^\mu(-Q_f \sin^2 \Theta_W + (1 - \gamma_5)\frac{I_{3f}}{2})
$$

may be included by finite renormalizations [15]

$$
\begin{aligned}
G_\mu &\to \rho_f G_\mu \\
\sin^2 \Theta_W &\to \kappa_f \sin^2 \Theta_W,
\end{aligned}
\tag{29}
$$

where $\rho_f = 1 + \Delta\rho_{se} + \Delta\rho_{f,vertex}$ and $\kappa_f = 1 + \Delta\kappa_{se} + \Delta\kappa_{f,vertex}$. The potentially large self-energy contributions (se) are universal. The analogues of Eq. (9) for $\Delta\rho$ and $\Delta\kappa$ read

$$\Delta\rho_{se} = \Delta\rho + \Delta\rho_{se,rem} \tag{30}$$

$$\Delta\kappa_{se} = \frac{c_W^2}{s_W^2}\Delta\rho + \Delta\kappa_{se,rem}$$

with $\Delta\rho$ defined in Eq. (10). The vertex contributions are (if $f \neq b$) relatively small (but not negligible) and flavor dependent. We may define effective $\sin^2\Theta$'s by

$$\sin^2\Theta_f = \kappa_f \sin^2\Theta_W \tag{31}$$

$$\simeq (1 + \Delta\kappa_{f,vertex})\sin^2\bar{\Theta}$$

$$\sin^2\bar{\Theta} = (1 + \Delta\kappa_{se})\sin^2\Theta_W$$

where $\sin^2\bar{\Theta}$ is a flavor independent auxiliary quantity. The $\sin^2\Theta$'s can be defined directly in terms of the input parameters α, G_μ and M_Z by generalizing

$$\sqrt{2}G_\mu M_Z^2 \cos^2\Theta_W \sin^2\Theta_W = \frac{\pi\alpha}{(1 - \Delta r)} \tag{32}$$

to

$$\sqrt{2}G_\mu M_Z^2 \cos^2\bar{\Theta} \sin^2\bar{\Theta} = \frac{\pi\alpha}{(1 - \Delta\bar{r})} \tag{33}$$

if we include the self-energy contributions only, or,

$$\sqrt{2}G_\mu M_Z^2 \cos^2\Theta_f \sin^2\Theta_f = \frac{\pi\alpha}{(1 - \Delta r_f)} \tag{34}$$

if we include the vertex corrections as well. For the Δr's we find

$$\Delta\bar{r} = \Delta r + \frac{\bar{c}^2 - \bar{s}^2}{\bar{c}^2}\Delta\kappa_{se} \tag{35}$$

$$\Delta r_f = \Delta\bar{r} + \frac{\bar{c}^2 - \bar{s}^2}{\bar{c}^2}\Delta\kappa_{f,vertex}.$$

Using Eq. (30) we obtain

$$\Delta\bar{r} = \Delta\alpha - \Delta\rho + \Delta\bar{r}_{rem}. \tag{36}$$

In all cases the $\Delta\alpha$ term is identical. The leading heavy top dependence is given by

$$\Delta\bar{r}^{top} = \frac{\sqrt{2}G_\mu M_W^2}{16\pi^2}\left\{-3\frac{m_t^2}{M_W^2}\right\} \tag{37}$$

$$\Delta r_b^{top} = \frac{\sqrt{2}G_\mu M_W^2}{16\pi^2}\left\{-\frac{1 + s_W^2}{c_W^2}\frac{m_t^2}{M_W^2}\right\}.$$

Except from extra top contributions from the $Z b\bar{b}$-vertex all heavy particle effects are universal i.e. $\Delta r_{f\neq b}^{top} = \Delta\bar{r}^{top}$ and $\Delta r_f^{Higgs} = \Delta\bar{r}^{Higgs}$.

Figure 5 exhibits the different behavior as a function of m_t of the various Δr's.

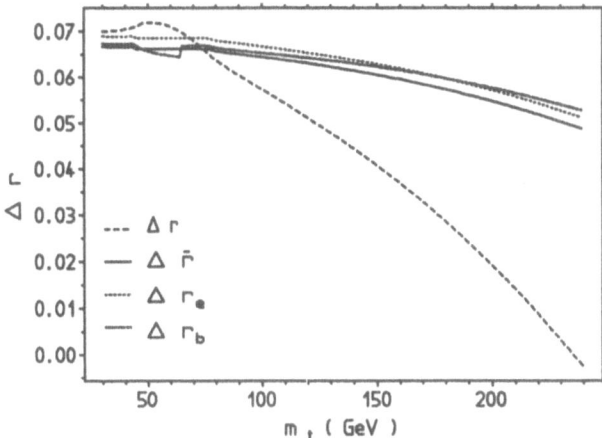

Figure 5. Δr's defined in Eqs. (32-34) as functions of the top mass ($M_Z = 91\ GeV$, $m_H = 100\ GeV$)

For the proper resummation of the large higher terms we obtain

$$\frac{1}{1 - \Delta r_f} \rightarrow \frac{1 - \Delta \rho_G}{1 - \Delta \alpha} + \Delta r_{f,rem}.$$

Again we may estimate the theoretical uncertainty by evaluating the remainder $\Delta r_{f,rem}$ in different parametrizations. Numerical results are given in Tab. 3.

Table 3. Scheme dependence ($M_Z = 91.17$ GeV, $m_H = 100$ GeV, $m_t = 60$ GeV)

Input Parameters	Δr_e	$\sin^2 \Theta_e$	ρ_e
α, M_W, M_Z	0.06743	0.23410	0.99749
α, G_μ, M_Z	0.06789	0.23427	0.99805
α, G_μ, M_W	0.06840	0.23445	0.99796
G_μ, M_W, M_Z	0.06696	0.23393	0.99795
$\alpha, G_\mu, \sin^2 \Theta_{\nu_\mu e}$	0.06795	0.23429	0.99775

Input Parameters	Δr_b	$\sin^2 \Theta_b$	ρ_b
α, M_W, M_Z	0.06409	0.23290	1.01844
α, G_μ, M_Z	0.06455	0.23306	1.01749
α, G_μ, M_W	0.06508	0.23325	1.01744
G_μ, M_W, M_Z	0.06360	0.23272	1.01745
$\alpha, G_\mu, \sin^2 \Theta_{\nu_\mu e}$	0.06461	0.23308	1.01731

From Tab. 3 we obtain

$$\Delta r_e = 0.06773 \pm \begin{matrix} 0.00067 \\ 0.00077 \end{matrix} \qquad \Delta r_b = 0.06439 \pm \begin{matrix} 0.00069 \\ 0.00079 \end{matrix}$$

or $\sin^2 \Theta_f \simeq 0.0003$. This uncertainty is smaller than the one obtained in Eq. (24) for the W mass prediction.

The corrected widths and asymmetries can be obtained using Eqs. (25) and (28) together with the $\sin^2 \Theta_f$'s given in the Tab. 3. In case of hadronic final states a QCD correction factor must be added.

CONCLUSIONS

One-loop predictions of measurable quantities are approximations which depend on the choice of the physical input parameters (to be chosen in a self-consistent way). The differences obtained signal missing higher order terms which must be interpreted as a theoretical uncertainty.

In order to reach the precision required at LEP/SLC, leading higher order effects must be included, thereby the scheme dependence is substantially reduced (by about a factor 5). The remaining uncertainties turn out to be small enough such that they do not obscure measurable effects.

A conservative estimate of the higher order uncertainties obtained by studying the scheme dependence of typical observables yields

$$\delta(\Delta r)^{higher-order} \simeq \delta(\Delta r)^{hadronic} \simeq 0.001 \qquad (38)$$

which amounts to a uncertainty of $\delta M_W \simeq 20$ MeV in the W mass and $\delta \sin^2 \Theta_W \simeq 0.0004$ in $\sin^2 \Theta_W$.

We have not considered the scheme dependence of the QCD corrections to the heavy top contribution. We refer to J. Kühn's contribution [16] for a discussion of

Table 4. Sensitivity of Δr, Eqs. (32-34), to top and Higgs effects versus experimental precision for various measurements. The top contribution $\delta(\Delta r)^{top}$ is given for $m_t = 200$ GeV, $M_Z = 92$ GeV and $\sin^2 \Theta_W = 0.23$. For $\delta(\Delta r)^{Higgs}$ we give the shift obtained if we change m_H from 0 to 1 TeV.

	Error	$\delta \Delta r^{exp}$	$\delta \Delta r^{theo}$	$\delta \Delta r^{top}$	$\delta \Delta r^{Higgs}$
M_W	100 MeV	0.0056	0.0017	-0.0494	0.0156
$A_{FB}^{\mu\mu}$	0.025	0.0043	0.0012	-0.0136	0.0067
A_{LR}	0.003	0.0013	0.0012	-0.0136	0.0067
$A_{FB,pol}^{b\bar{b}}$	0.027	0.0028	0.0012	-0.0028	0.0067

the QCD corrections. The uncertainties from this source are $\delta(\Delta r)^{top} = 0.0005$ and $\delta(\Delta r_e)^{top} = 0.00015$ for $m_t < 150\ GeV$.

Table 4 summarizes expected experimental uncertainies compared with effects obtained from virtual top quark or Higgs boson effects. The theoretical uncertainty includes the hadronic error, the higher-order uncertainty and the uncertainty in the perturbative QCD correction of the top contribution, added in quadrature.

In should be kept in mind that full two-loop calculations still could yield unexpected answers. From experience with QCD, we know that surprises concerning the size of next order results are possible in non-abelian gauge theories.

ACKNOWLEDGEMENTS

I wish to thank Norman Dombey for the invitation to this interesting meeting and for the very kind hospitality at University of Sussex. I am very grateful to Jochem Fleischer for help in preparing the tables ond figures for W-pair production and to Chris Fasano for reading the manuscript.

REFERENCES

1. "Physics with LEP", CERN 86-02 (1986), Eds. J. Ellis and R. Peccei;
   "ECFA Workshop on LEP 200", CERN/ECFA 87-08 (1987),
   Eds. A. Böhm and W. Hoogland

2. A. Sirlin, Phys. Rev. D 22 (1980) 971

3. A. Denner, these Proceedings
   M. Böhm, A. Denner, T. Sack, W. Beenakker, F. Berends, H. Kuijf,
   Nucl. Phys. B 304 (1988) 463

4. J. Fleischer, F. Jegerlehner, M. Zrałek Z. Phys. C 42 (1989) 409;
   F. Jegerlehner, in "Radiative Corrections for e^+e^--Colliders", Ed. J. H. Kühn,
   Springer Berlin 1989

5. B. Grządkowski, Z. Hioki, Phys. Lett. 197 (1987) 213;
   B. Grządkowski, Z. Hioki, H. J. Kühn, Phys. Lett. 205 (1988) 388;
   Z. Hioki, Preprint TOKUSHIMA 88-01, Tokushima (1988)

6. F. Jegerlehner, in "Testing of the Standard Model ",
   Eds. M. Zrałek, R. Mańka, World Scientific Publ., Singapore, 1988

7. M. Veltman, Nucl. Phys. B 123 (1977) 89

8. M. Consoli, W. Hollik, F. Jegerlehner, Phys. Lett. 227 (1989) 167; see also:
   W. Hollik, Preprint DESY 88-1988, to appear in: Fortschritte der Physik

9. J. J. van der Bij, F. Hoogeveen, Nucl. Phys. B 283 (1987) 477

10. F. Jegerlehner, Z. Phys. C 32 (1986) 425

11. B. W. Lynn, D. Kennedy, C. Verzegnassi, contribution to
    "Electroweak radiative corrections at LEP energies", CERN-EP/87-70, p. 40;
    D. C. Kennedy, B. W. Lynn, SLAC-PUB 4039 (1986, revised 1988)

12. G. Passarino, R. Pittau, Phys. Lett. 228 (1989) 89

13. F. Jegerlehner, Z. Phys. C 32 (1986) 425;
    W. Hollik, H.-J. Timme, Z. Phys. C 33 (1986) 125

14. F. Jegerlehner, Z. Phys. C 32 (1986) 195;
    H. Burkhardt, F. Jegerlehner, G. Penso, C. Vrezegnassi,
    Z. Phys. C 43 (1989) 497

15. A. Sirlin, W. J. Marciano, Nucl. Phys. B 189 (1981) 442;
    A. A. Akhundov, D. Yu. Bardin, T. Riemann, Nucl. Phys. B 276 (1986) 1

16. J. Kühn, these Proceedings;
    A. Djouadi and C. Verzegnassi, Phys. Lett. 195B (1987) 265 ;
    A. Djouadi, Nuovo Cimento 100 A (1988) 357;
    B. A. Kniehl, J . H. Kühn and R. G. Stuart, Phys. Lett. 214 (1988) 621;
    D. Yu. Bardin, A. V. Chizov, Dubna preprint E2-89-525 (1989)

LIGHT QUARK LOOPS

H.Burkhardt

University of SIEGEN

1 Introduction

An important subset of virtual electroweak corrections is due to closed fermion loops. In particular the correction to the photon self energy due to the known fermions is rather large and equivalent to a change of the QED fine structure constant α from its low energy value to the LEP energy scale by about 6 %. It can be calculated exactly in the case of leptons.

For light quarks instead, masses are not unambiguously defined and QCD corrections large. Fortunately there is a possibility of avoiding sizeable theoretical uncertainties from the light quark contribution: The hadronic part of $\Pi_{\gamma\gamma}$ can be expressed as a dispersion integral using the experimentally measured cross section of e^+e^- annihilation into hadrons. The uncertainty in the hadronic contribution from light quarks is then determined by experimental errors.

Only little new information on e^+e^- annihilation into hadrons at low energies has become available since the 1988 re-evaluation of the dispersion integral that has been described in detail in reference [1].

The current experimental knowledge with some outlook into the future will be reviewed in the first part of my talk. Practical aspects connected to the implementation of these correction into computer programs for LEP/SLC physics will conclude this conference contribution.

2 Review of experimental errors, outlook

The real part of the renormalized, gauge invariant hadronic vacuum polarization contribution to the transverse component of the photon self energy is obtained from the principal value dispersion integral :

$$Re\Pi_{\gamma\gamma\ had} \;=\; \frac{\alpha s}{3\pi}\, P \int_{m_\pi^2}^\infty \frac{R(s')}{s'(s'-s)}\, ds'$$

R is the cross section for e^+e^- annihilation into hadrons through one photon exchange, normalized to the point like QED cross section for lepton pair production. Figure 1 shows as solid line the parametrization of R used in the integration. The parametrization follows the average value of all published data. Some selected data points with their statistical errors and a QCD calculation for R have been included in the picture. Resonances have

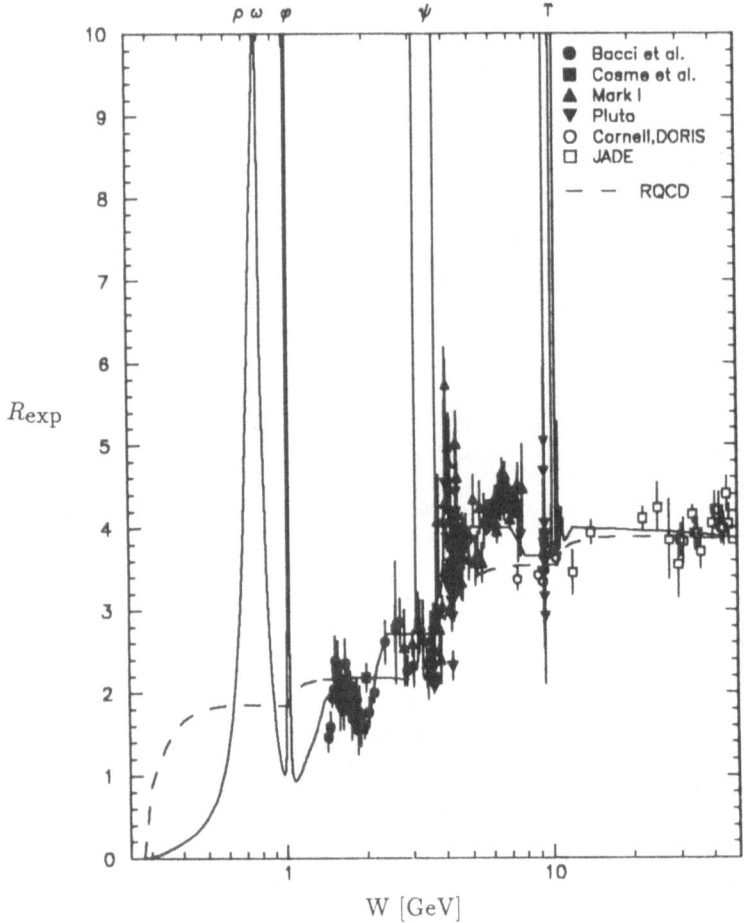

Figure 1. Parametrization of R including Resonances up to W = 50 GeV

been parametrized by Breit-Wigner forms using the leptonic widths from reference [3] (with the exception of the ρ where an effective leptonic widths of 0.5 ± 0.3 keV has been used) and the continuum through a set of straight lines. The whole integration and error calculation can be performed analytically for all space- and timelike values of four momentum transfer. For the error calculation, the continuum contribution has been subdivided into four broad regions. Rather conservative error estimates have been assigned to each of these regions.

The various hadronic contributions to $\Pi_{\gamma\gamma}$ at the mass of the Z are summarized in table 1. The uncertainty is dominated by the contribution from low energies (the result for W = 0 - 12 GeV alone is 1.64 ± 0.08). The parametrization for energies above W = 40 GeV follows second order QCD with $\alpha_s = 0.12$. The assumed relative error in R for energies above 12 GeV is 3 % and largely covers the experimental uncertainty in the knowledge of α_s.

Table 1. Hadronic (u,d,s,c,b) contributions to $\Delta\alpha$ for $W = 92$ GeV

W range	rel.error	$\Delta\alpha$ in %
1.0 - 2.3 GeV	20 %	0.20 ± 0.039
2.3 - 9.0 GeV	10 %	0.72 ± 0.072
9.0 - 12.0 GeV	10 %	0.17 ± 0.017
12.0 - ∞	3 %	1.24 ± 0.037
all resonances	3 %	0.55 ± 0.018
total	3 %	2.88 ± 0.09

Let us look now, what could be expected to change in future. Some still preliminary and unfortunately unpublished results from DM2 in the energy range $W = 1.3$ - 2.2 GeV and from Crystal Ball for $W = 5.0$ - 7.4 GeV should help to reduce the error in the important low energy region [4]. The preliminary DM2 data agrees well in average with the parametrization used here and the relative error in the energy range $W = 1$ - 2.3 GeV could probably be lowered to 10 %. (The low energy contribution in table 1 would now be 0.20 ± 0.020 % but the overall error for all energies would still be close to 0.09 %).

A reduction of the overall error would need a clarification of the energy range from $W = 5$ to 7 GeV, where so far only results from the MARK I collaboration have been published [5]. The old MARK I results seem to be systematically rather high (parametrized by R=4 $\pm$10%).

The DM2 and Crystal Ball results together could probably help to reduce the overall relative error from currently 3 to maybe 2 %. The Crystal Ball results are at the same time expected to lie below the Mark I results and therefore to reduce also the central value (well within the present overall error).

On the longer term, improvements could come from the sophisticated detectors proposed for future e^+e^- charm, τ and beauty factories.

3 Practical implementation

The photon self energy has to be known for some applications (Bhabha scattering) over a wide range of four momentum transfer. For the lepton case, the contribution to the photon self energy can be written as [2]:

$$\Pi_{\gamma\gamma}(s) = \frac{\alpha}{3\pi} \sum_f N_c Q_f^2 P(s, m_f) \quad ; \quad P(s, m) = \frac{1}{3} - \frac{1 + 2m^2}{s} F(s, m, m)$$

with

$$F(s, m_1, m_2) = 1 + \left(\frac{m_1^2 - m_2^2}{s} - \frac{m_1^2 + m_2^2}{m_1^2 - m_2^2}\right) \log \frac{m_1}{m_2} + \frac{RT}{s} \log \frac{R - T}{R + T}$$

and

$$R = \sqrt{s - (m_1 + m_2)^2} \qquad T = \sqrt{s - (m_1 - m_2)^2}$$

R,T can be real or imaginary. N_c is the number of colors (3 for quarks, 1 for leptons). $\Pi_{\gamma\gamma}$ has an imaginary part if $s > (m_1 + m_2)^2$ through $\log(-|x|) = \log|x| + i\pi$. Having $m_1 = m_2$ in the photon self energy, the F-function can be simplified to

$$F(s, m, m) = 2 + \beta \log \frac{\beta - 1}{\beta + 1} \qquad \beta = \sqrt{1 - \frac{4m^2}{s}}$$

For $|s| \gg m^2$ (in the limit $|\beta| \to 1$) we get

$$P_{asymp.}(s, m) = -\frac{5}{3} + \log\left(-\frac{s}{m^2}\right)$$

Numerically we find (assuming always $M_Z = 92$ GeV):

from the full formula

$$
\begin{array}{lll}
t = & -M_Z^2 & P(t, m_\tau) = 6.2213 \\
s = & M_Z^2 & P(s, m_\tau) = 6.2168 - 0.99999915\, i\pi
\end{array}
$$

from the asymptotic formula

$$
\begin{array}{lll}
t = & -M_Z^2 & P(t, m_\tau) = 6.2191 \\
s = & M_Z^2 & P(s, m_\tau) = 6.2191 - i\pi
\end{array}
$$

We see that the approximation is correct to better than $4 \cdot 10^{-4}$. Due to their smaller masses, the e, μ are even better approximated. We conclude, that the asymptotic formula is sufficient to calculate the e, μ, τ vacuum polarization in the s-channel at Z-energies.

Luminosity determination relies on small angle Bhabha scattering that is dominated by t-channel exchange with much lower four momentum transfer. At Z-energy and 20 mrad scattering angle for example, we have $t = -\frac{1}{2}(1 - \cos\theta)s \approx -\frac{s\theta^2}{4} \approx -0.01^2 s \approx -1$ GeV^2. The results from the full and asymptotic formula for P(s,m) are now very different:

$$P(-(0.01\, M_Z)^2, m_\tau) = 0.0608 \quad and \quad P_{asymp.}(-(0.01\, M_Z)^2, m_\tau) = -2.824$$

A number of Bhabha programs (for example OLDBAB [7] and BHLUMI [8]) used incorrectly the asymptotic formula in all cases in a function called REPI. This resulted in an underestimate of the Bhabha cross section by about 0.5 % at small scattering angles.

The dispersion integral result for the hadronic contribution to the photon self energy is well approximated by [6] :

$$\Pi_{\gamma\gamma}(s) \approx \Pi_{\gamma\gamma}(-t) \approx A + B\log(1 + C\, s)$$

Parametrization for $\Pi_{\gamma\gamma}$ from u,d,s,c,b quarks			
$\sqrt{|s|}$ [GeV]	A	B	C
0.0 - 0.3	0.00000	0.00835	1.000
0.3 - 3.0	0.00000	0.00238	3.927
3.0 - 100.0	0.00165	0.00299	1.000
100.0 - ∞	0.00221	0.00203	1.000

For the imaginary part we have $Im\Pi_h(s) = \frac{\alpha}{3} R(s)$ where R(s) is the parametrization of the normalized hadronic cross section as used in the dispersion integral. Figure 2 shows the result of the full calculation of $-Re\Pi_h$ and (as broken line) the parametrization for a broad range of time- and spacelike momentum transfers. Within the plot resolution, the parametrization shows perfect overlap for spacelike momentum transfers. For timelike momentum transfers the parametrization should be only used for sufficiently large values of s far from resonances (some GeV above Υ threshold).

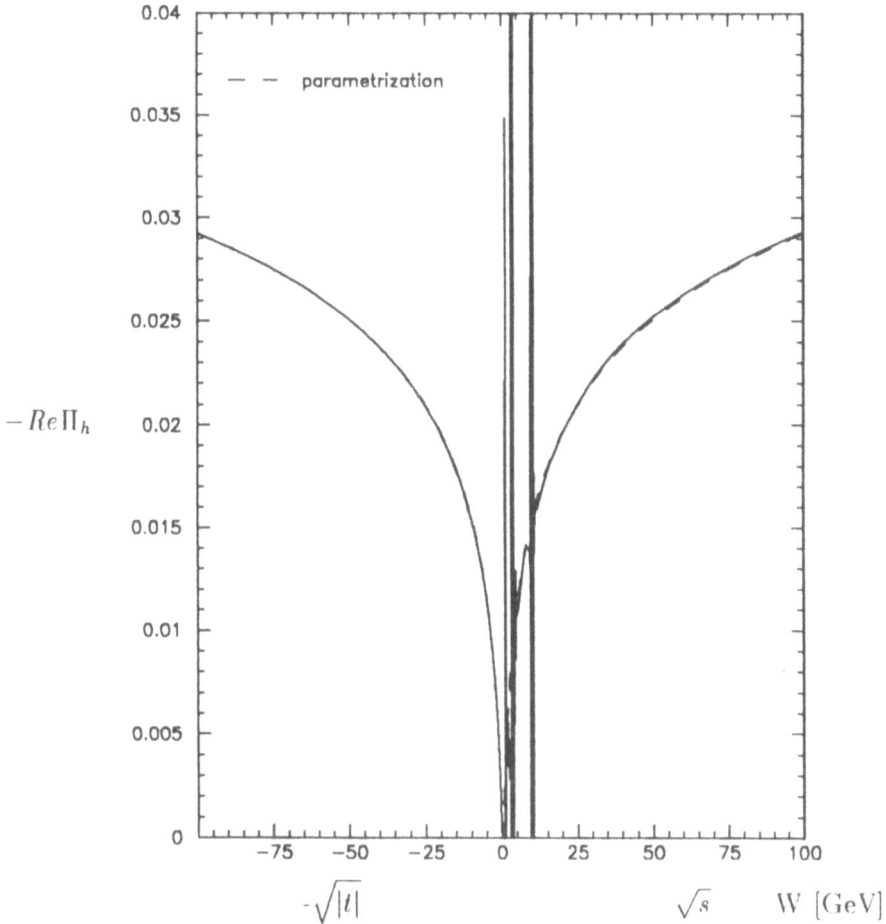

Figure 2. Result and Parametrization for time- and spacelike momentum transfers

Some programs use instead the same formula for quarks and leptons. Adjusting quark masses, this can lead to the correct result, but only in a very restricted energy range.

In the limit s → 0 we get

$$-\Pi_{\gamma\gamma h}(s) = \frac{\alpha s}{5\pi} \sum_f \frac{Q_f^2}{m_f^2} = const \cdot s$$

which can be tuned to agree with the parametrization $0.00835 \log(1 + |s|) \approx 0.00835\, s$ if the lightest quark masses are chosen to be not lighter than about 200 MeV. For $s \gg m_f^2$ we have

$$-\Pi_{\gamma\gamma h}(s) = \frac{\alpha}{\pi} \sum_f Q_f^2 \log s \;-\; \frac{\alpha}{\pi} \sum_f Q_f^2 (\frac{5}{3} + \log m_f^2)$$

We see, that the factor in front of the logarithm has a constant value that cannot be adjusted through quark masses:

$$\frac{\alpha}{\pi} \sum_f Q_f^2 = \frac{11\alpha}{9\pi} = 0.00284$$

The value is a few percent lower than the B = 0.00299 in the parametrization. This is due to QCD corrections $(1 + \frac{\alpha_s}{\pi} \approx 1.031$ at Z energies) that have been neglected in the free field theory. If we fix the s,c,b masses at 0.5, 1.5 and 4.5 GeV, then we have to lower the u,d masses to about 37 MeV in order to get the right correction for energies around M_Z (this was done in the program BABAMC [9] and resulted in an overestimation of the Bhabha cross-section at small angles by 0.7 %).

Apparently there is no satisfactory set of fixed quark masses, that allows to parametrize the hadronic contribution to $\Pi_{\gamma\gamma}$ over a broad region of momentum transfers. Moreover using fixed quark masses one has to worry about QCD corrections, that instead are automatically included in the dispersion relation approach. For reference, the following table gives the result of the sum of the lepton and hadron contribution.

It has been calculated using the full expression including all mass terms for leptons and the parametrization of the dispersion integration for hadrons.

Table 2.

Results for the e, μ, τ + u,d,s,c,b vacuum polarization		
s, t [GeV^2]	$Re\ \Pi_{\gamma\gamma}$	$Im\ \Pi_{\gamma\gamma}$
$-(92.0)^2$	0.06024	0.00000
$-(50.0)^2$	0.05375	0.00000
$-(10.0)^2$	0.03677	0.00000
$-(1.0)^2$	0.01653	0.00000
$(2.3)^2$	0.02222	-0.01143
$(6.0)^2$	0.03073	-0.01690
$(10.0)^2$	0.03647	-0.01618
$(92.0)^2$	0.06024	-0.01665

References

[1] H.Burkhardt, F.Jegerlehner, G.Penso, C.Verzegnassi in Polarization at LEP, G.Alexander et al. Ed., CERN 88-06 Vol I, 145-157, subm. to Z.Phys.C

[2] W.F.L.Hollik DESY 88-188 and M. Böhm, W. Hollik, H. Spiesberger, Fortschr. Phys. 34 (1986) 687

[3] Review of Particle Properties, Phys.Lett.B204(1988)

[4] R.Baldini, private communication, W. Lockman, private communication

[5] J.L.Siegrist et al (MarkI),Phys.Rev.D26(1982)969

[6] H.Burkhardt, TASSO Note 192 (1981) and DESY F35-82-03 (1982)

[7] F.A.Berends and R.Kleiss, Nucl.Phys. B228(1983) 537

[8] S.Jadach and B.F.L. Ward, unpublished

[9] F.A.Berends, W.Hollik and R.Kleiss, Nucl.Phys B304(1988) 712

ONE-LOOP FLAVOUR-CHANGING NEUTRAL CURRENTS

IN THE STANDARD MODEL

A. Barroso

Departamento de Física, Universidade de Lisboa
Ed. C1, Piso 4, Campo Grande 1700 Lisboa, Portugal

Abstract : The flavour-changing amplitudes $\gamma q_i q_j$ and $Z q_i q_j$ are computed for off-shell particles and without approximations. The calculations are done in the 't Hooft-Feynman gauge, and the on-shell renormalization scheme is used. The results for on-shell quarks and for the gluon and photon vertices are compared with the ones derived using the usual approximation of neglecting the external quark masses. This approximation is shown to be unreliable.

INTRODUCTION

In the Glashow-Weinberg-Salam (GWS) model[1] the flavour-changing neutral currents (FCNC), which couple to the gluon, photon or the Z-boson, arise at one-loop level, as a result of quark mixing.

A small survey of this subject ought to be started with the classical paper of Glashow Illiopoulos and Maiani,[2] which has pointed out that FCNC are suppressed, roughly by terms of the order $\Delta m_l^2/M^2_w$, where Δm_l^2 stands for the largest difference of squared masses of the quarks inside the loop. After this important work, the study of FCNC was used to derive constraints for the mass of the charmed quark,[3,4] and for the mixing angles[5] in the Kobayashi-Maskawa (KM) matrix.

Up to the end of the seventies all calculations of the FCNC vertices were done using four approximations, namely: i) external quarks on-shell; ii) neglecting external quark masses; iii) setting the external quark momenta equal to zero; and, iv) considering only the leading terms in m_l/M_w. Then, in 1980 and 1981, Ma and Pramudita[6] and Inami and Lim[7] repeated these calculations removing the fourth approximation. In the following year, Deshpande et al.[8] gave the first exact calculation of the off-shell $\gamma q_i q_j$ vertex, followed, one year later, by a similar calculation of the $g q_i q_j$ vertex by Chia[9]. The corresponding results for the $Z q_i q_j$ vertex were obtained by me and one of my students and have just been published[10].

In this talk I briefly review our calculation and compare some of its results with the ones derived using the approximations that I have already mentioned. Before I start, let me point out that this is an important topic, not only for the study of B-meson physics, but it is also relevant for e^+e^--physics, where rare Z decays can give a top mass dependent rate (e.g. $Z \longrightarrow b\bar{s} + \bar{b}s$), or even a top production mechanism below the $t\bar{t}$-threshold (e.g. $Z \longrightarrow t\bar{c} + \bar{t}c$)[11].

FLAVOUR-CHANGING $g q_i q_j$, $\gamma q_i q_j$ AND $Z q_i q_j$ VERTICES

Our calculations were performed in the 't Hooft-Feynman gauge. For up-type quarks in the vertex, the matrix element V_{ji} represents $U_{jl}U^{\dagger}{}_{li}$ and, for down-type quarks, it represents $U^{\dagger}{}_{jl}U_{li}$, where U is the KM matrix. Summation over the internal quark index l is assumed. So, for the FC vertices and due to the unitarity of the matrix U, terms, where V_{ji} multiplies factors which are independent of the internal quark mass, are zero and will be dropped. Dimensional regularization is used to deal with divergent diagrams, and I use the notation

$$\zeta = 2/\varepsilon - \gamma + \ln(4\pi) - \ln(M_W/\mu)^2 , \tag{1}$$

where $\varepsilon = 4 - d$, and d stands for the dimension of momentum space, γ is the Euler constant and μ is some arbitrary mass.

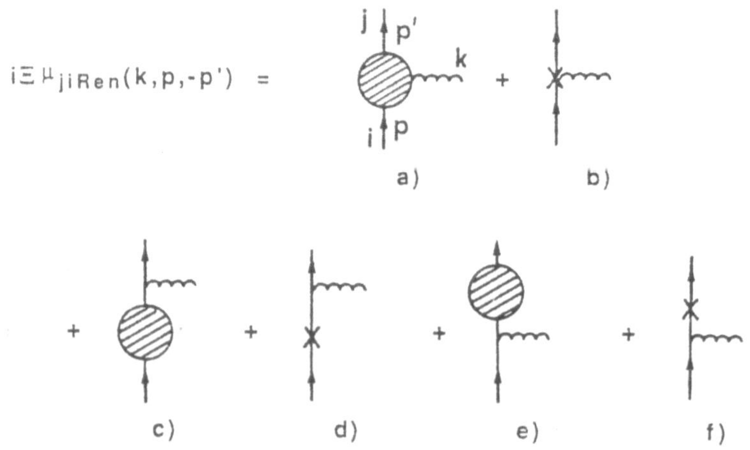

Fig. 1 - The one-loop FC effective vertex.

For any of the neutral gauge bosons, the corresponding renormalized flavour-changing vertex, $i\,\Xi^{\mu}_{jiRen}$, is represented in fig. 1. The diagrams a) and b) give the renormalized proper vertex, $i\Lambda^{\mu}_{jiRen}$, while the corrections to the external legs, $i\Omega^{\mu}_{jiRen}$, are represented by the diagrams c), d), e) and f). For these it is immediately to write :

$$\Omega^{\mu}_{jiRen} = X\,\frac{\not{p}+m_j}{p^2-m_j^2}\,(-\Sigma_{ji}(p)+\sigma_{ji}(p))\,,\tag{2}$$

and

$$\Omega^{\mu}_{jiRen} = (-\Sigma_{ji}(p')+\sigma_{ji}(p'))\,\frac{\not{p}'+m_i}{p'^2-m_i^2}\,X,\tag{3}$$

with

$$X = \begin{pmatrix} e \\ -g/\cos\theta_W \end{pmatrix}\left[\begin{pmatrix} Q \\ T_3-Q\sin^2\theta_W \end{pmatrix}\gamma^{\mu}\gamma_L + \begin{pmatrix} Q \\ -Q\sin^2\theta_W \end{pmatrix}\gamma^{\mu}\gamma_R\right],\tag{4}$$

where Q denotes the charge, in units of e>0, and T_3 is the third component of the weak isospin of the quarks in the external legs. In eq. (4) the upper line corresponds to the gamma vertex and the lower one to the Z vertex. The results for the gluon vertex are similar to the ones for γ replacing e by g_s and Q by the SU(3) generators $\lambda_a/2$. The flavour-changing self-energy and its counterterm are denoted Σ_{ji} and σ_{ji} respectively.

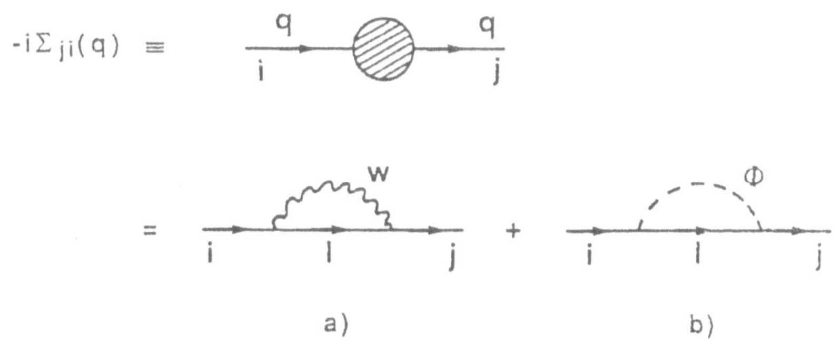

Fig. 2 - The one-loop FC quark self-energy.

At one-loop order, the unrenormalized FC quark self-energy, $-i\Sigma_{ji}$, shown in fig. 2, receives contributions from loops with charged gauge bosons, diagram a), and from loops with charged unphysical scalars, $\Phi^{\pm}$, diagram b). If q denotes the momentum of the external lines, m_l the mass of the internal quark and m_i and m_j the masses of the external quarks, the result is

$$\Sigma_{ji} = (g^2/16\,\pi^2)\,V_{ji}\,[D\,\slashed{q}\,\gamma_L + E\,\slashed{q}\,\gamma_R + F\,(m_i\,\gamma_R + m_j\,\gamma_L)]\,, \tag{5}$$

where D, E and F are the following functions of q^2 :

$$D = [\,1 + (m_l/M_W)^2/2\,]\int_0^1 dx\,(1-x)\,\ln\delta - \zeta\,(m_l/M_W)^2/4\,, \tag{6a}$$

$$E = (m_i m_j/M_W^2)/2\int_0^1 dx\,(1-x)\,\ln\delta\,, \tag{6b}$$

$$F = (m_l/M_W)^2/2\,[\,\zeta - \int_0^1 dx\,\ln\delta\,]\,, \tag{6c}$$

with

$$\delta = 1 - x + (m_l/M_W)^2\,x - (q^2/M^2_W)\,x\,(1-x)\,. \tag{7}$$

The divergent terms in D and F, proportional to the squared mass of the internal quark, are due to the coupling of the charged unphysical scalar, in diagram b). Those stemming from diagram a), which only contributes to the function D, cancel upon summation over the internal quark flavour.

The structure of the counterterm, $i\sigma_{ji}$, for the FC quark self-energy, depends only on the quark wave-function renormalization. In the free Lagrangian, the bare left-handed ($L^\circ_k = \gamma_L \Phi^\circ_k$) and right-handed ($R^\circ_k$) quark fields are scaled by a matrix in the flavour space, which deviates from the identity at order g^2, i. e.,

$$L^\circ_k = \sqrt{Z_{Lki}}\,L_i = (\,\delta_{ki} + 1/2\,\delta Z_{Lki}\,)\,L_i\,. \tag{8a}$$

$$R^\circ_k = \sqrt{Z_{Rki}}\,R_i = (\,\delta_{ki} + 1/2\,\delta Z_{Rki}\,)\,R_i\,. \tag{8b}$$

where L_i and R_i are the renormalized fields. In terms of the renormalized quantities, the Lagrangian acquires an extra term, $\overline{\Phi}_j\,\sigma_{ji}\,\Phi_i$, given by

$$\sigma_{ji} = a_{ji}\,\slashed{q}\,\gamma_L + b_{ji}\,\slashed{q}\,\gamma_R + c_{ji}\,\gamma_L + d_{ji}\,\gamma_R\,. \tag{9}$$

with

$$a_{ji} = 1/2 \, (\, \delta Z_L + \delta Z^{\dagger}_L \,)_{ji} \, , \tag{10a}$$

$$b_{ji} = 1/2 \, (\, \delta Z_R + \delta Z^{\dagger}_R \,)_{ji} \, , \tag{10b}$$

$$c_{ji} = - \, 1/2 \, (\, m_j \, \delta Z_L + m_i \, \delta Z^{\dagger}_R \,)_{ji} \, , \tag{10c}$$

$$d_{ji} = - \, 1/2 \, (\, m_i \, \delta Z^{\dagger}_L + m_j \, \delta Z_R \,)_{ji} \, . \tag{10d}$$

Notice that σ_{ji} obeys the relation

$$\sigma_{ji} = \gamma_0 \, \sigma^{\dagger}_{ji} \, \gamma_0^{\dagger} \, , \tag{11}$$

imposed by the hermiticity of the Lagrangian. Since we are only interested in the FC counterterm ($i \neq j$), I have omitted the mass renormalization constant $\delta m_k = m^0_k - m_k$. The renormalized FC quark self-energy, $-i \Sigma_{jiRen}$, is obtained adding the counterterm to the unrenormalized self-energy, i. e.,

$$- \, i \, \Sigma_{jiRen} = - \, i \, \Sigma_{ji} + i \, \sigma_{ji} \, . \tag{12}$$

The values of the constants a_{ji}, b_{ji}, c_{ji}, and d_{ji} of eq.(9) are now determined by the renormalization conditions imposed upon $-i \Sigma_{jiRen}$. In the on-shell renormalization scheme these conditions are

$$\overline{u}_j \, (q) \, \Sigma_{jiRen} \, (q) \, |_{q = m_j} = 0 \, , \tag{13a}$$

where no summation over j is implied, and u_j is a solution of the Dirac equation with mass m_j. Notice that eq.(13a) is only imposed upon the non-absorptive parts of the self-energy, i. e., only the real parts of D, E and F will appear in the renormalization condition. From eq.(13a), and using the property (11), it is easy to obtain

$$\Sigma_{jiRen} \, (q) \, u_i \, (q) \, |_{q = m_i} = 0 \, . \tag{13b}$$

Separating the terms in γ_L from those in γ_R, in eqs.(13a) and (13b), we obtained[10]

$$a_{ji} = C' \, a; \quad b_{ji} = C' \, b; \quad c_{ji} = C' \, c \quad \text{and} \quad d_{ji} = C' \, d \, , \tag{14}$$

with

213

$$a = [\, m^2_j\, D(m^2_j) - m^2_i\, D(m^2_i) + m_i\, m_j\, (E(m^2_j) - E(m^2_i))$$

$$+ (m^2_i + m^2_j)\, (F(m^2_j) - F(m^2_i))\,] \,/\, [\, m^2_j - m^2_i\,], \tag{15a}$$

$$b = [\, m^2_j\, E(m^2_j) - m^2_i\, E(m^2_i) + m_i\, m_j\, (D(m^2_j) - D(m^2_i))$$

$$+ 2\, m_i\, m_j\, (F(m^2_j) - F(m^2_i))\,] \,/\, [\, m^2_j - m^2_i\,], \tag{15b}$$

$$c = m_j\, [\, m^2_i\, (D(m^2_i) - D(m^2_j)) + m_i\, m_j\, (E(m^2_i) - E(m^2_j))$$

$$- 2\, m^2_i\, F(m^2_j) + (m^2_i + m^2_j)\, F(m^2_i)\,] \,/\, [\, m^2_j - m^2_i\,], \tag{15c}$$

$$d = m_i\, [\, m^2_j\, (D(m^2_i) - D(m^2_j)) + m_i\, m_j\, (E(m^2_i) - E(m^2_j))$$

$$+ 2\, m^2_j\, F(m^2_i) - (m^2_i + m^2_j)\, F(m^2_j)\,] \,/\, [\, m^2_j - m^2_i\,] \tag{15d}$$

and

$$C' = \frac{g^2}{16\,\pi^2}\, V_{ji}\,. \tag{16}$$

Using these results in eq.(12) it is easy to see that the renormalized self-energy is independent of ζ, and thus finite.

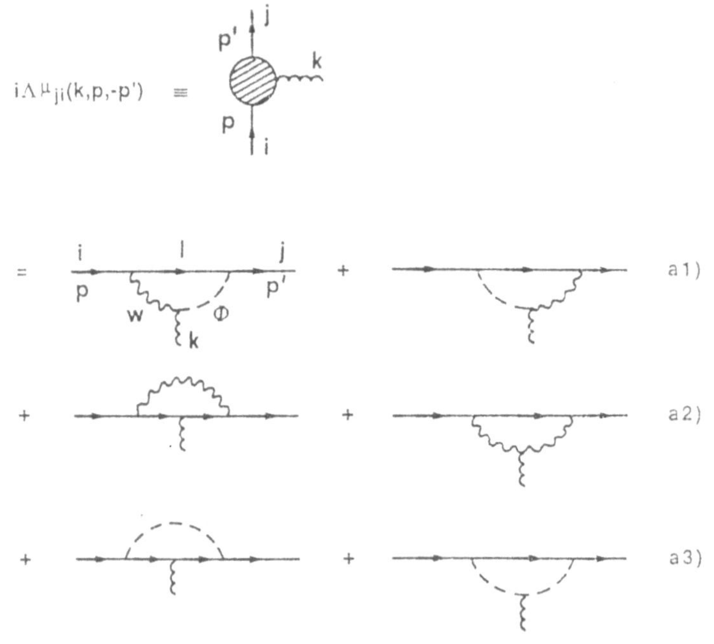

Fig. 3 - The one-loop FC proper vertex.

In fig.3, I show the one-loop diagrams that contribute to the proper vertex $i\Lambda^{\mu}_{ji}$. Since gluons do not couple to the W-boson, only two of these diagrams contribute to the gq_iq_j vertex. The diagrams where the neutral gauge boson couples to the internal quark are infinite: the divergent terms from the one with the W boson inside the loop cancel upon summation over the internal quark flavour, whereas those from the diagram with the charged unphysical scalar, $\Phi^{\pm}$, are proportional to m^2_i, and do not cancel. The divergent term is of the form $\gamma^{\mu}\gamma_L$. The renormalized proper vertex, $i\Lambda^{\mu}_{jiRen}$, is obtained adding the contribution from the counterterm, $i\lambda^{\mu}_{ji}$. The latter is generated following the same prescription as for the self-energy counterterm, i. e., in the neutral boson-quark interaction term in the Lagrangian, we perform a scaling of the bare fields and of the coupling constants. For the FC counterterm, at one-loop order, there are no contributions from the gluon, γ or Z field renormalization nor from the renormalization of the couplings, and the result is

$$i\,\lambda^{\mu}_{ji} = -i \left(\begin{matrix} e \\ -g/\cos\theta_W \end{matrix} \right) \left[\left(\begin{matrix} Q \\ T_3 - Q\sin^2\theta_W \end{matrix} \right) a_{ji}\, \gamma^{\mu}\,\gamma_L \right.$$

$$\left. + \left(\begin{matrix} Q \\ -Q\sin^2\theta_W \end{matrix} \right) b_{ji}\, \gamma^{\mu}\,\gamma_R \right], \tag{17}$$

where a_{ji} and b_{ji} are the same as in the self-energy counterterm. Notice that, among all the possible terms, $i\lambda^{\mu}_{ji}$ only includes those in $\gamma^{\mu}\gamma_L$ and in $\gamma^{\mu}\gamma_R$. Furthermore, their values are fixed without requiring any additional renormalization condition.

The relation between the counterterm for the proper vertex and the counterterm for the self-energy can be obtained explicitly, using the WST identities which we[10] have also derived. I won't explain the details here but, let me simply quote the relationships obtained for the counterterms, which are :

$$k_{\nu}\,\lambda^{\nu}_{ji}\,(k,p,-p') = -e\,Q\,[\,\sigma_{ji}(p') - \sigma_{ji}(p)\,] \tag{18a}$$

and

$$k_{\nu}\,\lambda^{\nu}_{ji}\,(k,p,-p') = i\,M_Z\,\lambda_{ji}\,(k,p,-p')$$

$$- g/\cos\theta_W\,[\,\sigma_{ji}(p')\,(Q\sin^2\theta_W - T_3\,\gamma_L) - (Q\sin^2\theta_W - T_3\,\gamma_R)\,\sigma_{ji}(p)\,] \tag{18b}$$

for the $\gamma q_i q_j$ and $Z q_i q_j$ vertices, respectively. In the previous equation, $i\lambda_{ji}$ is the counterterm for the $\Phi_Z q_i q_j$ FC proper vertex, namely:

$$i\lambda_{ji} = g\, T_3/M_W\, (d_{ji}\, \gamma_R - c_{ji}\, \gamma_L)\,. \tag{19}$$

It is interesting to point out that, the WST identities can be used to determine directly the value of the proper vertex counterterm without computing the self-energy. In fact, the renormalization conditions, eqs.(13a) and (13b), applied to the one-loop WST identities, give

$$k_\nu \Lambda^\nu_{jiRen}\,|_{\text{on-shell}} = 0 \tag{20a}$$

and

$$k_\nu \Lambda^\nu_{jiRen}\,|_{\text{on-shell}} = i\, M_Z\, \Lambda_{jiRen}\,|_{\text{on-shell}}\,, \tag{20b}$$

for the photon and Z vertices, respectively. From these equations one can obtain the value of the counterterms $i\lambda^\nu_{ji}$ and $i\lambda_{ji}$.

The final result for $i\,\Xi^\mu_{jiRen}$, can be written in the form

$$\Xi^\mu_{jiRen} = C\,\{\,[\,A_1\, \not{p}\, p^\mu + A_2\, \not{p}\, k^\mu + A_3\, \not{k}\, p^\mu + A_4\, \not{k}\, k^\mu + A_5\, \not{k}\, \gamma^\mu\, \not{p}$$
$$+ A_6\, \gamma^\mu + A_7\, k^2\, \gamma^\mu + A_8\, p.k\, \gamma^\mu + A_9\, p^2\, \gamma^\mu + A_{10}\, \gamma^\mu\, \not{p}$$
$$+ A_{11}\, \gamma^\mu\, \not{k} + A_{12}\, p^\mu + A_{13}\, k^\mu\,]\, \gamma_L + [\,A_r \rightarrow B_r\,]\, \gamma_R\,\}, \tag{21}$$

with

$$C = \left\{ \begin{array}{c} e \\ -g/\cos\theta_W \end{array} \right\} \frac{g^2}{16\,\pi^2}\, V_{ji}\,, \tag{22}$$

where the upper line corresponds to the $\gamma q_i q_j$ vertex, and the lower one to the $Z q_i q_j$ vertex. The incoming momenta corresponding to the neutral gauge boson and to the quarks, q_i and q_j, are denoted by k, p, and $-p' = -(k+p)$, respectively. The coefficients A_r and B_r (r=1,...,13) are scalar functions of the momenta of the external particles, and they depend on the masses and quantum numbers of the external and internal quarks. They are listed in table 1 of ref. 10 and there is no need to repeat them here. However, notice that, in the table, the entries to be multiplied by $(m_i/M_W)^2$ are the ones labelled $i\Lambda^\mu_{ji}(c)$ and not $i\Lambda^\mu_{ji}(a)$ as it is wrongly written in the table's caption.

ON-SHELL EFFECTIVE VERTEX

For on-shell quarks, the general form of $i\,\Xi^\mu_{jiRen}$ reduces to a sum of terms proportional to $k^\mu \gamma_{L,R}$, $p^\mu \gamma_{L,R}$ and $\gamma^\mu \gamma_{L,R}$. Alternatively, using Gordon's identities, one can write

$$\Xi^\mu_{jiRen}\big|_{\text{on-shell}} = C\,[\,(a_1 k^\mu + a_2 \gamma^\mu + a_3\, i\,\sigma^{\mu\nu} k_\nu)\, \gamma_L + (a_i \longrightarrow b_i)\,\gamma_R\,],\quad (23)$$

with

$$a_1 = m_i\,(B_2 - B_4) + m_j A_4 + A_{11} + A_{13} - m_j/2\,A_3$$

$$- 1/2\,A_{12} - m_i/2\,(B_1 - B_3 - 2\,B_5)\,,\qquad\qquad (24a)$$

$$a_2 = m_i\,m_j/2\,B_1 + (m^2_j - m^2_i)/2\,A_3 + m_j/2\,A_{12} + m^2_i/2\,A_1$$

$$+ m_j/2\,B_{12} + m^2_i\,A_9 + m_i\,B_{10} + A_6 + k^2\,A_7$$

$$+ (m^2_j - m^2_i - k^2)/2\,A_8\,,\qquad\qquad (24b)$$

$$a_3 = [\,-\,m_i\,(B_1 - B_3 - 2B_5) - m_j\,A_3 - 2\,A_{11} - A_{12}\,]\,/2\,,\qquad (24c)$$

and similar equations for b_i ($i=1,2,3$), but interchanging the A's and B's.

Since we are working in the on-shell renormalization scheme, the contributions from the renormalized corrections to the external legs are zero. Thus, only diagrams a) and b) in fig.1, i. e., the renormalized proper vertex, contribute to the on-shell effective vertex. However, evaluating all the diagrams in fig.1 for on-shell external quarks it is easy to see that the contribution of the unrenormalized diagrams is finite, and thus, the same must happen with the contribution from the counterterms. Moreover, this contribution, given by graphs b), d) and f), adds up to zero, which means that the effective vertex, for on-shell quarks is not renormalized, i. e., it can be obtained simply by summing up the one-loop diagrams a), c) and e), in fig.1. In fact, the counterterms for the corrections to the external legs, shown in graphs d) and f), are

$$\omega^\mu_{ji} = X\,\frac{\not p + m_j}{p^2 - m^2_j}\,\sigma_{ji}(p) + \sigma_{ji}(p')\,\frac{\not p' + m_i}{p'^2 - m^2_i}\,X\,.\qquad (25)$$

If we replace σ_{ji} by its expression, given in eq.(9), and add the contribution of the counterterm for the proper vertex, given by eq.(17), we obtain an expression in terms of the constants a_{ji}, b_{ji}, c_{ji} and d_{ji}. For on-shell quarks, this is easily seen to be zero, independently of the explicit values of those constants, which are fixed by the renormalization conditions. Hence, the cancellation of the counterterms follows directly from their structure, irrespective of the renormalization scheme which has been chosen. This means that such a result arises from the symmetry of the Lagrangian, and, in this sense, it is a fundamental result in the renormalization of the theory.

In the renormalization scheme that we are using, some counterterms are generated by renormalizing the fields, with the sole purpose of absorbing the divergencies in all one-particle-irreducible (1PI) Green functions. Such is the origin of the $i\lambda^v_{ji}$ counterterm that renormalizes the FC proper vertex, and of the counterterm $i\sigma_{ji}$ for the FC self-energy. Field renormalization is required for a formal proof of the renormalizability, based on the analysis of the generating functional of the 1PI Green functions (e. g. ref. 12). However, it can be ignored, if we are simply interested in the calculation of S-matrix elements. In fact, the Green functions may remain divergent, provided that the S-matrix elements, which are the physically meaningful predictions of a field theory, are finite. Such a renormalization programme has been proposed by Marciano and Sirlin[13]. There, the counterterms are generated only from the renormalization of the parameters, which is necessary for the consistency of the theory, and there is no renormalization of the fields. So, if the diagonalization of the mass matrices, both in the quark sector and in the neutral gauge boson sector, is achieved rotating the fields through bare mixing angles, off-diagonal counterterms do not arise,[14,15] and the counterterm Lagrangian has the same structure as the tree level one. In particular, counterterms for both the FC quark self-energies and the FC vertices do not exist. Hence, the previous result now follows trivially.

APPLICATION TO b —> s γ AND b —> s g

Let me apply the previous results to the transitions b —> s γ and b —> s g. It is clear that the transition amplitudes can be written in the form

$$T = \left\{ \begin{array}{l} e\,\bar{u}(p')\,\Gamma^\mu\,u(p) \\[2ex] g_s\,\bar{u}(p')\,\dfrac{\lambda_a}{2}\,\Gamma^\mu\,u(p) \end{array} \right. \tag{26}$$

where

$$\Gamma^\mu = \frac{G_F}{2\sqrt{2}\,\pi^2} \sum_l U^\dagger_{sl} U_{lb} [(k^2 \gamma^\mu - \slashed{k}\,k^\mu)$$

$$(F_{1L}\,\gamma_L + F_{1R}\,\gamma_R) - i\,\sigma^{\mu\nu} k_\nu M_W (F_{2L}\,\gamma_L + F_{2R}\,\gamma_R)] \tag{27}$$

and $F_{iL,R}$ are dimensionless form factors which are different for the gluon and photon emissions. I am considering that the quarks are on-shell and I have re-expressed the previous result in such a way that the validity of the on-shell WST identity is obvious. In fact, from eq. (23) and noting that, now, the momentum k is outgoing, it is easy to obtain the relations

$$F_{2L} = a_3/M_W \quad ; \qquad\qquad F_{2R} = b_3/M_W \tag{28a}$$

$$F_{1L} = \frac{M_W^2}{m_i^2 - m_j^2}(m_j a_1 + m_i b_1) \tag{28b}$$

$$F_{1R} = \frac{M_W^2}{m_i^2 - m_j^2}(m_i a_1 + m_j b_1) \tag{28c}$$

Hermiticity of the effective current implies

$$a_1(m_i, m_j) = -b^*_1(m_j, m_i) \tag{29a}$$

$$a_3(m_i, m_j) = b^*_3(m_j, m_i) . \tag{29b}$$

On the other hand, looking at eqs (24) and using table 1 of our paper[10], it is simple to conclude that, to a very good approximation, a_1 and a_3 are directly proportional to m_j while b_1 and b_3 are proportional to m_i. This observation and the use of relations (29), neglecting the small imaginary part due to the phase in the KM matrix, leads to the conclusion :

$$F_{2L} \approx \frac{m_s}{M_W} F_2 \quad ; \qquad\qquad F_{2R} \approx \frac{m_b}{M_W} F_2 \tag{30a}$$

$$F_{1L} \approx \frac{M_W^2}{m_b^2 - m_s^2}(m_s^2 X - m_b^2 X) \equiv F_1 \tag{30b}$$

and

$$F_{1R} \approx \frac{M_W^2}{m_b^2 - m_s^2} (m_s m_b X - m_s m_b X) = 0 .$$ (30c)

Within this approximation, there are two form factors : a charge form factor, F_1, and a dipole form factor F_2. I have checked that the ratio F_{2R}/F_{2L} differs from m_b/m_s by 2% for a light quark in the loop ($m_l = 0.1$ GeV) and the accuracy of the approximation increases with m_l. On the other hand, F_{1R}/F_{1L} is of the order 10^{-4} for a light quark and decreases by two orders of magnitude when m_l goes from 1 GeV to 200 GeV.

In fig. 4 the full curves show the F_1 form factors as a function of the mass of the quark in the loop. The dashed curve corresponds to the gluon form factor and

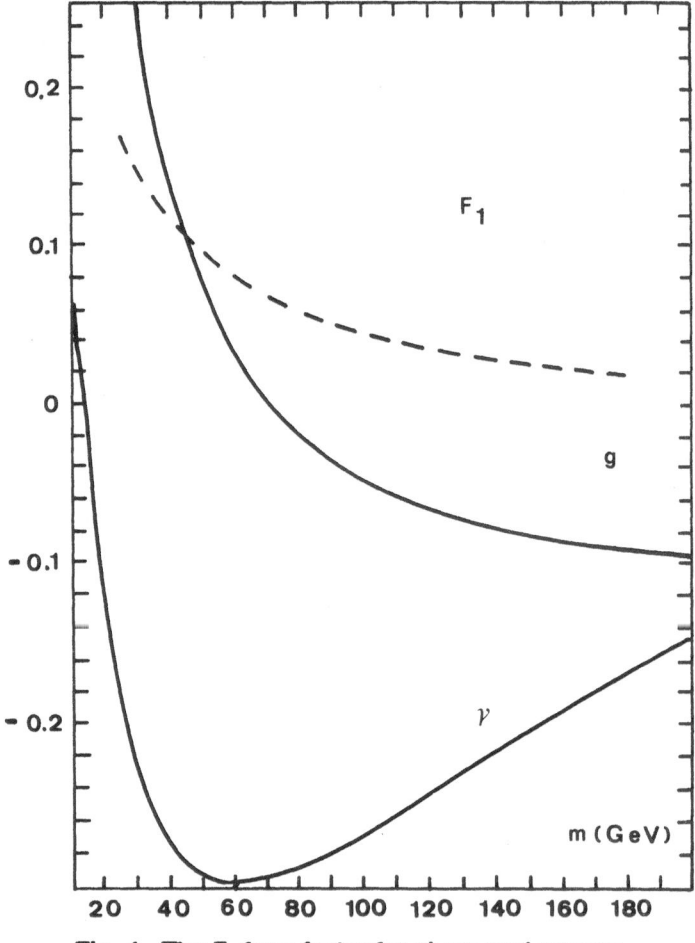

Fig. 4 - The F_1 form factor for gluon and gamma.

was obtained by Hou[16] "ignoring external masses and momenta whenever possible". The comparison with our exact result is puzzling. This is even more so for the F_2 form factor displayed in fig. 5. Again, the dashed curve is due to Hou and was derived using the same approximation.

If $m_b = m_s$ the terms in γ_5 cancel and the F_2 form factor gives, at $k^2 = 0$, a contribution to the magnetic moment. This ought to be finite even for a massless fermion in the loop as it is indeed in the case of the electron magnetic moment. It is clear that our exact results have this behaviour.

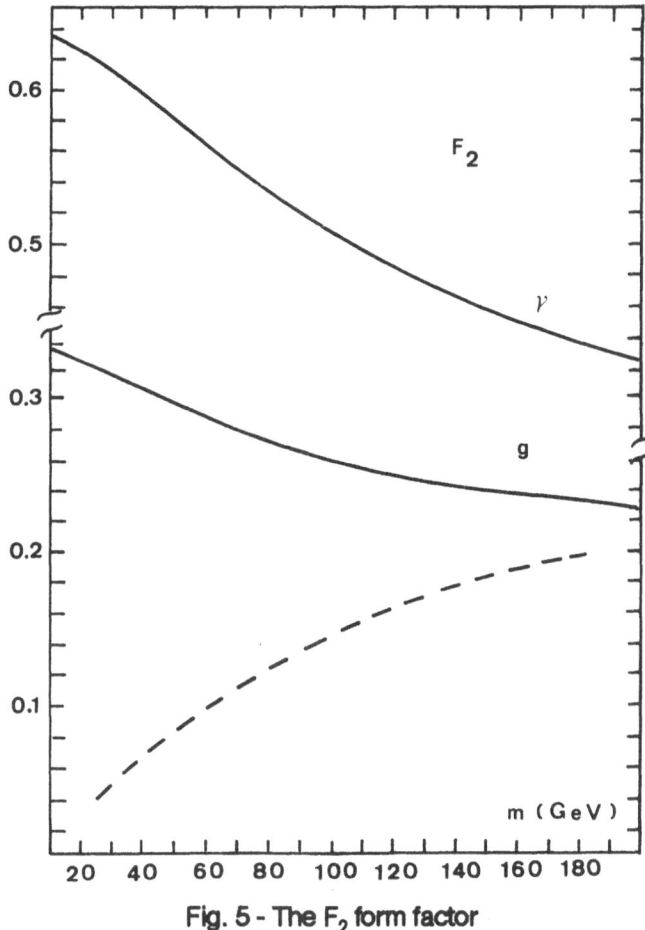

Fig. 5 - The F_2 form factor

Let me now see why the approximation fails so badly. If we return to eqs. (24), set $m_i = m_j = 0$ and use the values of A_i and B_i given in ref.10, the result is :

$$\Xi^{\mu}_{jiRen|on\text{-}shell} = e \frac{g^2}{16\pi^2} \sum_I V_{ji} (-\frac{1}{2} Q) \gamma^{\mu} \gamma_L = 0 . \tag{31}$$

This means that, ignoring, "whenever possible", the external quark masses, before evaluating the diagrams, is not a consistent approximation. Because it is so widely used I would like to emphasize this conclusion.

Perhaps, this could have been a good place to end this talk. However, allow me a few more minutes to show one application of the form factors just derived. From eq. (27) it is straightforward to derive the width for the b —> s γ transition, i. e.,

$$\Gamma(b \longrightarrow s\ \gamma) = \frac{G_F^2 m_b^5}{192\ \pi^3}\ \frac{6\ \alpha}{\pi}\ \left|\ \sum_i U_{is}^* U_{ib} F_2(m_i)\ \right|^2. \tag{32}$$

Then, the branching ratio, Br[B —> K γ], is

$$Br\ [B \longrightarrow k\ \gamma] = \frac{\Gamma(b \longrightarrow s\ \gamma)}{\Gamma(b \longrightarrow (u,c)\ l\nu)}\ Br\ [B \longrightarrow l\nu\ X]$$

$$= \frac{\frac{6\ \alpha}{\pi}\ \left|\ \sum_i U_{is}^* U_{ib} F_2(m_i)\ \right|^2}{[\ U_{ub}^2 + 0.5\ U_{cb}^2\ ] \times 0.88}\ 0.12 \tag{33}$$

and this I plot in fig. 6 as a function of the top quark mass. As one can see, the branching ratio rises from 10^{-4} to 10^{-5} when m_t varies from 40 GeV till 200 GeV. In the previous result, 0.5 is a phase space factor and 0.88 is a QCD correction factor to semileptonic B decay[16]. Large QCD corrections to the b —> s γ decay have also been discussed (e. g. ref. 17) but these are not my present concern.

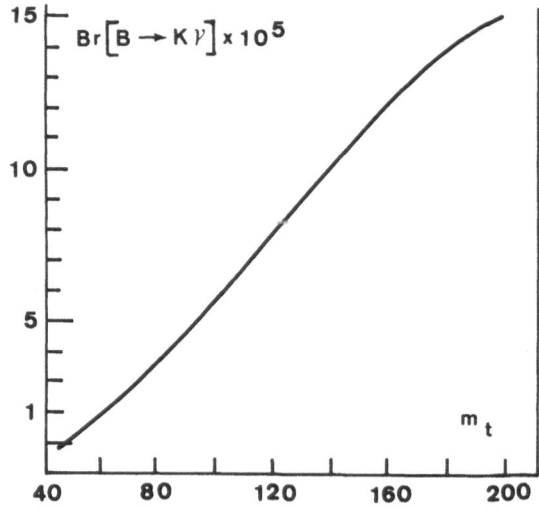

Fig. 6 - The branching ratio B —> K γ as a function of m_t.

To conclude, let me advertise our exact one-loop amplitudes for the flavour-changing Zq_iq_j and γq_iq_j effective vertices, for off-shell external particles, in the HF gauge. Its use in several FC processes should be very easy and the results can be substantially different from those derived neglecting the masses and the momenta of the external quarks.

REFERENCES

1. S. L. Glashow, Nucl. Phys. **22** (1961) 579.
   S. Weinberg, Phys. Rev. Lett. **19** (1967) 1264.
   A. Salam, in "Elementary Particle Theory: Relativistic Groups and Analyticity", Nobel Symposium nº8 N. Svartholm ed. (Almqvist and Wiksell, Stockholm, 1968), p.367.
2. S. L. Glashow, J. Illiopoulos, L. Maiani, Phys. Rev. **D2** (1970) 1285.
3. M. K. Gaillard, B. W. Lee, Phys. Rev. **D10** (1974) 897.
   M. K. Gaillard, B. W. Lee, R. E. Shrock, Phys. Rev. **D13** (1976)2674.
4. A. I. Vainshtein, I. B. Khriplovich, JETP Letters **18** (1973) 83.
   E. Ma, Phys. Rev. **D9** (1974) 3103.
5. R. E. Shrock, M. B. Voloshin, Phys. Lett. **87B** (1979) 375.
6. E. Ma and A. Pramudita, Phys. Rev. **D22** (1980) 214.
7. T. Inami and C. S. Lim, Progr. Theor. Phys. **65** (1981) 297.
8. N. G. Deshpande and G. Eilam, Phys. Rev. **D26** (1982) 2463.
   N. G. Deshpande and M. Nazerimonfared, Nucl. Phys. **B213** (1983) 390.
9. S. P. Chia, Phys. Lett. **130B** (1983) 315.
10. J. M. Soares and A. Barroso, Phys. Rev.**D39** (1989) 1973.
11. A. Axelrod, Nucl. Phys. **B209** (1982) 349.
    K. Hikasa, Phys. Lett. **149B** (1984) 221.
    G. Eilam, Phys. Rev. **D28** (1983) 1202.
12. C. Itzykson and J. B. Zuber, "Quantum Field Theory", McGraw-Hill, USA, 1980.
13. W. J. Marciano and A. Sirlin, Nucl. Phys. **B39** (1975) 303.
14. A. Sirlin, Phys. Rev. **D22** (1980) 971.
15. J. M. Soares, "The renormalization of the flavor-changing neutral currents", Master's thesis, Universidade de Lisboa, 1988.
16. W.-S. Hou, Nucl. Phys. **B308** (1988) 561.
17. B. Grinstein, R Springer and M. Wise, Phys. Lett **202B** (1988) 138.

GENERALIZED RADIATIVE CORRECTIONS FOR HADRONIC TARGETS

C. de Calan[a], H. Navelet[b] and J. Picard[c]

a) Centre de Physique Théorique, Ecole Polytechnique
   PALAISEAU, 91128-F
b) DPT, CEA-IRF, CEN-Saclay, GIF-SUR-YVETTE, 91191-France
c) DPHN-HE, CEA-IRF, CEN-Saclay, GIF-SUR-YVETTE, 91191-France

INTRODUCTION

After the pionnering work of Schwinger in the '40s for potential scattering, the most elaborate and widely used radiative correction procedure for electro-magnetic scattering on hadronic targets remains that given in the '60s by Y.S. Tsai[1] for structureless spin zero bosons. However, with the enhanced accuracy of new experimental data - the error is currently down to 1-2 % on electron scattering cross sections -, and the improvements in theoretical models and computing power, some of the approximations acceptable then may be questioned today. The error on the computed radiative correction due to these approximations is difficult to estimate because "[Publications] present only the result and very few intermediary steps of the calculation" as Källén and Sabry remarked sadly 35 years ago[2], let alone unstated approximations, overlooked divergences and *ad hoc* compensations "guessed on intuitive physical grounds". Moreover, when the target is scattered into several components, a procedure to radiatively correct coincidence experiments is still wanting.

The generalization of radiative correction to many particle processes and coincidence experiments was first given by Yennie, Frautschi and Suura[3]. Unfortunately, its implementation by Meister & Yennie[4] holds flawed approximations which were shown by Mo and Tsai[5] to give absurd results for heavy, high Z, targets. Another procedure to correct inelastic electron scattering with coincidence detection by de Calan and Fuchs[6] also leads to wrong results for high Z targets.

Considering all the pitfalls encountered by our forerunners, we decided to tackle the whole problem anew. The detailed account of our calculation will be given in a forthcoming report readable, we hope, by experimentalists as well. Here, instead, we will just outline what we did, comment on our differences with previous works and compare the different results.

GENERAL FRAMEWORK AND CALCULATION

With the availability and speed of today's computers we did not, as Tsai, MY[4] and CF[6] had to do in the '60s, neglect terms of order unity or non logarithmic terms. Actually, besides assuming a minimal electro-magnetic interaction (structureless particles) and a constant Born amplitude, no approximation was made except in the calculation of the relatively small spin contribution to which we will return. A careful and thorough calculation of the Feynman integrals seems to be essential to obtain correct results whatever the target nucleus and the particle energies.

Using the charge conservation hypothesis, the radiative correction to a given process involving n particles can be reduced[3] to a sum of radiative corrections to all $n(n-1)/2$ participating pairs (i,j). The total radiative correction δ to order α ($\simeq 1/137$) is written:

$$\delta = VP + \Sigma_{i<j} (\hat{B}_{ij} + ReB_{ij} + S_{ij}),$$

where VP is the vacuum polarisation contribution to be included when the Born amplitude involves a photon exchange. The indexes i,j run of course over all charged particles exclusively. $\hat{B}_{ij}$ and B_{ij} are the sum of all real and virtual processes respectively, and S_{ij} the spin-convection[4] contribution which we calculated for spin 1/2 particles only.

More specifically:
i) Assuming charge renormalisation, the vacuum polarisation term VP - only the closed electron loop is significant here[1] - is readily taken off the shelf (from Jauch & Rohrlich, Theory of Photons and Electrons).
ii) The real processes contribution $\hat{B}_{ij}$ is a sum of three integrals of the type:

$$\hat{I}_{ij} = \int_{-1}^{+1} dx \int_{\Omega} d^3k \; \frac{1}{k^o} \frac{1}{[k\,p_x]^2}.$$

The domain Ω, such that $|\vec{k}| \le k_m$ the maximum energy of the bremsstrahlung photon, is most conveniently expressed in the Lorentz frame where $\vec{k}$ is isotropic; and $p_x = \frac{1+x}{2}p_i + \frac{1-x}{2}p_j$ where $p_{i,j}$ are the four-vectors of the pair considered.

iii) The virtual processes term B_{ij} of which only the real part ReB_{ij} will be relevant, is a sum of three of each integrals:

$$J_{ij} = \int \frac{d^4k}{(k^2 - 2kp_i\theta_i + i\varepsilon)(k^2 + 2kp_j\theta_j + i\varepsilon)}$$

and

$$I_{ij} = \int \frac{d^4k}{k^2 + i\varepsilon} \frac{1}{(k^2 - 2kp_i\theta_i + i\varepsilon)(k^2 + 2kp_j\theta_j + i\varepsilon)}$$

where $\theta = \mp 1$ if the particle enters or leaves the Feynman graph describing the reaction. The three ultra-violet divergences from the J_{ij} cancel in the B_{ij} sum. The integral parametrisation of the integrands leads to $x \in [-1,+1]$ integrals on $p'_x = \frac{1+x}{2}p_i\theta_i - \frac{1-x}{2}p_j\theta_j$, where p'^2_x has two zeros within the integration interval $[-1,+1]$ only when $\theta_i = \theta_j$.

Although $\hat{B}_{ij}$ and B_{ij} are separately infrared divergent and must be handled with care, their sum is finite[3]. The finite part of the corresponding integrals were first calculated as in YFS, i.e. giving a fictitious mass to the photon; then using the dimensional regularisation method to check an unsuspected singularity to be discussed shortly; and, finally, each analytical result was cross-checked by numerical computation of the integrals using CERN standard DGAUSS and CAUCHY's routines.

The calculation of the finite part from $\hat{I}_{ij}$, is straightforward - albeit quite tedious - and its contribution numerically the most important. Incidentally, the result given by Mork and Olsen[7] is wrong by a factor two, and the "correcting" footnote by Maximon[8] compounds several errors or misprints.

The infrared finite sum of integrals J_{ij} is much simpler. The values of the finite parts from the J_{ij} are somewhat smaller than those of I_{ij} but, although they were altogether neglected by Tsai[1], they are still relevant to the accuracy we are aiming at.

Extracting the finite part from I_{ij} gives no major problem except for an extra singularity when $\theta_i = \theta_j$ which introduces an additional term

$\pi^2 p_i p_j [(p_i p_j)^2 - p_i^2 p_j^2]^{-1/2}$, (ref. 3, C-15), particularly troublesome for a low energy pair. This singularity was probably overlooked by Tsai. Fortunately, as can be shown using Coulomb wavefunctions and will be shown on more general grounds in a forthcoming paper[9], this singularity contributes to naught.

iv) In addition to the above so called[4] convection-convection, i.e. spin-independent, term and main contribution to the radiative correction, the spins introduce small additional terms that we estimated for spin 1/2 particles. The only significant one is the spin-convection interference term, S_{ij}, for the calculation of which the commutator $[\not{p}_i, \not{p}_j]$ must be factorized somehow. This commutator we approximated by $p_i p_j - m_i m_j$, where $m_{i,j}$ are the particle masses, instead of $p_i p_j$ as MY[4] did which is again particularly poor for a low energy pair. The spin-spin term could lead to new UV divergences. However it does not for pure QED where it is negligible, and was neglected here for all cases.

RESULTS

Table 1 gives two examples of radiative corrections to electron elastic scattering on Hydrogen and Calcium targets for conditions described in the caption. The results are broken down into partial contributions from the Z^0, Z^1 and Z^2 terms where Z is the charge of the target and recoil particles. To be consistent with MT[5], all comparisons are to be made excluding the spin contribution. Our difference with Mo & Tsai is mainly due to the Z^0 term and, to a lesser extent, to the Z^1 term. This difference decreases when the spins are taken into account but remains significant. The Z^2 coefficient is obviously wrong in MY[4], particularly so for the Calcium target.

In Table 2 our radiative correction results for the inelastic reactions $p(e,e'\pi^+)n$ and $^{208}Pb(e,e'p)^{207}Tl$ are compared to those without spin effects of de Calan & Fuchs[6]. As for Meister and Yennie, their calculation does not fare too badly for the light target, but utterly fails for the heavy one.

Figures 1 and 2 show the behaviour of radiative corrections to the $^{48}Ca(e,e')^{48}Ca$ and $p(e,e'\pi^+)n$ reactions for different scattering conditions (see captions) on a wide range of incident energies. Our results with spins (continuous curves) or without spins (dashed curves) are compared respectively to the ones from MT[5] and CF[6]. Users will bear in

Table 1. Radiative corrections to electron elastic scattering

		Z^0	Z^1	Z^2	Total
	(no spin	-.2555	-.0231	-.0024	-.2809
(1)	(with spin	-.2227	-.0227	-.0019	-.2473
	(MT[5]	-.2138	-.0035	-.0022	-.2195
	(MY[4]	-.2138	-.0034	+.0008	-.2164
	(no spin	-.2515	-.0336	-.0007	-.2858
(2)	(with spin	-.2195	-.0313	-.0007	-.2515
	(MT[5]	-.2109	-.0054	-.0007	-.2170
	(MY[4]	-.2109	-.0051	+.9654	+.7494

Table 1. (1) refers to the reaction $p(e,e')p$ at incident energy $E_e =$ 2.206 Gev, detection angle $\theta_{e'} = 16$ degrees and detected energy resolution $\Delta E_{e'} = .01 E_{e'}$. (2) refers to $^{48}Ca(e,e')^{48}Ca$ at $E_e = .5$ GeV, $\theta_{e'} = 60$ degrees and $\Delta E_{e'} = .01 E_{e'}$. Z^0 is the correction term from the (e,e') pair, Z^2 the one from the (target, recoil) pair and Z^1 the contribution from all remaining pairs.

Table 2. Radiative corrections to electron inelastic scattering

		(e,e')	(others)	Total
	(no spin	-.2176	-.0512	-.2688
(1)	(with spin	-.1833	-.0474	-.2306
	(CF[6]	-.2498	-.0522	-.3020
	(no spin	-.1125	-.0017	-.1142
(2)	(with spin	-.0829	-.0593	-.1422
	(CF[6]	-.0845	5.4327	5.3482

Table 2. (1) refers to the reaction $p(e,e'\pi^+)n$ at incident energy $E_e = 1$ GeV, detected electron energy $E_{e'} = E_e/2$, pion detection angle $\theta_\pi = 90$ degrees, and energy resolutions $\Delta E_{e'} = .01 E_{e'}$ and $\Delta E_\pi = .03 E_\pi$. (2) refers to $^{208}Pb(e,e'p)^{207}Tl$ at $E_e = 1$ GeV, $E_{e'} = E_e/3$, $\theta_{e'} = 30$ degrees, $\Delta E_{e'} = .01 E_{e'}$ and $\Delta E_p = .03 E_p$. Under the (e,e') headline are the contributions from this pair only, whereas the contributions from the other pairs are compounded in the second column.

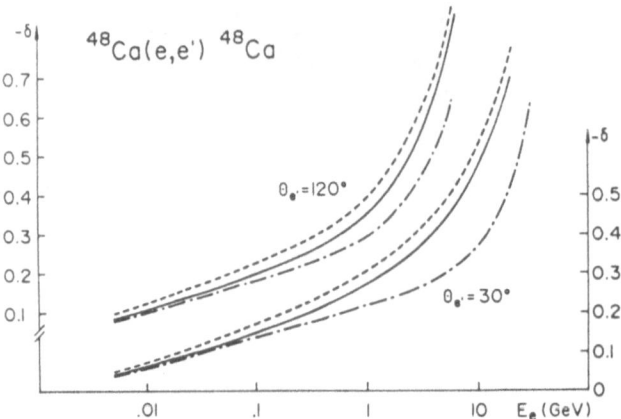

Fig. 1. The radiative correction to electron elastic scattering on ^{48}Ca is shown for a range of incident electron energies E_e at two scattering angles $\theta_{e'}$= 30 and 120 degrees, read respectively on the right and left scale. Both have detected energy resolution $\Delta E_{e'}$=.01$E_{e'}$. Our results with spins (continuous curves) and without spins (dashed curves) are compared to those from Mo & Tsai[5] (dot-dash curves).

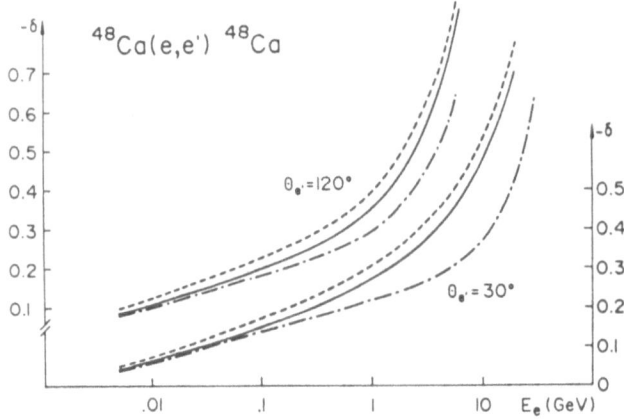

Fig. 2. The radiative correction to the reaction $p(e,e'\pi')n$ is shown for electron incident energies between .350 and 9 GeV with conditions i) $E_{e'}$= E_e/2 and θ_π= 20 degrees, on the right scale and ii) $E_{e'}$= E_e/3 and θ_π= 90 degrees, on the left scale; both have detected energy resolutions $\Delta E_{e'}$= .01$E_{e'}$ and ΔE_π = .03E_π. Drawing conventions are the same as in Fig. 1 where the dot-dash curves are now those of de Calan & Fuchs[6].

mind that the radiative correction δ remains sensible only inasmuch as $e^{\delta} \propto$ 1+δ, i.e. $\delta \leq .4$, beyond which higher orders of the cross section should be included.

CONCLUSION

Y.S. Tsai neglected the spin effects and his results differs from ours, considered without the spin-convection term S_{ij}, by some 40 % for 1 GeV electrons scattering at 30 degrees on ^{48}Ca (Fig. 1). This difference has no single obvious source and seems to be due only to its various approximations. The mathematical reduction of one result to the other could hopefully be done but would probably require as much work as was needed for each original calculation, due to the large number of functional relationships between dilogarithms (Spence functions).

The significant difference between our results and previous ones vindicate the effort put into a careful reappraisal of this time-honored problem plaguing charged particle scattering analysis. Furthermore, it may call for reconsideration of nuclear charge radii obtained through electron scattering with poor radiative correction.

ACKNOWLEDGEMENTS

We are grateful to Professor R. Balian for his interest in this old problem and to M. Bergère, J. Bros, D. Iagolnitzer, J. Lascoux and P. Vernin for helpful discussions throughout the course of this work.

References

1) Y.S. Tsai, Phys. Rev. 120(1960)269, 122(1961)1898 and SLAC-PUB-848, 1971.
2) G. Källén and A. Sabry, Dan. Mat. Fys. Medd., 29-17, 1955.
3) D.R. Yennie, S.C. Frautschi and H. Suura, Ann. of Phys. 13(1961)379.
4) N. Meister and D.R. Yennie, Phys. Rev. 130(1963)1210.
5) L.W. Mo and Y.S. Tsai, Rev. Mod. Phys. 41(1969)205.
6) C. de Calan and G. Fuchs, Il Nuovo Cim. 38(1965)1594 and 41(1966)A286.
7) K. Mork and H. Olsen, Phys. Rev. B140(1965)1661.
8) L.C. Maximon, Rev. Mod. Phys. 41(1969)193.
9) J. Bros and H. Navelet, in preparation.

REPORT ON WORKING GROUP A: RENORMALIZATION SCHEMES FOR ELECTRO-WEAK RADIATIVE CORRECTIONS

M. Böhm

Physikalisches Institut der Universität Würzburg
Am Hubland, D-8700 Würzburg, Federal Republic of Germany

Members of the working group

W.F.L. Hollik, M. Igarashi, F. Jegerlehner,
T.Riemann, B.F.L.Ward

Abstract

In this report we discuss questions which are connected with the renormalization of the electroweak standard model. Renormalization schemes are classified with respect to the physical renormalization conditions and parameters, the extent of explicit renormalization and other criteria. Four-fermion reactions are used as an illustrative example. The transformations between renormalization schemes are explained and the inclusion of higher order corrections is discussed.

1 Introduction

Predictions of the standard model of the electroweak interactions including radiative corrections now exist for many reactions. These results have been obtained in different renormalization schemes making use of different sets of parameters, gauges , renormalization conditions, field renormalizations.

Therefore the question of renormalization scheme dependence has to be adressed and indeed was discussed in the contributions [1] to this working group. In this report some remarks and conclusions on these topics will be added. Hopefully they help to clarify some points which have caused some misunderstanding and troubles in the past.

Radiative Corrections, Edited by N. Dombey and
F. Boudjema, Plenum Press, New York, 1990

2 Why and how renormalization

In order to discuss the meaning of renormalization let us begin with a simple situation: a theory described by a dynamical variable φ (e.g. a particle coordinate, a field), some parameter g and a Lagrangian $\mathcal{L}(\varphi, g)$. For the moment we also assume that we are able to calculate exactly some measurable quantities $\sigma_1(g), \ldots, \sigma_N(g)$. From an exact measurement M_1 of σ_1 it is possible to fix g and thus to obtain numerical predictions for $\sigma_2, \ldots, \sigma_N$. These can be compared to the corresponding measured quantities $M_2, \ldots, M_N$ yielding $N - 1$ tests of the theory. This is an admittedly overidealized description of the way physics is performed.

Now let us consider a reparametrization of the theory. We introduce a new parameter g_R by putting $g = g(g_R)$ then $\sigma_i = \sigma_i(g(g_{(R)})) = \sigma_i'(g_R)$. Under this reparametrization which may be extended to $\varphi = \varphi(\varphi_R)$ we have complete equivalence of relations between physical results. In this sense there is strict reparametrization independence of the exact calculations.

This is no longer the case if we perform approximate evaluations of the theory. Then the (φ, g)-scheme is in general not equivalent to the (φ_R, g_R)-scheme, the two may differ in higher order terms. As a consequence in approximate calculations there is no strict reparametrization invariance.

The case of relativistic quantum field theories needs some further refinement. Since higher order calculations may lead to divergent expressions, a regularization scheme like a momentum cutoff, Pauli-Villars or dimensional regularization is required. All these regularization schemes introduce a parameter Δ, where $\Delta \to \infty$ corresponds to vanishing regularization. As a consequence we have $\sigma_i = \sigma_i(g, \Delta)$ and if we determine g via σ_1 from M_1: $g = \sigma_1^{(-1)}(M_1, \Delta)$ we obtain a value for the "bare" coupling g which is cutoff dependent and therefore has no direct physical meaning. Inserting this in $\sigma_2(g, \Delta) = \sigma_2(\sigma_1^{(-1)}(M_1, \Delta), \Delta)$ gives a σ_2, which is independent of Δ. As a result bare quantities are regularization dependent, the relation between physical quantities is not. Let us clearly state here that it is possible to work with the bare quantities but that they lack a simple physical interpretation.

A convenient way to overcome this problem is to perform a reparametrization $g = g(g_R, \Delta)$, called renormalization, such that the renormalized g_R is finite in the limit $\Delta \to \infty$ and such that it has a simple physical meaning. We can do this also for the fields $\varphi = \varphi_R \cdot Z_\varphi^{-1/2}(\Delta, g)$ such that φ_R is finite and has a convenient normalization. The renormalization transformation $g \to g_R$, $\varphi \to \varphi_R$ has to fulfill two conditions: it must absorb the divergences and has to define the meaning of physical parameters. These things together define a Renormalization Scheme. As observed above for the general case exact results are RS independent, approximate results are not.

2.1 General classification of renormalization schemes

Our general discussion of renormalization in quantum field theories has shown that the freedom of reparametrization of the theory may be used to introduce convenient parameters and to absorb the divergences in the calculation of S-matrix elements or Green functions. A possible classification of RSs is based on the extent of these reparametrizations:

1. A first step consists in introducing suitable parameters and fields in the classical Lagrangian. This step is usually performed in the electroweak standard model. The original Lagrangian, which can be constructed from the gauge symmetry $SU(2)_L \times U(1)_Y$ with the gauge fields W_μ^a, B_μ, the complex scalar field Φ and left-handed fermion doublets Ψ_L, right-handed fermion singlets Ψ_R, contains the gauge couplings g_2, g_1, the bilinear and quadrilinear Φ couplings μ^2, λ and the Yukawa couplings g_y:

$$\mathcal{L}(W_\mu^a, B_\mu, \Phi, \psi_L, \psi_R; g_2, g_1, \mu^2, \lambda; g_y). \tag{1}$$

Choosing the unitary gauge and introducing the photon field A_μ, the charged fields $W_\mu^\pm$, and the Z-boson field Z_μ^0 with masses $M_W; M_Z$, the electric charge e, the physical Higgs field η, the Higgs mass M_H and the fermion masses m_f the physical content of the standard model can be directly seen. The corresponding Lagrangian

$$\mathcal{L}(W_\mu^\pm, A_\mu, Z_\mu^0; \eta; \psi_L, \psi_R; e, M_W, M_Z, M_H, m_f) \tag{2}$$

is indeed the starting point for many calculations of physical processes. Other convenient parameters are the μ-decay constant G_F and the weak mixing angle s_W^2. They are related to the set above in lowest order in the following way:

$$
\begin{aligned}
G_F &= \frac{e^2}{4\sqrt{2} M_W^2 (1 - \frac{M_W^2}{M_Z^2})}, \\
s_W^2 &= 1 - \frac{M_W^2}{M_Z^2}.
\end{aligned}
\tag{3}
$$

They can replace M_W or/and M_Z. Therefore RSs for the electroweak standard model may in a first step be distinguished by the parameters and fields which are used for the formulation of the theory.

As we have mentioned above it is possible to work with bare parameters and fields i.e. without renormalization. This avoids the explicit construction of a renormalization procedure at the price that the parameters are cutoff dependent and surely have no physical meaning. But the relations between physical quantities are finite and by taking as many input data as required to "fix" the free parameters one is able to obtain predictions for other physical processes. This procedure was used e.g. by [3].

2. A next step in the extent of renormalization consists in a renormalization of the bare parameters $(e_B, \ldots, s_{W,B}^2)$:

$$
\begin{aligned}
e_B &= e + \delta e = e(1 + \frac{\delta e}{e}) \\
M_{B,i}^2 &= M_i^2 + \delta M_i^2 = M_i^2(1 + \frac{\delta M_i^2}{M_i^2}) \\
m_{B,f}^2 &= m_f^2 + \delta m_f^2,
\end{aligned}
\tag{4}
$$

respectively

$$
\begin{aligned}
G_{F,B} &= G_F + \delta G_F = G_F(1 + \frac{\delta G_F}{G_F}) \\
s_{W,B}^+ &= s_W^2(1 + \frac{\delta s_W^2}{s_W^2}),
\end{aligned}
\tag{5}
$$

where $e, \ldots, s_W^2$ are the physical parameters. The quantities $\delta e, \delta M_i^2, \delta G_F, \delta s_W^2$ absorb possible divergences, their finite parts have to be adjusted in such a way that the corresponding physical quantities are defined including radiative corrections i.e. the quantum corrections to the electric charge, the masses etc.. The relations (3) induce equations between $\delta G_F, \delta s_W^2$ and $\delta e, \delta M_i^2$:

$$
\begin{aligned}
\frac{\delta G_F}{G_F} &= 2\frac{\delta e}{e} - \frac{\delta M_W^2}{M_W^2} - \frac{M_W^2}{M_Z^2 - M_W^2}\left(\frac{\delta M_Z^2}{M_Z^2} - \frac{\delta M_W^2}{M_W^2}\right), \\
\frac{\delta s_W^2}{s_W^2} &= \frac{M_W^2}{M_Z^2 - M_W^2}\left(\frac{\delta M_Z^2}{M_Z^2} - \frac{\delta M_W^2}{M_W^2}\right).
\end{aligned}
\tag{6}
$$

The strong inner structure of the standard model as a non-Abelian gauge theory with spontaneous symmetry breaking guarantees that with the help of $\delta e, \delta M_i^2, \delta m_f^2$ all S-matrix elements are finite and correctly calculated if in addition wave function renormalization i.e. the correct normalization of the in/outgoing states is performed.

3. If in addition to S-matrix elements also the Green function shall be finite and suitably normalized, field renormalization is required: $\varphi_B = Z_2^{-1/2}\varphi$. This procedure can be performed in a minimal sense: the number of field renormalization constants is chosen such that finiteness of the Green functions is obtained. In gauge theories this means that one field renormalization constant is introduced for every multiplet of gauge bosons, fermions, scalars i.e. for every irreducible representation of the gauge group. Consequently the renormalized propagators are finite but not all of them have residue one. Such a RS exists for the standard model using e and the masses M_W, M_Z, M_H, m_f as physical parameters as well as on-shell renormalization conditions: $\delta M_i^2 = \Sigma_i(k^2 = M_i^2)$ [4][5].

4. Complete field renormalization [6] works with renormalization matrices $\varphi_{i,B} = (Z_2^{-1/2})_{ij}\varphi_j$ for the multiplets. This allows to renormalize all residues in the diagonal propagators to one and to prescribe values for the non-diagonal parts of the propagators i.e. the mixing energies. A convenient choice is to forbid mixing for on-shell particles. In this case with complete field renormalization no additional wave function renormalization is required. On the other hand many renormalization constants have to be calculated and used in the calculation of radiative corrections.

5. Let us finally remark that renormalization must also be extended to the unphysical parts of the Green functions e.g. the longitudinal gauge boson propagators, the ghost propagators, the propagators of the unphysical Higgs fields and the mixing between gauge bosons and unphysical Higgses. A convenient renormalization of this sector can be constructed by requiring a simple structure for the renormalized generalized Ward identities [5].

2.2 What is a good renormalization scheme?

We have seen that there exist many possibilities to define a RS for the standard model. RSs may differ in the parameters they use and their precise definition, in the extent of renormalization, in the choice of gauge fixing conditions etc.. Therefore we present some criteria for the choice of RSs:

- A good RS should respect the structure of the theory. Since gauge theories are constructed obeying local gauge invariance a good RS should conserve the symmetry relations of the theory, especially the Ward identities.

- The standard model describes besides e^+e^- annihilation into fermion-antifermion pairs many interesting reactions, e.g. $e^+e^- \to W^+W^-$, $WW \to WW$. A good RS should allow to treat all of them in the same way. It should not be tailored only for the four-fermion processes.

- Besides 1-loop calculations sooner or later also higher loop corrections will become important. A good RS should be able to treat those without great modifications.

- The most important differences between RSs come from different choices of parameters. A good RS should use well-defined parameters with a clear physical meaning, if possible quantities which can be directly measured with a high accuracy.

- A RS should be as simple as possible for a consistent scheme and not more complicated than really necessary.

- The predictions for physical quantities should depend minimally on cryptic parameters. In the case of the SM these are presently the top quark mass and the Higgs mass.

Let us end this discussion with two further remarks. The parameters the theory is formulated with are not necessarily identical with the quantities which can be measured with the best accuracy. However, for the determination of the physical parameters the best existing experimental data should be used. Of course this will change with time. Some years ago we used α, G_F, s_W^2 the latter determined from ν- scattering. Meanwhile M_Z is determined with an accuracy better than 0.1% allowing a more precise input.

Our last remark concerns effective parametrizations. They may be useful to simplify the notation and the expressions for cross sections etc., but are not necessarily related to the question of RS dependence.

2.3   An example: four-fermion processes

Let us illustrate parameter renormalization by considering e^+e^-- annihilation into fermion pairs. As the set of relevant parameters in this sector of the standard model we choose e, M_Z, M_W; $c_W^2 = 1 - s_W^2 = M_W^2/M_Z^2$ is a shorthand notation. In lowest order we have $\gamma-$, $Z-$, W-exchange diagrams with the corresponding couplings: $\gamma_\mu Q_f$; $\gamma_\mu(I_f^3 - s_W^2 Q_f), \gamma_\mu \gamma_5 I_f^3; \gamma_\mu(1 - \gamma_5)/\sqrt{2}s_W$. To 1-loop order we have gauge-boson self-energies $\Sigma_\gamma, \Sigma_Z, \Sigma_{\gamma Z}, \Sigma_W$, vertex $F_{V,A}$ and box corrections. The parameter renormalizations are written as $\delta e, \delta M_Z, \delta M_W$. This leads us to the following corrected expressions:

$$\mathcal{M}_\gamma = \frac{e^2}{k^2}\gamma_\mu Q_i \otimes \gamma^\mu Q_f \left[1 + 2\frac{\delta e}{e} - \frac{\Sigma_\gamma(k^2)}{k^2} - 2\frac{s_W}{c_W}\frac{\Sigma_{\gamma Z}(k^2)}{k^2 - M_Z^2}\right]$$

$$+ \frac{e^2}{k^2}\left\{\gamma_\mu\left(F_V^e(k^2) + \frac{I_e^3}{2 s_W c_W}\frac{\Sigma_{\gamma Z}(k^2)}{k^2 - M_Z^2}\right)\right. \tag{7}$$

$$\left. - \gamma_\mu \gamma_5 \left(F_A^e(k^2) + \frac{I_e^3}{2 s_W c_W}\frac{\Sigma_{\gamma Z}(k^2)}{k^2 - M_Z^2}\right)\right\}$$

$$+ (e \leftrightarrow f) + \text{box contributions}$$

and similarly for $\mathcal{M}_Z, \mathcal{M}_W$ (see the contribution of W. Hollik to this workshop). All these matrix elements $\mathcal{M}_\gamma$, $\mathcal{M}_Z$, $\mathcal{M}_W$ contain universal expressions, which are independent of the flavour of the external fermions. They are written in square brackets. These universal contributions can be used to define the charge renormalization

$$2\frac{\delta e}{e} = \frac{\Sigma_\gamma(k^2)}{k^2}\Big|_{k^2=0} - 2\frac{s_W}{c_W}\frac{\Sigma_{\gamma Z}(0)}{M_Z^2}, \tag{8}$$

Z- and W-mass renormalization:

$$\delta M_Z^2 = Re\Sigma_Z(M_Z^2),$$
$$\delta M_W^2 = Re\Sigma_W(M_W^2).$$

Since these are all parameter renormalizations in the four-fermion sector they should be able to make finite $\mathcal{M}_\gamma, \ldots, \mathcal{M}_W$ as well. That this indeed is the case can be seen by an explicit calculation. The situation is non-trivial since the Z- and W-boson self energies contain besides constant divergent terms which are compensated by mass renormalization also divergent terms proportional to k^2. These infinities are also compensated which means that a partial field renormalization is contained in the prescription. The divergences of the vertex corrections contained in F_V, F_A are compensated by the divergent parts of $\Sigma_{\gamma Z}(0)/M_Z^2$. This is a consequence of the corresponding Ward identity.

Inserting the results for δe, δM_Z, δM_W we can rewrite eq. (7) and identify the following renormalized, finite gauge boson self energies ($\Pi_\gamma = \Sigma_\gamma(k^2)/k^2$),

$$\hat{\Sigma}^\gamma(k^2) = \Sigma^\gamma(k^2) - \Pi^\gamma(0) \cdot k^2$$

$$\hat{\Sigma}^{\gamma Z}(k^2) = \Sigma^{\gamma Z}(k^2) - \Sigma^{\gamma Z}(0) - k^2 \frac{c_W s_W}{c_W^2 - s_W^2}\left(\delta Z_2^Z - \delta Z_2^\gamma\right)$$

$$= \Sigma^{\gamma Z}(k^2) - \Sigma^{\gamma Z}(0) + k^2\left\{2\frac{\Sigma^{\gamma Z}(0)}{M_Z^2} - \frac{c_W}{s_W}\left(\frac{\delta M_Z^2}{M_Z^2} - \frac{\delta M_W^2}{M_W^2}\right)\right\}$$

$$\hat{\Sigma}^Z(k^2) = \Sigma^Z(k^2) - \delta M_Z^2 + \delta Z_2^Z(k^2 - M_Z^2),$$

$$\hat{\Sigma}^W(k^2) = \Sigma^W(k^2) - \delta M_W^2 + \delta Z_2^W(k^2 - M_W^2),$$

where we have introduced the abbreviations:

$$\delta Z_2^Z = -\Pi^\gamma(0) - 2\frac{c_W^2 - s_W^2}{s_W c_W}\frac{\Sigma^{\gamma Z}(0)}{M_Z^2} + \frac{c_W^2 - s_W^2}{s_W^2}\left(\frac{\delta M_Z^2}{M_Z^2} - \frac{\delta M_W^2}{M_W^2}\right),$$

$$\delta Z_2^W = -\Pi^\gamma(0) - 2\frac{c_W}{s_W}\frac{\Sigma^{\gamma Z}(0)}{M_Z^2} + \frac{c_W^2}{s_W^2}\left(\frac{\delta M_Z^2}{M_Z^2} - \frac{\delta M_W^2}{M_W^2}\right).$$

These expressions are doubly subtracted i.e. they contain besides the mass also the field renormalization. A comparison shows that they are identical with the result of an explicit on-shell renormalization procedure of the standard model with minimal field renormalization[5]. This demonstrates that within the framework of the parameters e, M_W, M_Z, M_H, m_f this RS is quite natural. It was proposed by[4] and worked out in[5] and meanwhile applied to many electroweak processes. Let us mention: $e^+e^- \to \mu^+\mu^-$, e^+e^-, $q\bar{q}$, $\gamma\gamma$, γZ, ZZ, WW, $\nu\bar{\nu}\gamma$; deep inelastic scattering and of course to μ-decay and ν_e-scattering as well as Z and W-decays.

The parameters of this scheme are usually determined from α, M_Z and G_F as numerical inputs. Using G_F instead of M_W has the advantage of absorbing big logarithms and reducing the top mass dependence of the corrected results.

2.4 Changing the RS

In the last section we have presented parts of the on-shell RS with minimal field renormalization. Besides this scheme several others have been worked out using different gauges [2], different physical parameters and different prescriptions for the renormalization of the fields. Of course this causes inconveniences as one has the problem to compare different schemes and to know how to switch from one scheme to another. But in principle it is good to have results from calculations in different schemes. Since the calculations of RCs are technically a complicated, often formidable task, independent calculations are needed as checks for the correctness of these results. Moreover, since calculations in different schemes differ by higher order contributions a comparison between results in different schemes may allow estimates of these neglected higher order corrections.

Let us point out, however, that for a consistent general test of the SM it is a bonus to have calculations in one and the same scheme for all the interesting processes. It may also be dangerous to propose special schemes for special processes since then the inner coherence in the tests of the SM may be lost.

We want to illustrate the change of a RS not in general terms but with a practical example, the transformation from the on-shell scheme with e, M_Z, M_W as parameters to a scheme using instead e, M_Z, G_μ. For this purpose we recapitulate the definitions of M_W and G_μ. M_W usually is defined as the position of the pole of the W-boson propagator. Consequently, the corrections δM_W^2 to M_W^2 in the on-shell scheme are defined as:

$$\delta M_W^2 = \Sigma_W(M_W^2). \tag{9}$$

In this scheme G_F is not an independent quantity but calculated as function of α, M_Z, M_W and the other parameters of the SM. We write in lowest order:

$$G_{F,B} = \frac{e_B^2}{4\sqrt{2}} \cdot \frac{1}{M_{W,B}^2(1 - M_{W,B}^2/M_{Z,B}^2)}. \tag{10}$$

We determine δG_F in the (e, M_Z, M_W)-scheme by evaluating the radiative corrections $(\delta_{QED}, \delta_\Sigma, \delta_V, \delta_{box}$ for e.m., self energies, weak vertex, box corrections) to μ-decay and expressing the result in the form:

$$
\begin{aligned}
\tau_\mu^{-1} &= \frac{G_F^2 m_\mu^5}{192\pi^3}\left(1 - 8\frac{m_e^2}{m_\mu^2}\right)\left[1 + \underbrace{\frac{\alpha}{2\pi}\left(1 + \frac{2\alpha}{3\pi}\ln\frac{m_\mu}{m_e}\right)\left(\frac{25}{9} - \pi^2\right)}_{\delta_{QED}}\right] \\
&= \left(\frac{e^2}{4\sqrt{2}M_W^2\left(1 - \frac{M_W^2}{M_Z^2}\right)}\right)^2 \frac{m_\mu^5}{192\pi^3}\left(1 - 8\frac{m_e^2}{m_\mu^2}\right) \cdot \\
&\quad \cdot \left[1 + \delta_{QED} + \underbrace{\delta_\Sigma + \delta_V + \delta_{box}}_{2\Delta r(\alpha, M_Z, M_W, M_H, m_f)}\right]
\end{aligned}
\tag{11}
$$

leading to

$$G_F = \frac{e^2}{4\sqrt{2}M_W^2(1 - \frac{M_W^2}{M_Z^2})} \cdot \frac{1}{1 - \Delta r} = G_{F,B}(1 + \frac{\delta G_F}{G_F}). \tag{12}$$

In the (e, M_Z, G_F) scheme G_F is defined by (11), M_W is the dependent quantity and obtained in lowest order by inverting eq. (10):

$$M_{W,B}^2 = M_{Z,B}^2(\frac{1}{2} + \frac{1}{2}\sqrt{1 - \frac{4\pi\alpha_B}{\sqrt{2}G_{F,B}M_{Z,B}^2}}). \tag{13}$$

The equation between the corrections δe, δM_Z, δG_F can be obtained by using eq. (6)

$$\frac{\delta G_F}{G_F} = 2\frac{\delta e}{e} + \frac{\delta M_W^2}{M_W^2} \cdot \frac{-M_Z^2 + 2M_W^2}{M_Z^2 - M_W^2} - \frac{\delta M_Z^2}{M_Z^2}\frac{M_W^2}{M_Z^2 - M_W^2}. \tag{14}$$

Solving this eq. for $\delta M_W^2/M_W^2$ and expressing in the resulting expressions M_W^2 with help of eq. (13) by G_F leads from the G_F-scheme to the M_W-scheme.

Special care has to be taken if RSs are compared where the "same" quantities are used but defined differently. An example for this situation is the weak mixing angle s_W^2. This may be defined in one scheme from the masses of the W- and Z-bosons, in another scheme by the formula of ν-scattering. Both schemes use s_W^2, but are different.

2.5 Higher order corrections

Many reactions have been calculated in the SM including $O(\alpha)$ electroweak corrections. Naively one would expect that the order of magnitude of these corrections is given by $\frac{\alpha}{\pi} \approx 0.25\%$. This, however, is not true since there are several mechanisms which generate large logarithms like $ln\frac{M_Z^2}{m_e^2} \approx 24$ or $ln\frac{\Delta E}{E} \approx 2-3$ which multiply each other and $\frac{\alpha}{\pi}$. The resulting corrections may reach the order of magnitude of the Born cross sections. For the case that the top quark is heavy, power type corrections proportional to $(\frac{m_t}{M_W})^2$ times $\frac{\alpha}{\pi}$ occur. If one is interested in percent accuracies or, as is the case for the study of the Z-peak in e^+e^--annihilation, in per mille accuracies higher order corrections are needed. The calculation of the complete $\mathcal{O}(\alpha^2)$ corrections in the electroweak standard model is a really complicated task and presently beyond our possibilities. Up to now only certain higher order corrections with a simple structure can be reliably evaluated. Fortunately these are just those which are numerically important since they generate the large terms mentioned above. They are:

- Infrared parts: the emission of a soft photon leads to the $ln\frac{\Delta E}{E}$ terms. Multiple soft photon radiation can be treated using the exponentiation prescriptions of [11].

- Mass singularities from collinear photons: $ln\frac{s}{m_e^2}$-terms occur from the emission of hard photons which are emitted almost parallel to one of the charged particles of the process. The leading log-terms $(\frac{\alpha}{\pi}ln\frac{s}{m_e^2})^n$ can be summed to all orders using renormalization group techniques [10] or the structure function approach[8]. Some results for the next to leading logs $(\frac{\alpha}{\pi})^n(ln\frac{s}{m_e^2})^{n-1}$ exist[7]. In order to check these approaches it is very important to have some explicit 2-loop calculations of these effects[7].

- The gauge boson self energies also contain big logarithms and in the case of heavy quarks powers in the quark mass. The leading log contributions are summed by using the Dyson series for the propagator:

$$\frac{1}{k^2 - M^2} - \frac{\sum(k^2)}{(k^2 - M^2)^2} + \ldots \Rightarrow \frac{1}{k^2 - M^2 + \sum(k^2)}. \tag{15}$$

The higher order heavy quark mass effects have been analyzed by[12].

For e^+e^- annihilation into fermion-antifermion pairs the inclusion of these leading higher order contributions has been worked out. Using the on-shell scheme with G_F as input quantity, the effects for physics at the Z-resonance have been discussed in many details and found to be important.

Different RSs for $\mathcal{O}(\alpha)$ corrections lead to different results when these $\mathcal{O}(\alpha)$ corrections are manipulated in order to include the leading higher order effects. This can be seen when the $\star$-scheme, which was proposed by[9], is compared to the on-shell scheme. For a consistent result it is necessary to compensate the effects of those terms in the $\star$-scheme which have been treated differently. After this has been done satisfying numerical agreement between both schemes is obtained. A question which is not yet answered is the following: Certain building blocks of the $\star$-scheme are not gauge invariant by themselves, but are used to generate higher order effects. Consequently it is not guaranteed that the resulting expressions are gauge invariant. Up to now also the contributions of heavy fermions including higher order effects are not treated correctly in the $\star$-scheme.

3 Conclusions

The renormalization of the electroweak standard model to 1-loop order meanwhile is well-understood. There exist several renormalization schemes using either the unitary gauge or renormalizable gauges like the 't Hooft-Feynman gauge. The calculations are done with minimal field renormalization of complete matrix field renormalizations. Differences exist in the use of physical parameters. In the on-shell scheme these are the electric charge and the masses of the gauge bosons, the Higgs and the fermions. Other schemes use instead the Fermi constant G_F and/or the weak mixing angle. The $\star$-scheme is also based on those quantities but they are defined in a different way. These schemes have been applied to the calculation of radiative corrections to many reactions. In almost all cases agreement (at least numerically) was achieved between independent calculations. Also the transformation from one RS to a different one is under control for 1-loop calculations.

The scheme which was worked out in most details and which found the greatest number of applications is the on-shell scheme with minimal field renormalization and e, M_Z, G_F as basic numerical input quantities.

The situation is not as good in what concerns higher order corrections. Up to now there exist no calculations doing systematic and complete 2-loop electroweak corrections. First 2-loop results can be found for gauge boson self-energies, initial state radiation and some other applications. If one is concerned only in the numerical accuracy of the results for four-fermion processes the situation is much better. There exist techniques to handle the numerically important higher order corrections: soft photons, collinear photons, leading logs in self-energies, heavy quark effects. Anything else is at best in status nascendi and needs a lot of systematic and practical work.

Acknowledgment: The author would like to thank the organizers for the pleasant atmosphere at this workshop.

References

[1] See the contributions in these proceedings by W.F.L. Hollik; M. Igarashi; F. Jegerlehner; T. Riemann; B.F.L. Ward.

[2] D.Yu. Bardin, P.CH. Christova, O.M. Federenko, Nucl. Phys. B **175** (1980) 435; B **197** (1982) 1.

[3] G. Passarino, M. Veltman, Nucl. Phys. B **160** (1979) 151; M. Consoli, M. Veltman, Nuc. Phys. B **160** (1979) 208.

[4] A. Sirlin, Phys. Rev. D **22** (1980) 971; W.J. Marciano, A. Sirlin, Phys. Rev. D **22** (1980) 2695; A. Sirlin, W.J. Marciano, Nucl. Phys. B **189** (1981) 442.

[5] M. Böhm, W. Hollik, H. Spiesberger, DESY **84-027** (1984); Fortschr. Phys. **34** (1986) 687.

[6] K.I. Aoki, Z. Hioki, R. Kawabe, M. Konuma, T. Muta, Progr. Theor. Phys. **64** (1980) 707; **65** (1981) 1001; Suppl. Progr. Theor. Phys. **73** (1982) 1.

[7] W. Beenakker, F.A. Berends, W.L. van Neerven, in: Proceedings of the Ringberg Workshop on "Radiative Corrections", ed. J.H. Kühn (1989).

[8] O. Nicrosini, L. Trentadue, Z. Phys. C **39** (1988) 479; G. Montagna, O. Nicrosini, L. Trentadue CERN-TH 5445789 (1989).

[9] D.C. Kennedy, B.W. Lynn, SLAC-PUB 4039 (1986, rev. 1988).

[10] G. Altarelli, in Physics at LEP, CERN 86-01 (1986).

[11] D.R. Yennie, S.C. Frautschi, H. Suura, Ann. of Phys. 13 (1961) 379.

[12] M. Consoli, W. Hollik, F. Jegerlehner, CERN-TH 5395/89 (1989).

RADIATIVE CORRECTIONS - AN EXPERIMENTALIST'S VIEW *

P. RANKIN

University of Colorado, Boulder, CO 80309-390

ABSTRACT

This paper reviews our current understanding of the effects of radiative corrections on the experimental measurements which will be made at the SLC and at LEP. It discusses how the shape of the Z^0 resonance is modified by initial state radiation, and considers how the Z^0 self energy corrections allow a probe of physics above the Z^0 mass scale. In particular, the sensitivity to the top quark mass of the various asymmetry measurements which may be made at the Z^0 pole is discussed.

Introduction

Experiments at the Z^0 pole aim first to define the Standard Model as precisely as possible and then to test it as stringently as possible. The achievement of both of these goals requires an understanding of how to extract fundamental parameters from the experimental data and how to relate many measurements to each other (including those made by other types of experiments at different energy scales).

The most basic version of the Standard Model requires three experimental measurements to be used to renormalize the bare parameters of the theory (the SU(2) coupling strength, the U(1) coupling strength, and the vacuum expectation value of the Higgs field). The measurements used to test the theory may not be made at the same energy scale as the measurements used to renormalize the bare parameters. It is usually necessary to "run" the couplings (include the effects of diagrams dependent on the energy) to the appropriate energy to make accurate

---

*Work supported by the Department of Energy, contract DE–0286ER-40253.

predictions. The changes in the input parameters due to this running reflect the particle content of the theory. Since we do not know exactly what this particle content is (or the exact masses of all the particles which we think exist) when we make a fourth measurement we are free to adjust quantities such as the top quark mass[1] to make our predictions agree with experiment. This means that we require four measurements to constrain the theory and a fifth one to test it. The fourth parameter is often called ρ and is usually thought of as being related to the Higgs structure of the model, but more generally it can be considered to be related to isospin splitting.

The fact that the self energies of the gauge bosons get contributions from particles too heavy to be produced directly by the decay of a particular gauge boson makes the comparison of theory to experiment depend on assumptions made about the physics beyond the Z^0 mass scale. Such measurements therefore probe mass scales above those directly accessible experimentally. For example, as we will see, the predictions of asymmetries at the Z^0 pole depend on the top quark mass, measuring an asymmetry therefore can be interpreted as measuring the mass of the top quark.[2] However, to begin in earnest our studies of the Standard Model we must first measure as accurately as possible the mass of the Z^0 resonance. This is already measured well enough to make it sensible to use it as a defining parameter of the Standard Model, the current world average for the Z^0 mass is 91.10 ± 0.05 GeV/c^2 (the world average Z^0 width is 2.58 ± 0.08 GeV/c^2).[3]

Measurement of the Z^0 Mass

The main contribution to the experimental error in determining the Z^0 mass comes from the determination of the exact energy at which measurements of the e^+e^- annihilation cross-section are made. The initial aim at both SLC and LEP is to know the absolute center of mass energy to about 40 MeV. This sets the initial scale to which it is desirable to control theoretical errors in the extraction of the Z^0 mass from the line shape to be of the order of a few MeV/c^2 .

The mass of the Z^0 is defined to correspond to the center of mass energy at which the real part of the Z^0 self energy is zero. This does not correspond to the peak of the resonance cross-section as observed by experiments because of the effects of initial state photon radiation. At machine center-of-mass energies which are nominally greater than the Z^0 mass photons radiated before the electrons and positrons collide result in the collision being more likely to produce a Z^0; the observed cross-section is therefore higher than the lowest order prediction. Conversely, at energies below the Z^0 pole the reverse occurs leading to a lower cross-section than predicted. The net result is that the resonance peak lies above the pole and that the cross-section falls off more slowly with increasing energy above the peak than it does with decreasing energy below the peak.

To calculate the energy separation between the pole and peak to an accuracy of a few MeV it is essential to allow for the effects of radiating multiple photons in calculations of the affect of initial state radiation. These effects can be parameterized in terms of a quantity[4]

$$\beta = t = \frac{2\alpha}{\pi}\left(\ln\left(\frac{s}{m_e^2}\right) - 1\right).$$

This is about 0.108 at SLC/LEP energies which implies that the effect of first order corrections is about 10%. A calculation which is to be accurate to better than 1% must therefore include at least second order corrections (t^2 terms).

Applying only first-order corrections (single photon radiated by the initial state) leads to an overestimate of the energy separation of the resonance pole and the observed resonance peak. It also leads to an underestimate of the observed peak cross-section. Many analytic calculations exist which can be used to study the effects of higher order corrections and they agree very well with each other. One effect which the calculations must include is the energy dependence of the Z^0 width (the imaginary part of the Z^0 self energy). If this effect is included, the resonance peak occurs about 94 MeV above the Z^0 pole for a Z^0 mass of 91 GeV/c^2 and a width of 2.5 GeV/c^2.[5]

As well as predicting the energy separation of the peak of the Z^0 resonance from the Z^0 pole, the analytic calculations also relate the observed peak cross-section to the cross-section at the pole (before the inclusion of the effects of initial state radiation) and relate the resonance width to the energies at which the observed cross-section has fallen off to half the peak cross-section . For hadrons, the observed peak cross-section is about 0.74 of the cross-section before initial state radiation and the energies at which the cross-section has half of its peak value are raised by about 60 MeV (below) and 430 MeV (above) relative to their original positions. This means that the separation of the half maxima positions is increased by a factor of about 1.14.

Detailed calculations have been done which include the effects of photon-Z^0 interference and show that the Z^0 line shape differs slightly for the muon channel compared to hadron channels. For muons, the half maxima are raised in energy by about 50 MeV (below) and 450 MeV (above) so their separation is increased by a factor of 1.15 (the muon line shape appears to be about 25 MeV wider than the hadronic line shape). The peak cross-section (which occurs about 1 MeV lower than it does for hadrons) is a slightly larger fraction (about 0.745) of the uncorrected cross-section. This enhancement of the muon cross-section relative to the hadron cross section must be allowed for when measuring branching fractions.[6]

The extremely good agreement between higher order analytic calculations means that they provide a standard Z line shape for a particular choice of input parameters (usually the Z^0 mass, width, and cross-section).[7] Monte Carlo programs which allow for the effects of initial state radiation when generating events can be compared to these analytic standards. If no cuts are applied to the generated events (that is, if the experiments are assumed to have perfect acceptance), the generated Z^0 line shapes agree very well with the standard provided (as one might expect) the Monte Carloes include higher order effects. The five Monte Carloes which satisfy this criterion are DYMU2,[8] EXPOSTAR,[9] YFS2/KORALZ3,[10] MOE10,[11] and ZBATCH(ZSHAPE).[12] The reader interested in a more detailed description of these Monte Carloes, and in their current status is urged to refer to the summary reports of Ronald Kleiss.[13] The ZBATCH and EXPOSTAR Monte Carloes should be considered to be "semi-analytic", they do not explicitly generate events, but are useful for studying effects such as those due to changes in the input parameters. These programs are in good agreement with each other and the standard calculations except, perhaps, in overall normalization. The differences of about 1% between the results of the various programs can mostly be traced to the use of slightly different routines to calculate the electroweak radiative corrections.[14]

Table 1.

Error in extracting the Z^0 mass and width as a function of a linear normalization error.

Cross-Section Measured 2 GeV below the Pole (As a fraction of the true cross-section)	ΔM_z (MeV)	$\Delta \Gamma$ (MeV)
0.8	192	-11
0.9	98	4
0.98	20	2
0.99	10	1
1.00	0.0	0
1.01	-10	-2
1.02	-20	-4
1.10	-99	-28
1.20	-195	-77

Table 2.

Error in extracting the Z^0 mass and width as a function of a normalization error depending on distance from Z^0 peak.

Cross-Section Measured $\pm$ 2 GeV from the Pole (As a fraction of the true cross-section)	ΔM_z (MeV)	$\Delta \Gamma$ (MeV)
0.80	-28	-370
0.90	-14	-190
1.00	0	0
1.10	16	200
1.20	33	412

Applying experimental cuts to the generated events increases the level of disagreement between different programs. They begin to disagree not only over the absolute normalization but also over relative point to point normalizations. However, the predicted relative change in acceptance when scanning across the peak differs by less than 1% between ZBATCH and KORALZ[15] even when studying effects as far as 5 GeV away from the Z^0 pole.

These disagreements can be put in perspective by studying the effect of normalization (scale) errors on the value of the Z^0 mass and width extracted from the data (the overall normalization error will also include normalization errors due to the luminosity measurement). Additive errors in normalization do not affect the mass measurement, multiplicative errors do not change the width, but the peak cross-section is sensitive to any error. Relative errors in normalization can be broadly divided into those which scale with the beam energy and those which scale with the absolute energy difference between the Z^0 peak position and the machine energy.

First, consider an error which has a linear dependence on the machine energy (for convenience, the normalization is assumed to be correct at the peak of the resonance). Table 1 shows the difference between the true Z^0 mass and width and the values found by fitting the data (using MINUIT) if such a normalization error exists. Second, consider an error which depends linearly on the energy difference (Table 2). In the first case if the cross-section is underestimated below the peak it is overestimated above it, in the second case it would again be underestimated above the peak. The mass measurement requires that errors depending linearly on the machine energy be kept to less than a percent over typical scan ranges of 5 to 10 GeV. The width measurement is less affected by linear errors than the mass (the effects each side of the peak compensate each other somewhat), but is more sensitive to errors which affect the relative ratio of the peak cross-section to cross-sections in the wings. A width measurement to 40 MeV/c^2 requires that these errors be kept to below a percent.

The luminosity measurements at the SLC and LEP rely on measurements of the Bhabha cross-section which are made at small enough angles that the contribution from s-channel Z^0 exchange is either negligible or calculable with negligible error. Over the past few months, much work has gone into studying the problems associated with Bhabha Monte Carloes, and as a result there has been a steady improvement over the last few months in the accuracy with which absolute cross-sections can be predicted.[16]

One source of discrepancies was the use of different routines to calculate the effects due to the vacuum polarization (Δr) of the photon. The accuracy with which the vacuum polarization can be calculated is limited by the "hadronic uncertainty". This uncertainty arises from the use of low energy data to evaluate the contribution of light quarks to the vacuum polarization of the photon. This contribution cannot be accurately calculated but must be estimated using dispersion relations. The estimate is limited in accuracy by the errors on the low energy measurements.[17] A very careful analysis (and re-analysis) of the existing data was made recently by Burkhart,Penso, Jegerlehner and Verzegnassi.[18] Their code should now be considered the standard for use in analyses. They find the contribution to Δr due to the known quarks (u,d,s,c,b) to be 0.0288 ± 0.0009. It has recently been pointed out[19] that the calculations also need to include a correction due to the hadronic interactions of heavy quarks. These effects, while small (they change Δr by about 2×10^{-3}) are of about the same order as the light quark uncertainty. The hadronic uncertainty, and the strong interaction contribution can be put into perspective by considering that if α_{em},G_F,and M_z are kept fixed, the change in Δr corresponds to about a 25 MeV/c^2 uncertainty in predictions of the W mass.

The effect of higher order (multiple photon) corrections to the cross-section must be either included or shown to be negligible. I think that it is still open to debate whether or not these corrections are of the order of a percent or not.[20] Arguments based on the fact that the effective $\ln Q^2$ for running at SLC/LEP is not very different from that at PEP/PETRA ($\ln Q^2$ is 24.2 at SLC and about 21.1 at PETRA) contend that if the Monte Carloes worked for PEP and PETRA they are fine for use at the SLC. However, it is known that first order Bhabha Monte Carloes needed exponentiation to fit the observed acollinearity distributions at PEP/PETRA.

The alignment of the inner (small angle) monitors used to measure the luminosity must be very precise. Since the Bhabha rate is so strongly dependent on alignment these detectors are sensitive to changes in the position of the interaction point. To decrease the sensitivity to longitudinal motion the acceptances of each side of the detector can be made slightly different. However, an asymmetric acceptance can make the detectors more sensitive to higher order corrections. This comes from the contribution due to events in which a hard photon is radiated. When the angular acceptance is the same for both tracks from an event, the correction due to the radiation of hard photons has been shown to be small, and higher order effects to essentially cancel. However, in the asymmetric case, there is an added correction due to events where "hard" radiation results in the acceptance of an event which would otherwise have been rejected. The number of events gained due to changes in the angle of tracks close to the inner edge of the detector is not exactly balanced by tracks moving outside the acceptance at the outer edge. How hard is "hard" (and therefore the probability such a photon will be radiated) depends on the event and how important the effect is depends on the details of the detector. The potential importance of higher order corrections increases with increasing asymmetry in the acceptance, and as the angles the detector operates at become smaller. The corrections due to second order (or higher) diagrams can be of the order of a percent and may therefore be significant sources of systematic errors.

It seems clear that higher order Monte Carloes are needed - at least to study such problems and check that higher order corrections can be safely neglected! It has been suggested[15] that exponentiation will not solve all the problems that potentially exist with the lower order Monte Carloes. However, rigorous approaches to multiple photon generation - such as those based on Yennie, Frautschi, and Suura's theory (YFS)[21] should work (at least in principle).[22] Programs currently under development include BHLUMI[23] and an extension of the MOE program. These programs can also avoid the need to divide photons up into "soft", where the final state is treated as a two-body, and "hard", where it is treated as three body. This eliminates potential errors in the normalization introduced because the exact value found for the total cross-section can be sensitive to the choice of photon energy used to separate the two types of events.

To summarize, experimentally the Z^0 mass can be measured to better than 50 MeV/c^2 fairly easily. The Z^0 width is expected to be known to within 30 MeV/c^2 and the aim is to measure the Z^0 cross-section to one percent. The width and cross-section can be predicted if we assume we know the particles to which the Z^0 can decay, and if we know the couplings of these particles to the Z^0. The next section discusses how well we can make these (and other) predictions.

Standard Model Parameters

Measuring the Z^0 line shape will comprise the first phase of the LEP/SLC experimental programs. The second phase of these programs to study Z^0 physics will consist of precision tests of the theory. The predicted couplings of particles to the Z^0 will be compared to their measured value. These tests will check the internal consistency of the model and probe for contributions to the self energies of the vector bosons from particles too heavy to be produced directly in Z^0 decays.

The coupling of particles to the Z^0 is determined by the sum of the squares of the axial(a) and vector(v) couplings. The partial widths for quarks must be multiplied by an factor of about 1.05 (corresponding to $\alpha_s = 0.14$ at PEP/PETRA energies)[24] to allow for the effects of final state gluon radiation. (Final state photon radiation has a negligible effect). The partial width to b-quarks is decreased relative to the d and s quark partial widths due to the effects of diagrams sensitive to the mass difference between the b-quark and its (presumed) partner the t-quark (these effects increase in importance as the limits on top quark mass are raised).[25] Table 3 gives the values of the couplings for the known fermion types assuming the effective value of $\sin^2\theta_w$ at the Z^0 pole is 0.233 (which corresponds to choosing the Z^0 mass to be 91.14GeV/c^2, and the Higgs and the top quark masses to be 100GeV/c^2).

The ratio of the bare weak and electromagnetic couplings gives the tangent of the Weinberg angle (θ_w). This angle can be defined in physical terms using the masses of the W and Z^0 gauge bosons as,

$$\sin^2\theta_w = 1 - \frac{M_w^2}{M_z^2}.$$

253

Table 3.

Partial Widths for $\sin^2 \theta_w(M_z) = 0.233$

STATE	$a^2 + v^2$	$\Gamma(\text{MeV}/c^2)$
ν	$1 + 1 = 2$	164
e	$(-1)^2 + (-1 + 4\sin^2 \theta_w)^2 = 1.0046$	82
u	$(1)^2 + (1 - (8/3)\sin^2 \theta_w)^2 = 1.144$	$93 \times 3 \times 1.05$
d	$(-1)^2 + (-1 + (4/3)\sin^2 \theta_w)^2 = 1.48$	$120 \times 3 \times 1.05$
all		2458

This definition, sometimes referred to as the on-mass-shell, or the Marciano-Sirlin definition is very convenient. Since the physical values of the masses are used, the effects of all diagrams (of any order) contributing to this quantity are automatically included. Implicit in this definition is the assumption that the Higgs structure of the model is the simplest possible (that is, that there are only two Higgs doublets, and only one physical Higgs particle remains after spontaneous symmetry breaking gives mass to the W and Z^0 gauge bosons). Currently, the value of $\sin^2 \theta_w$ found by direct measurements at $p\bar{p}$ colliders is 0.226 ± 0.009.[26] A much more accurate value of $\sin^2 \theta_w$ can, in principal, be derived in terms of the Z^0 mass, α_{em}, and G_F. The only problem with this method is the unknown value of the top quark mass which introduces an uncertainty into the calculation.[27] Table 4 compares two calculations of $\sin^2 \theta_w$ as a function of the top quark mass for a fixed value of M_z and the Higgs mass.[28]

The on-mass-shell definition of $\sin^2 \theta_w$ only gives the approximate value of the quantity relevant to discussions of particle couplings to the Z^0. The bare parameters of the theory are renormalized by relating them to physical quantities measured at particular energy scales. This renormalization only fixes the values of these parameters to lowest order when they are used to predict the results of measurements which involve different momentum transfers. The closer the renormalization conditions are to the conditions the new measurements are made at, the better the predictions will be. The apparent dependence of the predictions on the choice of renormalization measurements (or of "renormalization scheme") is removed by including higher order diagrams into the calculations. It is clearly preferable to work with quantities that include a more complete set of the diagrams instead of an approximation if one wants to test the theory (otherwise discrepancies between prediction and experiment lack significance).

Table 4.

A comparison of the value of $\sin^2\theta_w$
for various top quark masses.

The Z^0 mass is assumed to be 91.0 GeV/c^2 , the Higgs mass to be 100.0 GeV/c^2

Top Quark Mass (GeV)	Hollik	Lynn
60	0.2367	0.2369
100	0.2318	0.2320
150	0.2261	0.2263
180	0.2221	0.2222
200	0.2191	0.2193

The desire to be able to make accurate predictions (for physics at the Z^0 pole, and for other experiments) has lead to the development of "running" coupling constants which give the effective value of $\sin^2\theta_w$ ($\sin^2\theta_w$ corrected for the conditions under which the measurement is made). One such example is the quantity $\sin_*^2\theta$ which was proposed by Bryan Lynn.[29] The calculation of these quantities depends on the particle content of the model and on the masses of those particles. It is therefore not possible to make a unique prediction for the value of $\sin_*^2\theta$ at the Z^0 pole. However, once we make one measurement of this quantity the particle content needed to give the measured value must be consistent with that needed by other measurements. Once we can no longer adjust the particle content to make all measurements consistent with each other, we have disproved the model.

The effective value of $\sin^2\theta_w$ at the Z^0 pole ($\sin_*^2\theta$ (M_z)) is less sensitive to the exact value of the top quark mass than $\sin^2\theta_w$ is. The differences between $\sin_*^2\theta$ and $\sin^2\theta_w$ are shown graphically in Figure 1. The importance of using $\sin_*^2\theta$ in calculations at the Z^0 pole increases with increasing top quark mass. Table 5 compares the agreement between recent theoretical calculations of the effective value of $\sin^2\theta_w$ at the Z^0 pole as a function of top quark mass.

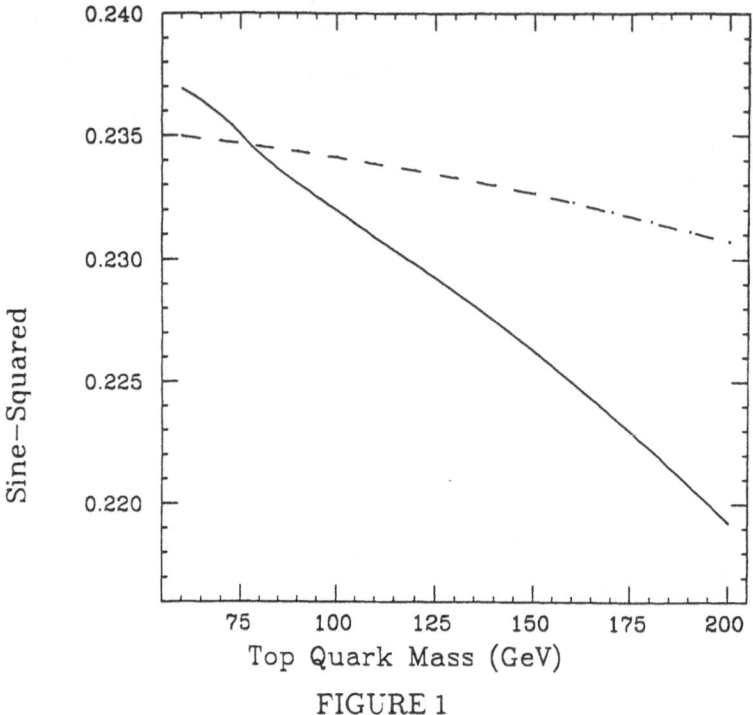

FIGURE 1

A comparison of the values of $\sin^2\theta_w$ (solid line) and $\sin^2_*\theta\,(M_z)$ (dashed line) as a function of top quark mass. For this figure, and for figures 2,3,5, and 6, M_z is assumed to be 91 GeV/c^2 , and the Higgs mass is assumed to be 100 GeV/c^2 .

Table 5.

A comparison of the value of $\sin^2_*\theta_w(M_z)$

for various top quark masses.

The Z^0 mass is assumed to be 91.0 GeV/c^2 , the Higgs mass to be 100.0 GeV/c^2

Top Quark Mass (GeV)	Hollik	Lynn
60	0.2350	0.2350
100	0.2341	0.2341
150	0.2326	0.2326
180	0.2315	0.2315
200	0.2307	0.2307

As can be seen although the calculations are independent of each other and differ slightly in the details of how and which higher order effects are included, they are essentially identical when measurements at the Z^0 are considered. The sensitivity needed for measurements of $\sin^2_* \theta$ in different processes is set by the size of the effects induced by changing the input parameters. If everything else is kept fixed; a 45 MeV/c² increase in the Z^0 mass decreases $\sin^2_* \theta$ by 0.0003,[30] increasing the top quark mass from 60 to 180 GeV/c² decreases $\sin^2_* \theta$ by 0.0035, and while decreasing the Higgs mass from 100 to 10 GeV/c² decreases $\sin^2_* \theta$ by 0.001, raising the Higgs mass to 400 GeV/c² increases $\sin^2_* \theta$ by only 0.0007.

Final state photon radiation increases the Z^0 width by less than 5 MeV/c² . Final state gluon radiation (which only modifies quark final states) has a more significant effect. The observed partial width to hadrons ($\Gamma_{h,obs}$) is given by

$$\Gamma_{h,obs} = \Gamma_h(1 + \frac{\alpha_s}{\pi} + 1.41(\frac{\alpha_s}{\pi})^2 + 64.8(\frac{\alpha_s}{\pi})^3......)$$
$$= \Gamma_h(1 + 0.04 + 0.0024 + 0.0046 + ...)$$

where Γ_h is the predicted width before the inclusion of final state effects. The analysis of D'Agostini, de Boer, and Grindhammer gives the overall factor for the increase in hadronic widths due to gluon radiation to be 1.046 ± 0.005.[23] The expansion in powers of α_s ceases to be well behaved at the fourth term (the co-efficient of the fourth term is larger than that of the third). Technically, this can be treated by truncating the expansion at the third term. It has been argued (for example by Kuhn[19] that it is in any case inconsistent to include the fourth term in the expansion without also adding in the effects of lower order diagrams sensitive to the mass splitting between the top and bottom quark which have been neglected. These latter effects are gaining in importance as the lower limit on the mass of the top quark keeps increasing. The b-quark partial width decreases as the top quark mass increases. However the effects are fairly small, changing the top quark mass from 80 to 200 GeV/c² has about a 10 MeV/c² affect on the Z^0 width.

Comparisons of predicted widths need to be done with some care. Some authors include QCD effects for example and some do not and the definition of the width may vary slightly from author to author. When like is compared to like however, the agreement between different calculations is much better than the anticipated experimental accuracy.

Studying the total cross-section will provide LEP/SLC experiments with a first glimpse at how well branching fractions and couplings agree with Standard Model predictions. The total cross-section however is not particularly sensitive to the physics beyond the Z^0 mass scale. In order to really probe the Standard Model (and beyond) measurements must be made of particle asymmetries at the Z^0 pole.

Asymmetry measurements with unpolarized beams

The most straightforward way to measure the vector coupling constant for a particular fermion final state is to measure the forward-backward asymmetry for Z^0 decays into that state. If the beams are unpolarized, the forward-backward asymmetry (A_{FB}) at the Z^0 pole is given by

$$A_{FB}^{f\bar{f}}(M_{Z^0}) = \frac{3a_e v_e a_f v_f}{(a_e^2 + v_e^2)(a_f^2 + v_f^2)},$$

where a_e, v_e, a_f and v_f represent the axial and vector couplings of the electron and the fermion of interest to the Z^0.

The sensitivity (S) of the forward-backward asymmetry to $\sin_*^2 \theta$ (the effective value of $\sin^2 \theta_w$ at the Z^0 pole) depends both on the value of $\sin_*^2 \theta$ and on the value of the asymmetry (which depends on $\sin_*^2 \theta$ and the fermion type).

$$S_f = A_{FB}^{f\bar{f}} * s_f$$
$$s_f = \frac{4}{v_e} - \frac{4Q_f}{v_f} - \frac{(8v_e - 8v_f Q_f)}{(1 + v_e^2)(1 + v_f^2)}$$

Q_f is the electric charge carried by the final state fermion. Table 6 compares the sensitivity of measurements of the muon forward-backward asymmetry to measurements of the b-quark asymmetry for selected values of $\sin_*^2 \theta$. A reasonable upper limit for $\sin_*^2 \theta$ is about 0.235. This corresponds to a Higgs mass of 500 GeV/c^2 , a Z^0 mass of 91 GeV/c^2 and a top quark mass of 80 GeV/c^2 . Raising the Z^0 mass or the top quark mass lowers $\sin_*^2 \theta$, decreasing the Higgs mass also lowers $\sin_*^2 \theta$ [31] The error on $\sin_*^2 \theta$ is given by

$$\delta \sin_*^2 \theta = \delta A_{FB}^{f\bar{f}}/S_f$$

As can be seen the muon asymmetry rapidly looses sensitivity as the value of $\sin_*^2 \theta$ approaches 0.25.

258

Table 6.

A comparison of the sensitivity of the unpolarized

forward-backward asymmetry for muons and for b quarks

for various values of $\sin_*^2 \theta$

$\sin_*^2 \theta$	Muons $A_{FB}^{\mu\bar{\mu}} * s_\mu = S_\mu$	b-quark $A_{FB}^{b\bar{b}} * s_b = S_b$
0.225	0.029× 78 = 2.31	0.139× 39=5.50
0.23	0.019× 99 = 1.87	0.112 ×50=5.61
0.235	0.011× 132=1.42	0.084 × 67=5.61
0.24	0.005 × 199=0.95	0.006 × 100=5.60

Figures 2 and 3 show the muon and b-quark asymmetries as a function of center of mass energy for a Z^0 mass of 91 GeV/c^2 , a Higgs mass of 100 GeV/c^2 and top quark masses of 60 and 180 GeV/c^2 . As one moves away from the pole these asymmetries change rapidly. Away from the Z^0 pole the dominant contribution to the asymmetry comes from the interference between photon exchange and Z^0 exchange, which depends on the axial couplings. However, at the Z^0 pole there is no interference and the asymmetry depends on the vector couplings of particles to the Z^0. The vector couplings are functions of $\sin_*^2 \theta$ (M$_z$) so it is these which one really wants to measure. The strong energy dependence requires close monitoring of the machine energy at which data is taken. It also makes these asymmetries sensitive to the details of initial state radiative corrections which effectively change the energy of the collision (and therefore the value of the asymmetry which is measured). The need to get angular correlations correct complicates the evaluation of initial state effects since not only the energy loss but also the momentum change due to radiated photons must be calculated correctly.

The Monte Carloes[8-11] which agree so well with each other and with analytic calculations when predicting the Z^0 line shape do not agree so well over predictions of the forward-backward asymmetry.[15] The muon forward-backward asymmetry is a particularly stringent test of Monte Carloes. Until recently no standard was available to check these Monte Carloes against. However, the recent development of CALASY,[32] a semi-analytic Monte Carlo (which plays the same role in forward-

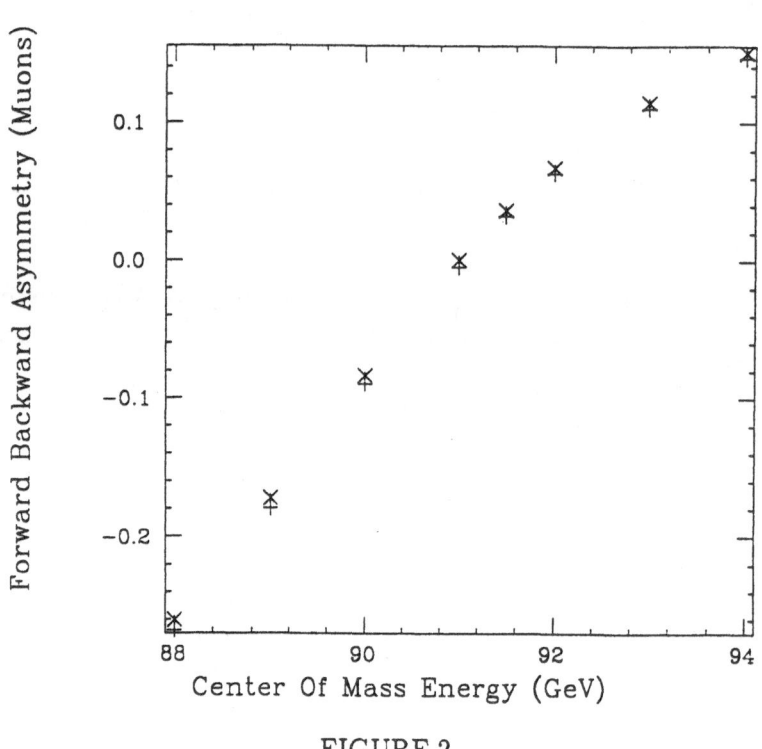

FIGURE 2

The muon forward-backward asymmetry as a function of center-of-mass energy for two top quark masses . The lower set of points corresponds to a 60 GeV/c^2 top quark mass, the upper to a 180 GeV/c^2 top quark mass. The vertical lines correspond to the error on the Monte Carlo simulation (the Monte Carlo EXPOSTAR was used).

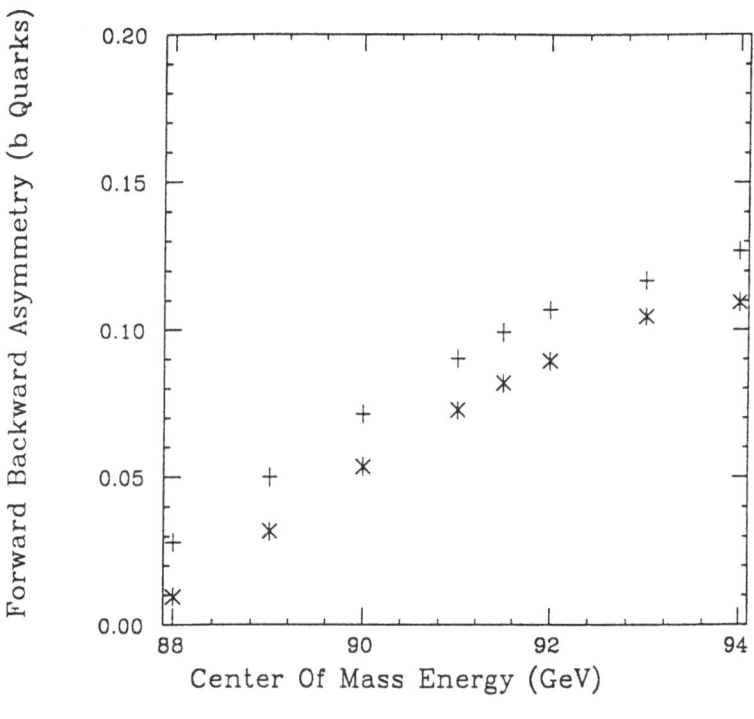

FIGURE 3

The b-quark forward-backward asymmetry as a function of center-of-mass energy for two top quark masses. The lower set of points corresponds to a 60 GeV/c^2 top quark mass, the upper to a 180 GeV/c^2 top quark mass. The vertical lines correspond to the error on the Monte Carlo simulation (the Monte Carlo EXPOSTAR was used).

backward asymmetry measurements that ZBATCH and EXPOSTAR do in line shape measurements) will help.[33] Initial-state radiative corrections also cause the flattening out of the energy dependence of the asymmetry above the Z^0 pole, due to the tendency for the electrons and positrons to radiate onto the pole.

Both the lack of sensitivity of the muon asymmetry to $\sin^2_* \theta$ and the particularly strong energy dependence of this asymmetry make the b quark asymmetry appear to be a better choice to use when extracting the value of $\sin^2_* \theta$. There is also an apparent statistical advantage since the Z^0 decays more frequently to b-quarks than to muons by about a factor of 4.6. However, the statistical advantage in using b-quarks is diluted by the efficiency for detecting b-quarks (and assigning them the correct charge). The other advantages are partially diminished by the fact that the interpretation of the b quark measurement is complicated by the occurrence of b-quark mixing, and by strong interaction effects. It has been estimated[34] that an integrated luminosity of 200 pb^{-1} will lead to a measurement of the muon asymmetry with an error of 0.0035 and a b-quark asymmetry with an error of 0.005. The error on the muon asymmetry measurement converts to an error on the measurement of $\sin^2_* \theta$ of 0.0025, while the b-quark measurement yields an error of 0.0009 (for $\sin^2_* \theta$ 0.235).

Polarization

The cross-section at the Z^0 pole is relatively insensitive to changes in (for example) the top quark mass. This is because changes in the partial widths are largely compensated for by the change in the total width of the resonance. (This cancellation would be perfect if effects due to photon-Z^0 interference could be ignored.) This insensitivity is shown in Figure 4 which gives the change in the muon cross-section at the pole as a function of the Z^0 mass when the top quark mass is changed from 60 to 180 GeV/c^2. The effects on the right and left-handed cross-sections considered individually are not so small but since the changes have opposite sign they compensate for each other. However, the almost perfect cancellation seen for a 92 GeV/c^2 Z^0 (the standard used in many studies) is not quite so good for lighter or heavier Z^0 masses. For a 91.1 GeV/c^2 Z^0, the change in the visible cross-section (hadrons, taus and muons) corresponding to this change in top quark masses is

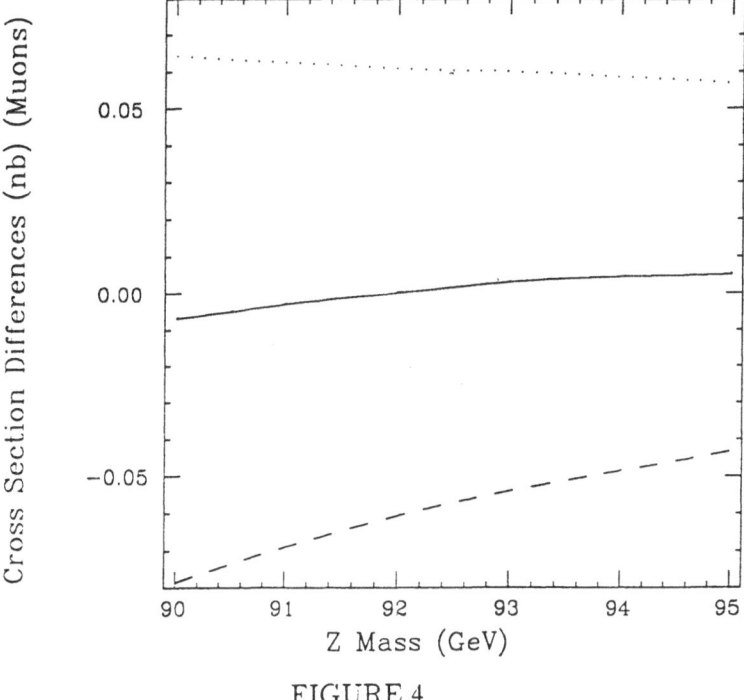

FIGURE 4

The difference (in nanobarns) in the unpolarized(solid line), left(dotted line), and right(dashed line) handed cross-sections for Z^0 decays to muons resulting from changing the top quark mass from 60 to 180 GeV/c^2 .

about 0.07% (which is experimentally unobservable). The right-handed and left-handed cross-sections change by -3% and 2.5% respectively. Polarizing the beams (or one of the beams) producing the Z^0 is desirable because it allows the isolation of the left and right-handed components of the cross-section.

The sensitivity to electroweak effects of the right and left-handed components of the total cross-section can be exploited by studying the left-right asymmetry, A_{LR}, which measures the difference between them. On the Z^0 pole, this asymmetry depends only on the couplings of electrons to the Z^0 and is given by,

$$A_{LR}(M_{Z^0}) = \frac{2Pa_ev_e}{a_e^2 + v_e^2},$$

where P is the magnitude of the polarization.

The sensitivity of this asymmetry to $\sin_*^2 \theta$ is not dependent on the value of $\sin_*^2 \theta$ (it is always 8). Figure 5 shows clearly how sensitive A_{LR} is to changes in

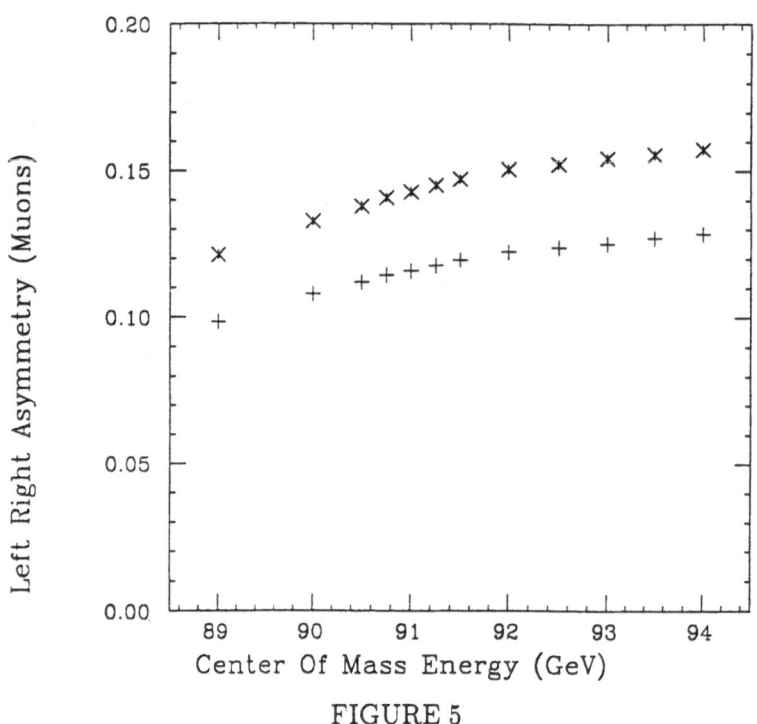

FIGURE 5

The left-right polarization asymmetry for muons as a function of center of mass energy for two top quark masses. The lower set of points corresponds to a 60 GeV/c^2 top quark mass, the upper to a 180 GeV/c^2 top quark mass. The vertical lines correspond to the error on the Monte Carlo simulation (the Monte Carlo EXPOSTAR was used).

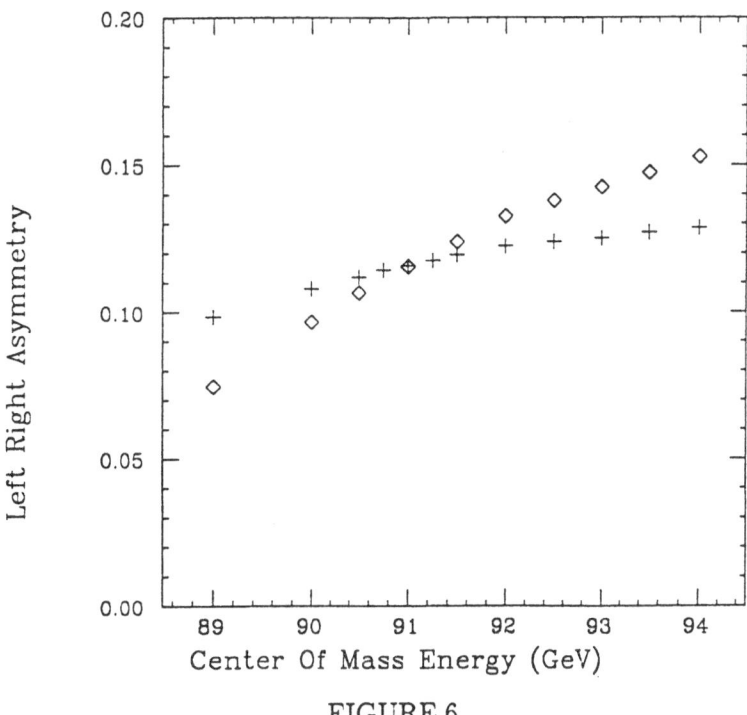

FIGURE 6

The left-right polarization asymmetry for the muon final state (crosses) and the b-quark final state (diamonds), as a function of center of mass energy.

the top quark mass. The energy dependence of this asymmetry is fairly weak so that it is comparatively insensitive to the effects of initial state photon radiation. The asymmetry at the Z^0 pole does not depend on the final state (Figure 6) which means that all events can be used to measure it. A discussion of the insensitivity of this measurement to final state effects and detector acceptances can be found in the paper by Lynn,[35] who has long been a champion of the advantages of this measurement.

The prospects for polarizing the electron beam at the SLC are very good. It is expected that interesting measurements of this asymmetry should be available in 1991. Combined with all the other measurements of electroweak parameters which will then be available it should be possible to test the Standard Model to the limits of our current understanding.

Summary and Acknowledgements

The prospects over the period of the next few years for a substantial improvement in the rigor with which the Standard Model has been tested are extremely good. After a long period of waiting for data from the SLC and LEP the early results have already revealed much information. The future, with the prospects of significant improvements in statistical errors and the possibility of polarized beams, promises to be very exciting.

The author is grateful for support as an Alfred P. Sloan Research Fellow, and for support by the U.S. Department of Energy as an Outstanding Junior Investigator. This paper has benefited from the comments of Jim Smith. I would like to thank the organizers of the conference for an enjoyable and informative experience at Brighton. I particularly liked being able to drink tea during the afternoon breaks.

REFERENCES

1. When the top mass is measured we could instead adjust the Higgs mass or the Higgs structure of the model.

2. This interpretation assumes that no unknown particles are contributing to the running.

3. "A Precision Measurement of the Mass and Width of the Z^0 Resonance at the Fermilab Tevatron",CDF Collaboration, PRL63, 720(1989); "Measurements of Z^0 Boson Resonance Parameters in e^+e^- Annihilation", M2 Collaboration, PRL63, 2173(1989); "Determination of the Number of Light Neutrino Species", ALEPH Collaboration, CERN-EP/89-132; "Measurement of the Mass and Width of the Z^0 Particle from Multihadronic Final States Produced in e^+e^- Annihilations", DELPHI Collaboration,CERN-EP/89-134; "A Determination of the Properties of the Neutral Intermediate Vector Boson Z^0", L3 collaboration, L3 preprint/001; "Measurement of the Z^0 Mass and Width with the OPAL detector at LEP", OPAL collaboration, CERN-EP/89-133.

4. The β parameter can be traced back to a paper by M.Greco in PL 56B,367 (1975). The value of the parameter at the low energies the calculations were then being applied at was 0.007, and since the James Bond movies were popular the name β was chosen.

5. Detailed formulae describing these effects can be found in R.N. Cahn, "Analytic forms for the e^+e^- annihilation cross-section near the Z^0 including initial state radiation", Phys.Rev. D36:2666,1987. Another useful treatment can be found in D.Y. Bardin et al," Energy dependent width effects in Z^0 line shape", Phys.Lett.B206:539,1988. An apparent disagreement as to the size of the energy dependent width effect between these two treatments can be traced to a difference in the form they assume for the Breit-Wigner when the width is constant. See also the "Z line Shape" by D.Y. Bardin et al. in the second CERN yellow report on LEP Z^0 physics, this contains an exhaustive list of references.

6. For more details see W.Beenakker et al, "Rules of thumb for the Z^0 line shape", University of Leiden preprint, October 1989.

7. One may or may not be free to choose all three of these variables. Various assumptions may be made which allow the width and/or the cross-section to be calculated within the framework of the Standard Model.

8. J.E. Campagne and R.Zitoun, University of Paris preprint LPNHE-88.08.

9. D.C. Kennedy at al, SLAC-PUB-4128(1988).

10. S.Jadach and B.F.L. Ward, Comp.Phys.Commun. ,1989,in press.

11. G. Bonvicinni and L.Trentadue, preprint UM-HE-88-36.

12. See G. Burgers contribution to Polarization at LEP, CERN 88-06.

13. For example his contributions to the Rinberg Workshop on Electroweak Radiative Corrections, and his contribution to the second CERN yellow report on LEP physics.

14. As of July 1989 the differences between the different libraries of routines to calculate electroweak corrections (Hollik,Stuart) were being resolved, and a standard library was coming into existence. A note of caution should however be sounded: agreement over results when particular terms are included is not the same as including the effects of all diagrams which contribute.

15. R. Kleiss, Status report on Electroweak Monte Carloes, Proceedings of the Rinberg conference on Electroweak Radiative Corrections and compare to the latest LEP status report.

16. B.F.L. Ward, private communication.

17. Further experiments at low energies could reduce this error significantly.

18. H.Burkhart et al, in Polarization at LEP, vol 1 (p 145-157).

19. J. Kuhn, proceedings of the Rinberg Conference on Electroweak Radiative Corrections and this conference.

20. Recent discussions (September 1989) on the status of Bhabha Monte Carloes with Bennie Ward suggest that these effects are now understood at the 1% level

21. D.R. Yennie, S.C. Frautschi and H. Suura, Annals of Phys. 13(1961)369.

22. S. Jadach and B.F.L. Ward," Exclusive exponentiation in the Monte Carlo Yennie, Frautschi, and Suura approach",TPJU 19/89, UTHEP 89-0703.

23. This approach, due to Jadach and Ward is based on YFS theory, (see article to appear in Phys.Rev.D40, Dec. 1st issue,1989) the program philosophy is the same as that of KORALZ3.

24. G. D'Agostini et al, " Determination of α_s and the Z Mass from measurements of the Total Hadronic Cross-Section", Desy 89-057.

25. D. Y. Bardin et al, "Z line shape", contribution to the second CERN yellow report on LEP100 Z physics.

26. P. Langacker," The implications of recent M_z, M_w and neutral current measurements for the top quark mass", UPR-0400T, September 1989.

27. When the W mass is much better measured we can use it to predict the top mass, or vice versa. The current uncertainty in the W mass is too large for interesting limits on the top mass to be obtained.

28. This comparison, and that of table 5 uses the versions of code supplied by W.Hollik and B.Lynn to the author. They do not rely on previously published values. The small difference can be attributed to slightly different treatments by the two authors of b/t quark effects.

29. There was some discussion at the conference about the gauge invariance of some of the running coupling schemes, in particular the $\sin^2_* \theta$ scheme. The problems raised with the original formulation of this scheme (D. Kennedy and B. W. Lynn, Nucl. Phys. B322:1,1989.) have been addressed by Lynn in SU-ITP-867 (Aug 1989) which has been submitted to Phys.Lett. Similar calculations have been developed by W. Hollik, the interested reader is recommended to read his DESY report 88-188, for a complete discussion of this issue.

30. $\sin^2 \theta_w$ is decreased by a similar amount. The direction of the change is different than one might expect because increasing the Z^0 mass by 45 MeV/c^2 corresponds to an increase of 50 MeV/c^2 in the W mass if everything else is kept fixed.

31. Increasing the Higgs mass to 1000 GeV would raise $\sin^2_* \theta$. Calculations give 0.0236, but they are unreliable for such a high Higgs mass.

32. D. Y. Bardin et al; contribution to the second CERN yellow report on LEP100 physics.

33. For the current (September 1989) status of these programs see D. Bardin et al," On Some New Analytic Calculations for the Process $e^+e^- \rightarrow f\bar{f} +$ (nγ)",CERN-TH.5434/89.

34. J. Drees, Proceedings of the Rinberg Conference on Electroweak radiative corrections.

35. B.W. Lynn,"High Precision Tests of Electroweak Physics on the Z^0 resonance", contribution to report on CERN yellow report on Polarization at LEP.

THE DYMU2 EVENT GENERATOR

J.E. Campagne and R. Zitoun

LPNHE, Universités Pierre et Marie Curie et Paris VII

Abstract

We present briefly the DYMU2 event generator [1,2].

1 Introduction

The strategy for event generation is a copy of the Drell-Yan process and of parton-parton scattering in hadron-hadron collisions. First one keeps up with the initial bremsstrahlung on each incoming line, then an elementary process is simulated. For the point of view of formulae, we use structure functions for photon energy spectra, order $O(\alpha)$ for photon angular distributions, and the Born cross section stands for the elementary process.

2 Initial state radiation

It is assumed that the incoming electron and positron are dressed objects, i.e. that they only have fractions of initial energy x_+ and x_- at the annihilation vertex; x_+ and x_- are distributed according to the structure function $D_e(x_\pm, s)$ describe in appendix, assuming that $E_{\gamma\pm} = (1 - x_\pm)\sqrt{s}/2 = k_\pm E_{beam}$.

The remaining energy fraction $1 - x_+$ and $1 - x_-$ should normally be radiated off by photons (including an infinite number of soft ones) and in some cases by real lepton pairs. In our generator, we simplify this pattern and attribute the radiated energy to only one photon for each incoming lepton. The angular distribution of the photons is given by the $O(\alpha)$ one:

$$f_+(\mathbf{k}) = \frac{\alpha}{2\pi^2 k}\left(1 - k\frac{\delta_+}{2}\right)^2 \left[\frac{1}{\delta_+\delta_-} - \frac{2m_e^2}{s}\frac{1-k}{1+(1-k)^2}\left(\frac{1}{\delta_+^2} + \frac{1}{\delta_-^2}\right)\right]$$

which is the probability that the e^+ emit a photon with reduced momentum $\mathbf{k}_+$ (modulus k_+ and angles Ω_γ^+ with respect to the incoming positron velocity $\mathbf{v}$)[1], for the electron $\mathbf{v}$ is changed to $-\mathbf{v}$.

---

[1] $\delta_\pm = 1 \mp v\cos\theta_\gamma$

3 Annihilation probability

After radiation, the e^+e^- system has an energy $\sqrt{s'}$ (where $s' = (e^+ + e^- - k_+ - k_-)^2$) less than the incoming energy $\sqrt{s}$. As the annihilation probability is a function of s', a hit and miss procedure is made on the value of s' according to a Born cross-section $\sigma^{(0)}(s')$ described below. The two step process of photon and s' generation is very efficient around the Z^0 peak but deteriorates off the peak. Let us note that we truncate the QED peak of $\sigma^{(0)}$ at low s' in order to speed up the generation. This results in generating less events than one should for $\sqrt{s'} \lesssim 5\text{GeV}$, however those events are normally few and, due to there very collimated topology, will be cut out for experimental analysis. An improvement of the production of events with low squared mass [2] is foreseen to help people searching for rare processes.

The Born cross section used is the G_μ improved one given by G.Burgers [3]. It is written as [3]

$$\sigma^{(0)}(s) = \frac{8}{3\pi s} G_\mu^2 M_Z{}^4 s_w{}^4 c_w{}^4 \left[1 + \frac{2C_v^2 s(s - M_Z{}^2) + s^2(C_v^2 + C_a^2)^2}{(s - M_Z{}^2)^2 + s^2 (\Gamma_Z/M_Z)^2} \right]$$

and realizes a very good approximation of full standard model calculation done in ZBATCH (at the permil level for $\sqrt{s}$ between 88GeV and 100GeV and for top mass between 50GeV up to 150GeV). The muon angular distribution is generated in the annihilation reference system according to the Born differential cross section, the z-axis being the virtual e^+ line of flight.

4 Final state radiation

The case of final state radiation is more delicate than initial state as it involves structure function at a scale known after (contrary to before) radiation. In the first order event generator MUS-TRAAL [4] a proper generation of final state kinematics is done: the (single) photon energy and direction are generate according to the first order spectra and the muon pair is produced on the mass shell using redefined Born expressions. As we do not know at present time how to generate one photon on each final line, we have kept for DYMU2 the same strategy and formulae, except that the photon spectrum is changed to the exponentiated spectrum $F_\mu(y, ys')$ equal to $F_e(x, s)$ where the electron mass has been replaced by the muonic one (see ref [2]). More explicitly we have used

$$F_\mu(y, ys) = \left(1 + \frac{\alpha}{\pi}(\frac{\pi^2}{3} - \frac{1}{2}) + \frac{3}{4}\beta_\mu(s) \right) \times \beta_\mu(s)(1 - y)^{\beta_\mu(s)-1} - \frac{\beta_\mu(ys)}{2}(1 + y) \ ,$$

where we have not included a y-dependence of $\beta_\mu(s)$ in the Gribov-Lipatov solution, the $O(\beta_\mu^2)$ hard corrections, and other next-to-leading terms. We know anyway that the final state corrections contribute only a little to the total cross section (.17% or 3pb). However we must keep in mind that improving final state radiation improves at the same times accollinearity distribution.

5 Miscellaneous

5.1 Interference

In the present version of the program there is no initial/final state interference. We want to mention that several authors have shown that for loose cuts the interference terms could be

---

[2]First we will change the generation of $x_\pm$ to take into account more accurately the QED peak with no loss in speed, see results in section 6.

[3]We use an linear s-dependent width.

neglected at the .001 level. In addition, a severe cut on the total radiated energy only reflects in a slight change of the angular distribution around $\cos\theta_\mu = \pm 1$; this change will be difficult to observe, due to fiducial cuts in the beam region (see section 6).

5.2 Input parameters

DYMU2 uses as basic inputs the Z^0 mass (M_Z) and width (Γ_Z) and $\sin^2\theta_W$. The final state charged fermion (electrons are not considered) are specified by the value of their color, weak isospin, charge and mass.

The user may change parameters in the Block Data statement linked with different approximations in the initial/final state bremsstrahlung:

- ID2: takes into account (1) or not (0) the non-singlet e^+e^- real pair corrections to the initial structure function $D_e(x,s)$ (it changes β_e into η', see appendix);

- ID3: does the same as ID2 for the singlet correction. One should keep in mind that this correction is in competition with the two photon physics which is not considered here, however these corrections give low muon mass squared which can be eliminate by muon energy cut;

- IEXPO: switch on (1) or off (0) the exponentiation of the initial soft region;

- FINEXP: switch on (1) or off (-1) the final state radiation, and if (1) then the exponentiated spectrum is used. The value (0) should not be used;

- XK0: is only relevent when one uses the IEXPO=0 option, and gives as in MUSTRAAL the separation between soft and hard photon;

- POIDS: when set to 1 events are generated with the weight 1, and when set to 0 the weight is different from 1 (i.e not as in reality).

the other flags are not user's ones.

5.3 The USER subroutine

After each generated event the USER subroutine is called by the main program RADCOR, the common VECLAB being filled with final state particles four momenta. The first call may be used for user's initialisation (for exemple HBOOK). This routine is called a last time at the end of the run with common RESULT filled with cross sections and asymmetries. This subroutine is very usefull for producing differential cross sections.

5.4 DYMU2 & sons

By modifications of charge, isospin, color and mass one can generate $q\bar{q}$ final states (without flavour mixing for the moment).

A version for polarised τ production decaying to $\pi\nu$ can be turned on by setting parameter TAU to 1. Other τ decays are under implementation.

6 Results and discussions

The results we present in this section concern the reaction $e^+e^- \rightarrow \mu^+\mu^-$ obtained using as physical parameters $M_Z = 92\text{GeV}$, $\Gamma_Z = 2.562\text{GeV}$ and $\sin^2\theta_W = .2296$. Each result will be presented without considering the interference term except at the end of this section where a comparison with Greco et al [5] is performed.

Table 1

Comparison between ZBATCH and DYMU2 predictions for the total cross section (nb) with for input parameters: $M_Z = 92\text{GeV}$, $\Gamma_Z = 2.562\text{GeV}$ and $\sin^2\theta_W = 0.2296$; for ZBATCH the width and $\sin^2\theta_W$ are calculated giving M_Z and $m_t = 60\text{GeV}$ and $m_H = 100\text{GeV}$. We have switch off the final state bremsstrahlung from DYMU2.

$\sqrt{s}$ (GeV)	82	88	92	94	100
ZBATCH	0.047	0.153	1.446	0.632	0.134
DYMU2	0.049	0.154	1.446	0.635	0.136

Table 2

Comparison between COMPACT and DYMU2 where final state radiation are in and we have made some cuts: $M_{\mu\mu} \geq 20\%\sqrt{s}$ and $\theta_\mu \epsilon [10°, 170°]$. Quoted errors are purely statistical ones for 100,000 events. The input parameters are $M_Z = 92\text{GeV}$, $\Gamma_Z = 2.562\text{GeV}$ and $\sin^2\theta_W = 0.2296$.

$\sqrt{s}$ (GeV)		88	92	94	96
COMPACT	σ (nb)	0.143	1.408	0.614	0.286
DYMU2	σ (nb)	0.147(4)	1.409(4)	0.614(4)	0.289(4)
COMPACT	A_{FB}	-0.365	0.001	0.121	0.181
DYMU2	A_{FB}	-0.366(4)	-0.002(4)	0.117(4)	0.180(4)

Table 3

Comparison between COMPACT and various calculous as F.A.Berends et al and L.Trentadue et al for the $A_{FB}(10^{-2})$. For convenience we have used an expression of the axial coupling where $\sin^2\theta_W$ is replaced by G_μ and the width has no s-dependence. The input parameters are $M_Z = 93\text{GeV}$, $\Gamma_Z = 2.5\text{GeV}$ and $\sin^2\theta_W = 0.23$.

$\sqrt{s}$ (GeV)	82	92.5	93	93.5	100
F.Berends et al	-53.34	-3.64	0.29	3.76	18.85
L.Trentadue et al	-53.96	-3.69	0.28	3.78	18.79
COMPACT	-53.95	-3.67	0.28	3.77	20.31

The use of the $O(\alpha^2)$ corrected photonic spectrum and the G_μ-scheme for weak corrections allows DYMU2 to reproduce perfectly enough (table 1) the right total cross section given by ZBATCH (when DYMU2 final state bremsstrahlung is switched off [4]). Unfortunately experimentalists needs cross sections with cuts (imposed by the acceptance of the detector for example) which are not provided by ZBATCH. So, we have proposed an ansatz (COMPACT integration program) described in references [2,6,7] to incorporate such cuts in a compact expression. The agreement is quite good (table 2). Extensive comparisons with more accurate semi-analytical formulae are under way by other authors [8] and preliminary results are encouraging.

The A_{FB} variable is also provided by COMPACT (table 3) which have been recently checked with F.A.Berends et al and L.Trentadue et al calculous [9]. Then, the table 2 also shows good agreement between DYMU2 and COMPACT when applying cuts.

In figs. 1.(a,b,c) and 2.(a,b,c) we present the $\mu^+\mu^-$ acollinearity angle ζ and the total photonic energy E_γ for three beam energies: $\sqrt{s}$ =80, 92, 100GeV. It is noticeable that we now keep up with the famous k_o parameter used at first order as no more gap in the photonic spectrum is visible in fig.3. Comparison with results in reference [2] shows that DYMU2 has been improved to properly generate very hard photons (i.e. very low muon mass squared) than it did. One also can see the contribution of the final state bremsstrahlung on the peak compared to 12GeV below or 8GeV above. One has not to conclude that the total size of final state bremsstrahlung is important because one forgets virtual corrections which compensates practically real ones to give the well known $3\alpha/4\pi$ factor. But anyway these plots reveal that any cuts on ζ or E_γ on the peak of the resonance will suppress more the final state radiation than the initial one.

The original part of DYMU2 is the staightforward photon generation where the function $f_\pm$ governs the angular distribution (fig. 4). Its backward peak allows us to mimic the double initial bremsstrahlung (see FPAIR program results [10]) as we generate one photon per initial line. The dynamics of these two photons can be appreciated by the two scatter plots:

- $(X_1 = k^{max}/\sum_{i=+,-} k^i, Y_1 = k^{max}/E_{beam})$ figs. 5.(a,b,c);
- $(X_2 = \sum_{i=+,-} k_z^i/E_{beam}, Y_2 = \sum_{i=+,-} k^i/2E_{beam})$ figs. 6.(a,b,c),

where k^{max} is the energy of the most energetic photon and k_z is the component along the beam axis of the photon momentum. These plots are completly new compared with first order event generator as MUSTRAAL [4], where the (X_1, Y_1) plot is one-dimensional and the (X_2, Y_2) is truncated at $Y_2 = 1/2$ and is practically a simple "V" due to $f_\pm$ forward-backward peaks. One also notices on these plots the influence of final state radiation.

On figs. 7.(a,b,c) we present ratio between the muon angular distribution given by DYMU2 and Greco et al formula (soft exponentiation performed from coherent states formalism) where in both we have cut the total radiated energy at 1GeV. The flatness of the ration has been recently confirmed by Bardin et al (exact first order calculation on $d\sigma/d\Omega_\mu$ and soft exponentiation) for all energy cuts and especially in the case where no cut is applied. Then the muon angular distribution is shown at $\sqrt{s} = 80, 92, 100$GeV on figs. 8.(a,b,c). These differential cross sections are not affected by final state bremsstrahlung as foreseen when one neglects interference terms.

In the previous discussion, we assumed that interference between initial and final state radiation is negligible. It is true on the peak as long as one does not apply severe cuts on total radiated energy i.e. E_γ^{max} =Max($\sum E_\gamma$) $\geq$ 1GeV. For first generation of experiments, this remark is irrelevant because, during the scanning phase of the Z^0, the statistical errors on σ and A_{FB} will be of the order of 5% (off the peak) and 3% (on) so no special care has to be taken. In addition, first order result of S.Jadach et al [11] shows that interference term contribution is of the order of 10^{-3} on A_{FB}. Nevertheless one can see on figs. 9.(a,b,c) the effect of interference

---

[4]We know that final state bremsstrahlung contributes for 3pb to the total cross section.

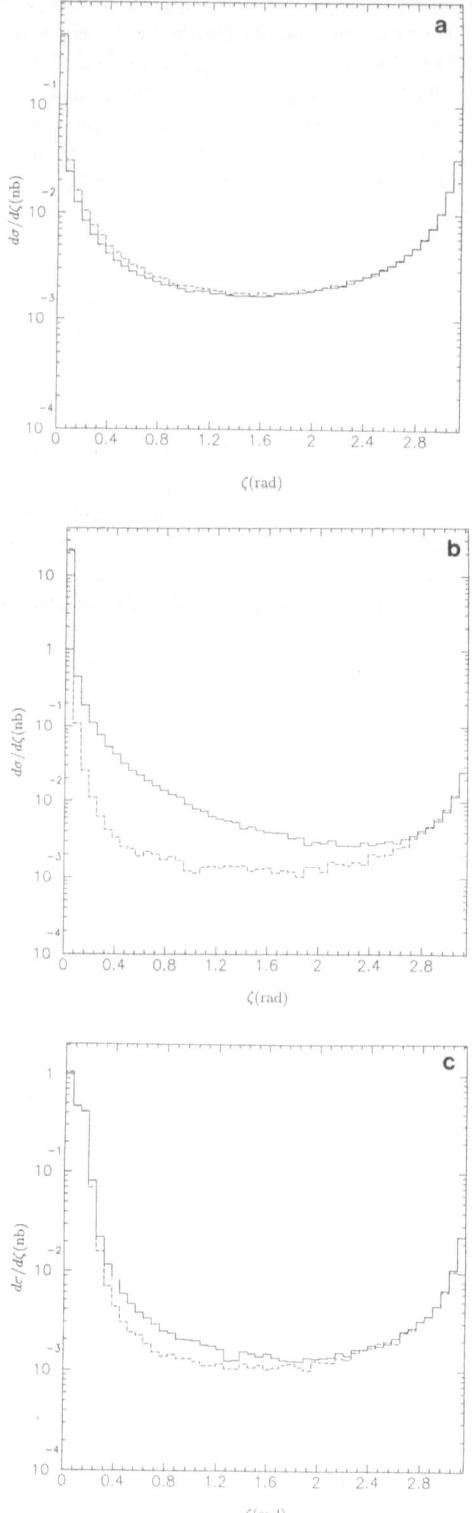

Figs.1. Distribution $d\sigma/d\zeta$ at energies $\sqrt{s}$ =80 (a), 92 (b), 100GeV (c) given by DYMU2 with initial and final state bremsstrahlung together (—), and with initial one alone (- - -).

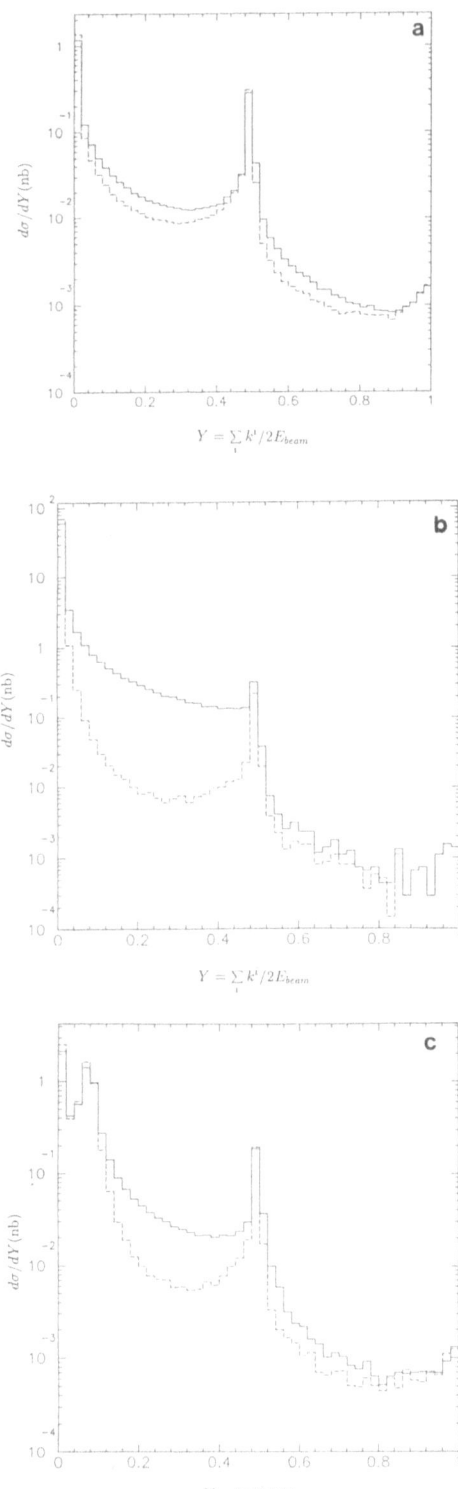

Figs.2. Distribution $d\sigma/dY$ where $Y = \sum_i k^i/2E_{beam}$ at energies $\sqrt{s}$ =80 (a), 92 (b), 100GeV (c) given by DYMU2. The full curve gives the predictions with initial and final state bremsstrahlung together ($i = +, -, f$), and the dashed curve with initial one alone ($i = +, -$).

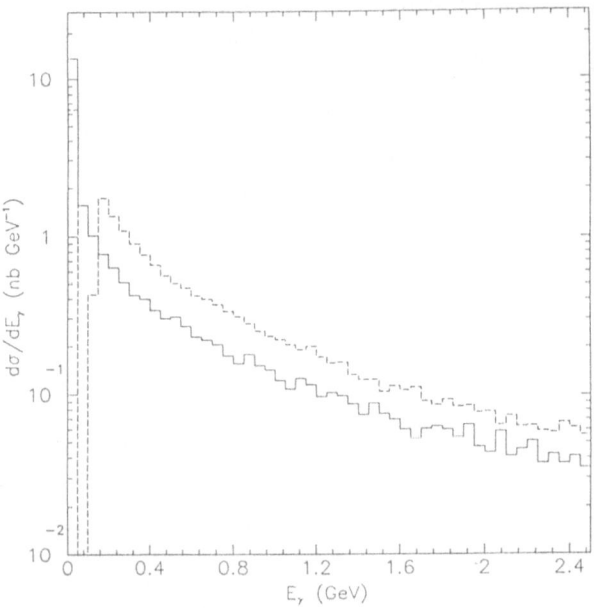

Fig.3. Photonic energy spectrum at $\sqrt{s} = M_Z$. The full curve is obtained with DYMU2. The dashed curve is obtained with MUSTRAAL [4]. We have taken M_Z =93GeV, Γ_Z =2.5GeV and $\sin^2 \theta_W$ =0.223.

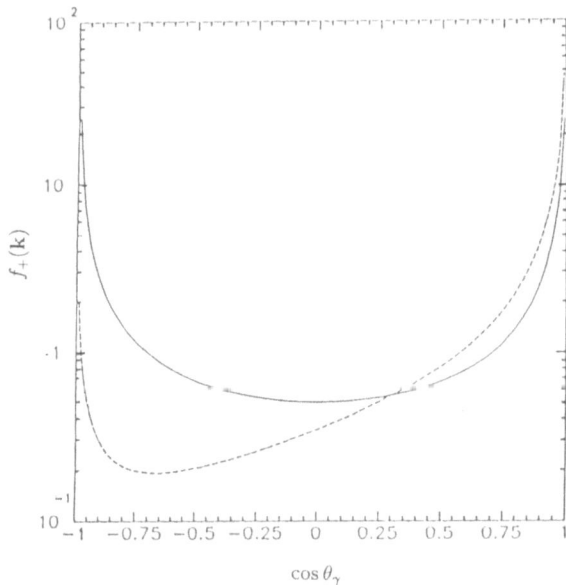

Fig.4. The function $f_+(\mathbf{k})$ (see definition in section 2) for $k = 0.1$ (—) and $k = 0.8$ (- - -).

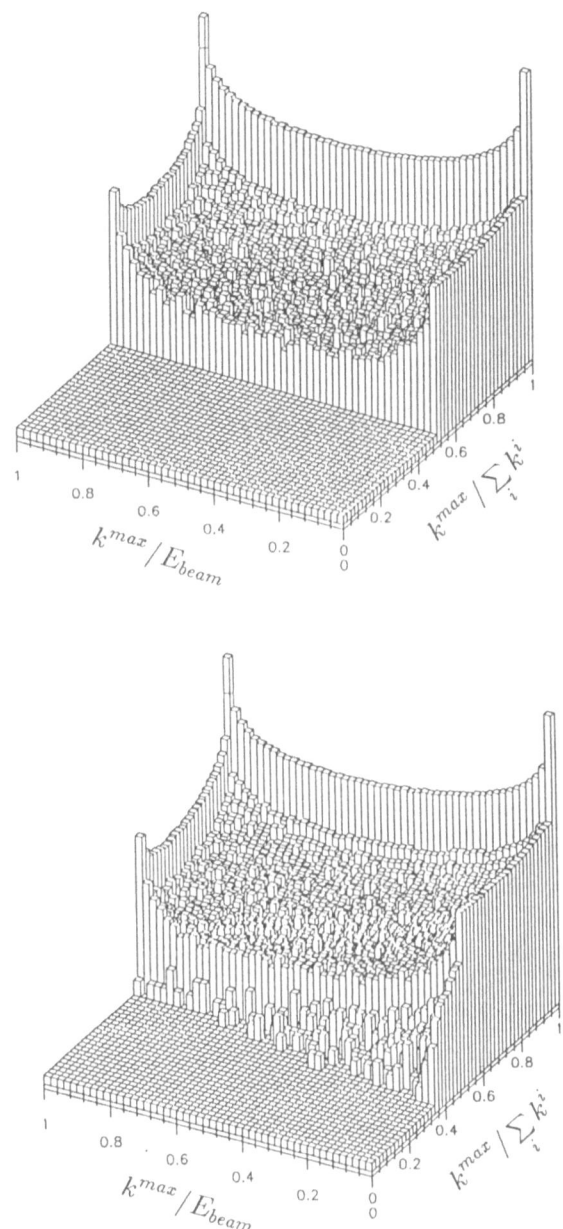

Fig.5. For $\sqrt{s}=80$ (a), 92 (b) and 100GeV (c) using DYMU2 we present $(k^{max}/\sum_i k^i, k^{max}/E_{beam})$ lego-plots where the top one just takes initial bremsstrahlung into account $(i=+,-)$, and the bottom one includes final radiation $(i=+,-,f)$.

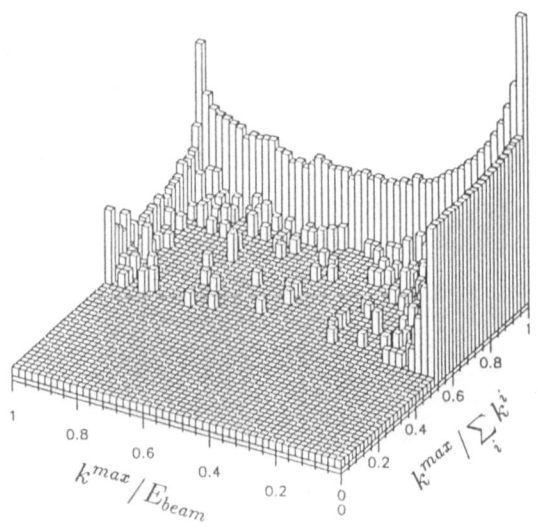

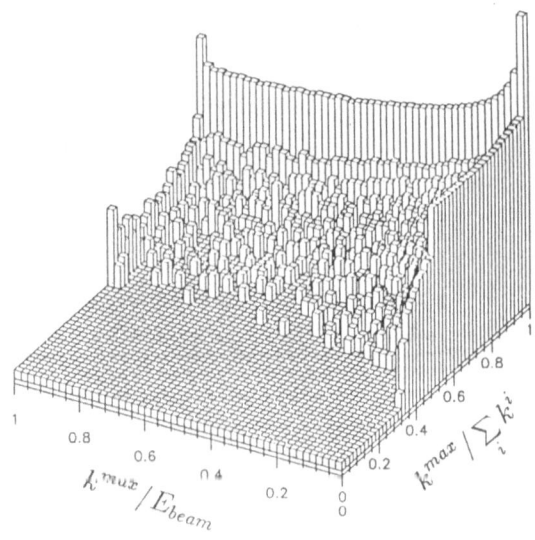

Fig.5b

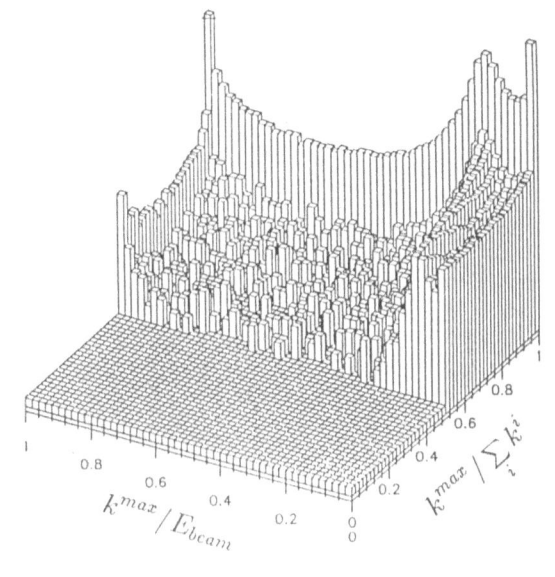

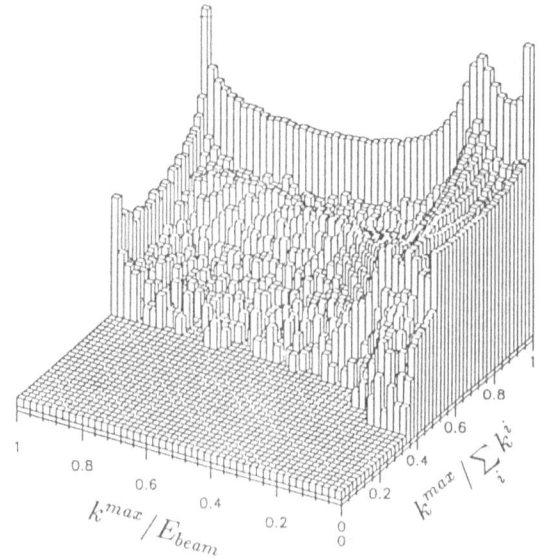

Fig.5c

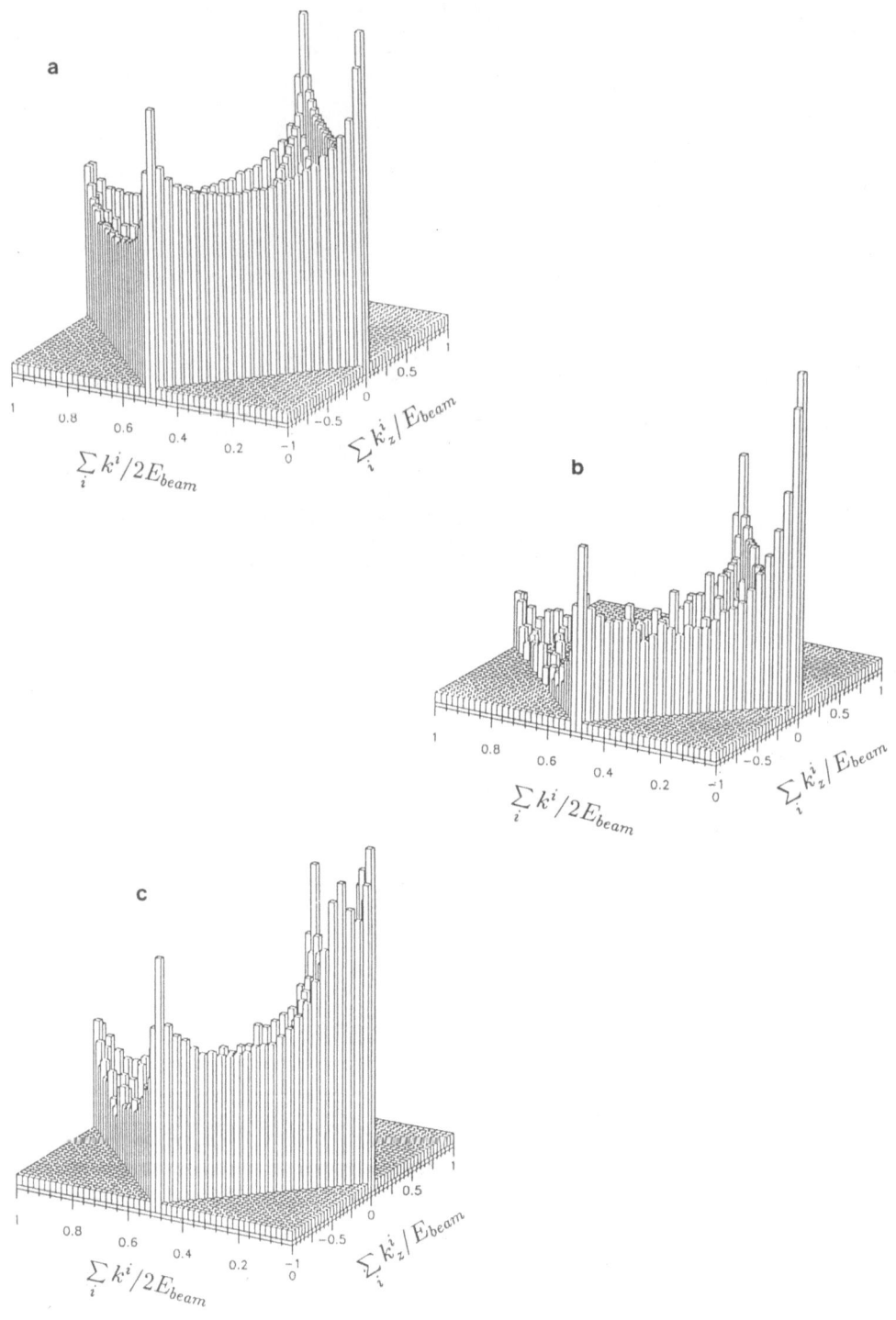

Fig.6. For $\sqrt{s}$ =80 (a), 92 (b) and 100GeV (c) using DYMU2 we present $(\sum_i k_z^i/E_{beam}, \sum_i k^i/2E_{beam})$ lego-plots where only the initial bremsstrahlung $(i = +, -)$ is taken into account.

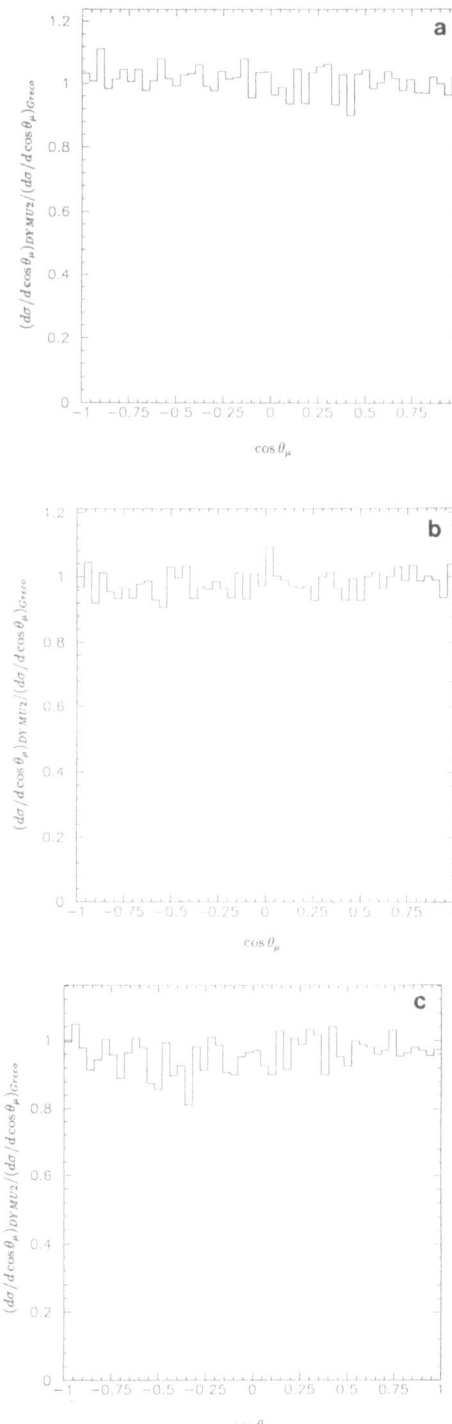

Fig.7. Ratio DYMU2 over Greco et al formula [5] for the muon angular distribution at $\sqrt{s}$ =90 (a), 92 (b) and 94GeV (c). Interference terms have been switch off.

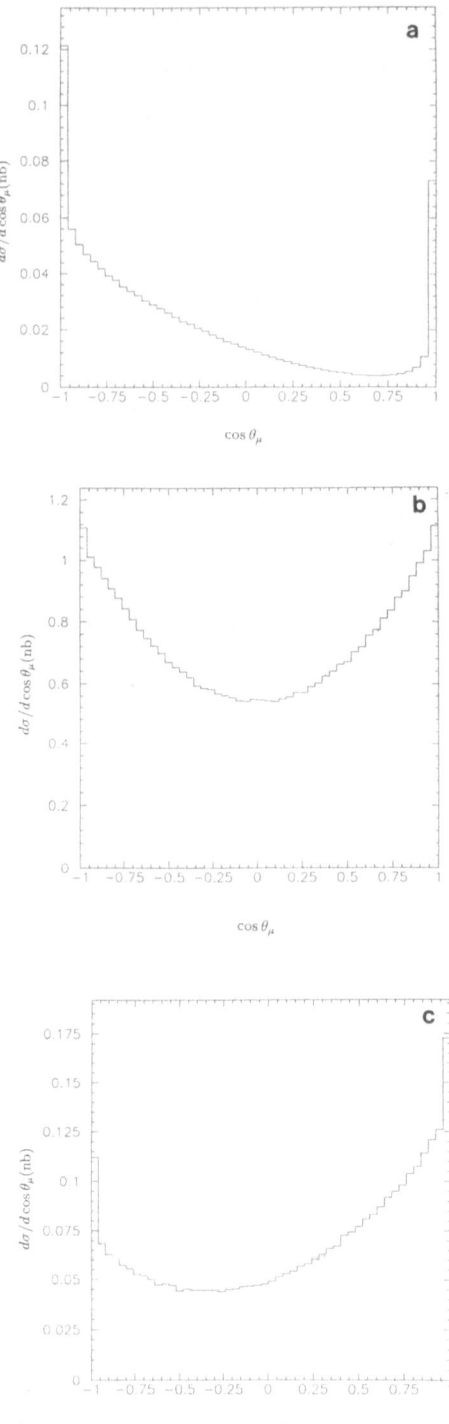

Fig.8. Distribution $d\sigma/d\cos\theta_{\mu^+}$ at energies $\sqrt{s}$ =80 (a), 92 (b), 100GeV (c) given by DYMU2 with initial and final state bremsstrahlung together. No visible difference appears when switching off the final radiation.

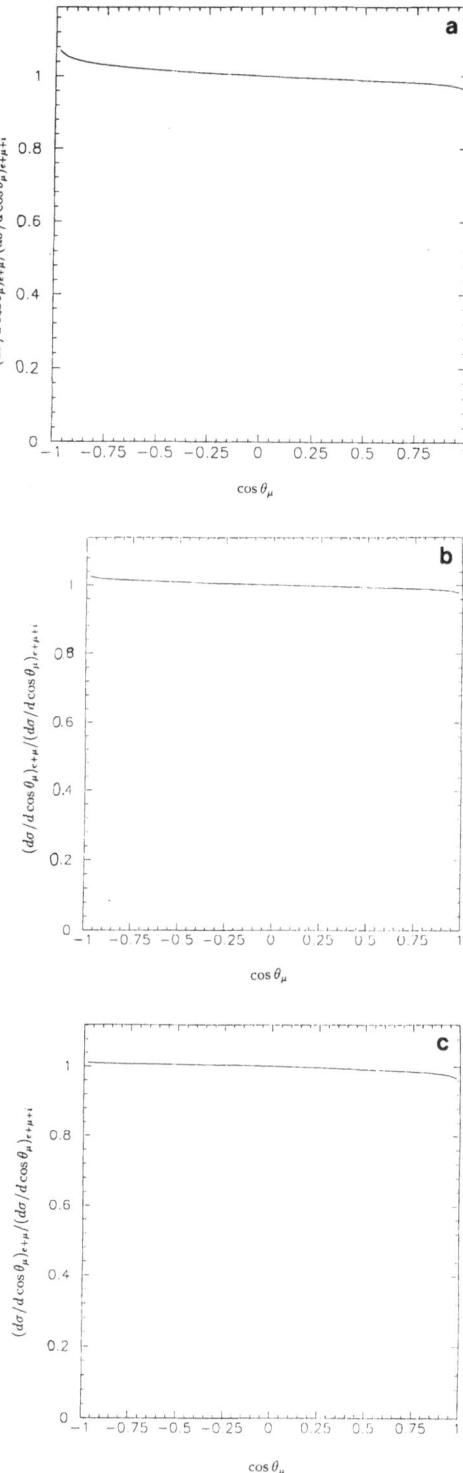

Fig.9. Ratio "without interference terms" $(e+\mu)$ over "with interference terms" $(e+\mu+i)$ for the muon angular distribution given by Greco et al formula [5], at various energies ($\sqrt{s}$ =90 (a), 92 (b) and 94GeV (c)) arround the pic, and for a total radiated energy less than 1GeV.

terms given by Greco et al formula, where interference contribution comes from exponentiated soft approximation. The distortion on the muon angular distribution is visible in the $\cos\theta_\mu = \pm 1$ region. However as E_γ^{max} increases, this effect vanishes. Moreover, it mainly concers beam pipe region ($\theta \lesssim 15°$) in which no observation will be available.

APPENDIX: The DYMU2 structure function

The authors had derived [2,12,13] from exact $O(\alpha^2)$QED corrections to the initial state of e^+e^- annihilation [14] the structure function $D_e(x,s)$ used in DYMU2. The result can be written

$$
\begin{aligned}
D_e(x,s) &= \frac{\exp\left\{\frac{\eta'}{8}(3 - 4\gamma_E)\right\}}{\Gamma\left(1 + \frac{\eta'}{2}\right)} \frac{\eta'}{2}(1-x)^{\frac{\eta'}{2}-1} - (1+x)\frac{\eta'}{4} + \frac{A(x)}{16}\eta'^2 \\
&+ \frac{\alpha\eta'}{4\pi}\left\{\left(2A + B' - \frac{3}{2}A_0 - \frac{1}{24}\right)\delta(1-x) + 2A(x) + B(x) + (1+x)A_0 \right. \\
&\left. - \frac{13}{18}P_{ee} + \frac{1}{6}P_{ee} * P_{ee} - \frac{1}{3}(1-x) - \frac{1}{6}(1+x)\ln x\right\}
\end{aligned}
$$

with the K-factor used in Drell-Yan formalism becoming

$$
K = 1 + \frac{\alpha}{\pi}A_0 + \frac{\alpha\eta'}{2\pi}\left(B - B' - \frac{A}{3} + \frac{5}{24}\right) \ .
$$

The η' variable allows to include the non-singlet corrections and is defined as:

$$
\eta' = \beta_e + \frac{\beta_e^2}{12} = -6\ln(1 - \frac{\beta_e}{6}) \ .
$$

References

[1] J.E.Campagne and R.Zitoun, Electromagnetic radiative corrections at LEP-SLC energies for experimentalists, LPNHEP 88.08 to be published in Z.Phys C Part.andFields.

[2] J.E.Campagne, Ph.D thesis, LPNHEP 89.02.

[3] G.Burgers, The shape and size of the Z resonance, Polarisation at LEP Vol.1 p121.

[4] F.A.Berends, R.Kleiss and S.Jadach, Comp.Phys.Comm 29 (1983) 185.

[5] M.Greco, G.Pancheri-Srivastava, Y Srivastava, Nucl Phys B171 (1980) 118; Nucl.Phys. B197 (1982) 543.

[6] J.E.Campagne and R.Zitoun, QED corrections to the forward backward asymmetry at LEP energies, LPNHEP 89.01 to be published in Phys. Lett. B.

[7] J.E.Campagne and R.Zitoun, Compact formula allowing for experimental cuts, CERN Yellow Book in preparation.

[8] D.Bardin et al, private communication.

[9] A_{FB} working group for the CERN Yellow Book in preparation.

[10] S.van der MARCK, FPAIR event generator, private communication.

[11] S.Jadach and Z.Wąs, Suppression of QED interference contribution to the charge asymmetry at the Z^0 resonance, CERN-TH.5127/88.

[12] J.E.Campagne and R.Zitoun, An expression of the electron structure function in QED, LPNHE-PARIS, LPNHEP-88.06

[13] J.E.Campagne and R.Zitoun, DYMU2 structure function, CERN Yellow Book in preparation.

[14] F.A.Berends, W.L.van Neerven and G.J.H.Burgers, Nucl.Phys **B297** (1987) 429; Phys.Lett. **B185** (1987) 395.

STRUCTURE FUNCTIONS APPROACH
TO ELECTROWEAK RADIATIVE CORRECTIONS

Luca Trentadue

Dipartimento di Fisica
Universitá di Parma
Parma, Italy

and

Istituto Nazionale di Fisica Nucleare
Gruppo Collegato di Parma
Sezione di Milano

Abstract

The structure function formalism to describe the electromagnetic radiative corrections in e^+e^- colliders is briefly discussed. Specific applications of the method to physical quantities are listed. Some more recent uses of the structure function method are also mentioned.

Introduction

An evaluation of electromagnetic radiative corrections to leptonic processes is required for precision tests of the electroweak theory around the Z^0 at the colliders LEP and SLC [1].

Starting from the first developments of the electron-positron accelerators this subject has been actively investigated [2]. The formalism of the structure functions, widely applied to describe the interactions of partons within the Quantum Chromodynamics (QCD), has been used [3-5] for a new series of applications also in the framework of the Quantum Electrodynamics (QED). Structure function evolution equations have been applied to describe the interactions of fermions within vector theories [6]. The more complex case of the interaction within non-Abelian gauge theories has been also considered [7]. At high energies, particularly around peaked resonances, the structure function approach is particularly useful since the infra-red and collinear singularity structure of the radiative corrections dominates the dynamics of the reaction [2,3-5].

Radiative Corrections, Edited by N. Dombey and
F. Boudjema, Plenum Press, New York, 1990

General Formalism

The probability of finding within a given electron state a an electron or a photon b at a given scale or virtualness $k^2 = s$ with a fraction of longitudinal momentum $x = \frac{k_l}{E}$ with k_l the longitudinal momentum of the electron and $E = \sqrt{s}/2$ can be given by defining the distribution $D_{ab}(x, s)$. Evolution equations can be defined for these distributions. The evolution equations can be written with the electron and positron distributions that satisfy the 'master' equation:

$$D_{ee}(x, s) = D(x, s) = \delta(1 - x) + \int_{m_e^2}^{s} \frac{dk^2}{k^2} \frac{\alpha(k^2)}{2\pi} \int_x^1 \frac{dz}{z} P^{NS}(z) D_{ee}(\frac{x}{z}, k^2) \quad (1)$$

with

$$\alpha(k^2) = \frac{\alpha}{1 - \frac{\alpha}{3\pi} \ln(k^2/m_e^2)}$$

where

$$P^{NS}(z) = P(z) = \frac{1 + z^2}{1 - z} - \delta(1 - z) \int_0^1 dx \frac{1 + x^2}{1 - x} \quad (2)$$

is the regularized $electron \rightarrow electron + photon$ vertex where the first term represents the 'real' photon radiation and the second the 'virtual' corrections. The $\delta(1 - x)$ represents a Born source term within the evolution equation and the regularization is obtained with the inclusion of the 'virtual' self-energy type contributions.

The physics involved in the problem of the radiative corrections to the initial state in $e^+ e^-$ annihilation is related to the emission of quanta from the annihilating electron and positron states and to their effect on the size and shape of the resonance peak. In the production of a resonance of mass M with colliding beams of total centre-of-mass energy $\sqrt{s} = 2E$. For $\sqrt{s} > M$, in the energy region above the resonance mass, this has the effect of decreasing the total center-of-mass energy to M. These states, therefore, effectively contribute to the resonance production. As a result a radiative tail arises on the right-hand side of the resonance peak and, the area under the peak not being affected by radiative corrections, a lowering of the maximum of the peak follows [2].

For an electron or positron state the radiative corrections are characterized in the perturbative series by terms of the type $(\alpha/\pi)^n \ln^p(s/m_e^2) \ln^q(E/\lambda)$ with m_e the electron mass and λ a scale on which there is a sizeable variation of the cross-section. In our case $\lambda = \delta E$ is the difference $\sqrt{s} - M$, i.e. the energy radiated away by the annihilating states. At the mass of the Z^0 $L = \ln(s/m_e^2) \simeq 24$ and the possibility that also the soft logarithms $l = \ln(E/\delta E)$ are also large implies that the effective expansion parameter $(\alpha/\pi)Ll$ is large too, and therefore these terms in the perturbative series must be taken into account and summed.

The process we are considering is the $e^+ e^-$ annihilation into the Z^0. According to a QCD analogy this scattering might be seen as a Drell-Yan process and, taking into account the well known theorems on the factorization of the mass and infra-red singularities [8], its cross-section can be written in the following form :

$$\sigma(s) = \int dx_1 \int dx_2 \, D_{e^-}(x_1, s) \, D_{e^+}(x_2, s) \, \sigma_0(x_1 x_2 s) \quad (3)$$

$D_{e^-(+)}(x, s)$ represents the electron (positron) structure function. $\sigma_0(x_1 x_2 s)$ represents the resonance-cross section at the reduced energy $x_1 x_2 s = s'$.

Various solutions can be obtained for $D(x, s)$ [3-5,9]. By using second order results for the electron form factor [10] and by iterating eq.(1) one has that[5]:

$$D_{e-(+)}^{NS}(x, s) = \frac{\beta}{2}(1 - x)^{\frac{\beta}{2}-1}\Delta^{\frac{1}{2}} - (1 + x)\frac{\alpha L}{2\pi} + (1 + x)\frac{\alpha}{2\pi}$$

$$+ \frac{1}{2}(\frac{\alpha L}{2\pi})^2[(1 + x)(-4\ln(1 - x) + 3\ln x) - \frac{4}{1 - x}\ln x - 5 - x]$$

$$- \frac{1}{2}(\frac{\alpha}{2\pi})^2(2L - 1)[(1 + x)(-4\ln(1 - x) + 3\ln x) - \frac{4}{1 - x}\ln x - 5 - x]$$

$$(4)$$

where $\beta = \frac{2\alpha}{\pi}(L - 1)$ and Δ is given by the expression [3,5] :

$$\Delta = 1 + \frac{\alpha}{\pi}(\frac{3}{2}L + \frac{\pi^2}{3} - 2) + (\frac{\alpha}{\pi})^2[(\frac{9}{8} - 2\zeta(2))L^2 + (-\frac{45}{16} \tag{5}$$

$$+ \frac{11}{2}\zeta(2) + 3\zeta(3))L - \frac{6}{5}(\zeta(2))^2 - \frac{9}{2}\zeta(3) - 6\zeta(2)\ln 2 + \frac{3}{8}\zeta(2) + \frac{57}{12}].$$

By substituting the result for $D_{e-(e+)}(x, s)$ into eq.(1) we have by defining $(1 - \chi)s = s'$ that the cross-section becomes:

$$\sigma(s) = \int d\chi \, \sigma_0((1 - \chi)s) \, H(\chi, s) \tag{6}$$

with

$$H(\chi, s) = \Delta \left(\beta\chi^{\beta-1} - \frac{\beta^2\chi^\beta}{4}\right) - \frac{\beta\chi^{\frac{\beta}{2}}\Delta_{(1)}^{\frac{1}{2}}}{4}\left[(2 - \chi)(1 + (1 - \chi)^{-\frac{\beta}{2}})\right.$$

$$\left. - \frac{\beta}{(2 + \beta)}\chi(1 - (1 - \chi)^{-\frac{\beta}{2}})\right] + \frac{\beta^2}{16}\left[(\chi - 2)(4\ln\chi - 3\ln(1 - \chi))\right.$$

$$\left. - \frac{4}{\chi}\ln(1 - \chi) - 6 + \chi + 2\chi - (2 - \chi)\ln(1 - \chi)\right],$$

where $\Delta_{(1)} = 1 + \frac{\alpha}{\pi}(\frac{3}{2}L + \frac{\pi^2}{3} - 2)$.

In the expression above not only the dominant $(\alpha L)^n$ terms are summed but also the less dominant $\alpha^n L^m$ with $n \geq m \geq 0$ are taken into account up to $n = 2$. The first term $\frac{\beta}{2}(1 - x)^{\frac{\beta}{2}-1}$ corresponds to the soft photon approximation. The terms proportional to $(1 + x)$ and the last term contain contributions which modify the result accounting for emission of hard photons. The Δ factor contains terms of the type $(\frac{\alpha}{\pi})^n L^m, n \geq m \geq 0$.

This expression differs [5] from the one obtained in ref.[3] for the inclusion of the terms $(\frac{\alpha}{\pi})^2 L^2, (\frac{\alpha}{\pi})^2 L, (\frac{\alpha}{\pi})^2 constant$ in the factor Δ . It agrees with the result obtained in ref. [11] by an exact $O(\alpha^2)$ calculation apart from terms that are relevant only within the 'hard' $x \to 0$ limit.

These facts all show that, $O(\alpha^3)$ corrections being really negligible, the accuracy reached in the cross-section in eq.(6) is below the one per cent level.

Also final state corrections [12] together with initial-final states interference, are needed to reproduce physical quantities such as acollinearities and asymmetries. In ref.[13] the general case of including the electromagnetic radiative corrections to the entire process has been considered. To this purpose the structure function formalism developed in [3-5] has been used and extended to both initial and final states.

The generalization of eq.(6) to take into account also final state radiation is, for the factorized part of the cross-section,

$$\sigma_f(s) = \int dx_1 dx_2 dy_1 dy_2 \, \sigma_0(s') \, D_e(x_1, s) D_e(x_2, s) D_\mu(y_1, s'') D_\mu(y_2, s'') \qquad (7)$$

where $s' = x_1 x_2 s$ and $s'' = y_1 y_2 s'$, with $x_{1,2}$ and $y_{1,2}$ fractions of longitudinal momentum of the electrons and muons respectively. $D_\mu(x, s)$ is the structure function for the muon, obtained from $D_e(x, s)$ with the substitution $m_e \to m_\mu$. Note that the scale s'' at which D_μ are evaluated is the invariant mass squared of the final real muon pair. By making the substitution $1 - x = x_1 x_2$ and $1 - y = y_1 y_2$ and recalling eq.(6) for the initial state radiator one has:

$$\sigma_f(s) = \int_0^{r_{max}} dx \, \sigma_0(s') \, H_e(x, s) \, F_\mu(r_{max} - x, s') \qquad (8)$$

where the upper limit r_{max} is the maximum fraction of radiation emitted and can be properly choosen and $F_\mu(z, s) = \int_0^z dx \, H_\mu(x, (1 - x)s)$. The final state radiation kernel $H_\mu(x, s)$ is defined as:

$$H_\mu(x, s) = \int_{1-x}^1 \frac{dy}{y} \, D_\mu(y, s) \, D_\mu\left(\frac{1 - x}{y}, s\right).$$

H_μ contains the same set of contributions as $H_e(x, s)$. eq.(8) is in a factorized form. σ_0 corresponds to the Born cross-section. To take into account also the non-factorizable corrections having photon lines connecting the initial with the final state, box diagrams and initial-final state interference contributions should be also included. One has:

$$\sigma(s) = \int_0^{r_{max}} dx \, \sigma_k(s') \, H_e(x, s) \, F_\mu(r_{max} - x, s') \qquad (9)$$

where $\sigma_k(s)$ is the effective integration kernel as defined above and given by [13]

$$\sigma_k(s) = \sigma_0(s) + \sigma_{box}(s) + \sigma_{int}(s).$$

Recently Aversa and Greco [14] by using the structure function formalism and the expression of the total cross section in eq.(9) have been able to derive an analytical solution of the same equation by explicitly including box and interference contributions.

Transverse Momentum Structure Functions

In order to deal with the problem of taking into account also the angles and the transverse momentum of the emitted photons, some extended distributions can be defined. Structure functions in eq.(1) can be generalized to take into account also the transverse degrees of freedom [15]. The evolution equation for $D(x, p_t; s)$ in the case of space-like kinematics, which is appropriate for the annihilation process, has the form:

$$D(x, p_t; s) = \delta(1 - x)\delta^{(2)}(p_t) + \frac{\alpha}{2\pi} \int_{m^2}^s \frac{dk^2}{k^2 + m^2} \int_x^1 \frac{dz}{z} P(z)$$

$$\int \frac{d^2 q_t}{\pi} \delta((1 - z)k^2 + z(1 - z)m^2 - q_t^2) D(\frac{x}{z}, p_t - \frac{x}{z}q_t; k^2). \tag{10}$$

$D(x, p_t; s)$ represents the probability of finding inside a parent electron, at the scale s, an electron with fraction of longitudinal momentum x and transverse momentum p_t with respect to the initial beam direction. eq.(10) can be solved by iterating the first term on the r.h.s., analogously to what can be done for the integrated $D(x, s)$ distribution. With one iteration of the source term we have:

$$D(x, p_t; s) = \delta(1 - x)\delta^{(2)}(p_t)$$

$$+ \frac{\alpha}{2\pi} P(x) \frac{1}{\pi} \frac{1}{p_t^2 + (1 - x)^2 m^2} \Theta\left((1 - x)s - p_t^2\right) + O\left(\frac{p_t^2}{E_\gamma^2}\right), \tag{11}$$

where Θ is the step function. As for the x-dependent distributions, we define a p_t-dependent radiator $H(x, p_t; s)$ as:

$$H(x, p_t; s) = \int_{1-x}^1 \frac{dz}{z} \int d^2 k_t D_{e^-}(z, k_t; s) D_{e^+}(\frac{1 - x}{z}, p_t - k_t; s),$$

where the indices e^- and e^+ label radiation from the electron and the positron respectively. The expression for $H(x, p_t; s)$ is, at order α,

$$H^{(\alpha)}(x, p_t; s) = \delta(x)\delta^{(2)}(p_t) + \frac{\alpha}{2\pi} P(1 - x)$$

$$\frac{1}{\pi} \left[\frac{1}{p_{te^+}^2 + x^2 m^2} + \frac{1}{p_{te^-}^2 + x^2 m^2}\right] \Theta\left(xs - p_t^2\right) + O\left(\frac{p_t^2}{E_\gamma^2}\right). \tag{12}$$

An angle-dependent radiator $H^{(\alpha)}(x, \cos\theta; s)$ can be also defined. The transverse momentum and the angle of the emitted photon, measured with respect to the incoming electron, are linked by the relations

$$p_{te^-}^2 = 2EE_\gamma(1 - \cos\theta) \quad p_{te^+}^2 = 2EE_\gamma(1 + \cos\theta).$$

One obtains, for the angle-dependent radiator, the expression:

$$H^{(\alpha)}(x, \cos\theta; s) = \frac{\alpha}{\pi} \frac{1 + (1 - x)^2}{x} \frac{1}{1 + \frac{4m^2}{s} - \cos^2\theta} + O\left(\frac{m^2}{s}\right).$$

$H(x, \cos\theta; s)$ can be rewritten as [16]:

$$H^{(\alpha)}(x, \cos\theta; s) = \frac{\alpha}{2\pi} \frac{1}{x} \left[2\frac{1 + (1 - x)^2}{1 + \frac{4m^2}{s} - \cos^2\theta} - x^2\right] + O\left(\frac{m^2}{s}\right).$$

It has been proposed [17] that the number of neutrino families is a quantity that can be determined by counting photons produced in electron-positron annihilation around the Z^0 resonance peak.

As for the total cross-section case this formalism has the main advantage that it is able to take into account and resum, in a compact and straightforward way, initial as well as final state configurations that contain soft and collinear radiation.

The neutrino counting problem is related to different final states, i.e. those containing one or more isolated photons [17]. These, radiated only by the initial electrons, are detected. Those configurations define a quantity which is less inclusive than the total cross-section and, for the neutrino counting problem, states with one or more radiated photons should also be accounted for.

The problem is to determine the energy and the angle of emission of photons radiated by initial electrons and positrons accompanying the production of Z^0's. When a Z^0 decays into a $\nu\bar{\nu}$ pair, the photon is the only observed state. In the average event, however, together with this process also hard photons radiated in the very forward direction and soft ones radiated all over the solid angle are likely to be produced. The former photons are lost in the beam pipe or rejected according to the geometrical set-up of the apparatus; the second ones will not be observed for energies below the detector threshold.

Radiative corrections to $e^+e^- \rightarrow \gamma\nu\bar{\nu}$ have been evaluated with standard Feynman-diagram tecniques [18,19] to the order α. The corresponding cross-section can be also computed by using the method of the structure functions for the initial fermions [20].

By using the radiator obtained in eq.(12), the cross-section for the process under consideration is

$$\frac{d^2\sigma_0}{dx\,dy} = H^{(\alpha)}(x, y; s)\, \sigma_0\left((1-x)s\right), \tag{13}$$

where σ_0 is the "reduced" cross-section for the process $e^+e^- \rightarrow Z, W \rightarrow \nu\bar{\nu}$:

$$\sigma_0(s) = \frac{G_F^2 s}{12\pi} \left(2 + \frac{N_\nu\left(g_v^2 + g_a^2\right) + 2\left(g_v + g_a\right)\left[1 - \frac{s}{M_Z^2}\right]}{\left[1 - \frac{s}{M_Z^2}\right]^2 + \Gamma_Z^2/M_Z^2} \right). \tag{14}$$

In order to analyze the various contributions to the radiative corrections, let us first consider the virtual and soft ones. One has [20] the soft spectrum:

$$\frac{d\sigma^{soft}}{dx} = H_{\theta_{min}}^{(\alpha)}(x, s) \int_0^{x_{min}} d\xi\, H(\xi, s)\, \sigma_0\left((1-x)(1-\xi)s\right). \tag{15}$$

eq.(15) describes the x spectrum of a real photon accompanied by soft and virtual radiation. The differential spectrum is given by:

$$\frac{d\sigma}{dx_1\,dx_2\,dy_1\,dy_2} = H^{(\alpha)}(x_1, y_1; s)H^{(\alpha)}(x_2, y_2; (1-x_1)s)\sigma_0((1-x_1)(1-x_2)s).$$

By substituting to $H^{(\alpha)}(x, y; s)$ the matrix element as given in eq.(12), the double bremsstrahlung matrix element is recovered.

It is necessary to distinguish two kinds of hard corrections: a) hard photons lost in the pipe and b) hard photons parallel (within an apparatus-dependent resolution angle) to the observed one. We have [20]

$$\frac{d\sigma^{pipe}}{dx} = H^{(\alpha)}_{\theta_{min}}(x;s)\cdot$$

$$\cdot \int_{\frac{x_{min}}{\sqrt{1-x}}}^{\sqrt{1-x}} dx_1 H^{pipe}(x_1;(1-x)s)\sigma_0\left((1-x_1)(1-x)s\right), \qquad (16)$$

where H^{pipe} is given by

$$H^{pipe}(x,s) \equiv 2 \cdot \int_{\cos\theta_{veto}}^{1-\frac{m^2}{s}} dy \, H(x,y;s) = \Delta(s)\,\beta_v\,x^{\beta_v-1} - \frac{1}{2}\beta_v(2-x)$$

$$+ \frac{1}{8}\beta_v^2\left[(2-x)\left[3\ln(1-x)-4\ln x\right] - \frac{4\ln(1-x)}{x} - 6 + x\right],$$

with $\beta_v = \frac{2\alpha}{\pi}\ln\left(\frac{s(1-c_v)}{m^2}\right)$ and $c_v = \cos\theta_v$.

$$\frac{d\sigma^{par}}{dx} = 2 \cdot \int_0^{c_m} dy H^{(\alpha)}(x,y;s)$$

$$\int_{\frac{x_{min}}{\sqrt{1-x}}}^{\sqrt{1-x}} dx_1 H^{(\alpha)}_{par}(x_1,y;(1-x)s)\sigma_0((1-x)(1-x_1)s). \qquad (17)$$

The spectrum of the observed photon $\frac{d\sigma}{dx}$ is given by the following sum:

$$\frac{d\sigma}{dx} = \frac{d\sigma^{soft}}{dx} + \frac{d\sigma^{pipe}}{dx} + \frac{d\sigma^{par}}{dx},$$

FORWARD-BACKWARD ASYMMETRIES

Let us confine ourselves to the process $e^+e^- \to \gamma, Z^0 \to \mu^+\mu^-$. The forward-backward asymmetry is defined as

$$A_0(s) = \frac{\sigma_0^{(+)}(s) - \sigma_0^{(-)}(s)}{\sigma_0(s)},$$

where $\sigma_0^{(+)}(s)$ and $\sigma_0^{(-)}(s)$ are the forward and backward hemisphere cross sections respectively, and $\sigma_0(s)$ is the total cross-section.

The calculation of the QED radiative corrections is reduced to the evaluation of the structure function for the initial states [3-5]. In particular the corrected cross-section can be written as a convolution of a "bare" cross-section and a "radiator" H

The structure function approach can be generalized to include final state radiative corrections too [13]. The corrected cross-section can be written in the form [21]

$$\sigma(s) = \int_0^{\varepsilon} dx H_{\varepsilon}(x,s)F_{\mu}\left(\varepsilon - x,(1-x)s\right)\sigma_0\left((1-x)s\right) + \int d\Omega_{\mu}\frac{d\sigma_B}{d\Omega_{\mu}},$$

where $\frac{d\sigma_B}{d\Omega_\mu}$ is the box and interference contribution (see for example ref.[12]). H_e and F_μ represent initial and final state radiation respectively [13].

As for the line shape case the factorized form is justified [21]. This result has been recently confirmed by an independent calculation [22].

By using the structure functions formalism it has been observed that initial-final interference contributions diagrams become important if photon-energy cuts are considered [21,23]. The contribution of the interference is, on the contrary, negligible when the radiation is inclusively integrated [21,23].

Monte Carlo

Structure functions formalism has been applied to develop a Monte Carlo code for electromagnetic processes [24]. The structure and the results of the corresponding algorithm are discussed in detail elsewhere in these proceedings by G. Bonvicini will not be mentioned here. It is only worth mentioning here that at the basis of algorithm are the main features of the structure function approach as described above. The radiatively corrected cross-sections in the Monte Carlo have the simple, factorized structure given by the convolution of a radiator with the corresponding Born cross-section. p_t-dependent structure functions can be used to describe both the longitudinal and the transverse degrees of freedom in the electron evolution. Multi-photon exponentiation and next-to-leading factorization are naturally implemented [25]. A different approach is discussed by R. Zitoun in these proceedings and further arguments concerning structure function Monte Carlo are in the contribution by R. Kleiss.

Further Applications

The structure function approach to radiative corrections has shown the ability to deal with the infrared structure of the electromagnetic radiation in a simple and effective way. Among the possible further applications of this same formalism is the inclusion of the effects due to the interference between initial and final states [26]. The relevance of such extension of the structure function method is related to the evaluation, within this same formalism, of the radiative Bhabha process [27].

Recently a general method has been proposed in ref.[28] to exponentiate leading and next-to leading logarithms to all orders of the perturbation theory. The techniques developed in ref.[28] can be applied to study the problem of the electromagnetic radiative corrections with an independent method. This approach allows to evaluate the structure of the non-dominant logarithmic terms arising from the mass singularities.

The e^+e^- annihilation cross section can be seen as an electrodynamical Drell-Yan process where the electrons, surrounded by the photon radiation, play the role of the annihilating partons. The cross section for e^+e^- annihilation is written as

$$\frac{d\sigma}{ds'} = \frac{1}{s}\sigma_0(s')W(s,\tau)$$

where s is the invariant energy square, $\tau = s'/s$ and $W(s,\tau)$ is the radiator containing radiation contributions to the elementary Born cross-section $\sigma_0(s')$. The total cross-section is given by $\sigma = \int ds'\, d\sigma/ds'$

The perturbative expansion for $W(s,\tau)$ shows contributions of the type

$$\frac{\alpha^n}{(1-\tau)}ln^{k-1}(1-\tau)L^n, \quad (m+k \le 2n)$$

with $L = ln(s/m_e^2)$ and m_e the electron mass and terms of the form

$$\alpha^n L^m f(1 - \tau), \quad (m \le n)$$

with $f(1 - \tau)$ an integrable function for $\tau \to 1$. The contributions above can be classified as collinear-soft and non-soft respectively. If $W_N(s) = \int d\tau \tau^{N-1} W(s, \tau)$ one has that [29]:

$$lnW_N^{IR}(s) = -\int \frac{d^3q}{4\pi\omega_q} \left(\frac{p_1}{p_1 \cdot q} - \frac{p_2}{p_2 \cdot q}\right)^2 A(\alpha(q_t^2))$$

$$[(1 - 2\frac{\omega_q}{\sqrt{s}})^{N-1} - 1]\theta(\sqrt{s}/2 - \omega_q)$$

where $\alpha(q^2)$ is the QED running coupling constant, p_1 and p_2 are the electron and positron momenta and the superscript IR means that takes into account *leading* and *next − to − leading* collinear-soft contributions. When only the radiation of photons from the interacting leptons is considered we have $A(\alpha) = \alpha/\pi$ with α being the fine structure constant and the usual result of soft photon exponentiation is recovered [2,3,4,5]. When also the production of real and virtual fermion pairs of mass m_f^2 is taken into account one can show [29] that the pair production contributions do exponentiate. At finite order the coefficients of the various perturbative terms coincide with the ones given in [11], since by expanding them up to the second order in α one gets for $lnW_N(s)$

$$\left(\frac{\alpha}{\pi}\right)^2 \int_0^1 dz \frac{z^{N-1} - 1}{1 - z}\theta((1 - z)^2 s - m_f^2)[\frac{1}{3}ln^2\frac{(1 - z)^2 s}{m_f^2} - \frac{10}{9}ln\frac{(1 - z)^2 s}{m_f^2}] + O(\alpha^3)$$

The pair production contribution differs qualitatively from the pure photonic term by showing an explicit logarithmic $(\alpha ln(1 - z))^n$ dependence that cannot be reproduced by the ordinary renormalization group equations [28,3-5,11].

Acknowledgements

I would like to thank Norman Dombey, Fawzi Boudjema and Claudio Verzegnassi for the help and support given to me and for all the efforts to organize this meeting. I would also like to thank G. Bonvicini, S. Catani and O. Nicrosini for many useful discussions and comments.

References

[1] G. Altarelli, in "Physics at LEP", CERN - Yellow Report, 86-02, J. Ellis and R. Peccei Editors, Geneva, February 1986; F. Gilman, SLAC-PUB-4002, June 1986, Talk presented at the Seventh Vanderbilt Conference on High Energy Physics, Nashville, Tennessee, May 15-17, 1986.

[2] E. Etim, G. Pancheri and B. Touschek, Nuovo Cimento 51B (1967) 276; G. Bonneau and F. Martin, Nucl. Phys. B27 (1971) 381; M. Greco, G. Pancheri-Srivastava and Y. Srivastava, Nucl. Phys. B101 (1975) 11, B171 (1980), 118; J. D. Jackson and D. L. Scharre, Nucl. Instruments and Methods 128 (1975) 13; F. A. Berends and R. Kleiss, Nucl. Phys. B178 (1981) 141; V. Baier, V. S. Fadin, V. Khoze and E. A. Kuraev, Phys. Rep. 78, 294 (1981); Y. S. Tsai, SLAC-PUB-3129 (1983), Presented at the Asia Pacific Conference, Singapore, June 12-18 1983; V. S. Fadin and V. A. Khoze, Yad. Fiz., 47 (1988) 1693.

[3] E. A. Kuraev and V. S. Fadin, Yad. Fiz. 41, 753 (1985) [Sov. J. Nucl. Phys. 41 (3), 1985, 466].

[4] G. Altarelli and G. Martinelli, in "Physics at LEP", CERN-Yellow Report, 86-02, J. Ellis and R. Peccei Editors, Geneva, February 1986.

[5] O. Nicrosini and L. Trentadue - Phys. Lett. 196B (1987) 551.

[6] L. Lipatov, Yad. Fiz. 20, 181 (1974)[Sov. J. of Nucl. Phys. 20, 94 (1975)]; V. Baier, V. S. Fadin and V. A. Khoze, Nucl. Phys. B65 (1973) 381; J. Kogut and L. Susskind, Phys. Rev. D9 (1974) 693, 3391.

[7] G. Altarelli and G. Parisi, Nucl. Phys. B126 (1977) 298.

[8] T. Kinoshita and J. Math. Phys., 3, 650 (1962); T. D. Lee and M. Nauenberg, Phys. Rev. 133 (1964) 1549.

[9] V. Gribov and L. Lipatov, Yad. Fiz. 15 (1972), 781, 1218 [Sov. J. Nucl. Phys, 15 (1972), 938, 675].

[10] R. Barbieri, J. A. Mignaco and E. Remiddi, Nuovo Cimento 11A (1972) 824.

[11] F. A. Berends, G. J. H. Burgers, and W. L. van Neerven, Phys. Lett. 185B (1987) 395; G. J. Burgers, Phys. Lett. 164B (1985) 167; F. A. Berends, G. Burgers, W. Hollik and W. L. van Neerven, Phys. Lett. B203 (1988) 177; J. P. Alexander, G. Bonvicini, P. S. Drell and R. Frey, Phys. Rev. D37 (1988).

[12] M. Greco, G. Pancheri and Y. Srivastava , Nucl. Phys. B101 (1975) 11, B171 (1980) 118, B197 (1982) 543; M. Greco - "Physics at LEP", CERN - Yellow Report, 86-02, J. Ellis and R. Peccei Editors, Geneva, February 1986; F. A. Berends, R. Kleiss and S. Jadach, Nucl. Phys. B202 (1982) 63.

[13] O. Nicrosini and L. Trentadue, Zeitsch. Phys. C39 (1988) 479.

[14] F. Aversa and M. Greco, LNF-Preprint 89/125 1989; M. Greco Proceedings of the "Workshop on QED Structure Functions", Ann Arbor, May 22-25, 1989.

[15] A. Bassetto, M. Ciafaloni and G. Marchesini, Phys. Rep. 100 (4) 1983; O. Nicrosini, L. Trentadue, Phys. Lett. B231 (1989) 487.

[16] G. Bonneau and F. Martin, Nucl. Phys. B27 (1971) 381; see also F. A. Berends and R. Kleiss, Nucl. Phys. B260 (1985) 32.

[17] A. D. Dolgov, L. B. Okun and V. I. Zacharov, Nucl. Phys. B41 (1972) 197; V. S. Fadin and V. Khoze, ZhETF, Pis. Red. 17, 8, 438 (1973); E. Ma and J. Okada, Phys. Rev. Lett. 41 (1978) 287; K. J. F. Gaemers, R. Gastmans and F. M. Renard, Phys. Rev. D19 (1979) 1605; G. Barbiellini, B. Richter and J. L. Siegrist, Phys. Lett. 106B (1981) 414.

[18] M. Igarashi and N. Nakazawa, Nucl. Phys. B288 (1987) 301.

[19] F. A. Berends, G. J. H. Burgers, C. Mana, M. Martinez and W. L. van Neerven, Nucl. Phys. B301 (1988) 583.

[20] O. Nicrosini and L. Trentadue, Nucl. Phys. B318 (1989) 1.

[21] O. Nicrosini, L. Trentadue, Proceedings of the "Ringberg Workshop on Electroweak Radiative Corrections", Ringberg Castle, Germany, April 3-7, 1989, J.H. Kuehn ed., Springer Verlag 1989; G. Montagna, O. Nicrosini and L. Trentadue, Phys. Lett. B231 (1989), 492.

[22] D. Bardin, M. Bilenky, A. Chizov, A. Sazonov, Yu. Sedykh, T. Riemann and M. Sachwitz, Phys. Lett. B229 (1989) 405.

[23] J. E. Campagne and R. Zitoun, LPNHE preprints 88-06, 88-08 (1988).

[24] G. Bonvicini and L. Trentadue, Nucl. Phys. B323 (1989) 253.

[25] G. Bonvicini, these proceedings.

[26] O. Nicrosini, Proceedings of the "Workshop on QED Structure Functions", Ann Arbor, May 22-25, 1989

[27] F. Aversa, G. Bonvicini, M. Greco, O. Nicrosini, L. Trentadue, work in progress.

[28] S. Catani and L. Trentadue, Phys. Lett. 217B (1989) 539; Nucl. Phys. B327 (1989) 353.

[29] S. Catani, L. Trentadue, Pis'ma Zh. Eksp. Teor. Fiz., 51, 1990, 72 [JEPT] Letters 51(2) 83 (1990)] and Proceedings of the "Workshop on QED Structure Functions", Ann Arbor, May 22-25, 1989.

COHERENT STATES AND STRUCTURE FUNCTIONS IN QED

Mario Greco

INFN - Laboratori Nazionali di Frascati
P.O. Box 13
00044 Frascati, Italy

ABSTRACT

The methods of coherent states and of structure functions in QED are considered in detail for a precise evaluation of the radiative effects at LEP/SLC. They are explicitly shown to give identical results for the exponentiated infrared factors and the $O(\alpha)$ terms corresponding to the initial and final state radiation. Furthermore interference effetcs from initial and final state radiation are introduced within the formalism of structure functions in a way which is suitable for Monte Carlo applications. The final formulae improve to $O(0.1\%)$ the evaluation of the e.m. effects.

The important role played by QED radiative corrections at LEP/SLC energies for precision tests of the standard model is well known[1,2]. The understanding and the detailed description of the multiphoton effects closely follow the pioneering work of Touschek and collaborators[3-5] at Frascati, more than twenty years ago, when a similar problem of precision had to be faced in testing Quantum Electrodynamics at "large" transverse momenta, i.e. at $Q^2 \sim 0$ (1 GeV^2).

The strategy was very clear: (i) to sum to all orders the large double and single soft and collinear logarithms of perturbation theory; (ii) to calculate exactly to one loop all remaining terms of $O(\alpha)$. The discovery of the J/ψ and other narrow resonances introduced further complications into the problem, due to the very nature of the resonant process and the subtle interference effects between initial and final leptonic states. Those were successfully described[6] along the same lines, in particular by exploiting the technique of the coherent states, introduced earlier[4,7] with the aim of having a realistic QED S-matrix.

The study[8] of genuine weak radiative effects at LEP/SLC energies clearly indicated the necessity of controlling e.m. corrections to a high degree of accuracy. Then the earlier analytical results were generalized to the case of Z_0 production to exact one-loop accuracy[9,10] and to all orders[9] in the leading logarithmic approximation, reaching a level of $O(1\%)$ precision. More recently initial state radiation effects have been evaluated[11] to two loops, pushing the theoretical accuracy - for the line shape measurements - to $O(0.1\%)$.

Radiative Corrections, Edited by N. Dombey and
F. Boudjema, Plenum Press, New York, 1990

To this aim the method of the structure functions[12], extended also to final states[13] in the reaction $e^+e^- \to \mu^+\mu^-$, has been shown to be quite powerful, suggesting a systematic approach to other processes as, for example, Bhabha scattering. However, only numerical solutions have been obtained so far in the case of a resonant cross section, leaving unclear the connection to the previous analytical approach of the problem. Furthermore the full extension of this technique to leptonic final states clearly demands an appropriate treatment of initial - and final - state interference effects which, on the other hand, are described to all orders in the coherent state formalism.

In this talk I will address the above questions, discussing analytical solutions of the structure functions method, including initial-final state interference effects, which are explicitly introduced in the formalism. More in detail I will explicitly show that the method of structure functions coincides with the coherent state approach with an accuracy of $O(1\%)$, giving explicitly in addition the $O(\alpha^2)$ corrections needed to improve further the theoretical precision.

The basic formula which describes the reaction $e^+e^- \to \mu^+\mu^-$ in the approach of structure functions without interference effects is the following[13]

$$d\sigma(s) = \int_0^\varepsilon dx \, d\sigma_0 \, ((1-x)\,s) \, H_e(x, s) \, F_\mu(\varepsilon - x, (1-x)\,s),$$ (1)

where $\varepsilon \equiv \Delta E/E$ is the energy resolution, and the initial and final state radiation kernels are given in terms of the electron and muon structure functions as

$$H_e(x, s) = \int_{1-x}^1 \frac{dz}{z} D_e(z, s) D_e\left(\frac{1-x}{z}, s\right),$$ (2)

$$F_\mu(x, s) = \int_0^x dy \, H_\mu(y, (1-y)\,s),$$ (3)

with $H_\mu(x, s)$ defined as in eq. (2). Eq. (1) describes the factorizable corrections only, corresponding to real and virtual photon emission from the initial and final legs, with no relative interference. By taking into account the effect of the soft radiation to all orders and of the hard one up to o (α^2) one obtains[13]

$$H_e(x, s) = \Delta_e(s) \, \beta_e \, x^{\beta_e - 1} - \frac{1}{2} \beta_e (2-x)$$

$$+ \frac{1}{8} \beta_e^2 \left\{ (2-x) [3 \ln(1-x) - 4 \ln x] - 4 \frac{\ln(1-x)}{x} + x - 6 \right\},$$ (4)

with $\beta_e = \frac{2\alpha}{\pi}(L_e - 1), L_e = \ln\left(\frac{s}{m_e^2}\right)$ and

$$\Delta_e(s) = 1 + \frac{\alpha}{\pi}\left[\frac{3}{2}L_e + 2(\zeta(2) - 1)\right] + \left(\frac{\alpha}{\pi}\right)^2 \left\{\left[\frac{9}{8} - 2\zeta(2)\right]L_e^2\right.$$

$$\left. + \left[3\zeta(3) + \frac{11}{2}\zeta(2) - \frac{45}{16}\right]L_e + \left[-\frac{6}{5}\zeta^2(2) - \frac{9}{2}\zeta(3) - 6\zeta(2)\ln 2 + \frac{3}{8}\zeta(2) + \frac{57}{12}\right]\right\} \quad (5)$$

$$\equiv 1 + \frac{\alpha}{\pi}\Delta_e^{(1)} + \left(\frac{\alpha}{\pi}\right)^2 \Delta_e^{(2)}$$

By insertion of (4) in eqs. (2) and (1) one easily obtains

$$\sigma(s) = \int_0^\varepsilon dx\ \sigma_0(s(1-x))\{\Delta_e(s)\Delta_\mu(s)\beta_e x^{\beta_e - 1}(\varepsilon - x)^{\beta_\mu} + R(x, \ldots)\} \quad (6)$$

where the first term in the r.h.s. of eq. ((6) is proportional to the leading soft contribution, while R (x, ...) give further correction terms of order (β^2) and β_ε in the final cross section. We will assume the fractional energy resolution $\varepsilon \equiv \Delta E/E$ of order $10^{-1} - 10^{-2}$.

By splitting the Born cross section $\sigma_0(s)$ as $\sigma_0 = \sigma_0^{QED} + \sigma_0^{INT} + \sigma_0^{RES}$, with

$$\sigma_0^{QED}(s) = A\frac{1}{s}$$

$$\sigma_0^{INT}(s) = B\ \mathrm{Re}\left\{\frac{1}{s - M^2 + i\Gamma M}\right\} \quad (7)$$

$$\sigma_0^{RES}(s) = C\frac{s}{(s - M^2)^2 + \Gamma^2 M^2}$$

the corresponding radiatively corrected cross sections are found[14] to be

$$\sigma^{QED}(s) = \sigma_0^{QED}(s)\Delta_e(s)\Delta_\mu(s)\varepsilon^{\beta_e + \beta_\mu} + \ldots \quad (8)$$

$$\sigma^{INT}(s) = \sigma_0^{INT}(s)\Delta_e(s)\Delta_\mu(s)\varepsilon^{\beta_\mu}\frac{\Gamma(1 + \beta_e)\Gamma(1 + \beta_\mu)}{\Gamma(1 + \beta_e + \beta_\mu)}\frac{1}{\cos\delta_R}$$

$$\cdot\ \mathrm{Re}\left\{e^{i\delta_R}\left[\frac{\varepsilon}{1 + \left(\frac{\varepsilon s}{M\Gamma}\right)\sin\delta_R\ e^{i\delta_R}}\right]^{\beta_e}\right\} + \quad (9)$$

$$\sigma^{RES}(s) = \sigma_0^{RES}(s)\Delta_e(s)\Delta_\mu(s)\varepsilon^{\beta_\mu}\frac{\Gamma(1 + \beta_e)\Gamma(1 + \beta_\mu)}{\Gamma(1 + \beta_e + \beta_\mu)}$$

$$\cdot\ \left|\frac{\varepsilon}{1 + \left(\frac{\varepsilon s}{M\Gamma}\right)\sin\delta_R\ e^{i\delta_R}}\right|(\cos\beta_e\ \phi - \cot\delta_R\ \sin\beta_e\ \phi) + \ldots \quad (10)$$

where $(M_R^2 - s)^{-1} \equiv \dfrac{\sin \delta_R \, e^{i\delta_R}}{M\Gamma}$, $\tan \delta_R = \dfrac{M\Gamma}{(M^2 - s)}$, $\phi = \arctan \left[\dfrac{\varepsilon s + M^2 - s}{M\Gamma}\right] - \arctan \left[\dfrac{M^2 - s}{M\Gamma}\right]$

and the dots indicate next to leading terms corresponding to R (x, ...) in eq. (6).

Comparing with the analogous expressions obtained in the framework of coherent states[2,9], one finds that exponentiated infrared and the $O(\alpha)$ factors coincide with those in eqs. (8-10). On the other hand extra terms are contained in eqs. (8-10) of $O(\beta^2)$ - in particular the factor $-\beta_e^2 \, (\phi^2/2)$ in eq. (13) arising from the expansion of $\cos \phi\beta_e$ - and $O(\varepsilon\beta)$, not written explicitly[15].

Concerning the remaining terms, including initial-final states interference, box diagrams, etc. they can also be included to $O(\alpha)$ in the approach of structure functions. However a general treatment of interference effects, to all orders, can be obtained as follows. For pure QED processes the simple rescaling[16] $s \to s \, (t/u)$, where s,t and u are the Mandelstam variables, gives the usual result

$$d\sigma^{QED} \approx d\sigma_0^{QED} \, e^{\beta_e + \beta_\mu + 2\beta_{int}}, \tag{11}$$

where $\beta_{int} = (4\alpha/\pi) \ln \mathrm{tg}\ \theta/2$. However this simple rule does not work for a resonant proces. Then, more generally, a full account of all radiative effects can be simply obtained by replacing the Born cross section $d\sigma_0 \, [(1 - x) s]$ in eq. (1) by $d\sigma_0 \, [(1 - x) s] \cdot K\,(x)$, where the K-factor includes all not-factorizable corrections, which can be determined to all orders for the soft contribution and to $O(\alpha)$ for the nonleading terms.

Indeed one obtains[17] the soft contributions, corresponding to initial-final state interference, the following expressions for the $K^{(i)}\,(x)$ factors

$$K^{QED}(x) = \frac{(\beta_e + \beta_{int})}{\beta_e} \, [x \, (\varepsilon - x)]^{\beta_{int}}$$

$$K^{INT}(x) = \frac{(\beta_e + \beta_{int})}{\beta_e} \left[\frac{s x}{s \, (1 - x) - M_R^2} \, (\varepsilon - x)\right]^{\beta_{int}} \tag{12}$$

$$K^{RES}(x) = \frac{(\beta_e + 2\beta_{int})}{\beta_e} \left|\frac{s x}{s \, (1 - x) - M_R^2}\right|^{2\beta_{int}} .$$

The above result is based on the observation[6] that in the pure QED process the virtual matrix element M_V^{QED} scales as $\{\lambda^2/s\}^{(\beta_e + \beta_\mu + 2\beta_{int})/4}$, while for a resonant process one has $M_V^{RES} \sim \{\lambda^2/s\}^{(\beta_e + \beta_\mu)/4} \cdot \{\lambda^2/(s - M_R^2)\}^{\beta_{int}/2}$, where λ is the minimum energy cutoff. Eqs. (12) allow us to obtain the complete analytical solution within the method of the structure functions, generalizing the results of ref. [14].

Then, after the substitution $d\sigma_0^{(i)} \, [(1 - x)s] \to d\sigma_0^{(i)} \, [(1 - x) s] \cdot K^{(i)}(x)$ in eq. (1), using eq. (12), and following ref. [14], one then finds[17] for the leading terms - corresponding to the resummation of the soft contributions:

$$d\sigma^{QED}(s) = d\sigma_0^{QED}(s) \left\{ \Delta_e(s)\,\Delta_\mu(s)\,\varepsilon^{\bar{\beta}_e + \bar{\beta}_\mu} \right.$$

$$\left. \cdot \frac{\Gamma(1+\bar{\beta}_e)\,\Gamma(1+\bar{\beta}_\mu)}{\Gamma(1+\bar{\beta}_e+\bar{\beta}_\mu)}\, {}_2F_1(1,\bar{\beta}_e; 1+\bar{\beta}_e+\bar{\beta}_\mu; \varepsilon) + \ldots \right\},$$

(13)

$$d\sigma^{INT}(s) = d\sigma_0^{INT}(s)\,\Delta_e(s)\,\Delta_\mu(s)\,\varepsilon^{\bar{\beta}_\mu}\,\frac{\Gamma(1+\bar{\beta}_e)\,\Gamma(1+\bar{\beta}_\mu)}{\Gamma(1+\bar{\beta}_e+\bar{\beta}_\mu)}\,\frac{1}{\cos\delta_R} \cdot$$

$$\cdot \,\mathrm{Re}\left\{ e^{i\delta_R}\left[\frac{\varepsilon}{1+\left(\frac{\varepsilon s}{M\Gamma}\right)\sin\delta_R\,e^{i\delta_R}}\right]^{\beta_e}\left[\frac{\varepsilon}{\varepsilon+\left(\frac{M\Gamma}{s}\right)\frac{e^{-i\delta_R}}{\sin\delta_R}}\right]^{\beta_{int}} \right\} + \ldots,$$

(14)

$$d\sigma^{RES}(s) = d\sigma_0^{RES}(s)\,\Delta_e(s)\,\Delta_\mu(s)\,\varepsilon^{\beta_\mu}\,\frac{\Gamma(1+\bar{\bar{\beta}}_e)\,\Gamma(1+\beta_\mu)}{\Gamma(1+\bar{\bar{\beta}}_\varepsilon+\beta_\mu)}\cdot\left|\frac{\varepsilon}{\varepsilon+\left(\frac{M\Gamma}{s}\right)\frac{e^{-i\delta_R}}{\sin\delta_R}}\right|^{2\beta_{int}}\cdot$$

$$\left|\frac{\varepsilon}{1+\left(\frac{\varepsilon s}{M\Gamma}\right)\sin\delta_R\,e^{i\delta_R}}\right|^{\beta_e}(\cos\beta_e\phi - \cot\delta_R\sin\beta_e\phi) + \ldots,$$

(15)

where $\bar{\beta}_{e,\mu} = \beta_{e,\mu}+\beta_{int}$, $\bar{\bar{\beta}}_e = \beta_e + 2\beta_{int}$.

A few comments are in order here. First of all the main β_{int}-dependence in eqs. (13-15) appears through exponentiated factors, which coincide with those found in refs. [6, 9], using the method of coherent states. Of course they also reproduce the exact one-loop calculations[9, 10]. Furthermore the β_{int}-dependence drops out completely in the limiting case of narrow resonance production ($\Gamma \ll \Delta\omega$), as for example the J/ψ.

Physically this can be understood through the observation that the initial and final state can no longer interfere since the long time delay ($\tau \sim 1/\Gamma$) due to the resonance formation and decay.

Grouping together all non-infrared factors coming from $\Delta_e(s)$, $\Delta_\mu(s)$ and the Γ-functions in eqs. (13-15), as well as those coming from the non-soft terms of the electron and muon radiators, one can then write, as in ref. [14],

$$\frac{d\sigma}{d\Omega} = \sum_i^{QED, INT, RES} \frac{d\sigma_0^{(i)}}{d\Omega}\left\{ C_{infra}^{(i)}\,(1+\bar{C}_F^{(i)}) + C_F^{'(i)} \right\}$$

(16)

where the infrared factors $C_{infra}^{(i)}$ are simply obtained from eqs. (8-10), and

$$\bar{C}_F^{(QED)} = \left(\frac{\alpha}{\pi}\right)\left[\Delta_e^{(1)} + \Delta_\mu^{(1)}\right] - \beta_\mu \varepsilon$$

$$+ \left(\frac{\alpha}{\pi}\right)^2\left[\Delta_e^{(2)} + \Delta_\mu^{(2)} + \Delta_e^{(1)}\Delta_\mu^{(1)}\right] - \frac{\pi^2}{6}\bar{\beta}_e\bar{\beta}_\mu - \frac{1}{4}\beta_e\bar{\beta}_e\varepsilon^{1-\bar{\beta}_e} \qquad (17)$$

$$\bar{C}_F^{(INT)} = \left(\frac{\alpha}{\pi}\right)\left[\Delta_e^{(1)} + \Delta_\mu^{(1)}\right] - \beta_\mu \varepsilon + \left(\frac{\alpha}{\pi}\right)^2\left[\Delta_e^{(2)} + \Delta_\mu^{(2)} + \Delta_e^{(1)}\Delta_\mu^{(1)}\right] - \frac{\pi^2}{6}\bar{\beta}_e\bar{\beta}_\mu$$

$$- \bar{\beta}_e \frac{\cos\phi\,(\beta_e + 1) + \tan\delta_R \sin\phi\,(\beta_e + 1)}{\cos\phi\beta_e + \tan\delta_R \sin\phi\beta_e}\left|\frac{\varepsilon}{1 + \dfrac{\varepsilon s}{M_R^2 - s}}\right| \qquad (18)$$

$$- \frac{1}{4}\beta_e\bar{\beta}_e \frac{\cos\phi + \tan\delta_R \sin\phi}{\cos\phi\beta_e + \tan\delta_R \sin\phi\beta_e}\left|\frac{\varepsilon}{1 + \dfrac{\varepsilon s}{M_R^2 - s}}\right|^{1-\beta_e}$$

$$\bar{C}_F^{(RES)} = \left(\frac{\alpha}{\pi}\right)\left[\Delta_e^{(1)} + \Delta_\mu^{(1)}\right] - \beta_\mu \varepsilon + \left(\frac{\alpha}{\pi}\right)^2\left[\Delta_e^{(2)} + \Delta_\mu^{(2)} + \Delta_e^{(1)}\Delta_\mu^{(1)}\right] - \frac{\pi^2}{6}\bar{\beta}_e\beta_\mu$$

$$- 2\bar{\beta}_e \frac{\cos\phi\,(\beta_e + 1) - \cot\delta_R \sin\phi\,(\beta_e + 1)}{\cos\phi\,\beta_e - \cot\delta_R \sin\phi\,\beta_e}\left|\frac{\varepsilon}{1 + \dfrac{\varepsilon s}{M_R^2 - s}}\right| \qquad (19)$$

$$- \frac{1}{4}\beta_e\bar{\beta}_e \frac{\cos\phi - \cot\delta_R \sin\phi}{\cos\phi\,\beta_e - \cot\delta_R \sin\phi\,\beta_e}\left|\frac{\varepsilon}{1 + \dfrac{\varepsilon s}{M_R^2 - s}}\right|^{1-\beta_e} .$$

Finally the factors $C'^{(i)}_F$ contain other $O(\alpha)$ finite terms, coming from bremsstrahlung and box diagrams, odd in the exchange $\theta \leftrightarrow \pi - \theta$, etc., and can be obtained from refs. [2,14].

The above equations represent our final result, which describes the radiative correction factors to an accuracy better than (1%). The effect of the new terms of $O(\beta^2, \varepsilon\beta)$ in eqs. (18) is shown[14] in figs. (1, 2), where we plot the ratios

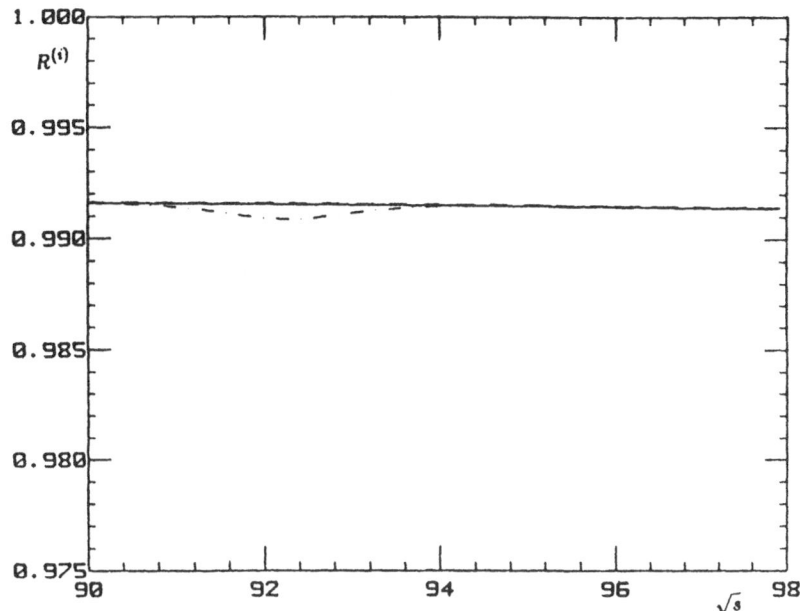

Fig. 1 Ratio $R^{(i)} = \frac{d\sigma^{(i)}[1+o(\alpha)+o(\beta\epsilon)+o(\alpha^2)]}{d\sigma^{(i)}[1+o(\alpha)]}$ for $e^+e^- \to \mu^+\mu^-$, with i=QED (solid line), INT (dashed line), RES (dotted-dashed line) for $\epsilon = 0.01$. The values of M and Γ are taken to be 92 Gev and 2.6 Gev respectively.

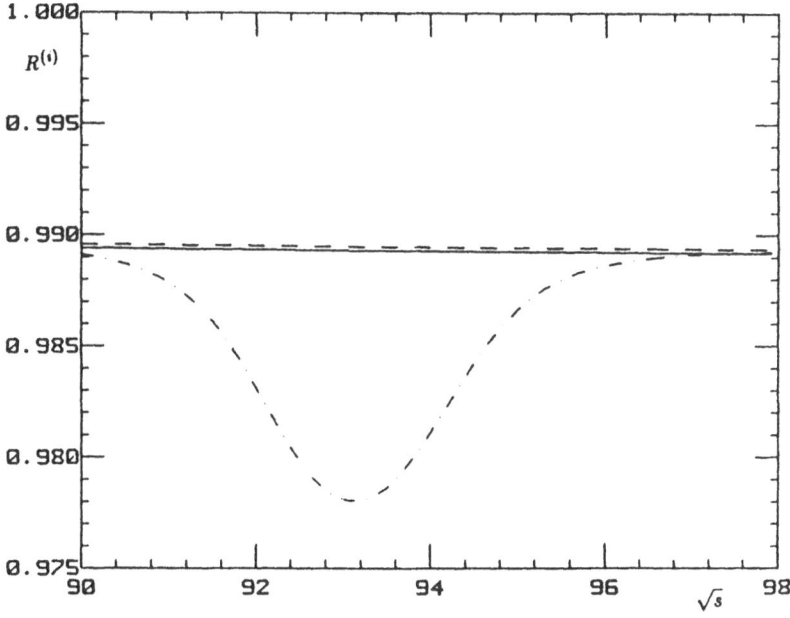

Fig. 2 Same as in fig. (1) for $\epsilon = 0.05$.

$$R^{(i)} = \frac{d\sigma^{(i)}\left[1 + O(\alpha) + 1 + O(\beta\varepsilon) + O(\alpha^2)\right]}{d\sigma^{(i)}\left[1 + O(\alpha)\right]}$$

for i = QED, INT, RES, for $\varepsilon = 0.01$ and $\varepsilon = 0.05$.

Notice that the factors $C^{(i)}_{infra}$ do not appear in the ratios $R^{(i)}$. We have taken the scattering angle $\theta = \pi/2$ in the factors $C^{(i)}_F$ to authomatically cancel the box diagram and other non factorizable contributions.

As is clear from figs. (1,2) the QED and INT corrections are practically constant in the resonant region to a value of about -0.01, while the RES correction is modulated essentially by the term $1 - (\beta_e^2\phi^2/2)$, with an extra factor of $O(0.01 - 0.02)$.

The extension of our results to the case of the Z line shape is straightforward. It simply corresponds to take the limit $\beta_\mu \beta_{int} = 0$, $\Delta_\mu = 1$ and $\varepsilon = 1 - (4\mu^2/s)$ in eq. (15). Then one simply obtains for $s \approx M_R^2$

$$\sigma(s) \approx \sigma_0(s) \left|\frac{M_R^2 - s}{M_R^2}\right|^{\beta_e} \frac{\sin(1 - \beta_e)\delta_R}{\sin \delta_R} \frac{\pi\beta_e}{\sin \pi\beta_e} \Delta_e(s) \tag{20}$$

which agrees with refs [6, 9, 18] up to constant factors of $O(\beta^2)$.

To conclude, we have explicitly shown that the approach of the coherent states and that of the structure functions offer two complementary methods in QED to achieve the theoretical accuracy required for precision measurements at LEP/SLC energies. They give identical results for the exponentiated and finite $O(\alpha)$ factors relative to the initial and final states radiation. Moreover we have also shown how to include the interference effects coming from initial and final state radiation within the approach of structure functions in QED. The additional $O(\alpha^2)$ corrections improve the theoretical accuracy to 0 (0.1%). The overall picture provides simple analytical formulae which can be easily extended to e^+e^- reactions other than $e^+e^- \rightarrow \mu^+\mu^-$, and in addition the method is suitable for MonteCarlo applications.

REFERENCES

[1] For recent reviews of LEP/SLC physics see, for example: Physics at LEP, edited by J. Ellis and R. Peccei, CERN 86-02 (1986);
Polarization at LEP, edited by G. Alexander et al., CERN 88-02 (1988).

[2] For a review of radiative corrections at LEP/SLC see, for example, M. Greco, La Rivista del Nuovo Cimento, Vol. 11, n. 5 (1988).

[3] E. Etim, G. Pancheri and B. Touschek, Nuovo Cimento 51B, 276 (1967).

[4] M. Greco and G. Rossi, Nuovo Cimento 50, 168 [1967].

[5] G. Pancheri, Nuovo Cimento 60, 321 (1969).

[6] M. Greco, G. Pancheri and Y. Srivastava, Nucl. Phys. B101, 234 (1975).

[7] V. Chung, Phys. Rev. B140, 1110 (1965).

[8] G. Passarino and M. Veltman, Nucl. Phys. B160, 151 (1979);
M. Consoli, Nucl. Phys. B160, 268 (1979).

[9] M. Greco, G. Pancheri and Y. Srivastava, Nulc. Phys. B171, 118 (1980).

[10] F.A. Berends, R. Kleiss and S. Jadach, Nucl. Phys. B202, 63 (1982);
M. Bohm and W. Hollik, Nucl. Phys. B204, 45 (1982).

[11] F.A. Berends, W.L. van Neerven and G.J.H. Burgers, Nucl. Phys. B297, 429 (1988).

[12] E.A. Kuraev and V.S. Fadin, Sov. J. Nucl. Phys. 41, 41, 466 (1985);
G. Altarelli and G. Martinelli, in Physics at LEP, ref. (1);
O. Nicrosini and L. Trentadue, Phys. Lett. 196B, 551 (1987).

[13] O. Nicrosini and L. Trentadue, Z. Phys. C39, 479 (1988).

[14] F. Aversa and M. Greco, LNF-89/025 (PT), May 1989.

[15] Soft corrections of $O(\beta^2)$ to the coherent state method are also discused by H. Spiesberger and M. Böhm Würzburg preprint, 1989.

[16] O. Nicrosini, talk given at the "Workshop on QED Structure Functions", Ann Arbor, May 1989.

[17] M. Greco, LNF-89/042 (PT), June 1989.

[18] R.N. Cahn, Phys. Rev. D36, 2666 (1987).

A MULTIPHOTON QED GENERATOR *

Giovanni Bonvicini [†]

University of Michigan, Ann Arbor MI 48109

Abstract

A method to calculate higher order radiative corrections with an explicit representation of the photon variables is presented here. The method can correctly exponentiate 2-particle QED processes up to next-to-leading factorization. Possible future improvements are discussed.

1 Introduction

Presently and in the foreseable future advancements in particle physics are expected to come from high statistics precision measurements or via detection of exceedingly rare events. Both types of experiments require powerful predictions of the behavior of known physics phenomena, for which analytic calculations in standard perturbation theory are inadequate. Much R&D on QED techniques has been done in recent years to achieve the needed precision (less than 1%) that the LEP experiments require.

A couple of relatively original approaches to the problem [1,2] suggest that QED processes can be calculated with high precision in an algorithmic way. There are a few relevant advantages in such an approach. The final formulation of the matrix element is more compact, calculable in a reasonable amount of computer time and conceptually simpler. Improving the calculation one order higher is simpler, and involves essentially the calculation of a new kernel for the algorithm. Both of these algorithms are naturally based on exponentiation and near-Poissonian multiplicity distributions for the photon number. This solves a problem of positive-definiteness known as "the cutoff problem" [2,3], which forces the photon energy to be above a minimum value which might be larger than one of the experimental resolutions. Results from such algorithms are usually more precise than the normal perturbation result because of the summation of large higher order leading and non leading logarithms.

---

*Work supported by the Department of Energy, contract DE-AC02-84ER01112.
[†]Now at CERN, European Laboratory for Particle Physics, CH-1211 Geneva 23, Switzerland.

Radiative Corrections, Edited by N. Dombey and
F. Boudjema, Plenum Press, New York, 1990

The algorithm of Ref. [1] makes itself preferred for a more rigorous formulation, while the algorithm of Ref. [2] is simpler, faster and does not require explicit calculations of higher order distributions. Development is under way for both of them, but the day in which we will have an algorithm capable of describing 4-particle processes (such as the e^+e^- and $\mu^+\mu^-$ final states) up to second order is still not in sight. This paper aims at giving a simple explanation of the features of these algorithms, in a very general way (Section 2). Section 3 is dedicated to the description of the Monte Carlo (MC) developed by L. Trentadue and myself, with the inclusion of final state photons (the latter feature has never been published before). Section 4 discusses the differences between this MC and traditionally exponentiated ones and gives a brief outlook of possible improvements of this method.

2 Multiphoton QED generators

A relativistic electron travels accompanied by a cloud of tightly bound photons some of which will be made real by a Q^2 exchange with an external potential.

The physical picture of parton beams is quite familiar in QCD and in the last few years has been used by everyone in QED for LEP physics purposes [4]. Starting with a parton beam approximation, one gets right away a predictions for QED corrections which includes multiple photon effects reasonably well. In our experiments, the Q^2 is always large enough to generate visible higher order effects, both in inclusive and exclusive observables.

Radiative corrections in QED are a series in three parameters, the fine structure constant α, the collinear logarithm $\log(s/m^2) = L$, where s is the Mandelstam invariant and m the particle mass, and the infrared logarithm $\log(k_0)$, where k_0 is the fractional energy resolution of the experiment [1]. Double leading approximation means that the most dominant terms, which are of the form $\alpha^n L^n l^n$, are included to all orders. This physically implies Poissonian photon multiplicity and uncorrelated photon emission. Improving from the pure double leading approximation is called next-to-leading factorization, which involves small deviations from Poissonian statistics, from "unitarity" (a Poissonian distribution sums to one) and correlated photon emission.

Probability densities are positive definite as opposed to positive (real) and negative (virtual) amplitudes calculated with Feynman diagrams. The fact that positive and negative terms merge to all orders into a positive-definite Poissonian statistics for the photon multiplicity [5] is the base of our method. The result is true for soft photons only, but we will see in the next section that it can be made true for two hard collinear photons as well.

The probability that an annihilation event will produce n photons from the initial state is

$$P(n) = \frac{\bar{n}^n e^{-\bar{n}}}{n!}, \qquad \bar{n} = \frac{\alpha}{\pi}(L-1)(-l-\frac{3}{4}). \tag{1}$$

When photons are of energy comparable to the beam energy, the phase space available for extra photons will be reduced. Therefore the first terms that we encounter in the perturbation series which deviate from Poissonian statistics (we exclude here non unitary terms) will be negative.

---

[1] We have argued previously [2] that since a Montecarlo is used both inclusively and exclusively the cutoff k_0 must be lower than all the physical scales in the experiment, ie acollinearity, energy and momentum resolution, beam energy spread and resonance width.

The major advantage in using an algorithm to calculate radiative corrections is that the computer will do automathically a convolution of many terms (a few 10^3 for double radiation from a continuous line, such as in the process $e^+e^- \to X$) while iteratively calculating the QED evolution. Some more advantages in the algorithmic approach are the following.

One, higher order corrections involve polilogarithms and multiple integrals of them. They have the general form [6]

$$S_{m,n}(x) = \int_0^1 dy \frac{\log^{m-1} y \log^n(1-xy)}{y}, \tag{2}$$

and look nasty, specially to an experimentalist. However, the MC method is *designed* to calculate nasty integrals. Solving an integral over all or part of the phase space so that one can put the result in the MC could be appropriately called "double working". The "ultimate MC" should be so ultimate as to calculate its Spence functions by itself. In some cases, like the MCs PYMU2 [7] and EXPOSTAR [8], summation over the photon multiplicity and integration over the transverse variables lose information on the process and lessen the power and meaning of the MC calculation.

Two, in the approach predicated here we always generate one photon at a time, excluding one branching in which the two final state fermions are generated. In other words, we start with a system with 4−momentum $(p_1 + p_2) = (\sqrt{s}, \vec{0})$, degrade it into a photon with 3−momentum $\vec{k}$ and a massive system P recoiling against it, and iterate this procedure. The transition will be written as

$$d^n\sigma \propto |M|^2 \frac{d^3\vec{P}}{E_P}\frac{d^3\vec{k}}{k}\delta^4(p_1 + p_2 - k - P) \tag{3}$$

and integration over $\vec{P}$ gives

$$d^n\sigma \propto |M|^2 \frac{d^3\vec{k}}{k}\delta(M_P^2 - s(1 - k/E)), \tag{4}$$

where E is the beam energy, which is a probability density in the photon variables only and contains no phase space factors to all orders.

Three, an algorithmic approach is self-adjusting. At every step variables and parameters of the evolution can be readjusted to improve the approximation of the method. Whether one can, this way, generate the exact matrix element is *the* crucial question for the future of this kind of algorithm and its applicability to QCD, and so far it has not been proven or disproven. "Linear combination of factorized kernels" [2] seems to be a good definition of the methods [1,2] and limitations that affect structure function MC (see last section) do not hold for it.

3 The MOE algorithm

Initially based on QED structure functions (SF) this algorithm [2] conserves the probabilistic approach but departs from its origin in the method of iteration and in the use of specialized kernels.

Evolution equations for the SF $D(x, s)$ can be started at the scale $Q^2 = m_e^2$ and evolved to the annihilation scale $Q^2 = s$,

---

[2] As dubbed by R. Kleiss.

$$D(x,s) = \delta(1-x) + \frac{\alpha}{2\pi}\int_{m^2}^{s}\frac{ds'}{s'}\int_{x}^{1}\frac{dz}{z}P(z)D(\frac{x}{z},s').\qquad(5)$$

$P(z)$ is the regularized $q \to q + g$ AP vertex [9] given by

$$P(z) = \frac{1+z^2}{1-z} - \delta(1-z)\int_{0}^{1}dt\frac{1+t^2}{1-t}.\qquad(6)$$

A practical method of solution in QED is the iteration to order n

$$\begin{aligned}
D(x,s) = \quad & \delta(1-x) \\
+ \quad & \frac{\alpha}{2\pi}(\int ds'/s'\,P(x) \\
+ \quad & \frac{\alpha}{2\pi}(\int ds''/s''\int\frac{dz}{z}P(\frac{x}{z})P(z) \\
+ \quad & \frac{\alpha}{2\pi}(\int ds'''/s'''\int\frac{dz}{z}\int\frac{dy}{y}P(\frac{x}{yz})P(z)P(y) + ...))).
\end{aligned}\qquad(7)$$

Integration in x above and below $x_0 = 1 - k_0$, and neglection of terms correspondent to the sea electrons, yields

$$\begin{aligned}
B(0) &= 1 - \bar{n} + \bar{n}^2/2 - \bar{n}^3/6 + ... &&= e^{-\bar{n}}, \\
B(1) &= \bar{n} - \bar{n}^2 + \bar{n}^3/2 + ... &&= \bar{n}B(0), \\
B(2) &= \bar{n}^2/2 - \bar{n}^3/2 + ... &&= (\bar{n}^2/2)B(0), \qquad etc.,
\end{aligned}\qquad(8)$$

with

$$\bar{n} = \frac{\alpha}{\pi}L(-l-3/4)\qquad(9)$$

The $B(n)$ are the branching ratios into n photons of energy greater than the cutoff at a given scale s. The substitution $L \to L - 1$ achieves agreement with the exact first order calculation.

Having in hand the multiplicity distribution, we now cast all multiple emission terms in the form of an explicit probability density [10]

$$\int\frac{dz_1}{z_1}...\frac{dz_n}{z_n}P(z_1)...P(z_n)P(\frac{x}{\Pi_i z_i}) = \frac{d}{dx}\int_{\Omega}dz_1...dz_n\,P(z_1)...P(z_n),\qquad(10)$$

where the integration volume Ω is defined by

$$z_i < x_0, \qquad \Pi_i z_i > x.\qquad(11)$$

The proof can be found in Ref. [2]. Eq. (10) shows that the difficult multiple integrals in Eq. (7) can always be cast into a convolution of elementary kernels to be done via MC, as stated in the previous Section. In the case of two photons the steps are

1. Generate z_1, $P(z)$ as defined in Eq. (6) and $0 < z < x_0$.

2. Repeat the first step for z_2.

3. Let $x = z_1 z_2$, $k_1 = 1 - z_1$ and $k_2 = z_1(1 - z_2)$.

The second photon is allowed by the evolution to have an energy $z_1 k_0 < k_2 < z_1$, part of which is below the cutoff. As the phase space for the second photon is

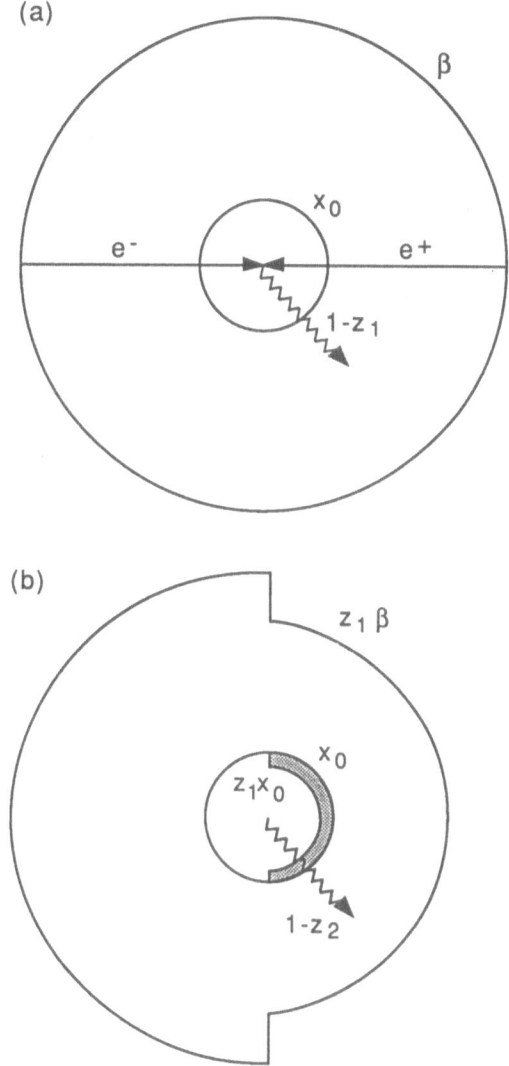

Figure 1. Evolution of the cutoff in the approximation that the electron and positron evolve independently. a) Phase space allowed to the first photon. b) phase space allowed to the second photon.

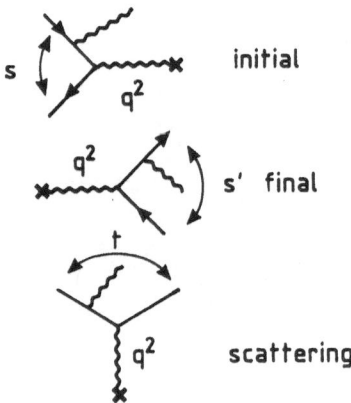

Figure 2. The initial, final, and scattering radiators.

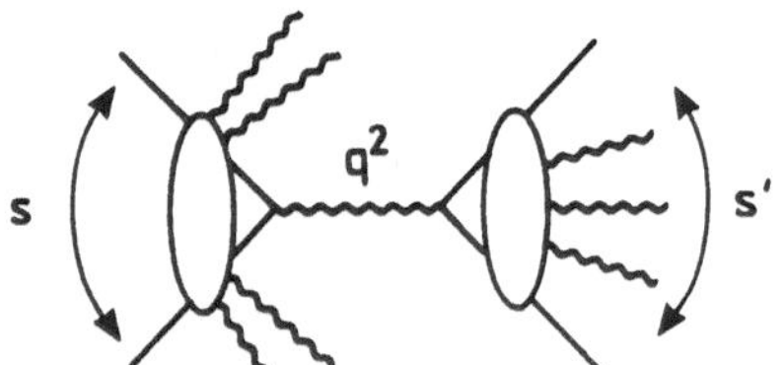

Figure 3. The process $e^+e^- \to \mu^+\mu^-$ as modeled in MOE.

reduced, the cutoff is moved inward to maintain a Poissonian statistics (Fig. 1). The probability of radiating a photon k_1 and a second photon k_2 below the cutoff is given by the integral of the $P(z)$ over the shaded area in Fig.1 and corresponds to a probability exchanged between the one and two photon amplitudes

$$(\frac{\alpha L}{2\pi})^2 \int_{1-x_0}^{1-z_1 x_0} \frac{2dz_2}{1-z_2} \frac{1+z_1^2}{1-z_1} = (\frac{\alpha L}{\pi})^2 \frac{1}{2} \log z_1 \frac{1+z_1^2}{1-z_1} \quad , \tag{12}$$

which has to be compared with those non-Poissonian $\alpha^2 L^2$ terms where one real photon plus one soft-virtual photon are radiated. The two expressions are identical and the virtual-virtual (0 real photons) and real-real (2 real photons) terms are easily proven to be factorizable in the $\alpha^2 L^2$ terms [11]. We conclude that our variable cutoff method achieves factorization of the L-dominant terms. As known from QCD, the AP evolution equations do achieve factorization of the $\alpha^n L^n$ terms, which translates into a variable cutoff when one generates the explicit photon variables.

Inclusion of the transverse degrees of freedom cannot follow the Bassetto Ciafaloni Marchesini [12] prescription for QCD. Use of independent kernels $D(x,s)G(p_T)$ is very effective in the soft-collinear region, but generates unacceptable distortions in the large angle, large energy region. The solution is simply to use the exact first-order radiation kernel. The radiation kernel from an external leg can be written compactly for 2-particle evolution, where i is the radiating leg and j the spectator leg (see Fig. 2)

$$P(r_i, r_j) = \frac{(1 \pm r_i)^2 + (1 \pm r_j)^2}{r_i(r_i + r_j)} - \frac{m^2}{r_i^2}(1 \pm (r_i + r_j)) \tag{13}$$

The r_i are the invariant masses of the photon with the ith leg, divided by the 4-product of the two on-mass shell legs

$$r_i = \frac{p_i \cdot k}{p_i \cdot p_j}, \tag{14}$$

and their correspondence with the longitudinal variable z is

$$z = 1 - \frac{1}{2}(r_i + r_j)C, \qquad C = \frac{p_i \cdot p_j}{\max(p_i \cdot p_j, q^2)} \tag{15}$$

which can be written more explicitly as follows

$$\begin{aligned} z &= & 1 - (k/E), & \text{initial,} \\ z &= & 1 - (k/E), & \text{final,} \\ z &= & 1 - (k/E)\cos\theta, & \text{scattering.} \end{aligned} \tag{16}$$

The $\pm$ in Eq. (13) become $+$ when the evolution is timelike (indeces 3 and 4), and $-$ when the evolution is spacelike (indeces 1 and 2), ie for initial state we get all $-$, for final state all $+$, and for scattering with an external potential a $+$ and a $-$ (Fig. 2).

Once the first photon is gone, one of the legs will have 4-momentum $(p_i - k)$ and $p^\mu p_\mu \neq m^2$. The "mass" of the particle is evolved accordingly and in higher order, when one or both the legs at a later step n in the evolution are off shell, the variables $(r_i)_n$ and C are defined in a similar way

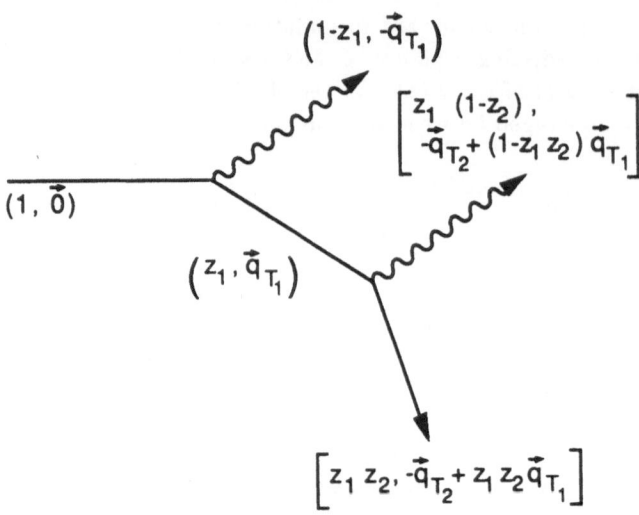

Figure 4. Evolution of a single leg.

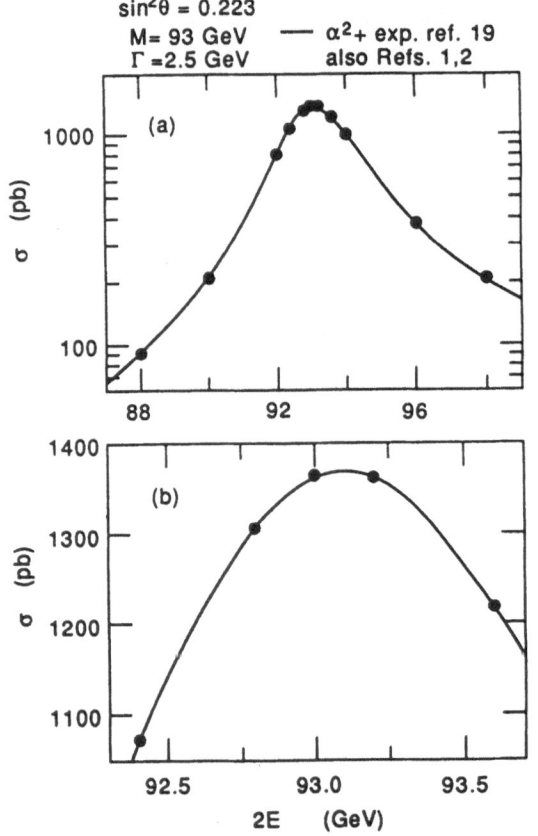

Figure 5. Radiative corrections to the Z line shape in the $\mu^+\mu^-$ channel as predicted by the Monte Carlo (data points), and by an analytic exact $O(\alpha^2)$ exponentiated calculation from Ref. [11]. The Monte Carlo was run with 10^5 events per point, a cutoff k_0 of $1/10^5$ and $s' > .1s$.

$$(r_i)_n = \frac{(p_i)_n \cdot k_n}{(p_i)_n \cdot (p_j)_n}, \qquad C = \frac{(p_i)_n \cdot (p_j)_n}{\max((p_i)_{n-1} \cdot (p_j)_{n-1}, \ (p_i)_n \cdot (p_j)_n)}. \tag{17}$$

The 4−momentum δ introduces the following set of recursive equations (Fig.4)

$$x_i = \ \Pi^i_{j=1} z_j, \qquad k_i \ = \prod^{i-1}_{j=1} z_j(1 - z_i), \tag{18}$$

$$p_{Ti} = \ q_{Ti} + x_i p_{Ti-1}, \qquad k_{Ti} \ = -q_{Ti} + (1 - x_i)p_{Ti-1}. \tag{19}$$

In the model described here, the initial and final state radiators are independent (Fig.3). This is true in the case of a very tall and narrow resonance. Results on the accuracy of the initial state radiator are described in detail in Ref. [2] and will be only summarized here. Fig. 5 shows a comparison between an analytic calculation [11] and the MC calculation of the line shape. They agree within .1% near the peak and within .3% several GeV above the peak. Fig. 6 shows that for one photon inclusive cases the agreement with an exponentiated calculation [13] is of order 1%.

In the particular case of final state evolution (Fig. 2b and 3), the evolution can proceed either forward or backward. Were the extra pairs of leptons to be considered, the evolution would have been tree-like and would have had to proceed forward. In our case we proceed backward, so as to avoid calculation of the Sudakov form factor [14]. We start from the $\mu^+\mu^-$ rest frame, with the μ on the $z-$axis, and assume $s' = q^2$ to sample the Poissonian. The total 4-momentum is increased by the addition of a photon generated according to the kernel of Eq. (13), and subsequently rescaled by means of a boost with

$$\vec{\beta} = \frac{\vec{k}}{2E - k}, \tag{20}$$

to comply with conservation laws. This procedure is not exact, but the approximation is of order $\alpha/\pi(m_\mu^2/s)$, which is negligible. The evolution then proceeds according to Eqs. (17) and (18).

Once the backward evolution is over, one has to readjust the statistics. If s' becomes small, the amplitude for photon emission goes to zero (Eq.(1)), so the too many photons generated by the assumption $s' = q^2$ are accepted with weight

$$W(n) = \frac{B(n)_{s'}}{B(n)_{q^2}} e^{-\bar{n}_{q^2}} = (\frac{\bar{n}_{s'}}{\bar{n}_{q^2}})^n e^{-n_{s'}}, \tag{21}$$

which is bound between 0 and 1 and effectively reproduces the exact statistics.

A sample distribution is shown in Fig. 7. In the approximation discussed here, the initial and final state are incoherent radiators (that would be true in the limit of a very narrow resonance) and the effect of the two energy losses can be separated to all orders. The MC reproduces correctly the zero amplitude in the small dimuon mass region.

This section therefore explains all the branching contained in MOE (July 1989 version) with the exclusion of the one generating the two fermions.

4   Effective beams, discussion and conclusions

We have left for this last Section the delicate argument of generating the $\mu^+\mu^-$ 4-momenta. Kleiss [3] has discussed the problem in detail. His conclusions can be summarized as follows:

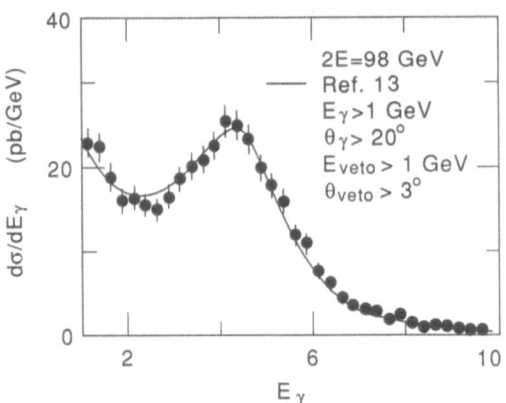

Figure 6. Radiative corrections to the $\gamma\nu\nu$ photon energy spectrum as predicted by the Monte Carlo (data points), and by an analytic exponentiated calculation. Photons are required to have an energy greater than 1 GeV, θ greater than 20 degrees and are vetoed if another photon of energy greater than 1 GeV has θ greater than 3 degrees.

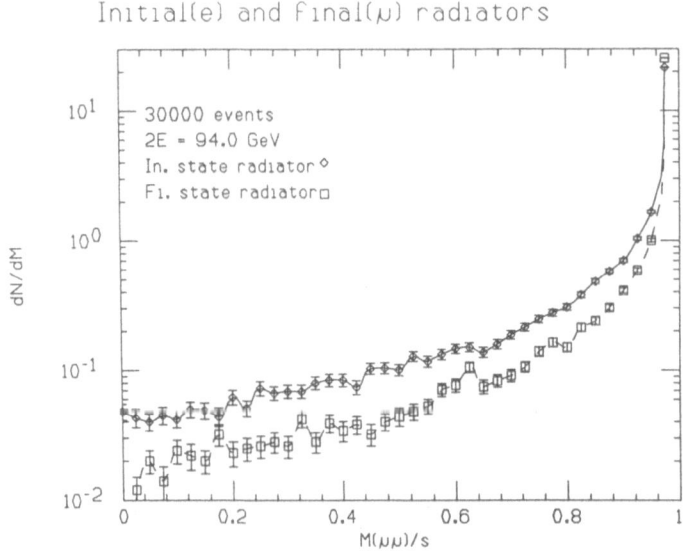

Figure 7. Energy loss spectrum from the initial and final state radiator.

- Factorization holds for single bremsstrahlung, with exact splitting between a radiation kernel and a modified Born term containing the substitution

$$\sigma(s, t, u) \rightarrow \sigma(s, s', t, t', u, u'),$$

where the arguments are the Mandelstam invariants without and with radiation.

- Factorization fails for double bremsstrahlung.

- Once a photon has been generated, correct generation of the $\mu^+\mu^-$ variables can be achieved by sampling an AP kernel $g \rightarrow q + \bar{q}$ with unequal probabilities

$$\frac{d\sigma}{dz} = Az^2 + B(1-z)^2, \qquad z = (\frac{1 - \cos\theta}{2}), \tag{22}$$

where the A and B coefficients are related to the beam energies in the $\mu^+\mu^-$ rest frame.

The conclusions listed above are consistent with a large body of literature about the properties of factorization in QCD, but are conveniently cast in algorithmic language. There are two disturbing aspects, that factorization does not hold in second order and that the kernels assume unequal probabilities even if there is a massive photon decaying into two fermions, like at PEP/PETRA. Unequal probability kernels are related to parity violating splittings, such as $Z \rightarrow f\bar{f}$, and should not appear in a parity conserving splitting. Since the two weights A and B in Eq. (22) have to be calculated from the kinematical distributions, they could rapidly become complicated in higher order and undermine the simplicity of this method, in the same way as phase space factors could be a problem.

"Effective beams" are directions in space such that, after generation of the initial state radiation, the $\mu^+\mu^-$ can be generated around them with uniform ϕ distribution and θ distribution as predicted by the Born term. As such, they are an important part of our original hypothesis, which is the possibility of building a higher order MC based on successive iterations of simple kernels. Implicitly, factorization as stated in the evolution equations [9], by Yennie, Frautschi and Suura [15] or in a coherent states formulation [5] assumes the "effective beams" to be the beamline to all orders.

Here we take a different approach. For sake of simplicity, we discuss the decay $\gamma \rightarrow f\bar{f}$, which is parity conserving. The extension of our method to a Z decay is straightforward. The Born term is

$$\sigma(s, t, u) \propto \frac{t^2 + u^2}{s^2} = z^2 + (1-z)^2, \tag{23}$$

and evolves as follows in the case of single emission [16]

$$\sigma(s, s', t, t', u, u') \propto \frac{t^2 + t'^2 + u^2 + u'^2}{ss'}. \tag{24}$$

After one photon has gone from the initial state we obtain two possible configurations by subtracting it randomly from either leg, and we get one virtual and one real beam, which are back-to-back in the system recoiling against the photon. Two different rotations after the same boost (Eq. (20)) have to be applied to align these beams with the z axis, depending on which leg has lost the photon, and the $\cos\theta$ variables in these two different systems get labels 1 and 2 depending on whether the e^+ or the e^- was left on shell. We assume these as effective beams, generate either the μ^+ or the μ^- angle first, with 50% probability, and make the other recoil.

We have the equality for the 4-momenta

$$p_1 + p_2 = p_3 + p_4 + k, \tag{25}$$

the equalities between Mandelstam invariants and 4-momenta

$$t^2 = (p_1 - p_3)^4 = (p_2 - (p_4 + k))^4 = ((p_2 - k) - p_4)^4$$
$$t'^2 = (p_2 - p_4)^4 = (p_1 - (p_3 + k))^4 = ((p_1 - k) - p_3)^4$$
$$u^2 = (p_1 - p_4)^4 = (p_2 - (p_3 + k))^4 = ((p_2 - k) - p_3)^4 \tag{26}$$
$$u'^2 = (p_2 - p_3)^4 = (p_1 - (p_4 + k))^4 = ((p_1 - k) - p_4)^4$$

and the equalities between integrated decay probabilites

$$\int d(\cos\theta_1) t^2 = \int d(\cos\theta_1) u^2 = \int d(\cos\theta_2) t'^2 = \int d(\cos\theta_2) u'^2,$$
$$\int d(\cos\theta_2) t^2 = \int d(\cos\theta_2) u^2 = \int d(\cos\theta_1) t'^2 = \int d(\cos\theta_1) u'^2. \tag{27}$$

From Eqs. (24), (26) and (27) it appears that the $\mu^+\mu^-$ kernel stays the same, with unitary weights, as long as the μ^+ and the μ^- are generated first with 50% probability. This means 4 possible branchings (which initial leg went off shell, and which fermion will be generated) for one photon. Since the proof of Ref. [3] holds for normal factorization, in which the "effective beam" is always the beamline, it does not hold for this kind of algorithm. Proving or disproving that the second order can be cast into a finite combination of "effective beams", the normalization of which is determined by simple branchings, remains the next very crucial step about this method and its usefulness to predict more complicated processes such as Bhabha scattering and multijet topologies in quantum chromodynamics.

In conclusion, we have presented a simple algorithm capable of reproducing the next-to-leading result for 2-particle radiators in QED, including photon multiplicities and transverse degrees of freedom. I thank R. Kleiss, K. Riles and L. Trentadue for useful discussions.

References

[1] S. Jadach and B.F.L. Ward, PRD 38: 2897, 1989.

[2] G. Bonvicini and L. Trentadue, NPB 323: 253, 1989.

[3] R. Kleiss, CERN-TH 5439-89.

[4] The first application of evolution equations to e^+e^- annihilation and to this day the most accurate (because of implementation of pair radiation) is: E.A.Kuraev and V.S.Fadin, SJNP 41(3): 466, 1985.

[5] E. Etim, G. Pancheri and B. Touschek, Nuovo Cim. 51 (1967) 276.

[6] R. Barbieri, J. Mignaco and E. Remiddi, NC 11A: 824, 1972.

[7] J. E. Campagne and R. Zitoun, these Proceedings.

[8] D. Kennedy *et al.*, SLAC-PUB-4128.

[9] G. Altarelli and G. Parisi, NPB 126: 298, 1977.

[10] C.J.Everett and D.E.Cashwell, LA-9721-MS.

[11] F.A. Berends, G.J.H. Burgers and W.L. Van Neerven, NPB 297: 429, 1988.

[12] A. Bassetto, M. Ciafaloni and G. Marchesini, NPB 163: 477, 1980.

[13] O.Nicrosini and L. Trentadue, NPB 318: 1, 1989.

[14] V.V. Sudakov, ZHETF 30: 187, 1956.

[15] D.R. Yennie *et al.*, Ann. Phys. 13: 379, 1961.

[16] F.A. Berends and R. Kleiss, NPB 177: 237, 1981.

EXCLUSIVE EXPONENTIATION IN THE MONTE CARLO
YENNIE-FRAUTSCHI-SUURA APPROACH

Stanisław Jadach

Institute of Physics, Jagellonian University
30-059 Kraków, ul. Reymonta 4, Poland
and
CERN, Geneva, Switzerland

B. F. L. Ward

Department of Physics and Astronomy
The University of Tennessee, Knoxville, Tennessee 37996-1200, U.S.A.
and
Stanford Linear Accelerator Center
Stanford University, Stanford, California 94309, U.S.A.

ABSTRACT

A closer look is taken into various types of the so-called "exponentiation" procedures in QED. In particular the question of the difference between the common (ad
hoc) inclusive exponentiation and the exclusive Yennie-Frautschi-Suura exponentiation is examined. The discussion is limited to the initial state bremsstrahlung in e^+e^-
annihilation. Numerical results from the YFS2 Monte Carlo are used to illustrate
the discussion.

1. Introduction

The procedure of summing up to infinite order the infrared divergent contributions in QED, so called "exponentiation", is known to be an efficient way of including
the higher order QED effects in the perturbative calculations. Typically, the first order perturbative calculation combined with the exponentiation procedure may give
a result quite close to the complete unexponentiated second order result [1].

The common exponentiation procedure, say for the first order calculation, consists
in combining in an ad hoc way two types of the calculations. The conventional
calculation provides the energy spectrum of the single hard photon. On the other

hand, from the classical papers on the exponentiation [2,3] one knows the distribution of the *total* energy of many soft photons. The two distributions are interpolated at some intermediate energy point and the resulting formula is used for predicting (successfully) the fermion pair effective mass spectrum and the total cross section.

The modern detectors can control, however, the topology of the experimental events (photon multiplicities and four-momenta) to a large extent and the above exponentiation procedure is not sufficient because it only provides a rather limited information on the inclusive spectra and the total event rate.

A closer look at the classical work on the exponentiation [2] reveals that the exclusive multi-photon distributions are well defined but they were not exploited for practical calculations due to a necessity of integration over the multibody Lorentz-invariant phase space.

In the recent works we have developed technical means for integrating the relevant phase space integrals using the Monte Carlo method [4,5,6]. The related computer programs simulate directly the multiple real photon emission - something what is very valuable for the present experiments in $e^{\pm}$ annihilation.

In the talk we shall shortly comment on the relation of the common inclusive exponentiation to the exclusive one. Some examples of the numerical results will be given and the applicability limits of both methods will be indicated.

The discussion will be mainly related to the YFS2 Monte Carlo [6] for the initial state bremsstrahlung in the electron-positron annihilation which is also a part of the KORALZ program [7]. This talk together with that of ref. [8] may be treated as an introduction to more technical papers on the exclusive exponentiation like [5,6].

The outline of the talk is the following: In the first part we shall recall what is the common (ad hoc) exponentiation procedure and we shall point out what are the conceptual and numerical uncertainties related to this procedure. Then, in the second part, we describe briefly the Yennie-Frautschi-Suura exponentiation scheme, and its implementation in the Monte Carlo, and we show how the above uncertainties can be resolved or better understood. Section 4 contains some summary remarks.

2. Common exponentiation – unanswerable questions

In the following we shall recall what is usually meant as an exponentiation procedure (we call it inclusive or ad hoc exponentiation) on the textbook example of exponentiating the Bonneau-Martin, $O(\alpha)$, expression for the total cross section in the annihilation process

$$e^{+}(p_1) + e^{-}(p_2) \rightarrow \mu^{+}(q_1) + \mu^{-}(q_1)(+\gamma(k)).$$

The Bonneau-Martin formula reads as follows

$$\sigma = \int\limits_{0}^{1} dv\, \rho_{\mathrm{BM}}(v)\, \sigma^{\mathrm{Born}}(s(1-v)), \tag{2.1}$$

where

$$\rho_{\mathrm{BM}}(v) = \delta(k)(1 + \gamma \ln \varepsilon + \delta_{VS}^{(1)}) + \gamma \Theta(v - \varepsilon)\frac{1 + (1 - v)^2}{2v}$$

$$\gamma = 2\left(\frac{\alpha}{\pi}\right)(L - 1), \quad L = \ln\frac{s}{m_e^2}, \quad \delta_{VS}^{(1)} = \left(\frac{\alpha}{\pi}\right)\left(\frac{3}{2}L - 2 + \frac{1}{3}\pi^2\right). \tag{2.2}$$

and $v = 1 - (q_1 + q_2)^2/(p_1 + p_2)^2 = 1 - s'/s$. In the single photon case v is simply a fraction of the beam energy carried by the photon, $v = 2k^0/\sqrt{s}$.* What is the usual ad hoc "exponentiation" procedure in the case of the above formula? It amounts to the following, essentially by hand, modification of the photon spectrum

$$\rho_{KF} = \gamma v^{\gamma-1}(1 + \delta_{VS}^{(1)}) + \delta_H^{(1)}(v)), \qquad \delta_H^{(1)}(v) = \gamma(-1 + \frac{1}{2}v). \tag{2.3}$$

There is definitely some freedom in the way it is done and we have followed here the prescription of ref. [10].

Let us examine more closely the question: *What is the meaning of the above procedure?* The exponentiated photon distribution $\rho(v)$ has the following two properties:

1. The original Bonneau-Martin $\rho(v)$ distribution (2.2) and the exponentiated one (2.3) coincide, up to terms of $O(\gamma^2)$, for hard photon, $v \sim 1$.

2. The main modification is made in soft photon region, $v \sim 0$, and it is done in such a way that, the cross section $\sigma(v < \varepsilon)$, when expanded up to $O(\gamma)$, remains the same.

There are convincing reasons for such a modification for $v << 0$. It is well known that in the region $v < \varepsilon << 1$ one has to sum up the contributions from the infinite number of soft virtual and real photons and, furthermore, it is known quite precisely [2] what are the resulting differential and integrated cross sections $d\sigma/dv$ and $\sigma(v < \varepsilon)$.†

Summarizing, the exponentiated formula (2.3) is well founded (A) in the region $v \sim 1$ where it becomes

$$\rho_{hard} = \gamma \frac{1 + (1 - v)^2}{2v} \tag{2.4}$$

and represents the *exclusive* single real photon bremsstrahlung contribution (as obtained directly from the Feynman rules) and (B) in the soft region $v << 1$

$$\rho_{soft} = \gamma v^{\gamma-1}(1 + \delta_{VS}^{(1)}), \tag{2.5}$$

where it represents the *inclusive* cross section summed/integrated over the infinite number of real and virtual soft photons with the total energy constrained to some small value: $\frac{2}{\sqrt{s}}\sum_{i=1}^{n} k_i^0 = v$, $v << 1$.

Our main observation is the following: *the usual exponentiation procedure is essentially the interpolation between the two expressions: exclusive hard of eq. (2.4) and inclusive soft of eq. (2.5).* Strictly speaking interpolation is made by ad hoc recipe following some sort of "continuity principle".*

---

* In the multiple photon case $v \sim \sum_i 2k_i^0/\sqrt{s}$ as long as there are no two anti-collinear hard photons.

† The necessity of such a modification is also signaled by the fact that the Bonneau-Martin cross section $\sigma(v < \varepsilon)$, in the limit $\varepsilon \to 0$ is ill defined – it becomes negative.

* Of course, such an ansatz can be quite successful. Its validity may be checked experimentally.

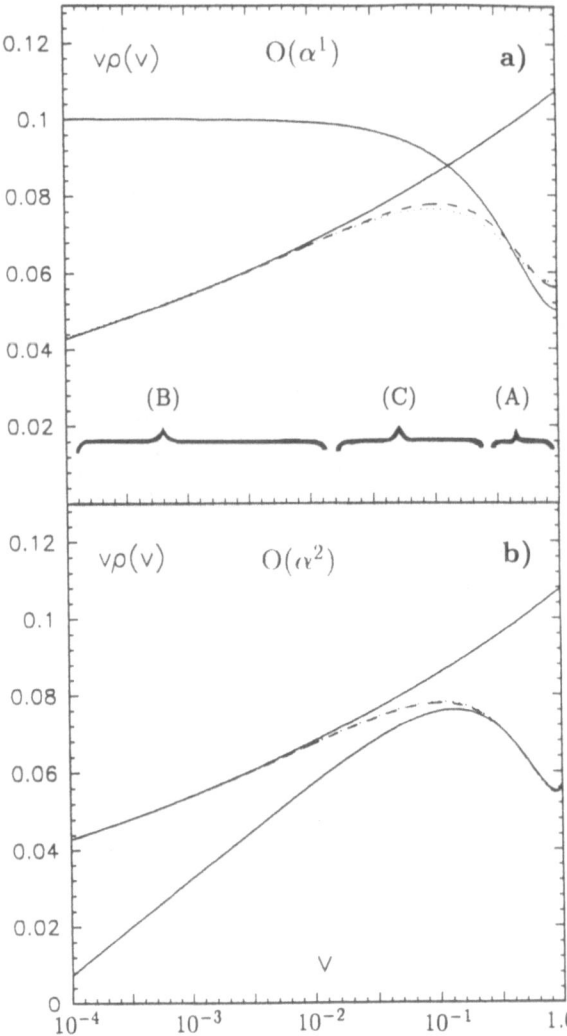

Fig. 1 *The various types of the distribution $v\rho(v)$. The two cases (a) and (b) represent $O(\alpha)$ and $O(\alpha^2)$ results correspondingly. The solid curve rising towards $v = 1$ represents $v\rho_{soft}(v)$, see for example eq. (2.5), and the other solid curve represents $v\rho_{hard}(v)$, see eq. (2.4). The dashed curve shows the distribution $v\rho(v)$ exponentiated/interpolated according to ref. [10] (see also eq. (2.3)) and the dotted curve shows exponentiation/interpolation according to ref. [6]. Note that the area under the "soft" curve $v\rho_{soft}(v)$ is roughly equal to unity.*

In the Fig. 1 (a) it is shown how the above interpolation works in practice. We plot the two source distributions $v\rho_{soft}(v)$ and $v\rho_{hard}(v)$ and the interpolation $v\rho_{KF}(v)$ as a function of $\ln v$ at $s = M_Z^2$. As we see, the exponentiated distribution coincides with $v\rho_{hard}(v)$ in the region (A), $0.3 < v < 1$, and with $v\rho_{soft}(v)$ in the region (B), $0 < v < 0.002$. The interpolation takes place within the interval (C) $0.02 < v < 0.3$.

The exponentiated distribution may be quite uncertain in the transition region (C). Furthermore, it is not clear what is the precise meaning of the result of the interpolation. This is because we interpolate among the quantities which represent rather different things: in the soft region (B) the contribution from many photons and in the hard region (A) the "Born-like" contribution from one photon only. The other (soft) photons disappear somehow in between, in the intermediate region (C). The related questions are: How are the many soft photons distributed over the multiphoton phase space? Are their momenta limited somehow? Do the resulting cross sections and distributions depend on the limits imposed on the (additional) soft photons? These question are quite relevant experimentally since modern detectors can see directly and indirectly photons down to $100 MeV$ and because the precision level in the LEP experiments will reach the level of $\delta\sigma/\sigma \simeq 10^{-3}$.

We postpone answering these questions to the next Section and we shall elaborate further on some interesting aspects of the "common exponentiation". More specifically, the usual way of getting some hint on the uncertainty resulting from the interpolation among soft and hard formulas is to take another "ad hoc" interpolation or to go to a higher perturbative order. Figs. 1 (a) and (b) illustrate both cases.

In Fig. 1 (a) we show what is the result of switching to another $O(\alpha)$ exponentiation/interpolation recipe. The dotted curve is obtained following the exponentiation prescription of ref. [6][*]

$$\rho_{YFS}(v) = \frac{e^{-C\gamma}}{\Gamma(1+\gamma)} e^{\delta_{\rm YFS}} \gamma v^{\gamma-1} (1 + \Delta_S^{(1)} + \Delta_H^{(1)}(v)), \quad \delta_{\rm YFS} = \left(\frac{\alpha}{\pi}\right)\left(\frac{1}{2}L - 1 + \frac{\pi^2}{3}\right),$$

$$\Delta_S^{(1)} = \left(\frac{\alpha}{\pi}\right)(L-1), \quad \Delta_H^{(1)}(v) = v\left(-1 + \frac{1}{2}v\right), \quad C = 0.57721566...$$
$$(2.6)$$

As we see in Fig. 1 (a) the above change of the type of exponentiation affects the distribution $d\sigma/dv$ around $v \sim 0.2$ by 2-3%.

The advantage of this new (also ad hoc) exponentiation is that the soft photon limit ($v << 1$) of the new formula (2.6) coincides precisely with the *exact* expression resulting from the soft photon resummation, as given in ref. [2][†] ,

$$d\sigma/dv = \sigma_{\rm Born}(s)e^{\delta_{\rm YFS}} \int \frac{dx}{2\pi} \exp\left[ixvW + \int \frac{d^3k}{k^0} \tilde{S}(p_1, p_2, k)\left(e^{-izk} - 1\right)\right]$$

$$= \sigma_{\rm Born}(s)e^{\gamma \ln \epsilon + \delta_{\rm YFS}} \sum_{n=0}^{\infty} \frac{1}{n!} \prod_{i=1}^{n} \int_{k_i^0 \geq \epsilon W} \frac{d^3 k_i}{k_i^0} \tilde{S}(p_1, p_2, k_i)\delta^4(vW - \sum_{i=1}^{n} k_i^0) \quad (2.7)$$

$$= \sigma_{\rm Born}(s)\frac{e^{-C\gamma}}{\Gamma(1+\gamma)} e^{\delta_{\rm YFS}} \gamma v^{\gamma-1}, \quad W = \frac{1}{2}\sqrt{s},$$

where

---

[*] The main difference is that $\gamma v^{\gamma-1}$ multiplies now the hard photon correction $\Delta_H(v)$.

[†] We set here $\Delta_S = 0$ for simplicity.

$$\delta_{\text{YFS}} = 2\alpha B + \tilde{B}$$

$$= \text{Re}\frac{i\alpha}{4\pi^2} \int \frac{d^4k}{k^2}\left(\frac{2p_1 - k}{k^2 - 2p_1 \cdot k} + \frac{2p_2 + k}{k^2 + 2p_2 \cdot k}\right)^2 + \int_{k^0 < W} \frac{d^3k}{k^0}\tilde{S}(p_1, p_2, k), \qquad (2.8)$$

$$\tilde{S}(p_1, p_2, k) = -\frac{\alpha}{4\pi^2}\left(\frac{p_1}{p_1 k} - \frac{p_2}{p_2 k}\right)^2.$$

We would like to stress that the eqs. (2.7) are exact identities and we refer the reader to refs. [6] and [8] for formal proofs – in particular on the transition from the Mellin type inclusive representation to the exclusive representation with the explicit integration/summation over soft photons.

In Fig. 1 (b) we show what happens if we go to $O(\alpha^2)$ exponentiation, i.e. interpolation. The starting "building blocks" for exponentiation are the "soft distribution"[*]

$$\rho_{soft}(v) = \frac{e^{-C\gamma}}{\Gamma(1 + \gamma)}e^{\delta_{\text{YFS}}}\,\gamma v^{\gamma - 1}(1 + \Delta_S^{(1)} + \Delta_S^{(2)}), \quad \Delta_S^{(2)} = \frac{1}{2}\left(\frac{\alpha}{\pi}\right)^2 L^2 \qquad (2.9)$$

which includes summation over infinite number of soft photons and the "hard distribution"

$$\rho_{hard} = \gamma\frac{1 + (1 - v)^2}{2v} + \left(\frac{\alpha}{\pi}\right)^2 L^2\left(-\frac{1 + (1 - v)^2}{v}\ln(1 - v)\right.$$

$$\left. + (2 + v)(\frac{1}{2}\ln(1 - v) - 2\ln v) - \frac{5}{2} - \frac{1}{2}(1 - v) + \frac{3}{v} + \frac{4}{v}\ln v\right), \qquad (2.10)$$

representing the second order (double bremsstrahlung) result without any exponentiation.

Before we interpolate the two distributions, let us note the remarkable difference with the $O(\alpha)$ case. The "hard photon" curve does not cross the "soft" one any more. In fact the "hard" one goes to $-\infty$ at $v \to 0$. This is related to the fact that, in the $O(\alpha^2)$, at fixed v, there are two types of contributions to $\rho(v)$: one from two real hard photons and another from one real and one virtual photon. The latter one (as noted in ref. [6]) becomes negative for $v \to 0$.

Now we "exponentiate" the above $O(\alpha^2)$ raw distributions (2.9) and (2.10), that is, we simply *interpolate* them. We do it either with the help of the formula (2.3) where we append $\delta_{VS}^{(1)}$ and $\delta_H^{(1)}$, see ref. [10,12], with the following $O(\alpha^2)$ contributions

$$\delta_{VS}^{(2)} = \left(\frac{\alpha}{\pi}\right)^2 L^2\left(\frac{9}{8} - \frac{1}{3}\pi^2\right), \qquad \delta_H^{(2)}(v) = \left(\frac{\alpha}{\pi}\right)^2 L^2\left(-\frac{1 + (1 - v)^2}{v}\ln(1 - v)\right.$$

$$\left. + (2 + v)\left(\frac{1}{2}\ln(1 - v) - 2\ln v\right) - \frac{5}{2} - \frac{1}{2}(1 - v)\right), \qquad (2.11)$$

or with the help of eq. (2.6) where we add $\Delta_S^{(2)}$ of eq. (2.9) and

$$\Delta_H^{(2)}(v) = \left(\frac{\alpha}{\pi}\right)L\left(-\frac{1}{4}(4 - 6v + 3v^2)\ln(1 - v) - v\right), \qquad (2.12)$$

see ref. [6]. The difference among the two types of the exponentiation/interpolation is now rather small, of order 0.2% at most.

---

[*] In these expressions we omitted some subleading, numerically unimportant, terms as in refs. [11,12].

Let us conclude by recalling main deficiencies of the common exponentiation:

1. There is no general and systematical framework of the type "calculations order by order" for this type of exponentiation – it is a matter of inventing recipes.

2. The phase space for real (soft) photons remains unspecified.

3. Finally, one should keep in mind that the common exponentiation is defined for some specific choice of the "principal" kinematical variable ($v = 1 - s'/s$ in the case presented above). It has to be repeated from the scratch if one decides to switch to another variable.[‡]

3. Yennie-Frautschi-Suura exclusive exponentiation

The aim of the following part of the talk is to explain what is the Yennie-Frautschi-Suura exclusive exponentiation. We shall show that the questions which could not be answered in the case of the common (ad hoc) exponentiation can be here either answered or shown to be not relevant. Let us indicate immediately what are the answers to the most disturbing questions from the former Section: The question "How many of the soft photon are there and where are they?" is unswered in the Yennie-Frautschi-Suura exclusive exponentiation as follows: there are no events without soft photons – soft real photons are always there; there is no sharp distinction in a form of cut-off between soft and hard real photons – the important and relevant distinction (in a given perturbative order) is between the part of the phase space where the actual differential cross section is or is not sufficiently precise with respect to the (experimental) precision requirements in mind. Furthermore, this type of exponentiation offers a definite prescription how, by going to higher orders, to improve in a systematical way on the precision of the calculation and, since Monte Carlo method is used, there is no need to rely on some specific choice of a "principal variable" for exponentiation nor on some specific kinematical cut-off's.

In the following we shall present the main aspects of the Yennie-Frautschi-Suura exponentiation. Only some most important aspects of the Monte Carlo phase space integration algorithm will be mentioned.

3.1 GENERAL PROPERTIES

Let us take as a starting point of discussion the Yennie-Frautschi-Suura formula for the entire perturbative series in a form suitable for the Monte-Carlo integration[♮]

$$
\sigma = \int \frac{d^3 q_1}{q_1^0} \frac{d^3 q_2}{q_2^0} \sum_{n=0}^{\infty} \frac{1}{n!} \prod_{i=1}^{n} \int_{k^0 < \epsilon W} \frac{d^3 k_i}{k_i^0} \tilde{S}(p_1, p_2, k_i) \delta^4 \left(p_1 + p_2 - q_1 - q_2 - \sum_{i=1}^{n} k_i \right)
$$

$$
e^{\gamma \ln \epsilon + \delta_{YFS}} \left\{ \tilde{\beta}_0(p_1, p_2, q_1, q_2) + \sum_{l=1}^{n} \frac{\tilde{\beta}_1(p_1, p_2, q_1, q_2, k_l)}{\tilde{S}(k_l)} \right.
$$

$$
\left. + \frac{1}{2} \sum_{\substack{l,j=1 \\ l \neq j}}^{n} \frac{\tilde{\beta}_2(p_1, p_2, q_1, q_2, k_l, k_j)}{\tilde{S}(k_l) \tilde{S}(k_j)} + ... \right\}.
$$

(3.1)

The other, Mellin-type representation, similar to that in eq. (2.7) used in the original YFS paper [2] and in many other papers on exponentiation [13] is algebraically and numerically equivalent to the above representation and we refer the reader to ref. [6] and [8] for further comments on the transition from one to another formulation.

---

‡ For example to sum of photon energies, upper cut on energy of any photon or a combination of the angular and energy cut-off.

♮ Here and in the rest of the paper we limit ourself to the initial state bremsstrahlung only.

Let us summarize on the most important properties of the above multi-distribution:

1. Eq. (3.1) represents the entire perturbative series (no approximations) reorganized in such a way that all infrared divergences cancel in the form-factor $\exp(\gamma \ln \varepsilon + \delta_{YFS})$.

2. The total cross section and any measurable quantity do not depend on the infrared cut-off (dummy) parameter ε which may be set arbitrarily low.

3. In the Monte Carlo real photons are explicitly generated above the energy threshold εW all over the entire phase space, i.e., many hard photons are generated as well. The average multiplicity of generated photons is approximately $\gamma \ln \frac{1}{\varepsilon}$.

4. The functions $\tilde{\beta}_i$ are infrared free and are calculable from the Feynman diagrams order by order. There is certain amount of freedom in defining $\tilde{\beta}'s$ related to the way of extracting the factors $\tilde{S}(p_1, p_2, k)$ from the differential distributions provided by the Feynman rules.[°]

5. In the N-th perturbative order the sum of $\tilde{\beta}'s$ gets truncated and we deal in practice with $\tilde{\beta}_i, i = 1, 2, ...N$. This we call the N-th order exclusive YFS exponentiation. In particular, in the lowest order exclusive exponentiation[•] we deal only with $\tilde{\beta}_0$.

The (truncated) formula (3.1) answers immediately the principal question of the previous Section: *Where are soft/hard real photons and how they are distributed in the phase space.* As we have noted already, the real photons are distributed over the complete phase space and, as we see, the eq. (3.1) defines the differential cross section everywhere. (The Monte Carlo is merely an exact, up to a statistical error, method of numerical integration over the phase space.) The limitation is hidden somewhere else. As usual in the perturbative expansion, due to the truncation in series of $\tilde{\beta}$'s, the differential cross section may be not precise enough. If it is true then one has to go to higher order and include more $\tilde{\beta}$'s in the expansion. In the following we shall discuss this point in more detail.

We would like to stress that the truncation of the series in $\tilde{\beta}$'s and the truncation of the perturbative expansion for each $\tilde{\beta}$ is the normal perturbative procedure. In fact we adhere to the following (conservative) algorithm of the perturbative calculation:

{1} Start with lowest order calculation $n = 0$,

{2} Calculate the quantity of interest $X^{(n)}$ in order n and $n + 1$,

{3} If $|X^{(n+1)} - X^{(n)}| < \delta$ where δ is required (experimental) precision then finish calculation else go to point {2}.

In practice such an algorithm would require to perform $O(\alpha^3)$ calculation of the QED calculation around Z and, in our opinion, such a calculation should be done at least in some approximation, to estimate the "theoretical precision" of the $O(\alpha^2)$ calculation [14].

[°] As long as the complete sum over $\tilde{\beta}$'s is kept this freedom does not influence the integrated and/or differential cross sections at all. This is analogous to the first step in the renormalization group technique – the scheme dependence comes when the series gets truncated.

[•] The results may now slightly depend on the way of defining $\tilde{\beta}$ functions.

3.2 ZERO ORDER EXPONENTIATION

The zero-order exponentiation, with $\bar{\beta}_0$ alone, is the case when only the most infrared-singular part is isolated and summed up to infinite order. As was proven in ref. [2] the virtual photon divergences sum up to $e^{\alpha B}$ and the real photon leading singularity is proportional to $\prod_{i=1}^{n} \bar{S}(k_i)$. The function $\bar{\beta}_0$ is what remains after factorization of these singularities

$$\rho_n(p_i, q_i, k_1, k_2, ..., k_n) = \prod_{i=1}^{n} \bar{S}(p_1, p_2, k_i) e^{2\alpha \mathrm{Re} B} \bar{\beta}_0(p_1, p_2, q_1, q_2). \tag{3.2}$$

It should be stressed that for any number n of real photons we get the same residue $\bar{\beta}_0$! Of course, it is calculable most easily for $n = 0$. In general, $\bar{\beta}_0$ is an expansion in a series of the coupling constant and in the zero-order case it coincides simply with the Born angular distribution

$$\bar{\beta}_0^{(0)}(p_1, p_2, q_1, q_2) = \frac{d\sigma^{\mathrm{Born}}}{d\Omega} \left(\mathcal{R}p_1, \mathcal{R}p_2, \mathcal{R}q_1, \mathcal{R}q_2 \right) = \frac{d\sigma^{\mathrm{Born}}}{d\Omega} \left(\theta_\mathcal{R}, s(1-v) \right). \tag{3.3}$$

The only complication is that, strictly speaking, $d\sigma^{\mathrm{Born}}/d\Omega$ is defined within the 2-body phase space $p_1 + p_2 - q_1 - q_2 = 0$ while $\bar{\beta}_0$, see eq. (3.2) has to be also defined for $p_1 + p_2 - q_1 - q_2 \neq 0$, see also the discussion in ref. [2]. In eq. (3.3) we do it by means of mapping $p_i \to \mathcal{R}p_i$, $q_i \to \mathcal{R}q_i$ such that $\mathcal{R}p_1 + \mathcal{R}p_2 - \mathcal{R}q_1 - \mathcal{R}q_2 = 0$. (The angle $\theta_\mathcal{R}$ is calculated using the reduced momenta.) This so called "reduction procedure" has to obey two basic rules: (i) it has to become an identity transformation when $p_1 + p_2 - q_1 - q_2 \to 0$ and (ii) it should be the same for arbitrary configuration of the photons which sum up to the same total four-momentum $K = \sum_{i=1}^{n} k_i = p_1 + p_2 - q_1 - q_2$.$^\nabla$ In eq. (3.3) we have implicitly imposed the additional condition $(\mathcal{R}p_1 + \mathcal{R}p_2)^2 = (q_1 + q_2)^2 = s(1-v)$ anticipating the resonance formation in the $s-$channel. The reduction is done in practice by means of boosts, rotations and rescaling of the momenta, see ref. [2] for more details. Summarizing, the zero-order version of the eq. (3.1) reads†

$$\sigma = e^{\gamma \ln \varepsilon + \delta_{YFS}} \int dv \frac{d\Omega_q}{4\pi} \sum_{n=0}^{\infty} \frac{1}{n!} \prod_{i=1}^{n} \int_{k^0 < \varepsilon W} \frac{d^3 k_i}{k_i^0} \bar{S}(p_1, p_2, k_i)$$

$$\delta^4 \left(v - \frac{(p_1 + p_2 - \sum k_i)^2}{s} \right) \frac{d\sigma^{\mathrm{Born}}}{d\Omega} \left(\theta_\mathcal{R}, s(1-v) \right). \tag{3.4}$$

The above zero-order exclusive exponentiation is of little practical use and we presented it in order to discuss some aspects of the inclusive exponentiation on the maximally simple example. The main deficiency of the zero-order exponentiation is twofold: (i) The distribution of the real photons is a simple product of infrared factors $\prod \bar{S}(k_i)$ – something which could be wrong if there is at least one hard photon, (ii) The virtual correction of the type $1 + \delta_S$ is missing. The total cross section may be imprecise by about 5% due to these defects. This will be corrected in the $O(\alpha)$ case.

∇ In particular the reduction knows nothing about the number of photons.

$\dagger$ Note that a distribution very close to this one is generated in the early stage in the YFS2 Monte Carlo – the contributions due to $\bar{\beta}_1$ and $\bar{\beta}_2$ are added by rejection.

3.3 First and Second Order Exponentiation

In the $O(\alpha)$ case the $\tilde{\beta}_0$ is calculated (up to a normalization constant) as follows

$$\tilde{\beta}_0^{(1)} = |e^{-2\alpha B}\mathcal{M}(\mathcal{O}(\alpha))|^2 = (1 - 2\alpha\mathrm{Re}B)(1 + 2\mathrm{Re}F_1)|\mathcal{M}_{\mathrm{Born}}|^2$$
$$= (1 - 2\alpha\mathrm{Re}B + 2\mathrm{Re}F_1)|\mathcal{M}_{\mathrm{Born}}|^2 = (1 + \Delta_S^{(1)})|\mathcal{M}_{\mathrm{Born}}|^2 \tag{3.5}$$

where $\mathcal{M}(\mathcal{O}(\alpha)) = (1 + F_1)\mathcal{M}_{\mathrm{Born}}$ is the one loop result obtained directly from the Feynman rules (Born + vertex correction). The reduction procedure is understood to be applied here similarly as in the previous zero order case. We expect, now, that the differential cross section will be improved in the limit $v \to 0$ due to the $1 + \Delta_S^{(1)}$ factor.

What about the cases with hard photons? In the case of $O(\alpha)$ exponentiation we include in (3.1) the $\tilde{\beta}_1$ term. Its definition results, see ref. [2], from the careful analysis of the distributions obtained from the Feynman rules. It is quite easy, however, to deduce its definition indirectly as follows: If in the sum

$$d\sigma_n \sim \tilde{\beta}_0 \prod_{i=1}^{n} \tilde{S}(k_i) + \sum_{l=1}^{n} \tilde{\beta}_1(k_l) \prod_{i \neq l} \tilde{S}(k_i), \tag{3.6}$$

all photons are soft, $k_i^0 << \sqrt{s}/2$, then the second sum is negligible because $\tilde{\beta}_1(k)$ is not singular in the infrared limit $k^0 \to 0$. If one photon, say $l = L$, is hard, $k_L^0 \sim \sqrt{s}/2$, then in the second sum only this term, $l = L$, is non-negligible and we can rewrite (3.6) as follows

$$\tilde{S}(k_1)...\tilde{S}(k_{L-1})\left(\tilde{\beta}_0(k_L)\tilde{S}(k_L) + \tilde{\beta}_1(k_L)\right)...\tilde{S}(k_n) = \tilde{S}(k_1)...\tilde{S}(k_{L-1})\rho_1(k_L)...\tilde{S}(k_n),$$

where the $\rho_1(k)$ should obviously coincide with the conventional single bremsstrahlung matrix element squared. The above implies the definition of the $\tilde{\beta}_1(k)$, see also refs. [2,6,15,5],

$$\tilde{\beta}_1^{(1)}(k) = \rho_1^{(1)}(k) - \tilde{S}(k)\tilde{\beta}_0^{(0)}. \tag{3.7}$$

Again, as is the case of $\tilde{\beta}_0$, the definition of $\tilde{\beta}_1$ must *necessarily* include the reduction procedure which eliminates from the four-momentum balance all photons but one.[†] It should be stressed that in the above definition one uses the lowest order[*] $\tilde{\beta}_0^{(0)}$ while in the first term of the sum (3.6) one uses rather the $O(\alpha)$ $\tilde{\beta}_1^{(1)}$. To summarize: *the $\tilde{\beta}_1$ contribution is negligible if all photons are soft (compared with beam momentum) and its role is to correct the distribution of the events with one hard and (possibly) many soft photons.*

To proceed from the $O(\alpha)$ to $O(\alpha^2)$ exponentiation is quite straightforward. The $\tilde{\beta}_0$ is calculated now as follows

$$\tilde{\beta}_0^{(2)} = |e^{-2\alpha B}\mathcal{M}(\mathcal{O}(\alpha^2))|^2 = (1 - 2\alpha\mathrm{Re}B + \frac{1}{2}(2\alpha\mathrm{Re}B)^2)(1 + 2\mathrm{Re}F_1)|\mathcal{M}_{\mathrm{Born}}|^2$$
$$= (1 + \Delta_S^{(1)} + \Delta_S^{(2)})|\mathcal{M}_{\mathrm{Born}}|^2 \tag{3.8}$$

where

[†] The reduction procedure for $\tilde{\beta}_1(k_l)$ which preserves the momentum k_l and eliminates the other ones has to fulfil another important constraint [6]: it has to preserve the value of $\tilde{S}(p_1, p_2, k_l) \simeq 1/k_{l,T}^2$, otherwise there is no guarantee that $\tilde{\beta}_1$ is small in the region $E_{beam} > k_l^0 > m_e$.

[*] Otherwise $\tilde{\beta}_1$ would be infrared singular at $k^0 \to 0$.

$$\Delta_S^{(2)} = \frac{1}{2}\left(\frac{\alpha}{\pi}\right)^2 L^2 + \left(\frac{\alpha}{\pi}\right)^2 \left\{ -\left(-\frac{13}{16} + \frac{3}{2}\zeta(2) - 3\zeta(3) \right)L \right.$$
$$\left. -\frac{16}{5}\zeta(2)^2 + \frac{51}{8}\zeta(2) + \frac{13}{4} - \frac{9}{2}\zeta(3) - 6\zeta(2)\ln(2) \right\} \tag{3.9}$$

is the exact $O(\alpha^2)$ contribution calculated using the $O(\alpha^2)$ vertex correction F_1 of ref. [1]. Note that the resulting $\Delta_S^{(2)}$ is remarkably shorter then the original $O(\alpha^2)$ vertex function F_1.

The $O(\alpha^2)$ $\tilde{\beta}_1$ is obtained from

$$\tilde{\beta}_1^{(2)}(k) = \rho_1^{(2)}(k) - \tilde{S}(k)\tilde{\beta}_0^{(1)}. \tag{3.10}$$

where one substitutes $O(\alpha^2)$ distribution $\rho_1^{(2)}$ which includes contributions from one real and up to one virtual photons[‡] and the $O(\alpha)$ $\tilde{\beta}_0^{(1)}$. The new object in $O(\alpha^2)$ exponentiation is now

$$\tilde{\beta}_2^{(2)}(k_1, k_2) = \rho_1^{(2)}(k_1, k_2) - \tilde{S}(k_1)\tilde{\beta}_1^{(1)}(k_2) - \tilde{S}(k_2)\tilde{\beta}_1^{(1)}(k_1) - \tilde{S}(k_1)\tilde{S}(k_2)\tilde{\beta}_0^{(0)} \tag{3.11}$$

which is constructed using the tree level distributions only[♮] (no virtual photons). The use of reduction procedure which eliminates all but two photons is understood implicitly.

3.4 COMPARISONS, NUMERICAL RESULTS

We are now in position to compare the ad hoc inclusive and YFS exclusive exponentiations and to answer in a more detail some "unanswerable" questions from the first part of our talk, illustrating the discussion with numerical Monte Carlo results.

As already indicated, the question "Where are the real soft photons?" referring to the matching, in the ad hoc exponentiation, the exclusive ρ_{hard} of region (A) and the inclusive ρ_{soft} of region (B) is in the present Yennie-Frautschi-Suura exclusive exponentiation *not really relevant* because there is no exclusive region (A) – hard photon can always be accompanied in this case with some real soft photons. There is also no mysterious transition region (C) any more.[◊] Nevertheless the distribution $v\rho(v)$ which was the object of common exponentiation in the previous Section can be obtained from the Monte Carlo representing the exclusive Yennie-Frautschi-Suura exponentiation and we may check that indeed the Monte Carlo result is rather close to the result of the ad hoc exponentiation/interpolation procedure. In Fig. 2 compare the $O(\alpha)$ and $O(\alpha^2)$ results from the Monte Carlo program YFS2 [6] with the "exponentiated distributions" from the previous Section. The agreement is very good and the analogous excellent agreement on the total cross section, see ref. [6], is a direct consequence of the agreement for $\rho(v)$. Let us also note that once the exponentiation-type Monte Carlo is set up then we may easily produce "exponentiated" distribution not only for v but for any other variable (acollinearity etc.).

---

‡ The distribution $\rho_1^{(2)}$ is not available in the literature and in [6] we used the leading/subleading approximation. The contributions of $O((\alpha/\pi)^2)$ are neglected in this way.

♮ In ref. [6] we use the leading/subleading logarithmic approximation for $\rho_2^{(2)}$.

◊ The role of interpolation by "continuity principle" is now, to some extent, played by the more complicated but better defined "reduction procedure". As it was already explained, it is a natural consequence of the process of extracting the infrared real photon singularities.

It is also clear that in the presented exclusive exponentiation there is *no sharp distinction*, for instance in a form of a cut-off, among the hard and soft real photons. The only distinction in the phase space is such that in some part of it the YFS matrix element of eq. (3.1) is less precise and in another one more precise than the actual (experimental) precision requirement. The precision depends on the actual perturbative order of the calculation and we shall try to illustrate this point by looking into $O(\alpha^0)$, $O(\alpha^1)$ and $O(\alpha^2)$ results and their differences.

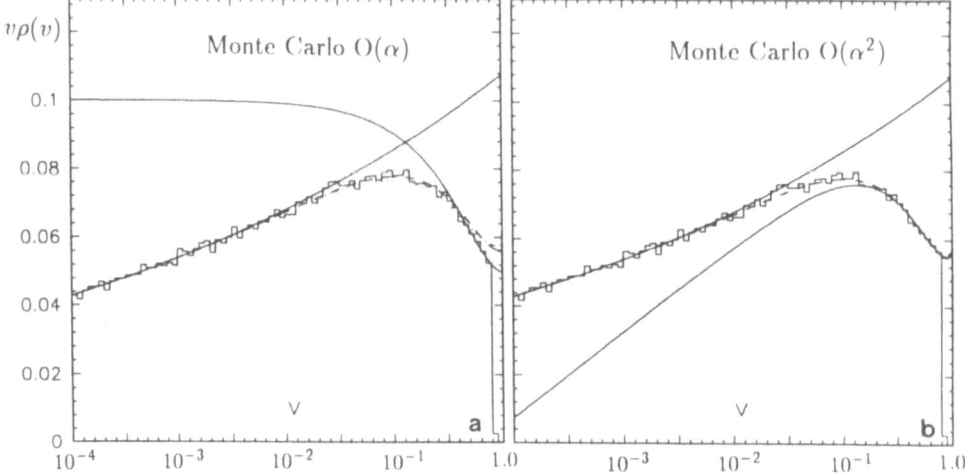

Fig. 2 *The histogram represents the distribution* $v\rho_{MC}(v) = (d\sigma/d\ln v)/\sigma^{Born}(s(1 - v))$, *from the Monte Carlo sample of* $4 \cdot 10^5$ *events produced with the help of the YFS2 program [6]. The continuous lines are the same as in Fig. 1 (b) and represent the* $O(\alpha^2)$ *"common exponentiation". This distribution depends very little on* $\sqrt{s}$. *We have used here* $\sqrt{s} = 40GeV$ *and we cut* $v < 0.8$.

In Fig. 3 a) we present the $O(\alpha^0)$ and $O(\alpha^1)$ Monte carlo results for $v\rho(v)$ and their difference. The difference between the two is most visible at $v > 0.1$ and it comes from the fact that the soft photon approximation is not very adequate the real hard photon – it could be wrong by the factor two (at most). In technical terms, the $O(\alpha^0)$ calculation does not include the $\bar{\beta}_1$ contribution. It should be stressed however that the difference persists at $v \to 0$ limit and this is due to the lack of the virtual correction Δ_1 in the $O(\alpha^0)$ version of the $\bar{\beta}_0$. For these reason the differences between $O(\alpha^0)$ and $O(\alpha^1)$ total cross sections may be of order 5%. This can be actually seen directly in Fig. 3 where we indicate a rectangle which corresponds roughly to 1% of the integrated cross section. [•]

In Fig. 3b) we examine the analogous difference between $O(\alpha^1)$ and $O(\alpha^2)$ Monte Carlo result for $v\rho(v)$. The difference is, as expected, much smaller. According to

---

• We say roughly because $\rho(v)$ is weighted with the Born cross section $\sigma(s(1 - v))$.

what was said in the beginning of this Section the difference between $O(\alpha^1)$ and $O(\alpha^2)$ may be regarded as a precision estimate of the $O(\alpha)$ calculation. In Fig. 3 b) we see that, as far as the integrated cross section is concerned, the precision of the $O(\alpha)$ exponentiated calculation is below 1% level. To estimate the precision of the $O(\alpha^2)$ calculation we would need the $O(\alpha^3)$ Monte Carlo calculation which is not yet available (although some attempts were made in this direction [14]). One may only guess that it is of order $\sim 0.1\%$. In the following part of this Section we shall show the result of some additional Monte Carlo exercises which demonstrate more clearly where the higher order contributions are located in the phase space.

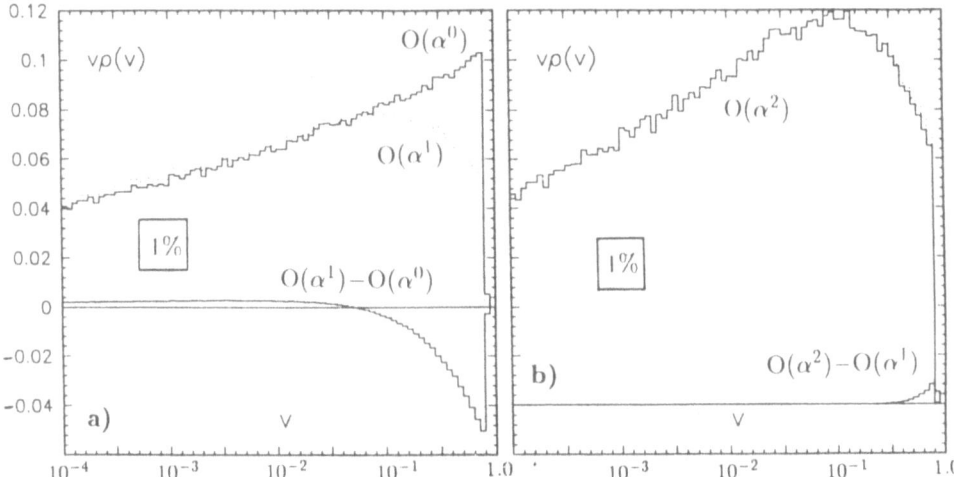

Fig. 3 *The distribution $v\rho_{MC}(v) = (d\sigma/d\ln v)/\sigma^{\mathrm{Born}}(s(1-v))$, from the Monte Carlo sample of $4 \cdot 10^5$ events produced with the help of the YFS2 program [6]. In the case (a) the upper solid line shows $O(\alpha^0)$ result and the dashed line represents $O(\alpha^1)$ result. The lower solid line shows the difference of the previous two. The case (b) includes the corresponding $O(\alpha^1)$ and $O(\alpha^2)$ results. We have used here $\sqrt{s} = 40GeV$ and we cut $v < 0.8$.*

In the $O(\alpha)$, we truncate the series on the $\tilde{\beta}_1$ term and the rates of events with two hard photons may be distorted – some sort of "double counting" of $\tilde{\beta}_1$ occurs. If we go to $O(\alpha^2)$ then the inclusion of $\tilde{\beta}_2$ improves the two hard photon distribution (the problem remains still for three hard photons). As a result, in Fig. 3 b) we find a pronounced peak in the plot of the difference between $O(\alpha^2)$ and $O(\alpha)$ at high v. It comes mainly from $\tilde{\beta}_2$ i.e. from the situations when we have two hard photons in the sample. Let us check this conjecture with the Monte Carlo exercise.

In Fig. 4 we plot the energy distribution of the photon which is not the hardest one but the second in the energy order. It means that if this photon has an energy $E_\gamma^{(2)} = 0.1 E_{beam}$ then in fact we have two photons with the energy $E_\gamma \geq 0.1 E_{beam}$.

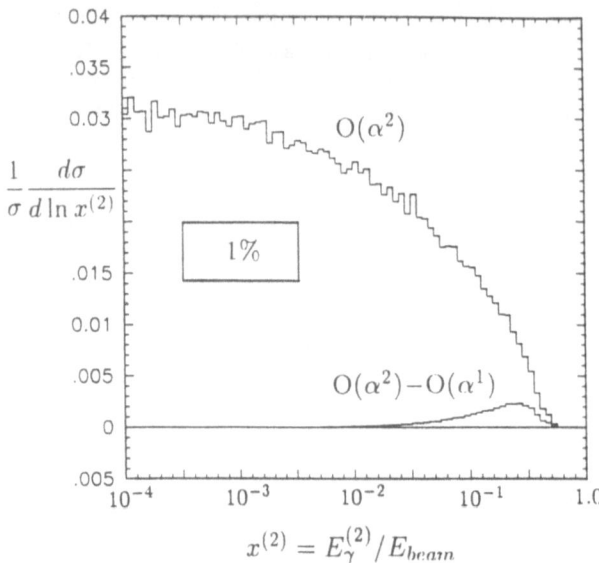

$$x^{(2)} = E_\gamma^{(2)}/E_{beam}$$

Fig. 4 *The probability distribution of the energy of the second photon in the energy order. Plotted is $O(\alpha^2)$ result and the difference with the $O(\alpha^1)$ result.*

The difference between the $O(\alpha^2)$ and $O(\alpha^1)$ results for this distribution is also shown in the Fig. 4 and we see that, as expected, it is well concentrated at the high v region.

One could argue that the higher order effect (in this case the $O(\alpha^2)$) could be at least partly eliminated by cutting off both in the Monte Carlo and the experiment the region with the cut of the type $E_\gamma^{(2)} < 0.1 E_{beam}$ which eliminates events with the simultaneous emission of two hard photons. This however may appear not to be really necessary because the gain in the precision is not substantial and it may also happen that the change of the cross section due to such a cut is in fact of order of the actual experimental precision requirement anyway.

In order to illustrate this point let us plot the change of the total cross section due to elimination of the events with three hard photons by means of the cut $E_\gamma^{(3)}/E_{beam} < x^{(3)}$. (Such a cut would hopefully eliminate some of $O(\alpha^3)$ effects.) As we see in the Fig. 5 in the vicinity of the Z resonance it does not really matter if we introduce such a cut or not because it affect the total cross section below 0.1%, even if we use $x_{min}^{(3)} \simeq 10^{-2}$. This is a simple reflection of the fact that events with many hard photons are more and more rare when we go to higher number of hard photons. In this case the third photon in the energy order has a distribution which is rather strongly suppressed at high energy end of its spectrum. This can be seen in Fig. 6 where we plot the energy distribution for the first three photons in the energy order. Of course, such a study is not really complete before we are able to calculate the $O(\alpha^3)$ corrections. and examine the difference between the $O(\alpha^3)$ and $O(\alpha^2)$ results.

4. Summary

We have discussed extensively what is the difference between the common inclusive exponentiation and the exclusive Yennie-Frautschi-Suura exponentiation. The discussion concentrated around the problem of the phase space limits for soft real photons and on the question where the higher order corrections are located in the phase space.

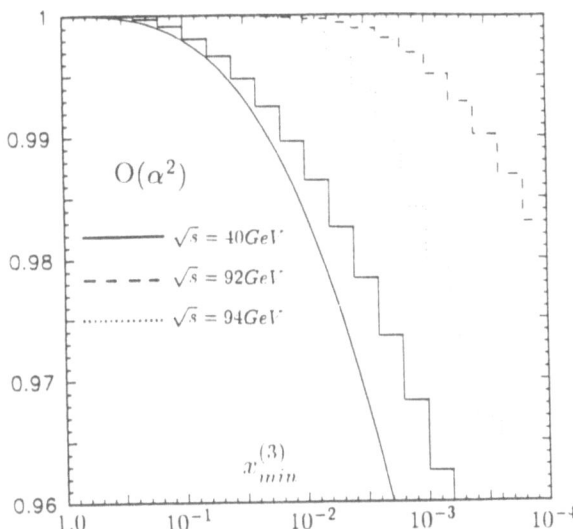

Fig. 5 *The Monte Carlo result for the probability of getting three hard photons with the energy above $x_{min}^{(3)} E_{beam}$ limit. This probability is plotted as a function of $\ln x_{min}^{(3)}$ at $\sqrt{s} = 40 GeV, M_Z, M_Z \pm 2 GeV$. The continuous smooth line represents a simple estimate $(\gamma^3/6)|\ln x_{min}^{(3)}|^3$.*

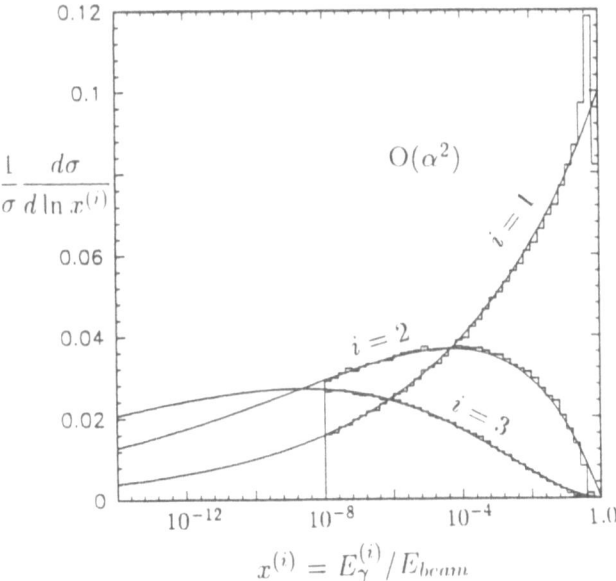

Fig. 6 *The energy distribution of the photon which has respectively the highest energy $E_\gamma^{(1)}$, then the second and third highest energy $E_\gamma^{(i)}, i = 2, 3, ...$. Plotted are the probability distributions $dp/d\ln x^{(i)}$ where, $x^{(i)} = E_\gamma^{(i)}/E_{beam}$, at $\sqrt{s} = 40 GeV$. The continuous lines represent a simple analytical estimate $p_i(x) = e^{\gamma \ln x}|\gamma \ln x|^{i-1}/(i-1)!$ Monte Carlo samples of $4 \cdot 10^5$ event were generated with $\varepsilon = 10^{-8}$ and $v < 0.8$.*

Acknowledgments

The authors are grateful to SLAC and CERN for their hospitality and support, especially to the MKII,SLD,ALEPH and DELPHI collaborations at these laboratories.The authors acknowledge useful discussions with Dr.Z. Was.This work was partly supported by grant no. CPB.01.09 of the Polish Ministry of Education and by the U.S. Department of Energy,contracts DE-AS05-76ER03956 and DE-AC03-76SF00515.The authors thank the Organizers for giving them the opportunity to lecture at this very stimulating Workshop.

REFERENCES

1. F. A. Berends and W. L. Van Neerven and G. J. H. Burgers, Nucl. Phys. **B297** (1988) 429.

2. D. R. Yennie, S. C. Frautschi and H. Suura, Annals of Phys. **13** (1961) 379.

3. K. T. Mahanthappa, Phys. Rev. **126** (329) 1962.

4. S. Jadach, "Yennie-Frautschi-Suura soft photons in the Monte Carlo generators", preprint of Max-Plack-Institut, München, MPI-PAE/PTh 6/87 (1987).

5. S. Jadach and B. F. L. Ward, SLAC-PUB-4543 (1988), Phys. Rev. **D38** (1988) 2897.

6. S. Jadach and B.F.L. Ward, " YFS2 –the second order ¦monte Carlo for fermion pair production at LEP/SLC with the initial state radiation of two hard and multiple soft photons", TPJU-15/88, SLAC-PUB-4834 report, to appear in Computer Phys. Commun.

7. S. Jadach, Z. Wąs, R. G. S. Stuart, B. F. L. Ward and W. Hollik, "KORALZ the Monte Carlo program for τ and μ pair production processes at LEP/SLC", unpublished, may be obtained from Z. Wąs: WASM @ CERNVM.

8. S. Jadach and B.F.L. Ward, in Proceedings of the Ringberg-workshop, April 1989, CERN report TH.5399/89.

9. G. Bonneau and F. Martin, Nucl. Phys. **177** (1988) 56.

10. E. A. Kuraiev and V. S. Fadin, Sov. J. Nucl. Phys. **41** (1985) 466.

11. J. P. Alexander, G. Bonvicini, P. S. Drell and R. Frey, "Radiative Corrections to the Z^0 resonance", SLAC-PUB-4376, Phys. Rev. **D37** (1988) 56.

12. G. Burgers, "The shape and size of the Z resonance", in " Polarization at LEP", CERN report 88-06, eds. J. Ellis and R. Peccei, CERN, Geneva, (1988).

13. M. Greco, G. Pancheri and Y. N. Srivastava, Nucl. Phys. **B101** (1975) 234; Phys. Lett. **56B** (1975) 367.

14. S. Jadach and M. Skrzypek, Jagellonian University preprint TPJU-3/89, January 1080.

15. B. F. L. Ward, Phys. Rev. **D36** (1987) 939; Acta Phys. Pol. **B19** (1988) 465.

ON THE TREATMENT OF E.W.R.C. IN M.C. EVENT
GENERATORS FOR LEP *

M. Martinez and R. Miquel

Laboratori de Física d'Altes Energies
Universitat Autònoma de Barcelona
E-08193 Bellaterra (Barcelona) Spain

C. Mana

CERN, CH-1211 Geneva, Switzerland

Abstract

The aim of this work is twofold: first the discussion of which is the status of the theoretical steps towards the precise-enough M.C. event generator, and second some prospects on how to deal with the event generators available at present. Therefore a simple analysis of the size and effect of the relevant corrections to be taken into account to fulfill the foreseen accuracy needed at LEP energies is presented. Additionally, a review of the different ways of implementing them developed so far and available at present in the form of M.C. event generators is presented. The intrinsic differences are commented and the real range of validity of each implementation is discussed.

1 Introduction

The foreseen accuracy of experimental measurements at LEP has motivated in the last years a vast theoretical effort in order to provide experimenters the precise predictions of the Standard Model necessary to extract clear information out of the data.

The necessity of including higher order effects in the calculations when predictions providing a systematic uncertainty of few percent are required has become clear not just to theorists but also to experimentalist who, mostly involved in the role of preparing the whole experimental set-up, in general do not have the necessary background to understand the technical aspects of the problem.

In the mean time, different groups of theorists have been working in trying to clarify the proper way of solving the problems. Different approaches have been developed and released to the experimentalists

Short time ago, the increasing multiplicity of M.C. event generators (which actually should have been contemplated just as intermediate steps of the work, in spite that they were often released as "the final version"), together with the use of them without a clear knowledge of their contents and natural limitations originated an alarming number of results of simulations leading, in general, to predictions which differed among themselves for more than the claimed

---

*Presented by M.Martinez

accuracy of each generator. Fortunately, this situation has been changing in the last months and at this moment, encouraging agreement between different calculations is being reported for most of the interesting processes for LEP energies.

The purpose of this paper is twofold: first the clarification of which is the status of the theoretical steps towards the precise-enough M.C. event generator, and second some prospects on how to deal with the use the M.C. event generators available at present.

The outline of the paper is as follows. In the next section we discuss an operative way of getting close to the problem: analyzing the evolution of M.C. calculations in the previous generation of accelerators (PETRA-PEP). Section 3 analyzes the obvious new needs for LEP energies. In section 4 the treatment the so-called oblique corrections is considered. Section 5 is devoted to the discussion of the different ways of including the infrared QED corrections. Section 6 deals with the way of implementing a reasonable calculation of the Z_0 width. In section 7 the common features and differences as well as the obvious limitations of each one of the methods reviewed are discussed. Finally an overview of the status of the questions and some prospects for experimentalists involved in the use of the existing M.C. event generators are presented.

2 The PETRA-PEP Monte Carlo event generators

The previous step from which the evolution of M.C. event generators towards the LEP requirements has started to develop is the set of programs which have been extensively used for the data analysis in PETRA and PEP. These calculations have been carefully confronted with data by all the experiments and therefore can be considered as accurate-enough for the energies in which they where working. The evolution of M.C. calculations to adapt their characteristics to the PETRA-PEP necessities can be schematized in the following steps:

A) The starting point is the use of programs consisting in just tree level calculations including just QED diagrams for two body final state processes. For some processes intended to be used to measure E.W. effects (μ pairs and τ pairs) some additional diagrams are taken into account: basically only the interference of QED diagrams with Z_0 ones at the tree level (customarily included just as a correction factor to the QED matrix element squared). At this level no hard photons are taken into account. For the calculation of the total cross section this is a reasonable approach already for processes such as Bhabha scattering (low Q^2) but can give just around a 10% accuracy for processes such as μ pair production (high Q^2) since:

  i.- Weak corrections (even the ones due to the missing tree level diagrams) are very small in any case.

  ii.- QED corrections (if no cut on additional photons or particle acollinearity are applied) are rather small for Bhabha but sizeable for μ pairs. As this point will be of paramount importance, let us discuss it in more detail: as it is well known, the cross section for the process $e^+e^- \longrightarrow \mu^+\mu^-$ (which we will use as an example for simplicity) corrected with one loop initial state radiation and assuming that the dependence of σ_0 on s is smooth, can be written in the following way:

$$\sigma_1(s) = \sigma_0(s)(1 + \delta_1 + \beta \ln x_M) \tag{1}$$

where

$$\beta \equiv \frac{2\alpha}{\pi}(\ln \frac{s}{m_e^2} - 1) \qquad (2)$$

and

$$\delta_1 \equiv \frac{3}{4}\beta + \frac{\alpha}{\pi}(\frac{\pi^2}{3} - \frac{1}{2}) \qquad (3)$$

x_M represents the maximum photon energy allowed by the detection cuts. If no photon cut or acollinearity cut is applied, then $x_M \to 1$ and the correction is:

$$\delta_1 \sim 8\% \qquad\qquad (\sqrt{s} \sim 40 GeV) \qquad (4)$$

In this expression, the real photon contribution has been integrated over the photon energy and its infrared divergency has been cancelled with the one due to the virtual photon correction, leading to the term $\beta \ln x_M$ being δ_1 the remaining correction. Let's realize that since the Q^2 is very small for Bhabha events, the total correction is also small:

$$\delta_1 \sim 0.06\% \qquad (5)$$

Nevertheless, since the presence of real bremsstrahlung photons is clearly detected in all channels affecting therefore all the differential cross sections, it becomes obvious that radiative corrections (which at the beginning of PETRA-PEP running where understood by many experimentalists just as the number of detectable *real photons*) should be taken into account somehow.

B) The following step was the development of computational methods to handle radiative processes at the tree level:

   i.- The use of helicity amplitudes to compute processes involving many Feynman diagrams (such as the radiative ones) [1].

   ii.- The application of more elaborated Monte Carlo techniques to handle the collinear peaks of the multidifferential bremsstrahlung cross sections [2]. Each real photon of energy E_k emitted by a charged particle p introduces in the matrix element M a factor

$$\frac{1}{E_k(1 - \beta_e \cos \theta_{pk})} \qquad\qquad \beta_e = \sqrt{1 - \frac{m^2}{p_0^2}} \qquad (6)$$

in such a way that if the photon direction is nearly parallel to p (collinear) then the cross section peaks. The dependence in E_k leads to the infrared divergence which can be regularized for instance just by taking a minimum photon energy ("hard photon").

These two tools allow the construction of M.C. event generators for studying specifically final states such as $e^+e^-\gamma$, $\mu^+\mu^-\gamma$, and so on. As experimentally the detection of a photon requires that it has an energy larger than a certain amount determined by the detector resolution, a natural definition of "hard photon" comes up.

C) The next step attacked is the construction of a M.C. event generator combining the radiative and the non-radiative cross section for a given process. This is obtained by calculating all the $O(\alpha)$ QED corrections to the Born QED diagrams ("reduced QED corrections"). When computing the real photon part, it turns out to have an infrared divergence which cancels out exactly with the one showing up in the interference of the tree-level diagrams with the virtual photon ones. To handle the problem, the solution customarily adopted is splitting the real photon integral into two parts:

- a part in which the photon energy is smaller than a certain limit x_0. The photon energy integral is computed analytically in the limit $x \longrightarrow 0$ (soft photon) and the result is added to the one of the virtual part cancelling out the infrared divergences.
- the part in which the photon energy is larger than the soft limit (hard photon). The cross section is treated as a radiative cross section at the tree level.

For instance, the $O(\alpha)$ QED corrections due to initial state radiation can be written as

$$
\begin{aligned}
\sigma_1(s) \;=\; & \sigma_0(s)(1 + \delta_1 + \beta \ln x_0) \\
& + \int_{x_0}^{x_M} \sigma_0(s(1-x))\beta(\frac{1}{x} - 1 + \frac{x}{2})dx
\end{aligned}
\tag{7}
$$

As actually the splitting of the real photon integral into soft and hard parts is arbitrary, the total result has to be independent on the value of the soft-hard limit (provided the s dependence of σ_0 is smooth enough).

D) For the processes allowing more accurate E.W. measurements ($\mu^+\mu^-$ and $\tau^+\tau^-$) further developments required for the pushing experimental accuracy were: including in the calculations all the Z_0 diagrams at the tree level and furthermore affected as well by the photonic corrections ("full QED corrections").

This was essentially the state of the art in M.C. event generators used at PETRA-PEP before the requirements for LEP-SLC energies where taken into account. In spite that the complete one loop E.W. corrections for processes such as $\mu^+\mu^-$ existed already [3], they were not available in the form of M.C. event generators.

3 Requirements for a precise-enough M.C. for LEP-SLC

The first obvious realization about the requirements to obtain reliable M.C. event generators for LEP energies was that there was no strong reason to expect all the non-photonic one loop corrections to be negligible and therefore, for completeness they all were included in the M.C. event generators (though most of the corrections are really small in a given renormalization scheme and can be safely neglected). To do so, it was necessary to study in detail the one loop renormalization of the Standard Model choosing a clear and handy renormalization scheme such as the "on shell" one [4]. The calculations have been generally done in the Feynman gauge mainly because of the trivial form of the boson propagators it allows, in spite of the existence of unphysical fields which have to be taken into account in the diagrammatics. Nevertheless, some people have been using other gauges (unitary, $\xi - gauge$) allowing thus stringent comparisons.

This lead to a set of M.C. event generators which in principle were thought to be the Standard for LEP-SLC energies. They were orthodox theoretically speaking since they contained a full one-loop expansion of the Glashow-Weinberg-Salam theory for a given process, and thus they were by definition gauge-independent and divergency-free. In one hand, their contents where clear and reproducible (in the sense that since they were strictly full one-loop E.W. M.C. event generators, one knew exactly which Feynman diagrams were included and which not and one could compare exactly between different calculations using the same renormalization scheme) and in other hand nobody believed that a calculation of the full two-loop E.W. correction (which would have preserved the orthodoxy) was feasible.

At that moment the development of resummation of higher order terms for only "isolated" parts of the corrections was seen by many people with strong caution since it represented a loss of theoretical orthodoxy and clarity in favour of phenomenological requirements producing potentially just effective approaches.

The corrections to the Born terms can be classified as "weak" (non-photonic) and QED (photonic) ones. This division is sensible since the second corrections form a gauge invariant subset and helps to clarify the origin and effect of the corrections.

3.1 Weak corrections

The weak ones, thought were not the dominant ones at PETRA energies and can be estimated not to be the dominant ones either at LEP energies, are the most interesting ones since they contain the gauge theory character of the Standard Model, the dependence on unknown parameters as the Top and the Higgs masses and are also sensitives to "new physics".

In the "on shell" scheme and using the Feynman gauge, the most important ones for all the relevant observables turn out to be the oblique ones, essentially the Z_0 self energy which amounts up to 10% few GeV under and above M_Z. The oblique corrections at the one loop level do not depend on detection cuts. In any case it is obvious that if accuracies of about 1% are to be reached, higher order terms of these corrections have to be taken into account.

3.2 QED Corrections

The second ones, namely the QED corrections, do not have any special theoretical interest, but one shall recall that in the case of PETRA generators were the dominant ones and their size was strongly dependent on the detection cuts applied. Anyway, if no photon cut was applied, they were expected to amount less than 10%. This fact will not hold for LEP energies because of the following: near the Z_0 peak the QED corrections become large because the emission of initial state photons decreases the effective Q^2 in the Z_0 propagator. We can estimate numerically the effect of the presence of the resonance assuming that the Z_0 shape makes an effective cut in the maximum photonic energy of

$$x_M \sim \frac{\frac{\Gamma_Z}{2}}{E_b} = \frac{\Gamma_Z}{M_Z} \tag{8}$$

in such a way that the infrared term of the corrections becomes

$$\beta \ln \frac{\Gamma_Z}{M_Z} \sim -40\% \tag{9}$$

That is, at LEP energies, independently of the detection cuts (which may still enhance the size of the infrared correction) we will have always extremely important QED corrections due to the neighbourhood to the Z_0 resonance. This enormous size of the QED one loop corrections implies that:

1. We cannot expect in any case higher order QED corrections to be small and therefore we have to take them into account to obtain reasonable accuracies.

2. As the bulk of these QED corrections comes from the infrared parts of the initial state radiation correction, and in the soft limit this part factorizes from the rest of the cross section, we shall keep in mind that the interplay among photonic and non-photonic corrections is important (if we do not take it into account we can be overestimating seriously the effect of oblique corrections).

Additionally some technical problems arise in the M.C. implementation of the soft-hard separation technique making the "pure one loop" calculation unpractical:

a) x_0 cannot be taken large because in this technique for the soft part we are neglecting the photon energy, and this can be a quite bad approximation since if we take $x_0 = 0.01$ for instance (what will match an experimental energy resolution of .5 GeV) then the actual dependence of σ_0 would give

$$\sigma_0(M_Z^2) - \sigma_0(M_Z^2(1 - 0.01)) \sim 0.1 \ \sigma_0(M_Z^2) \tag{10}$$

in such a way that to assure a variation of around 1 per mil we should actually take x_0 of about 0.001. In practice, this energy dependence of the soft cross section near the resonance can be taken into account approximately to correct it for the value of x_0 used and many M.C. do include this correction already.

b) x_0 cannot be taken small because as we have seen, the soft cross section may become negative and therefore it is not possible to use it as a P.D.F. in a M.C. procedure.

3.3   Additionnal remarks

A last important remark in order is the following: in spite that the calculations discussed so far are intended to give us the cross section up to $O(\alpha^3)$ they are not exactly $O(\alpha^3)$ calculations. The reason is that actually the Z_0 width we are putting into the denominator of the Z_0 propagator is just an $O(\alpha)$ calculation.

Therefore, while we are off-resonance, $((s - M_Z)^2 >> M_Z \Gamma_Z)$ the real part of the propagator dominates and then we have effectively an $O(\alpha)$ corrected calculation but when we are sitting on top of the resonance $((s - M_Z)^2 \to 0)$ this is no more true, and we end up with an $O(1)$ corrected cross section. This means that if we want our M.C. to be accurate enough on top of the Z_0 resonance, one of the things we need is including into the Z_0 propagator a more accurate calculation of the Z_0 width than the simple tree level one.

Therefore, the study of the complete E.W. one-loop calculation at LEP energies shows that there are essentially four aspects which have to be studied carefully to build M.C. event generators to fulfill the accuracy requirements of LEP experiments, namely:

a) Higher order OBLIQUE corrections

b) Higher order QED corrections

c) Interplay of photonic and non-photonic corrections

d) A more accurate evaluation of the Z_0 width

4 Oblique corrections

The oblique corrections are the ones to the boson propagators. As they are universal (in the sense that they don't depend on the specific process studied), they can be computed once for ever and included as basic building blocks of the theory as we will see now.

The corrected boson propagators can be related to the proper self energies through the Dyson equations (fig.1)

$$
\begin{aligned}
G_\gamma &= D_\gamma - D_\gamma \hat{\Sigma}^\gamma G_\gamma - D_\gamma \hat{\Sigma}^{\gamma Z} G_{\gamma Z} \\
G_Z &= D_Z - D_Z \hat{\Sigma}^Z G_Z - D_Z \hat{\Sigma}^{\gamma Z} G_{\gamma Z} \\
G_{\gamma Z} &= -D_Z \hat{\Sigma}^Z G_{\gamma Z} - D_Z \hat{\Sigma}^{\gamma Z} G_\gamma \\
G_W &= D_W - D_W \hat{\Sigma}^W G_W
\end{aligned}
\tag{11}
$$

where $\hat{\Sigma}^\gamma$ is the proper photon self energy, $\hat{\Sigma}^{\gamma Z}$ is the proper $\gamma - Z_0$ mixing, $\hat{\Sigma}^Z$ is the proper Z_0 self energy and

$$
D_\gamma = \frac{1}{s} \quad D_Z = \frac{1}{s - M_Z^2} \quad D_W = \frac{1}{s - M_W^2}
\tag{12}
$$

are the tree-level propagators. The solution of these equations is the following:

$$
\begin{aligned}
G_\gamma &= \frac{1}{s + \hat{\Sigma}^\gamma(s) - \frac{(\hat{\Sigma}^{\gamma Z}(s))^2}{s - M_Z^2 + \hat{\Sigma}^Z(s)}} \\
G_Z &= \frac{1}{s - M_Z^2 + \hat{\Sigma}^Z(s) - \frac{(\hat{\Sigma}^{\gamma Z}(s))^2}{s + \hat{\Sigma}^\gamma(s)}} \\
G_{\gamma Z} &= -\frac{\hat{\Sigma}^{\gamma Z}(s)}{(s + \hat{\Sigma}^\gamma(s))(s - M_Z^2 + \hat{\Sigma}^Z(s)) - (\hat{\Sigma}^{\gamma Z}(s))^2} \\
G_W &= \frac{1}{s - M_W^2 + \hat{\Sigma}^W(s)}
\end{aligned}
\tag{13}
$$

in such a way that we can interpret the resulting propagators as a summation of all the 1PI contributions to the vector bosons self energies to all orders (see figure). If originally the self energies are calculated at the one-loop level, this implies adding up all the leading log contributions to the self energies:

$$
\alpha^n \ln^n \frac{s}{m^2}
\tag{14}
$$

which, in fact, are just really important for the fermionic loops contribution.

The direct use of the result we have obtained from Dyson equations has a couple of drawbacks: the first is that the appearance of $(\hat{\Sigma}^{\gamma Z}(s))^2$ in the denominator introduces higher order terms which affect the definition of the boson masses through the renormalization conditions. The second is that the resummation produced by Dyson equations leads to gauge dependent results which have to be treated with caution. As we will clarify later, the vector boson contributions to the boson self-energies can be pointed out as the responsible for this non-invariance. As in the Feynman gauge their size is essentially negligible, we can affirm that the numerical effect on the final results due to the gauge dependence of these equations would be very small, so that in principle we could just forget about the problem.

Different methods have been suggested so far to implement this resummation in practice, essentially they follow three different ideas, namely

a) neglecting the effect of $(\hat{\Sigma}^{\gamma Z}(s))^2$ in the denominator and also forgetting about gauge non-invariance. This is done in the "propagator substitution method".

b) trying to absorb the result of Dyson equations in the definition of a set of universal running coupling constants. In this absorption, the necessary terms to preserve gauge-invariance have to be resummed as well in the proper way. This is the basic idea of the STAR scheme.

c) trying to absorb the result of Dyson equations in a set of forms factors preserving gauge-invariance. This has been recently claimed to be achieved by the Dubna-Zeuthen [5] collaboration.

4.1 Propagator substitution

If we just want to keep terms of $O(\alpha)$ in the denominator of the propagator with the leading-log terms resummed to all orders, we can neglect the terms

$$- (\hat{\Sigma}^{\gamma Z})^2 \tag{15}$$

and therefore we end up with the following rule

$$
\begin{aligned}
\frac{1}{s} &\rightarrow \frac{1}{s + \hat{\Sigma}^{\gamma}(s)} \\
\frac{1}{s - M_Z^2} &\rightarrow \frac{1}{s - M_Z^2 + \hat{\Sigma}^Z(s)} \\
\frac{1}{s - M_W^2} &\rightarrow \frac{1}{s - M_W^2 + \hat{\Sigma}^W(s)}
\end{aligned}
\tag{16}
$$

which can be applied in any of the occurrences of the boson propagators (at the tree level and in the corrections themselves). This is the sense in which we can affirm that these corrections are universal. This simple substitution rule, introduces automatically the interplay of oblique corrections with the rest in the proper way.

The approximation made is sensible since the size of $\hat{\Sigma}^{\gamma Z}$ is rather small. Nevertheless, it has been shown that the situation changes sizeably if the Top mass is large (say > 150 GeV) and then, it becomes necessary the use of the full expression (eqs. 13) of the propagators to match the foreseen experimental needs.

348

4.2 The "improved Born" amplitude

A more serious attempt to take benefit of the universality of these oblique corrections can be obtained by studying specifically the four fermion interaction in the following way: The result of Dyson eq. can be seen also as a rediagonalization of the neutral boson mass matrix after radiative corrections mixing Z_0 and photons have been included

$$(D_{\mu\nu})^{-1} = i g_{\mu\nu} \begin{pmatrix} s + \hat{\Sigma}^\gamma(s) & \hat{\Sigma}^{\gamma Z}(s) \\ \hat{\Sigma}^{\gamma Z}(s) & s - M_Z^2 + \hat{\Sigma}^Z(s) \end{pmatrix} \tag{17}$$

Neglecting in the self-energies all the imaginary parts (but, of course, that of the Z_0 one) which can be shown to be small and replacing their full s dependence by just a simple linear one like

$$\hat{\Sigma}^\gamma(s) = s \hat{\Pi}^\gamma(M_Z^2)$$

$$\hat{\Sigma}^{\gamma Z}(s) = s \hat{\Pi}^{\gamma Z}(M_Z^2) \tag{18}$$

being

$$\hat{\Pi}^\gamma(M_Z^2) \equiv Re(\frac{\hat{\Sigma}^\gamma(s)}{s})|_{s=M_Z^2}$$

$$\hat{\Pi}^{\gamma Z}(M_Z^2) \equiv Re(\frac{\hat{\Sigma}^{\gamma Z}(s)}{s})|_{s=M_Z^2}$$

$$\hat{\Pi}^Z(M_Z^2) \equiv \frac{\partial}{\partial s} Re[\hat{\Sigma}^Z(s) - \frac{(\hat{\Sigma}^{\gamma Z}(s))^2}{s + \hat{\Sigma}^\gamma(s)}]|_{s=M_Z^2} \tag{19}$$

which turns out to be a rather good approximation, the four-fermion amplitude can be written as

$$M_{NC} = \frac{e^2}{1 + \hat{\Pi}^\gamma(M_Z^2)} Q_e Q_f \frac{1}{s} \gamma_\mu \otimes \gamma^\mu$$

$$+ \frac{e^2}{4 s_W^2 c_W^2} \frac{1}{1 + \hat{\Pi}^Z(M_Z^2)} \frac{1}{s - M_Z^2 + i \frac{s}{M_Z} \Gamma_Z}$$

$$\gamma_\mu[I_3^e - 2 \bar{s}_W^2 Q_e - I_3^e \gamma_5] \otimes \gamma^\mu [I_3^f - 2 \bar{s}_W^2 Q_f - I_3^f \gamma_5] \tag{20}$$

where Γ_Z denotes the physical width. This expression can be rewritten as

$$M_{NC} = Q_e Q_f \frac{e^2(M_Z^2)}{s} \gamma_\mu \otimes \gamma^\mu$$

$$+ \frac{G_\mu}{\sqrt{2}} \kappa \frac{M_Z^2}{s - M_Z^2 + i \frac{s}{M_Z} \Gamma_Z} \gamma_\mu[I_3^e - 2 \bar{s}_W^2 Q_e - I_3^e \gamma_5] \otimes \gamma_\mu [I_3^f - 2 \bar{s}_W^2 Q_f - I_3^f \gamma_5] \tag{21}$$

that is, by using a sort of "running" parameters $e(M_Z^2)$ and $\bar{s}_W^2$ defined as

$$e^2(M_Z^2) \equiv \frac{e^2}{1 + \hat{\Pi}^\gamma(M_Z^2)}$$

$$\bar{s}_W^2 \equiv s_W^2 - s_W c_W \frac{\hat{\Pi}^{\gamma Z}(M_Z^2)}{1 + \hat{\Pi}^\gamma(M_Z^2)} \qquad (22)$$

we are able to write an "improved" Born-like matrix element squared. The correction factor κ deviates from 1 only for a heavy top

$$\kappa \simeq 1 + \frac{3\alpha}{16\pi s_W^2} \frac{m_t^2}{M_W^2} \qquad (23)$$

and also in practice

$$\bar{s}_W^2 \approx s_W^2 + \frac{\alpha(M_Z^2)}{4\pi} \frac{3c_W^2}{4s_W^2} \frac{m_t^2}{M_W^2} \qquad (24)$$

in such a way that the residual terms of the full expressions are of the same size as the vertex and box corrections to be included next.

The basic conclusion of this approach is that a good implementation of the oblique corrections resummation near the Z_0 pole can be obtained already just by using in the tree level expressions

- i.- The running electric charge $e^2(s)$ in the photon diagrams.

- ii.- G_μ plus an "energy-dependent" width in the Z_0 diagrams.

4.3 The STAR scheme

One out of the most elaborated ways of treatment of the "oblique" corrections is the so-called STAR scheme [6], which can be regarded actually as a new renormalization scheme with running parameters. It takes full benefit of the universality of oblique corrections by including them already as basic building blocks of the theory that can be used to write the Feynman diagram amplitudes. This allows a fast evaluation of the effect of oblique corrections in any observable by using the following procedure: write the observable at the tree level in terms of "bare" parameters and include the relevant QED corrections; then just substitute the "bare" parameters by "starred" ones.

Basically, the idea of the method is as follows: First one starts with a renormalized lagrangian writing down the four-fermion amplitude in terms of "bare" parameters:

$$M_{NC} \quad - \quad e_0^2 G_{AA} Q Q' + e_0^2 [Q(\frac{I'_3 - s_0^2 Q'}{s_0 c_0}) + Q'(\frac{I_3 - s_0^2 Q}{s_0 c_0})] G_{ZA}$$

$$+ \quad e_0^2 (\frac{I'_3 - s_0^2 Q'}{s_0 c_0})(\frac{I_3 - s_0^2 Q}{s_0 c_0}) G_{ZZ} \qquad (25)$$

Then one uses Dyson eqs. to obtain the full boson propagators G_{AA}, G_{ZZ} and G_{WW} and Z_0 mixing as it was done at the beginning of the section. The main difference at this point is that since we are still dealing with "bare" quantities, the expression cannot be used as it is because we do not know what e_0, s_0 are.

As we mentioned before, there are some theoretical problems which arise as a result of using Dyson equations to obtain the full propagators in this way, namely

a) It is easy to realize that since the vector boson contributions to the self energies are not gauge invariant (fig.2), actually the results for the full boson propagators are not gauge invariant either.

b) When computing Π_{ZA} (the $\gamma - Z$ mixing) we find terms proportional to M_W^2 which do not vanish when $Q^2 = 0$. This implies that there is a $\gamma - Z$ mixing term at $Q^2 = 0$, which means that the neutral sector has been misdiagonalized since we want the photon to have no weak component at all at $Q^2 = 0$. (This didn't happen in the previous method because there renormalized quantities are used, and one of the renormalization conditions is precisely that Π_{ZA} has to vanish for $Q^2 = 0$).

These problems come clearly from the non-abelian nature of the theory and to solve them we will have to perform a rediagonalization of the neutral sector. The solution of both problems can be found realizing that the gauge dependence of the boson self energies cancels out at the one loop level with part of the weak vertex corrections[1] (fig.3). We know that every vertex in the theory will pick up corrections which can be treated as modifications of the coupling constants:

$$g_0 \rightarrow g_0(1 + g_0^2 \Gamma) \tag{26}$$

We can split Γ into two parts: $\Gamma = \Gamma' + \tilde{\Gamma}$, the former one being gauge-dependent and the second one gauge-independent. Since the first one cancels with a similar part appearing in the oblique corrections, which are universal, it must be universal as well. The other way around, any process-dependent part of Γ has to be gauge-independent, because there is nothing to cancel with in the oblique corrections. Therefore we can include this part of the vertex correction by defining a new universal coupling constant

$$\tilde{g} \equiv g_0(1 + g_0^2 \Gamma') \tag{27}$$

and rediagonalize the mass matrix with this new coupling. The proper treatment of this fact before using Dyson eq. ensures that problem "a" disappears and moreover, requiring that the new π_{ZA} must be purely transverse (proportional to Q^2), we isolate the actual part Γ' of Γ needed to fulfill requirement "b", namely

$$\Gamma' = \frac{-\Pi_{ZA}^L}{\tilde{g} g_0' M_{W_0}^2} \tag{28}$$

After this has been done, we can write a four-fermion amplitude which has on it all the vacuum polarization resummed properly, preserving gauge invariance and decoupling properly the oblique part from the remaining one:

---

[1]This has been claimed by Lynn and Kennedy in spite that many people affirms that also some parts of the W boxes are needed to cancellate the gauge dependence. This doesn't really affect the numerical results as commneted above and, apparently, even if this is true, some minor changes would restore the rigorous validity of the method we are discussing [7]

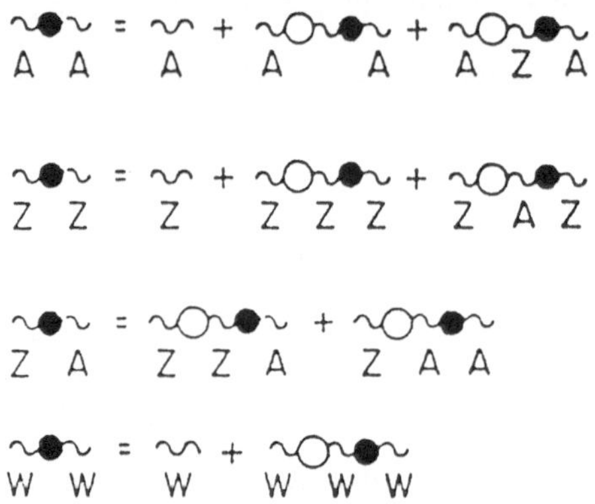

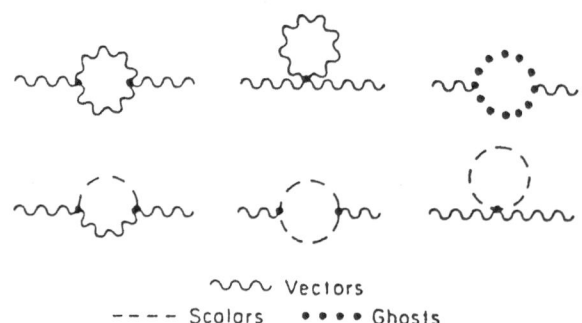

Figure 1. Graphic representation of Dyson equations. The empty bubble stands for the 1 particle irreducible (1PI) self energy.

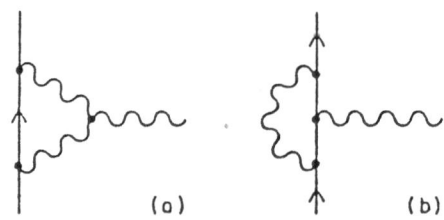

⌇⌇ Vectors
---- Scalars •••• Ghosts

Figure 2. Boson contributions to the boson self energies.

(a)   (b)

Figure 3. Vertex diagrams which are gauge dependent.

$$M_{NC} = G_{ZZ}\frac{\tilde{e}^2}{\tilde{s}^2\tilde{c}^2}[I_3 - Q(\tilde{s}^2 - \tilde{s}\tilde{c}\frac{[ZA]}{[A]})][I'_3 - Q'(\tilde{s}^2 - \tilde{s}\tilde{c}\frac{[ZA]}{[A]})]$$
$$+ \tilde{e}^2\frac{QQ'}{[A]} \tag{29}$$

We can rewrite this amplitude in terms of finite parameters in the following way

$$M_{NC} = \frac{e_*^2}{q^2(1 - iIm\Pi'^*_{AA})}$$
$$+ \frac{e_*^2}{s_*^2c_*^2}\frac{[I_3 - Q(s_*^2 - is_*c_*Im\Pi'^*_{ZA})][I_3' - Q'(s_*^2 - is_*c_*Im\Pi'^*_{ZA}]}{q^2(1 + (Im\Pi'^*_{AA})^2)(1 + \frac{\lambda}{s_*^2c_*^2}) + \frac{e_*^2}{4\sqrt{2}s_*^2c_*^2G_{\mu*}\rho^*} - i\sqrt{s}\Gamma_Z^*} \tag{30}$$

where e_*^2, s_*^2, $G_{\mu*}$ and ρ_* are finite and renormalization-scheme independent (since they depend only on "bare" quantities) functions of Q^2. Namely

$$\frac{1}{e_*^2} \equiv \frac{1}{e_0^2} - Re(\Pi'_{QQ} + 2\Gamma')$$

$$s_*^2 \equiv Re\left(\frac{\frac{1}{g_0^2} - (\Pi'_{3Q} + 2\Gamma')}{\frac{1}{e_0^2} - (\Pi'_{QQ} + 2\Gamma')}\right)$$

$$\frac{1}{4\sqrt{2}G_{\mu*}} \equiv \frac{M_{W_0}^2}{g_0^2} - Re(\Pi_{11} - \Pi_{3Q}^T - 2\Pi_{3Q}^L)$$

$$\frac{1}{\rho_*} \equiv \frac{1}{\rho_0} - 4\sqrt{2}G_{\mu*}Re(\Pi_{33} - \Pi_{11}) \tag{31}$$

being the rest of "starred" quantities just calculable functions of the four basic "starred" parameters. If we assume that $\rho_0 = 1$, that is only Higgs doublets, then ρ_* becomes also just a function of the remaining three "starred" functions, which must be considered then as the only fundamental ones. If we find a well-defined procedure of fixing their value at a given Q^2 we will have a well defined quantity. To do so, we use equation (29) to compute the QED amplitude in the Thompson limit, and from here we can just define

$$e_*^2(0) = 4\pi\alpha \tag{32}$$

In the same manner, to fix a value for s_*^2 we can impose that the real part of the Z_0 propagator vanishes at $q^2 = -M_Z^2$ and then we have

$$s_*^2c_*^2|_Z = \frac{e_*^2}{4\sqrt{2}M_Z^2G_{\mu*}\rho_*}\bigg|_Z \quad \frac{1}{1 + (Im\Pi'^*_{AA})^2}\bigg|_Z \quad - \lambda|_Z \tag{33}$$

And finally, the value of $G_{\mu*}$ is fixed by comparing the muon lifetime predicted using the equations we have obtained with the one deducted from the experiment (after subtracting and adding the proper corrections).

Therefore, we end up with a rigorous and well defined procedure of including all the oblique corrections in the calculation of 4-fermion amplitudes. This procedure, through allowing the inclusion of these effects everywhere in our calculation by means of the proper use of "starred" quantities, makes automatic the correct factorisation of these corrections with the rest and, also the inclusion of the oblique corrections in the Z_0 width.

If we neglect the imaginary parts (but the ones of the Z_0 self energy)

$$
\begin{aligned}
Im\Pi'^*_{AA} &\longrightarrow 0 \\
Im\Pi'^*_{ZA} &\longrightarrow 0 \\
\lambda &\longrightarrow 0
\end{aligned}
\tag{34}
$$

then we end up with

$$
M_{NC} = \frac{e_*^2}{q^2} + \frac{e_*^2}{s_*^2 c_*^2} \frac{[I_3 - Q s_*^2][I_3' - Q' s_*^2]}{q^2 + \frac{e_*^2}{4\sqrt{2}s_*^2 c_*^2 G_{\mu_*} \rho^*} - i\sqrt{s}\Gamma_Z^*}
\tag{35}
$$

being then

$$
M_Z^2 = \frac{e_*^2}{4\sqrt{2}s_*^2 c_*^2 G_{\mu_*} \rho^*}
\tag{36}
$$

the same relation we have at the tree level and

$$
\sqrt{s}\Gamma_Z^* = Im\Pi_{ZZ}^*
\tag{37}
$$

so that what we have is just a Born-like matrix element but in terms of "starred" parameters. As, in fact, near the peak we have

$$
\begin{aligned}
\rho^* &\simeq 1 \\
G_{\mu*} &\simeq G_\mu
\end{aligned}
\tag{38}
$$

the result we obtain consists in modifying the Born amplitude by substituting

i.- the "starred" electric charge $e_*^2(s)$ in the photon diagrams.

ii.- G_μ plus an "energy-dependent" "starred" width in the Z_0 diagrams.

which is numerically equivalent to the substitutions obtained in the previous sections.

For the M.C. implementation, the complete set of "starred" functions is available and for processes in which the Q^2 in the boson propagator lines changes from event to event (for instance radiative processes) the fact that their behaviour is smooth can be used to spare CPU time by using just tabulations and interpolation.

5 Infrared corrections

The infrared corrections are the dominant QED ones as argued before and the fact that their size depends so strongly on experimental detection conditions, makes them a complicated problem to treat in practice making unavoidable the intensive use of M.C. techniques.

A systematic and rigorous study of these corrections was done already many years ago (Y.F.S.) [8] and actually practical implementations of exponentiation techniques have been used already extensively for the study of the QED radiative corrections to resonance widths [9]. Nevertheless the clarification of the proper way of applying these techniques to the problem we are discussing is rather recent and we will try to describe just the different approaches developed for this purpose.

5.1 The "inductive" approach

The way higher order QED corrections have to be included can be studied in an inductive way analyzing which new terms show up when two loop corrections are calculated. This has been done for the initial state QED corrections for μ pair production [10] and the result can be written in the following way:

$$
\begin{aligned}
\sigma_z(s) &= \sigma_0(s)\{1 + \delta_1 + \delta_2 + \beta \ln x_0 + \delta_1 \beta \ln x_0 + \frac{1}{2}\beta^2 \ln^2 x_0\} \\
&+ \int_{x_0}^1 \sigma_0(s')[\beta\{\frac{1}{x}(1 + \delta_1 + \beta \ln x) - 1 + \frac{x}{2}\} + \delta_2^H]dx
\end{aligned}
\tag{39}
$$

where

$$
\begin{aligned}
\delta_2 &\equiv O(\alpha^2) \text{ correction independent of } x_0 \\
\delta_2^H &\equiv O(\alpha^2) \text{ non-leading hard correction} \\
s' &\equiv s(1 - x)
\end{aligned}
\tag{40}
$$

We can estimate the effect of the new corrections as we did for the one loop one near the resonance:

$$
\sigma_2(s) \sim \sigma_0\{1 + \delta_1 + \delta_2 + \beta \ln(\frac{\Gamma_Z}{M_Z}) + \delta_1 \beta \ln(\frac{\Gamma_Z}{M_Z}) + \frac{1}{2}\beta^2 \ln^2(\frac{\Gamma_Z}{M_Z})\}
\tag{41}
$$

$$
\begin{aligned}
\delta_1 &= 9\% \\
\delta_2 &= -0.5\% \\
\beta \ln(\frac{\Gamma_Z}{M_Z}) &= -40\% \\
\delta_1 \beta \ln(\frac{\Gamma_Z}{M_Z}) &= -3.6\% \\
\frac{1}{2}\beta^2 \ln^2(\frac{\Gamma_Z}{M_Z}) &= 8\%
\end{aligned}
\tag{42}
$$

and it is obvious that while the non-infrared corrections δ_1, δ_2 fall rapidly (allowing us to expect the next term to be of the order of a per mil), the infrared ones do it slowly (the next-order terms can be estimated to be of the order of a percent). Moreover, looking at the soft part, we can see that the infrared terms exhibit a clear structure of the kind:

$$
\sigma_0(s)\{1 + \delta_1 + \delta_2\}(1 + \beta \ln x_0 + \frac{1}{2}\beta^2 \ln^2 x_0 + ...) = \sigma_0(s)(1 + \delta_1 + \delta_2)x_0^\beta
\tag{43}
$$

The same can be seen concerning the hard part, namely, we can induce that the effect of resumming all the Sudakov logs will give us an expression like

$$
\int_{x_0}^1 \sigma_0(s')[\beta(x^{\beta-1}(1 + \delta_1 + \delta_2) - 1 + \frac{x}{2}) + \delta_2^H]dx
\tag{44}
$$

The sum of these two parts gives us an expression of the total exponentiated cross-section including up to $O(\alpha^2)$ corrections exactly (the $O(\alpha)$ one can be obtained by just dropping all the δ_2 terms).

Lets realize that this expression of the cross section fixes up all the problems of QED corrections we warned at the beginning:

i.- in one hand, it has resummed all the large infrared corrections.

ii.- additionally, the problem with the range of x_0 values has disappeared: the soft cross section is defined positive for any value of x_0 (even for $x_0 = 0$ which would be the most precise election)

The problem of this method is that only information about the total energy loss due to photons is available. The rest has been implicitly integrated. The variable x represents essentially the photon energy sum if no transverse component is produced. Therefore, all we can expect to reproduce properly is the behaviour of the total integrated cross section.

Nevertheless, a reasonable approximation of the effect of photons in the differential cross section can be obtained by modifying a one-loop M.C. event generator with the following transformation:

i.- in the soft part, eliminate the infrared terms $\beta \ln x_0$ from the corrections and multiply the cross section by x_0^β.

ii.- in the hard photon part: substitute the piece $\beta \frac{1}{x}$ by $\beta x^{\beta-1}(1 + \delta_1)$.

Doing so, we are neglecting the effect of all the photons carrying an energy smaller that x_0 (but this is not a problem since we can take x_0 as small as we want) and additionally we are assuming that when the energy radiated is above this cut-off, then it is carried by just one photon. This is clearly an approximation since eq.(44) is adding the contribution from the radiation of an infinite number of photons but, however, when the radiated energy is large enough it tends to be carried by just one or at most two energetic photons. So the approximation is very good in this case.

5.2 The Structure Functions approach

A different way of attacking the problem is by using the Structure Functions approach [11] like in QCD. The effect of initial state radiation can be explained with the following picture: real electrons are particles "dressed" by photon radiation in such a way that the interacting "partons" in e^+e^- collisions in practice carry just a fraction of the real particle energy, being the rest lost in form of radiation when a collision takes place. The probability distribution of finding an electron with fractional energy

$$z = 1 - \sum_i \frac{E_{\gamma_i}}{\sqrt{s}/2} \tag{45}$$

at a momentum transfer squared $Q^2 - s$ is given by the Structure Function

$$D_e = (z, s) \tag{46}$$

in such a way that in analogy with the Drell-Yang processes in QCD we can write the total cross section as

$$\sigma(s) = \int^1 dz_1 dz_2 D_e(z_1, s) D_e(z_2, s) \sigma_0(s z_1 z_2) \tag{47}$$

In this formula all the radiative corrections due to initial state radiation (virtual and real) are accounted for in the Structure Function expression, and all the other effects, like oblique corrections, have to be included explicitly in σ_0.

The calculation of the Structure Function is obviously process independent and can be done by using Altarelli-Parisi evolution equations:

$$D_e(z,s) = \delta(1-z) + \int_{m_e^2}^{s} \frac{ds'}{s'} \frac{\alpha(s')}{2\pi} \int_z^1 \frac{da}{a} P(a) D_e(\frac{z}{a}, s') \tag{48}$$

being

$$\alpha(s') \;=\; \frac{\alpha}{1 - \frac{\alpha}{3\pi}\ln(\frac{s'}{m_e^2})} \quad \text{QED running coupling constant}$$

$$P(z) \;=\; \frac{1+z^2}{1-z} - \delta(1-z)\int_0^1 dx \frac{1+x^2}{1-x} \quad \text{Kernel} \tag{49}$$

This last expression is the so-called regularized vertex function. Its first term takes care of real photon emission (is just the well-known Weiszacker-Williams photon spectrum) whereas the second one reflects the contribution coming from virtual photons.

This equation can be solved analytically for all orders in the soft limit (the interesting one since we know that the bulk of photonic corrections comes from soft photonic radiation) leading to the Gribov-Lipatov [12] solution:

$$D_e(z,s) \;=\; \frac{\exp\{\frac{\eta}{4}(\frac{3}{2} - 2\gamma_E)\}}{\Gamma(\frac{\eta}{2})}(1-z)^{\frac{\eta}{2}-1}$$

$$\eta \;\equiv\; -6\ln(1 - \frac{\alpha}{3\pi}\ln(\frac{s}{m_e^2})) \qquad \gamma_E \equiv \text{Euler constant} \tag{50}$$

In the soft limit (that is, neglecting the s dependence of σ_0) this leads to

$$\sigma(s) = \sigma_0(s)\frac{x_0^\eta}{\Gamma(\eta+1)} \exp\{\frac{\eta}{4}(\frac{3}{2} - 2\gamma_E)\} \tag{51}$$

which expanded in powers of α up to $O(\alpha^2)$ gives:

$$\sigma(s) \;=\; \sigma_0(s)[1 + \frac{\alpha}{\pi}L(2l + \frac{3}{2}) + (\frac{\alpha}{\pi})^2 L^2(2l^2 + \frac{10}{3}l + \frac{11}{8} - \frac{\pi^2}{3})]$$

$$L \;\equiv\; \ln(\frac{s}{m_e^2}) \qquad l = \ln x_0 \tag{52}$$

This result is to be compared with the one coming form the complete $O(\alpha^2)$ calculation, which with the same notation can be written as

$$\sigma(s) \;=\; \sigma_0(s)1 + \frac{\alpha}{\pi}\{L(2l + \frac{3}{2}) - 2l + \frac{\pi^2}{3} - 2\}$$

$$+ \;(\frac{\alpha}{\pi})^2\{L^2(2l^2 + 3l + \frac{9}{8} - 2\zeta(2)) + L \quad (...) + \quad (...)\} \tag{53}$$

revealing that the Structure Functions expansion reproduces exactly the $O(\alpha^2)$ calculation in the leading-log approximation. Moreover it accounts exactly for the $O(\frac{\alpha}{\pi}L)$ terms and very approximately for terms with $(\frac{\alpha}{\pi}L)^2$ with l powers smaller than 2. It can be seen that eq.(51) reproduces also the soft photon exponentiated equation at the leading-log level:

$$\sigma(s) \simeq \sigma_0(s) x_0^{\frac{2\alpha}{\pi}L}(1 + \frac{\alpha}{\pi}\frac{3}{2}L) \tag{54}$$

The Gribov-Lipatov equation includes all the leading-logs in the soft photon limit but, of course not for the hard photon case. Additionally the two equations above show differences due to the next-to-leading terms some of which can be included with the simple substitution rule $L \longrightarrow L - 1$. Actually, the Gribov-Lipatov solution can be manipulated following a simple procedure to include all the relevant terms we would like to have [11] The basic idea is:

i.- Take the evolution equation (48) and substitute in the right part the lowest order expression of

$$D_e(z, s) = \delta(1 - z) \tag{55}$$

to obtain the $O(\alpha)$ solution (which now will show a soft photon part as well a a hard photon one).

ii.- by comparing with the $O(\alpha)$ Gribov-Lipatov solution, the missing leading-log terms due to hard photons can be detected and included directly in the Gribov-Lipatov solution.

iii.- moreover, non-leading terms can be included by just substituting in the expression of the cross section (47) the $D_e(z, s)$ we want to complete and comparing the result with the exact one-loop calculation to look for the missing terms we want to pick up.

Of course, this can be done for $O(\alpha^2)$ as well and in any case, the infrared terms are obtained from the Structure Functions evolution equation whereas the rest of terms are obtained by comparison with the result of a complete Feynman diagram calculation.

The Structure Functions method, demonstrates that the induction made in the previous method about the behaviour of infrared corrections is correct because the leading-log terms show up naturally in the same way. In other hand, to reach the same accuracy attainable with the previous method, it needs a further elaboration of the calculational results obtained there and, as it has been described here, it gives no real additional information on the photon energy. Nevertheless, for M.C. purposes it provides a clear and easy way of including the effect of initial state radiation since eq. (47) is directly implementable in a M.C. event generator.

Eq. (47) can be used to compute the total cross section as well as to take into account distribution effects unsensitive to transverse components (let's realize that transverse information has been implicitly integrated in the Structure Functions that we are using). To account for transverse effects, two possible ways have been proposed:

(a) ONE-PHOTON: The total cross section is computed using eq.(47) but when the amount of energy radiated is larger than a certain x_0, the one-hard-photon exponentiated cross section (the same one as in the previous method) is used instead of the soft one. In this way, the total cross section is correct and the main distributions show the approximation of assigning the total photonic cross section to just one photon.

(b) MULTI-PHOTON: The total cross section is computed also using eq.(47) but when the amount of energy is larger than a certain x_0, a multiphoton production algorithm is applied [13]. The basic idea is the following: when the photonic energy is above a certain quantity, use is made of an iteration of the evolution equation (48) to obtain an expansion in α rather than an exponentiated solution like the Gribov-Lipatov one. For instance, iterated to third order this gives

$$
\begin{aligned}
D_e(z,s) \;=\;& \delta(1-z) \\
&+ \frac{\alpha}{2\pi} \int \frac{ds'}{s'} P(z) \\
&+ (\frac{\alpha}{2\pi})^2 \int \frac{ds''}{s''} \int \frac{da}{a} P(\frac{z}{a}) P(a) \\
&+ (\frac{\alpha}{2\pi})^3 \int \frac{ds'''}{s'''} \int \frac{da}{a} \int \frac{db}{b} P(\frac{z}{ba}) P(a) P(b)
\end{aligned}
\tag{56}
$$

This provides the following basic ingredients to treat multiphoton production:

i.- the number of produced photons is computed using a Poissonian distribution

$$
\begin{aligned}
B(0) &= 1 - \bar{n} + \frac{\bar{n}^2}{2} - \frac{\bar{n}^3}{6} + ... = e^{-\bar{n}} \\
B(1) &= \bar{n} B(0) \\
B(2) &= \frac{\bar{n}^2}{2} B(0)... \\
\bar{n} &= \frac{\alpha}{\pi} L(-l - \frac{3}{4})
\end{aligned}
\tag{57}
$$

which in fact can be directly deduced from the iterated solution of the evolution equation integrating over z and dropping non-dominant terms.

ii.- the longitudinal energy for each photon is generated using the Weiszaker-Williams expression that shows up as the Kernel in the evolution equation.

iii.- to obtain the transverse energy of each photon, the Structure Function keeping information on the transverse momentum is used $D_e(z, p_T, s)$.

$$
\sigma(s) = \int dz_1 dz_2 d^2 p_{T_1} d^2 p_{T_2} \, \sigma_0(s') D_e(z_1, p_{T_1}, s) D_e(z_2, p_{T_2}, s)
\tag{58}
$$

This function verifies the evolution equation

$$
\begin{aligned}
\frac{\partial D_e(z_1, p_T, s')}{\partial s'} \;=\;& \frac{\alpha}{2\pi} \frac{1}{s' - m_e^2} \int_x^1 \frac{da}{a} P(a) \int \frac{d^2 q_T}{\pi} \delta \\
& D_e(\frac{z}{a}, p_T - \frac{z}{a} q_T, s')
\end{aligned}
\tag{59}
$$

where δ guarantees a correct kinematic splitting at the vertex which iterated up to $O(\alpha)$ gives

$$D_e(z_1, p_T, s) = B(0)\delta(1-z)\delta^2(p_T)$$

$$+ \quad B(1)P(z)\frac{1}{p_T^2 + (1-z)^2 m_e^2} + 0(\frac{p_T^2}{E_\gamma^2}) \tag{60}$$

showing, therefore, the approximated photon p_T distribution.

5.3 Y.F.S.

The most elaborated and rigorous way of attacking the problem is based upon the early work of Yennie, Frautschi and Suura [14]. A careful study of the infrared terms leads to the following general master equation

$$\sigma = \sum_{n=0}^{\infty} \frac{1}{n!} \int \frac{d^3 q_1}{q_1^0} \frac{d^3 q_2}{q_2^0} (\prod_{i=1}^{n} \frac{d^3 k_i}{k_i^0} \tilde{S}(p_1, p_2, k_i)\theta(\frac{2k_i^0}{\sqrt{s}} - \epsilon))$$

$$\delta^4(p_1 + p_2 - q_1 - q_2 - \sum_{i=1}^{n} k_i)$$

$$exp\{2\alpha ReB(p_1, p_2) + \int \frac{d^3 k}{k_0} \tilde{S}(p_1, p_2, k)\theta(\epsilon - \frac{2k^0}{\sqrt{s}})\}$$

$$[\tilde{\beta}_0(p_1, p_2, q_1, q_2) + \sum_{l=1}^{n} \frac{\tilde{\beta}_1(p_1, p_2, q_1, q_2, k_l)}{\tilde{S}(k_l)} + \sum_{l,j=1 \atop l \neq j}^{n} \frac{\tilde{\beta}_2(p_1, p_2, q_1, q_2, k_l, k_j)}{\tilde{S}(k_l)\tilde{S}(k_j)}] \tag{61}$$

where

$$2\alpha B(p_1, p_2) = \frac{i\alpha}{4\pi^2} \int \frac{d^4 k}{k^2 - m_\gamma^2} (\frac{2p_1 - k}{k^2 - 2p_1 k} + \frac{2p_2 + k}{k^2 + 2p_2 k})^2 \Rightarrow \text{virtual photon}$$

$$\tilde{S}(p_1, p_2, k) = -\frac{\alpha}{4\pi^2} (\frac{p_1}{p_1 k} - \frac{p_2}{p_2 k})^2 \Rightarrow \text{real photon} \tag{62}$$

which are the universal parts due to radiation, and the $\tilde{\beta}$ functions contain all the non-infrared parts and are, obviously, process-dependent. Their expression depends on the order of perturbation expansion one wants to keep in the non-infrared part and this is related directly with the number of hard photons one wants to reproduce correctly. If one hard photon is to be seen, one needs the use of a complete $O(\alpha)$ QED Feynman diagram calculation to compute the functions $\tilde{\beta}_0$ and $\tilde{\beta}_1$ which in this case will represent

- $\tilde{\beta}_0$: is the complete $O(\alpha)$ virtual+soft photon part of the cross section after eliminating the virtual infrared contribution.

- $\tilde{\beta}_1$: is the complete $O(\alpha)$ real photon part of the calculation after subtracting explicitly the infrared part coming from the factorisation of the infrared factor $\tilde{S}(p_1, p_2, k)$ times $\tilde{\beta}_0$ at the Born level.

This can be called "first order exponentiation" and, obviously, if two real photons have to be seen, a second order exponentiation is needed for which the master equation is the same but $\tilde{\beta}_0$, $\tilde{\beta}_1$ and $\tilde{\beta}_2$ have to be computed using an $O(\alpha^2)$ calculation.

From the master equation we can learn two things:

1. how the infrared terms do factorize and appear exponentiated in the soft photon approximation

2. how the differential cross section for multiphoton production giving the full photon four-momenta information shows up naturally in the soft limit and can be corrected to produce hard photons by computing properly the $\tilde{\beta}$ functions.

It can be easily seen that the master equation predicts a Poissonian distribution of the number of photons if only infrared terms are taken into account (that is, only $\tilde{\beta}_0$ is kept and the soft limit is taken), which is in very good agreement with the one obtained in the previous method.

In practice, the M.C. implementation of this method is rather involved since the rigorous treatment of the multiphotonic phase space and four-momentum conservation requires a quite complicated procedure [15].

6 The Z_0 width

As we saw already, at least the next order corrections to the Z_0 width should be included in a calculation in order not to loose accuracy when sitting on the peak.

The Z_0 width is given at the tree level by the imaginary part of the one loop Z_0 self energy $Im\hat{\Sigma}^Z_{(1)}(s)$. If we use the full boson propagators, we can be sure that some higher order terms are already included, like

$$-\frac{(\hat{\Sigma}^{\gamma Z}(s))^2}{s + \hat{\Sigma}^\gamma(s)} \tag{63}$$

Besides of that, when using the STAR scheme, the fact that the "starred" width is a function of the "starred" basic parameters, guarantees that resummed oblique corrections to the Z_0 width have been automatically included.

In any case, what is still missing is the two loop irreducible part of the self energy $Im\hat{\Sigma}^Z_{(2)}(s)$. To include these corrections in a reasonable way, we can use the fact that at the Z_0 pole we know that

$$
\begin{aligned}
Im\hat{\Sigma}^Z_{(1)}(M_Z^2) &= M_Z\Gamma_Z^{(0)} \\
Im\hat{\Sigma}^Z_{(2)}(M_Z^2) &= M_Z(\Gamma_Z^{(1)} - \Gamma_Z^{(0)})
\end{aligned}
\tag{64}
$$

and additionally the fact that the imaginary part of the self energy behaves around the the peak as

$$\hat{\Sigma}^Z(s) \propto s \qquad s > 0 \tag{65}$$

Therefore a reasonable way of including the relevant corrections to the width is simply by taking

$$Im\hat{\Sigma}^Z(s) \simeq \hat{\Sigma}^Z_{(1)}(M_Z^2) + \frac{s}{M_Z^2}M_Z(\Gamma_Z^{(1)} - \Gamma_Z^{(0)}) \tag{66}$$

where $\Gamma_Z^{(1)}$ includes the Z_0 three body decay width as well as strong corrections (which are responsible for most of the uncertainty) and the electroweak ones (but the ones already included as the Z-gamma mixing and the oblique ones) and has been calculated already [16]. The difference between using the STAR scheme for the inclusion of oblique corrections or not in this calculation is of $O(\alpha^3)$ and can be neglected in front of the hadronic uncertainties.

7 Comparison

We are going to point out the obvious limitations and advantages of the different methods reviewed here for the treatment of the questions of OBLIQUE corrections resummation and infrared correction EXPONENTIATION and multiphoton production.

Concerning the first point it is obvious that the first method is just an approximation, which can be rather good if we are working in a region in which $\hat{\Sigma}^Z$ is by far larger than $\hat{\Sigma}^{\gamma Z}$ and is very easy to implement in any already existing calculation. The "improved BORN amplitude" is an step ahead which contains actually all the relevant elements to ensure an accuracy in this part of the calculation well below the hadronic uncertainties. Finally, the STAR scheme is the most rigorous way of dealing with the problem, out of the three reviewed here, besides being the most complicated one, though the difference with the previous method can be smaller than the uncertainties coming from other parts of the calculations. The implementation of both methods in an existing calculation is not difficult provided you have the STAR functions already available. It is important to point out that since in all the methods the basic ingredients are the one loop self energies, they are seriously in trouble if the Top mass turns out to be large. In this case some correction factors should be included or even better, the result of a two loop self energy calculation.

Concerning the second point, it is obvious that only the infrared terms in the soft photon limit do factorize rigorously and thus can be exponentiated. This result is taken into account in the same manner by all the three methods presented. A complete $O(\alpha^2)$ calculation was needed not just to corroborate in an inductive way this fact, but also to provide all the methods the source to obtain the missing relevant terms which cannot be deduced directly from the study of the infrared parts.

We can first classify all the methods in two groups: the ones producing only one hard photon, and these producing multiphotons. From the first group, the main differences between the "inductive" approach and the Structure Functions one provided the proper $D_e(z,s)$ is taken to include non-leading effects, is that the second method gives always the right answer for the longitudinal components (even in the soft photon case) whereas the first one will always show a gap in distributions such as acollinearities. For the hard part, the distributions will be the same by construction and the results you can expect from both methods for the total cross section are essentially the same [7].

The second group is obviously superior to the first one for the study of differential distributions affected by multiphotons though the results for the total cross section must be essentially the same. As a drawback, they tend to be somewhat slower because of the multiphoton production procedure in spite that this can be adjusted by means of the arbitrary soft-hard energy limit, which actually acts as a regulator on the number of photons produced. Both treatments of multiphoton events can be expected to give roughly the same answer being the only relevant differences the following:

- In one hand, the treatment of multiphoton event looks more rigorously done in the Y.F.S. method for two reasons: first, the complete photon four-momentum distribution is known taking, thus, care of correlations among photons, and second, the "master" formula provides the correct way of including a correct calculation up to any order in perturbation theory, that is, in spite that just a Poissonian distribution of the number of photons or a simple photon four-momenta spectrum can be utilized to approximate the differential cross section in a M.C. implementation, we can use afterwards the "master" equation to correct for the approximation used and thus obtain the right result up to the desired perturbation order. This cannot be done rigorously in the Structure Functions multiphoton approach reviewed here because it uses a Poissonian distribution which cannot be corrected for the non-infrared effects and this affects the photon-by-photon longitudinal and transverse variables, which are the only ones that we can control. Additionally in the generation of the photon transverse momentum as reviewed here $0(\frac{p_\gamma^2}{E_\gamma^2})$ terms are neglected and this can be a rather bad approximation when hard transverse photons are studied.

- In other hand, the Structure Functions approach, as mentioned before, gives always the right longitudinal distributions regardless on the soft-hard limit taken to regulate the number of produced photons. In the Y.F.S. method, a precise acollinearity distribution can be obtained only by lowering the soft-hard limit, thus producing many undetectable photons which represent a waste of computing time.

8 Conclusions

The relevant modifications to the previous generation of M.C. event generators (PETRA - PEP) needed "a priori" for working at LEP energies have been discussed. The different methods developed so far to implement them have been briefly reviewed showing their main features and also pointing out their similarities and differences.

We cannot see the methods reviewed here as the final answer to the problem of perturbative expansions in the E.W. theory, but rather as an effective treatment of the relevant questions needed at present to reach the requested accuracy in LEP-SLC experiments. In any case, they all constitute an step ahead in the clarification of the real meaning of perturbation theory and its confrontation with the experiment.

The main conclusion is that if we want to get around 1% accurate theoretical predictions for processes dominated by the Z_0 peak influence we have to check that the M.C. event generator that we are using, is at least an $O(\alpha)$ corrected calculation and that the authors have foreseen some treatment of all the following fundamental aspects:

a) Resummation of OBLIQUE corrections

b) Exponentiation of infrared QED corrections

c) Interplay of photonic and non-photonic corrections

d) A more accurate evaluation of the Z_0 width than the simple tree level one

For most of the applications any combination of the methods reviewed here can produce already the required accuracy. A nice example on this fact is the excitingly good agreement between the $\nu\bar{\nu}\gamma$ M.C. event generator programs NNGG03 (Mana-Martinez-Miquel)[17] and KORALZ (adapted for Colas-Mirabito-Wąs [18] for neutrino production) which have been written independently and using completely different techniques:

NNGG03 is a program based upon a complete $O(\alpha)$ QED corrected $\nu\bar{\nu}\gamma$ calculation which uses the STAR scheme for the oblique corrections, the "inductive" method for the treatment of infrared QED corrections and includes the Z_0 width in the way explained here.

KORALZ is a program based upon a complete $O(\alpha)$ QED corrected $\nu\bar{\nu}$ calculation which uses the "propagator substitution" method for the oblique corrections, the Y.F.S. scheme to treat the infrared QED corrections (including also leading and next-to-leading logarithmic terms of the $O(\alpha^2)$ QED corrections) and to produce multiphoton events and includes also corrections to the Z_0 width.

For some applications, however, the use of one technique instead of another may be of paramount importance. You can realize that twofold:

- if possible IN ADVANCE, because it is easy to understand that some techniques presented here, by construction, are not designed to give accurate results in the special observable you might be interested in (this is just matter of information), or

- A POSTERIORI, because after debugging the programs and checking that they give the same answer in the observables they are supposed to do it, you get different results for the observable that you are calculating.

In this last case, a M.C. implementation combining the most accurate techniques for each of the problems discussed here should be the starting point to try to understand whether still further neglected corrections should be taken into account or not.

References

1 . P. de Causmaker et al., Nucl. Phys. **B206**,53 (1982)
   G.R. Ferrar and F. Neri, Phys. Lett. **130B**,109 (1983)
   F.A. Berends et al. (CALKUL Collaboration), DESY 83-125 (1983)
   R. Kleiss and W.J. Stirling, Nucl. Phys.**B262**,235 (1985)
   P.H. Daverveldt, Ph. D. Thesis, Leiden Univ. (1985)
2 . F.A. Berends and R. Kleiss, Nucl. Phys. **B228**,537 (1983)
   S. Jadach and Z. Was, Acta Phys. Pol **B15**,1151 (1984)
3 . G. Passarino and M. Veltman, Nucl. Phys.**B160**,151 (1979)
4 . M. Boehm, H Spiesberger and W. Hollik, DESY 84-027 (1984)
   R.G. Stuart, Ph. D. thesis, RAL-T-008 (1985)
   W. Hollik, DESY 88-188 (1988) and references therein
5 . D. Bardin and T. Riemann, these proceeedings
6 . D.C. Kennedy and B.W. Lynn, SLAC-PUB-4039 (Rev) (1988)
7 . R. Miquel, Ph. D. Thesis, Universitat Autònoma de Barcelona, UAB-LFAE 89-02 (1989)
8 . D.R. Yennie, S.C. Frautschi and H. Suura, An. of Phys. **13** (1961) 379
9 . W. Buchmueller and S. Cooper in "High Energy Electron Positron Physics", eds. A. Ali and P. Soeding (World Scientific, Singapore, 1988) and references therein
10 . F.A. Berends, G.J.H. Burgers and W.L. van Neerven, CERN-TH.4772/87 (1987)
11. E.A. Kuraev and V.S. Fadin, Sov. J. Nucl. Phys. **41** (1985) 466
   O. Nicrosini and L. Trentadue, Phys. Lett. **B196** (1987) 551
12. V. Gribov and L. Lipatov, Sov. J. Nucl Phys. **15** (1972) 438, 675
13. G. Bonvicini and L. Trentadue, University of Michigan preprint, UM-HE-88-36 (1988)
14. S. Jadach and W.F.L. Ward, Phys. Rev. D **38**,2897 (1988)
15. S. Jadach, Max Planck Institut preprint, MPI-PAE/PTh 6/87 (1987)
16. W. Wetzel in CERN 86-02 (1986)
17. C. Mana, M. Martinez and R. Miquel, in preparation
18. P. Colas, L. Mirabito and Z. Was, CEN-Saclay preprint DphPE 89-10 (1989)

FPAIR, AN EVENT GENERATOR FOR FERMION PAIR PRODUCTION AT LEP/SLC ENERGIES

S.C. van der Marck

Instituut-Lorentz, University of Leiden
P.o.b. 9506, 2300 RA Leiden, The Netherlands

We want to present a new Monte Carlo program that treats the higher order corrections to the process $e^+e^- \longrightarrow f\bar{f}$ ($f \neq e^-$). The need for a treatment of such corrections will be taken for granted. A fully analytical calculation of the corrections is discussed elsewhere in these proceedings [1]. Also other Monte Carlo programs are discussed [2]. Here we introduce a Monte Carlo program that does not use the structure function formalism, but that calculates bremsstrahlung matrix elements up to second order. In this Monte Carlo the following corrections have been incorporated.

- $\mathcal{O}(\alpha)$ and $\mathcal{O}(\alpha^2)$ QED corrections on the initial state,

- $\mathcal{O}(\alpha)$ QED corrections on the initial state,

- one loop weak corrections.

Not (yet) included are

- interference between initial and final state corrections,

- soft photon exponentiation.

The weak corrections are incorporated by using the library of the program ZSHAPE (see ref. 3). The final state corrections are treated using a modified version of the algorithm given in [4].

The initial state corrections however are the more difficult ones from this purely Monte Carlo point of view. The contribution from these corrections is split into several parts:

$$\sigma = \sigma_{0\gamma} + \sigma_{1\gamma} + \sigma_{2\gamma}. \tag{1}$$

Here $\sigma_{i\gamma}$ corresponds to the part where i photons have an energy larger than some value k_0 and can therefore in principle be detected. We will call these photons hard photons. The no hard photon part ($\sigma_{0\gamma}$) and the one hard photon part ($\sigma_{1\gamma}$) one has known how to handle for quite some time now. The form of the virtual and soft corrections has been taken from ref. 5. The part where two photons can become hard is the difficult part. This part is incorporated by calculating the matrixelements of figure 1 without approximations. The implementation of these matrix elements has been checked with collinear and soft limits.

This approach bears in itself a potential problem, which is generally called the k_0 problem. The problem is that none of the contributions $\sigma_{i\gamma}$ may become negative, which would prohibit a probabilistic treatment such as the Monte Carlo treatment is. However if one lowers the value of k_0 the contribution $\sigma_{1\gamma}$ will at some point become negative as it has a logarithmic dependence on k_0:

$$\sigma_{1\gamma} \sim 1 + \frac{2\alpha}{\pi}\left(\log\frac{s}{m_e^2} - 1\right)\log\frac{k_0}{E_{beam}} + \dots \tag{2}$$

Radiative Corrections, Edited by N. Dombey and
F. Boudjema, Plenum Press, New York, 1990

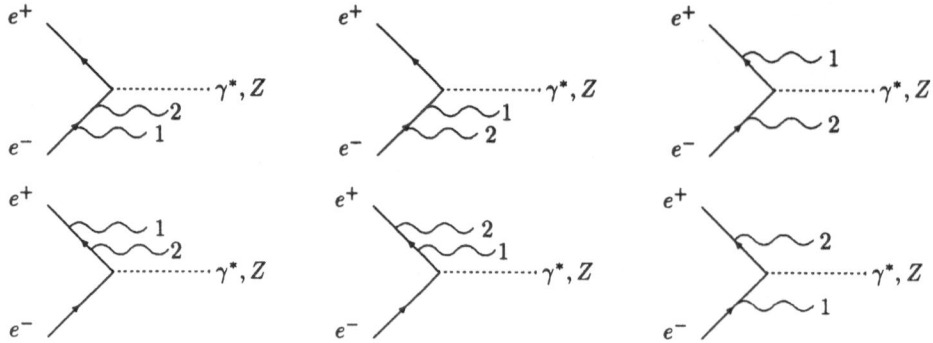

Figure 1. The Feynman diagrams for double bremsstrahlung from the initial state. The two bremsstrahlung photons have been labelled by 1 and 2 respectively, the dashed line represents the virtual photon or Z that decays into a fermion pair.

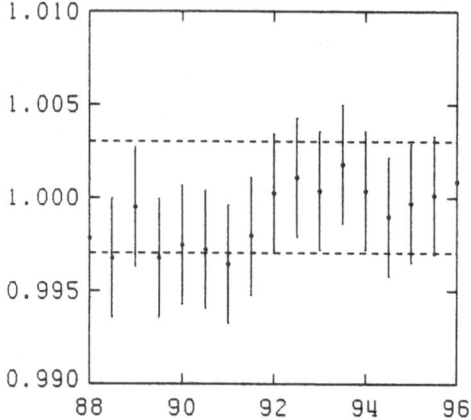

Figure 2. The ratio of the FPAIR result and the ZSHAPE result.

A rough estimate reveals that the lower bound on k_0 approximately is

$$k_{0,min} = 10^{-4} E_{beam}.$$

This value seems to be low enough for all practical purposes. The other contributions do not become negative at any point, for $\sigma_{0\gamma}$ has a $\log^2(k_0/E_{beam})$ dependence, whereas $\sigma_{2\gamma}$ corresponds to an in principle detectable cross section.

At this point we may insert a brief remark on the use of the program. All the fermion masses are of course input parameters, but as we use the library of the program ZSHAPE, quantities as Γ_Z, $\sin^2 \vartheta_W$ and m_W are calculated, once the input parameters m_Z, m_{top} and m_H have been specified. The program then determines the boundaries of the phase space and generates events in phase space without any restrictions on any variables.

Having introduced the program thus far, we would like to check the results of the program and conclude by giving some numerical results. The first thing we want to do is to check the result for the total cross section. This we do by plotting the ratio of the FPAIR result and the result of the program ZSHAPE, which can be used to serve as a benchmark for event generators. In figure 2 this ratio is plotted over an energy range that includes the Z peak. For the input parameters we used $m_Z = 92$ GeV, $m_H = 100$ GeV, $m_{top} = 60$ GeV, which leads to $\sin^2 \vartheta_W = 0.2296$ and $\Gamma_Z = 2.562$ GeV. The particular process we did these runs for

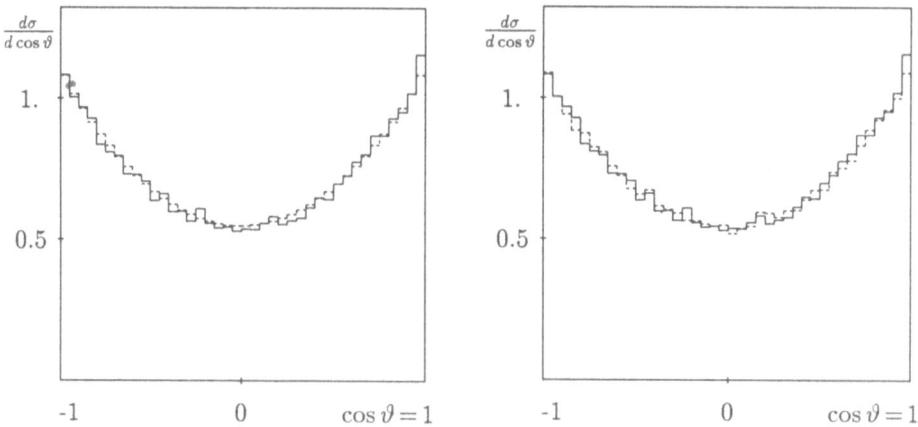

Figure 3. The distribution of $\cos\vartheta$ for $\sqrt{s} = 92$ GeV by the programs FPAIR (solid lines, left and right), DYMU3 (dashed line, right) and MUCUTCOS (dashed line, left).

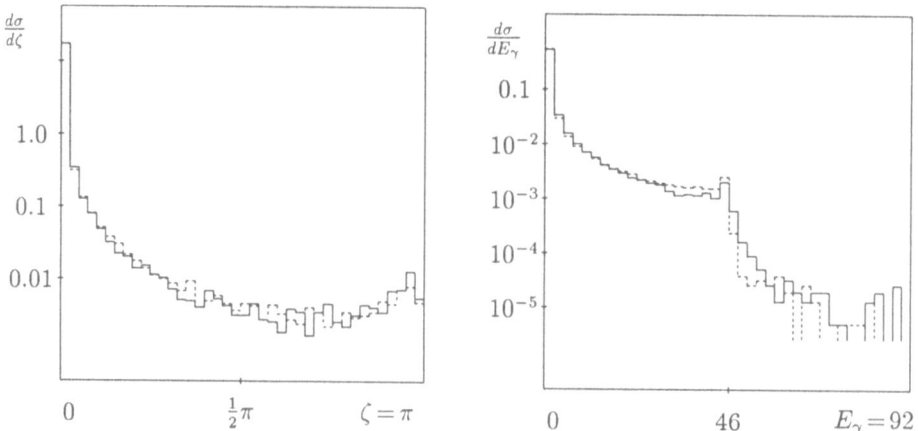

Figure 4. The distributions of E_γ and ζ at $\sqrt{s} = 92$ GeV by the programs FPAIR (solid line) and DYMU3 (dashed line).

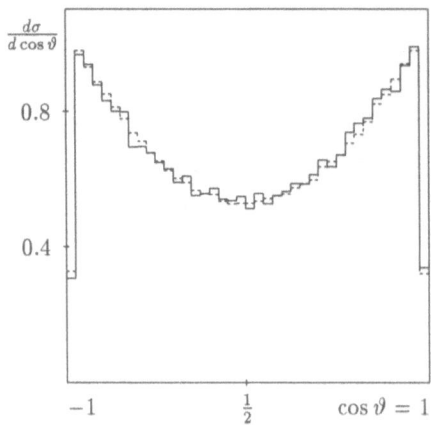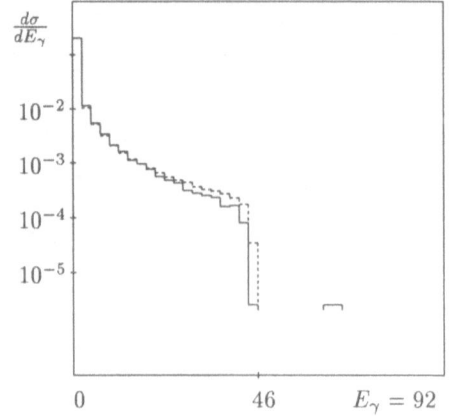

Figure 5. The distributions of $\cos\vartheta$ and E_γ as obtained by the programs FPAIR (solid line) and DYMU3 (dashed line), using the cuts as defined in the text. $\sqrt{s} = 92$ GeV.

was muon pair production, so $f = \mu^-$. These parameters have been used for the other figures as well. The numbers from the program ZSHAPE have been obtained using the option where exponentiation is switched off, to make an honest comparison possible. The difference with the exponentiated version can be up to 1%. In this figure two dashed lines have been plotted to mark 0.3 % deviation from the ideal result, which is 1. Concluding that this looks to be OK, we can proceed with differential cross sections.

Differential cross sections we will compare with another Monte Carlo program, namely DYMU3 [6]. We will consider the distributions of

$$
\begin{aligned}
\cos\vartheta &= \cos\angle(q_+, p_+) \ , \\
\zeta &= \pi - \angle(q_+, q_-) \ , \\
E_\gamma &= \sum_i k_i^0 \ .
\end{aligned}
$$

The last variable is the total energy carried away by photons. For the $\cos\vartheta$ distribution a comparison with an analytical calculation is possible. In [7] a program MUCUTCOS is presented that calculates the $\cos\vartheta$ distribution using analytical formulae. In this program the weak corrections are approximated by using the so called G_μ representation, which should be well enough for our purposes here.

First we plot these distributions when no cuts, but for $\sqrt{s'} > 1$ GeV, are applied. It is good to remind ourselves that, as has been pointed out in [5], we need a rather strict cut on s' to get rid of the contribution of the so called two photon processes. The cut $\sqrt{s'} > 1$ GeV is not nearly strict enough for that purpose. However for the moment we only want to compare two theoretical calculations, so we may apply any cut we want. Taking $\sqrt{s'} > 1$ GeV can be taken in order to be able to treat the final state fermion, which was taken to be the muon, as massless. The results are shown in figures 3 and 4.

The agreement, for as far as the can be seen from these histograms, seems to be there. Only in the E_γ distribution the two programs differ slightly in the hard photon area. This however need not to be a surprise, for describing the hard photon part well was one of the reasons for the chosen approach in writing this Monte Carlo. That these differences integrate to a negligible amount for the total cross section is the statement that the structure function approach is good for inclusive quantities, up to and including the next to leading log terms of order α^2.

Finally we can apply some cuts, for which we will take

$$
\begin{aligned}
15^\circ < \vartheta &< 165^\circ \\
\zeta &< 10^\circ \\
\sqrt{s'} &> 10 \text{ GeV}
\end{aligned}
\tag{3}
$$

The results of the runs with these cuts are shown in figure 5. We do not plot the ζ distribution any more, since this variable has been restricted to a small interval by our cuts. Again the results of the two programs compare well, but for the hard photon part in the E_γ distribution.

References

[1] Talk given by F.A. Berends.
[2] Talks given by G. Bonvicini, J.-E. Campagne and Z. Was.
[3] Report Line Shape working group to the LEP workshop 1989, to be published.
[4] R. Kleiss, Phys. Lett. B180 (1986) 400.
[5] F.A. Berends, G.J.H. Burgers and W.L. van Neerven, Nucl. Phys. B297 (1988) 429.
[6] A later version, dated July 25th 1989, of the program DYMU2, presented in
   J.-E. Campagne, Ph.D. thesis, Paris 1989.
   J.-E. Campagne and R. Zitoun, Z. Phys. C43 (1989) 469.
[7] D. Bardin, L. Vertogradov, Yu. Sedykh and T. Riemann, CERN preprint TH.5434/89.

HIGHER ORDER CORRECTIONS TO THE $e^+e^- \longrightarrow \nu\bar{\nu}\gamma$ REACTION*

C. Mana

CERN, CH-1211 Geneva, Switzerland

M. Martinez and R. Miquel

Laboratori de Física d'Altes Energies
Universitat Autònoma de Barcelona
E-08193 Bellaterra (Barcelona) Spain

Abstract

Radiative corrections to radiative neutrino production are discussed. One loop corrections are briefly reviewed. After emphasizing the importance of higher order QED corrections, their implementation using the Structure Functions approach is described. Next, the inclusion of the main weak corrections using the 'star' scheme is explained. Finally some results are presented, showing effects in the range 5–6% with respect to the pure one loop corrected cross-section. Also some comparisons with another calculation featuring multiphoton generation are shown, and good agreement is found between them, both for the total cross-section and the photonic distributions.

1 Introduction

Radiative neutrino production has been proposed long time ago ([1]) as a clean way of measuring the number of generations within the Standard Model framework. The first calculations were done in the lowest order approximation and using the contact approximation for the W diagrams ([2]). Not until recently complete tree level calculations have been performed ([3],[4]) and also one-loop radiative corrections have been taken into account ([4]-[6]). Since their effect has been found to be large (a common feature of radiative corrections to four fermion neutral current processes around the Z^0 peak), we have undertaken the task of including the main higher order (i.e., more than one-loop) radiative corrections.

Our goal is to achieve a precision better than 1% in the estimation of the total cross-section. This is of the order of the experimental needs, because, although the effect due to an extra generation would be much larger (around 25%), other reactions leading to single photon final states (radiative production of sneutrinos, photinos, etc.) have cross-sections in the per cent range. On the other hand, the error in the determination of the absolute luminosity will be the limiting experimental error, and it will not be smaller than 1%. Therefore, this is a sensible choice for the planned accuracy.

---

*Presented by R. Miquel

Table 1. Comparison between different approximations for the tree level cross-section.

$\sqrt{s}$(GeV)	σ_2^0(pb)	σ_{GGR}^0(pb)	σ_4^0(pb)	σ_5^0(pb)
98	126.5(3)	129.5(3)	128.7(3)	128.7(3)
150	5.80(3)	9.97(3)	7.74(5)	7.79(5)

In section 2 we will review quickly the exact lowest order calculation. Next we explain with no technical details the one-loop corrections in section 3. After justifying the need for higher order corrections, we present in section 4 the two implementations we have done of the treatment of higher order QED corrections : one based in the 'inductive' approach ([7]) and another one based on the Structure Functions approach ([8]). Section 5 deals with the inclusion of the main higher order weak corrections by means of the 'star' scheme ([9]). Also the need for the inclusion of one loop corrections to the imaginary part of the Z^0 self energy is emphasized there. Finally, section 6 contains the results of a comparison done between our Monte Carlo program built along the lines explained in the previous sections (NNGG03) and another one based on the Yennie-Frautschi-Suura approach ([10]). These results have been taken from ref. [11]. The conclusions are given in section 7.

2   Lowest order calculation

The diagrams entering the tree level matrix element squared are those depicted in fig. 1. The result for the different helicity amplitudes can be found in ref. [4]. We have taken these results expressing the matrix element squared in terms of α, $\sin^2\theta_W$ and M_Z, i.e. not using the tree level relation between those quantities and G_μ to get an 'improved' tree level result which is neither the true tree level result nor the one obtained after the propagator (also called oblique) corrections are included. In Table 1 we can see the results for the integrated cross-section when making different approximations to the exact calculation. The meaning of the different cross-sections is as follows :

- σ_2^0 is the cross-section taking into account only the two Z^0 diagrams.

- σ_{GGR}^0 is the cross-section obtained neglecting the last diagram in fig. 1 and taking the limit $M_W^2 \to \infty$ in the other W diagrams. It corresponds to the classical result of Gaemers, Gastmans and Renard (GGR) presented in [2].

- σ_4^0 is the result obtained neglecting the last diagram but taking into account without any approximation the one-W diagrams.

- σ_5^0 is, finally, the total exact result without any approximation.

We have taken $M_Z = 93$ GeV, $\sin^2\theta_W = 0.230$, $E_\gamma > 1$ GeV and $|\cos\theta_\gamma| < \cos 15°$.
The main conclusions that we can extract from Table 1 are the following :

- The contribution from the diagram with two W propagators is negligibly small in all the energy range of LEP/SLC.

- The GGR calculation is essentially good enough at LEP1/SLC.

- At LEP1/SLC the bulk of the cross-section is due to the Z diagrams. Hence, we will only need to worry about radiative corrections to the Z diagrams.

In fig. 2 we show the W contribution, to be understood as the difference between the results of σ_5 and σ_2, as a function of the center of mass energy. We see that for energies between M_Z and 110 GeV it lies between 1% and 4%. This confirms our previous assumption. Therefore, in the following we will only compute radiative corrections to the Z diagrams.

3 One-loop corrections

As costumarily done, we divide the one-loop QED corrections into virtual ones and real ones. The first are due to the interference between the tree level diagrams and the diagrams shown in fig. 3. The calculation has been performed using the reduction techniques developed by Passarino and Veltman ([12]). The result can be found in [4]-[6].

Concerning the real photon corrections, we have to distinguish between the soft photon corrections and the hard photon ones. The first ones come from the diagrams of fig. 4 where one of the photons has an energy smaller than $x_0 E_b$, with x_0 an arbitrary (small : around $10^{-2} - 10^{-3}$) number. These corrections are computed analytically neglecting the soft photon kinematical effects and combined with the virtual photon ones to give an infrared finite cross-section depending on x_0.

The hard photon contribution cancels this dependence on x_0 by including events with may look like single photon events because of any of the following reasons:

- One of the photons have fractional energy larger than the minimum detectable photon energy, x_D, and the other one has $x_0 < x < x_D$.

- For this second photon, $x > x_D$ but it goes too low angle and, then, it cannot be detected.

- The two photons are almost collinear and are seen as a single object.

We have computed this cross-section with the spinor techniques explained, for instance, in ref. [6]. This has proved to be a powerful tool when dealing with calculations involving a lot of diagrams or a lot of particles.

Finally, we have also considered the more important weak corrections, namely (in the 'on-shell' renormalization scheme ([13])) the Z^0 self energy. This correction is taken into account just performing the following substitution wherever appears the Z propagator :

$$\frac{1}{q^2 - M_Z^2 + i\Gamma_Z M_Z} \longrightarrow \frac{1}{q^2 - M_Z^2 + Re\Pi_{ZZ}(q^2) + iIm\Pi_{ZZ}(q^2)} \tag{1}$$

where Π_{ZZ} is the Z^0 self-energy given by the diagram of figure 5. A discussion of the treatment of the propagator corrections is given in section 5. At the moment, we just apply (1) to every propagator appearing both in the soft and hard parts. We neglect the non-leading weak corrections ([14]).

The total one-loop corrected cross-section can be expressed as :

$$\begin{aligned}
\sigma_{\nu\bar{\nu}\gamma(\gamma)} &= \sigma_{V+S}^{ew} + \sigma_{\nu\bar{\nu}\gamma\gamma}^{ew} \\
&\equiv \sigma_{\nu\bar{\nu}\gamma}^{ew}(1 + \delta^{V+S}) + \sigma_{\nu\bar{\nu}\gamma\gamma}^{ew}
\end{aligned} \tag{2}$$

where the ew superscript means that the Z^0 propagator has been corrected.

For the results given in this section we have used the following set of cuts to define the single photon event :

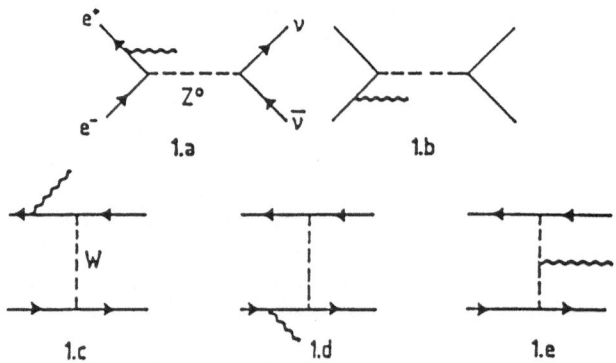

Figure 1. Feynman diagrams of the process $e^+e^- \rightarrow \nu\bar{\nu}\gamma$ in the Born approximation.

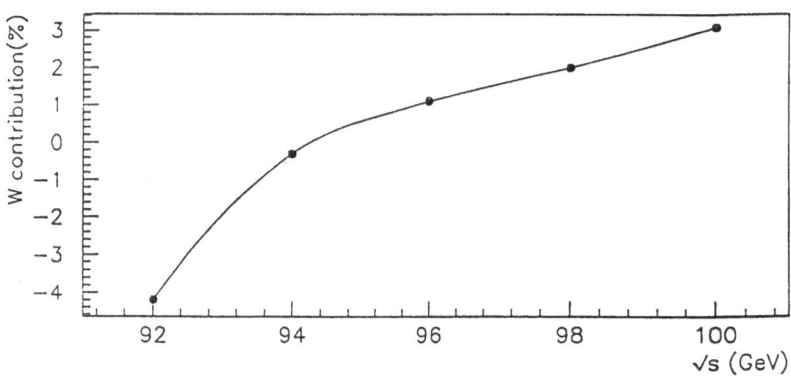

Figure 2. Contribution of the W diagrams squared and of their interference with the Z diagrams. $M_Z = 92$ GeV

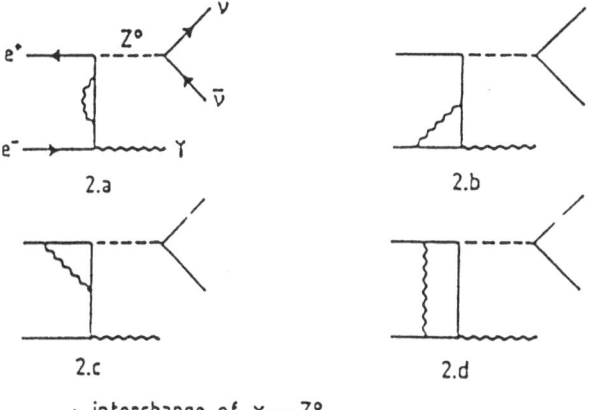

Figure 3. Diagrams contributing to the QED virtual corrections.

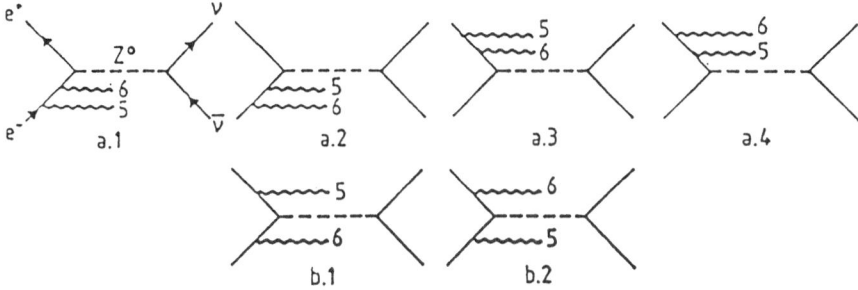

Figure 4. Real photon diagrams contributing to the one loop QED radiative corrections.

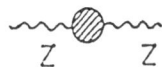

Figure 5. Z^0 self energy diagram.

- Minimum photon energy : $E_D \equiv x_D E_b = 1$ GeV.

- Minimum detection angle : $\theta_D = 15°$.

- Veto angle : $\theta_V = 2.4°$.

- Two photon separation angle : $\theta_{RES} = 1°$.

and the following set of input parameters for the calculation :

- The fine structure function : $\alpha \simeq 1/137$.

- The muon decay constant : $G_\mu = 1.16637 \times 10^{-5}$ GeV^{-2}.

- The Z^0 mass : $M_Z = 93$ GeV.

- The Higgs boson mass : $M_H = 100$ GeV.

- The top quark mass : $m_t = 35$ GeV.

We use the value of G_μ to fix the value of M_W through the one-loop relation

$$\frac{G_\mu}{\sqrt{2}} = \frac{\pi \alpha}{2 s_\theta^2 M_W^2 (1 - \Delta r)} \tag{3}$$

We need the top and Higgs masses because they enter the calculations of the Z^0 self energy and of Δr. Finally, one has to choose a mass for the heavy charged leptons which are in the same $SU(2)$ doublet that the extra neutrinos, since they enter the calculation of the Z^0 self energy. We take for all of them $M_{HEAVY} = 100$ GeV. However, the numbers presented here are for $N_\nu = 3$. Furthermore, we have taken the soft-hard separation limit as $k_0 \equiv x_0 E_b = 0.2$ GeV.

Table 2 contains the results for the different corrections at two center of mass energies. In this table, we have defined $\delta_{Z^0} \equiv \frac{\sigma_{\nu\bar{\nu}\gamma}^{ew} - \sigma^0}{\sigma^0}$. We can see there that the QED corrections are extremely large at the Z peak : adding to δ^{V+S} the hard correction $\delta^H \equiv \frac{\sigma_{\nu\bar{\nu}\gamma\gamma}^{ew}}{\sigma_0}$, we find a total QED correction $\delta^{QED} \equiv \delta^{V+S} + \delta^H = -42.6\%$ Also the Z^0 propagator corrections are sizable. Comparing the first row (tree level calculation) with the last one (final result after the one-loop calculation) we see a huge effect when running at the Z^0 peak ($\sqrt{s} = 93$ GeV) : around -23%. This is almost the difference in the cross-section when adding a new generation (around 25%). At the higher energy, $\sqrt{s} = 100$ GeV, the total effect in the integrated cross-section is rather small ($\sim 3.5\%$). However, it is much more important in the photon distributions. We have shown in fig. 6 b), c) and d) a comparison between the distributions for the photon energy, the cosine of the polar angle and the tranverse momentum obtained with the tree level and the one-loop calculations for the two center of mass energies studied before. We can see that the energy distribution at $\sqrt{s} = 100$ GeV is greatly modified by the one-loop corrections, making the spectrum to be softer as a consequence of having the possibility of radiating an extra photon. Then, if an experimental cut is made in the photon energy in order to reduce the background, the actual cross-section will change a lot. So, even at this higher energy, radiative corrections are important. A similar effect can be seen in the tranverse momentum distribution.

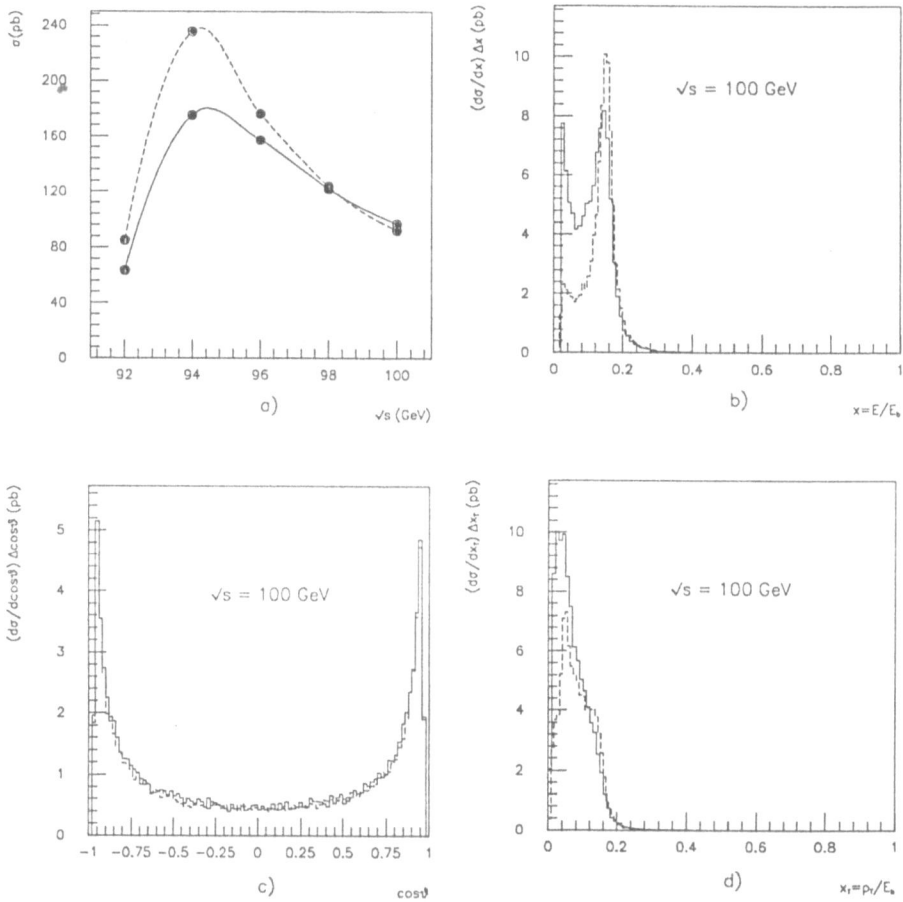

Figure 6. Comparison between tree level and one-loop results :
a) Integrated cross-section.
b) Differential cross section with respect to the photon energy.
c) Differential cross section with respect to the photon polar angle.
d) Differential cross section with respect to the photon transverse momentum.
Solid line : corrected cross-section.
Dashed line : lowest order.

Table 2. Effect of the one-loop corrections. See in the main text the definitions of the quantities appearing in the table.

$\sqrt{s}(\text{GeV})$	93	100
$\sigma^0(\text{pb})$	77.9 ± 0.7	94.2 ± 0.5
δ^{V+S}	-62.5%	-57.3%
δ_{Z^0}	20.0%	15.3%
$\sigma^{ew}_{V+S}(\text{pb})$	44.8 ± 0.4	54.6 ± 0.3
$\sigma^{ew}_{\nu\bar{\nu}\gamma\gamma}(\text{pb})$	15.5 ± 0.3	42.8 ± 0.3
$\sigma_{\nu\bar{\nu}\gamma(\gamma)}(\text{pb})$	60.2 ± 0.4	97.5 ± 0.4

4 Higher order QED corrections

The need for higher order corrections can be summarized in the following points which can be read off the one-loop results :

- The QED initial state corrections are really large in absolute value, around -40%. Therefore, we need higher order initial state QED corrections.

- The oblique corrections are large, around 15% : we need also higher order oblique corrections.

- From the previous two points we can extract that the effect of considering one-loop QED corrections combined with one-loop oblique corrections will be around -6%. Then we have to consider carefully the QED-oblique interplay.

Let's comment briefly why initial state QED corrections are large around the Z peak. We can write the cross-section for a general $e^+e^- \to f\bar{f}$ ($f \neq e$) process including one-loop initial state QED corrections as :

$$\sigma_1(s) = \sigma_0(s)(1 + \delta_1 + \beta \ln x_0) + \int_{x_0}^{1} \beta(\frac{1}{x} - 1 + \frac{x}{2})\sigma_0(s')dx \qquad (4)$$

where the first part is the soft cross-section and the rest the hard one; x is the photon energy in units of the beam energy and $s' = s(1 - x)$; x_0 is the soft-hard separation cut; δ_1 is the part of the soft radiative corrections which is independent of x_0, its value being

$$\delta_1 = \frac{\alpha}{\pi}(\frac{3}{2} \ln \frac{s}{m_e^2} + \frac{\pi^2}{3} - 2) = \frac{3}{4}\beta + \frac{\alpha}{\pi}(\frac{\pi^2}{3} - \frac{1}{2}) ; \qquad (5)$$

β is defined as

$$\beta \equiv \frac{2\alpha}{\pi}(\ln \frac{s}{m_e^2} - 1) \qquad (6)$$

and it can be regarded as an effective coupling constant for bremsstrahlung. At LEP energies it is large : $\beta \simeq 0.11$. This is one of the reasons causing the QED radiative corrections being so large at the Z_0 peak.

The other major reason has to do with the fact that we are close to a (relatively) narrow resonance as the Z^0 one. Since σ_0 is falling rapidly when far from the pole, the upper limit of the integral in (4) is effectively cut to

$$x_M \sim \frac{\Gamma_Z/2}{E_b} \tag{7}$$

where Γ_Z is the Z^0 width and E_b, the beam energy. And then we can approximate

$$\sigma_1(s) \sim \sigma_0(s)(1 + \delta_1 + \beta \ln \Gamma_Z/M_Z) \tag{8}$$

Putting some standard values for the Z width and mass, we find $\beta \ln \Gamma_Z/M_Z \sim -40\%$!

In the case of final state radiation or in absence of any resonance, the restriction in x_M would disappear and we would end up with no large logs.

There is still another, more technical reason for needing higher order QED corrections. We know that the soft-hard limit, x_0, has no physical content and that any physical result from a Monte-Carlo event generator has to be independent of the value of x_0 in a reasonable range. Since in the soft part we are neglecting the photon energy and the cross-section is a steep function of s, we cannot choose a large value for x_0 if we want a reasonable accuracy. For instance, at the Z peak and for $x_0 = 0.01$ (a standard value) we find

$$\frac{\sigma_0(M_Z^2) - \sigma_0(M_Z^2(1 - 0.01))}{\sigma_0(M_Z^2)} \sim 10\% \tag{9}$$

while for $x_0 = 0.001$ the difference is roughly one per mil. Then, we can conclude that we need to choose $x_0 \sim 0.001$ or lower. Actually, this can be taken into account easily modifying the soft-photon integral. However, choosing a large value for x_0 also affects the accuracy with which some distributions are reproduced. This is specially true for the acollinearity distribution.

On the other hand, if we look at equation (4) we can see that putting x_0 too small would cause the soft part to become negative. Since we have to interpret it as a probability density function, this is unacceptable. We have to put x_0 around 0.001 or higher. So, clearly, we have a problem also here, since there is almost no window left for x_0. We will see how this problem is also solved when the exponentiation is performed.

After seeing that one-loop corrections are not enough, the first idea has to be to try and compute two loop initial state QED corrections. This has been done in ref. [7] with the following result :

$$\begin{aligned}
\sigma_2(s) &= \sigma_0(s)(1 + \delta_1 + \delta_2 + \beta \ln x_0 + \delta_1 \beta \ln x_0 + \frac{1}{2}\beta^2 \ln^2 x_0) \\
&+ \int_{x_0}^1 dx \left[\beta\left(\frac{1}{x} - 1 + \frac{x}{2}\right)(1 + \delta_1 + \beta \ln x) + \delta_2^H\right] \sigma_0(s')
\end{aligned} \tag{10}$$

where δ_2 is the $O(\alpha^2)$ correction independent of x_0 and δ_2^H is the $O(\alpha^2)$ non-leading hard correction, the leading one being the one which is proportional to $1/x$. Now x is defined in such a way that $s' = s(1 - x)$ is the q^2 in the boson propagator. The explicit expressions for δ_2 and δ_2^H can be found in [7].

If we look carefully at this last equation, it will become clear that it looks like an $O(\alpha^2)$ expansion of the following expression :

$$
\begin{aligned}
\sigma_{obs}(s) &= \sigma_0(s)(1 + \delta_1 + \delta_2)x_0^\beta \\
&+ \int_{x_0}^1 dx \left[\beta x^{\beta-1}(1 + \delta_1 + \delta_2) - \frac{\beta}{2}(2 - x)(1 + \delta_1 + \beta \ln x) + \delta_2^H\right] \sigma_0(s') \quad (11)
\end{aligned}
$$

which we will take as our exponentiated formula.

With this treatment we have solved the two problems that we had mentioned before :

- The large corrections coming from the x_0-dependent terms in the initial state corrections have been summed up to all orders. Therefore, the overall precision of the calculation for the initial state QED corrections is around one per mil, the expected size of δ_3.

- The problem with the value of x_0 has also disappeared. Now we can choose x_0 much smaller than 0.001 without any problem, since the soft part never becomes negative. Actually, we could choose $x_0 = 0$, and this is indeed done in the Structure Functions approach, which will be presented later.

This approach can be implemented very easily in any one-loop Monte Carlo calculation. We have done it following these steps (it has to be noted that, in our problem, σ_0 will be the cross-section for radiative neutrino production, i.e., already with one hard photon) :

- In the virtual + soft photon corrections part : substitute $1 + \delta_1 + \beta \ln x_0$ by $(1 + \delta_1)x_0^\beta$.

- In the hard photon corrections part : substitute the term containing the piece $\beta\frac{1}{x}$ by $\beta\frac{1}{x}x^\beta(1+\delta_1)$ Since it is not easy to isolate explicitly from our hard photon cross-section the $\frac{1}{x}$ term, we have just multiplyed everything by $x^\beta(1+\delta_1)$. This makes us to include also the corrections to the hardest piece in an exponentiated way which may not be rigorous. However, we do know that the β^2 piece is taken into account correctly, so the difference with the rigorous treatment will be at most in the β^3 term. But in all the calculation we have neglected those terms, which could amount around 0.1%.

Since we are mainly interested in the total cross-section of the process as a measurement of the number of neutrino generations, this approximation will likely be enough. However, if we want to simulate our process accurately in order to take into account detector effects, for example, we have to worry also about how well do we reproduce the differential cross-sections. In this approach we will neglect all the effects of photons carrying a fractional energy smaller than x_0. This is not a serious problem since now, with the exponentiated formula (11), we can choose x_0 as small as we want, so that the effect will be really negligible. When the energy radiated is above this cut-off, we assume that is carried away by just one photon (besides the visible one). This is clearly an approximation, since what we are doing in (11) is adding up the contribution from the radiation of an infinite number of photons. However, when the radiated energy is large enough, it tends to be taken by just one hard photon or at most two. So our approximation is very good in this case. It has been shown in ref. [11] that the probability of having three photons or more of a noticeable energy is very small around the Z peak. Their effects in the total cross-section are accounted for, but their effects in the distributions are taken into account in an approximate way. This effect cannot produce differences in the observable cross-section at a level larger than some parts in 10^{-3}. Results in section 6 will confirm this statement.

The second approach follows the ideas of Kuraev and Fadin and Nicrosini and Trentadue ([8]) of using the Stucture Functions formalism. In this approach the initial state electrons are given a structure of electrons and photons. If $D_e(z,s)$ is the probability of finding an electron with fractional momentum z at center of mass energy s, we have

$$\sigma(s) = \int_{z_1^m}^1 dz_1 \int_{z_2^m}^1 dz_2 D_e(z_1,s)D_e(z_2,s)\sigma_0(sz_1z_2) \tag{12}$$

The lower limit is the minimum fractional energy needed to create a pair of final state fermions.

The function $D_e(z,s)$ satisfies the Altarelli-Parisi equation ([15]). It can be solved to all order in the soft photon limit (large z) ([16]). And then the hard photon corrections up to order β^2 can be added by looking at the explicit iterative second order solution. Finally, the second order virtual corrections can be added by comparing with a complete second order calculation like that of [7]. The final result reads ([8]) :

$$\begin{aligned}
D_e(z,s) &= \frac{1}{2}\beta(1-z)^{\frac{\beta}{2}-1}\Delta' - \frac{1}{4}(1+z)\beta \\
&+ \frac{1}{32}\beta^2\left[(1+z)[3\ln z - 4\ln(1-z)] - \frac{4}{1-z}\ln z - 5 - z\right]
\end{aligned} \tag{13}$$

where Δ' includes the virtual corrections up to $O(\alpha^2)$: $\Delta' = 1 + \frac{\delta_1}{2} + \cdots$. This leads to the following expression for the cross-section

$$\begin{aligned}
\sigma(s) &= \int_0^1 dx H(x,s)\sigma_0(s(1-x)) \\
H(x,s) &= \Delta \cdot \beta x^{\beta-1} - \frac{1}{2}\beta(2-x) + \frac{1}{8}\beta^2\left\{(2-x)[3\ln(1-x) - 4\ln x]\right. \\
&\left. - 4\frac{\ln(1-x)}{x} - 6 + x\right\} + O(\beta^3)
\end{aligned} \tag{14}$$

where $\Delta = \Delta'^2 - \frac{\beta^2\pi^2}{24}$. Comparing the result obtained for $H(x,s)$ with the integrand in the hard part of (11), we see that they are very similar, so that we would expect the two methods to give very similar results. We will see in the following that this is indeed the case.

We will use the expression with the Structure Functions (12) for the calculation of the total cross-section. However when the total amount of energy radiated is larger than a certain x_0 cut, we will use *only for the differential cross-section* the exact hard cross-section computed previously corrected with the replacement explained before for the $\frac{1}{x}$ piece. In this way, we are sure that we get the right answer for both the total cross-section and the main distributions. Let's see how we do it in practice :

- First, we generate the fractional energy of the electrons in the collision z_1 and z_2 according to an approximated distribution function $\tilde{D}_e(z,s) = \frac{\beta}{2}(1-x)^{\beta/2-1}$, and compute the new center of mass energy for the annihilation $s' = sz_1z_2$.

- Then we generate the phase-space variables for the two neutrinos and the visible photon according to the differential tree level cross-section at center of mass energy s' and boost them to the lab reference frame.

- Next, we compute the weight of the event as

$$W = \frac{D_e(z_1, s)}{\tilde{D}_e(z_1, s)} \frac{D_e(z_2, s)}{\tilde{D}_e(z_2, s)} \tag{15}$$

Computing the mean of those weights and multiplying by the integral of the approximant we will find the total cross-section and with the help of a rejection algorithm we will have a sample of unweighted events.

- Once we have an unweighted event, we compute the amount of energy taken by the non-visible photons : $x = 1 - z_1 z_2$. If $x < x_0$, we are through and go to next event; if $x > x_0$, we throw away this event and, instead, we generate an unweighted event according to the modified hard differential cross-section. Then, we go to next event.

This second implementation is supposed to give more accurate results than the first one for the distributions. As mentioned before, with the Structure Functions approach we have more kinematical information than with the 'inductive' approach : we know how the radiated energy is shared between the two initial state particles. Furthermore, all the effects of *collinear* radiation are taken into account in a precise way. With our approach of taking the exact second order matrix element for the hard photon radiation we are sure that we also take into account transverse radiation in a sensible way. For the total cross-sections, they are supposed to give equivalent results. Actually, we will see that this is indeed the case.

Now we are going to present the results of the contribution of the higher order QED corrections. Throughout this section we are going to use the following set of cuts to define a single photon event :

- Minimum energy : $E_D \equiv x_D E_b = 1$ GeV.

- Minimum polar angle : $\theta_D = 15°$.

- veto angle : $\theta_V = \theta_D = 15°$.

- Two photon separation angle : $\theta_{RES} = 1°$.

And the following input parameters for the calculation :

- $M_Z = 92$ GeV, $M_H = 100$ GeV, $m_t = 60$ GeV, $M_{HEAVY} = 100$ GeV.

And the standard values for α and G_μ. M_{HEAVY} is the mass of the leptons which are in the doublet with the extra neutrinos. It enters the calculation of the self energy of the Z, for instance. We have taken the soft-hard cut-off as $k_0 \equiv x_0 E_b = 0.2$ GeV for the pure one-loop calculation and 0.002 GeV for the other ones. Of course, the result is independent of the value chosen.

First of all in Table 3 we show a comparison between the pure one-loop result for the integrated cross-section (σ^1), the cross-section obtained after adding the higher order QED corrections via the implementation of the approach of Berends et al. (σ^I), and after adding the higher order QED corrections with the Structure Functions approach (σ^{SF}). In all the cases the only weak correction included is the modification of the Z^0 propagator as indicated by equation (1).

The main conclusions from this table are two : firstly, there is very good agreement between the two ways we have tried for including higher order QED corrections in this region.

Table 3. Comparison between one-loop and exponentiated cross-sections.

$\sqrt{s}$(GeV)	92	94	100
σ^1(pb)	63.7 ± 0.2	175.4 ± 0.6	96.7 ± 0.4
σ^I(pb)	67.5 ± 0.3	184.5 ± 0.7	95.6 ± 0.5
σ^{SF}(pb)	67.6 ± 0.4	185.1 ± 1.3	94.8 ± 0.7

The results are always compatible and the differences of the central values are between 0.1 and 0.8%. On the other hand, the corrections with respect to the pure $O(\alpha)$ calculation are sizable, especially at the pole. They range from $+6.0\%$ at $\sqrt{s} = 92$ GeV to -1.1% at $\sqrt{s} = 100$ GeV. Comparing with table 2 where a comparison between tree level and one-loop results is made, we realize that higher order corrections go in the opposite way the order α ones go. They tend to compensate the overstimation of the radiative effects done by the first order corrections.

5 Higher order oblique corrections

We have chosen to use the 'star' scheme developped by Kennedy and Lynn ([9]) to include in our calculation the oblique corrections[1]. Here we will give just a rough idea of the procedure. The details can be found in ref. [6].

We can write a general four-fermion neutral current process matrix element with the one-loop oblique corrections included through Dyson equations (which means that the leading-log terms are included to all orders) as :

$$M_{NC} = e_0^2 \frac{QQ'}{q^2 + \Pi_{AA}} + \frac{e_0^2}{s_0^2 c_0^2} \frac{\left[I_3 - Q(s_0^2 - s_0 c_0 \frac{\Pi_{ZA}}{q^2 + \Pi_{AA}})\right]\left[I_3' - Q'(s_0^2 - s_0 c_0 \frac{\Pi_{ZA}}{q^2 + \Pi_{AA}})\right]}{q^2 - \frac{e_0^2}{4\sqrt{2}s_0^2 c_0^2 G_{\mu 0}\rho_0} + Re\Pi_{ZZ} + iIm\Pi_{ZZ}} \quad (16)$$

Now defining a set of finite functions (we do not write the imaginary parts)

$$e_*^2(q^2) = \frac{e_0^2}{1 - \frac{\Pi_{AA}(q^2)}{q^2}}$$

$$s_*^2(q^2) = s_0^2 - s_0 c_0 \frac{\Pi_{ZA}(q^2)}{q^2 + \Pi_{AA}(q^2)}$$

$$\frac{1}{4\sqrt{2}G_{\mu*}}(q^2) = \frac{1}{4\sqrt{2}G_{\mu 0}} - \frac{s_0}{c_0}\Pi_{WW}(q^2) + \frac{s_0 c_0}{e_0}\Pi_{ZA}(q^2) + \frac{s_0^2}{e_0^2}\Pi_{AA}(q^2)$$

$$\frac{1}{\rho_*(q^2)} = \frac{1}{\rho_0} - 4\sqrt{2}G_{\mu*}(q^2)\left(\Pi_{ZZ}(q^2) - \Pi_{WW}(q^2)\right) \quad (17)$$

---

[1] In this Workshop there has been some discussion about the gauge-invariance of the 'star' treatment([17]). The main conclusions seem to be that the treatment is gauge-dependent in a formal sense, but that this fact is completely irrelevat numerically, because the pieces missing to achieve gauge-invariance are very small in the t'Hooft-Feynman gauge in which the authors of ref. [9] work. For a more detailed discussion see, for instance, [6]

we obtain :

$$M_{NC} = \frac{e_*^2(q^2)QQ'}{q^2} + \frac{e_*^2}{s_*^2 c_*^2} \frac{[I_3 - Qs_*^2(q^2)][I_3' - Q's_*^2(q^2)]}{q^2 - \frac{e_*^2}{4\sqrt{2}s_*^2 c_*^2 G_{\mu_* \rho_*}}(q^2) + iM_Z \Gamma_Z^*(q^2)}$$

(18)

We have reabsorbed all the oblique corrections in four finite, universal functions using the Dyson equations. This treatment has some advantadges :

- The matrix element looks like the tree level one.

- All the leading-log terms of the oblique corrections are summed up to all orders and the corrections appear automatically in the Z width, for instance. The 'starred' functions being universal, the running coupling constants which appear in the numerator of the neutral current matrix element are the same that appear in the definition of Γ_Z^* and, thus, they cancel at the pole showing that the total cross-section is insensitive to the oblique corrections. Other schemes for computing radiative corrections would have obscured this fact.

- We can now take the new matrix element as an effective Born amplitude and apply to it QED corrections. In this way, we assure that the oblique corrections are not overstimated. This would happen if QED corrections were not factorized with respect to the oblique ones. Since, as we have seen, QED corrections are extremely large near the Z pole, this would have implied a large error in the oblique corrections.

One can also absorb the non-abelian vertex corrections (which are also universal) into the 'starred' functions ([6]). Then we will have also these corrections factorized with respect to the QED ones. However, the effect is almost negligible, as we will show later on.

Another important point concerning higher order oblique corrections is that of the corrections to $Im\Pi_{ZZ}$ around the pole, as pointed out for the first time by Wentzel ([18]). If we look at the expression for the matrix element for the Z^0 current, we see that at the Z peak, $q^2 = M_Z^2$, we just have

$$M^{NC} \sim \frac{\alpha A}{iM_Z \Gamma_Z^*(M_Z^2)}$$

(19)

where the A in the numerator means something without factors of α in the first approximation. Then the whole numerator is of order α. Since Γ_Z^* is also an $O(\alpha)$ quantity, the matrix element (19) is of order 1. This means that an order α correction to Γ_Z^*, which is equivalent to an order α^2 contribution to $Im\Pi_{ZZ}$, is just an order α correction to the matrix element, and, hence, it is probably needed to reach the accuracy we are seeking.

The fact that in front of the 'starred' width we have the 'starred' coupling constants means that we are including the oblique corrections also in the width, as we have pointed out previously. However, these are not the only corrections that have to be taken into account. There are also the direct corrections shown in the diagram of fig. 7. This kind of diagrams have not been computed at the moment. Nevertheless, we do not need to know their contribution at any q^2, but only when $q^2 \simeq M_Z^2$. We know that exactly at the Z pole the following equations hold

$$\alpha Im\Pi_{ZZ}^{(1)}(M_Z^2) = M_Z \Gamma_Z^{(0)}$$

(20)

$$\alpha^2 Im\Pi_{ZZ}^{(2)}(M_Z^2) = M_Z(\Gamma_Z^{(1)} - \Gamma_Z^{(0)})$$

(21)

Figure 7. Example of one loop corrections to $Im\Pi_{ZZ}$.

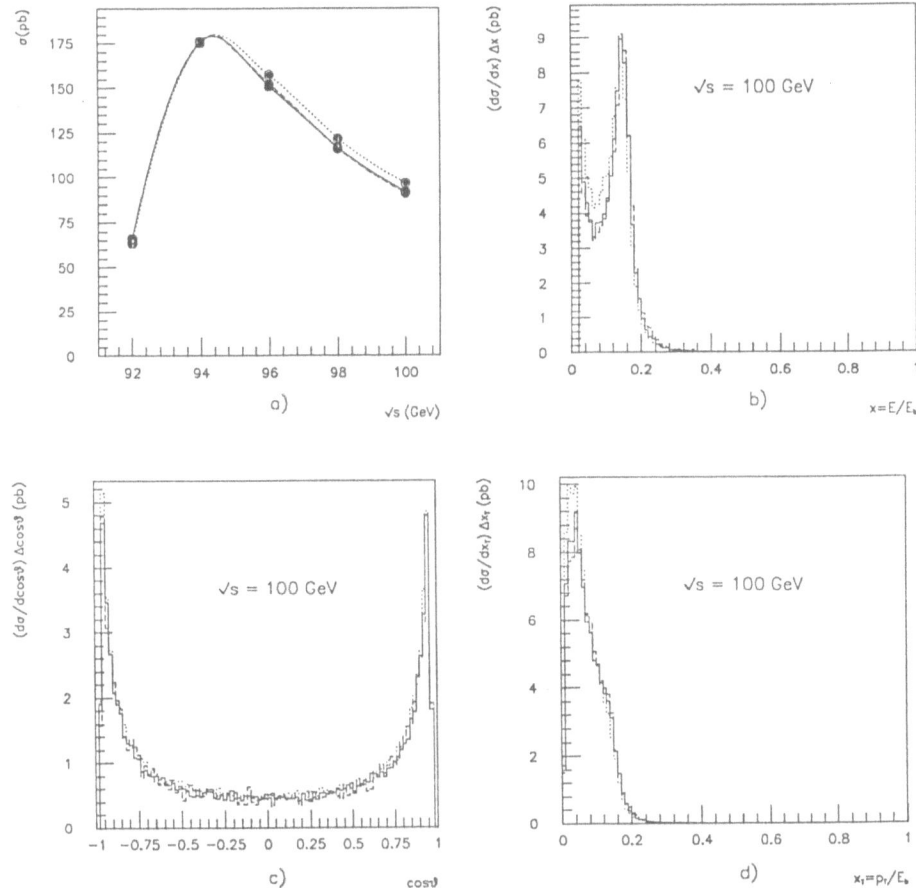

Figure 8. Comparison between pure one-loop results and the ones obtained including higher order corrections :

a) Integrated cross-section.

b) Differential cross section with respect to the photon energy.

c) Differential cross section with respect to the photon polar angle.

d) Differential cross section with respect to the photon transverse momentum.

Solid line : Higher order result with inductive exponentiation.

Dashed line : Higher order result with structure functions exponentiation.

Dotted line : One-loop result.

Table 4. Comparison between the cross-sections without 'star' scheme (σ^I), with 'star' scheme unmodified (σ^{I*}) and with 'star' scheme modified by us ($\hat{\sigma}^I$).

$\sqrt{s}$(GeV)	92	94	100
σ^I(pb)	67.5 ± 0.3	184.5 ± 0.7	95.6 ± 0.5
σ^{I*}(pb)	66.4 ± 0.3	175.9 ± 0.8	93.3 ± 0.5
$\hat{\sigma}^I$(pb)	66.0 ± 0.3	175.6 ± 0.7	92.4 ± 0.5

where $\alpha Im\Pi_{ZZ}^{(1)}$ is the one-loop self-energy, $\alpha^2 Im\Pi_{ZZ}^{(2)}$ is the second order contribution to this self energy, $\Gamma_Z^{(0)}$ is the tree level width and $\Gamma_Z^{(1)}$ is the one-loop corrected Z^0 width, computed in ref. [19]. It includes the complete electroweak corrections together with the QCD corrections to the hadronic branching ratios. This last effect turns out to be the one which is numerically the most important.

We also know that, to good accuracy, the behaviour of the imaginary part of the Z self energy around the peak is

$$Im\Pi_{ZZ}(s) \propto s. \tag{22}$$

Then, a good approximation for the two-loop contributions to $Im\Pi_{ZZ}$ will be

$$Im\Pi_{ZZ}(s) \simeq \alpha Im\Pi_{ZZ}^{(1)}(s) + \frac{s}{M_Z^2} M_Z(\Gamma_Z^{(1)} - \Gamma_Z^{(0)}) \tag{23}$$

This will be a very good approximation near the peak. Far away from it, it will not be so good, but then we do not need such a precision for the imaginary part, since the real part will no longer be small and will dominate the behaviour of the propagator.

Now, we are going to discuss the effect of putting the weak corrections using the 'star' scheme and the effect of the modification we have made to it. We see in Table 4 the cross-section with exponentiation, σ^I from Table 3, and the cross-sections with inductive exponentiation and 'star' scheme without (σ^{I*}) and with our modifications to include the non-abelian vertices ($\hat{\sigma}^I$). Note that since we are including in the Z width the QCD corrections, we need to fix the value of the strong coupling constant, α_s. We have taken $\alpha_s(M_Z^2) = 0.12$.

We see that the largest effect due to the inclusion of the 'star' treatment is at $\sqrt{s} = 94$ GeV, i.e., when we recover the Z peak due to the emission of the visible photon. This effect is important : around -4.6%. Below and above this energy the correction is decreasing up to around -2%. This kind of behaviour is reasonable since most of the effect is due to the inclusion of the one-loop corrections to the width and this is important when the Z^0 is on-shell. The correction due to the inclusion of the non-abelian vertices is almost negligible : it is always less that 1%.

Combining the higher order QED with the higher order weak corrections (with the 'star' scheme modified), we can find the total correction with respect to the one-loop cross-section. It is shown in Table 5 and in figure 8 a).

The QED correction dominates at the peak and hence the total correction is positive. Slightly above the peak, the two corrections are almost of the same size and opposite signs,

Table 5. Percentual total corrections due to higher order effects.

$\sqrt{s}$(GeV)	92	94	100
δ(%)	3.6	0.1	-4.4

Table 6. Integrated cross-sections (pb) with at least one photon with $E_\gamma > 0.5$ GeV and $15° < \theta_\gamma < 165°$

$\sqrt{s}$(GeV)	90	92	94	99	110
KORL03	38.9 ± 0.2	143.7 ± 1.0	245.1 ± 1.2	117.0 ± 0.6	46.3 ± 0.3
NNGG03	38.4 ± 0.1	144.9 ± 0.2	245.4 ± 0.5	117.6 ± 0.1	46.0 ± 0.1

so that the global correction is almost zero. We cannot see any particular reason for this and, therefore, we think it is completely casual. Finally, above the peak both corrections are negative giving a large total effect.

In fig. 8 b), c) and d) we can see the effect of all the higher order corrections altogether. As we see they do not modify very much the photon distributions, although some effects can be seen in the photon and transverse momentum distributions.

6   Comparison with other calculations

A thorough comparison has been performed in ref. [11] between our Monte Carlo calculation with the exponentiation a la Berends et al. and the version of KORL03 ([10]) modified by Colas, Mirabito and Wąs for the neutrino counting process. All the numbers in this section are taken from ref. [11].

The program in ref. [11] includes multiphoton radiation following the Yennie, Frautschi, Suura approach ([20]) and the oblique radiative corrections have been computed by Stuart ([21]). The neutrino version adds exactly the contributions of the one W diagrams and neglects the two W one.

The results have been obtained with the following set of parameters : $M_Z = 92$ GeV, $m_{top} = 60$ GeV, $M_H = 100$ GeV and $N_\nu = 3$. Furthermore only the events with at least one photon with $E_\gamma > 0.5$ GeV and $15° < \theta_\gamma < 165°$ are considered.

The agreement in the integrated cross-section has been found to be very good, better than 1% in the region of interest for the neutrino counting experiment, i.e., at and slightly above the Z peak. This is shown in Table 6 and in fig. 9 a) and b).

It has to be noted the fact that KORL03 has larger errors. The reason is that to have a large sample of events a lot of CPU time is needed, since most of the events will not have any hard photon. This is not the case with our program, which has always a photon from the tree level.

Also the main distributions have been compared in ref. [11] and the agreement has been found satisfactory. In fig. 10 a), b) and c) we show the distribution of energy, polar angle (defined in such a way that it is flat for a bremsstrahlung photon) and tranverse momentum

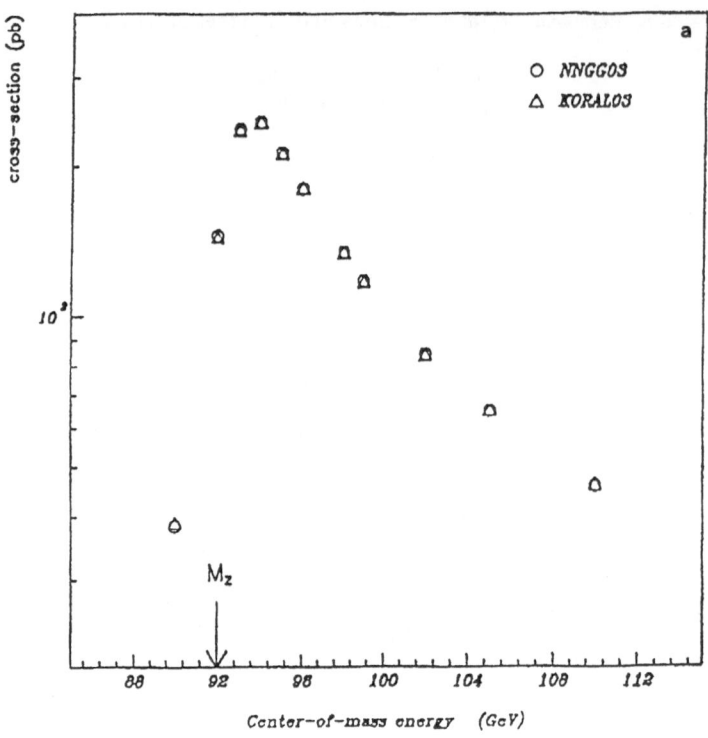

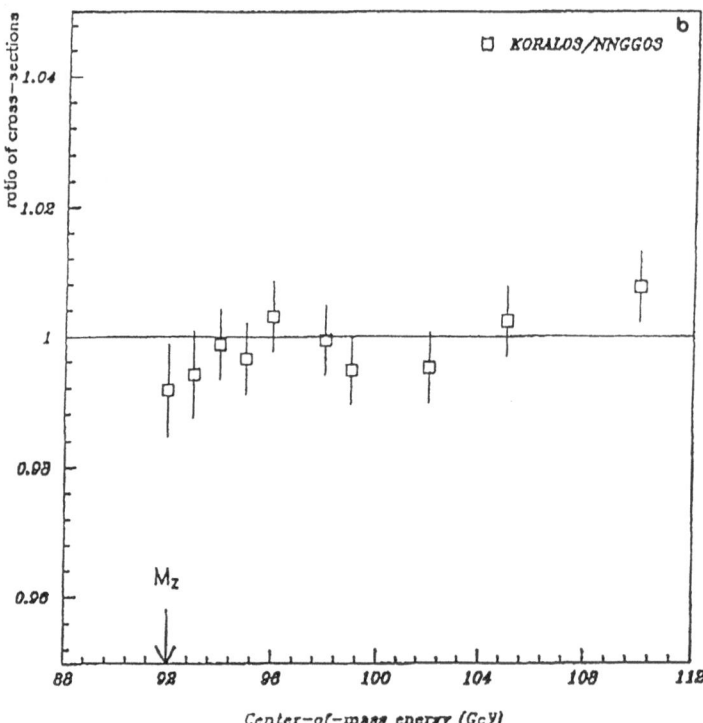

Figure 9. Comparison between our calculation and the one with KORALZ:
a) Integrated cross-section.
b) Ratio of cross-sections.

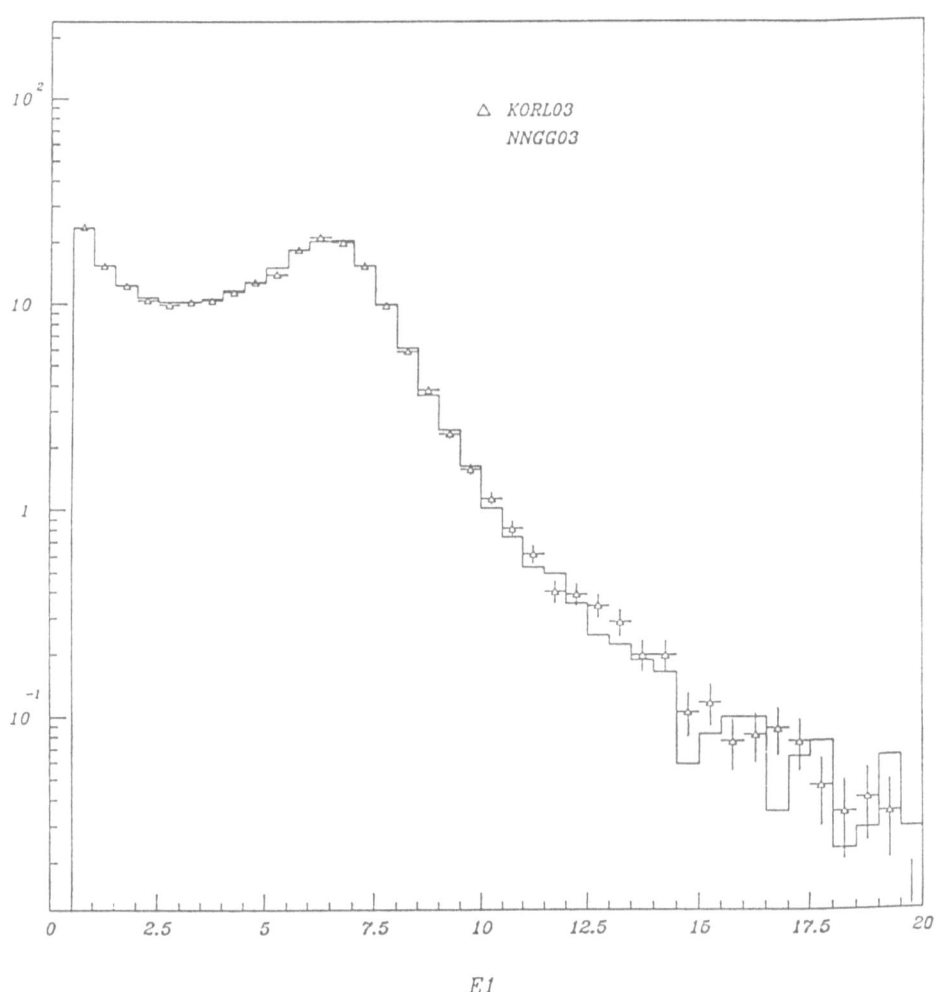

Figure 10. Comparison between our calculation and the one with KORALZ:
a) E_γ distribution.

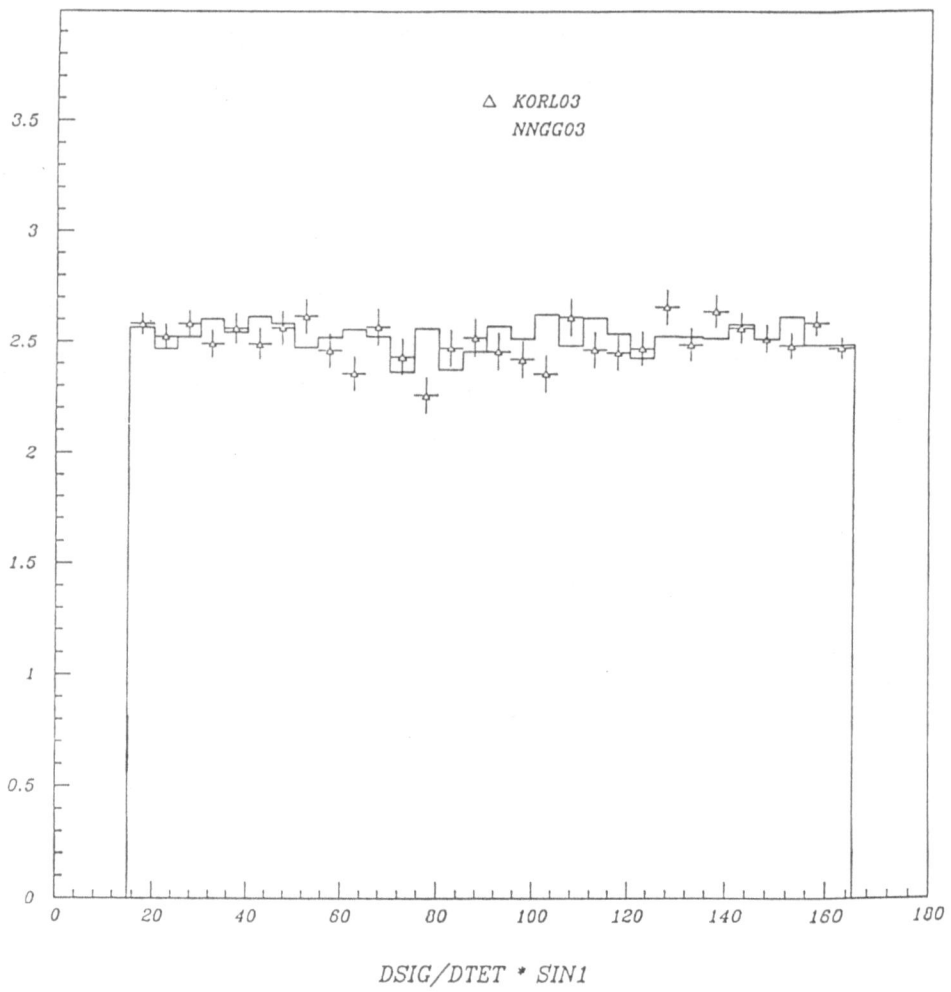

Figure 10. Comparison between our calculation and the one with KORALZ:
b) $\cos\theta_\gamma$ distribution.

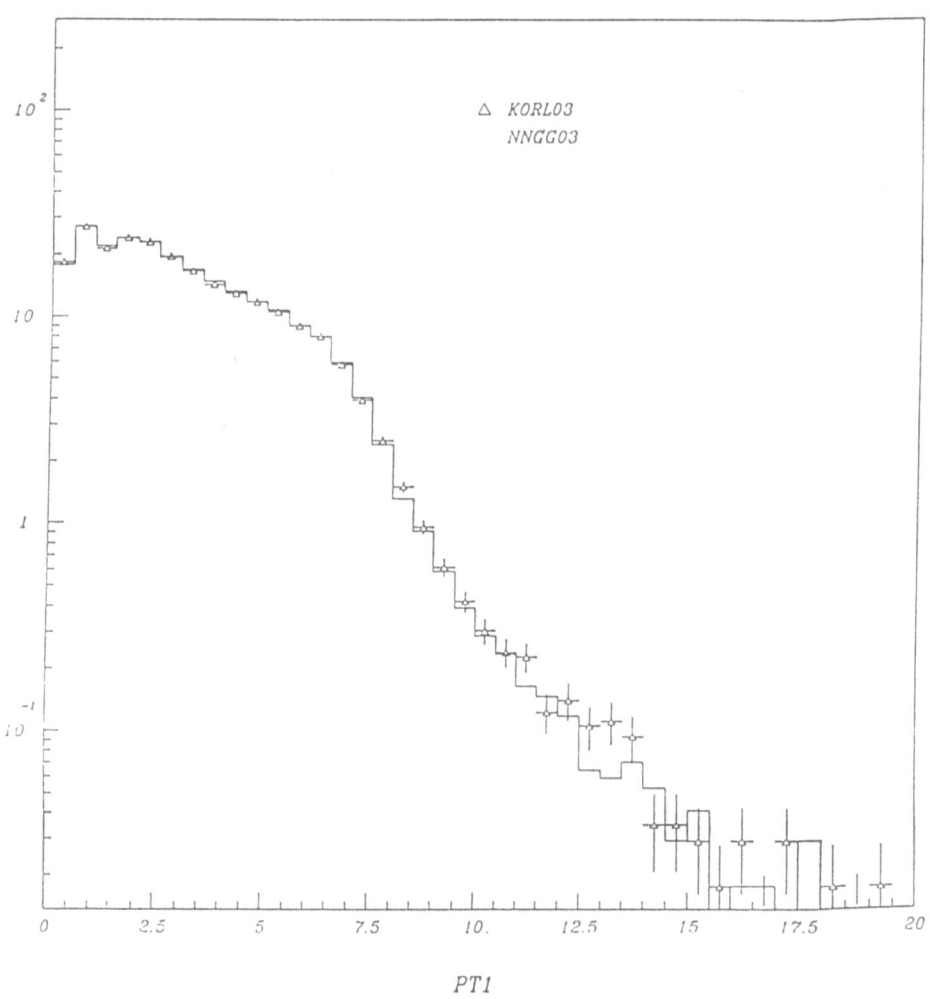

Figure 10. Comparison between our calculation and the one with KORALZ:
c) $p_{T\gamma}$ distribution.

for the hardest photon at a center of mass energy of 99 GeV. No noticeable differences between the two programs can be seen. The same has been checked at $\sqrt{s} = M_Z = 92$ GeV and at $\sqrt{s} = 110$ GeV.

7 Summary and conclusions

We have presented a calculation of $e^+e^- \to \nu\bar{\nu}\gamma$ including :

- Exact tree level calculation.
- Complete $O(\alpha)$ QED corrections to the Z diagrams.
- Leading QED corrections exponentiated.
- Oblique corrections and non-abelian vertices included via 'star' scheme.
- Corrections to $Im\Pi_{ZZ}$.

The calculation has been implemented as a Monte Carlo event generator of unweighted events. The one-loop corrections give a $\sim -25\%$ effect, while higher order corrections altogether amount something around $\pm 5\%$, depending on the center of mass energy.

Results from a comparison with KORL03 done in ref. [11] have been discussed showing a very good agreement (better than 1%) both in cross-section and main distributions in the region of experimental interest. This is a powerful check of both calculations.

We can conclude that the overall precision of our calculation is of the order of 1%, if not better. Hence, it is precise enough for the experimental needs.

Acknowledgements

We thank very much the authors of ref. [11] for allowing us to take their results concerning the comparison between KORL03 and NNGG03.

References

1 . A.D. Dolgov, L.B. Okun and V.I. Zakharov, Nucl. Phys. **B41** (1972) 197
   E. Ma and J. Okada, Phys. Rev. Lett. **41** (1978) 287
2 . K.J.F. Gaemers, R. Gastmans and F.M. Renard, Phys. Rev. **D19** (1979) 1605
3 . M. Martinez, unpublished (1985)
   L. Bento, J.C. Romao and A. Barroso, Phys. Rev. **D33** (1986) 148
4 . F.A. Berends, G.J.H. Burgers, C. Mana, M. Martinez and W.L. van Neerven, Nucl. Phys. **B301** (1988) 583
5 . M. Igarashi and N. Nakazawa, Nucl. Phys. **B288** (1987) 301
6 . R. Miquel, Ph. D. Thesis, Universitat Autònoma de Barcelona, UAB-LFAE 89-02 (1989)
7 . F.A. Berends, G.J.H. Burgers and W.L. van Neerven, CERN-TH.4772/87 (1987)
8 . E.A. Kuraev and V.S. Fadin, Sov. J. Nucl. Phys. **41** (1985) 466
   O. Nicrosini and L. Trentadue, Phys. Lett. **B196** (1987) 551
9 . D.C. Kennedy and B.W. Lynn, SLAC–PUB–4030 (Rev) (1988)
10. S. Jadach, R.G. Stuart, B.F.L. Ward and Z. Wąs, KORL03 Monte-Carlo, unpublished
11. P. Colas, L. Mirabito, Z. Wąs, CEN-Saclay preprint DphPE 89-10 (1989)
12. G. Passarino and M. Veltman, Nucl. Phys. **B160** (1979) 151
13. W.F.L. Hollik, DESY 88-188 (1988)
14. M. Böhm and Th. Sack, Z. Phys. **C 35** (1987) 119
15. G. Altarelli and G. Parisi, Nucl. Phys. **B126** (1977) 298
16. V. Gribov and L. Lipatov, Sov. J. Nucl Phys. **15** (1972) 438, 675
17. A. Sirlin, in these Proceedings. 18. W. Wentzel in CERN 86-02 (1986)
19. W. Beenakker and W. Hollik, Z. Phys **C 40** (1988) 141
20. D.R. Yennie, S.C. Frautschi and H. Suura, An. of Phys. **13** (1961) 379
21. R.G. Stuart, unpublished

THE QED RADIATIVE CORRECTIONS FOR A_{pol} AT LEP[*]

Stanisław Jadach

Institute of Physics, Jagellonian University
30-059 Kraków, ul. Reymonta 4, Poland
and
Zbigniew Wąs[†]

Max Planck Institut für Physik und Astrophysik
D-8000 München 40, Föhringer Ring 6, FRG

Abstract. In this paper we discuss the effects of the QED corrections on the τ polarization measurement at LEP. We discuss also how τ polarization A_{pol} can be applied for precision tests of the Standard Model. The observables which can be constructed for this purpose and different classes of radiative corrections which affect these quantities are carefully reviewed. We compare the size of the radiative corrections with the size of expected statistical and systematic errors for A_{pol} measurement.

1. Introduction

The experiments at SLC and LEP will offer an opportunity for precision tests of the electroweak sector of the Standard Model (SM) [1]. The general principle of these tests is the following: using, for example, α, G_μ and M_Z as an input one calculates the values of some precisely measurable quantities and compares them with experimental results. If the accuracy of the experiment is good enough to match the size of $O(\alpha)$ electroweak corrections then the agreement constitutes positive experimental evidence for the validity of the SM beyond the tree level. On the other hand, any sizeable discrepancy would indicate new physics. In addition to the left-right asymmetry A_{LR}, which offers the best sensitivity but requires polarized beams, and the forward-backward asymmetry A_{FB}, the τ spin polarization asymmetry A_{pol} in the reaction $e^+e^- \to \tau^+\tau^-$ is of particular interest.

In a sense the above three observables are complementary, they all measure couplings of leptons to the Z. From the experimental point of view they do it, however, in a different way and thus will be influenced differently by systematic errors. On the other hand, they are testing slightly different aspects of the model. For instance, they involve coupling constants of leptons (μ, e, τ) from various families and therefore check lepton universality.

---

[*] Presented by Z. Wąs
[†] Permanent adress: Institute of Nuclear Physics, Cracow ul. Kawiory 26a, Poland

Radiative Corrections, Edited by N. Dombey and
F. Boudjema, Plenum Press, New York, 1990

The pure electroweak radiative corrections, which are of main interest, are in fact not the only ones present in the data and also not the most sizeable. In the unified electroweak model pure QED corrections and QCD (hadronic) corrections entering typically through vacuum polarization corrections form other important groups of radiative effects. The one loop pure electroweak corrections to forward-backward asymmetry A_{FB}, to initial state polarization asymmetry A_{LR}, which are the same as to the integrated final state polarization asymmetry $A_{\rm pol}$, were calculated by several groups [2] and are gradually being combined with QCD/hadronic corrections. In these calculations the long distance QED effects (bremsstrahlung) are usually neglected, having in mind that this will be done somehow later.

In this paper we concentrate on the τ polarization asymmetry $A_{\rm pol}$, which can be measured in τ pair production and the subsequent decay process. This observable was considered [3,4] as an important data point in precision tests of the SM model. It remained, however, in the shadow of the forward-backward asymmetry A_{FB} for muons which is experimentally much easier to measure. It appears, however, that although A_{FB} can be measured more easily and with a smaller statistical error it is less sensitive to the $\tau - Z$ vector coupling constant and thus to $\sin^2 \theta_W$. It is also subject to large QED corrections [5] and other problems resulting from its rapid dependence on the center of mass energy (CMS) $\sqrt{s}$ across the Z resonance. The τ polarization asymmetry $A_{\rm pol}$ benefits, similarly to A_{LR}, from the same property of the linear dependence on the vector coupling constant to the Z. It varies less strongly with $\sqrt{s}$ than A_{FB}. It is not sensitive to the initial/final state bremsstrahlung interference, similar to A_{LR} [6] and A_{FB} [7] (to some extent).

The clear disadvantage of $A_{\rm pol}$ is that it has to be measured using τ decay distributions and since this can be done only for some decay modes a substantial loss in statistics occurs. Systematic uncertainties arising from a misidentification of the τ decay modes may also degrade the value of $A_{\rm pol}$ as a promising observable. In the following we shall mainly try to answer the question how pure QED effects (hard bremsstrahlung) influence the τ polarization measurement.

The layout of the paper is the following: In Section 2 we collect the basic facts on the τ polarization asymmetry $A_{\rm pol}$. We introduce the notation and in particular we define the experimental measurables related to $A_{\rm pol}$. Then we show the dependence of $A_{\rm pol}$ on the center of mass energy and the scattering angle. We concentrate mainly on the one decay mode $\tau \to \pi\nu$ which will be used most extensively as τ polarimeter. In the discussion we shall use Born differential distributions both for production and for decay processes, sometimes limiting ourselves to Z exchange only, we will also present analytical expressions which preserve many properties of the Born distributions but incorporate leading QED corrections as well. The dependence on the $Z - e$, $Z - \tau$ and $W - \tau$ couplings will be shown explicitly. We then construct two π energy asymmetries which respectively measure Z couplings to leptons (universality must be assumed) and Z couplings to τ.

In Section 3 we concentrate on QED effects. We show how various types of the bremsstrahlung, i.e. photon emission from incoming beams, outgoing τ's and decay products, influence the $A_{\rm pol}$ measurement. The numerical results are presented.

Section 4 contains the summary and concluding remarks. We compare QED effects with the size of an error in measurement of $A_{\rm pol}$. Finally we will check how the size of the experimental errors matches the weak effects entering through top and Higgs mass dependence.

2. Basic facts on the τ polarization asymmetry

The Born differential distribution for τ pair production $e^+e^- \to \tau^+\tau^-$ including the dependence on the τ^- longitudinal polarization p reads as follows

$$\frac{d\sigma_{\text{Born}}}{d\cos\theta}(s,\cos\theta;p) = (1+\cos^2\theta)F_0(s) + 2\cos\theta F_1(s)$$
$$+p[(1+\cos^2\theta)F_2(s) + 2\cos\theta F_3(s)]. \tag{2.1}$$

with the four formfactors

$$F_0(s) = \frac{\pi\alpha^2}{2s}(q_e^2 q_\tau^2 + 2\text{Re}\chi(s)q_e q_\tau v_e v_\tau + |\chi(s)|^2(v_e^2+a_e^2)(v_\tau^2+a_\tau^2)),$$

$$F_1(s) = \frac{\pi\alpha^2}{2s}(\quad 2\text{Re}\chi(s)q_e q_\tau a_e a_\tau + |\chi(s)|^2 \, 2v_e a_e \, 2v_\tau a_\tau),$$

$$F_2(s) = \frac{\pi\alpha^2}{2s}(\quad 2\text{Re}\chi(s)q_e q_\tau v_e a_\tau + |\chi(s)|^2(v_e^2+a_e^2)\, 2v_\tau a_\tau),$$

$$F_3(s) = \frac{\pi\alpha^2}{2s}(\quad 2\text{Re}\chi(s)q_e q_\tau a_e v_\tau + |\chi(s)|^2 \, 2v_e a_e \, (v_\tau^2+a_\tau^2)),$$

$$\tag{2.2}$$

and

$$\chi(s) = \frac{s}{s - M_Z^2 + is\Gamma_Z/M_Z}$$

The $q_e, v_e, a_e, q_\tau, v_\tau, a_\tau$ are the Z coupling constants to the electron and τ respectively.

The above formfactors are related directly to the Born cross section σ_{Born}, forward backward asymmetry A_{FB}, the τ polarization asymmetry A_{pol}, the combined polarization forward backward asymmetry $A_{\text{pol}}^{\text{FB}}$

$$\sigma_{\text{Born}}(s) = \frac{8}{3}F_0(s),$$

$$A_{\text{FB}}(s) = \frac{1}{\sigma_{\text{Born}}}\Big\{\sigma(\cos\theta>0) - \sigma(\cos\theta<0)\Big\} = \frac{3}{4}\frac{F_1(s)}{F_0(s)} \simeq \frac{3}{4}\frac{2v_e a_e}{v_e^2+a_e^2}\frac{2v_\tau a_\tau}{v_\tau^2+a_\tau^2},$$

$$A_{\text{pol}}(s) = -\frac{1}{\sigma_{\text{Born}}}\Big\{\sigma(p=+1) - \sigma(p=-1)\Big\} = -\frac{F_2(s)}{F_0(s)} \simeq -\frac{2v_\tau a_\tau}{v_\tau^2+a_\tau^2},$$

$$A_{\text{pol}}^{\text{FB}}(s) = -\frac{1}{\sigma_{\text{Born}}}\Big\{\sigma(\cos\theta>0,p=+1) - \sigma(\cos\theta>0,p=-1)$$

$$- \sigma(\cos\theta<0,p=+1) - \sigma(\cos\theta<0,p=-1)\Big\} = -\frac{3}{4}\frac{F_3(s)}{F_0(s)} \simeq -\frac{3}{4}\frac{2v_e a_e}{v_e^2+a_e^2},$$

$$\tag{2.3}$$

and the angular dependence of τ polarization asymmetry

$$A_{\text{pol}}^{\text{Born}}(s,\cos\theta) = -\frac{d\sigma^{\text{Born}}(\cos\theta,p=+1) - d\sigma^{\text{Born}}(\cos\theta,p=-1)}{2d\sigma^{\text{Born}}(\cos\theta,p=0)}$$

$$\tag{2.4}$$

$$= -\frac{(1+\cos^2\theta)F_2(s) + 2\cos\theta \, F_3(s)}{(1+\cos^2\theta)F_0(s) + 2\cos\theta \, F_1(s)}.$$

In eq. (2.3) we have also indicated the approximate relation to the coupling constants at the Z peak (γ exchange neglected).

The polarization p of the τ^- can be measured by looking into the energy distribution of a given decay product,

$$h(u) = \frac{1}{N}\frac{dN}{du}, \qquad u = \frac{2E_{\text{dec.prod.}}}{\sqrt{s}}, \qquad \int\limits_0^1 h(u)du = 1. \qquad (2.5)$$

Here we shall limit ourselves to $\tau \to \pi\nu, e\nu\bar{\nu}$ with the particular emphasis on π decay. The decay distributions read

$$h_\pi(p; u) = h_0^\pi(u) + ph_1^\pi(u) = 1 - p\omega(2u - 1),$$

$$h_e(p; u) = h_0^e(u) + ph_1^e(u) = \frac{1}{3}[(5 - 9u^2 + 4u^3) + p(1 - 9u^2 + 8u^3)],$$

$$\int\limits_0^1 h_0^{\pi,e}(u)du \equiv 1, \qquad \int\limits_0^1 h_1^{\pi,e}(u)du \equiv 0. \qquad (2.6)$$

The parameter

$$\omega = \frac{2g_V g_A}{g_V^2 + g_A^2} \qquad (2.7)$$

includes the coupling constants of $W^\pm$ boson to τ. The standard $V - A$ hypothesis corresponds to $\omega = -1$. We exclude scalar and tensor couplings from our considerations.

What is actually measured in the experiment is the double differential distribution in $\cos\theta$ and u which is obtained by appropriately combined eq. (2.1) and eq. (2.6)

$$\frac{d\sigma_{\text{Born}}}{d\cos\theta du}(s, \cos\theta, u) = (1 + \cos^2\theta)F_0(s)h_0(u) + 2\cos\theta F_1(s)h_0(u)$$

$$(1 + \cos^2\theta)F_2(s)h_1(u) + 2\cos\theta F_3(s)h_1(u). \qquad (2.8)$$

On the other hand the main aim of the experiments is to measure quantities

$$\mathcal{A}_{i,\,(i=e,\mu,\tau)} = 2v_i a_i/(v_i^2 + a_i^2). \qquad (2.9)$$

and perhaps also terms like ω which describe couplings of τ and τ decay products to W. In def. (2.9) we will use index $i = l$ when universality of the Z couplings to leptons is asummed.

3. QED bremsstrahlung effects

The main message of the QED corrections to A_{pol} close to Z can be stated as follows: (1) the initial state bremsstrahlung effect is negligible, (2) the final state bremsstrahlung has a sizeable influence on A_{pol}.

The direct influence of the initial state photon emission is negligible because A_{pol} (like A_{LR} depends weakly on $\sqrt{s}$ (contrary to A_{FB}) and the smearing of the center of the mass energy close to the top of the Z resonance is strongly cut off by the Z line shape. The final state bremsstrahlung does not change the polarization of the τ directly, the helicity non-conservation induced by photon emission being of order $\frac{\alpha}{\pi}A_{\text{pol}} \simeq 10^{-4}$. However since A_{pol} is measured from the slope of the energy distribution of a decay product therefore the softening of its energy spectrum due to either emission of a photon from the τ prior to its decay or from the τ decay product itself, influences the experimental estimate of A_{pol} quite significantly.

The first numerical estimate of the above QED effects was obtained in ref. [4] see also [8] with help of the $O(\alpha)$ QED Monte Carlo calculation. The ultimate answer on the size of the QED effects will always require the Monte Carlo type calculation because the QED effects are inherently dependent on the experimental cut-offs. Monte Carlo is also necessary for any reliable estimation of the systematical errors.

On the other hand it is possible, and highly desirable, to understand qualitatively and if possible quantitatively, the QED corrections using the simple semi-analytical calculation (analytical with a single numerical integration at most). This type of calculations can only be done for simplified or absent cut-offs. A good example is the second order QED result for the total cross section which [9] can be parametrized with the integral of the type

$$\sigma = \int_0^1 dv \, \rho(v) \, \sigma^{\text{Born}}(s(1-v)), \qquad (3.1)$$

where the $\rho(v)$ function can be found in refs. [10,11] and for the sake of completeness we include in Appendix A the formula used in this Section. It parametrizes the total cross section from the second order exponentiation-type Monte Carlo generator of ref. [11] to a precision of 0.1%.

In ref. [12] it was shown that a similar expression works remarkably well for the A_{FB} asymmetry in the presence of the first and second order initial state bremsstrahlung. In the following we shall present an extension of this result to the A_{pol} asymmetry. In this case we must find a simple way to calculate the energy distribution of the final state τ decay product, typically π e or μ. The dependence on the scatering angle is also of interest. The relevant formulas presented below are strictly speaking valid for the final state bremsstrahlung in the leading logarithmic approximation. We understand that the quality of this approximation has to be checked against the Monte Carlo which includes the nonleading terms. We expect, however, these approximations to be rather good for two reasons: they have proved to work very well for the initial state bremsstrahlung and because the final state bremsstrahlung effects are not large therefore the precision requirements are not very high. Leading-logarithmic approximation should be sufficient. The application range is limited to simple or absent cut-offs.

Let us consider the τ production and decay process. We shall speak in the following about the τ decay into $\pi\nu$ but all our discussion will apply for $e\nu\bar{\nu}$ and $\mu\nu\bar{\nu}$ decays as well.

Let us describe the whole τ production and decay process in terms of a chain of fragmentation processes, see Fig. 1. The e^+ and e^- beams, prior to their collision, lose some of their energy due to the initial state bremsstrahlung such that at the moment of the annihilation they create an object, $\gamma + Z$, of an effective mass $\sqrt{s(1-v)}$. The emitted photons are highly collimated with the incoming beams and therefore the rest frame

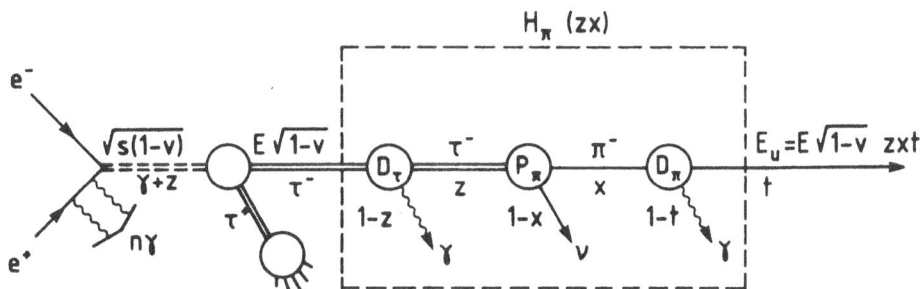

Figure 1. *The fragmentation chain for the semi-analytical estimate of the QED initial and final state bremsstrahlung effects.*

of the $\gamma + Z$ system is usually boosted with respect to the laboratory system[*] roughly along the beams. Let us call the rest frame of the $\gamma + Z$ system the "reduced frame". In the next step the τ pair is created. In the leading-logarithmic approximation, the emission of a photon from the τ may be considered separately as a fragmentation process, namely, the τ of an energy $E' = \frac{1}{2}\sqrt{s(1-v)}$ fragments (in the reduced frame) into a photon with the energy $E'' = E'(1-z)$ and a τ with the energy zE' according to the standard Altarelli-Parisi distribution $D(z)$, see Appendix A. The angular distribution of the τ (in the reduced frame) and its polarization is determined by the $d\sigma^{\mathrm{Born}}/d\cos\theta$. Let us concentrate on the fragmentation history of only one τ. It decays into π of energy $E''' = xE''$ and a neutrino and this process may be viewed, from the reduced frame, as a fragmentation process with the fragmentation function $h_\pi(x)$. Finally the decay pion may again emit a photon and its energy gets reduced once again to $E_\pi = tE'''$. The fragmentation function $D_\pi(t)$ which governs this process is given in Appendix A. It should be noted that the above three step fragmentation process in the final state does not change (in this leading-log picture) the total cross section, i.e. all fragmentation functions sum up to unity.

Let us discuss the initial state bremsstrahlung first: the polarized differential distribution in the collinear approximation is given then, following closely ref. [12], by the simple expression

$$\frac{d\sigma}{d\cos\theta}(s,\cos\theta;p) = \int\limits_0^1 dv\rho(v)\frac{d\sigma^{\mathrm{Born}}}{d\cos\theta}\big[s(1-v),\cos\theta;p\big]$$
$$= (1+\cos^2\theta)\tilde{F}_0(s,u) + 2\cos\theta\tilde{F}_1(s,u)$$
$$+ p\Big[(1+\cos^2\theta)\tilde{F}_2(s,u) + 2\cos\theta\tilde{F}_3(s,u)\Big], \tag{3.2}$$

where

$$\tilde{F}_i(s,u) = \int\limits_0^{1-u^2}\frac{dv}{\sqrt{1-v}}\rho(v)F_i\big(s(1-v)\big), \tag{3.3}$$

The asymmeties A_{pol} and A_{FB} can be obtained from the formula (2.3) provided that we replace the Born form-factors F_i with the convoluted form-factors $\tilde{F}_i$. From formula (3.2) we can learn that the effect of the initial state bremsstrahlung at the Z position on the polarization asymmetry A_{pol} is about a factor ~ 5 smaller than on the asymmetry A_{FB} (note also that in addition A_{pol} is about factor of 5 bigger than the Born prediction for A_{FB}). As we will see later it is in fact negligible experimentally.

In order to discuss the effect of the final state bremsstrahlung, which is much more important than the effect of the initial state emission, we need to know the distribution of the pion energy. In the collinear approximation it can be written as follows

$$\frac{d\sigma}{d\cos\theta du}(s,\cos\theta,u) = \int\limits_0^1 dv\rho(v)\frac{d\sigma^{\mathrm{Born}}}{d\cos\theta}\big[s(1-v),\cos\theta\big]\int\limits_0^1 dz D_\tau\Big(\frac{m_\tau^2}{s(1-v)},z\Big)$$
$$\int\limits_0^1 dx\, h_\pi\Big(A_{\mathrm{pol}}^{\mathrm{Born}}\big[s(1-v),\cos\theta\big],x\Big)\int\limits_0^1 dt\, D_\pi(t)\delta(u-zxt\sqrt{1-v}) \tag{3.4}$$

The pion energy $E_\pi = u\frac{1}{2}\sqrt{s} = zxt\sqrt{1-v}\frac{1}{2}\sqrt{s}$, in the above expression is defined in the "reduced frame". From now on we shall forget, however, that the energy E_π and

---

[*] As already mentioned on the top of Z one does not expect any hard initial state photons therefore this boost is not very big.

scattering angle θ in the above distribution are given in the "reduced frame" and we shall use it for the energy and angle measured in laboratory system. It was checked by comparing with the Monte Carlo in ref. [12] that this is a very good approximation (even far away from the Z peak) for the angular distribution and we shall check with the Monte Carlo later on, in this Section, that the same is true for the u-variable[†].

Most of the integrations in the above formula can be done analytically such that we are left with only one integral over the v-variable. We proceed as follows: the fragmentation functions for the two photonic emissions in the final state, one out of τ and another out of π, may be combined with the τ fragmentation function h_π into an effective single fragmentation function for the process $\tau \to \pi\nu(\gamma(\gamma))$ defined as follows

$$H_\pi(p;x) = H_0^\pi(x) + pH_1^\pi(x),$$

$$H_i^\pi(x) = \int_0^1 \frac{dz}{z} \int_0^1 \frac{dt}{t} D_\tau\left(\frac{m_\tau^2}{s(1-v)}, z\right) D_\pi(t)\, h_i^\pi\left(\frac{x}{zt}\right) \theta\left(1 - \frac{x}{zt}\right),$$

$$\int_0^1 H_0^\pi(x)dx = 1, \qquad \int_0^1 H_1^\pi(x)dx = 0.$$

(3.5)

Using the linearity property of the functions $h_\pi(x,p)$ and $H_\pi(x,p)$ with respect to polarization p we arrive at the compact formula

$$\frac{d\sigma}{d\cos\theta du}(s,\cos\theta,u) =$$

$$\int_0^{1-u^2} \frac{dv}{\sqrt{1-v}} \rho(v) \frac{d\sigma^{\text{Born}}}{d\cos\theta}\left[s(1-v),\cos\theta\right] H_\pi\left(-A_{\text{pol}}^{\text{Born}}\left[s(1-v),\cos\theta\right], \frac{u}{\sqrt{1-v}}\right)$$

(3.6)

The advantage of the above regrouping is that H^π can be calculated analytically and one is left effectively with only one integral over v which can be done (for a given s and $\cos\theta$) numerically rather easily. The expressions for the distribution H^π as well as the analogous functions $H^{e,\mu}$ for the leptonic decays are given in the appendix A.

Note that for a *semi-quantitative* discussion of the electroweak corrections the above formula can also be used, provided we substitute

$$\frac{d\sigma^{\text{Born}}}{d\cos\theta} \longrightarrow \frac{d\sigma^{\text{Born}}}{d\cos\theta} + \frac{d\sigma^{\text{EWC}}}{d\cos\theta},$$

where the last term contains non-photonic "pure electroweak" corrections[*] The asymmetries can then be calculated easily by additional integrations and/or fits made over the $\cos\theta$ and u dependence.

The eq. (3.6) can be rewriten in the Born-like maner, see eq. (2.8),

$$\frac{d\sigma}{d\cos\theta du}(s,\cos\theta,u) = (1+\cos^2\theta)W_0(s,u) + 2\cos\theta W_1(s,u)$$

$$+(1+\cos^2\theta)W_2(s,u) + 2\cos\theta W_3(s,u),$$

(3.7)

where

[†] Alternatively it is possible in the experiment to do a boost in the direction of the beams which maximizes thrust of the τ pair event and take the pion energy and the scattering angle in this frame.

[*] In fact one should also modify the s-channel propagators of γ and Z in both terms as well. The above procedure involves summation of the higher order QED corrections and to what extent it can be justified is a separate interesting problem in itself.

$$W_i(s, u) = \int\limits_0^{1-u^2} \frac{dv}{\sqrt{1-v}} \rho(v) F_i(s(1-v)) H_0^\pi\left(\frac{u}{\sqrt{1-v}}\right), \quad i = 1, 2,$$

$$W_i(s, u) = \int\limits_0^{1-u^2} \frac{dv}{\sqrt{1-v}} \rho(v) F_i(s(1-v)) H_1^\pi\left(\frac{u}{\sqrt{1-v}}\right), \quad i = 3, 4.$$

(3.8)

which may be easier for practical calculation because the integration over $\cos\theta$ can be done analytically.

4. Testing the Standard Electroweak Model

As we have seen in the former Section all the effects of the QED bremsstrahlung do not change the functional shape of the Born distribution (2.8) and it is well approximated after inclusion of bremsstrahlung by formula (3.7) which we recall here once again

$$\frac{d\sigma}{d\cos\theta du}(s, \cos\theta, u) = (1 + \cos^2\theta)W_0(s, u) + 2\cos\theta W_1(s, u)$$
$$(1 + \cos^2\theta)W_2(s, u) + 2\cos\theta W_3(s, u)$$

where the functions $W_i(s, u)$ include all types of energy smearing due to initial and final state QED bremsstrahlung and are defined in eq. (3.8). Note that $\int_0^1 W_{2,3}(s, u)du \equiv 0$.

In the experiment one shall fit the above distribution to the data and determine the coupling constants. We shall consider three possible sets of the assumptions (constraints) in the fit:

1. Assume lepton universality among the electron and the τ and also the pure $V - A$ couplings of the tau to $W^\pm$ boson

$$v_e = v_\tau, \quad a_e = a_\tau, \quad \omega = -1.$$

This is a one-parameter fit which determines roughly $v/a = 1 - 4\sin^2\theta_W$. Such a fit can be done for each decay channel independently. Most efficient will be the π decay mode. In this case we can eliminate the region of $\cos\theta < 0$ where τ is polarized only a little.

2. Assume pure $V - A$ but allow for different (non-universal) couplings of beam electrons and final state τ's. In this case one performs the two-parameter fit which determines v_τ/a_τ and v_e/a_e. Such a fit corresponds approximately to deducing v_τ/a_τ from A_{pol} and v_e/a_e from $A_{\text{pol}}^{\text{FB}}$ simultaneously. The electron and τ couplings will be found independently but with a larger error than the universal v/a in the previous method. The fit can be done for each decay channel independently, the π mode being expected the best.

3. Renounce $V - A$ assumption and universality. In this case there are three free parameters in the fit $\omega, v_e/a_e, v_\tau/a_\tau$. In this case the admixture of $V + A$ can be determined from the electron and muon energy spectrum. In practice this amounts to a simultaneous fit of the energy/angle distribution of the pion and of the electron/muon.

Let us concentrate here on points (1) and (2), more precisely on the measurement of the *universal* lepton's couplings to the Z and the measurement of the τ couplings when universality is not assumed.

First let us turn to the discussion of phenomenological consequences of eqs. (2.1), (2.6), (2.8). At the top of the Z resonance eq. (2.4) can be simplified:

$$A_{\mathrm{pol}}(\cos\theta) = \frac{\mathcal{A}_\tau + \frac{2\cos\theta}{1+\cos^2\theta}\mathcal{A}_e}{1 + \frac{2\cos\theta}{1+\cos^2\theta}\mathcal{A}_e\mathcal{A}_\tau}. \tag{4.1}$$

The τ's produced in the forward directions are strongly polarized, up to twice the mean polarization, whereas polarization is very small in the backward regions and approaches zero for $\cos\theta = -1$. The mean τ polarization is determined by couplings of Z to τ whereas the asymmetry in the polarization is due to the couplings of Z to the electron. We can easily check that the τ polarization given by the approximate eq. (4.1) differs only very little from the predictions of eq. (2.4) and that the τ polarization changes only very little with the CMS energy.

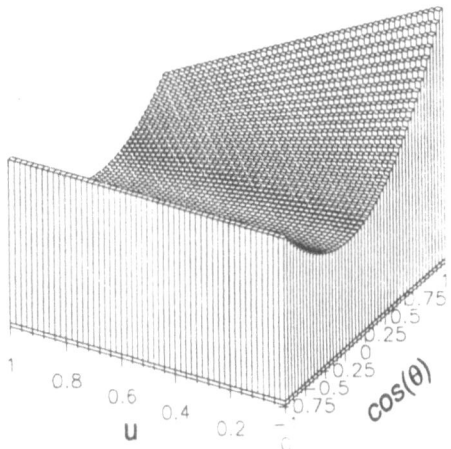

Figure 2. *The 2-dimensional distribution, described by the eq. (2.8) of the $\tau \to \pi$ decay, $\sqrt{s} = M_Z$, (to visualize better the effects we take: $M_Z = 93$ GeV, $\sin^2\theta_W = 0.222$, $\Gamma = 2.557 GeV$).*

In Fig. 2 we show the two-dimensional u, $\cos\theta$ distribution of the $\tau \to \pi$ decay. As we can see, the slope of the π energy spectrum can be used, at every angle θ separately to measure angular τ polarization distribution. For the purpose of the further discussion we limit ourselvesf to the more general asymmetries $A_\pi^{(1,2)}$, defined as follows

$$A_\pi^{(1,2)} = \frac{N^{(1,2)}(u_\pi > u_0) - N^{(1,2)}(u_\pi < u_0)}{N^{(1,2)}(u_\pi > u_0) + N^{(1,2)}(u_\pi < u_0)} \simeq \frac{1}{2}A_{pol}^{(1,2)} \tag{4.2}$$

where u_0 denotes the mean energy of the π while $N^{(1,2)}(u_\pi > u_0)$ and $N^{(1,2)}(u_\pi < u_0)$ denote respectively the number of the detected pions faster and slower than the mean energy. In the case 1) we take into account all π's accessible for the detection. In the case 2) we exclude the backward region ($\cos\theta < 0$) where the τ sample is nearly unpolarized. The mean τ polarization averaged over the region 1) $A_{pol}^{(1)}$ depends practically only on $\mathcal{A}_\tau$ whereas in the case 2), $A_{pol}^{(2)}$ depends on the combination of $\mathcal{A}_\tau$ and $\mathcal{A}_e$.

Numerical estimates of the QED effects and experimental errors

The first numerical estimate of the above QED effects was obtained in refs. [13,4], see also [8] with the help of the $O(\alpha)$ QED Monte Carlo calculation. The ultimate answer on the size of the QED effects will always require the Monte Carlo type calculation because the QED effects are inherently dependent on the experimental cut-offs. Monte Carlo is also necessary for any reliable estimation of the systematical errors.

The Monte Carlo program [14], which could in principle be used to perform such a high precision study already exists. However this study requires good knowledge of the detector properties. This will have probably to wait until after first data samples are analyzed, and some experience is gained how detectors work in practice.

Let us now tabulate estimates of the QED bremsstrahlung effects. We will use Z mass $M_Z = 91 GeV$, Z width $\Gamma = 2.557 GeV$ and $\sin^2 \theta_W = 0.232$. In the lowest possible level of approximation, when only Z exchange is included, the mean τ polarization is given by the following equations

$$2A_\pi^{(1)} = A_{\text{pol}}^{(1)} = \mathcal{A}_\tau,$$
$$2A_\pi^{(2)} = A_{\text{pol}}^{(2)} = \frac{7\mathcal{A}_l}{4 + 3\mathcal{A}_l^2}. \tag{4.3}$$

The direct QED corrections which modify the relation (4.3) between A_{pol} and the couplings of the τ lepton to the Z, originate from pure s-channel photon exchange and from bremsstrahlung in the initial and final states. As can be seen from Table 1,

Table 1 *Corrections to eq. (4.3) from purely photonic contribution to the Born approximation, from initial state bremsstrahlung and from final state bremsstrahlung calculated up to $O(\alpha)$. All corrections are calculated for $\sqrt{s} = M_Z$.*

A_{pol}	γ contribution to Born	Initial state brem.	Final state brem.
$A_{\text{pol}}^{(1)}$	−0.0009	−0.0021	−0.0002
$A_{\text{pol}}^{(2)}$	−0.0016	−0.0037	−0.0004

these direct QED corrections are small (compared to $A_{\text{pol}} \simeq 0.1 - 0.2$), they do not exceed 0.007. The reason is the following: The biggest correction due to the emission of hard photons in the initial state is strongly reduced because on the top of the Z the photon spectrum is damped by the resonance shape.

Final state bremsstrahlung is practically unable to modify the τ polarization directly. The main correction originates from the helicity-flip [13] contribution which is of $O(\frac{\alpha}{\pi} A_{\text{pol}})$. As in the case of the asymmetry A_{LR} [0] there is a rather strong cancellation between the corrections of boxes and initial-final state bremsstrahlung interference. This class of corrections is completely negligible even for relatively strong cut-offs.

QED bremsstrahlung distorts the relation (2.6) between the slope of π energy spectrum and A_{pol}. In fact this indirect effect of QED on the measurement of the τ polarization is significantly larger than the direct ones previously discussed. The emission of the hard photon in the initial or final state, reduces the invariant mass of the $\tau^\pm$ pair and thus the distributions of their decay products are deformed. Table 2 presents the deformation of the π's energy asymmetry and subsequent Born level estimate on A_{pol} in the $\tau \to \pi\nu$ decay due to hard photon emission. The energy of the photons emitted in the final state is usually larger than the energy of photons emitted in the initial state. The initial state bremsstrahlung correction is not only small by itself but is also cancelled partially by the correction to formula (4.3). The contribution of the photons emitted in the final state is the dominant one. The various contributions to the asymmetry in the π energy are given in Table 2[*].

Table 2 *Variation of the Born level estimates of $A_{\text{pol}}^{(1,2)}$, see formulae (2.6) and (4.2), obtained from A_π due to the QED bremsstrahlung: A) no cuts on π energy, B) π energy in the range $0.05\sqrt{s} - 0.45\sqrt{s}$. Numerical results obtained with the help of eq. (3.7) are given in brackets. The columns of the table corresponds respectively to Born, Born + initial bremsstrahlung, Born + initial + final bremsstrahlung and Born + initial + final + bremsstrahlung in τ decay.*

A) no cuts	Born	+ Initial	+ Final	+ In decay
$A_{\text{pol}}^{(1)} = 2A_\pi^{(1)}$	0.127 ± 0.002	0.129 ± 0.003	0.154 ± 0.003	0.164 ± 0.003
$A_{\text{pol}}^{(2)} = 2A_\pi^{(2)}$	0.232 ± 0.003	0.232 ± 0.004	0.260 ± 0.004	0.266 ± 0.004

B) cut on E_π (u)	Born	+ Initial	+ Final	+ In decay
$A_{\text{pol}}^{(1)} = 2.222A_\pi$	0.138 ± 0.002	0.138 ± 0.003	0.160 ± 0.003	0.171 ± 0.003
(semianalytically)	(0.142)	(0.141)	(0.163)	(0.173)
$A_{\text{pol}}^{(2)} = 2.222A_\pi$	0.245 ± 0.003	0.241 ± 0.004	0.269 ± 0.004	0.273 ± 0.004
(semianalytically)	(0.245)	(0.247)	(0.268)	(0.279)

We may summarize the discussion of QED corrections by saying that the indirect corrections in contrast to the direct ones, are sizeable and also depend on experimental cut-offs.

5. Summary and conclusions

Let us now compare expected statistical and systematic errors in the measurement of A_{pol} with the size of the QED effects. From the sample of $3 \cdot 10^6$ Z we get 10^5 τ pairs, and $2 \cdot 10^4$ τ's decaying into pions. This brings the statistical error in the measurement

---

[*] In our discussion we have assumed that the bremsstrahlung photons not only escape undetected but do not affect measurement of the π energy as well. In reality neither of this two is true and the final answer on the size of the bremsstrahlung effect requires study of the particular detector. In our case (B), results obtained with the help of the Monte Carlo [14], include only simple cut, which eliminate regions sensitive to the π mass corrections.

of $A_\pi^{1(2)}$ down to respectively 0.007 (0.01) and $A_{pol}^{(1,2)}$ down to 0.014 (0.02)[*] . Note that $A_{pol}^{(2)}$ is nearly a factor of 2 more sensitive to $\sin^2\theta_W$ than $A_{pol}^{(1)}$. From the results of the Tables 1,2 we may thus conclude that even though all QED corrections represent only $\frac{1}{4}(\frac{1}{3})\sigma$ correction for $A_{pol}^{1(2)}$, the correction for $A_\pi^{1(2)}$ which will be used to measure A_{pol}, can be as big as 2σ. This is not an effect which could deteriorate the quality of the observable, but definitely must be carefully taken into account. The dominant corrections originate from *final* state bremsstrahlung. On the other hand corrections for the leptonic modes are bigger, comparable to the polarization effect itself and they can cause a problem, the solution would however require careful study of the detector properties.

Usually corrections to the τ decay width are considered to be the most interesting corrections [17]. They can be absorbed into the an overall normalization: branching ratio. On the other hand, as it was shown in [18] the big leading-log terms determining the bulk of the bremsstrahlung in decay corrections to the A_π are free from the π structure effects (formfactors).

Finally let us comment on the interesting point of the electroweak effects. We have practically excluded this topic from our considerations because within the SM, predictions for integrated A_{pol} are the same as for A_{LR}. The point is that the 0.01 error in measurement of $\mathcal{A}_l$ (ie. error 0.02 for $A_{pol}^{(2)}$) matches in size the variation due to the top mass increase from 50 to $120 GeV$, see eg. [2].

We can summarize this talk as follows. The measurement of τ polarization asymmetry, A_{pol}, is expected to be a very good, possibly even the best precision test of the SM at LEP. Effects of the QED corrections are rather well understood. The detailed study of the QED corrections and experimental conditions is however not yet completed.

APPENDIX A

Ingredients for the semi-analytical calculations

In the practical applications in this paper we shall use for the initial state bremsstrahlung $\rho(v)$ distribution the simple formula of ref. [11], similar to that of ref. [10],

$$\rho(v) = \frac{e^{-C\gamma}}{\Gamma(1+\gamma)} e^{\delta_{YFS}} \gamma v^{\gamma-1}(1 + \delta_S + \delta_H(v)),$$

$$\delta_{YFS} = \left(\frac{\alpha}{\pi}\right)\left(\frac{1}{2}L - 1 + \frac{\pi^2}{3}\right), \quad \delta_S = \left(\frac{\alpha}{\pi}\right)(L-1) + \frac{1}{2}\left(\frac{\alpha}{\pi}\right)^2 L^2,$$

$$\delta_H(v) = v\left(-1 + \frac{1}{2}v\right) + \left(\frac{\alpha}{\pi}\right)L\left(-\frac{1}{4}(4 - 6v + 3v^2)\ln(1-v) - v\right) \tag{A1}$$

$$L = \ln\frac{s}{m_e^2}, \quad \gamma = 2\left(\frac{\alpha}{\pi}\right)(L-1), \quad C = 0.57721566...,$$

which is the second order sub-leading approximation with the proper resummation of the soft photons. The formula of this class was given for the first time in ref. [19] and it parametrizes the total cross section from the Monte Carlo of ref. [11] to within 0.1%. The above formula omits the effect of the production of the additional light fermion pair, which is known to be small and affecting the overall normalization of the cross section.

---

[*] As it was shown in [15] (see also [16]) the statistical error can be further decreased up to even 30% if one includes all τ decay modes for the analysis of A_{pol} and also exploits spin correlations of τ^+ and τ^- decay products. This ambitious program requires however much better, than it is at present, knowledge of the systematic errors for all τ decay modes.

Let us now show how to calculate the $H_i^{\pi,e,\mu}$ distributions by convoluting the τ fragmentation functions $h_i^{\pi,e,\mu}(x)$ with the bremsstrahlung (leading logarithmic) kernels

$$D_\tau\left(\frac{m_\tau^2}{s(1-v)},z\right) = \delta(1-z) + \gamma_\tau d_\tau(z), \quad \gamma_\tau = \frac{\alpha}{\pi}\left(\ln\frac{s(1-v)}{m_\tau^2} - 1\right),$$

$$d_\tau(z) = \left(\frac{1}{1-z}\right)_+ + \frac{3}{4}\delta(1-z) - \frac{1}{2}(1+z), \quad \int_0^1 d_\tau(z)dz = 0,$$

$$D_\pi\left(\frac{m_\pi^2}{m_\tau^2},t\right) = \delta(1-t) + \gamma_\pi d_\pi(t), \quad \gamma_\pi = \frac{\alpha}{\pi}\ln\frac{m_\tau^2}{m_\pi^2},$$

$$d_\pi(t) = \left(\frac{1}{1-z}\right)_+ + \delta(1-z) - 1, \quad \int_0^1 d_\pi(z)dz = 0,$$

(A2)

For our precision it is possible to neglect the term of order $\gamma_\tau\gamma_\pi$ and the double convolution gets replaced with the two single convolutions[†]

$$H_i(x) = \int_0^1 \frac{dz}{z} \int_0^1 \frac{dt}{t} D_\tau\left(\frac{m_\tau^2}{s(1-v)},z\right) D_\pi\left(\frac{m_\pi^2}{m_\tau^2},t\right) h_i\left(\frac{x}{zt}\right) \theta\left(1-\frac{x}{zt}\right)$$

$$= h_i(x) + \gamma_\tau \int_x^1 \frac{dz}{z} d_\tau(z) h_i\left(\frac{x}{z}\right) + \gamma_\pi \int_x^1 \frac{dz}{z} d_\pi(z) h_i\left(\frac{x}{z}\right)$$

For $\tau \to \pi$ decay we get

$$H_0^\pi(x) = 1 + \gamma_\tau\left[\ln(1-x) - \frac{1}{2}\ln x + \frac{1}{2}x + \frac{1}{4}\right]$$
$$+ \gamma_\pi\left[\ln(1-x) + 1\right]$$

$$H_1^\pi(x) = 2x - 1 + \gamma_\tau\left[(2x-1)\ln(1-x) - \frac{1}{2}(2x-1)\ln x + \frac{3}{4}\right]$$
$$+ \gamma_\pi\left[(2x-1)\ln(1-x) - 2x\ln x + 2x - 1\right]$$

(A3)

For $\tau \to \mu$ decay we have

$$H_0^\mu(x) = 2 - 6x^2 + 4x^3 + \frac{4}{9}\rho(-1 + 9x^2 - 8x^3)$$
$$+ (\gamma_\tau + \gamma_\mu)\left\{\ln(1-x)(2 - 6x^2 + 4x^3) - \ln(x)(1 - 6x^2 + 4x^3) - \frac{8}{3} + 4x^2 - x - \frac{1}{3}\right.$$
$$+ \frac{9}{4}\rho\left[\ln(1-x)(-1 + 9x^2 - 8x^3) - \ln(x)(-\frac{1}{2} + 9x^2 - 8x^3) + \frac{16}{3}x^3 - 8x^2 + 2x + \frac{2}{3}\right]\right\},$$

(A4)

[†] This corresponds to neglecting the simultaneous photon emission from the tau and the decay product.

$$H_1^\mu(x) = \frac{-2}{3} + 4x - 6x^2 + \frac{8}{3}x^3 + \frac{4}{9}\delta(1 - 12x + 27x^2 - 16x^3)$$
$$+ (\gamma_\tau + \gamma_\mu)\left\{ \ln(1-x)(-\frac{2}{3} + 4x - 6x^2 + \frac{8}{3}x^3) \right.$$
$$+ \frac{4}{9}\delta\left[\ln(1-x)(1 - 12x + 27x - 16x^3) \right.$$
$$\left. \left. - \ln(x)(\frac{1}{2} - 6x + 27x^2 - 16x^3) + \frac{32}{3}x^3 - 16x^2 + 7x - \frac{5}{7}\right]\right\}$$

(A5)

Here ρ and δ are the standard parameters see ref. [20] characterizing the τ coupling constants to $W^\pm$ boson.

REFERENCES

1. S. L. Glashow, Nucl. Phys. **22** (1961) 579;
   S. Weinberg, Phys. Rev. Lett. **19** (1967) 1264; Phys. Rev. **D5** (1972) 1412;
   A. Salam, in *Elementary Particle Theory*, ed. N. Svartholm, Stockholm, 1968, p 361.

2. W. Hollik, DESY report, DESY-88-188,December 1988, and references therein.

3. G. Altarelli et al., in Physics at LEP, CERN 86-02 (1986) Vol. 1 p. 1.

4. F. Dydak et al. ECFA Workshop on LEP 200, CERN 87-08 (1987) Vol. 1, p. 157.

5. R. Kleiss, Hard Bremsstrahlung in $e^+e^- \to \mu^+\mu-$ Matching Theory and Experiment, in Physics at LEP 86-02 Vol 1 p. 153.

6. S. Jadach, J. H. Kühn, R. G. Stuart, Z. Wąs, Z. Phys. **C38** (1988) 609; J. H. Kühn, R. G. Stuart, Phys. Lett. **B200** (1988) 360.

7. S. Jadach, Z. Wąs, Phys. Lett. **B219** (1989) 103.

8. F. Boillot and Z. Wąs, Z. Phys **C43** (1989) 109.

9. F. A. Berends and W. L. Van Neerven and G. J. H. Burgers, Nucl. Phys. **B297** (1988) 249.

10. G. Burgers, "The shape and size of the Z resonance", in " Polarization at LEP", CERN report 88-06, eds. J. Ellis and R. Peccei, CERN, Geneva, (1988).

11. S. Jadach and B.F.L. Ward, " YFS2 –the second order monte Carlo for fermion pair production at LEP/SLC with the initial state radiation of two hard and multiple soft photons", to appear in Computer Phys. Commun.

12. S. Jadach, Z. Wąs, "First and higher order noninterference QED radiative corrections to the charge asymmetry at the Z resonance", 1989 Munich preprint, MPI-PAE/PTh 33/89.

13. Z. Wąs, Acta Physica Polonica **B18** (1987) 12.

14. S. Jadach, B. F. L. Ward, Z. Wąs, R. G. S. Stuart and W. Hollik, "KORALZ the Monte Carlo program for τ and μ pair production processes at LEP/SLC", unpublished, may be obtained form Z. Was: WASM @ CERNVM.

15. C. A. Nelson, Tests for New Physics from τ spin correlation functions for $Z \to \tau^+\tau^- \to A^+B^-X$, Preprint, New York 1989, SUNY BING 1/30/89.

16. J. Kühn, F. Wagner, Nucl. Phys. **B236** (1984) 16.

17. W. Marciano, A. Sirlin, Phys. Rev. Lett. **61** (1988) 1815.

18. S. Banjere, Phys. Rev. **D34** (1986) 2080.

19. E. A. Kuraiev and V. S. Fadin, Sov. J. Nucl. Phys. **41** (1985) 466 .

20. Rewiev of Particle Properties, Phys. Lett. **B204** (1988) 1.

REVIEW OF THE SESSION ON EVENT GENERATORS

Ronald Kleiss

CERN, Geneva

I have been asked to give the summary talk on the session on Monte Carlo event generators. This is not a trivial task. An exhaustive review of all that was presented would not be very useful, since all that can be found elsewhere in these proceedings. Instead, I have chosen to address a few separate questions that were raised during this week's discussions. Since presenting a review/summary unavoidably calls for judgements and subjective statements, I will also make a number of those: of course, these are solely my personal point of view.

I shall start with briefly calling to mind the various talks that were given, ordered by speaker. Then I shall redo the same, ordered by topic. Subsequently I shall try to formulate a number of the more fundamental questions in the Monte Carlo / radiative corrections field, and then try to answer them (again, in a possibly biased and subjective way!).

A summary of the talks, ordered by speaker

Patricia Rankin has in fact provided us with the first description of the various radiative correction programs *in actual use!* By the time these proceedings are coming out, LEP itself will of course have many events, but certainly people have already learned a lot from what is going on at the SLC. Unfortunately, the Z mass was still kept semisecret in Brighton, but Patricia presented the strategy used at MARK II for extracting the Z mass with very satisfying accuracy from their modest event sample. She showed the results of comparisons between several Monte Carlo programs: ZBATCH, EXPOSTAR, KORALZ, MMGE92, and DYMU2. Also, she spent a good deal of time on the various well-known items of theoretical rather than programming interest: the influence of M_t, the use of $\sin^2 \theta_w$ versus $\sin^2 \bar{\theta}_w$, $\sin^2 \theta_w^*$, etcetera, and the expectations for $A_{FB}^{\mu\mu}$, A_{FB}^{bb}, and even A_{LR}, which SLC still has a better chance of measuring than LEP. A subject that she also discussed, and which I think needs a little bit more of attention than it has received so far, is the luminosity measurement: I basically agree with what she said, but since that is somewhat different from what she said at the Ringberg meeting in April, one can see that this question has not been settled completely.

Jean-Eric Campagne sketched for us the principles on which the DYMU2 Monte Carlo is constructed. In my opinion this is at present one of the nicest event generators, precisely for the reason that its structure is not very complicated, and can in fact be described completely in a short talk. He moreover presented comparisons of DYMU2 with various semianalytical predictions, which is very interesting since we start to be in a position (mainly owing to the efforts of Dimitri Bardin, Tord Riemann and their collaborators of Dubna and Zeuthen) where more than just the total cross section can be studied in this way. Comparisons such as these

Radiative Corrections, Edited by N. Dombey and
F. Boudjema, Plenum Press, New York, 1990

are still essential (I think) to gain the necessary confidence in the Monte Carlos that will be actually used in the data analysis.

Giovanni Bonvicini gave us a similar description of the strategy underlying the MOE event generator. This is a rather more complete and rigorous treatment of the QED structure functions than goes into DYMU2: however, the results of the two programs seem to be quite compatible (except, of course, that the photon multiplicity is not limited in MOE, and limited to 2 or 3 in DYMU2). He presented comparisons with ZBATCH for the line shape and the s' spectrum (s' is the invariant mass of the final-state fermion pair). Some time was spent in a discussion of the merits and possibilities of the use of 'effective' and 'virtual' beam momenta, and the question of 'Lorentz invariance' versus a 'well-defined muon mass'. These terms look like a lot of technical jargon, and in fact they are: the fact that people like him and me can get quite excited about conceptual problems like that only indicates that we are Monte Carlo fanatics, and not that these questions are numerically important. In fact, he presented convincing evidence that the uncertainties in the MOE result will be either of order $\mathcal{O}(\alpha^2)$ or $\mathcal{O}(m_\mu^2/s)$, at least close to the Z peak. However, two remarks are in order here. Although a version of MOE has been released which also generates the muon (or, more generally, fermion) momenta, it has not yet been widely used, and a critical comparison of for instance the muon angular distribution, in the same way as was done for DYMU2, is absolutely necessary. Secondly, the communication between different people working in the Monte Carlo field is often hampered by the fact that separate groups have built up their own particular jargon. A statement like 'our cross sections factorize because the algorithm is unitary' will be understood differently by different Montecarloists, if it is understood at all. I notice a clear need for people to get together and unify our field a little: in fact, that is one of the aims of this workshop. Finally, I think that we still need an explicit *formula* for the distributions as they are in fact generated by the MOE Monte Carlo (or, equivalently, a precise description of the way the event weights are computed). In that respect, I think the DYMU2 program is still superior to MOE since it provides very clear answers to such questions.

Steven van der Marck presented the latest addition to the field, the FPAIR Monte Carlo. That program does not, at present, include exponentiation but it rather aims at a precise description of hard bremsstrahlung up to 2 loops, which is not implemented in any of the other programs. In particular the collinear radiation is treated very carefully. At present the program should be good for hard-photon phenomenlogy and such: a version including exponentiation and also final-state radiation is under way. A nice thing was the detailed comparisons possible between FPAIR and DYMU2 that we saw: again, the fact that DYMU2 restricts itself to 2 initial-state photons makes it excellent for this kind of work.

Ramon Miquel described the NNGG02 program, which is a quite specialized Monte Carlo for the neutrino-counting process $e^+e^- \to \nu\bar{\nu}\gamma(\gamma,\cdots)$. A very nice feature is that exponentiation is implemented in both the 'ad-hoc' way (which is here called 'inductive exponentiation') and with structure functions: it should be possible to obtain a lot of insight on the usefulness of 'ad-hoc' exponentiation using this program. The agreement with KORALZ appears to be quite satisfactory, which indicates that neutrino counting, at least, is understood well enough for LEP purposes (leaving aside, of course, the issues of the possible backgrounds which neither of the programs can address). A point that comes to mind is the following: a 1% agreement between NNGG02 and KORALZ, and a 2% agreement between NNGG02 and MOE suggests a 2% agreement between MOE and KORALZ. Is that in fact the case? It would be interesting to know but has not been studied as far as I know.

Staszek Jadach gave an exceptionally clear presentation of the beautiful YFS algorithm for 'exclusive exponentiation', and its relation to 'common', 'inductive', or 'ad-hoc' exponentiation (all these things mean the same, another example of random jargon development). His description of the YFS algorithm, as well as possibilities for checking it and making comparisons with other results, are again tremendously helped by the fact that analytical results for the s' spectrum contributions from the $\tilde{\beta}_0$, $\tilde{\beta}_1$, and $\tilde{\beta}_2$ exist. There are some ambiguities left due to $\tilde{\beta}_{3,4,...}$ and the '$\mathcal{R}$-procedure', but these are again minor. In my opinion, the YFS algorithm represents one of the major recent advances in electroweak-Monte-Carlo technology.

Sbyszek Wąs described the application of the above (and, of course, a lot more) to the study of τ polarization at LEP: he reviewed its principles, and the influence of QED radiative corrections on this very important measurement. Unfortunately, since no competing Monte Carlos exist for this process, comparisons are hard to make, but the semianalytical results from CALASY indicate that a very good job has been done.

Mario Greco presented the latest views on the connection between the use of structure functions and exponentiation using the coherent-state approach. He argues that also the interference between initial- and final-state radiation can be accomodated in this scheme, basically by the use of a K factor (stated in another way, by assigning an additional weight to the Monte Carlo events). In my personal opinion, this subject is still not understood well enough, and in particular it did not become clear to me how his results (valid for soft photons) should be applied to not-so-soft photon bremsstrahlung. I think that a lot of work is still needed on this problem, which may well become the most difficult one in the near future. It has been argued (mainly by Staszek Jadach and Sbyszek Wąs) that this interference can be neglected at the Z peak, for not too tight cuts, but this only marks the limits of the area where we can afford our lack of understanding, and will not avail us whenever we move out of it.

Manuel Martinez presented an appealing overview of the ingredients that are nowadays going into radiative corrections programs. It appears especially from his talk that we have indeed come a long way since PETRA and PEP! Although he did not dwell on any specific results, it can be said that our level of sophistication is finally approaching the one needed for reliable and precise LEP phenomenology — just in time.

A summary of the talks, ordered by subject

Let me now discuss the various physics subjects, and the status we can consider ourselves to be in. I shall also mention what I feel to be the most appropriate programs. Of course, what is the 'best' Monte Carlo depends both on your personal taste and what precisely you want to use it for, but a number of recommendations seem to be fair enough.

- **The line shape:** This is in good shape! We have at our disposal very accurate results from different groups that all agree with each other. The program KORALZ has been considerably improved by including the electroweak library due to Wolfgang Hollik and Wim Beenakker. Still, in my opinion the fact that now two differring results can be obtained from KORALZ depending on which library is used is a weakness rather than an asset: the authors should quickly decide which one is correct *and drop the wrong one!* A new ZBATCH-type program, called ZSHAPE, has been introduced, which is better from the accuracy (and programming) point of view: I would like to recommend it for replacing ZBATCH. In addition, the Dubna-Zeuthen group has released their set of semianalytical programs MUCUT/MUCUTCOS/DIZET/ZBIZON, which provide us with the line shape, and the angular distribution up to first order — a very impressive result. Another nice point is that KORALZ, DYMU2 and FPAIR now all support final-state fermions other than muons as well.

- **Fermion angular distribution and asymmetries:** we are getting there! The agreement between the various programs has improved a lot in the last few months, and moreover the new semianalytical results prove their invaluability already. As I have said above, MOE needs a good amount of checking and comparing before we can claim that it also gets the A_{FB} right.

409

- **Neutrino counting:** this seems to have been essentially settled as far as the signal is concerned. A study of the backgrounds is still under way: in the LEP reports we can expect an up-to-date review of those. In all, my feeling is that we will be able to count neutrinos with confidence, using KORALZ, NNGG02, or MOE, which should all agree to sufficient precision. Some more sophisticated background simulations may still be necessary, but that will only be a matter of time.

- **Small-angle Bhabha scattering:** especially in the last few weeks there has been very good, and very needed, improvement. The various programs OLDBAB, BABAMC, HOWLEEG, and BHLUMI now all agree to a level that will probably be sufficient. For symmetric cuts the $\mathcal{O}(\alpha)$ corrections are now agreed to be quite small, and exponentiation does not change that noticeably. Comparing the present status with what it was during the Ringberg meeting, we can be quite satisfied.

- **Large-angle Bhabha scattering:** a sad story! No single piece of real improvement has been reported here, or elsewhere. I think that this subject has not been receiving the attention that it deserves: this is basically due to the fact that it is not an easy subject, and can not be discussed at all without relying on the computer (note that even a thing like 'total cross section' cannot be defined for Bhabha scattering), and it has been quite impopular with a number of theorists for that reason. It would be sad to see interesting measurements of the electroweak paramaters from this process become impossible because of lack of theoretical input: at present, we have only BABAMC, and this is only good to $\mathcal{O}(\alpha)$ which we know from the muon/fermion case to be really insufficient at LEP.

A problem with the star scheme?

The thing that I want to discuss now has, in itself, not much to do with Monte Carlos, but rather with the underlying electroweak effects. It has been stressed several times during this workshop that there may be conceptual problems involved in the use of the so-called 'starred-scheme' introduced by Bryan Lynn and his coworkers. In particular it has been argued that this scheme violates gauge invariance. I think the best way to understand this is the following. In schemes like the 'starred-scheme' one tries to incorporate as much of the higher-order effects as possible into 'effective' or 'running' parameters such as a running QED coupling constant $\alpha(Q^2)$ or an effective $\alpha(M_Z^2)$, or into $\sin^2 \bar{\theta}_w$ or $\sin^2 \theta_w^*$. Most importantly, the photon and Z self-energy diagrams due to fermion loops are resummed to all orders: these diagrams, which indeed contain most of the large logarithms, are gauge-invariant. In the 'starred-scheme' one also resums the 'nonabelian' photon and Z self-energy graphs where a W^+W^- pair occurs in the loop. It can simply be seen that these diagrams are not gauge-invariant by cutting the s-channel diagram for $e^+e^- \to Z \to \mu^+\mu^-$ which such a Z self-energy insertion in two: as is well known, this gives us the imaginary part of the diagram, and, by the optical theorem, also the total cross section for $e^+e^- \to Z \to W^+W^-$, corresponding to just one Feynman diagram. It is common knowledge that this does *not* give the correct result for this total cross section: one has to add also the s-channel diagram with γ exchange, and the t-channel diagram with ν_e exchange, in order to get the correct gauge-invariant answer. Therefore, the imaginary part of the W^+W^- loop diagram for the Z vacuum polarization (and *a fortiori* the whole vacuum polarization) is not by itself gauge-invariant: a gauge invariant correction only results if we add to it the $\gamma - Z$ mixing graphs, the γ self-energy due to W's, the triangle graphs with γW^+W^- and ZW^+W^- couplings, and also the W^+W^- box graph. It is clearly not easy to include all those diagrams into an all-orders resummation! In the 'starred-scheme' part of the triangle and box graphs are included in the resummation, but it is generally agreed by most people that this does not make the result gauge-invariant.

Fortunately, one always has to pick a particular gauge when performing a computation. The authors of the 'starred-scheme' have chosen to use the 't Hooft-Feynman gauge: in this gauge, the results of the WW boxes, the WW loops and and triangle graphs are all comfortably small, at least at LEP energies. Resumming part of this small effect to all orders, while keeping another part of these small terms to first order only, will of course result in a small error! In fact, the error is obviously (small)^2 since the first order can be done exactly. So, using this lucky gauge, there is not really any need to worry about the gauge-noninvariance. In a (for this discussion) unlucky gauge, like the unitary gauge that is used by the Dubna-Zeuthen people, the WW loops and boxes, and the triangle graphs, are all individually *divergent*, and clearly it would be impossible to perform the above partial summation.

Some things come to mind, however. In the first place, if the effect of the W's is so small anyway, why bother to resum it at all? In fact this is advocated by people like Wolfgang Hollik: it is probably better from a consistency point of view to resum only the fermion loop contributions which are large, but gauge-invariant, and do the W loops only to first order. Secondly, the 'starred-scheme' is an attempt at an *effective* Lagrangian for the standard model, which should as much as possible be independent of a particular process, and therefore useful in many different processes. Unfortunately, most of the processes relevant at LEP have already been computed, so the advantages of having an effective Lagrangian there are not so clear. At higher energies (LEP200? LHC/SSC? CLIC?) many things have not been done yet, and an effective Lagrangian might be useful there — but at those high energies (where $\log(s/M_W^2)$ will be large just like $\log(s/m_f^2)$) the WW loops will also be large, even in the lucky 't Hooft-Feynman gauge, as well as the boxes and triangles! In that case the gauge-noninvariance problem pops up again, and we cannot be sure to be resumming in the right way in the 'starred-scheme'.

My personal view is that it is not particularly dangerous to use the 'starred-scheme', but there is no particular merit in it, either. Claims that this is the way in which electroweak radiative corrections must be computed are not, at present, justified.

Optimizing the k_0 problem?

In Patricia Rankins talk an interesting suggestion was made concerning the so-called k_0-dependence problem. For a discussion of this problem I would like to refer to the proceedings of the Ringberg meeting, but I want to make a remark on it here. It is well known that in many event generators there is a dependence of the result for the cross section on the value chosen for k_0, the cutoff between what one wants to call 'hard' bremsstrahlung and what one wants to call 'soft' bremsstrahlung. Ideally, the resulting cross section should not depend on this arbitrary cutoff. For too large values of k_0, however, unacceptable simplifications are made in the description of bremsstrahlung photons that are actually not so soft: for too low values of k_0, negative intermediate results for cross sections occur, which invalidate the Monte Carlo approach. If one is lucky, there is an intermediate region where one can choose any (small enough) k_0 value without running into positivity problems and without changing the resulting cross section, but in many cases there is always a slight dependence. It was argued that one could in fact try to 'optimize' the result by choosing such a k_0 as to give a cross section closer to the 'exact' one, that is, the result one would obtain using full exponentiation. I think this is **very** dangerous! The k_0 sensitivity is really there in the program: it is due either to the physical assumptions you put in by truncating the QED perturbation series, or (hopefully not, but possibly) to bugs in the program. **The 'best' choice for k_0 is always the smallest possible one.** This is in contrast to the QCD case, where for instance one can choose the Q^2 scale in a process to improve the convergence of the perturbation series: this Q^2 scale is a thing that has physical interpretation (although different people may differ on what it is), but the k_0 parameter is, and should always be taken te be, an in principle arbitrary number occurring somewhere in an algorithm, which puts a limit on the validity of the Monte Carlo approach. I think that playing around with the limitations of our knowledge, in order to mimick physical behaviour, is an extremely bad idea. An example: it is known that the Z peak is reduced by about 40% in first order QED: higher orders change this from -40% into

-30%, i.e. they contribute about +10% to the cross section. If one would like to mimick this by changing k_0, a value would be necessary that is either so very small that many negative weights would occur (which we can't handle), or it would have to be very close to 1, which is also clearly unacceptable. For other distributions, different k_0 would be necessary, and so on.

Some overall questions, and putative answers

I will now discuss some questions of a more general nature which were raised (not for the first time) during this week, and give you you my opinion on them, for what it's worth.

In the first place, Guido Altarelli has presented his view of how the radiative corrections are to be studied. Roughly speaking, his position is that the experimentalists should try to remove from their data the well-known and so-called 'uninteresting' QED effects, and then more or less hand ot over to the theorists, who would derive 'interesting' consequences like the value of M_t or the existence of a Z' boson. Clearly an experimentalist could possibly take the opposite stance. Instead, what I think is necessary (and, fortunately, happening) is that the experimentalists should appreciate theoretical issues, and vice versa. During the last few years, more and more of the theoretical uncertainties and ambiguities have become common knowlegde to a degree that is really impressive to me. The obligations between theorists and experimentalists are those of mutual education, not merely the transmission of Monte Carlos one way and 'cleaned-up' data the other way.

The second issue that was raised is the old one of how the data should be presented. In particular Friedrich Dydak has again made a strong point for the use of 'canonical cuts': **at least once** the 'raw' data should be made available, under a set of cuts that is as simple as possible, and common to at least the four LEP experiments. Of course, this is a politically tricky demand to make of four different experimental groups that all pride themselves on doing particular things better than the other groups, but for the sake of avoiding confusion like the one that arose at PETRA over the measurement of $\sin^2 \theta$ it seems to be essential that these suggestions (with which I agree wholeheartedly) be taken very seriously. The second-best approach would of course be to publish both 'uncorrected' and 'corrected' data together, as has become in fact the practice at PETRA where often the 'corrected' data and the corrections that were applied are stated more or less clearly. Of course it will always be difficult to extract information from plots of (often) quite complex experimentally determined quantities, but at the end of the day we would really like to see for ourselves that all the experiments were in fact compatible in their measurements rather than in the published value for, say, M_Z.

Another important point was the organization of the analysis. Should there be 'authorized' and 'definitive' Monte Carlos, to go for instance into the CERN program library? This would be of course a desirable situation, in which everybody would know quite unambiguously what programs were used in the data analysis of every experiment. Unfortunately, at this moment that seems to me not a viable approach. For one thing, most programs are still being developed: there are open issues, like the initial-final-state radiation interference, for which simply no definitive treatment exists yet. Also, programs that do not have continuous and enthousiastic support from the authors themselves tend to fossilize and die, as has happened to ZBATCH and, to a lesser extent, EXPOSTAR, which is mainly kept alive through the dedicated efforts of Alain Blondel. Also, simply stating 'we used such-and-such event generators' is not sufficient: a values for quantities like $\alpha_s(M_Z^2)$, M_t and M_H will always have to be used as parameters, and therefore mentioned in the presentation as well. In addition, many programs support many different modes (like the KEYRAD options in KORALZ) which are equally important and have to be specified. At this moment, the only feasible thing to do seems to perform the analysis, publish the results, and *store the program that was used, in the way it was used, and not to touch that version!* Probably, in about a year from now, most programs will attain something like a definitive status and can be put into the CERN program libarary as authorized versions — but by that time, many exciting LEP results will already have been published.

This finishes my talk. On the whole, I have got the impression that nothwithstanding the many issues that still have to be resolved, we are in quite good shape for the start of LEP, at least as far as the electroweak software is concerned. I would like to thank the organizers of the Brighton conference for supplying us with the stimulating environment in which we have been able to assess our situation at the eve of the exciting period which, I hope, we shall be entering.

TESTS OF QCD IN E^+E^- PHYSICS

Z. Kunszt* and P. Nason

Institute of Theoretical Physics
ETH, Höngg, Zürich
Switzerland

INTRODUCTION

QCD[1] has been tested in numerous experiments at PETRA, PEP and TRIS-TAN [2] [3] [4] [5] [6] [7] [8] [9]. The new experiments at LEP are expected to provide substantial improvements as a result of the higher energy, the much higher luminosity and the better detector resolutions and efficiencies.

In the reactions of e^+e^- annihilation the initial state is completely known. This is a considerable advantage over deep inelastic scattering or hard hadronic collisions, where the input structure functions cannot be calculated with perturbative methods. Therefore in e^+e^- annihilation into hadrons there are a number of quantities, like the total cross section and various infrared safe jet variables (thrust, spherocity, energy-energy correlations $etc.$), which can be calculated in perturbative QCD in terms of a *single parameter* $\Lambda_{\overline{MS}}$. The dependence on the long distance properties of the theory is expected to be negligibly small. They appear only as power corrections which become negligible at high enough energies. Measuring any of these quantities we also obtain a measurement of $\Lambda_{\overline{MS}}$ †. We are forced to consider infrared safe quantities since the perturbative QCD predictions are valid for quark and gluon final states while in the experiments hadron properties are measured. This is a major difficulty in testing QCD, which in general does not arise when the precision tests of the electroweak theory is considered.

An additional difficulty is that the coupling constant α_S is not very small. In many cases we are unable to calculate physical quantities in the necessary order of the perturbative expansion, required to meet the experimental accuracy. Therefore, we must assess also the theoretical uncertainties, due to unkown higher order corrections.

If we consider quantities which are defined in terms of a single large scale, then finite order perturbation theory is the appropriate method to use. If two or more widely different scales are involved, large logarithmic factors can arise which might destroy the convergence of the perturbative expansion. Fortunately in several important cases these large logarithms can be resummed in all order of perturbation theory, providing us with an improved perturbative treatment. For example, such a resummation is implemented in shower Monte Carlo programs [11] [12] [13], where the leading collinear and soft logarithms are correctly taken into account. Techniques are

---

*Presented by Z. Kunszt.

†For a recent summary on the measurements of α_S see for example Ref.[10].

Radiative Corrections, Edited by N. Dombey and
F. Boudjema. Plenum Press, New York, 1990

also available for calculating QCD effects in the back-to-back region (the so called Sudakov effects) [14], where the resummation of double and single logarithmic terms is carried out. The shower Monte Carlo models, however, which are well suited for the description of many detailed properties of two jet prodcution such as the hadron multiplicity distributions, the Q^2 evolution of the fragmentation functions *etc.*, provide us only with the pole approximation for the infrared safe jet variables. In principle, it is possible to have a Monte Carlo program which includes the complete next to leading order QCD corrections, however, such a Monte Carlo program is not yet available.

In this talk I shall consider only the simplest QCD predictions for LEP[15]

1. The total hadronic width.

2. Three jet dominated quantities, such as thrust, oblateness, heavy jet mass, energy-energy correlation, jet clusters and inclusive jet cross sections.

3. Four and five jet dominated quantities.

I note that there is an other important simple test of QCD namely the measurement of the Q^2 evolution of the fragmentation functions which I shall not consider here. Their measurement at LEP will provide us high Q^2 values such that the effects of the standard QCD evolution equations can be tested.

THE TOTAL HADRONIC WIDTH

At lower energy measurements the total hadronic width is usually normalized to the lowest order uncorrected muonic cross section. For LEP one can discuss directly the hadronic width of the Z boson which appears in the formulae of the line shape, and of the hadronic cross section. In leading order of perturbation theory assuming massless quarks the hadronic width is

$$\Gamma_h^0 = \frac{G_F M_Z^3}{24\pi\sqrt{2}} 3 \sum_i (a_i^2 + v_i^2)$$

(1)

The vector and axial coupling of fermions to the Z boson are given in Table 1.

Table 1. Vector and axial coupling of the Z boson to fermions.

f	v_f	a_f
ν	1	1
$e\,\mu\,\tau$	$\left(-1 + 4\sin^2(\theta_W)\right)$	-1
$u\,c\,t$	$\left(1 - \frac{8}{3}\sin^2(\theta_W)\right)$	1
$d\,s\,b$	$\left(-1 + \frac{4}{3}\sin^2(\theta_W)\right)$	-1

There are three kinds of corrections to the hadronic width that we must consider: the strong corrections, the mass corrections, and the electroweak corrections.

The effect of the electroweak corrections is a small: less than a part in a thousand.

The strong corrections to R are given by the formula:

$$\frac{\Gamma_h^0 + (\delta\Gamma_h)^S}{\Gamma_h^0} = 1 + \frac{\alpha_S}{\pi} + c_2\left(\frac{\alpha_S}{\pi}\right)^2 + c_3\left(\frac{\alpha_S}{\pi}\right)^3$$

$$= 1 + \frac{\alpha_S}{\pi} + (1.986 - .115 n_f) \left(\frac{\alpha_S}{\pi}\right)^2$$

$$+ (70.986 - 1.2 n_f - .005 n_f^2) \left(\frac{\alpha_S}{\pi}\right)^3 + \tilde{c}_2 \left(\frac{\alpha_S}{\pi}\right)^2 + \tilde{c}_3 \left(\frac{\alpha_S}{\pi}\right)^3. \quad (2)$$

Here and in what follows α_S stands for the running coupling in the $\overline{\text{MS}}$ scheme with five flavours, evaluated at the mass of the Z, $\alpha_S^{(5)}(M_Z^2)$. There are certain type of corrections which do not appear in the case of pure vector coupling. They are deonoted by $\tilde{c}_2$ and $\tilde{c}_3$. These contributions are given by the classes of graphs depicted in Fig. 1, and they are the only terms which receive different contributions from the axial and

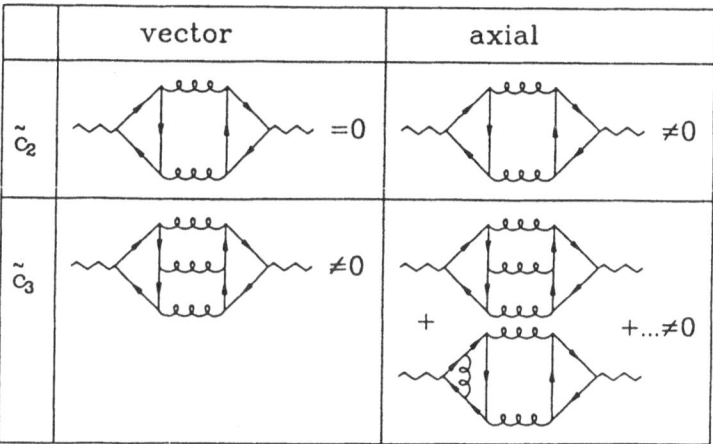

	vector	axial
$\tilde{c}_2$	$=0$	$\neq 0$
$\tilde{c}_3$	$\neq 0$	$+ \quad + ... \neq 0$

Fig. 1. The diagrams that contribute differently to the axial and to the vector part of the Z coupling. The curly lines represent gluons, and the jigsaw lines denote the weak current.

vector part of the Z coupling. The vector contribution was calculated to order α_S^2 by different groups[16], and is by now a well established result. The contribution of order α_S^3, including the vector contribution of $\tilde{c}_3$, was obtained recently by one group[17]. The term $\tilde{c}_2$ was calculated in Ref.[18]. It has the form

$$\tilde{c}_2 = -\frac{a_b^2}{\sum_q (v_q^2 + a_q^2)} \frac{1}{3} I_2 \left(\frac{M_Z}{2 M_t}\right) \quad (3)$$

where I_2 has the approximate expression

$$I_2 \left(\frac{M_Z}{2 M_t}\right) = 9.25 - 1.037 \left(\frac{M_Z}{2 M_t}\right)^2 - .632 \left(\frac{M_Z}{2 M_t}\right)^4 - 6 \log \left(\frac{M_z}{M_t}\right). \quad (4)$$

It has been found that $I_2/3$ varies between 2 and 5 for M_t between 60 and 250 GeV.

The term $\tilde{c}_3$ has the form

$$\tilde{c}_3 = 1.679 \frac{\left(\sum_q v_q\right)^2}{\sum_q (v_q^2 + a_q^2)} + \frac{\left(\sum_q a_q\right)^2}{\sum_q (v_q^2 + a_q^2)} I_3 \left(\frac{M_Z}{2 M_t}\right). \quad (5)$$

A complete calculation of I_3 does not exist; although some partial results are available*.

---
*See Kühn's talk in this volume.

Quark mass corrections to the hadronic width of the Z boson have been calculated long time ago. The vector part was found by Schwinger[19] in QED (the correspondig QCD result is obtained by simply changing the coupling and introducing an appropriate color factor.) Results for more general cases, which include the vector and axial current, are given in Ref.[20], and are consistent with other determinations[21]. They all agree with Schwinger's result for the vector part.

Here we only present the small mass limit which can be easily obtained from Ref.[20], and is adequate for the bottom contribution. If a top quark with mass less than $M_Z/2$ were found, one should refer to the exact result. Expanding in m^2/S the contribution of a single massive quark to the vector and axial part of the width we get

$$\frac{\Gamma_{f\bar{f}}^V(m)}{\Gamma_{f\bar{f}}^V(0)} = 1 + \frac{\alpha_S}{\pi} 3 \frac{4m^2}{S} \tag{6}$$

$$\frac{\Gamma_{f\bar{f}}^A(m)}{\Gamma_{f\bar{f}}^A(0)} = 1 - \frac{3}{2}\frac{4m^2}{S} - \frac{\alpha_S}{\pi}\frac{3}{2}\frac{4m^2}{S}(1 + 2\log(m^2/s)). \tag{7}$$

The logarithmic term is a running mass effect. Hence it can not appear in the vector part since the term proportional to m^2/S is absent in zero order in α_S .

It is instructive to give the numerical evaluation of all the corrections available the hadronic width. We have obtained

$$\Gamma_h = \frac{G_F M_Z^3}{24\pi\sqrt{2}} 3 \sum_q (a_q^2 + v_q^2)(1 + \delta_f) \times$$

$$\left\{ 1 \right.$$

$$+ \frac{\alpha_S}{\pi}(1 + .449\alpha_S + 6.57\alpha_S^2) \qquad = .035 + .00173 + .002778$$

$$- \left(\frac{\alpha_S}{\pi}\right)^2 \frac{a_b^2}{\sum_q(a_q^2+v_q^2)} \frac{I_2\left(\frac{M_Z}{2M_t}\right)}{3} \qquad = -1.8\,10^{-4} \frac{I_2\left(\frac{M_Z}{2M_t}\right)}{3}$$

$$+ \left(\frac{\alpha_S}{\pi}\right)^3 \frac{1.679\left(\sum_q v_q\right)^2 + a_b^2 I_3\left(\frac{M_Z}{2M_t}\right)}{\sum_q(a_q^2+v_q^2)} \qquad = 1.89\,10^{-5} + 6.46\,10^{-6} I_3\left(\frac{M_Z}{2M_t}\right)$$

$$- \frac{4m_b^2}{M_Z^2} \frac{3}{2} \frac{a_b^2}{\sum_q(a_q^2+v_q^2)} \qquad = -2.6\,10^{-3}$$

$$\left. + \frac{4m_b^2}{M_Z^2} \frac{\alpha_S}{\pi} \frac{3v_b^2 - \frac{3}{2}a_b^2\left(1 + 2\log\frac{m_b^2}{M_Z^2}\right)}{\sum_q(a_q^2+v_q^2)} \right\} \qquad = 1.02\,10^{-3} \tag{8}$$

where δ_f includes the effect of the electroweak corrections, and it is less than a part in a thousand. The numerical values in the formula are obtained with $\alpha_S = 0.11$, $M_Z = 91$GeVand $m_b = 5$GeV.

One may speculate that the difficulty that the second subleading QCD correction is larger than the first one, signals the breakdown of the perturbative expansion*. On the other hand, the first subleading correction in R is unusually small[†] in comparison with corrections obtained for other quantities. Hence one may also argue that this is the reason why the second subleading term is larger than the first one. It is very difficult either to support or to refuse these kind of speculations. Accepting nevertheless the latter interpretation, we can estimate the even higher order corrections.

---

*See West's talk in this volume.

[†]See the following section on jet production.

Let us assume somewhat arbitrarily that the third subleading correction has the form α_S^4 times a coefficient of order one, say between minus five and plus five. The coefficient of α_S^3 is indeed around two, so that this is reasonable. Assuming $\alpha_S = .12$, the corresponding correction to R would be in this case $\pm.1\%$, just within our goal of accuracy.

In closing we consider the methods proposed [22] to measure R at LEP. First of all, one can extract the hadronic width from the value of the total width, obtained from a fit to the line shape. Assuming [22] that one can get the total width with an error $\delta\Gamma = 50\,\mathrm{MeV}$, the corresponding uncertainty in the hadronic width is

$$\frac{\delta\Gamma_h}{\Gamma_h} = \frac{50}{1789} = 2.8\%. \tag{9}$$

Since the size of the radiative QCD corrections to Γ_h is small ($\delta\Gamma_h/\Gamma_h \approx \frac{\alpha_S}{\pi} = .035$), the corresponding uncertainty in the determination of α_S is large ($\approx 70\%$).

A second method is to extract the hadronic width from a direct measurement of the hadronic cross section. In the peaking approximation we obtain

$$\int \sigma(e^+e^- \to \text{hadrons})dE = \frac{6\pi^2}{M_Z^2}\frac{\Gamma_h\Gamma_{e^+e^-}}{\Gamma_{\text{TOT}}}. \tag{10}$$

In the right hand side of eq.(10) the hadronic width appears both in the numerator and in the denominator. Hence there is a reduced sensitivity to the strong corrections. In leading order in α_S one has

$$\frac{\Gamma_h}{\Gamma_{\text{TOT}}} = \frac{\Gamma_h^0}{\Gamma_{\text{TOT}}^0}\left(1 + \frac{\alpha_S}{\pi}\frac{\Gamma_{\text{leptons}}}{\Gamma_{\text{TOT}}}\right) = \frac{\Gamma_h^0}{\Gamma_{\text{TOT}}^0}\left(1 + .3\frac{\alpha_S}{\pi}\right). \tag{11}$$

This method is therefore applicable only if the absolute luminosity is very well measured, so as to compensate for the above mentioned effect.

Perhaps the most promising possibility is to consider the ratio of the hadronic events to the leptonic events. This quantity is directly related to the ratio of the hadronic width to the leptonic width. In order to get the value of α_S with a precision of 10%, one needs to measure this ratio with a precision of about .5%*. Whether we consider the total width, the total cross section, or the ratio of hadronic to muonic events, electroweak effects should be always taken appropriately into accounts.

QCD DESCRIPTION OF THREE JET PRODUCTION

The smallness of the QCD coupling constant α_S at high energy implies that the hadronic final states in e^+e^- annihilation can be classified into distinct topological signatures characterized by the formation of increased number of jets of hadrons[23][24][25].

The majority of the events have two jet structure with an angular distribution with respect to the beam axis $(1+\cos^2\theta)$ typical of spin $\frac{1}{2}$ quarks[26]. A small fraction of events have planar toplogy and are dominated by three jet production. The production of three jet events at PETRA was predicted by the theory and its observation gave a rather direct evidence for the existence of the gluons[23]. The spin of the gluon has been tested by measuring the angular distributions of the jets. For example, the observed angular distribution[27] in the Ellis-Karliner angle[28] $\bar{\theta}$ is in agreement with the behaviour of the emission of spin one gluons but it is in disagreement with the emission of spin zero gluons. To test the details of the three jet matrix element of perturbative QCD many other topological correlations have been studied.

---

*In Ref.[22] the more pessimistic value of 1.5% was assumed.

The applicability of perturbative QCD to the description of jet production[29] is based on a cancellation theorem (strong version of the KNL theorem[30][31]) which ensures that in appropriately defined physical quantities the contributions of soft and collinear parton emissions cancel[32]. Consequently, infrared safe quantities defined in terms of quarks and gluons can reliably be calculated in any given fixed order of perturbation theory. In the real world, however, we have hadrons. This difficulty is avoided by a technical assumption based on asymptotic freedom. We assume that the perturbative short distance partonic distributions of infrared safe quantities defined in terms of partons can be identified with the corresponding physical quantities defined in terms of hadrons. This assumption might be valid provided we average the result over resolution bins of width large enough for the perturbative result to be justifiable ("bin smearing"). In more physical terms, we assume that for infrared safe quantities, the confinement effects only limit the angular and energy resolution of our "quark and gluon detectors". It is expected that the non perturbative effects are power suppressed with some power of $1\text{GeV}/Q$ where Q is the typical physical scale of the problem in GeV.

In the approach of Sterman and Weinberg[29], the aim was to find an appropriate definition of jet which is well defined to all orders in perturbation theory. Introducing finite jet resolution parameters the averaging over the non-perturbative effects may be taken into account. One does not need to use the concept of a jet necessarily. There is a wide variety of infrared shape variables whose distributions are calculable in perturbative QCD[33][34] and which, after appropriate bin smearing, can be directly compared with the data. The only requirements for such quantities are that they should be infrared and collinear safe.

Theoretically, the most interesting infrared and collinear safe quantities are those which are dominated by the three jet configurations. In leading order of QCD, these quantities are of $\mathcal{O}(\alpha_S)$. Therefore $\Lambda_{\overline{\text{MS}}}$ can be extracted meaningfully from the data using these quantities only if we know the theoretical prediction at least to $\alpha_S{}^2$. The next to leading order corrections to the production of three partons have been calculated by Ellis Ross and Terrano (ERT)[35]. Their calculation is valid for quantities which do not depend upon the orientation of the final state. The analytic results of ERT are the input in the numerical evaluation of the physical cross sections. The integration of the ERT formulas is difficult. Distributions in infrared and collinear safe quantities are obtained as 5-dimensional intagrals over a sum of functions which are separately divergent. The divergences cancel statistically in the sum. The coefficient of the terms of order $\alpha_S{}^2$ in the distributions in infrared and collinear safe quantities can be calculated once and for all with the required accuracy. An improved treatement is reported in Ref. [15].

The ERT calculation has been checked by the completely numerical analysis of Vermaseren et al.[36] . They calculated the thrust distribution and their result have been in agreement with the evaluation of the thrust distribution based on the ERT formulae[37][38]. More importantly the analytic formulae of ERT have been positively confirmed by Fabricius et al. (FKSS)[39]. Hence it can be considered a well established result.

At the Z pole, strictly speaking, the ERT fromulas are not enough to give the complete $\mathcal{O}(\alpha_S{}^2)$ corrections. The diagram in Fig. 1 that contributes to the term $\bar{c}_2$ in the total width, also contributes, which is not known yet. In what follows we shall assume that such terms are small so they will be neglected.

There is a large variety of distributions of soft and collinear safe quantities which can be predicted by the theory. From a theoretical point of view, a good strategy for testing QCD at LEP, would be to measure a large number of such distributions, compare them to the theoretical predictions, and look at the general agreement (or disagreement) of the theory with the experimental results.

The strategy followed in many of the previous e^+e^- analysis was different. It was

420

attempted to implement directly the radiative corrections into Monte Carlo programs [40] [41] [42] (the so called matrix elements (ME) Monte Carlos), and to use appropriate fragmentation models in order to give a complete description of the events. In these studies first appropriate parton clusters were defined, in order to separate two, three and four clusters events. Each cluster is than hadronized according to a certain model. Whatever hadronization model was used, it was tuned to reproduce the bulk of the data, which is in the two jet regime. This approach has many advantages, especially from an experimental point of view. In view of the many other ambiguities at lower energies this procedure can be justified. It must be clearly stated, however, that certain features of this method is in contradiction whith the principles of perturbative QCD, and it is bound to fail at higher energies. Let us discuss in more detail where is the weakest point of this method. In the ME Monte Carlo programs the multiplicity of the partons is given by the starting configuration. It can only be two, three or four. Therefore the full hadronic final state multiplicity is generated by the hadronization model. Furthermore the virtuality of the initial partons are larger and larger as the incoming energy is increased. This is not valid in general. We know that perturbative effects, collinear splitting and soft emission of partons influence the multiplicity at the perturbative level, and non-perturbative hadronization effects come into play only when the virtuality of the partons is as low as the typical hadronic scale. Shower Monte Carlo programs[11][12] [43][13] are useful tools to examine this problem. In particular, they also provide a model of the two, three, and four parton production in the pole approximation which is accurate only near the collinear limit (and therefore less accurate at the parton level than ME Monte Carlos*). However, they also describe within the same approximation the production of more then four partons which are not included at all in the ME Monte Carlo programs. In the study of Ref.[15] it was found that the average number of partons produced perturbatively, i. e. before their virtuality is below 1 GeV is 6 at PETRA energy, while it is 9 at LEP. If 6 may be considered a good approximation to 4, 10 is not. When fragmentation models are tuned to fit the data in ME Monte Carlos, they are forced to include not only non-perturbative effects, but also some genuine higher order, high transverse momentum effects, which are not contained in the 3 and 4 parton matrix elements.

Since in the case of the three jet dominated quantities at LEP energies, neither the shower Monte Carlos nor the ME Monte Carlos are good enough to describe the data, we suggest that, instead, one should try to compare directly the parton level QCD prediction with the data. In this case first one should carefully estimate the uncertainties in the QCD predictions. Then one should separately study the size of the higher twist contributions by investigating the energy dependence of the shape variable distributions. Finally one should estimate the hadronization effects using for example the shower Monte Carlo models.

We have written a new efficient[45] integration program for the ERT formulae. We have modified the pole terms of the ERT formulae in order to achieve better convergence of the integrals. The program is a simple Monte Carlo integration with some important sampling at the parton level. Therefore any shape variable or jet variable distribution can be calculated with the help of a single program with the required accuracy. We have estimated theoretical errors due to even higher order QCD effects in two ways: by changing the subtraction scale μ in an appropriate range, and by a method which uses shower Monte Carlo programs [11] [12] [43] [42]. In this last method, (see Ref.[15]) the higher order corrections are estimated via the study of the sensitivity to Q_0 which denotes the scale at which the shower developement is stopped. The sensitivity to hadronization effects, however, has been estimated by switching on and off the hadronization which takes place after the shower is stopped. In this way we can relatively well separate the effects of the higher order corrections, and the effects of the hadronization.

We shall consider thrust, oblateness, energy-energy correlations, broad mass

---

*The so called NLO schower Monte Carlos [44] include the next to leading logarithmic corrections to the shower but they do not include $\mathcal{O}(\alpha_S{}^2)$ corrections to three and four parton production.

square (with respect to the thrust axis), jet cluster distributions. We shall also study inclusive jet production, where the infrared safe, calorimetric jet definition of Refs.[46,47] is used. The study of the calorimetric jet definition is motivated also by the fact that the bulk of the jet data at hadron colliders (Sp$\bar{\text{p}}$S, Tevatron) is presented with the help of some explicit jet definition. It is therefore important that the e^+e^- jet data are also analysed in terms of similarly defined jets. In this way we can establish a connection between the e^+e^- and $p\bar{p}$ jet data.

Thrust distribution

The weighted differential cross section in thrust, at the renormalization scale μ is given by

$$\frac{1}{\sigma_0}(1-T)\frac{d\sigma}{dT} = \frac{\alpha_S(\mu)}{2\pi}A_T(T)$$
$$+ \left(\frac{\alpha_S(\mu)}{2\pi}\right)^2\left[A_T(T)2\pi b_0\log(\mu^2/S) + B_T(T)\right] \quad (12)$$

where σ_0 is the leading order cross section of the reaction

$$e^+ + e^- \to \text{hadrons} , \quad (13)$$

and

$$b_0 = \frac{33 - 2n_{\text{f}}}{12\pi} . \quad (14)$$

A_T and B_T are universal functions, which only depend upon the perturbative structure of the theory. The weight $(1-T)$ was introduced for convenience to damp the fast growth of the distribution near the $T \to 1$, two jet region.

We have given explicitly the full dependence upon the subtraction scale μ in eq. (12). The physical quantity expressed by eq. (12) must be independent on μ, within the order of the accuracy of the equation. To order α_S^2 eq. (12) satisfies the renormalization group equation. The structure of the perturbative expansion does not tell us precisely at which scale μ should we evaluate our formula. However, a change of μ with less than an order of magnitude will generate a change of order α_S^3 in the value of the cross section section. Various schemes have been proposed, which give recepies for the best choice of the subtraction scale. The only thing we know for sure about the choice of μ is that it should be of the order of the scale involved in the problem. With this choice we avoid the presence of large terms of the order $(\alpha_S \log(\mu/S))^n$ to all order in the perturbative expansion.

The leading order contribution is known in analytic form[25] Our program generates three and four parton events with an appropriate weight (not necessarily positive). In order to obtain the function B_T, one calculates the thrust of each event, multiplies the weight by $1-T$, and adds it to the appropriate bin in T. The function B_T is shown in Fig. 2. We have also indicated in Fig. 2 a second histogram (barely visible in the case of thrust), which gives the calculated standard deviation for the corresponding bin. (For more details see Ref.[15].)

In Fig. 3 we plotted the predictions of QCD for LEP energy $M_Z = 91\text{GeV}$ at three different values of μ, $\mu = M_Z/4$, $M_Z/2$ and M_Z, and $\Lambda_{\overline{\text{MS}}} = .140\text{GeV}$ (this is the value we will assume throughout our discussion). The three curves for different values of μ give an estimate of the higher order effects. We can see that at LEP energy, this corresponds to an ambiguity of roughly 10%. We prefer to choose $M_Z/2$ as our central value, since the energy involved in the strong processes is always somewhat smaller than the total annihilation energy.

That the effect of hadronization at LEP can be studied using the shower Monte Carlo programs. In these programs, from an initial parton with virtuality Q, a shower

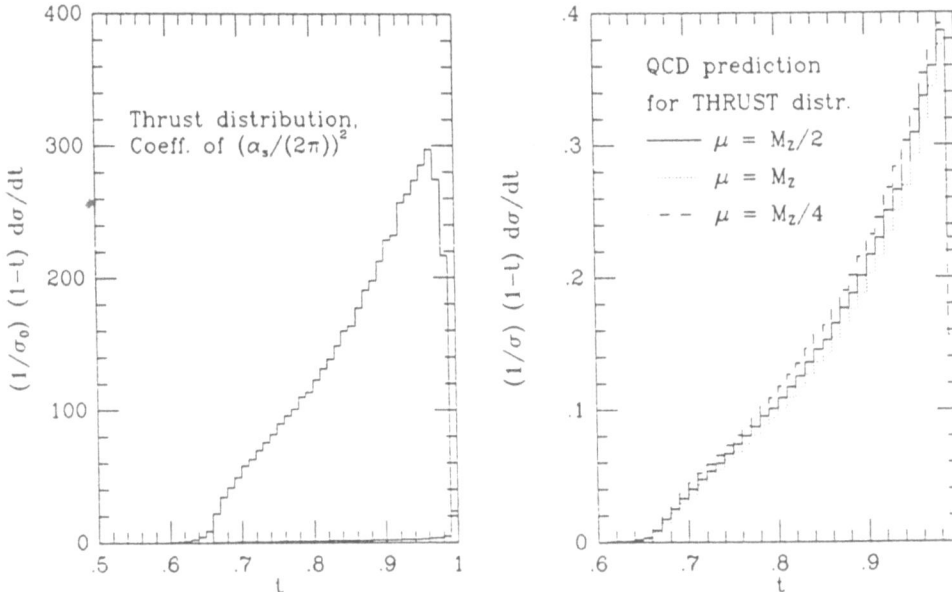

Fig. 2. Histogram for the function $B_t(t)$.

Fig. 3. Physical prediction for thrust distribution at LEP.

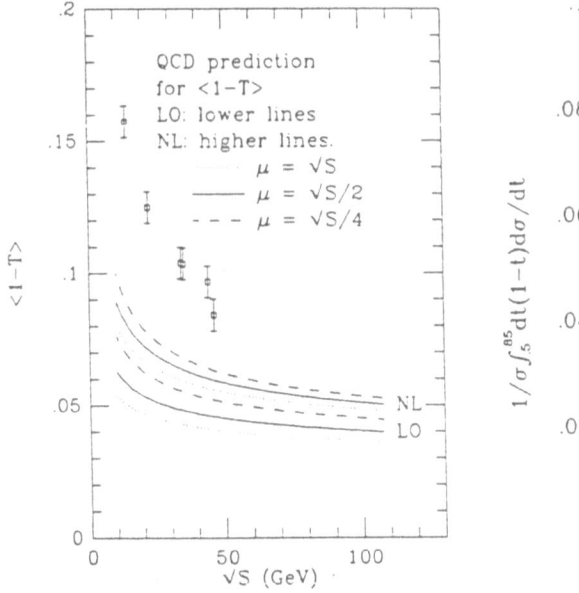

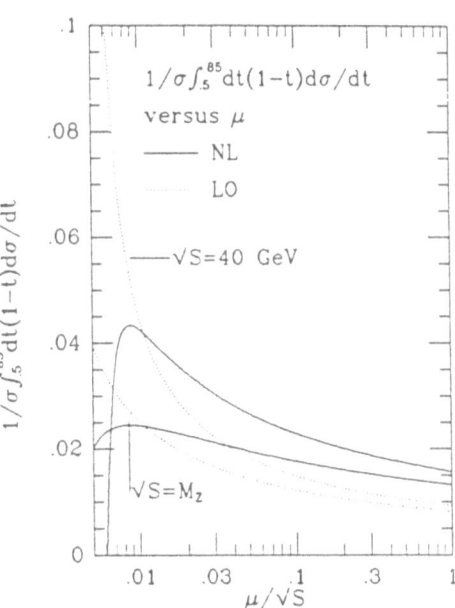

Fig. 4. $< 1 - t >$ and data points versus $\sqrt{S}$.

Fig. 5. Scale dependence of thrust at PETRA and LEP energies.

of partons of decreasing virtuality is developed. The shower is terminated at some small value Q_0 of the virtuality of the partons, and the hadronization is performed by a cluster algorithm or by string fragmentation. In order to simplify our discussion we shall only consider the average value of thrust $< 1 - T >$. The perturbative prediction for $< 1 - T >$ is given in Fig. 4, as a function of the annihilation energy, at three different values of the renormalization scale μ. We also plotted the same figures the data points (which are taken from Ref.[48]). The energy dependence of the data points and the difference between the theoretical and experimental values clearly indicates that at PETRA and PEP energies the higher twist contributions are very important. However, at LEP their effect might become marginal.

The effects of hadronization at LEP energy has been estimated by switching on and off the hadronization algorithm of HERWIG[15]. The corresponding effect for $< 1 - T >$ is about 3%. In order to estimate the higher order effects, one should calculate the change in the HERWIG prediction when the shower is stopped at the level of 4 partons, with respect to the result when the shower is developed down to the hadronization scale. This can be done with HERWIG by changing Q_0. The result for $< 1 - T >$ is about 16%.

We conclude that the ambiguity in the next to leading order perturbative QCD calculation due to higher order effects is larger than the size of the estimated hadronization effects. Therefore, we suggest that the thrust distribution as given on Fig. 3 and the value of $< 1 - T >$ given in Fig. 4 should be directly compared with the $e^+ e^-$ hadronic data (maybe after correcting for the b quark mass and the B decays). The Monte Carlo programs should only be used for unfolding the raw data. In Fig. 3, the shape of the $\sqrt{S}$ dependence of the perturbative QCD prediction for $< 1 - T >$ reflects the $\sqrt{S}$ dependence of α_S. Assuming $\Lambda^{(5)} = 0.14 \text{GeV}$. its value at LEP is predicted to be in the interval .05-.055. Comparing the leading order (LO) and the next-to-leading order (NL) prediction in the figure, we note that the K-factor of $< 1 - T >$ even at the Z-pole is rather large: $K = 1.35 - 1.4$ depending on the scale choice. We notice that the uncertainty in the theoretical prediction for $< 1 - T >$ is reflected directly in the uncertainty of α_S. Thus, considering only theoretical errors, one expects to be able to determine α_S within 10% from $< 1 - T >$. Such an accuarcy in the value of α_S can be translated into an $\approx 50\%$ accuarcy in the value of $\Lambda_{\overline{MS}}$.

The scale ambiguity may be even larger. A rather extreme example is given by the scale dependence of the integrated cross section $\int_{.5}^{.85} dT (1 - T) \frac{1}{\sigma} \frac{d\sigma}{dT}$. In Fig.5 we plotted its value as a function of $\mu / \sqrt{S}$ for $\sqrt{S} = 40 \text{and} 91 \text{GeV}$ in leading and next to leading order. The maximum of the next to leading order curves corresponds to the optimal scale μ_{opt} of the minimal sensitivity principle. It may appear somewhat disturbing that the maximum of the next to leading order curves is at the scale value less than 1 GeV. This is due to the large ($\approx 45\%$) positive correction. This does not mean that the the perturbative result has to be abandoned. Instead we should recall that at the next to leading order the MS principle gives a unique correspondence between the size of the correction and the "best" choice of the scale. If a physical cross section is parametrized as

$$d\sigma = C \alpha_S^n (1 + r \alpha_S))$$

then the MS principle gives the relation

$$\mu_{opt} = \sqrt{s} e^{-\frac{r}{2 n b_0} - \frac{1}{2(n+1) b_1}}$$

where b_0 is given by eq. (14) and $b_1 = \frac{153 - 19 n_f}{24 \pi^2}$.

At the next to leading order if the correction r is large and positive the optimal scale is small if the correction is large and negative the optimalization scale is large. We note, also, that for the optimalization it is significant what one chooses to calulate in the first place. Clearly there are always independent higher order effects that cannot be anticipated with the intuition, guesswork and prejudice needed for the application of the minimal sensitivity principle and alike.

The considerations of this section can be repeated for any other shape variable or three jet dominated quantity. It is important to explore also other shape variables, since in next to leading order different quantities might have rather different corrections.

Oblateness

Oblateness was used by the MARK - J experiment[49] in their jet analysis of the hadronic events.

The oblateness distribution can be given in the standard form similar to the case of thrust

$$
\frac{1}{\sigma_0} O \frac{d\sigma}{dO} = \frac{\alpha_S(\mu)}{2\pi} A_O(O)
$$
$$
+ \left(\frac{\alpha_S(\mu)}{2\pi}\right)^2 \left[A_O(O)2\pi b_0 \log(\mu^2/S) + B_O(O)\right] .
$$
(15)

In Fig. 6 we show the value, B_O as histograms. It is remarkable that the next to leading corrections are negative for all values of oblateness. (For more details see Ref.[15].) Fig. 7 shows the energy dependence of the average value of oblateness $< O >$ at the usual three different choices of the scale μ. At $\mu = \sqrt{S}$ we have found

$$
< O >= (1.29 \pm 0.002)\alpha_S(S)[1 - (4.3 \pm 0.15)\alpha_S(S)] .
$$
(16)

The large negative correction is an interesting prediction of the theory and the consistency of this prediction can be tested at LEP. It appears that from the study of the oblateness α_S can be extracted with $\pm 10\%$ accuracy.

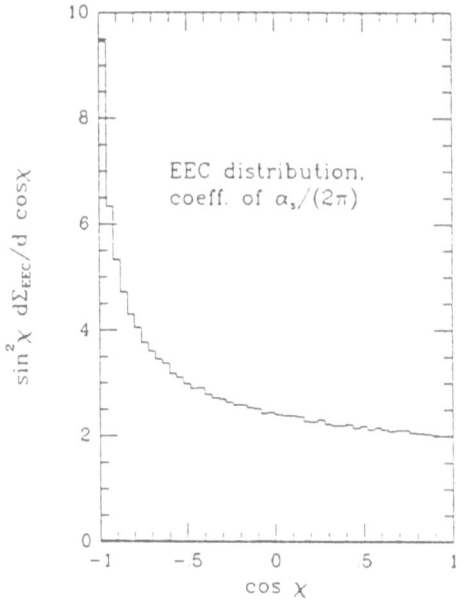

Fig. 6. Physical prediction for oblateness distribution at LEP.

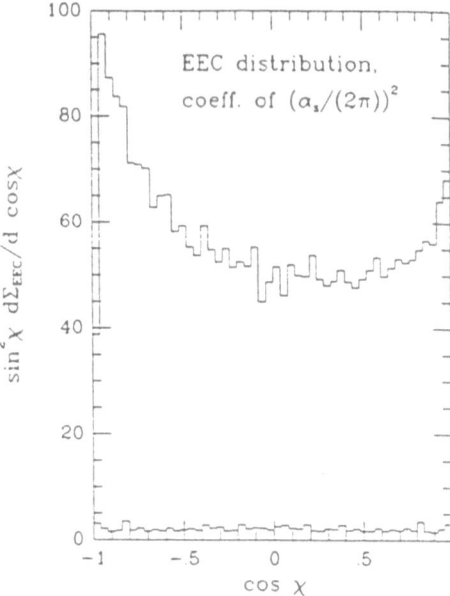

Fig. 7. Average values of oblateness versus the energy.

Energy-energy correlation

The energy energy correlation was proposed in Ref.[50], and several experimental groups have studied it[51,52,53]. As usual, we express the cross section as

$$\frac{\sin^2 \chi}{\sigma_0} \frac{d\Sigma}{d\cos\chi} = \frac{\alpha_S(\mu)}{2\pi} A_{\text{EEC}}(\cos\chi)$$

$$+ \left(\frac{\alpha_S(\mu)}{2\pi}\right)^2 \left[A_{\text{EEC}}(\cos\chi)2\pi b_0 \log(\mu^2/S) + B_{\text{EEC}}(\cos\chi)\right] + \mathcal{O}(\alpha_S^3). \quad (17)$$

The lowest order term is given by[50]. The histograms for A_{EEC} and B_{EEC} are plotted in Fig. 8 and 9. Two works[54][55] have been published the radiative corrections to the EEC distribution. We find a value for B_{EEC} which is systematically larger than Ref.[54] by about 20%. Also the authors of Ref.[55] claim to find larger values then those in Ref.[54]*

The asymmetry in the energy-energy correlation (AEEC) is usually defined as

$$\frac{d\Sigma_{\text{AEEC}}}{d\cos\chi}(\chi) = \frac{d\Sigma_{\text{EEC}}}{d\cos\chi}(180^\circ - \chi) - \frac{d\Sigma_{\text{EEC}}}{d\cos\chi}(\chi) \quad (18)$$

It has smaller radiative corrections then the EEC itself. From Fig. 8,9 it is apparent that the radiative correction term is more symmetric than the lowest order. It has been argued[55] that non-perturbative hadronization effects may also be much smaller for the asymmetry. This conclusion was based upon the sensitivity of the EEC to an energy resolution parameter ϵ. It was found that while the introduction of an energy resolution parameter modifies the energy-energy correlation to the order ϵ, the asymmetry is modified to order ϵ^2. From the HERWIG Monte Carlo study, we find that the influence of the parton shower developement is 16% for the EEC, and 1% for the AEEC, while the effect of hadronization is 1% for the EEC and -7% for the asymmetry, when the initial process is $q\bar{q}$. For the $q\bar{q}g$ initial process with a thrust cut, we find 12% and -15% effect of the shower developement on the EEC and on the AEEC respectively, while the hadronization effect is 1% for the EEC and -6% for the AEEC. We may conclude that the EEC seems to be fairly insensitive to hadronization, while the asymmetry has a modest sensitivity. On the other hand, shower developement seems to influence more the EEC, although the result for the asymmetry are inconclusive. From this comparison, we learn that it is very difficult to estimate effects for which we have a very poor understanding. There is also an experimental indication that non-perturbative effects may be smaller in the asymmetry than in the energy-energy correlation[57]. It is found that the energy dependence of the AEEC shows smaller power type corrections than the EEC. The LEP experiments will help in clarifying this issue.

In Fig. 10 we report the pure QCD prediction for the EEC distribution at LEP. Again we see that the width of the band of sensitivity to a scale change varies from 10 to 20%. For the asymmetry the scale sensitivity is much smaller, since the radiative correction itself is smaller, as can be seen in Fig. 11.

Heavy jet mass

The heavy jet mass as shape variable has been proposed in Ref.[58]. It is easier to implement in practice a slightly different definition of the heavy and light masses using thrust . Instead of minimizing the value of the sum of the invariant masses of two

---

*We also note that our result is in complete agreement with the recent new evaluation of Ref.[56].

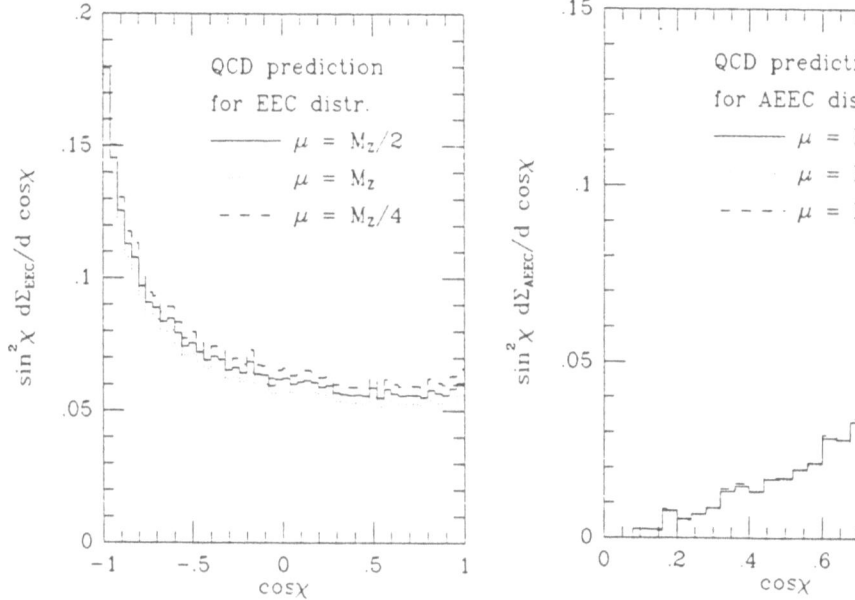

Fig. 8. Histogram for $A_{EEC}\cos(\chi)$.

Fig. 9. Histogram for $B_{EEC}\cos(\chi)$.

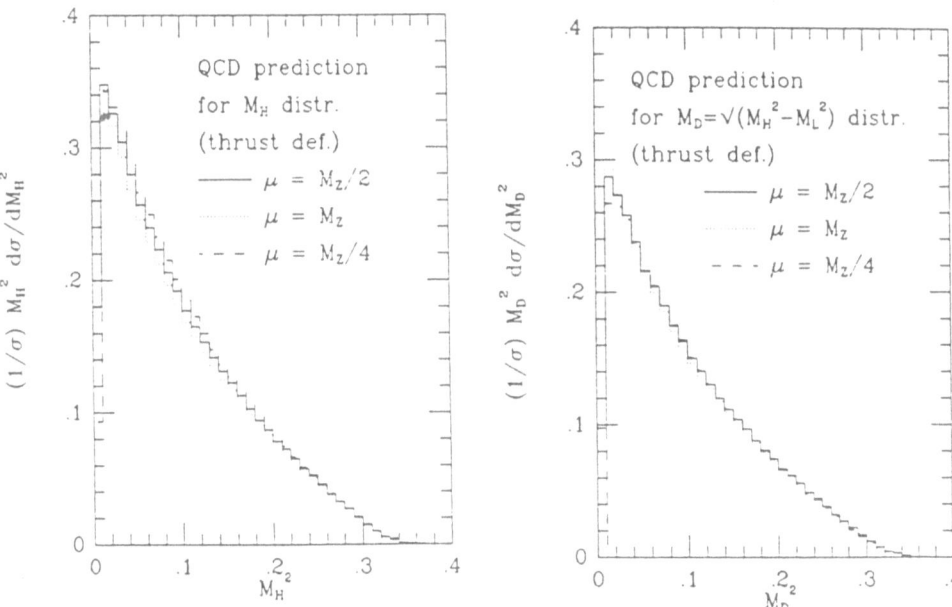

Fig. 10. Prediction for the EEC distribution.

Fig. 11. Prediction for the AEC distribution.

groups of final particles, one simply calculates the invariant masses $M_1^2(\vec{n}_T), M_2^2(\vec{n}_T)$ in the two hemispheres separated by the plane orthogonal to the thrust axis $\vec{n}_T$:

$$
\begin{aligned}
M_{H,t}^2 &= \max\left[M_1^2(\vec{n}_T), M_2^2(\vec{n}_T)\right] \\
M_{L,t}^2 &= \min\left[M_1^2(\vec{n}_T), M_2^2(\vec{n}_T)\right] .
\end{aligned}
\tag{19}
$$

To order α_S the two definitions give the same value for both the heavy and the light masses. The kinematical bounds for this definition will have again kinematical recession.

A further variation of the heavy mass, light mass shape variables are given by the differences

$$
M_{D,t}^2 = M_{H,t}^2 - M_{L,t}^2 .
\tag{20}
$$

The differential cross sections in heavy mass and light mass variables are parametrized in the usual form

$$
\begin{aligned}
\frac{1}{\sigma_0} M_I^2/S \frac{d\sigma}{d(M_I^2/S)} &= \frac{\alpha_S(\mu)}{2\pi} A_I(M_I^2/S) \\
&\quad + \left(\frac{\alpha_S(\mu)}{2\pi}\right)^2 \left[A_I(M_I^2/S) 2\pi b_0 \log(\mu^2/S) + B_I(M_I^2/S)\right]
\end{aligned}
\tag{21}
$$

where $I = (H,t), (D,t), (L,t)$.

This gives a good illustration how easy to proliferate the number of shape variables. Nevertheless we have chosen here to discuss all these variations. Our motivation is to show the sensitivity of the results to such modifications. It is important to keep in mind that the experimental definition in terms of hadrons must always match the definition in terms of partons. The values of all the B_I functions are given in Ref.[15] as well as the predictions for the physical cross sections at LEP.

In the leading order $1 - T$, $M_{H,t}^2/S$, and $M_{D,t}^2/S$ have the same values. The K-factors for these quantities, however, are significantly different. In Figs. 12-13 we plotted the physical cross sections as defined by eq.(21) at three different values of μ. Since the corrections changes their signs in all these quantities there is always a region where the sensitivity to the scale change is very small.

At $\mu = \sqrt{S}$ we obtain* for the average values

$$
\begin{aligned}
< 1 - T > &= 1.05 \frac{\alpha_S}{\pi} + (10.1 \pm 0.08)\left(\frac{\alpha_S}{\pi}\right)^2 \\
< M_{H,t}^2/S > &= 1.05 \frac{\alpha_S}{\pi} + (4.7 \pm 0.07)\left(\frac{\alpha_S}{\pi}\right)^2 \\
< M_{D,t}^2/S > &= 1.05 \frac{\alpha_S}{\pi} - (0.08 \pm 0.07)\left(\frac{\alpha_S}{\pi}\right)^2 .
\end{aligned}
\tag{22}
$$

It is remarkable that the corrections are so widely different. The energy dependence of the average values with three different scale choice is shown in Figs. 14-15. In all the four cases the sensitivity to the scale choice is negligible small.† From these figures we may conclude that all these quantities are good for measuring α_S . As long as we do not understand the physical reason why some of these corrections are

---

*We find different results than the ones in Ref.[59] and in Ref.[60].

†The sensitivity to the scale choice remains week as long the corrections are smaller than 20%.

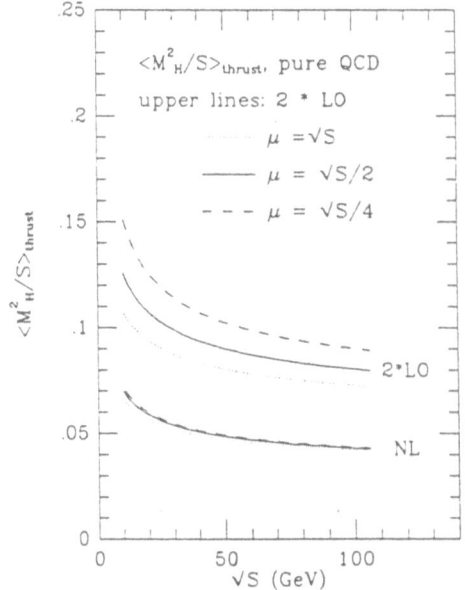

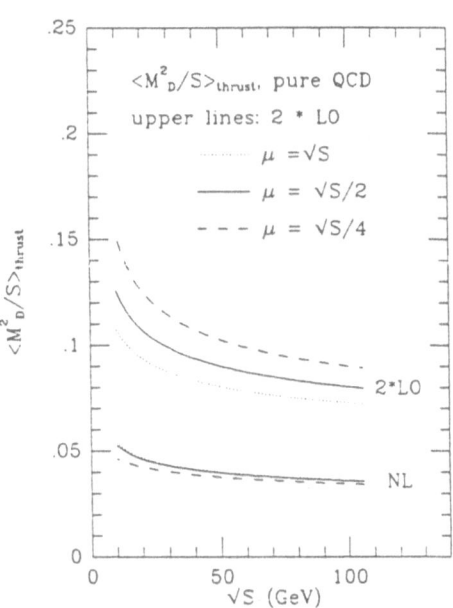

Fig. 12. Prediction for $M_{H,t}$ distribution.

Fig. 13. Prediction for $M_{D,t}$ distribution.

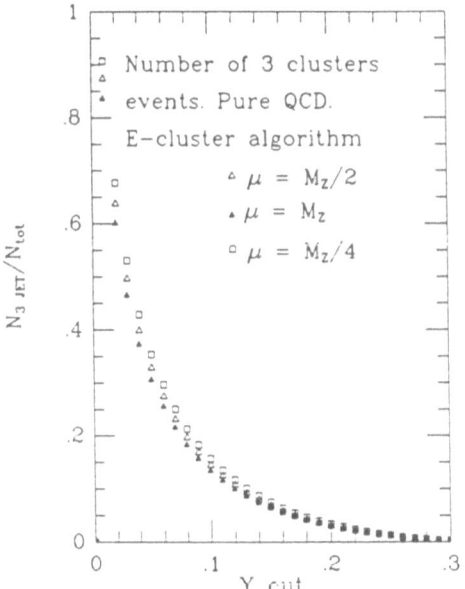

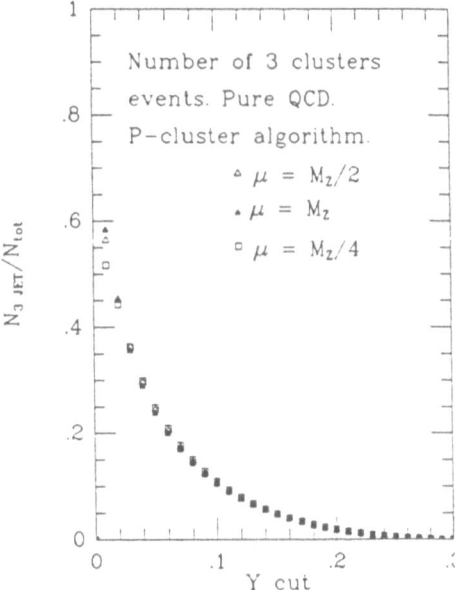

Fig. 14. Average value for $< M_{H,t} >^2$ distribution.

Fig. 15. Average value for $< M_{D,t} >^2$ distribution.

much smaller than the others we may think that they are very small just accidentally. Therefore our estimate on the even higher order corrections might be too optimistic. The study of the $< M^2_{H,t}/S >$ and $< M^2_{D,t}/S >$ with the shower Monte Carlo method confirms this conclusion. For $< M^2_{H,t}/S >$ the shower effect is large, (equal to 12%) for $q\bar{q}$ intial state, but small (2%) for $q\bar{q}g$ initial state. The hadronization effect, however, is negligible for both intial states. This is a remarkable result which deserves further attention. In the case of $< M^2_{D,t}/S >$ the shower effect is small (-1%) for $q\bar{q}$ intial state, but large (-15%) for $q\bar{q}g$ initial state. The hadronization effect is larger: -13% and -6%, respectively. Since the shower Monte Carlos partially include α_S^3 and higher order effects, this indicates that the small correction found for $< M^2_{D,t}/S >$ is most likely accidental. Nevertheless, it appears that $< M^2_{D,t}/S >$ might be a particlularly good quantity to measure α_S. We refer the reader to the recent analysis of CELLO[48] and TASSO[61].

Jet clusters

The existence of quark[26] and gluon jets has been proven first in e^+e^- annihilation. In these early studies various shape parameters were used to establish the result. In many later applications, various jet resolution parameters have also been used in conjunction with shape parameters. The problem of overlapping jets was avoided by the use of small jet resolution parameters. Clearly, good infrared safe jet finding algorithms give competitive methods to the shape variables. Moreover, with explicitely defined jets one can ask more detailed questions concerning jet properties.

Any jet analysis should start with some infrared safe jet definition. Let us consider first the criteria of a good jet definition[46][47]. (i) It is defined at any order of perturbation theory; (ii) it gives finite cross sections at any order of perturbation theory; (iii) it is insensitive to hadronization; (iv) it is simple to implement both experimentally and theoretically.

In e^+e^- annihilation simple, infrared safe jet finding algorithms are provided by the cluster algorithms based on an invariant mass cutoff. There is a certain amount freedom how we define these algorithms. Consequently, various versions appeared in the literature. Here we shall consider i) the E-cluster algorithm, ii) the P-cluster algorithm and iii) the E_0-cluster algorithm.

These algorithms are all equivalent to order α_S but give different results to α_S^2. The circumstance that we can use different algorithms is not an ambiguity of the theory. Instead it just reflects the fact that we have a certain amount of freedom in the definition of a jet. Hence a valid comparison between the theory and the experiment requires that the algorithm must be the same in terms of hadrons as in terms of partons.

The E-cluster jet finding algorithm is defined as follows. For all pairs of particles a and b of an event the scaled invariant mass squared $y_{ab} = (p_a + p_b)^2/S$ is calculated. Then the two particles with the smallest value of y_{ab} denoted as y_{ij} are replaced by a pseudo particle of four momentum

$$p^\mu_{ij} = p^\mu_i + p^\mu_j \tag{23}$$

if their scaled invariant mass is larger than some given threshold value y_{cut}. This procedure is repeated until all y_{ab} pair invariant masses are greater than y_{cut}. The resulting number of clusters is called the jet or cluster multiplicity of the event.

The P-cluster and E_0-cluster jet finding algorithms are variations of this algorithm. They use different prescriptions for calculating the four momenta of the pseudo particle. In the case of the P-cluster algorithm instead of eq.(23) the four momenta of the pseudo-particle are calculated as

$$\vec{p}_{ij} = \vec{p}_i + \vec{p}_j$$
$$E_{ij} = |\vec{p}_{ij}| \; , \tag{24}$$

Here we slightly depart from the definition of the P-cluster algorithm which is usually given in the literature. In the usual implementations, one scales all momenta at each step, in order to mantain energy conservation. We emphasize that these different definitions receive different radiative corrections at order α_S^2.

In the case of the E_0-cluster jet finding algorithm we have

$$E_{ij} = E_i + E_j$$
$$\vec{p}_{ij} = \frac{E_{ij}}{|\vec{p}_i + \vec{p}_j|} (\vec{p}_i + \vec{p}_j) \; . \tag{25}$$

The number of three jet events with jet resolution y_{cut} can be conveniently parametrized again in the form

$$\frac{\sigma_{3\text{jet}}(y_{\text{cut}})}{\sigma_0} = \frac{\alpha_S(\mu)}{2\pi} A_{n3}(y_{\text{cut}})$$

$$+ \left(\frac{\alpha_S(\mu)}{2\pi} \right)^2 \left[A_{n3}(y_{\text{cut}}) 2\pi b_0 \log(\mu^2/S) + B_{n3,\text{I}}(y_{\text{cut}}) \right] \; , \tag{26}$$

$$I = E, \; p \text{ and } E_0 \; ,$$

while the number of four jet events can be given as

$$\frac{\sigma_{4\text{jet}}(y_{\text{cut}})}{\sigma_0} = \left(\frac{\alpha_S(\mu)}{2\pi} \right)^2 B_{n4}(y_{\text{cut}}) \; . \tag{27}$$

The values of the functions A_{n3} and B_{n4} are the same for all the three algorithms since in the leading order no dependence on the algorithm can occur. The values of A_{n3}, $B_{n3,E}$, $B_{n3,p}$ and B_{n3,E_0} are given in Ref.[15]. In all the three cases the corrections are rather large (20%- 40 %). Using eq.(26) (but normalized to σ_{tot} and not to σ_0) we can easily calculate the fraction of three jet clusters as a function of the y_{cut}. The results are shown in Figs.16-18, for three different values of the renormalization scale. We can see that a change in the algorithm gives a definitely larger change in the prediction than the change in the renormalization scale μ. The former can be as high as 40% while the latter is equal to $\approx \pm 10\%$.

Assuming that in the number of the 3 jet clusters the hadronization effect will not be larger than in the case of the shape parameters, we can see that the consistency of the algorithms at the parton and hadron level is important.

It is not clear to us whether in the experimental studies, the matching between the theoretical and the experimental algorithms was rigorously required. The JADE experiment of Ref.[62] and the recent TASSO experiment of Ref.[64] used the E_0-cluster jet finding algorithm, with the practical modification that the cluster masses were calculated in term of the experimentally well defined visible energy, and not of the total energy

$$\tilde{y}_{ab} = \frac{E_a E_b (1 - \cos(\theta_{ab}))}{E_{\text{visible}}^2} \tag{28}$$

After correcting for E_{visible}, the data on three jet clusters can be directly compared with the perturbative QCD prediction calculated in the E_0-cluster jet finding algorithm.

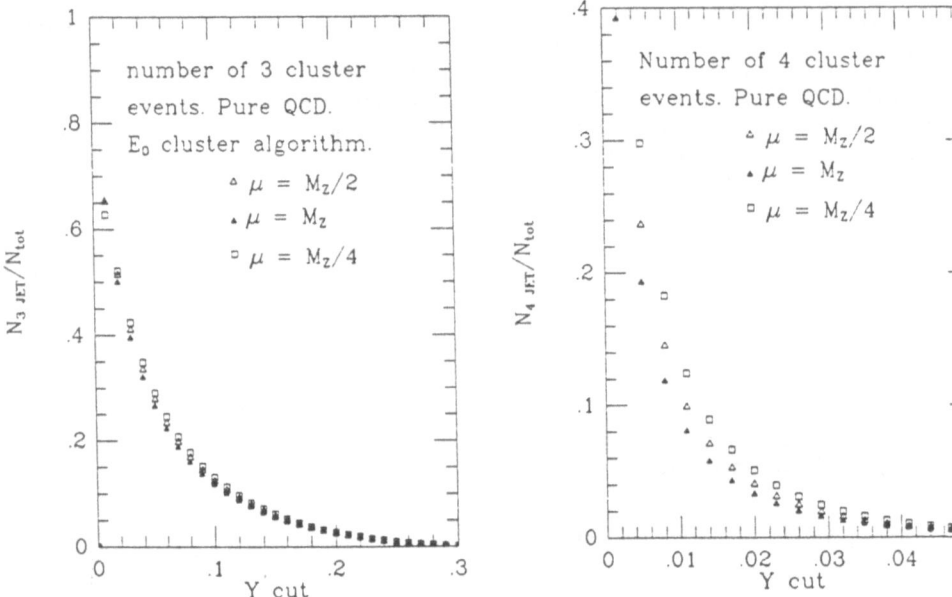

Fig. 16. Three cluster multiplicity. E-cluster algorithm.

Fig. 17. Three cluster multiplicity. P-cluster algorithm.

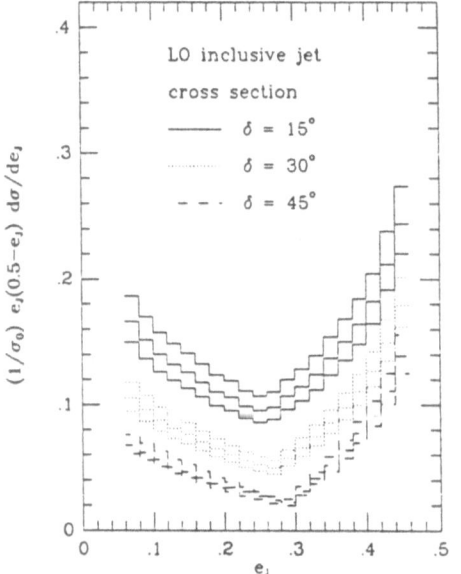

Fig. 18. Three cluster multiplicity. E_0-cluster algorithm.

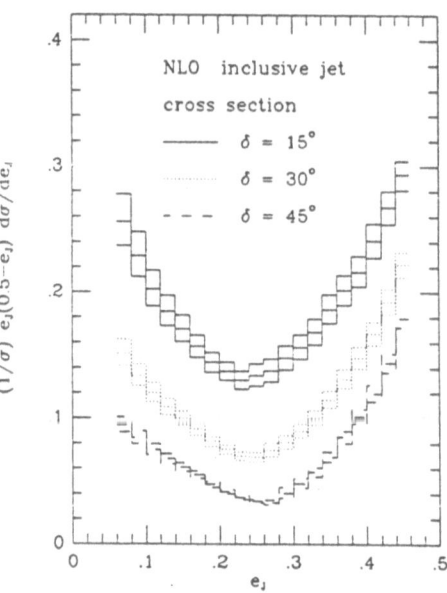

Fig. 19. Four cluster multiplicity.

Of course, Therefore, there is no "scheme" ambiguity in the theoretical prediction. There are instead different jet finding algorithms, and one has to make sure that the parton and hadron level jet finding algorithms being used are the same.

We note that the algorithms defined by eqs.(23) and (25) corresponds to the algorithms used in the recent study of the production of jet clusters in Ref.[63]. Our result is in agreement with the curves labelled as ERTE and ERTE_0 in Fig. (4.6) of this reference. Our defintion of the P-cluster jet finding algorithm is different from the ERT$\vec{p}$ algorithm of Ref.[63], since we do not rescale all the momenta to reestablish energy momentum conservation. Such a modification in the algorithm can lead to 10-20% differences.

In Fig. 19 we show the predicted fraction of the production of four jet clusters at three different values of the renormalization scale M_Z, $M_Z/2$ and $M_Z/4$. This quantity is known only to leading order therefore the scale ambiguity is large. This unsatisfactory situation can only be improved by calculating the next to leading order α_S^3 corrections.

Inclusive jet production

The significance of the jet analyses which use some explicit jet definition has considerably increased at the $Sp\bar{p}S$ and the Tevatron experiments [65] [66] [67]. The reasons for this are the reduced topological freedom (the discussion of boost invariant quantities is much simpler), the background of the so called underlying events and the clarity of the jet signal.

In high energy e^+e^- , ep, pp and $p\bar{p}$ collisions, the discovery of many interesting new physics signals depends on jet properties. In many future experiments the background suppression could be considerably improved if the the jet properties would be better known. Clearly, it is most important to achieve the best possible description of jet physics and to provide connection between the jet studies carried out at the various colliders. One of the simplest jet measure is the spectrum of inclusively produced jets. It is remarkable that this simple quantity has not been studied in e^+e^- annihilation.

In Refs.[68,15] we followed the calorimetric jet definition of Ref.[46] with some obvious modifications needed in the case of e^+e^- annihilation. This definition has been succesfully used by the CDF experiment[67]. We employed an algorithm[68] in which the total energy of the jets is always equal to the incoming energy and jets within jets are not counted as independent jets.

In the soft region the jet energy $E_J \approx 0$ and in the two jet region $E_J \approx \sqrt{S}/2$ the jet spectra is singular because of soft and collinear effects. One should therefore avoid this region. Alternatively, we can suppress the singular behaviour by multiplying the cross section with the weight $e_J(0.5 - e_J)$ where e_J is defined as

$$e_J = E_J/\sqrt{S} . \tag{29}$$

The weighted cross section can be conveniently parametrized in the usual form

$$\frac{1}{\sigma} e_J(0.5 - e_J) \frac{d\sigma}{de_J} = \frac{\alpha_S(\mu)}{2\pi} A_J(e_J)$$
$$+ \left(\frac{\alpha_S(\mu)}{2\pi}\right)^2 \left[A_J(e_J)2\pi b_0 \log(\mu^2/S) + B_J(e_J)\right] . \tag{30}$$

The scale invariant functions $A_J(e_J)$, $B_J(e_J)$ are tabulated in Ref.[15].

In order to exibit the expected size of the uncertaintes in the QCD prediction we plotted the cross section value in Fig. 20 at three different half-opening angles

$\delta = 15°$, $30°$ and $45°$ and at three different values of the renormalization scale $\mu = M_Z, M_Z/2, M_Z/4$. The corresponding Born cross sections are given on Fig. 21.

Comparing the curves on these two figures one can see that the K-factor is large $\approx 1.3 - 1.5\%$. We can also observe the expected reduced sensitivity to the scale choice in the next to leading order prediction. The spectrum has a stationary point around $e_J \approx 0.24$ and the corrections change slightly the shape of the spectrum, leading to a more symmetric behaviour. In the region around $e_J \approx 0.24$ the uncertainty due to a change in the renormalization scale is around 13 % . This part of the spectrum offers the best opportunity to extract α_S from such a measurement. There is a sizeable sensitivity to the change of the jet opening angle.

The shape of the jet spectrum, the dependence on the jet opening angle, the increased size of the jet rate with respect to the Born approximation and the relatively small sensitivity to the choice of the renormalization scale around $e_J \approx 0.24$ are all very interesting features of the QCD prediction. The study of inclusive jet prodution offers a novel method to test QCD at LEP.

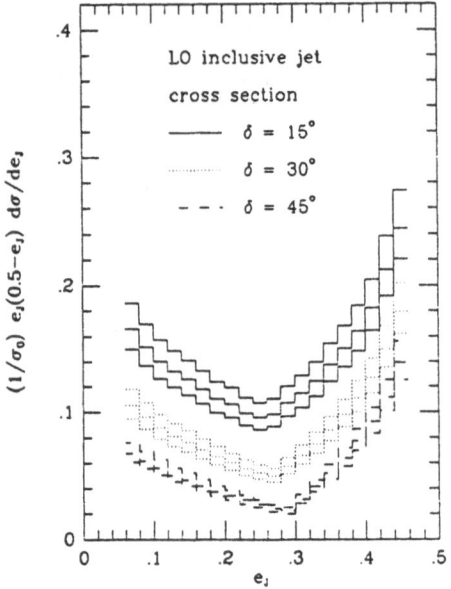

Fig. 20. Leading order prediction for the inclusive jet cross section at three different value of δ. The band corresponds to $\mu = M_Z, M_Z/2, M_Z/4$.

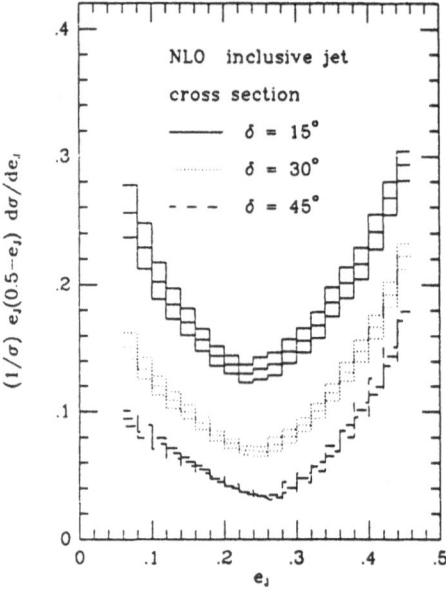

Fig. 21. Next to leading order prediction for the inclusive jet cross section at three different value of δ. The band corresponds to $\mu = M_Z, M_Z/2, M_Z/4$.

MULTI JET PRODUCTION

The production of four and five well separated jets have been clearly observed in the PETRA and PEP experiments [62] [64] [8]. Assuming a factor of 5 reduction in the rate for each additional emission of a well separated energetic jet and assuming that 10^6 hadronic events will be produced at LEP, even seven jet events will be observable. This is a rather exotic possibility since the study of seven jet events allows us to test the theory in effects of the order of α_S^5.

The exact leading order QCD matrix elements are available in the form of computer program up to five parton productions[70] [71]*. Testing QCD in multijet phenomena, however, can be only semi-quantitative, since the next to leading order corrections are not known. Similarly to the case of three jet dominated quantities, we expect that including the next to leading order corrections, significant modifications will occur. We recall also that the Born cross sections for producing higher and higher number of jets are more and more sensitive to the choice of the renormalization scale and to the choice of the jet resolution parameters (see for example the scale ambiguity in the four jet production rate depicted in Fig.19). Nevertheless, the study of the production of multijet events provides a very non trivial test of the theory.

Shower programs have also been used to calculate multijet rates [62] [13]. The shower Monte Carlo programs in the region of energetic, well separated jets are expected to give only a crude approximation. In spite of this, at PETRA energies, they have been found to be rather successfull in describing four and five jet event rates. It would be interesting to test the shower Monte Carlos with respect to the exact five parton matrix elements in this region.

Many four jet shape variables, which are non trivial only for four or more jet production, have been suggested in the literature. The most popular quantities are the acoplanarity[25], the number of four jet clusters (eq.(27)), the light jet mass, etc.[73]. All of them can be used to isolate four jet dominated events. In Ref.[71] Sterman-Weinberg type of jets with ϵ , δ jet resolution parameters and cluster algorithms with y_{cut} have been used to estimate the five jet rates.

An interesting opportunity in four jet studies is the possible direct test of the non-abelian gluon self coupling specific to QCD, appearing first in second or higher order in α_S . We have many indirect test of the non-abelian character of the theory, since the whole consitency of the perturbative approach relies on asymptotic freedom. Perhaps the strongest test is given by jet production at $Sp\bar{p}S$ and at Tevatron. The production rate of jets of transverse momenta up to ≈ 50 GeV at Tevatron is dominated by gluon initial states. It appears that it is impossible to describe the hadron collider phenomena without the contributions from gluon gluon scattering. Regardless of this it is interesting to test the presence of the gluon self coupling also in the properties of four or more jet production. It was suggested[72] to compare QCD with an abelian theory where the eight coloured gluons of QCD are replaced with abelian gluons. Their couplings are normalized in such a way that the three jet cross section remains the same ($\alpha_{abelian} = C_F \alpha_S$). As a consequence of this modification the fraction of the two quark two antiquark final states increases more than an order of magnitude. To obtain a measurable effect, various angular correlations (motivated by "gluon alignment"[74] and by the differences in the helicity structure[73][75]) have been considered. Unfortunately, the differences found in realistic Monte Carlo calculations are at best of the order of 20-30%. In view of fact that the contribution of the higher order corrections is expected to give modifications of similar size it might become difficult to draw firm conclusions from these considerations. Recently it has been suggested[76] that one should try to test directly the fraction of the two quark two antiquark final states by tagging $b-$quark jets. Assuming 0.4 tagging efficiency, the single b-tagging would be $\approx 20\%$ in QCD and $\approx 30\%$ in the abelian modification. Double b-tagging, however, would give $\approx 0.2\%$ in QCD and $\approx 2.4\%$ in the abelian theory. Therefore, double b-tagging would provide a really convincing test. Assuming 1% reconstructed four jet events in 10^6 hadronic events it appears that the double tag test can be carried out at LEP. Since the abelian modification gives rise to an order of magnitude increase in the fraction of double tag events, the unknown radiative higher order corrections can not change the conclusion of such an analysis.

Finally we note that parity violating correlations have also been considered in Ref.[77].

---

*Due to new technical development, if it is needed the exact matrix elements can be calculated for even higher number of jets[70].

References

[1] M. Gell-Mann, *Acta Physica Austriaca, Suppl.* **IX**(1072)733;
H. Fritzsch and M. Gell-Mann, XVI International Conference on High Energy Physics, Batavia, Vol. II p.135 (1972);
H. Fritzsch, M. Gell-Mann and H. Leutwyler, *Phys. Lett.* **47B** (1973) 365.

[2] P. Söding and G. Wolf, *Ann. Rev. Nucl. Part. Sci.* **31** (1981) 231.

[3] B. Adeva *et al.*, *Phys. Rep.* **109** (1984) 131.

[4] S. L. Wu, *Phys. Rep.* **107** (1984) 59.

[5] B. Naroska, *Phys. Rep.* **148** (1987) 67.

[6] W. de Boer, SLAC-PUB 4482, Proc. of the Warsaw Symposium on Elementary Particle Physics, Kazimierez, Ed. Z.Adjuk, Poland, (1987), p.503.

[7] Topaz Collaboration: M. Yamauchi *et al.*, KEK-Preprint-88-63, Presented at the Int. Conf. on High Energy Physics, Munich, Germany, Aug. 4-10, 1988;
H. Sagawa (KEK, Tsukuba), KEK Preprint 88-041 (1988).

[8] S. Bethke *et al.*, SLAC-PUB-4944 (1989). Submitted to Zeit. Phys. C.

[9] For a recent excellent review see for example:
A. Ali and F. Barreiro, in Advanced Sereis on Directions in High Energy Physics, Vol 1., editors: A. Ali and P. Soding, 1988. World Scientific (1988) 873 P.

[10] G. Altarelli, CERN-TH-5290/89, Feb 1989. 76pp. Submitted to Ann. Rev. Nucl. Part. Sci.

[11] G. Marchesini and B. R. Webber, *Nucl. Phys.* **B238** (1984) 1.

[12] G. Marchesini and B. R. Webber, Cavendish-HEP-88/7 (1988) 24pp. Updated version.

[13] T. Sjöstrand and M. Bengsston, *Comput. Phys. Commun.* **43** (1987) 367.

[14] J. C. Collins and D. E. Soper, *Nucl. Phys.* **B284** (1987) 253, and references therein.

[15] Z. Kunszt, P. Nason, G. Marchesini and R. B. Webber, ETH-PT/89-39 (1989).

[16] K. G. Cheterkyn, A. L. Kateev and F. V. Tkachov, *Phys. Lett.* **85B** (1979) 277;
M. Dine and J. Sarpinstein, *Phys. Rev. Lett.* **43** (1979) 668;
W. Celemaster and R. J. Gonsalves, *Phys. Rev. Lett.* **44** (1979) 560; *Phys. Rev.* **D21** (1980) 3112.

[17] S. G. Gorishny, A. L. Kateev and S. A. Larin, *Phys. Lett.* **212B** (1988) 238.

[18] B. A. Kniehl and J. H. Kühn, MPI-PAE/PTh 12/89.

[19] J. Schwinger, "Particles, Sources and Fields", vol II (Addison-Wesley 1973).

[20] L. J. Reinders, H. R. Rubistein and S. Yazaki, *Phys. Rep.* **127** (1985) 1.

[21] J. Jersák, E. Laerman, and P. M. Zerwas, *Phys. Rev.* **D25** (1982) 1218;
A. Djouadi, Nuov. Cim. **100A** (1987) 265.

[22] G. Altarelli, in J. Ellis,(Ed.) and R. Peccei (Ed.): Physics at Lep, Vol. 1, pp. 3 - 34.

[23] J. Ellis, M. K. Gaillard, G. G. Ross *Nucl. Phys.* **B111** (1976) 253, Erratum-*Nucl. Phys.* **B130** (1977) 516.

[24] A. M. Polyakov, *JETP Sov. Phys.* **32** (1971) 296 and *JETP Sov. Phys.* **33** (1971) 850.

[25] A. De Rujula, J. Ellis, E. G. Floratos and M. K. Gaillard, *Nucl. Phys.* **B138** (1978) 387.

[26] R. F. Schwitters, *et al.*, *Phys. Rev. Lett.* **35** (1975) 1320;
G. G. Hanson, *et al.*, *Phys. Rev. Lett.* **35** (1975) 1609.

[27] R. Brandelik, *et al.*, *Phys. Lett.* **97B** (1980) 453.

[28] J. Ellis and I. Karliner, *Nucl. Phys.* **148B** (1979) 141.

[29] G. Sterman and S. Weinberg, *Phys. Rev. Lett.* **39** (1977) 1436.

[30] T. Kinoshita, *J. Math. Phys.* **3** (1965) 56.

[31] T. D. Lee and M. Nauenberg *Phys. Rev.* **133** (1964) 1549

[32] G. Sterman, *Phys. Rev.* **D17** (1978) 2789; and *Phys. Rev.* **D17** (1978) 2773.

[33] H. Georgi and M. Machacek, *Phys. Rev. Lett.* **39** (1977) 1237.

[34] E. Fahri, *Phys. Rev. Lett.* **39** (1977) 1587.

[35] R. K. Ellis, D .A. Ross and A. E. Terrano, *Nucl. Phys.* **B178** (1981) 421.

[36] J. A. M. Vermaseren, K. J. F. Gaemers and S. J. Oldham, *Nucl. Phys.* **B187** (1981) 301.

[37] Z. Kunszt, *Phys. Lett.* **99B** (1981) 429, *ibid.* **107B** (1981) 123.

[38] R .K. Ellis and D. A. Ross, *Phys. Lett.* **106B** (1981) 88.

[39] K. Fabricius, G. Kramer, G. Schierholz and I. Schmitt, *Zeit. Phys.* **C11** (1981) 315.

[40] P. Hoyer, P. Osland, H. G. Sander, T. F. Walsh and P. Zerwas, *Nucl. Phys.* **28** (1983) 243.

[41] A. Ali, E. Pietarinen and J. Willrodt, DESY T-80-01, (1980).
A. Ali, E. Pietarinen, G. Kramer, J. Willrodt, *Phys. Lett.* **93B** (1980) 155.

[42] T. Sjöstrand, *Comput. Phys. Commun.* **28** (1983) 243.

[43] R. D. Field and S. Wolfram, *Nucl. Phys.* **B213** (1983) 65;
G. C. Fox and S. Wolfram, *Nucl. Phys.* **B168** (1980) 285;
T. D. Gottschalk, *Nucl. Phys.* **B239** (1984) 349.

[44] K. Kato and T. Munehisha, *Phys. Rev.* **D36** (1987) 61.

[45] Z. Kunszt and P. Nason, paper in preparation.

[46] S .D. Ellis, Z. Kunszt and D. E. Soper, *Phys. Rev.* **62** (1989) 726.

[47] D. E. Soper, Proceedings of the XIXth International Symposium on Multiparticle Dynamics, Arles, France, June, 1988.

[48] CELLO Collaboration: H. J. Behrend *et al.*, Desy report, DESY 89-019 (1989).

[49] MARK-J Collaboration: D. P. Barber *et al.*, *Phys. Rev. Lett.* **43** (1979) 830.

[50] C. Basham, L. Brown, S. Ellis and S. Love, *Phys. Rev. Lett.* **41** (1978) 1585, *Phys. Rev.* **D19** (1979) 2018 and *ibid.* **D24** (1981) 2382.

[51] W. Bartel *et al.*, *Zeit. Phys.* **C25** (1984) 231.

[52] TASSO Collaboration: W. Braunschweig *et al.*, *Zeit. Phys.* **C36** (1987) 349.

[53] CELLO Collaboration: H. J. Behrend *et al.*, *Zeit. Phys.* **C14** (1982) 95.

[54] D. G. Richards, W. J. Stirling and S. D. Ellis, *Nucl. Phys.* **B229** (1983) 317.

[55] A. Ali and F. Barreiro, *Nucl. Phys.* **B236** (1984) 269.

[56] N. K. Falck, G. Kramer, *Zeit. Phys.* **C42** (1989) 459.

[57] PLUTO collaboration: C. Berger *et al.*, *Zeit. Phys.* **C12** (1982) 297.

[58] T. Chandrahoman and L. Clavelli, *Nucl. Phys.* **B184** (1981) 365.

[59] L. Clavelli and D. Wyler, *Phys. Lett.* **103B** (1981) 383.

[60] J.del Peso, L. Labarga and F. Barreiro, DESY 88-108 (1988).

[61] TASSO Collaboration: W. Braunschweig *et al.*, DESY report 89-069 (1989).

[62] W. Bartel *et al.*, *Zeit. Phys.* **C33** (1986) 23.

[63] N. Magnussen, PhD Thesis, Internal Report DESY F22-89-01 (1989).

[64] TASSO Collaboration: W. Braunschweig *et al.*, *Phys. Lett.* **B214** (1988) 286.

[65] G. Arnison *et al.*, *Phys. Lett.* **123B** (1983) 115.

[66] M. Banner *et al.*, *Phys. Lett.* **118B** (1982) 203.

[67] J. Huth, Proceedings of the SSC Workshop on Calorimetry for SuperCollider, March 13-17, 1989.

[68] Z. Kunszt and P. Nason, ETH-PT/89-27 (1989).

[69] A. Ali, J. G. Körner, Z. Kunszt, G. Kramer, E. Pietarinen, G. Schierholz, and J. Willrodt, *Phys. Lett.* **82B** (1979) 285; *Nucl. Phys.* **B167** (1980) 454.

[70] F. A. Berends, W. T. Giele, H. Kuijf, Print-89-0055 (Leiden) (1988).

[71] N. K. Falck, D. Graudenz and G. Kramer, *Phys. Lett.* **220B** (1989) 299 see also DESY 89-046, (1989).

[72] K. J. F. Gaemers and J. A. M. Vermaseren, *Zeit. Phys.* **C7** (1980) 81.

[73] O. Nachtmann and A. Reiter, *Zeit. Phys.* **C16** (1982) 45.

[74] J. G. Körner, G. Schierholz and J. Willrodt, *Zeit. Phys.* **C7** (1980) 81.

[75] M. Bengsston and P. M. Zerwas, *Phys. Lett.* **B208** (1988) 306.

[76] Z. Fodor, ITP Budapest Report No.465 (1988).

[77] L. Clavelli G. v. Gehlen, *Phys. Rev.* **D27** (1983) 1495.

QCD AND PRECISION TESTS OF ELECTROWEAK INTERACTIONS

J.H. Kühn

Max-Planck-Institut für Physik und Astrophysik
– Werner-Heisenberg-Institut für Physik –
P.O.Box 42 12 12, Munich (Fed. Rep. Germany)

Abstract: The influence of hadronic interactions on precision tests of the electroweak theory is investigated: (i) In $\mathcal{O}(\alpha^2)$ the total cross section is affected by initial state radiation not only of photons but also of hadrons. (ii) Hadronic interactions influence the contribution of heavy quarks to the vacuum polarizations of γ, Z and W. (iii) QCD corrections reduce the forward backward asymmetry of heavy quarks. (iv) The $\mathcal{O}(\alpha_s^2)$ correction of the Z decay rate differs from the correction to R calculated for the photon induced reaction at lower energies. It will be shown that these corrections all have to be included in precision tests of the Standard model. Their magnitude is sufficiently well under control not to endanger the interpretation of these measurements in the forseeable future.

1. Introduction

Forthcoming experiments at electron positron colliders will probe the predictions of the standard model of electroweak interactions at a hitherto unrivalled level of precision. They will be sensitive to one of the most fundamental aspects of the theoretical formulation – quantum corrections as predicted within the perturbative treatment of a local quantum field theory.

Many of these experiments will – directly or indirectly – involve hadron physics:

i) Measurements requiring high statistics, like the determination of the left-right asymmetry A_{LR} will be performed with hadronic final states. The Z line shape will be explored with leptonic and hadronic final states. The restriction to leptonic final states is not necessarily a remedy of the problem of hadron physics: Around the Z peak hadronic and leptonic production rates are closely correlated through unitarity constraints.

ii) The forward backward asymmetry of quarks will be of interest for the determination of $\sin^2 \theta_W$ and could eventually even be sensitive to physics beyond the standard model.

iii) The influence of the hadronic vacuum polarization plays, furthermore, a crucial role for the evaluation of weak corrections to asymmetries and rates, involving quark and leptonic final states as well.

One thus has to assure that the final error on the quantities of interest is not dominated by "hadronic uncertainties", as is the case in neutrino nucleon scattering. There the systematic error in the present result [1] for the weak mixing angle $\sin^2 \theta_W = 0.233 \pm 0.003 \pm 0.005$ originates from the uncertainty in the choice of m_c — in other words from uncontrolled hadron physics.

In most of the calculations the hadronic system is substituted by free quarks. Occasionally perturbative QCD corrections are included. A notable exception is the hadronic part of Π_{AA}, the photon vacuum polarization arising from light quarks, which must be evaluated on the basis of experimental data via dispersion relations. The other self energies Π_{ZZ}, Π_{ZA} and Π_{WW} and in particular the parts induced by heavy quarks are calculated using the parton model, often even ignoring QCD corrections.

A systematic analysis in this field has to proceed in two directions: The applicability and the limitations of the parton model have to be examined and further calculations have to be performed, either based on perturbative QCD or on dispersion techniques.

This review will be concerned with several topics which demonstrate the close relation between hadron physics and electroweak precision measurements and which are of relevance for e^+e^- reactions at high energies. Chapter 2 will deal with (real and virtual) initial state radiation of hadrons. The evaluation of hadronic corrections to the vacuum polarization from heavy quarks in chapter 3 will be based on a combination of perturbative QCD and dispersion relations. The rest of the paper is devoted to reactions with hadronic final states. QCD corrections to the forward backward asymmetry and their influence on the determination of the weak mixing angle will be treated in chapter 4. Chapter 5 will be concerned with deviations from the parton model predictions for the total cross section $\sigma(e^+e^- \rightarrow hadrons)$ and for A_{LR} measured with hadronic final states. The emphasis will be on "non-universal" corrections to R which appear firstly in order (α_s^2) α_s^3 for the (axial) vector induced part of the rate and which originate from flavour singlet intermediate states. After a general discussion of the formalism and of vector current induced rates to $\mathcal{O}(\alpha_s^3)$ the results of our (two loop) calculation for the singlet part of axial current induced rates to $\mathcal{O}(\alpha_s^2)$ will be presented. For $b\bar{b}$ final states one obtains additional corrections with a top mass dependent coefficient between ≈ -1 and ≈ -5. The paper concludes with a discussion of the combined $\mathcal{O}(\alpha_s^2)$ and $\mathcal{O}(\alpha_s^3)$ corrections to the Z decay rate.

2. Hadronic initial state radiation*

Measurements of the total cross section for e^+e^- annihilation have reached a level of precision [3] where the influence of higher order radiative corrections is no longer negligible. Also the determination of the Z mass and width through the resonance line shape at future e^+e^- colliders will be influenced by radiative corrections and again the treatment to $\mathcal{O}(\alpha)$ is insufficient [4].

Radiative corrections are dominantly due to initial state radiation. To $\mathcal{O}(\alpha^2)$ purely photonic contributions as well as those from real and virtual leptonic and hadronic states are relevant. Photonic and leptonic corrections have been calculated in Refs. 5, 6, 7.

Hadronic corrections are known to contribute approximately 50% to the large logarithms that appear in the vacuum polarization $\Pi(q^2)$ for large q^2. A priori they might be as important in the high energy region for $\mathcal{O}(\alpha^2)$ vertex corrections as the aforementioned large leptonic terms. One might try to guess the characteristic scale for initial state radiation, which is presumably lower than s, thus reducing the relative importance of hadrons compared to electrons, or one might evaluate the lepton formulae with effective quark masses. However, a more rigorous treatment is at hand [2]. Just like $\Pi(q^2)$ also these $\mathcal{O}(\alpha^2)$ hadronic corrections are determined by the quantity $R(s) \equiv \sigma_{had}/\sigma_{point}$ measured in lower energy e^+e^- collisions, assuming that $R(s)$ approaches a constant value for large s.

In Ref. 2 an expression for the hadronic contribution to the virtual $\mathcal{O}(\alpha^2)$ corrections has been derived which is valid for arbitrary q^2. It becomes particularly simple in the large q^2 region. The information contained in $R(s)$ can be condensed in its asymptotic behaviour together with three moments which fix the coefficients of the $\ln^n q^2/m^2$ terms. A similar approach has been developed for real soft and hard hadron radiation which in the high energy region depends on the same moments. The formalism can be easily applied to the special case of lepton radiation and reproduces earlier results for virtual radiation [5, 6, 7] and the logarithmically enhanced terms from real radiation [5].

A special situation arises close to the Z peak. The Born cross section changes rapidly when the energy varies by $\Gamma_Z/2 \approx 1.3$ GeV. On the other hand, hadron production in e^+e^- annihilation has its threshold at $2m_\pi$, peaks at about 800 MeV and shows drastic variations up to 4 GeV. The "soft" approximation that can be justified for radiation of additional e^+e^- or $\mu^+\mu^-$ pairs is not applicable to this region. At that point one has to resort to numerical techniques. The numerical result will therefore also be compared to a simplified treatment based on ρ meson dominance which provides the bulk of real radiation. Finally the effect of hadronic and leptonic corrections on the Z line shape will be presented.

---

*This chapter is largely based on [2]

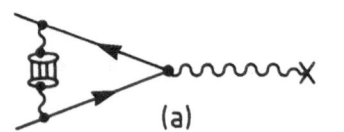
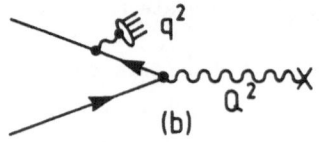

Fig. 1. Feynman diagram indicating the hadronic contribution to the $\mathcal{O}(\alpha^2)$ correction to the formfactor(a) and real hadron radiation(b).

1. Virtual Corrections

The $\mathcal{O}(\alpha^2)$ correction to the electron form factor which originates from the leptonic and hadronic vacuum polarization $F_{vac}^{(4)}$ is shown in Fig. 1. It is obtained from the $\mathcal{O}(\alpha)$ result by replacing the photon propagator by the $\mathcal{O}(\alpha)$ renormalized vacuum polarization insertion [8]

$$\frac{g_{\alpha\beta}}{q^2 - \lambda^2 + i\epsilon} \rightarrow \frac{g_{\alpha\delta}}{q^2 - \lambda^2 + i\epsilon}(q^2 g^{\delta\epsilon} - q^\delta q^\epsilon)\left(\frac{\alpha}{\pi}\right)\Pi^{(2)}(q^2)\frac{g_{\epsilon\beta}}{q^2 - \lambda^2 + i\epsilon} \tag{1}$$

where q denotes the momentum of the internal photon line and λ the photon mass introduced to regulate infrared singularities. $\Pi^{(2)}(q^2)$ is related to $R = \sigma_{had}/\sigma_{point}$ via

$$\Pi^{(2)}(q^2) = \frac{q^2}{3}\int_{4m_\pi^2}^\infty \frac{dq'^2}{q'^2}\frac{R(q'^2)}{q^2 - q'^2 + i\epsilon} \tag{2}$$

where the threshold of σ_{had} is located at $2m_\pi$. $\Pi^{(2)}(q^2)$ vanishes at $q^2 = 0$ together with its first derivative so that no infrared divergencies appear and the limit $\lambda^2 \rightarrow 0$ is legitimate. The photon propagator is thus effectively replaced as follows:

$$\frac{g_{\alpha\beta}}{q^2 + i\epsilon} \rightarrow \frac{\alpha}{3\pi}\int_{4m_\pi^2}^\infty \frac{dq'^2}{q'^2}R(q'^2)\frac{g_{\alpha\beta}}{q^2 - q'^2 + i\epsilon} \tag{3}$$

Exchanging the order of integration one finds

$$\left(\frac{\alpha}{\pi}\right)^2 F_{vac}^{(4)}(q^2) = \frac{\alpha}{3\pi}\int_{4m_\pi^2}^\infty \frac{dq'^2}{q'^2}R(q'^2)\frac{\alpha}{\pi}F^{(2)}(m_e^2, q'^2, q^2) \tag{4}$$

where $F^{(2)}(m_e^2, q'^2, q^2)$ denotes the vertex correction due to the exchange of a boson of mass $(q'^2)^{1/2}$, such that the rhs. can effectively be interpreted as originating from a superposition of vector masses of mass $(q^2)^{\frac{1}{2}}$. For the contribution from heavy leptons or hadrons we have $m_\pi \gg m_e$. Since we are furthermore only interested in the high energy limit $(q^2 \gg m_e^2)$, we may set $m_e = 0$ without encountering any mass singularities.

In this case the one loop vertex correction depends only on the ratio $u \equiv q^2/(q'^2 - i\epsilon)$ and is particularly simple [9, 10]

$$F^{(2)}(0, q'^2, q^2) \equiv \rho(u) = -\frac{7}{8} - \frac{1}{2u} + \left(\frac{3}{4} + \frac{1}{2u}\right)\ln(-u) - \frac{1}{2}\left(1 + \frac{1}{u}\right)^2(\zeta(2) - \text{Li}_2(1 + u)) \tag{5}$$

(Li$_2$ and Li$_3$ denote the di- and trilogarithms, $\zeta(2) = \pi^2/6$, and $\zeta(3) = 1.202\ldots$). $F_{vac}^{(4)}$ is then expressed in terms of R and ρ through the following equation

$$F_{vac}^{(4)}(q^2) = \frac{1}{3} \int_{4m_\pi^2}^{\infty} \frac{dq'^2}{q'^2} R(q'^2) \rho\left(\frac{q^2}{q'^2 - i\epsilon}\right) \tag{6}$$

which is the starting point of the subsequent discussion.

For a parametrization for $R(s)$ derived from data [11] eq. (6) can be evaluated numerically. One finds for the hadronic contribution to $F_{vac}^{(4)}(q^2)$:

$$F_{had}^{(4)} = -57.9 \quad \text{at} \quad \sqrt{q^2} = 93 \text{ GeV} \tag{7}$$

with an estimated error of about 5%. It is smaller than the corresponding contributions from electrons but larger than those from muons and τ's:

$$F_{electron}^{(4)} = -255.9 \quad F_{muon}^{(4)} = -29.5 \quad F_\tau^{(4)} = -2.2 \tag{8}$$

For practical applications it is, however, convenient to consider functions $R(s)$ which converge for large s towards an asymptotic value $R(\infty)$.

Defining moments R_n through*

$$R_n \equiv \int_0^1 \frac{dx}{x} \frac{\ln^n x}{n!} \left(R(4m_\pi^2/x) - R(\infty) \right) \tag{9}$$

$F_{vac}^{(4)}$ can be cast into the following form in the high energy limit

$$\begin{aligned}
\text{Re} F_{vac}^{(4)}(q^2) =& R(\infty)\left[-\frac{1}{36}L^3 + \frac{1}{8}L^2 + \left(\frac{\zeta(2)}{6} - \frac{7}{24}\right)L - \frac{\zeta(2)}{4} + \frac{5}{16}\right] \\
& + R_0\left[-\frac{1}{12}L^2 + \frac{1}{4}L + \frac{\zeta(2)}{6} - \frac{7}{24}\right] + R_1\left[-\frac{1}{6}L + \frac{1}{4}\right] - \frac{1}{6}R_2
\end{aligned} \tag{10}$$

where $L \equiv \ln q^2/4m_\pi^2$. With the parametrization of Ref. 11 one finds for hadrons

$$R(\infty) = 4.0 \quad R_0 = -8.31 \quad R_1 = 13.1 \quad R_2 = -15.6 \tag{11}$$

Eq. (10) leads to an excellent parametrization for $\sqrt{q^2} > 1$ GeV as may be seen from Fig. 2. For muons the moments can be calculated analytically so that the result of Refs. 5, 6 for the form factor contribution of muons is easily verified.

2. Real Corrections

Radiation of real hadrons can be calculated from

$$\frac{d^2\sigma}{dq^2 dQ^2} = \sigma_{Born}(Q^2)\left(\frac{\alpha}{\pi}\right)^2 R(q^2) f(s, q^2, Q^2) \tag{12}$$

---

*It should be noted that $R(\infty)$ and R_0 also determine the large q^2 behaviour of the vacuum polarization, $\Pi^{(2)}(q^2) = (R(\infty)\ln q^2/4m^2 + R_0)/3$.

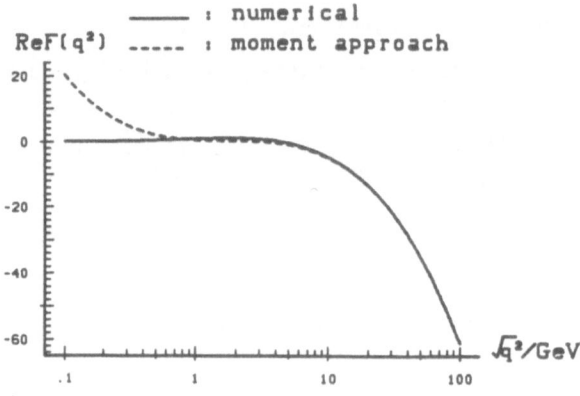

Fig. 2. $\mathcal{O}(\alpha^2)$ corrections to the electron form factor from hadrons (numerical integration and moment approach).

f denotes a kinematical factor, $\sigma_{Born}(Q^2)$ stands for the lowest order cross section for e^+e^- annihilation into any final state with invariant mass Q^2, and q^2 is the invariant mass of the radiated system (see Fig. 1b). Depending on the experimental setup, events with radiation of additional hadrons may be sorted out, e.g. in the case when Z decays into muon pairs and the soft hadrons are observed or when a natural cutoff on radiation is provided by running on a narrow resonance with width below $2m_\pi$ — such as a toponium resonance. In other circumstances these events will contribute to the total cross section, a situation characteristic for hadronic final states in the continuum.

The cross section for "soft" hadron radiation, which includes all events with radiated hadrons up to some energy $\Delta \ll \sqrt{s}$ is given by

$$\sigma_{soft}(s, \Delta) = \int_{4m^2}^{\Delta^2} dq^2 \int_{(\sqrt{s}-\Delta)^2}^{(\sqrt{s}-\sqrt{q^2})^2} dQ^2 \frac{d^2\sigma}{dq^2 dQ^2} \tag{13}$$

If at the same time $R(\Delta^2)$ is sufficiently close to $R(\infty)$ and the variation of $\sigma_{Born}(Q^2)$ for Q^2 between $(\sqrt{s}-\Delta)^2$ and $(\sqrt{s}-2m)^2$ is sufficiently small σ_{soft} may be expressed in terms of the moments R_n defined above ($\mathcal{L} \equiv \ln \Delta^2/m^2$),

$$\frac{\sigma_{soft}(s,\Delta)}{\sigma_{Born}(s)} = \left(\frac{\alpha}{\pi}\right)^2 \frac{4}{3}\left[R(\infty)\left(\frac{1}{24}\mathcal{L}^3 - \frac{\zeta(2)}{4}\mathcal{L} + \frac{\zeta(3)}{2}\right) + R_0\left(\frac{1}{8}\mathcal{L}^2 - \frac{\zeta(2)}{4}\right) + \frac{1}{4}R_1\mathcal{L} + \frac{1}{4}R_2\right] \tag{14}$$

For μ pairs this directly leads to the result given in Refs. 5, 7. When the virtual correction is added to the soft radiation the moment R_2 and the $\ln^3 s/4m^2$ term are cancelled ($L \equiv \ln s/4m^2$, $\ell \equiv \ln 2\Delta/\sqrt{s}$):

$$2\mathrm{Re}F_{vac}^{(4)}(s) + \frac{\sigma_{soft}(s,\Delta)}{\sigma_{Born}(s)} = \left(\frac{\alpha}{\pi}\right)^2 \left\{\left[R(\infty)\left(\frac{L^2}{2} - \zeta(2)\right) + R_0 L + R_1\right]\left(\frac{2}{3}\ell + \frac{1}{2}\right) + \right.$$
$$\left. \left(R(\infty)L + R_0\right)\left(\frac{2}{3}\ell^2 - \frac{7}{12}\right) + R(\infty)\left(\frac{4}{9}\ell^3 + \frac{2}{3}\zeta(3) + \frac{5}{8}\right)\right\} \tag{15}$$

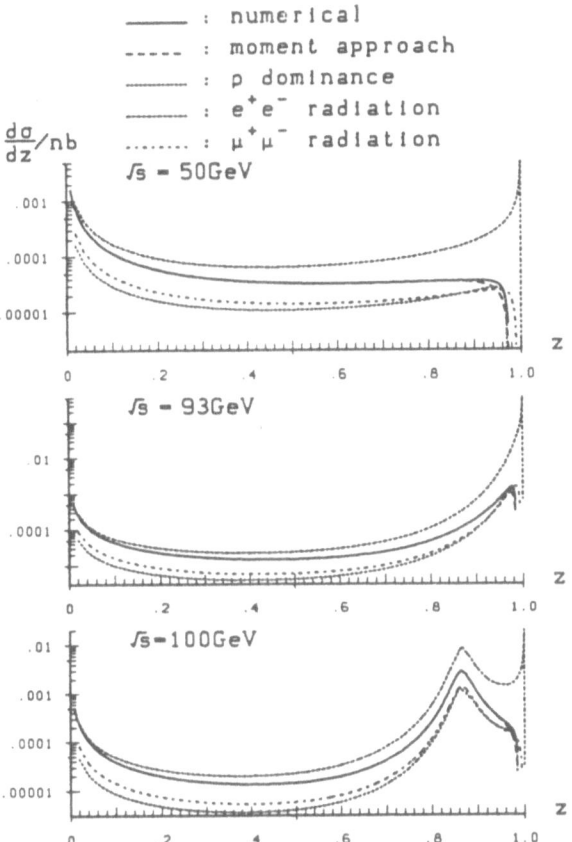

Fig. 3. Differential cross section for hard radiation of hadrons, electrons and muon pairs as a function of $z = Q^2/s$. The hadronic radiation is evaluated through numerical integration, using the moment approach and ρ meson dominance. σ_{Born} stands for $\sigma(e^+e^- \to \mu^+\mu^-)$.

This result is consistent with the expectation that the maximal power of the logarithm should be n in n-th order perturbation theory [12].

Under similar assumptions on the behaviour of R and for $z \equiv Q^2/s$ chosen such that $2m/\sqrt{s} \ll \Delta/\sqrt{s} \le (1 - \sqrt{z})$ one may also calculate the cross section for hard radiation. The relevant formulas can be found in [2]

All approximations discussed in the context of real radiation are strictly applicable to electron and muon pairs in the continuum and on the Z resonance and to hadron radiation in the continuum only. The discussion of real hadron radiation is more involved in the neighbourhood of the Z resonance, since $R(q^2)$ varies rapidly for $\sqrt{q^2}$ between $2m_\pi$ and 4 GeV. Within this energy range also σ_{Born} exhibits rapid variations and provides an effective cutoff on $\sqrt{q^2}$ which varies with $(M_Z - \sqrt{s})$ but is typically of order Γ_Z. In this case only the numerical evaluation of eq. (12) leads to an adequate result over the full range of z. It is, however, noteworthy that — as a consequence of the low effective

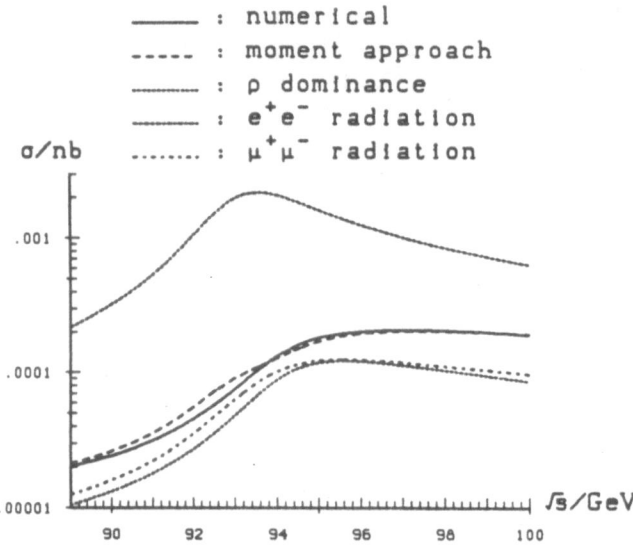

Fig. 4. Cross section for the radiation of hadron, electron and muon
pairs, as a function of $\sqrt{s}$, integrated between $z_{min} = 0.5$ and 1.

cutoff — the hadronic system is dominantly a ρ meson for $z \geq 0.95$. This is evident from
Fig. 3 where the differential distribution $d\sigma/dz$, evaluated with the complete hadronic
contribution, is compared with the one for which only the ρ meson has been taken into
account. This figure also shows radiation of electron and muon pairs. The cross section
is fairly small in the region $0.1 \leq z \leq 0.6$ and events with small z will be quite distinct
from the bulk of events close to $z = 1$. Hence the integrated cross section was evaluated
with a cut at $z_{min} = 0.5$, and the result is practically independent from the precise
choice of z_{min} as long as it is chosen between 0.1 and 0.6. In Fig. 4 the production cross
section is shown separately for radiation of hadrons, electron and muons pairs. The close
agreement between the numerical integration and the moment approach for the integrated
cross section is remarkable.

The combined effect of hadron and lepton radiation on the line shape is at the edge
of the planned experimental accuracy.

3. Hadronic corrections to Δr for large m_t

It is well known that a large splitting between top and bottom quark masses would
lead to a significant decrease of Δr, the quantity which summarizes the influence of
radiative corrections in the relation between m_W, m_Z, α and G_μ. For $m_t = 230$ GeV, a
value consistent with present experimental information, the additional piece originating

from a heavy top would amount to $\Delta r(tb) \approx -0.07$. In most calculations the hadronic interaction in the heavy quark sector is neglected and quarks are treated just like leptons. However, one expects the influence of hadronic interactions on these terms to be of order $\alpha_s/\pi \cdot 0.07 \approx 3 \cdot 10^{-3}$. They could therefore be larger than the uncertainty from the light hadrons' vacuum polarization [13] and comparable to the Higgs mass dependence, if m_H is varied e.g. from 10 to 100 GeV.

One may attempt to evaluate the influence of hadron physics on $\Delta r(tb)$ by calculating hadronic corrections to the various vacuum polarizations $\Pi_{AA}, \Pi_{ZA}, \Pi_{ZZ}$ and Π_{WW} entirely in the framework of perturbative QCD*. In this case the imaginary parts of the various $\Pi's$ are not described correctly in the threshold region and the quarkonium resonances $(t\bar{t})$ and $(t\bar{b})$ are not taken into account. Furthermore, the relation between the quark mass which appears in the calculation and the masses of the $(t, \bar{t})$- and T-mesons is not obvious.

In Ref. 17 another approach has been proposed. The imaginary part of the vacuum polarization is modelled in the region far above threshold using perturbative QCD, with the argument of the running coupling constant $\alpha_s(\mu^2)$ chosen alternatively as $\mu^2 = 4p_t^2$ or s. The resonance region has been evaluated with the resonance parameters (masses and couplings) for $(t\bar{t})$ and $(t\bar{b})$ bound states derived from a potential model [18]. The real part is then deduced via dispersion relations. The result of this approach is shown in Fig. 5.

Fig. 5a gives the hadronic correction to $\Delta r(tb)$ originating from the continuum region under three different assumptions on the behaviour of α_s: i) $\alpha_s = 0.12 = const.$, ii) $\mu^2 = s$ and iii) $\mu^2 = 4\vec{p}^2$, where $\vec{p}$ denotes the momentum of the top quark. The constituent quark mass $m_t = m_T - 400$ MeV is closely related to the mass of the top meson. General considerations suggest that the true answer should lie between the choices ii) and iii) (with slight preference for iii), such that the shaded band is indicative of present theoretical uncertainty.

For comparison also shown is the hadronic correction to $-\cos^2\theta_W/\sin^2\theta_W \Delta\rho(tb)$ which coincides for large m_t with $\Delta r(tb)$ but differs significantly below $m_t \approx 100$ GeV. Fig. 5b gives the contribution of $(t\bar{t})$ and $(t\bar{b})$ resonances to Δr and $-\cos^2\theta_W/\sin^2\theta_W \Delta\rho$, which has to be added.

The sum of these two pieces amounts to a change of $\Delta r(tb)$ by about $2 \cdot 10^{-3}$ for m_t between 70 and 200 GeV, well consistent with the order of magnitude estimate indicated above. It corresponds to a shift of the W mass (for fixed m_Z, G_μ, α) of about 25 MeV. The corresponding changes in the prediction for the left-right asymmetry A_{LR}, which are also studied in Ref. 17 are significantly smaller.

---

*For attempts in this direction see [14, 15, 16].

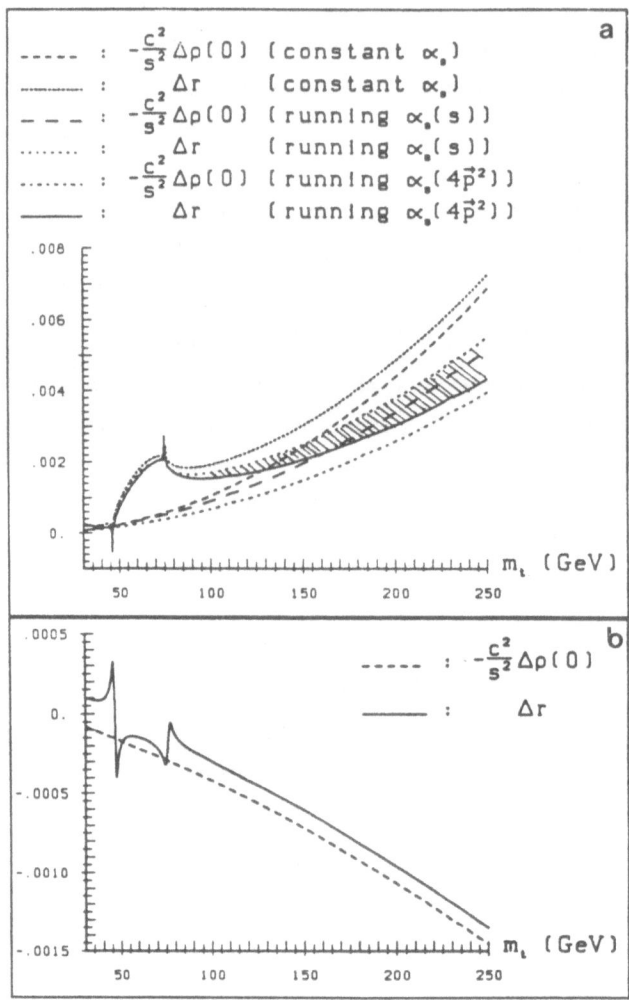

Fig. 5. QCD corrections for the th—doublet contribution to $\triangle r$ and to the leading term $\propto \triangle\rho(0)$. a) Continuum contribution for different choices of α_s. b) Resonance contribution.

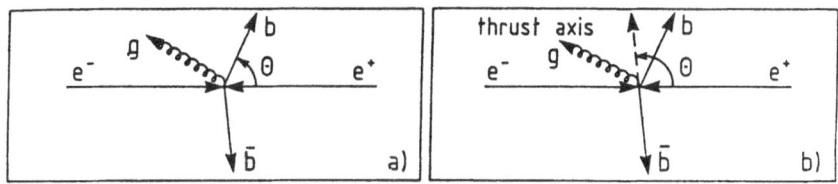

Fig. 6. Two different definitions of the scattering angle: a) through the quark direction, b) through to the thrust axis.

4. Forward-backward asymmetries with quark final states

The measurement of forward backward asymmetries with quark final states will constitute an important tool to test the validity of the Standard Model and to determine the relative strengths of vector vs. axial-vector couplings of the various fermion species, in particular of the electron. All formulas below apply to light (u, d, s) and heavy (c, b) quarks equally well. However, in practice tagging of charmed and bottom quarks will be easier [19, 20]. Thus the influence of QCD and of mass corrections has to be studied*. In Born approximation and on top of the Z the asymmetry is given by

$$A_{FB}^0 = \frac{3}{4} \frac{\sigma_F^0}{\sigma_V^0 + \sigma_A^0} \tag{16}$$

with

$$\sigma_F^0 = 2v_e a_e 2v_q a_q \beta; \qquad \sigma_V^0 = (v_e^2 + a_e^2)v_q^2(3 - \beta^2)/2; \qquad \sigma_A^0 = (v_e^2 + a_e^2)a_q^2\beta^3 \tag{17}$$

Once QCD corrections are included a careful definition of the scattering angle is mandatory. One may, for example, choose the direction of the b quark (Fig. 6a), or alternatively, the direction of the thrust axis (Fig. 6b). QCD corrections to A_{FB} for arbitrary quark mass, based on the first definition have been calculated in Ref. 22. For small $\mu^2 = 4m_q^2/s$ and on top of the Z the corrections to the various terms appearing in eq. (16) can be cast into the form

$$\sigma_i = \sigma_i^0\left(1 + c_i(\frac{\alpha_s}{\pi})\right) \tag{18}$$

with

$$c_F = \frac{2\pi}{3}\mu + \frac{2\mu^2}{3}\left[\frac{9}{2} + \frac{\pi^2}{8} + \frac{1}{8}(\log\frac{4}{\mu^2})^2 + \frac{3}{2}\log\frac{4}{\mu^2} - \frac{5}{2}\log 2\right] + \ldots \approx 0.348$$

$$c_V = 1 + 3\mu^2 + \ldots \approx 1.033 \tag{19}$$

$$c_A = 1 + 3\mu^2 \log\frac{4}{\mu^2} + \ldots \approx 1.197$$

---

*A more detailed treatment of these effects which includes also the impact of QED and electroweak corrections can be found in Ref. 21.

449

(numerical values for $\mu = 2 \times 4.8$ GeV/91 GeV). Note that c_F vanishes in the zero-mass limit. The correction to A_{FB} on top of the resonance then reads

$$A_B^{FB} \rightarrow A^{FB} \approx A_B^{FB}\left[1 - \frac{\alpha_s}{\pi}\left(1 - c_F + 3\mu^2 \frac{v_Q^2 + a_Q^2 \log\frac{4}{\mu^2}}{v_Q^2 + a_Q^2}\right)\right] \tag{20}$$

with the leading part due to the change of the normalization. The dominant effect is thus a reduction of the asymmetry by about 4% through a factor $(1 - \frac{\alpha_s}{\pi})$ for $\mu^2 = 0$; mass corrections lower the the coefficient of $\frac{\alpha_s}{\pi}$ from 1 to about 0.8.

Similar results are obtained if the thrust axis is chosen as reference direction. Defining two-jet-events through the requirement that the jet invariant masses are less than $\sqrt{y}E_{cm}$, one finds corrections to the two jet forward backward asymmetry which depend on y:

$$A_{FB}(2\text{jet}) = A_{FB}^0(1 - k_A(y)\frac{\alpha_s}{\pi}) \tag{21}$$

Although the corrections to the numerator and the denominator (with coefficients k_F and k_2) in eq. (16) become quite large for small y, the coefficient k_A is in general fairly small, typically below 0.5 for a reasonable range of the cutoff parameters [21] as shown in Tab. 1.

Tab. 1. *QCD corrections to 2-jet cross sections and forward-backward asymmetries* [21].

Reference axis : thrust or b-quark axis

	y	k_2	k_F	k_A
$m_c = 1.5$ GeV	0.02	−11.33	−11.83	−0.51
	0.04	−6.23	−6.71	−0.48
	0.08	−2.53	−2.99	−0.46
$m_b = 4.5$ GeV	0.02	−11.63	−12.17	−0.54
	0.04	−6.23	−6.71	−0.48
	0.08	−2.45	−2.84	−0.39

We note that the impact of uncertainties from higher order QCD corrections on the experimental determination of the weak couplings is fairly small: b quark asymmetries on top of the Z amount to about 0.09 for $M_Z \approx 91$ GeV. Taking the uncertainty from higher order QCD corrections as 1/3 of the leading $\mathcal{O}(\alpha_s)$ term this results in a "systematic" theoretical error in the prediction for A_{FB} of about 10^{-3} — well below the anticipated statistical error for 10^7 Z and about 1/3 of the uncertainty from $B\bar{B}$ mixing [which reduces the asymmetry by a factor $(1 - 2\bar{\chi})$], if $\delta\bar{\chi}/\bar{\chi} = 0.1$ is assumed. As illustrated in Fig. 7, the b-quark asymmetry will be an excellent tool to measure the weak mixing angle. In fact, apart from the indirect determination of $\sin\theta_W$ through α, G_μ and m_Z

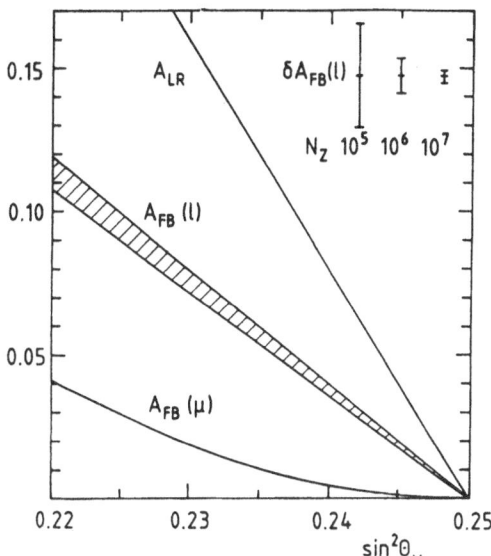

Fig. 7. Dependence of the left-right asymmetry A_{LR}, the forward-backward asymmetry of leptons from B-decays (including QCD corrections and mixing) $A_{FB}(\ell)$ and of μ-pairs $A_{FB}(\mu)$ on $\sin^2 \theta_W$. Also shown is the combined uncertainty from fragmentation, QCD and mixing [21].

it may well lead to the most precise value at LEPI — as long as longitudinally polarized beams are not available.

5. Non-universal QCD corrections

1. Influence of QCD corrections on A_{LR} and the Z line shape

In leading order of the electroweak coupling the total cross section for e^+e^- into hadrons can be cast into the following form

$$
\sigma_{\{L\}}^{\{R\}} = \frac{4\pi\alpha^2}{3s} \left[\frac{(v_e \mp a_e)^2}{y^2} \left| \frac{s}{s - M_Z^2 + iM_Z\Gamma_Z} \right|^2 (r^{(V,V)} + r^{(A,A)}) \right.
$$
$$
+ 2Q_e \frac{(v_e \mp a_e)}{y} \mathrm{Re}\left(\frac{s}{s - M_Z^2 + iM_Z\Gamma_Z} \right) r^{(em,V)} \tag{22}
$$
$$
\left. + Q_e^2 r^{(em,em)} \right]
$$

with

$$
Q_e = -1 \quad v_e = 2I_e^3 - 4Q_e \sin^2\theta_W \quad a_e = 2I_e^3 \quad y = 4\sin^2\theta_W \cos^2\theta_W \tag{23}
$$

R and L denote the electron beam polarization (positrons are assumed as unpolarized). The functions $r^{(i,j)}$ are defined as the transverse parts of $\sum_n \langle 0|J_\mu^i|hk \rangle \langle h|J_\nu^j|0 \rangle$ and are the natural generalizations of the familiar $R \equiv \sigma_{had}/\sigma_{point} = r^{(em,em)}$ to the currents of interest in the high energy region. In the massless parton model they are given by

$$r^{(V,V)} = 3\sum_q \frac{v_q^2}{y^2} \qquad r^{(A,A)} = 3\sum_q \frac{a_q^2}{y^2} \qquad r^{(em,em)} = 3\sum_q Q_q^2 \qquad r^{(em,V)} = 3\sum_q Q_q \frac{v_q}{y}$$

(24)

The left right asymmetry is evidently independent of the final state as long as the leading $|Z|^2$ term is considered. In the very moment when $Z - \gamma$ interference and the pure QED contribution are included, the question arises to which level of accuracy the predictions of the parton model for the line shape, the normalization and for A_{LR} can be maintained. The normalization of $r^{(i,j)}$ and thus of the cross section is obviously affected by QCD corrections. If these corrections were different for the different i,j and for different quark final states then the parton model predictions would be invalidated not only for the overall normalization, but also for the line shape and in particular for A_{LR} in a rather complicated and perhaps even uncontrolable way. A closely related problem arises if one considers the second term in eq. (23): If $r^{(em,V)}$ is real, $Z - \gamma$ interference vanishes on top of the Z resonance. If $r^{(em,V)}$ were complex, a non-negligible final state dependence of A_{LR} could arise at the peak.

In this chapter these questions will therefore be adressed in a rather general form. It will be investigated to which extent QCD corrections for $r^{(i,j)}$ are given by a universal factor and the leading non universal terms will be identified and calculated. It will be convenient to write these corrections in the form

$$r^{(i,j)} = r_{Born}^{(i,j)} \left(1 + c_1^{(i,j)}(\frac{\alpha_s}{\pi}) + c_2^{(i,j)}(\frac{\alpha_s}{\pi})^2 + c_3^{(i,j)}(\frac{\alpha_s}{\pi})^3 + \dots\right)$$

(25)

For clarity and completeness we stress that $r^{(ij)}$ contains in addition to the charge factors mass corrections which read

$$r_{Born}^{(i,j)} = r_{Born}^{(i,j)}\Big|_{\mu^2=0} \times \begin{cases} \beta(3 - \beta^2)/2 & \text{for } i,j = em \text{ or } V \\ \beta^3 & \text{for } i = j = A \end{cases}$$

(26)

with $\beta \equiv \sqrt{1 - \mu^2}$ and $\mu^2 \equiv 4m_q^2/s$.

2. *QCD corrections to $\mathcal{O}(\alpha_s)$ including mass terms*

$\mathcal{O}(\alpha_s)$ corrections including non vanishing quark masses have been calculated in Refs. 14,22.

The leading terms for small μ^2 are given by

$$c_1^{(i,j)} = \begin{cases} 1 + 3\mu^2 + \dots & \text{for } i,j = em \text{ or } V \\ 1 + 3\mu^2 \log(4/\mu^2) + \dots & \text{for } i = j = A \end{cases}$$

(27)

The coefficient $c_1^{(AA)}$ exceeds unity by $\approx +0.17$ for b quarks so that the mass correction in 1st order QCD is as important as the entire 2nd order of the Z width.

$c_2^{(ij)}$ and all higher corrections have been calculated for massless quarks only. In the remainder of this chapter which will be concerned with two or three loop amplitudes quarks in the final state will be considered as massless, thus implicitly assuming $m_Z < 2m_t$.

3. Vector currents up to $\mathcal{O}(\alpha_s^3)$

QCD corrections are identical to $\mathcal{O}(\alpha_s^2)$ for all vector current induced cross sections,

$$c_2^{(i,j)} = 1.986 - 0.115n_f \quad \text{for } i,j = em \text{ or } V \tag{28}$$

since the quark line that couples to the external current forms one closed loop in all Feynman diagrams under consideration.

As noted in Ref. 23 a new diagram with three gluons as intermediate states appears in third order. It contributes specifically for the flavour singlet configurations. The resulting terms can be considered separately and are proportional to the square of the singlet content of the respective current. It is thus convenient to split c_3 into singlet (S) and non-singlet (NS) parts

$$
\begin{aligned}
c_3^{(i,j)}(NS) &= 64.861 \quad \text{for } i,j = em \text{ or } V \\
c_3^{(em,em)}(S) &= -\frac{(\Sigma Q_q)^2}{3\Sigma Q_q^2}1.679 \\
c_3^{(V,V)}(S) &= -\frac{(\Sigma v_q)^2}{3\Sigma v_q^2}1.679 \\
c_3^{(V,em)}(S) &= -\frac{(\Sigma v_q)(\Sigma Q_q)}{3\Sigma v_q Q_q}1.679
\end{aligned}
\tag{29}
$$

Numerically one finds

$$
\begin{aligned}
c_3^{(em,em)}(S) &= -0.051 \\
c_3^{(V,V)}(S) &= -0.549 \\
c_3^{(V,em)}(S) &= +0.202
\end{aligned}
\tag{30}
$$

for 5 massless quarks and $\sin^2 \theta_W = 0.23$.*

The masses of real and virtual quarks are assumed to be zero in this calculation, i.e. the top quark contribution has been ignored. In principle also diagrams involving virtual top quarks should be included. However, power counting – combined with the requirements of gauge invariance – demonstrates that these contributions are suppressed by $(m_Z/2m_t)^2$ (consistent with the Appelquist Carazzone decoupling theorem) and can thus be neglected for large m_t. For $2m_t/M_Z = \mathcal{O}(1)$ no full calculation is at hand. Taking

---

*As a trivial consequence one observes that the QCD corrections to the W decay rate are simply given by the non-singlet corrections.

the opposite extreme $2m_t/M_Z \ll 1$ to estimate the potential influence of such terms, one finds again coefficients smaller than one

$$c_3^{(em,em)}(S) = -0.336$$
$$c_3^{(V,V)}(S) = -0.251 \tag{31}$$
$$c_3^{(em,V)}(S) = +0.351$$

which are completely negligible compared to $c_3(NS)$. Therefore one expects that the singlet contribution has no practical influence on c_3 also for arbitrary m_t between these limiting cases.**

4. Phase of $r^{(em,V)}$

Closely related arguments can be applied to the discussion of the phase of $r^{(em,V)}$ [25]. Whereas the other functions $r^{(i,j)}$ are evidently positive, neither the sign nor the phase of $r^{(em,V)}$ are fixed a priori. However, it will be shown in the following that the QCD induced phase is extremely small. For this purpose it will first be shown that $r^{(em,V)}$ is real in a fictitious $SU(n)$ invariant theory. $SU(n)$ breaking due to the large top mass will then introduce small corrections which are subsequently estimated.

Let us first consider a fictitious model with n mass degenerate quarks (applicable for $n_f = 5$ to the case $m_q^2 \ll s \ll m_t^2$ with $q \neq t$). One may then expand J^{em} and J^V in terms of $SU(n)$ currents j^i ($i = 1, \ldots n^2 - 1$; for J^{em} and J^V only the diagonal components are relevant) and the singlet current j^s with real coefficients a

$$J^{em} = \sum_i a_i^{em} j^i + a_s^{em} j^s$$
$$J^V = \sum_i a_i^V j^i + a_s^V j^s \tag{32}$$

One thus obtains

$$r^{(em,V)} \propto \sum_i a_i^{em} a_i^V \sum_h \langle 0|j^i|0\rangle\langle h|j^i|0\rangle + \sum a_s^{em} a_s^V \sum_h \langle 0|j^s|h\rangle\langle h|j^i|0\rangle$$
$$+ \sum_i a_i^{em} a_s^V \sum_h \langle 0|j^i|h\rangle\langle h|j^s|0\rangle + \sum_i a_s^{em} a_i^V \sum_n \langle 0|j^s|h\rangle\langle h|j^i|0\rangle \tag{33}$$

The last two terms vanish as a consequence of $SU(n)$ invariance, the first two terms are evidently real.

Even if $4m_t^2 > s$, virtual top contributions lead to a violation of $SU(n)$ invariance which leads to an a priori undetermined phase of $\sum_h \langle 0|j^s|h\rangle\langle h|j^i|0\rangle$. Such terms are induced by diagrams of the following type and may appear in third or higher order. The induced phase vanishes in the limit of $4m_t^2/s \to 0$, as discussed above and in the limit of $m_t^2/s \to \infty$ as a consequence of

---

**This qualitative argument is supported by the smallness of vector current induced part of the $Z \to ggg$ decay rate calculated in Ref. 24.

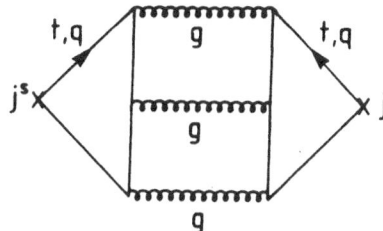

Fig. 8. Feynman diagram whose absorbtive part contributes to the non-vanishing phase of $r^{em,V}$ if the top contribution is included in the loops.

decoupling. For $4m_t^2/s \approx 1$ it is expected to be of order $(\alpha_s/\pi)^3$ with a non-enhanced coefficient. This leads to a tiny $Z - \gamma$ interference of order

$$\mathcal{O}\Big(\frac{\Gamma_Z}{M_Z}(\frac{\alpha_s}{\pi})^3\Big) \approx \mathcal{O}(10^{-6}) \tag{34}$$

5. *Corrections for axial currents*

Flavour non-singlet corrections to $r^{(AA)}$ in the massless limit are the same as those for the vector current. In one and two loop

$$\begin{aligned}
c_1^{(AA)} &= 1 \\
c_2^{(AA)} &= 1.986 - 0.115 n_f
\end{aligned} \tag{35}$$

Whereas induced flavour singlet contributions to $r^{(i,j)}$ arise for vector currents only in $\mathcal{O}(\alpha_s^3)$, they are present already in second order* for $r^{(AA)}$ as is evident from Fig. 9.

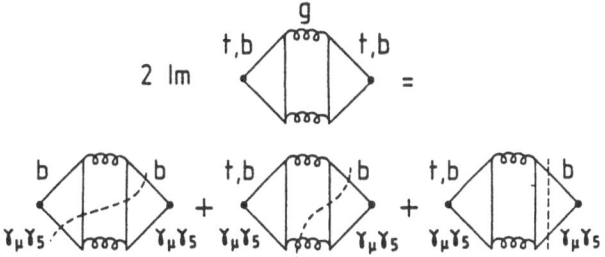

Fig. 9. Feynman diagrams pertinent to the $\mathcal{O}(\alpha_s^2)$ corrections calculated in eq. (36). Permutations are omitted.

The triangle anomaly originating from the t and b quark loops individually is compensated in the sum ($a_t = -a_b$!). Mass dependent terms, however, survive this cancellation.

---

*For vector currents this part is absent as a consequence of Furry's theorem [26].

In the limit of $m_t^2/s \to \infty$ the result increases logarithmically – a signal of the break down of anomaly cancellation if top is removed from the theory.

There is increasingly strong experimental evidence that $2m_t \gtrsim M_Z$. Therefore one expects that the coefficient of this $\mathcal{O}(\alpha_s^2)$ correction is sizeable in practice. The relevant contributions originate from two-, three- and four-particle intermediate states as shown in Fig. 9. (Two on-shell gluons are forbidden by Yang's theorem.) Each term is individually infrared and ultraviolet finite and can be interpreted as contribution to the two-, three- and four-jet cross section. A lengthy calculation [27] leads to the following correction terms which appear for Γ_A in addition to those discussed above:

$$\delta\Gamma_A(b\bar{b}) = \frac{1}{3}(\frac{\alpha_s}{\pi})^2 I((M_Z/2m_t)^2)\Gamma_A^{QPM} \tag{36}$$

with

$$\begin{aligned}
I(a) &= I_2(a) + I_3(a) + I_4 \\
&= +6l + 2A^2 - \frac{15}{4} \\
&\quad + \sqrt{\frac{1}{a} - 1}\left\{ 2\,\mathrm{Cl}_2(2A) + 2A(2l - 3) - \frac{1}{a}\Big[\mathrm{Cl}_2(2A) + 2A(l-1)\Big]\right\} \\
&\quad - \frac{1}{a}(4A^2 + 1) - \frac{1}{a^2}\Big(\mathrm{Cl}_3(2A) - \zeta(3) + A\,\mathrm{Cl}_2(2A)\Big) \\
&\approx -9.250 + 1.037a + 0.632a^2 + 6l
\end{aligned} \tag{37}$$

where $A = \arcsin\sqrt{a}$, $l = \ln 2\sqrt{a}$, $\zeta(3) = 1.202\ldots$ and Cl_i denotes Clausen's function of i-th order [28]. $2m_t \geq M_Z$ and $m_b = 0$ is assumed. I_2, I_3 and I_4 characterize the two, three and four parton configurations. The function $I_4 = \frac{\pi^2}{3} - \frac{15}{4}$ is independent of m_t, I_3 approaches a constant for large m_t and I_2 increases logarithmically. I_2, I_3 and I_4 are shown in fig. 2a, the overall correction $I/3$ in fig. 2b.

The magnitude of the additional terms is well comparable to the $\mathcal{O}(\alpha_s^2)$ part of the nonsinglet corrections (eq. (35)) and varies between -0.18 and -0.81% for m_t between $M_Z/2$ and 250 GeV and $\alpha_s/\pi = 0.04$.

It should be mentioned that also final states with light quarks (u, d, c, s) are individually affected by these considerations – however, the individual changes compensate in the total rate. Specifically, the effect on two and three parton configurations is opposite equal for u- and d-type quarks respectively and can be read directly from eq. (36). Additional terms in the cross section with the four partons $u\bar{u}\,u\bar{u}$ and $d\bar{d}\,d\bar{d}$ respectively are equal in sign and magnitude. They are compensated by a corresponding term from the "mixed" configuration $u\bar{u}\,d\bar{d}$ which appears with a relative factor -2.

The logarithmically increasing term in I_2 can be deduced directly from Refs. 29, 30. Consider a kinematical situation where the masses of both U and D quarks in the loop

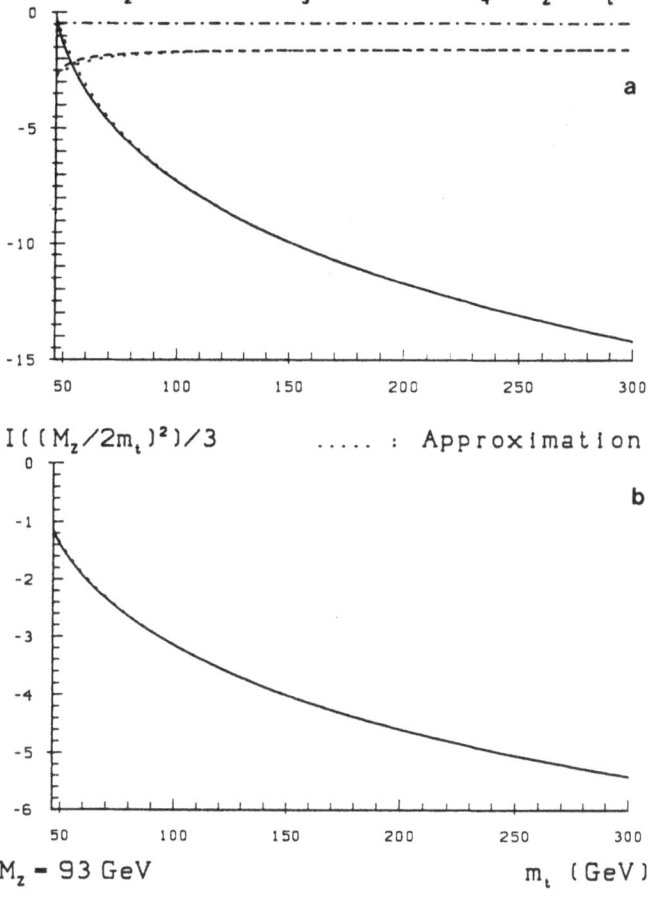

Fig. 10. The functions I_2, I_3, I_4 which determine the corrections for two, three and four parton final states. b) The function $I/3$ which determines the overall corrections.

are larger than $\sqrt{s}/2$ (as considered in Ref. 31) and where the quarks in the final state are massless. Then an effective flavour singlet piece is induced in the axial current

$$J_A = \frac{1}{2}(u\bar{u} - d\bar{d}) + \frac{1}{2}(u\bar{u} + d\bar{d})(\frac{\alpha_s}{\pi})^2 \frac{1}{6}\Big(I_2(s/(2m_U)^2) - I_2(s/(2m_D)^2)\Big) \qquad (38)$$

and

$$\frac{1}{6}\Big(I_2(s/(2m_U)^2) - I_2(s/(2m_D)^2)\Big) \xrightarrow[m_Q \to \infty]{} \ln\frac{m_U}{m_D} \qquad (39)$$

in accordance with Refs. 29, 30, 31.

Acknowledgements: I would like to thank Abdelhak Djouadi, Bernd Kniehl, Maria Krawczyk, Robin Stuart and Peter Zerwas for pleasant collaborations on various topics treated in this review.

REFERENCES

[1] U. Amaldi at al., *Phys. Rev.* **D 36** (1987) 1385.

[2] B.A. Kniehl, M. Krawczyk, J.H. Kühn and R.G. Stuart, *Phys. Lett.* **209 B** (1988) 337.

[3] W. deBoer, Proceedings of the Workshop on Radiative Corrections for e^+e^- Collisions, Springer Verlag, 1989, ed. J.H. Kühn.

[4] G. Altarelli et al., in "Physics at LEP", CERN Report 86-02 (1986).

[5] E.A. Kuraev and V.S. Fadin, Sov. J. Nucl. Phys. **41** (1985) 466.

[6] G.J.H. Burgers, *Phys. Lett.* **164 B** (1985) 167.

[7] F.A. Berends, G.J.H. Burgers and W.L. van Neerven, *Phys. Lett.* **185 B** (1987) 395; *Nucl. Phys.* **B 297** (1988) 429; Erratum, *Nucl. Phys.* **B 304** (1988) 92.

[8] R. Barbieri, J.A.Mignaco and E. Remiddi, *Nuovo Cimento* **11 A** (1972) 824,865.

[9] B. Grzadkowski, J.H. Kühn, P. Krawczyk and R.G. Stuart, *Nucl. Phys.* **B 281** (1987) 18.

[10] M. Böhm, H. Spiesberger and W. Hollik, *Fortsch. Phys.* **34** (1986) 687.

[11] H. Burkhardt, private communication and Tasso Note No.192.

[12] V.V. Sudakov, *Soviet. Phys. JETP* **3** (1956) 65.

[13] W. Burkhardt, Proceedings of the Workshop on Radiative Corrections for e^+e^- Collisions, Springer Verlag, 1989, ed. J.H. Kühn.

[14] T.H. Chang, K.J.F. Gaemers and W.L. van Neerven, *Nucl. Phys.* **B 202** (1982) 407.

[15] A. Djouadi, Preprint PM/87-53;
A. Djouadi and C. Verzegnassi, *Phys. Lett.* **195 B** (1987) 265.

[16] F. Jegerlehner, in "Proc. XI Int. School of TP", Szczyrk, Poland, World Scientific(1988).

[17] B.A. Kniehl, J.H. Kühn and R.G. Stuart, *Phys. Lett.* **214 B** (1988) 621.

[18] J.L. Richardson, *Phys. Lett.* **82 B** (1979) 272.

[19] J. Drees et al., in "Polarization at LEP" CERN 88-06, Vol. 1, pg.317;
J. Drees, Proceedings of the Workshop on Radiative Corrections for e^+e^- Collisions, Springer Verlag, 1989, ed. J.H. Kühn, and refs. therein.

[20] J.H. Kühn, P.M. Zerwas et al., Heavy Flavours at LEP, MPI-PAE/PTh 49/89, sect.2.2.

[21] A. Djouadi, J.H. Kühn and P.M. Zerwas, preprint MPI-PAE/PTh 48/89.

[22] J. Jersak, E. Laermann and P.M. Zerwas, *Phys. Lett.* **98 B** (1981) 363.

[23] S.G. Gorishny, A.L. Kataev and S.A. Larin, *Phys. Lett.* **212 B** (1988) 238.

[24] J. van der Bij, *Nucl. Phys.* **B 313** (1989) 237.

[25] S. Jadach, J.H. Kühn, R.G. Stuart and Z. Was, *Z. Phys.* **C 38** (1988) 609.

[26] W.H. Furry, *Phys. Rev.* **51** (1937) 125.

[27] B.A. Kniehl and J.H. Kühn, *Phys. Lett.* **224 B** (1989) 229; MPI-PAE/PTh 39/89.

[28] R. Lewin, Dilogarithms and Associated Functions, MacMillan, London

[29] S. Adler, *Phys. Rev.* **177** (1968) 2426.

[30] J. Collins, F. Wilczek and A. Zee, *Phys. Rev.* **D 18** (1978) 242.

[31] Y. Kizukuri et al., *Phys. Rev.* **D 23** (1981) 2095.

QCD CORRECTIONS TO HIGGS DECAY AND PRODUCTION AT LEPI/SLC *

A. Djouadi and F. Oualitsen

Institut Theor. Physik, RWTH Aachen, D-5100 Aachen, FRG

ABSTRACT

Strong interaction corrections to the decay rates and the production cross sections of the minimal standard model Higgs boson are discussed for the processes which are relevant to LEPI and SLC experiments.
A recent calculation of the QCD correction to the decays $H \longrightarrow \gamma\gamma$ and $Z \longrightarrow H\gamma$ in the limit of very large fermion masses is presented.

The major ingredient of the standard electroweak theory, the Higgs boson which is the remnant of the mechanism responsible for the spontaneous symmetry breaking of $SU(2)_L \times U(1)_Y$, is still waiting for experimental evidence. Its mass is not predicted by the model but theoretical considerations favor a lower limit of $O(10$ GeV$)$ and an upper limit of $O(1$ TeV$)$. This range is, typically, the one which will be covered by the new generation of e^+e^-, pp and $p\bar{p}$ machines.
If the Higgs is fairly light ($M_H \leq M_W$), the experiments at LEPI and SLC may already offer a very good opportunity to discover it. This possibility has been intensively reviewed in two big reports [1,2] and a third one [3] is in preparation.
A carefull study of the production cross sections and the decay rates of the Higgs boson must include radiative corrections and especially the QCD corrections which could give rise to sizeable contributions due to the large value of the strong interaction coupling constant.
In what follows, we shall discuss these corrections to the main processes relevant at LEPI/SLC. We shall restrict ourselves to the minimal case where the gauge symmetry is broken with one doublet of scalar fields, leading to a scalar Higgs particle whose mass, to be definite, will be assumed to lie in the range:

$$10 \text{ GeV} \leq M_H \leq 60 \text{ GeV} \tag{1}$$

---

*Supported in part by the W. German Bundesministerium für Forschung und Technologie.

Decay rates

1) $H \longrightarrow f\bar{f}$

The special feature of the Higgs boson is that it couples most strongly to heavy particles and thus tend to decay into the heaviest available ones. Since the decay into the gauge bosons is kinematically forbidden at the Z peak, the most important channel will be into a heavy fermion–antifermion pair. The rate is given by

$$\Gamma(H \to f\bar{f}) = \frac{N_c G_F m_f^2}{4\sqrt{2}\pi} M_H \, v^3 \tag{2}$$

where N_c is a color factor and $v^2 = 1 - 4m_f^2/M_H^2$. Because the top quark is expected to be rather heavy, the Higgs will preferencially decay into $\tau^+\tau^-$, $c\bar{c}$ and mostly $b\bar{b}$. For m_b=5 GeV the partial width goes from 1 Mev for $M_H \simeq 20$ GeV to 3 MeV for $M_H \simeq 60$ GeV (see the numbers given in [2]).

The QCD corrections to the decay width into a quark–antiquark pair (see Fig.1) have been computed almost ten years ago, first by Braaten and Leveille [4] and later in ref.[5]. Their result can be written in the form:

$$\Gamma^{(\alpha_s)} = \Gamma \left[1 + \frac{\alpha_s}{\pi} \Delta_H(v) \right] \tag{3}$$

The complete expression of $\Delta_H(v)$ is rather lengthy but we shall be interested here only in its behavior in the two limiting situations: $M_H \simeq 2m_q$ and $M_H \gg m_q$. In the former case it exhibits the well known singularity near threshold:

$$\Delta_H(v \to 0) \sim \frac{2\,\pi^2}{3\,v} - \frac{4}{3}$$

which invalids the perturbative calculation. As usual one has to sum all the terms $(\alpha_s/v)^n$ to make the prediction trustworthy. In the second case the correction Δ_H shows the rather unexpected singularity:

$$\Delta_H(v \to 1) \sim -4\ln\frac{m_q}{M_H} + 3$$

coming from the renormalization of the fermion–Higgs boson coupling. Indeed Kinoshita's theorem guarantees the absence of mass singularities only in the unrenormalized (one can also include mass renormalization) total decay width. This large logarithmic term needs to be summed to all orders of perturbation theory to make the prediction reliable. This gives a corrected rate (Λ is the QCD scale and we use 5 flavors) [4,5]:

$$\Gamma^{(\alpha_s)} = \Gamma \left(\frac{\ln 2m_q/\Lambda}{\ln M_H/\Lambda} \right)^{24/23} \left[1 + \frac{\alpha_s}{\pi} \left(\Delta_H(v) + 4\ln\frac{m_q}{M_H} \right) \right] \tag{4}$$

which shows that even after the summation of the leading logs the correction remains large for moderate values of the Higgs mass and lead to a decrease of the width of about $\sim 30\%$ for $M_H \simeq M_W$.

2) $H \longrightarrow \gamma\gamma$

Another decay mode of the Higgs boson is the decay into two photons. The coupling $H\gamma\gamma$ does not occur at the tree level but is induced through the triangle diagrams

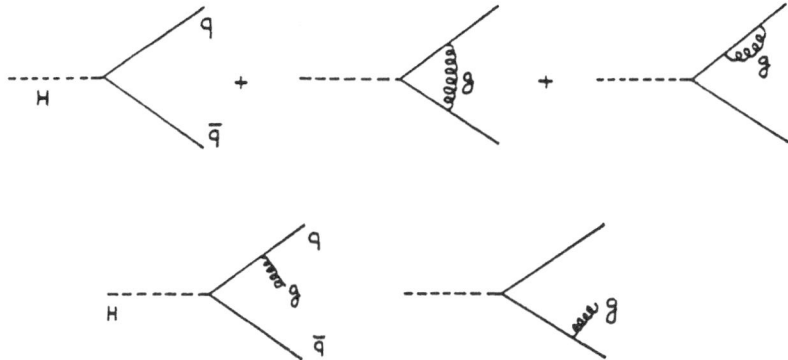

Fig. 1   Diagrams contributing to $H \longrightarrow q\bar{q}(g)$

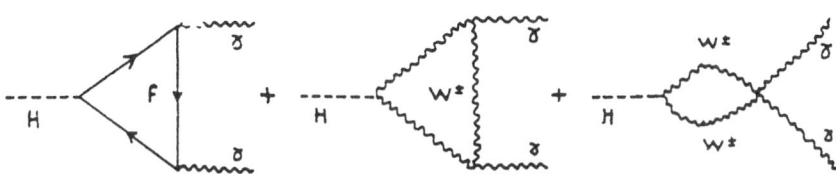

Fig. 2   Diagrams contributing to $H \longrightarrow \gamma\gamma$. Crossed diagrams are to be added.

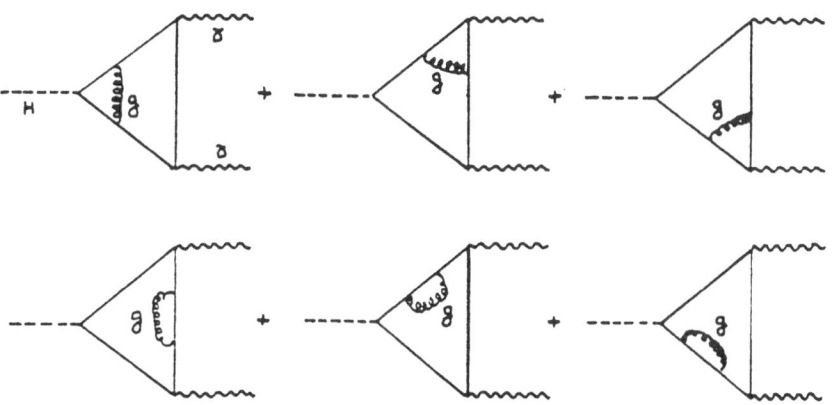

Fig. 3   Diagrams contributing to the QCD corrections for $H \longrightarrow \gamma\gamma$. Crossed diagrams and contreterms are to be added.

shown in Fig.2. The width is given by [6]:

$$\Gamma(H \longrightarrow \gamma\gamma) = \frac{\alpha^2 G_F}{8\sqrt{2}\pi^3} M_H^3 \left| \sum_F N_c Q_F^2 I_F + I_W \right|^2 \qquad (5)$$

Due to the proportionality of the $Hf\bar{f}$ coupling to the fermion masses, the light fermions contribution is negligibible (for $m_F < M_H/2$)[1], so that in the minimal standard model with three generations of fermions only the top quark and the W boson contribute. The expression of I_F and I_W can be approximated, for the Higgs mass range relevant at the Z peak, by

$$
\begin{aligned}
I_W &= -\frac{7}{4} \\
I_F &= \frac{1}{3}\left(1 + \frac{7}{120}\mu^2\right)
\end{aligned}
\qquad (6)
$$

where $\mu = M_H/m_F$. Some important comments are, at this stage, appropriate.

First of all eq.(6) shows a feature which can be refered to as "the decoupling of the decoupling theorem". Indeed, due to the fact that H couples proportionally to their masses, heavy particles need not to decouple from low energy processes. This reaction, thus, can be seen as counting the number of heavy fermions present in the theory.

The coefficients I_W and I_F have opposite signs. The two contributions interfere destructively but the bosonic one dominates in the the minimal standard model. However if there is one more generation of heavy fermions it would give a contribution $\frac{8}{9}$ which, when added to that of the top quark, will almost cancel the W contribution, making this decay invisible. So any significant change of these coefficients (due to radiative corrections for instance) will lead to a dramatically different situation.

One can also notice that the fermionic contribution is practically not affected by the ratio $\mu = M_H/m_F$ so that a fermion can be considered as heavy, already if $m_F > M_H/2$ (see also footnote). This property is only accidental and needs not to be true for the higher order corrections.

Finally, since the coefficients in eq.(6) are of O(1) the decay width is small. However this process might become very important for Higgs production (with mass around 100 GeV) with the advent of the new generation of linear e^+e^- machines. As a matter of fact, Blankenbecler and Drell have recently shown [8] that, under some realistic assumptions on the parameters of the new accelerators, one can produce roughly 100 H's per day in two photons fusion via Beamstrahlung!

Let us now turn to the discussion of QCD radiative corrections to this process. The reason why one should worry about them is that, as we have seen previously, they are very large for the decay into quarks. In addition, Visotsky [9] has shown that these corrections are also very large in the case of the decay of a toponium, say, into $H\gamma$ [10] which also proceeds via triangle diagrams.

To derive these corrections we computed the relevant six two–loop diagrams depicted in Fig.3 and the corresponding counterterms. We performed this calculation in the limit of large fermion mass but retained the leading M_H/m_Q term because, as we already mentionned, it might give sizeable contribtions in the range $M_H \sim m_Q$, contrary to the case of the one–loop approximation.

---

[1]Note that this contribution is peaked in a very small region around the threshold, but here, non–perturbative effects occur [7] and we shall not discuss them.

The technical details will be given elsewhere [11]. Here we only report our result and spend few comments on it. It is as follows:

$$I_Q^{\alpha_s} = \frac{1}{3}(1 + \frac{7}{120}\mu^2) \times \left[1 + \frac{\alpha_s}{\pi}\left(1 - \frac{1}{4}\mu^2\right)\right] \tag{7}$$

As it can be seen this correction is very small ($\leq 5\%$ for reasonnable values of α_s). The mass effects are no longer negligible but (if we trust this approximation) they are still small in the range of Higgs masses that we are interested in (we assume that $m_t > 70$ GeV).

Contrary to the case of the QCD correction to $H \to q\bar{q}$ the result does not contain mass singularities. This fact can be seen as follows. The counterterm renormalizing the Higgs–quark coupling is simply given by:

$$i\partial\Sigma(\hat{p})/\partial\hat{p}|_{\hat{p}=m} + i\Sigma(\hat{p} = m)$$

where Σ is the self energy of the fermion of momentum p and mass m. The first piece will compensate one of the wave function counterterms (the two other ones being canceled by the photon vertices couterterms by virtue of the Ward identity). The second piece is just a mass counterterm (non singular) which has to be multiplied by the one–loop result so that, practically, we are only dealing with mass renormalization and we already know that it does not introduce any singularity.

A final comment is that, one can can easily derive the correction of eq.(7) in the limit $M_H = 0$ (i.e without the μ^2 term). Indeed, the ITEP [12] group has proposed a low-energy theorem which relates the amplitude $H \to \gamma\gamma$ to the photon self–energy. Due to the fact that fermion masses are generated via spontaneous symmetry breaking, the net effect of emitting a massless Higgs from the vacuum polarization function of the photon is to replace the fermion mass m_F by $m_F(1 + (\sqrt{2}G_F)^{1/2}\varphi)$ in the expression of the latter (φ is the Higgs field). When including the QCD corrections to the photon self–energy one gets, after substitution, the result (7) without the μ term.

3) $H \longrightarrow gg$

As for the precedent case, this process proceeds via quark loops (see Fig.4). The decay width is found to be :

$$\Gamma(H \longrightarrow gg) = N_H^2 \frac{\alpha_s^2 G_F}{4\sqrt{2}\pi^3} M_H^3 |I_Q|^2 \tag{8}$$

where N_H is the number of quarks obeing $m_Q > M_H/2$ and I_Q is given by the second part of eq.(6). Because of the large value of α_s this decay is much more important than the two photons decay and is of comparable size as the decay into tau's and c's when M_H approaches M_Z (even for $N_H = 1$). Since there is now no W loops which interfere, this process acts truly as a "cash register" of heavy quarks (note that it could also get contributions from possibly new strong interacting particles).

Another reason why this process is very important is that, when turned arround, it is the best hope to discover the Higgs boson in pp collisions [13]. Indeed, if its mass is larger than 200 GeV, gluon fusion (together with massive gauge boson fusion) will be the most efficient way to produce it.

Because of the presence of the three–gluon coupling (which gives several additional diagrams), the computation of the QCD corrections to $H \to gg$ is much more complicated than for $\gamma\gamma$. This calculation is under way[14].

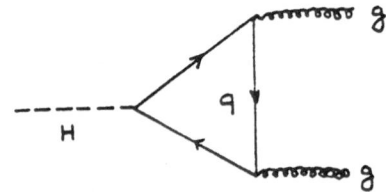

Fig. 4   Diagram contributing to $H \longrightarrow gg$.

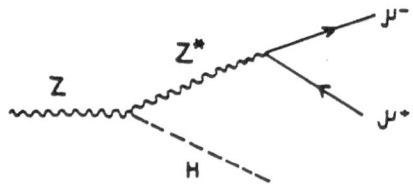

Fig. 5   Diagram contributing to $Z \longrightarrow H\mu^+\mu^-$.

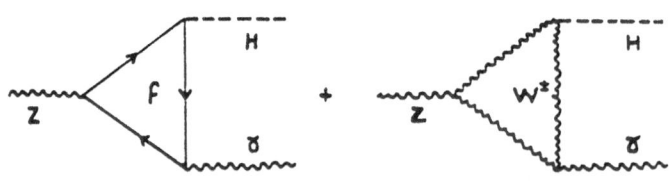

Fig. 6   Diagrams contributing to $Z \longrightarrow H\gamma$.

Production cross sections

1) $Z \longrightarrow H f \bar{f}$

Since the toponium has a too large mass to be produced at LEP/SLC (otherwise it would have acted like a Higgs factory [10]) the most promizing process for Higgs production at the Z resonance is:

$$e^+ e^- \longrightarrow Z \longrightarrow H \, Z^* \longrightarrow H f \bar{f}.$$

As shown in Fig.5 the real Z emits the Higgs and goes off–shell to decays into a fermion–antifermion pair. The rate is given in [15] and the number of $H \mu^+ \mu^-$ events for 10^7 Z goes from 660 for $M_H \simeq 10$ GeV to 10 events for $M_H \simeq 55$ GeV.

The channel $Z \longrightarrow H q \bar{q}$ is not very appealing to look for because of the huge hadronic jets background and the QCD corrections to the decay of the Z into the quark–antiquark pair might not be relevant.

2) $Z \longrightarrow H \gamma$

Another interesting mechanism to produce the Higgs boson is the decay $Z \longrightarrow H\gamma$ [16]. For a Higgs mass of 50 GeV it is comparable to the previous one and becomes dominant for heavier Higgs. As for the decay of the Higgs into two photons, it proceeds via triangle diagrams (Fig.6). The rate is given by :

$$\frac{\Gamma(Z \to H\gamma)}{\Gamma(Z \to \mu^+ \mu^-)} \simeq 2.5 \times 10^{-5} \left(\frac{E_\gamma}{M_Z} \right)^3 |A|^2 \tag{9}$$

with $A = A_F + A_W$. Since the axial part of the Z coupling to the virtual fermions does not contribute (the Higgs and the photon have opposite C–parity and thus only the vectorial part couples), the fermionic contribution is identical to that of $H \to \gamma\gamma$, after altering the electroweak charges appropriately, in the limit of a large fermion mass:

$$A_F \longrightarrow \frac{2}{3} N_c Q_F \frac{(T_F^{3L} - 2 Q_F \sin^2 \theta_W)}{\cos \theta_W} \tag{10}$$

The absence of the axial coupling of the Z and the smallness of the vectorial one will make all these contributions rather small: a heavy charged lepton gives a contribution of 0.045 and heavy quarks 0.27 and 0.31 for down type and up type quarks respectively (we use $\sin^2 \theta_W = 0.22$). In this limit, the QCD correction is the same as the one that we have obtained for $H \to \gamma\gamma$ i.e:

$$A_Q^{(\alpha_s)} = A_Q \times \left[1 + \frac{\alpha_s}{\pi} \right] \tag{11}$$

The expression of the bosonic loop may be found in [16]. It interferes destructively with the fermionic one but it is always dominating (at least 15 times larger than the contribution of a very heavy top) unless the number of new generations exceeds six.

Since the factor A is of order $O(1)$ the process $Z \to H\gamma$ has a small branching ratio (~ 10 events per 10^7 Z's for $M_H \sim 50$ GeV), but its clear signature may compensate for this inconvenient.

We have discussed the QCD radiative corrections to Higgs decay and production at the Z peak, in the framework of the minimal standard model. These corrections are large in the case of Higgs decay into a quark–antiquark pair but small for the decay into two photos and for Higgs production via the the process $Z \to H\gamma$.

Acknowledgements

We are grateful to P. M. Zerwas for suggesting this topic and for instructive discussions. A. D. would like to thank the organizers of the conference, especially F. Boudjema and N. Dombey, for their warm hospitality.

REFERENCES

[1] G. Barbiellini et al., DESY Report 79/27 (1979).

[2] H. Baer et al., CERN Yellow Book 86-02 (1986) 297.

[3] J. P. Franzini and P. Taxil, LEP Workshop, in preparation.

[4] E. Braaten and J. P. Leveille, Phys. Rev. D22 (1980), 715.

[5] N. Sakai, Phys. Rev. D22 (1980), 2220;
T. Inami and T. Kubota, Nucl. Phys. B179 (1981), 171.

[6] J. Ellis, M. K. Gaillard and D. V. Nanopoulos, Nucl. Phys. B106 (1976) 292;
M. K. Gaillard, Com. Nuc. Par. Phys. 8(1978) 31.

[7] M. Drees and H. Hikasa, CERN Preprint TH/89-5393 (1989).

[8] R. Blakenbecler and S. D. Drell, Phys. Rev. Lett. 61 (1988) 24.

[9] M. I. Visotsky, Phys. Lett 97B (1980), 159.

[10] F. Wilczek, Phys. Rev. Lett. 39 (1977) 1304.

[11] A. Djouadi and F. Oualitsen, Aachen Preprint, in preparation.

[12] A. I. Vainshtein et al., Sov. J. Nucl. Phys. 30 (1979) 711.

[13] H. Georgi et al., Phys. Rev. Lett. 40 (1978) 692.

[14] A. Djouadi, M. Spira and P. M. Zerwas, in preparation.

[15] J. D. Bjorken, Proc. 1976 SLAC Summer Inst., ed. M. Zipf, Stanford 1976.

[16] R. N. Cahn, M.S. Chanowitz and N. Fleishon, Phys. Lett. 82B (1979) 113.

ELECTROWEAK RADIATIVE CORRECTIONS IN DEEP
INELASTIC ELECTRON PROTON SCATTERING

W. Hollik

CERN, Geneva, Switzerland

Abstract

An overview is given on higher order electroweak effects in neutral and charged current processes in deep inelastic ep scattering at HERA energies in the framework of the Standard Model based on $SU(3)_C \times SU(2)_W \times U(1)$.

1 Introduction

The fundamental process in neutral and charged current deep inelastic lepton-proton scattering is the 4-fermion scattering between the lepton and the quarks inside the nucleon: $lq \rightarrow lq$. In lowest order this process is of pure electroweak origin, in the Standard Model mediated by the exchange of photons and weak Z^0 resp. $W^{\pm}$ bosons.

The electroweak Standard Model contains, besides fermion masses, quark mixing angles, and the mass of the Higgs scalar, three free parameters in the gauge sector. For a comparison between theory and experiment three independent experimental input data are required for fixing the SU(2) and U(1) gauge coupling constants g_2, g_1, and the vacuum expectation value v of the Higgs field. It is, however, more practical to deal with parameters such that each of them has a direct relation to a specific experiment. For deep inelastic scattering a natural choice is given by the electromagnetic fine structure constant α (Thomson scattering), and the masses of the Z and W bosons characterizing the photon, neutral current Z and charged current W exchange. The masses of the Higgs boson and the top quark enter the higher order calculations as additional free parameters unless a direct experimental determination of their values is available.

The existing calculations of electroweak radiative corrections for HERA processes [1-6] follow the lines of the on-shell scheme which treats the masses of the vector bosons M_Z and M_W in a symmetric way: the one-loop amplitudes for 4-fermion processes are expressed in terms of M_Z and M_W (besides α) as the basic quantities defined as the real parts of the pole positions in the Z and W propagators. The Fermi constant G_{μ}, measured from the muon lifetime, can then be calculated as a constraint

$$G_{\mu} = G(\alpha, M_Z, M_W, M_H, m_t)$$

Radiative Corrections, Edited by N. Dombey and
F. Boudjema, Plenum Press, New York, 1990

on the value of M_W after specifying the other quantities. In practice this constraint is used to determine the numerical value for M_W. This removes the dependence of the measurable quantities in deep inelastic scattering from explicit W mass measurements which will not reach the accuracy of the Fermi constant.

After the choice of three input data all other mesurable quantities are predictions according to the structure of the theory. The relations between the experimental quantities, however, are changed when higher order perturbative contributions are included. The electroweak Standard Model as a spontaneously broken gauge theory implies that the theoretical prediction for any observable quantity can (at least in principle) be calculated to an arbitrary order in perturbation theory. In practice the lowest order expressions are supplemented by the one-loop radiative corrections which can be improved by certain higher order leading terms where they tend to become large. For an adequate analysis of experiments the discussion of radiative corrections is an indispensable part:

- Experimental tests of the validity of radiative corrections in various types of reactions and in various kinematical domains are tests of the Standard model at the quantum level.

- For detecting signals from possible "new physics" the Standard Model predictions have to be known with the adequate accuracy.

- The special situation in ep collisions allows to probe the structure of the proton by measurements of the quark distributions, thereby testing also the QCD sector of the Standard Model. For this purpose reliable predictions for the electroweak sector are necessary.

- Radiative corrections can be large. This applies in particular to the conventional QED bremsstrahlung corrections. These QED contributions do not contain additional information about the weak interaction part of the complete theory, but they constitute the bulk of the radiative corrections in particular for deep inelastic processes at high momentum transfer. Their proper treatment in the data analysis is of basic importance for testing the QCD and the non-QED electroweak part of the Standard Model.

The radiative correction problem in deep inelastic ep scattering is therefore characterized by the interplay of the strong interaction between the quarks inside the nucleon and the electroweak interactions between the quarks and the incident leptons. The hadronic part as yet cannot be calculated from first principles and has to be treated as an additional experimental input entering the radiative corrections. On the other hand, this hadronic input cannot be extracted from experiments without knowledge of the size of the electroweak radiative corrections.

The present contribution will give an update of the situation of electroweak radiative corrections for neutral and charged current HERA processes. Recent comparisons between independent calculations show that the numerical results are now in agreement within about 1% and are therefore checked against each other with sufficient accuracy.

2 Neutral current deep inelastic scattering

2.1 Lowest order differential cross section

In the quark-parton model and also in the leading log approximation of QCD the fundamental processes may be considered to be the neutral current 4-fermion scattering processes between the electron and the quark constituents of the proton, as indicated in Figure 1.

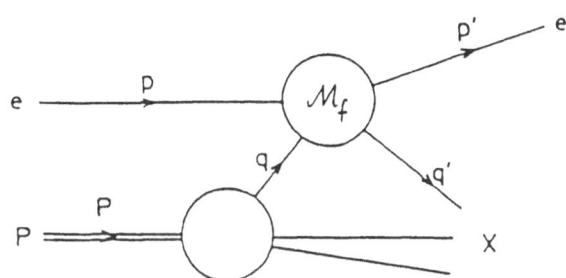

Figure 1. Deep inelastic scattering in the parton model

Defining the kinematical variables x, Q^2 by

$$Q^2 = -(p - p')^2 \, , \tag{1}$$
$$q = x \, P$$

the differential cross section for $e\,p \to e\,X$ in the parton picture is given by

$$\frac{d^2\sigma}{dx \, dQ^2} \sim \sum_f |\, \mathcal{M}_f \,|^2 \cdot q_f(x) \tag{2}$$

where $q_f(x)$ denotes the distribution function of the quark with flavor f inside the proton.

For left/right-handed polarized incident electrons the lowest order cross section is explicitly written with help of the dimensionless variable $y = Q^2/xS$, $S = (p + P)^2$ as follows:

$$\frac{d^2\sigma}{dx \, dQ^2}(e_{L,R}) = \frac{2\pi\alpha^2}{xQ^4} \left[\left(1 + (1 - y)^2\right) F_2^{L,R} + \left(1 - (1 - y)^2\right) x F_3^{L,R} \right] \tag{3}$$

The structure functions

$$F_2^{L,R} = \sum_f (xq_f + x\bar{q}_f) \cdot A_f^{L,R} \tag{4}$$
$$x F_3^{L,R} = \sum_f (xq_f - x\bar{q}_f) \cdot B_f^{L,R}$$

469

contain the quark (q_f) and anti-quark $(\bar{q}_f)$ distribution functions as well as coupling constants and propagators corresponding to photon and Z boson exchange:

$$A_f^{L,R} = Q_f^2 + 2Q_e Q_f (v_e \pm a_e) v_f \frac{Q^2}{Q^2 + M_Z^2} + (v_e \pm a_e)^2 (v_f^2 + a_f^2) \left(\frac{Q^2}{Q^2 + M_Z^2}\right)^2 \quad (5)$$

$$B_f^{L,R} = \pm 2Q_e Q_f (v_e \pm a_e) a_f \frac{Q^2}{Q^2 + M_Z^2} \pm 2(v_e + a_e) v_f a_f \left(\frac{Q^2}{Q^2 + M_Z^2}\right)^2$$

According to the γ, $\gamma - Z$ interference, and Z contributions the cross section can also be decomposed into

$$d^2\sigma = \sigma_\gamma + \sigma_{\gamma Z} + \sigma_Z . \quad (6)$$

In the range $Q^2 \geq 10^3$ GeV^2 the interference and pure Z terms become of non-negligible influence; at $Q^2 \simeq 10^4$ GeV^2 they are of the same size as the pure photon exchange contribution.

The parametrization of the lowest order cross section in terms of the vector and axial vector coupling constants $v_{e,f}$, $a_{e,f}$ is quite general and applies to any model where a vector boson with v and a couplings is exchanged together with the photon. In the Standard Model, however, these quantities are not independent. For a quantitative evaluation a well known ambiguity in the Born level presentation shows up: the coupling constants can either be expressed by means of (Q_f and I_3^f denote charge and weak isospin of the fermion f)

$$v_f = \frac{I_3^f - 2s_W^2 Q_f}{2s_W c_W} , \quad a_f = \frac{I_3^f}{2s_W c_W} , \quad (7)$$

with the mixing angle in the minimal model

$$s_W \equiv \sin\theta_W , \quad c_W \equiv \cos\theta_W = M_W / M_Z , \quad (8)$$

or in the following way with help of the Fermi constant:

$$a_f = \left(\frac{\sqrt{2} G_\mu M_Z^2}{4\pi\alpha}\right)^{1/2} I_3^f \quad (9)$$

$$v_f = \left(\frac{\sqrt{2} G_\mu M_Z^2}{4\pi\alpha}\right)^{1/2} (I_3^f - 2Q_f s_W^2).$$

The relation between these two parametrizations is established by identifying the lowest order expression for the muon lifetime with the effective Fermi model result:

$$\frac{G_\mu}{\sqrt{2}} = \frac{\pi\alpha}{2s_W^2 M_W^2} . \quad (10)$$

The numerical results obtained from these two parametrizations are different in general. Of course, after incorporating consistently the weak radiative corrections described next we obtain an unambigous result.

2.2 Classification of the one-loop corrections

2.2.1 QED corrections

This gauge invariant subclass consists of all photonic virtual and real corrections which are obtained from the underlying Born diagrams by attaching a single photon line to each charged fermion line in all possible ways (diagrams in Figure 2 together with virtual photon insertions in the external fermion lines). The infrared singularities are cancelled in the sum of the 2- and 3-body final states after integrating the real photons over their kinematically and experimentally (by cuts) allowed phase space.

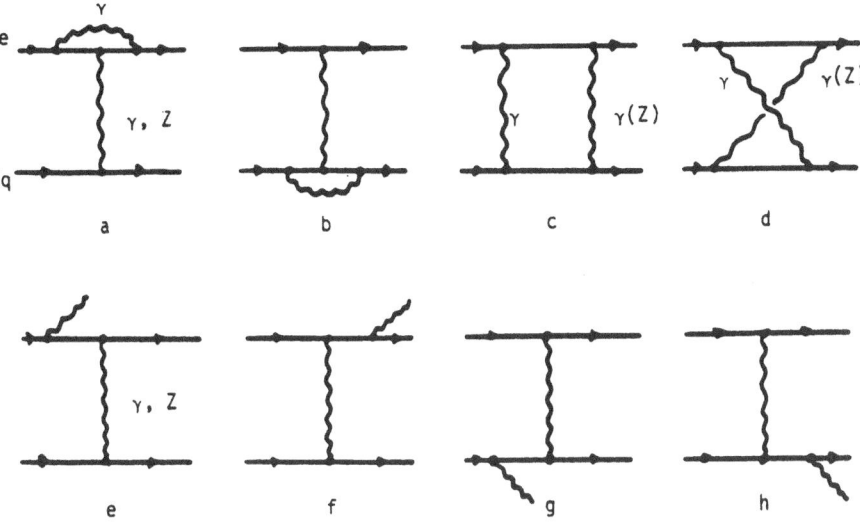

Figure 2 . QED corrections to $eq \rightarrow eq$

The QED corrections can be further subdivided into

- QED corrections at the lepton line (Figure 2 a,e,f);

- QED corrections at the quark line (Figure 2 b,g,h);

- the interference of bremsstrahlung from the lepton and the quark line (Figure 2 e-h) and the box diagrams (Figure 2 c,d) which connect the lepton and quark line by an extra virtual photon.

For the further discussion of the qualitatively different features it is important to note that each of these classes is infrared finite and gauge invariant.

2.2.2 Weak corrections

The remaining set of the infrared finite non-QED or weak corrections, schematically depicted in Figure 3, consists of the one-loop corrected diagonal gauge boson propagators, the non-diagonal propagator with $\gamma - Z$ mixing, the vertex corrections after splitting off the photonic terms [1] and the box diagrams with $Z - Z$ and $W - W$ exchange. Being independent of the experimental set up these contributions can be absorbed into an effective Born amplitude which is expressed in terms of Q^2 dependent vector and axial vector coupling constants. A comparison between different calculations is straightforward when it is ensured that the same physical input has been used for renormalization and computation of the cross section.

---

[1] these are always combined with the external fermion self energies

2.3   Existing calculations

For deep inelastic ep processes with $Q^2 \ll M_Z^2$ the most important electromagnetic corrections to the pure QED contribution (only γ exchange) have been discussed by Mo and Tsai [7] and later by Akhundov et al. [8].

QED corrections arising from real and virtual photons from the lepton line have been calculated in the leading logarithmic approximation by Kripfganz und Möhring [9], Beenakker et al. [10], and Blümlein [11]. The leading log (LL) contribution together with the fermionic part of the photon and Z self energies have already been treated earlier by Consoli and Greco [12]. The question of quark mass singularities according to QED bremsstrahlung from the quark line is discussed in [13] by extending the QCD Altarelli-Parisi equations for the Q^2 evolution of the quark distribution functions by the corresponding collinear QED terms.

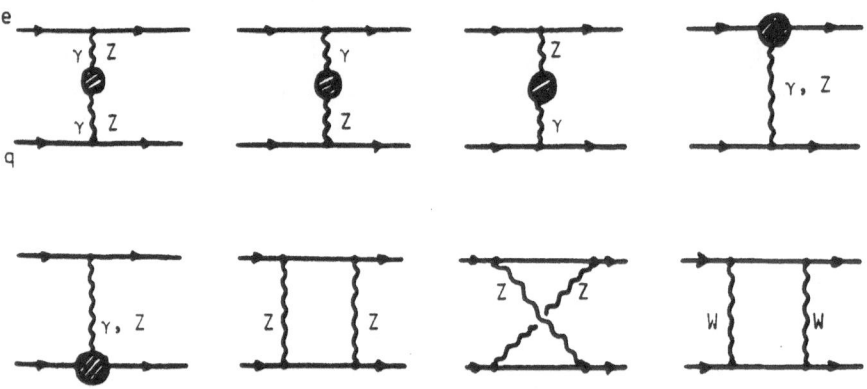

Figure 3 . Weak corrections to $eq \to eq$

Until now two independent calculations of the full $O(\alpha)$ corrections to the neutral current scattering process have been performed: the calculation by Bardin et al. [2,5] based on the on-shell scheme in the unitary gauge, and the calculation by Böhm and Spiesberger [3], also in the on-shell scheme but in the 't Hooft-Feynman gauge. Both calculations make use of analytical and numerical integration methods. All calculations in [2-6] treat the radiated photon as totally inclusive.

2.4   Discussion of the weak (non-QED) corrections

As outlined in the introduction, the concrete calculations of electroweak corrections for HERA processes have been performed by choosing α, M_W, M_Z as definition of a renormalized "physical" input for fixing the corresponding renormalized self-energies and vertex corrections at the one-loop level. This so-called "on-shell scheme" utilizes the weak mixing angle in the version [14]

$$s_W^2 = 1 - \frac{M_W^2}{M_Z^2}.$$ (11)

In practice, M_W as well as s_W^2 are determined from M_Z via the corrected form of the relation (10) [14]:

$$\frac{G_\mu}{\sqrt{2}} = \frac{\pi\alpha}{2s_W^2 c_W^2 M_Z^2} \cdot \frac{1}{1 - \Delta r}.$$ (12)

Since the correction term $\Delta r(\alpha, M_Z, M_W, M_H, m_t)$ depends also on the Higgs and top masses M_H, m_t, the s_W^2 determined in this way becomes dependent on the presently empirically unknown part of the theory. Note, however, that s_W^2 as derived from (12) for a given Z mass is not the effective mixing angle in the neutral current couplings. The coupling constants get additional contributions by $\gamma - Z$ mixing, vertex corrections, and heavy box diagrams. Also the dependence of the effective mixing angle on m_t and M_H is different from that of s_W^2.

The incorporation of the non-QED radiative corrections into 4-fermion amplitudes has been discussed in detail in several reports [15,16,17]. Here we concentrate only on the main results.

In a compact notation the weak corrections can be summarized in terms of an effective matrix element for the underlying parton subprocess.

photon exchange amplitude

The dressed photon amplitude can be written in the following way:

$$\mathcal{M}_\gamma = \frac{e^2}{1 - \hat{\Pi}_\gamma} \cdot \frac{1}{Q^2} \cdot [(Q_e + F_V^e)\gamma_\mu - F_A^e \gamma_\mu \gamma_5] \otimes \left[(Q_f + F_V^f)\gamma^\mu - F_A^f \gamma^\mu \gamma_5\right]$$ (13)

with Q^2 dependent vertex form factors $F_{V,A}$ and the vacuum polarization $\hat{\Pi}_\gamma(Q^2)$. Writing it in the denominator corresponds to the summation of the leading log terms from the light fermions in case of large Q^2. The bosonic content has to be understood as expanded to $O(\alpha)$. The typical size of these contributions at $Q^2 \simeq M_Z^2$ is:

$$\hat{\Pi}_\gamma \simeq 0.06,$$
$$F_{V,A} \simeq 0.001.$$

Aiming a precision of 1% the restriction to the fermionic vacuum polarization is sufficient. For the hadronic part a convenient parametrization has been given by Burkhardt et al. [18] which fits the results obtained from the updated dispersion integral over experimental $e^+ e^- \rightarrow hadrons$ data.

Z exchange amplitude

The dressed Z amplitude can be expressed with help of Q^2 dependent form factors ρ and κ in the following way:

$$\mathcal{M}_Z = \sqrt{2} M_Z^2 G_\mu \rho \frac{[I_3^e - 2Q_e s_W^2 \kappa_e)\gamma_\mu - I_3^e \gamma_\mu \gamma_5] \otimes \left[I_3^f - 2Q_f s_W^2 \kappa_f)\gamma^\mu - I_3^f \gamma^\mu \gamma_5\right]}{Q^2 + M_Z^2}.$$
(14)

The correction factors ρ, κ contain a universal (i.e. fermion type independent) part and a fermion dependent non-universal part:

$$\rho = 1 + (\Delta\rho)_{univ} + (\Delta\rho)_{non-univ}$$
$$\kappa_{e,f} = 1 + (\Delta\kappa)_{univ} + (\Delta\kappa)_{non-univ} .$$

The universal pieces can become large if the top is heavy, in the leading quadratic term given by

$$(\Delta\rho)_{univ} \simeq \frac{3G_\mu m_t^2}{8\pi^2\sqrt{2}} \equiv \Delta\rho \tag{15}$$

and

$$(\Delta\kappa)_{univ} \simeq \frac{c_W^2}{s_W^2} \Delta\rho \tag{16}$$

yielding

$$\kappa_{e,f} s_W^2 \simeq s_W^2 + c_W^2 \Delta\rho \tag{17}$$

where s_W^2 is the quantity derived from (12).

The non-universal parts of ρ and κ contain such (potentially) large terms only for heavy flavors ($f = b, t$); in all other cases they are practically independent of M_H, m_t and in size smaller than 1%. The form (14) does not yet contain the heavy box diagrams. They depend, besides on Q^2, also on x, but still have a Born-like structure and can easily be added. Numerically they do not give large corrections. For details we refer to the literature [3,5].

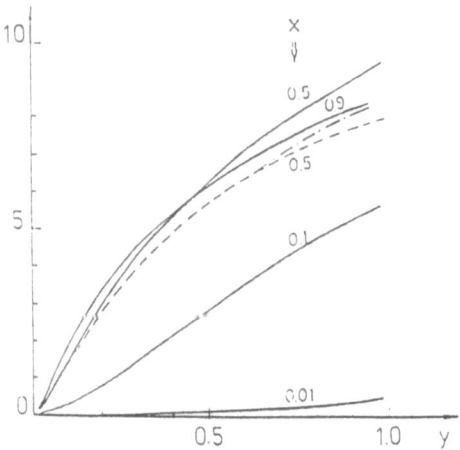

Figure 4. Weak corrections to $d^2\sigma/dxdy$ in % ,
quark densities from [19]
full curves from [5], dashed from [3];
dashed-dotted: from [5] and linearized in α.
$M_Z = 93$ GeV, $M_H = 100$ GeV, $m_t = 30$GeV, $s = 10^5$ GeV^2.

Figure 4 shows the correction to $d^2\sigma/dxdy$, relative to the Born cross section (3) with coupling constants (7) and s_W^2 from (12), calculated in [5], and for comparison the dashed curve from [3]. The difference between the two results is due to the resummation of $\hat{\Pi}_\gamma$ according to (13), as done in [5], whereas [3] contains $\hat{\Pi}_\gamma$ in linearized form. Linearizing the expressions of [5] to $O(\alpha)$ (dashed-dotted) yields agreement up to $y = 0.7$; but also at higher y the deviations are not bigger than about half a percent.

We want to conclude this discussion of the non-QED corrections with a practical recipe for obtaining an approximate cross section which incorporates all large corrections and may be sufficient for many purposes. The dominant corrections associated with $\hat{\Pi}_\gamma$ and $\Delta\rho$ (in case of a heavy top) can easily be included in the Born formulae (3)-(5) by the following prescription: replace

$$\alpha \rightarrow \alpha(Q^2) = \frac{\alpha}{1 - \hat{\Pi}_\gamma(Q^2)}$$

and use the coupling constants from (9) with

$$
\begin{aligned}
G_\mu &\rightarrow G_\mu(1 + \Delta\rho) \\
s_W^2 &\rightarrow s_W^2 + c_W^2\Delta\rho
\end{aligned}
$$

where $\Delta\rho$ is given in (15).

2.5 Discussion of the QED corrections

2.5.1 Kinematics

The infrared singular virtual photon corrections have to be combined with the real photon bremsstrahlung contributions in order to give a finite and physically sensible result. Since the radiative parton subprocess has an additional photon in the final state we have to specify in which variables the corrected cross section is expressed.

From an inclusive measurement of the final electron 4-momentum the variables

$$S = (p + P)^2, \quad Q^2 = -(p - p')^2, \quad U = -(p' - P)^2$$

respectively the dimensionless quantities

$$y = 1 - \frac{U}{S}, \quad x = \frac{Q^2}{S - U}$$

are used for measuring the distributions $d^2\sigma/dxdy$. For the description of the parton subprocess we have the variables

$$x' = \frac{s}{S}, \quad y' = \frac{Q'^2}{S}, \quad s' = (p' + q')^2$$

with

$$s = (p + q)^2, \quad Q'^2 = -(q' - q)^2.$$

Keeping x, y fixed, the inclusive radiatively corrected cross section is obtained from (full phase space)

$$\frac{d^2\sigma^B}{dxdy} = \int_x^1 dx' \int_0^y dy' \int_{s_1'}^{s_2'} ds' \frac{d^5\sigma}{dxdydx'dy'ds'}. \tag{18}$$

For the more complicated kinematical bounds on s' see e.g. ref [3].

In the classification of 2.2.1 it was emphasized that the subsets of corrections at the lepton and the hadron line as well as the interference between them are gauge invariant and allow a separate treatment.

2.5.2 Radiation from the lepton line

This subset of real photon emission from the electron (Figure 2 e,f) and the leptonic vertex correction (Figure 2 a) constitutes the bulk of all radiative corrections and represents the practically most important contribution: it can amount up to a 20 - 50% correction (and even larger at small x) with steep increase in the region of large Q^2. This behaviour can be understood from the following reasons:

- large logarithmic terms of the form

$$\frac{\alpha}{\pi} \log \frac{Q^2}{m_e^2}$$

  are present due to the radiation of photons collinear with the emitting lepton;

- the emission of a very energetic photon shifts the momentum in the propagator of the exchanged γ to a value which is essentially smaller than determined from the energy and momentum of the final electron (hard photon tail).

The second effect is primarily of kinematical nature; it can in principle be reduced in magnitude by applying suitable cuts to the emitted photon. For example, the restriction of the photon energy to $E_\gamma \leq 10$ GeV keeps the corrections below 20% [9].

The results for the relative corrections in % without phase space cuts are displayed in Figure 5: the exact $O(\alpha)$ calculation of [5] together with the LL approximation of [11] in 5a, and the updated results of [3] in 5b. All the various calculations, including also the LL approximation of [10], agree now pretty well at the percent level.

2.5.3 Radiation from the quark line

The QED corrections associated with the quark line (Figure 2 b,g,h) exhibit a different behaviour: they are essentially smaller in magnitude, typically a few percent, and are flat functions of Q^2 (Figure 6). The main reason for this difference is the absence of the hard photon tail in the photon propagator, which dominates the radiation from the lepton at large y.

However, another problem now comes up: the occurence of quark mass singularities of the type

$$\frac{\alpha}{\pi} Q_f^2 \log \frac{Q^2}{m_q^2} \tag{19}$$

which have their origin, quite in analogy to the leptonic case, in the radiation of photons collinear with the emitting quarks. They depend on quark masses which are physically not well understood parameters. Whereas in the collinear radiation from charged leptons the presence of mass singularities is a physical reality (since the electron mass is a physical parameter), quark mass singularities can never be isolated from the corresponding quark distribution functions. Their appearance is therefore

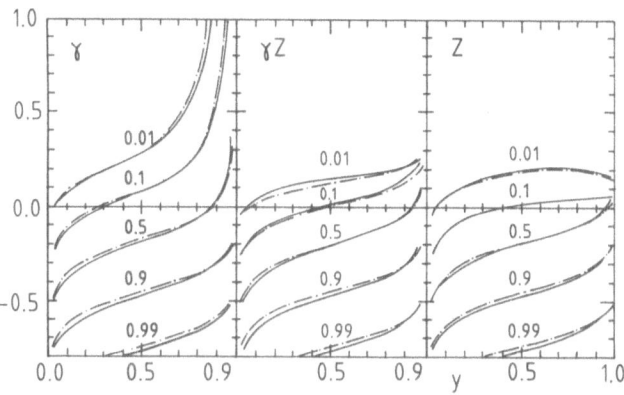

Figure 5a.
Leptonic QED corrections, relative to the γ, γZ, Z cross section of (6).
—·—·— from [5], ———— LL approximation [11]

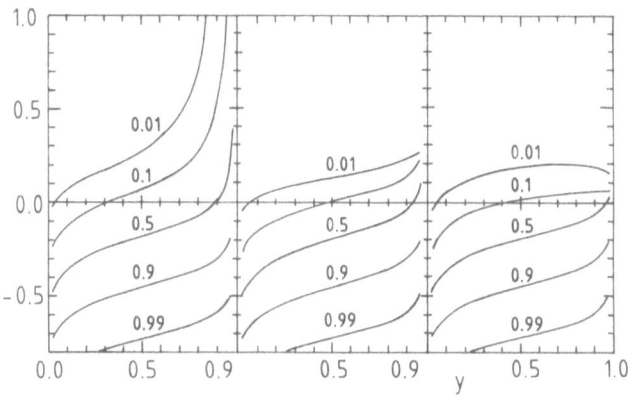

Figure 5b. updated results of [3]. Parameters as in Figure 4.

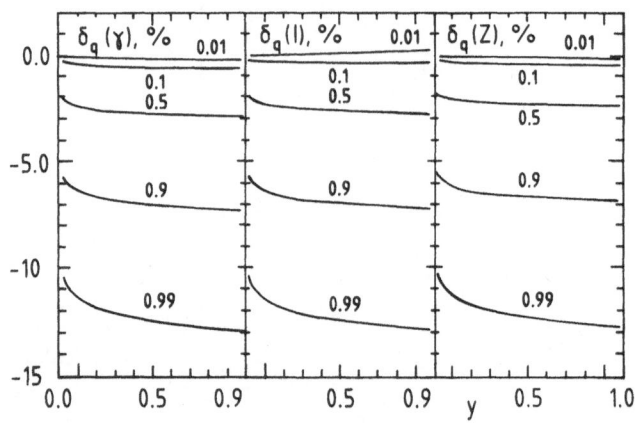

Figure 6a.
Hadronic QED corrections from [5], relative to the γ, γZ, Z cross section of (6).

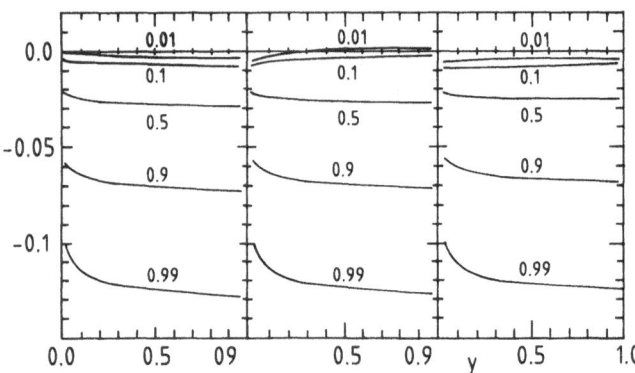

Figure 6b. Hadronic QED corrections from [3], updated. Parameters as in Figure 4.

an artifact which is related to a naive interpretation of the $q_f(x)$ in (2) as "bare" distribution functions.

In the naive parton model the hadronic part of the QED corrections δ_h^f modifies the lowest order cross section with bare parton distributions $q_f^0(x)$ as follows:

$$\frac{d^2\sigma}{dx\,dQ^2} = \sum_f \frac{d\sigma^f}{dQ^2} \cdot q_f^0(x) \left(1 + \delta_h^f(Q^2)\right) \ . \tag{20}$$

Any measurement of the bare $q_f^0(x)$ would have to include these QED corrections in the data analysis. On the other hand, it is more natural to define Q^2 dependent parton distributions

$$q_f(x,Q^2) = q_f^0(x) \cdot (1 + \delta_h^f(Q^2))$$

which absorb the singular leading log contributions. If expressed in terms of $q_f(x,Q_0^2)$, extracted from data at a given value of Q_0^2, the remnant correction for a process at another Q^2 will result in

$$\frac{d^2\sigma}{dx\,dQ^2} = \sum_f \frac{d\sigma^f}{dQ^2} \cdot q_f(x,Q_0^2) \left(1 + \delta_h^f(Q^2) - \delta_h^f(Q_0^2)\right) \tag{21}$$

which is now free of the quark mass singularities; instead of (19) the typical leading log correction terms are now of the form

$$\sim \frac{\alpha}{\pi} Q_f^2 \log \frac{Q^2}{Q_0^2} \ .$$

For the calculation of the cross section (21) parton distributions are required which have been determined without taking into account the QED corrections associated with the quark lines in the data analysis.

Q^2 evolution of the quark distributions

A more refined treatment of the Q^2 evolution of the parton distribution functions applies the method of the Altarelli-Parisi (A-P) equations, in analogy to QCD, in order to sum the leading log contributions to all orders. In the extensive literature concerning the strong interaction corrections for deep inelastic scattering it has been shown that the leading logarithmic corrections of QCD can be accounted for by replacing the bare parton distribution functions by Q^2 dependent ones where the Q^2 dependence is governed by the A-P equations, and the x dependence has to be taken from experiment.

For semileptonic processes at high Q^2 QED quark mass singularities show up in a similar way as in QCD calculations. The factorization property of these terms

$$\left(\frac{\alpha}{\pi} \log \frac{Q^2}{m^2}\right)^n$$

guarantees that they are process independent and can be absorbed into effective quark distributions as well. The absorbed terms then do not show up as explicit corrections (as it would be in eq. (20)) but introduce an additional Q^2 dependence of the $q_f(x,Q^2)$ which is of electromagnetic origin.

479

This dependence can be described by additional electromagnetic terms in the A-P equations:

$$\frac{dq_i}{dt}(x,t) = \frac{\alpha_s}{2\pi} \int_x^1 \frac{dy}{y} \left(\sum_j q_j(y,t) \cdot P_{q_iq_j}(\frac{x}{y}) + G(y,t) \cdot P_{q_jG}(\frac{x}{y}) \right) \qquad (22)$$
$$+ \frac{\alpha}{2\pi} \int_x^1 \frac{dy}{y} q_i(y,t) \cdot P_{qq}^\gamma(\frac{x}{y}) \cdot Q_{q_i}^2$$
$$+ O(\alpha_s^2)$$
$$+ O(\alpha_s \cdot \alpha)$$

for the quark (q_i) and gluon (G) distributions. In one-loop order the transition function P_{qq}^γ is given by

$$P_{qq}^\gamma = \frac{1+z^2}{(1-z)_+} + \frac{3}{2}\delta(z-1) \qquad (23)$$

which differs from the QCD kernels $P_{q_iq_j}$ only by a group factor.

The numerical solution of these equations [13] yields for the QED evolution of the valence quarks $q_V(x,Q^2)$ an increase of 1 - 2% at $Q^2 = 10^4 GeV^2$ relative to the experimental input $q_V(x,Q_0^2)$ at $Q_0^2 = 4\ GeV^2$ [13].

There are of course also other terms in the QED corrections at the quark lines which are not of the leading log behaviour and which depend on the specific process. These can be considered as explicit corrections appearing in the corrected cross section formula (21). In principle they influence the extraction of the quark distributions from experimental data; in practice, however, they are negligible if an experimental accuracy of 1% order is not accessible and a simplified discussion is sufficient.

2.5.4   Interference between lepton - hadron radiation

The separation into leptonic and hadronic parts of the QED corrections is not unique due to the presence of box diagrams (Figure 2 c,d) and interference between leptonic and quarkonic bremsstrahlung in Figure 2 (e,f)-(g,h). In contrast to the previous subset these terms have no analogon in the QCD corrections for deep inelastic scattering.

In the approximation for small fermion masses the box contributions can be cast into the conventional current-current interaction form

$$J_\mu^e \cdot J^{\mu,q} \cdot D_{\gamma,Z}(Q^2) \cdot F_{\gamma,Z}(Q^2,x) \qquad (24)$$

with currents $J_\mu^{e(q)}$ depending only on the electron (quark) electroweak coupling constants, propagators $D_{\gamma,Z}$ which carry the momentum Q^2 of the hard scattering process, and functions $F_{\gamma,Z}$ which depend only on the kinematical variables Q^2 and $xS = s$, the CMS energy squared of the parton subprocess. The real photon contributions can be expressed in a similar way. As a consequence, the total interference part is not affected by either lepton or quark masses and is therefore also free of mass singularities. Typically logarithms and dilogarithms of the form

$$\log(\frac{Q^2}{xS}), \quad Li_2(\frac{Q^2}{xS})$$

or with slightly more complicated arguments are the ingredients of the invariant functions $F_{\gamma,Z}$.

The non-universality of these terms (the $\gamma\gamma$ and γZ boxes are different in their non-infrared parts) and the explicit dependence also on the variable x forbid an absorption of these interference terms in the quark distribution functions. They have to be taken into account separately and show up as explicit corrections in the cross section.

Quantitatively these interference terms are well behaved: they do not exceed the 5% level in the corrections to the cross section for reasonable values of x, y, and are relatively flat functions of Q^2 (Figure 7). However, for some specific observables like the electron-positron charge asymmetry they can become of sizable and non-negligible magnitude.

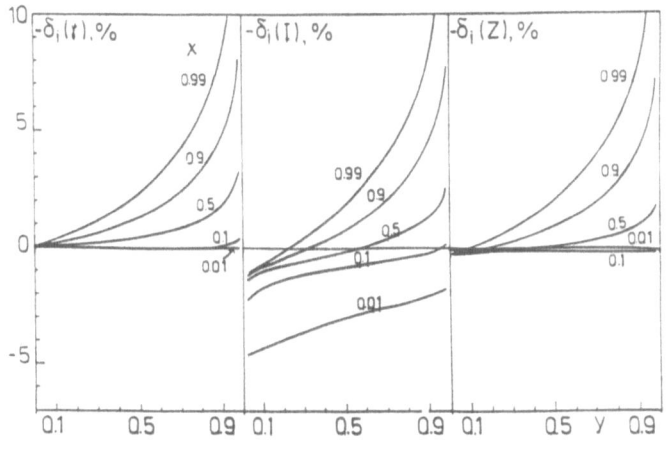

Figure 7.
The lepton-quark interference corrections relative to the γ, γZ, and Z cross section (6) (from [5])

A further problem appearing in the $O(\alpha)$ interference terms should also be mentioned: If the photon is produced at the lepton legs the correct choice for the momentum scale in the parton distributions is obviously $Q'^2 = -(q - q')^2$, whereas for photons produced at the quark legs we would guess $Q^2 = -(p - p')^2$ to be correct. For the interference terms there is no such simple choice. To give the correct answer it would be necessary to analyze the combined $O(\alpha\alpha_s)$ effects on the Q^2 dependence of the structure functions. For practical purposes, however, it is sufficient to know that several choices for the momentum scale in the q_f yield only differences which are numerically negligible.

3 Charged current deep inelastic scattering

3.1 Lowest order cross section

The charged current differential cross section for

$$e + p \rightarrow \nu_e + X$$

with left-handed polarized electrons can be written in two ways:

$$
\begin{aligned}
\frac{d^2\sigma}{dx dQ^2}(e_L) &= \frac{\pi\alpha^2}{2s_W^4} \cdot \frac{1}{(Q^2 + M_W^2)^2} \left[u + c + (1-y)^2(\bar{d} + \bar{s}) \right] \quad (25) \\
&= \frac{G_\mu^2}{\pi} \left(\frac{M_W^2}{Q^2 + M_W^2} \right)^2 \left[u + c + (1-y)^2(\bar{d} + \bar{s}) \right].
\end{aligned}
$$

The second form has the advantage that it is not changed by large radiative corrections associated with the electromagnetic α and with $\Delta\rho$. The only source for large corrections is the photon radiation collinear to the incoming electron.

3.2 Electroweak one-loop corrections

Complete $O(\alpha)$ electroweak radiative corrections to the charged current process at energy scales comparable to the W mass have been calculated by two different groups [4,6]; QED-like corrections in the LL approximation are also treated in [11].

According to the general classification of the one-loop corrections one could interpret all diagrams with an additional real or virtual photon as displayed in Figure 8 (in addition to the self energies of the external charged fermions) as "QED" corrections. In contrast to the neutral current process, however, such a subset is not gauge invariant. Moreover the appearance of the nonabelian γWW vertex in the real and virtual photonic diagrams, not present within the conventional QED but typical for a non-abelian gauge theory, indicates that the usual classification convenient for NC processes is not very sensible for the charged current case.

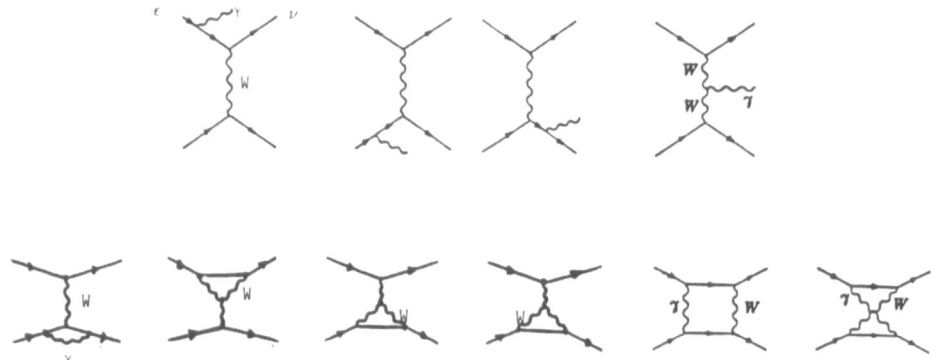

Figure 8. Real and virtual photon corrections to the CC parton process

A gauge invariant separation is only possible for the infrared- and mass-singular parts of the photonic corrections. If quark masses are used for regularization of the collinear photon-quark singularities the initial (f) and final (f') quarks appear in different combinations:

$$Q_f^2 \log \frac{Q^2}{m_f^2}, \quad Q_{f'}^2 \log \frac{Q^2}{m_{f'}^2}, \quad Q_e Q_f \log \frac{Q^2}{m_e m_f}, \quad Q_e Q_{f'} \log \frac{Q^2}{m_e m_{f'}}.$$

As shown in [6], all terms of the virtual contributions which contain the final quark masses cancel against the real photon contributions where the photon is collinear with the final quark, according to the KNL-theorem [20]. The only left-over singularities are the same universal parts as in the NC case. Those terms, depending on the initial quark masses, can therefore be absorbed in the universal parton densities $q_f(x, Q^2)$ as it was discussed in the previous section. The numerical results of Figure 9, however, contain the mass terms explicitly and use the density functions of [19] as input.

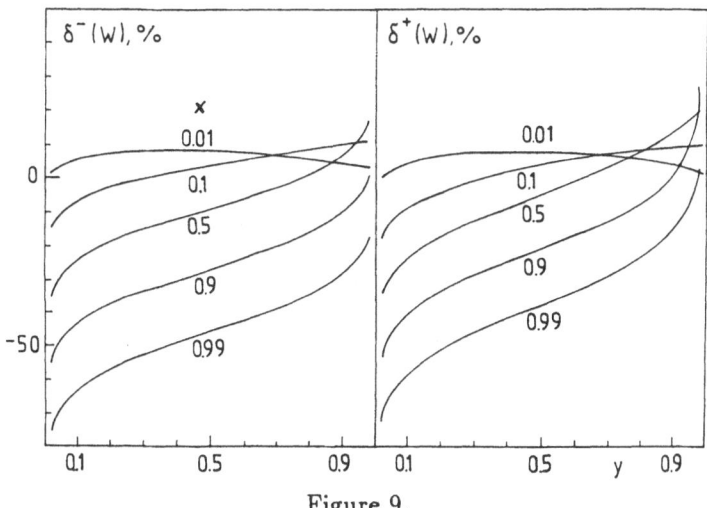

Figure 9.

Total electroweak corrections to the CC cross section (from [6]) for electrons ($-$) and positrons ($+$)

Since the separation of a QED-like subclass, in principle possible for the singular terms, is not unique we prefer to discuss the electroweak corrections to the CC cross section as total corrections. Figure 9 displays the total electroweak corrections in percent relative to the Born cross section (25) in the G_μ parametrization. They are dominated by the radiation from the electron leg. Differently from NC scattering there is no hard photon tail for $y \to 1$ since the cross section is a smooth function of Q^2. The sizeable negative corrections for large x are due to the restriction of the photon phase space to the soft-photon domain; they are qualitatively the same as in the NC case.

4 Conclusions

Radiative corrections in deep inelastic scattering processes are in general large at HERA energies.

In neutral current scattering the by far largest part consists of electromagnetic corrections (real and virtual bremsstrahlung) associated with the lepton line. Radiation from the quarks can be treated to its largest extent in terms of a QED induced Q^2 evolution of the parton distribution functions absorbing the leading logarithmic terms with quark mass singularities. These singular terms are universal and appear in the charged current process as well. The left-over non-universal parts from quark radiation and lepton-quark interference contribute at the level of a few percent.

Among the corrections not of the bremsstrahlung type the dominant contributions enter in terms of the running electromagnetic $\alpha(Q^2)$ in the photon exchange amplitude and the $\Delta\rho$ term with the m_t^2 increase in the Z exchange diagram. The charged current amplitude if expressed in terms of G_μ is not affected by these types of corrections. The other corrections are typically of order 1% or below.

In the charged current scattering process the conventional separation into QED and weak corrections is less sensible since such a separation is neither unique nor gauge invariant. For the inclusive cross section the corrections are smaller than in the NC case for large y; for small y and large x they are similar to the NC corrections, both dominated by soft-photon radiation.

In view of the large $O(\alpha)$ corrections the higher order QED contributions $\geq O(\alpha^2)$ play a non negligible role. Their most important part can be obtained by an exponentiation of the leading soft photon corrections and leads to an additional positive contribution in the (inclusive) cross section [5].

A further contribution to the NC cross section is expected in form of a radiative tail from the elastic peak. This elastic tail has not been discussed here since it is not of the deep-inelastic type. For totally inclusive measurements of the final lepton only, however, both contributions cannot be distinguished. A calculation of the elastic tail in the HERA range has been performed in [21]. For $x < 0.1$ and $y > 0.7$ it becomes of non-negligible influence.

Acknowledgement
I want to thank D. Bardin, F. Berends, J. Blümlein, M. Böhm, T. Riemann, H. Spiesberger for helpful discussions and for supplying me with information on the most recent results.

References

1. Proceedings of the DESY Workshop on "Physics with HERA", Hamburg 1987, ed. R. Peccei, Vol. 2, p. 577

2. D. Yu. Bardin, O.M. Fedorenko, N.M. Shumeiko, J. Phys. G : Nucl. Phys. 7 (1981) 1331;
   D.Yu. Bardin, C. Burdik, P. Ch. Christova, T. Riemann, Dubna Preprint E2-87-595 (1987)

3. M. Böhm, H. Spiesberger, Nucl. Phys. B 294 (1987) 1081

4. M. Böhm, H. Spiesberger, Nucl. Phys. B 304 (1988) 749

5. D. Yu. Bardin, C. Burdik, P. Ch. Christova, T. Riemann, Z. Phys. C 42 (1989) 679

6. D. Yu. Bardin, C. Burdik, P. Ch. Christova, T. Riemann, Dubna Preprint E2-89-145, Z. Phys. C (to appear)

7. L.W. Mo, Y.S. Tsai, Rev. Mod. Phys. 41 (1969) 205

8. A.A. Akhundov et al., Sov. J. Nucl. Phys. 26 (1977) 660

9. J. Kripfganz, H.J. Möhring, Z. Phys. C 38 (1988) 653

10. W. Beenakker, F.A. Berends, W.L. van Neerven, Proceedings of the Ringberg Workshop on "Radiative Corrections for e^+e^- Collisions", Ringberg 1989, ed. J.H. Kühn, p. 3

11. J. Blümlein, Berlin Zeuthen Preprints PHE 89-8, 89-10, and Proceedings of the International Conference "Frontiers in Elementary Particle Physics", Kazimierz 1989, eds. Z. Ajduk and S. Pokorski

12. M. Consoli, M. Greco, Nucl. Phys. B 186 (1981) 519

13. J. Kripfganz, H. Perlt, Z. Phys. C 41 (1988)319

14. A. Sirlin, Phys. Rev. D 22 (1980) 971

15. W. Hollik, DESY 88-188 (1988), Fortschr. Phys. (to appear)

16. M. Consoli, W. Hollik, F. Jegerlehner, CERN-TH.5527/89 (1989), to appear in the Proceedings of the Workshop on Z Physics at LEP, eds. G. Altarelli, R. Kleiss, C. Verzegnassi

17. D.Yu. Bardin, M.S. Bilenky, G.V. Mitselmakher, T. Riemann, Berlin-Zeuthen Preprint PHE 89-05 (1989), Z. Phys. C (to appear)

18. H. Burkhardt, F. Jegerlehner, G. Penso, C. Verzegnassi, in: "Polarization at LEP", CERN 88-06, eds. G. Alexander et al., Vol. 1, p. 145

19. D.W. Duke, J.F. Owens, Phys. Rev. D 30 (1984) 49

20. T. Kinoshita, J. Math. Phys. 3 (1962) 650;
T.D. Lee, M. Nauenberg, Phys. Rev. 133 (1964) 1549

21. A.A. Akhundov, D.Yu. Bardin, C. Burdik, P.Ch. Christova, L.V. Kalinovskaja, Dubna Preprint E2-89-405 (1989)

THE LARGE COEFFICIENT PROBLEM: CAN WE MAKE SENSE OUT OF QCD

PERTURBATION THEORY? [1]

Geoffrey B. West

Theoretical Division, T-8
Los Alamos National Laboratory
MS B285
Los Alamos, NM 87545

I INTRODUCTION

There is the possibility of an impending crisis looming on the horizon for QCD. The problem is that in many processes, large coefficients arise in the perturbation series expansion leading to serious uncertainties concerning its predictive power. Until recently most of the examples of such a phenomenon occurred in the calculation of decay rates. These were, by and large, either ignored or dismissed using possible scheme-dependence arguments as a way out. However, more recently a calculation of the 3-loop contribution to the total e^+e^- annihilation cross-section was performed which gave an enormous coefficient of the order of 50 times that of the 2-loop term[1]. If correct, this would imply that the 3-loop contribution actually exceeds that of the 2-loop! Thus, **from a conservative viewpoint, the validity of the perturbation series expansion as an estimate for the total e^+e^- cross-section is called into question.** Such a cautionary attitude should even be extended to the lowest order parton-model result, $\sum Q_i^2$; (Q_i being the charge of the ith quark species). Since this process has played a key rôle in the development and understanding of QCD and since, in many ways, it is one of the cleanest methods for extracting α_s (the conventional QCD fine structure constant) the problem can no longer be avoided. Furthermore, there is no reason to doubt (and, in fact, good reasons to believe) that this problem should occur in all physical processes. Coming to grips with it is, of course, not only important for testing QCD but also for extracting fundamental quantities such as α_s. Clearly one needs to understand the nature and origin of such large coefficients before one can confidently continue to use perturbative estimates. Such problems can be expected to occur universally so, using different methods to determine α_s, for example, will not circumvent the difficulty.

The purpose of this talk is to focus on these problems. I shall first review the experimental situation with some examples illustrating the problem. I shall then discuss various general

---

[1] Talk given at the Radiative Corrections Workshop held at the University of Sussex, Brighton, England, July 1989

components and properties of perturbation theory (such as renormalization and causality) before attempting to give a possible resolution of the problem.

Everyone is, of course, familiar with perturbation theory as a calculational tool; Feynman diagrams and their accompanying rules of computation are the stock-in-trade for all particle physicists, theorists and experimentalists alike. Indeed, the phenomenal success of quantum electrodynamics is surely one of the crowning achievements of modern physics. Its predictions are in remarkable agreement with experiment, in some cases to one part in 10^{12}. Calculationally, this success is rooted in perturbation theory which has become the cornerstone for understanding the consequences of any relativistic quantum field theory. In spite of this, perturbation theory has serious limitations beyond those associated, for example, with such questions as bound states, or spontaneous symmetry breaking. Already in 1952 Dyson[2] had pointed out in an elegant paper that the perturbation series could not be convergent. The argument is deceptively simple and very physical: imagine changing α to $-\alpha$ so that like charges now attract and opposite ones repel. Clearly the ground state of this new theory is quite different from that of ordinary QED since virtual pairs created in the vacuum now repel one another. Thus, perturbing around the original "trivial" vacuum of QED will clearly be insufficient to describe this new situation; the structure of the new theory and, in particular, of its vacuum cannot, therefore, be obtained by simply setting $\alpha = -\alpha$ in the Feynman perturbation series. Consequently perturbation theory must be non-analytic in α at $\alpha = 0$ signifying that the series has zero radius of convergence. An amusing historical note to this is that Dyson came to this conclusion[3] after having first claimed that the series was in fact convergent and that QED was therefore a closed book!

If the nature of the divergence of the series were such that it were asymptotic then the situation is, at least in principle, controllable. For, in such a case, as will be reviewed below, a good estimate for the sum of the series is obtained by keeping "only" the first π/α(≈ 400 in QED) terms. In practice, this means that, since α is so small, perturbation theory will, in fact, give an accurate estimate; only at absurdly high order do serious deviations begin to develop. Thus the fact that the series is actually divergent would be of no practical importance. Clearly, then, the nature of the divergence (i.e. whether, for example, it is asymptotic or not) is a potentially deep and important question. It presumably bears upon the question of the self-consistency of QED and whether it needs to be imbedded in a larger, possibly asymptotically free theory.

The second limitation is a practical one and was best expressed by Feynman[4] himself in 1959; he was concerned about developing an approximate algorithm for estimating higher order terms in the perturbation series without having to laboriously (and, to some extent, mindlessly) calculate Feynman diagrams. Of course, his concern is somewhat less of a problem these days given the advent of fast computers and sophisticated software, nevertheless, his remarks are worth repeating. To quote:

> "It seems that very little physical intuition has yet been developed in this subject. In nearly every case we are reduced to computing exactly the coefficient of some specific term. We have no way to get a general idea of the result to be expected. To make my view clearer, consider, for example, the anomalous electron moment, $[\frac{1}{2}(g - 2) = \alpha/2\pi - 0.328\alpha^2/\pi^2]$. We have no physical picture by which we can easily see that the correction is roughly $\alpha/2\pi$, in fact, we do not even know why the sign is positive (other than by computing it). In another field

we would not be content with the calculation of the second-order term to three significant figures without enough understanding to get a rational estimate of the order of magnitude of the third. We have been computing terms like a blind man exploring a new room, but soon we must develop some concept of this room as a whole, and to have some general idea of what is contained in it. As a specific challenge, is there any method of computing the anomalous moment of the electron which, on first rough approximation, gives a fair approximation of the α term and a crude one to α^2; and when improved, increases the accuracy of the α^2 term, yielding a rough estimate to α^3 and beyond?"

Although we shall not be able to meet Feynman's challenge directly, nevertheless the techniques discussed here do constitute the beginning of an answer.

Returning to the question at hand, namely QCD rather than QED, we should note that the difficulties there are exacerbated for at least two independent reasons: (i) since $\alpha_s \gg \alpha$, the problem of the divergence of the series is much more serious and (ii) there are explicit non-perturbative phenomena (instantons and the like) associated with new local minima of the action. The question of the interplay between other minima of the action beyond the trivial one and ordinary perturbation theory is a subtle one which we shall discuss below. Regardless, it is clear that the QCD situation is a serious one in that large coefficients can (and do) occur early in the expansion. In some cases, as reviewed immediately below, they occur ridiculously early. Ultimately a methodology based on understanding the nature of the series must be devised for handling them.

II REVIEW OF SOME EXAMPLES

Before discussing some explicit QCD examples, let us examine QED and use it to briefly discuss the question of scheme dependence. One of the best known QED series is that for (g-2) of the electron which was quoted above by Feynman; it reads

$$\frac{1}{2}g_e = 1 + 0.5(\alpha/\pi) - 0.328(\alpha/\pi)^2 + 1.183(\alpha/\pi)^3 + \cdots \tag{1}$$

which certainly looks like a well-behaved series. Recall that in deriving this equation a certain renormalization scheme has been implicitly used; for example, α could be defined through threshold Thompson scattering from (on-shell) electrons. It is via such a definition that we deduce from experiment that $\alpha \approx (137.03\cdots)^{-1}$. This is a natural definition especially for low-energy quantities such as g_e. One could, in principle, use other schemes associated with high energy QED phenomena where the corresponding α would be smaller. Generally speaking, different schemes are related by some polynomial relationship[5]: $\alpha' = \alpha + a_1\alpha^2 + a_2\alpha^3 + \cdots$ (the a_i being constants). Although, the final result for a physical quantity such as g_e does not depend on which scheme is chosen it is clearly not very sensible to choose one associated with a high energy process when dealing with low energy phenomena. In any case, in QED, since electrons are observable, there are "natural" schemes such as via Thompson scattering, which are the most appropriate ones for the definition of α. This is in contradistinction to QCD where there are no analogous "natural" schemes associated with experiments where quarks or gluons are real observables. Because of asymptotic freedom, however, these can be approximated in a variety of high energy experiments such as the total e^+e^- total cross-section measurement. In any case, as already stated, the final result represented, for example, by the sum of the perturbation series must be scheme invariant. On the other hand, to a finite order in perturbation theory, the result will, in general, be scheme-dependent and this is a major source of ambiguity (and confusion). There have been many attempts to define a "best" scheme appropriate to a particular process, however, none are totally satisfactory and all necessarily leave a residue of uncertainty. I shall not, in this talk, be much concerned with such problems especially since for any physical process, **scheme invariant quantities can be defined**[5,7] even in finite orders of perturbation theory. However, to illustrate the problem suppose we use a scheme where[6] $\alpha'/\pi = \alpha/\pi(1 - 10\alpha/\pi)$, then the series (1) reads

$$\frac{1}{2}g_e = 1 + 0.5(\alpha'/\pi) + 4.67(\alpha'/\pi)^2 + 94.61(\alpha'/\pi)^3 + \cdots \tag{2}$$

which now looks like a badly behaved series! This means that when dealing with large coefficients some care must be taken to express quantities in a scheme-invariant fashion. In this example, incidentally, $(\alpha')^{-1} \approx 140.29$ corresponding to an "inappropriate" α' defined at an energy scale significantly greater than the low energy scale of g_e.

Before moving onto QCD, it is worth mentioning one other example from QED and that is the decay of orthopositronium into three photons. The width for this process is given by[6]

$$\Gamma \approx \frac{2\alpha^6}{9\pi}m_e(\pi^2 - 9)[1 - 10.35(\frac{\alpha}{\pi}) + \cdots] \tag{3}$$

There are several points worth noting about this; first is the appearance of a large coefficient in the leading order correction; this receives its dominant contribution from the graph shown in fig.1 and is large in any reasonable scheme. Note, however, that, because the tree-graph contribution is $\sim \alpha^6$, this correction is rather sensitive to the scheme. Eq. (3) is one of the few (and possibly only) cases where the theoretical predictions of QED are in serious disagreement with experiment. In fact, there is a 5-standard deviation discrepancy which could be explained if the coefficient of the $(\alpha/\pi)^2$ term were of order 300! Scheme dependence has been evoked to explain this, but it could be a situation where true large coefficients are occurring. Notice, incidentally, the occurrence of the curious coefficient $(\pi^2 - 9)$; conceivably the fact that this almost vanishes (presumably accidentally) contributes to the sensitivity of this process.

Perhaps the first example of a large coefficient occurring in perturbative QCD is in the decay of the 0^{-+} heavy quark state η_B into two gluons[8]. Typical graphs are shown in fig. 2. The calculation yields

$$\frac{\Gamma(\eta_B \to 2g)}{\Gamma(\eta_B \to 2\gamma)} = \frac{2}{9Q_B^2}\left(\frac{\alpha_s}{\alpha}\right)\left[1 + 22.4\left(\frac{\alpha_s}{\pi}\right) + \cdots\right] \tag{4}$$

Bound states effects have been completely ignored and the initial valence quarks and final state gluons treated as if free. As in the orthopositronium case, the large coefficient (22.4) is scheme-dependent since the tree-graph contribution is $O(\alpha_s^2)$.

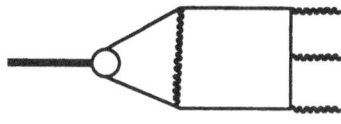

Fig. 1

Dominant contribution to the decay of orthopositronium into three photons

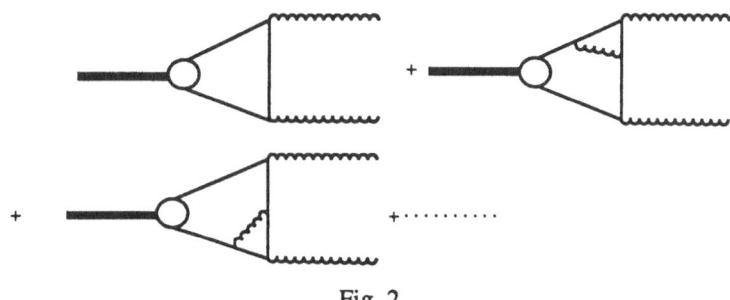

Fig. 2

Typical graphs contributing to the decay of paraquarkonium

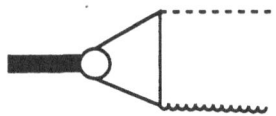

Fig. 3

Leading graph in the decay $Y \to h + \gamma$

Ruling out a light higgs has been a frustratingly difficult enterprise mostly because of its very weak coupling[9]. The best limits come from rare K and B decays, the higgs being detected via its decay into $\mu^+\mu^-$ pairs. This, therefore, requires accurate knowledge of the branching ratio into this mode which can lead to serious uncertainties. An elegant way of avoiding this problem is to use the decay $Y \to h + \gamma$ and to look for a single hard photon in which case, the decay mode of the higgs is irrelevant[10]. At tree level, the decay proceeds via the diagram shown in fig. 3. Ignoring bound state effects and treating the system as if it were positronium, one obtains

$$\frac{\Gamma(Y \to h\gamma)}{\Gamma(Y \to \mu^+\mu^-)} \approx \frac{G_F m_b^2}{\sqrt{2}\,\pi\alpha} \left(1 - \frac{m_h^2}{m_Y^2}\right) \tag{5}$$

In this ratio the crude bound state effects, which would be reflected by the wave-function at the origin in a non-relativistic loosely bound system, cancel. In such an approximation $m_Y \approx 2\,m_b$. In addition to inherently non-perturbative bound state corrections to this formula there are the usual QCD perturbative radiative corrections to the sub-process $b\bar{b} \to h\gamma$. A calculation of these leads to[11]

$$\Gamma(Y \to h\gamma) \approx \Gamma_0(Y \to h\gamma)\left[1 - a\left(\frac{\alpha_s}{\pi}\right)\right] \tag{6}$$

where, for $m_h \ll m_Y$, $a \approx 13$ again showing the appearance of a large coefficient. Taken at face value this formula would rule out [12] a higgs in the range 200 MeV$\lesssim m_h \lesssim$6GeV. However the correction here **reduces** the tree graph result by $\sim 80\%$ making the calculation suspect. When uncertainties about possible bound state corrections are folded in, one must certainly take this conclusion *cum grano salis*, especially since a conservative attitude is mandated when dealing with the existence of the higgs! Notice incidentally that unlike the previous formulae the expansion in eq. (6) begins with $(\alpha_s)^0$ so that the result should be scheme-invariant.

This process has already been alluded to in the introduction. The total cross-section (R) for e^+e^- annihilation into hadrons relative to that of $\mu^+\mu^-$ pairs is directly related to the absorptive part of the polarization tensor defined by

$$\Pi_{\mu\nu}(q) = i\int d^4x\, e^{iq\cdot x}\langle 0|T[j_\mu(x)j_\nu(0)]|0\rangle \tag{7}$$

$$- [q^2 g_{\mu\nu} - q_\mu q_\nu]\Pi(q^2) \tag{8}$$

where $j_\mu(x)$ is the electromagnetic current operator: $R = 12\pi \text{Im}\ \Pi$. If μ is an arbitrary renormalization scale parameter then the perturbative expansion of R reads

$$R\left[q^2/\mu^2, \alpha_s(\mu)\right] = \left(\sum Q_i^2\right)\sum_{n=0}^{\infty} \tilde{a}_n(q^2/\mu^2)(\alpha_s/\pi)^n. \tag{9}$$

In the $\overline{MS}$ scheme with 5 flavors, the coefficients have the following [1] values when $q^2 = |\mu^2|$:

$$\tilde{a}_o = \tilde{a}_1 = 1 \; ; \tilde{a}_2 = \mathbf{1.41} \text{ and } \tilde{a}_3 = \mathbf{64.9}. \tag{10}$$

The last of these is truly remarkable. To get some idea of its implications, note that if it is neglected then a comparison with data at $\sqrt{q^2} = 34\,GeV$ leads to $\alpha_s = 0.169$ corresponding to $\Lambda_{QCD} \approx 610\,MeV$. On the other hand, it if is included then $\alpha_s = 0.150$ corresponding to $\Lambda_{QCD} \approx 314\,MeV$, a reduction of 50%. To make this even more dramatic, Fleischer et al [13] have made a Padé approximant fit to the series. Though this should be taken *cum grano salis* it does illustrate just how serious things could become: they find $\tilde{a}_4 \approx 10^4$ leading to $\alpha_s \approx 0.114$ and $\Lambda_{QCD} \approx 60.3\,MeV$ and $\tilde{a}_5 \approx 1.5 \times 10^6$ leading to $\alpha_s \approx 0.086$ and $\Lambda_{QCD} \approx 6.3\,MeV$. Clearly, then, one needs to understand the nature of the series and its implicit non-perturbative character before one can begin to confidently extract quantities like α_s and $\sum Q_i^2$ from the data. In a sense one can compare this situation with that which existed in weak interactions prior to unification. At that time the Fermi theory gave an adequate (and reasonably accurate) description of low energy weak interaction phenomenology in spite of the fact that it was non-renormalizable so that eventually it would break down. Similarly, here, one might argue that, because of asymptotic freedom, perturbation theory should give an adequate estimate at truly infinite energies. However, at finite energies appropriate to present-day experiment, there are potentially large corrections due to the divergent nature of the series, even though the theory is renormalizable. Just as one had to wait for the development of a renormalizable theory of the weak interactions in order to control and consistently define the infinities in each term of the perturbative expansion, so, from a conservative point of view, one must await a similar procedure for dealing with the complete sum of the series before being confident that our predictions are consistently meaningful.

III   GENERAL PRINCIPLES, DEFINITIONS AND TECHNIQUES

Perturbation theory can be thought of as being generated by an expansion of the path integral around local minima of the action [S_m, say, where $\delta S/\delta A_\mu|_{s=s_m} = 0$]. Symbolically, then, Π can be thought of as having the following representation:

$$\Pi(q^2, g^2) \approx \sum_{m,n=0}^{\infty} A_{mn}(q^2)e^{-Sm/g^2}(g^2)^{n-\nu_m} \tag{11}$$

where g is the usual gauge coupling ($\alpha_s \equiv g^2/4\pi$) and the dependence on μ has been suppressed. The series with $m = 0$ defines ordinary perturbation theory as represented by the usual sum of Feynman graphs. The multi-instanton sectors and so forth, which are topologically separated from ordinary perturbation theory, are represented by the $m \neq 0$ series. As generated directly from the path integral, the expansion (11) is very sick: each coefficient $A_{mn}(q^2)$ is divergent as is each series summed over n. The first of these problems is conventionally dealt with via some renormalization scheme (and, if necessary, by some infrared cut-off) whereas the second disease is typically ignored. The challenge is to find a consistent scheme to control the divergence implicit in the sums. To attack this problem, I shall need to review some general properties of series expansions with emphasis on asymptotic expansion. However, before doing so I want to remind the reader of the constraints on Π rendered by renormalizability and causality. The point is that the latter dictates general analytic properties as a function of q^2 whereas the former tells us that q^2 and g^2 are not, in fact, independent variables. Thus, the general dependence on, and analytic structure, in g^2 is, in some sense, known.

Because j_μ is a composite operator there is an additional divergence above the usual multiplicative ones needed to renormalize QCD that must be cancelled to render the theory finite. This is the single $q\bar{q}$ intermediate state which gives a logarithmically divergent contribution to Π even in free field theory. On the other hand j_μ, being a conserved current, has no anomalous dimension. The renormalization group equation resulting from the invariance of Π to changes in the scale μ therefore reads (neglecting quark masses)[14]

$$\left[\mu \frac{\partial}{\partial \mu} + \beta(g) \frac{\partial}{\partial g} \right] \Pi \left[\frac{q^2}{\mu^2}, g^2(\mu) \right] = I[g^2(\mu)] \tag{12}$$

The presence of I reflects the composite nature of j_μ. Now, the general solution to this equation can be expressed as follows[14]:

$$\Pi \left(\frac{q^2}{\mu^2}, g^2 \right) = F \left[\frac{q^2}{\mu^2} e^{2K(g)} \right] + \phi(g^2) \tag{13}$$

where

$$K(g^2) \equiv \int^g \frac{dg'}{\beta(g')} \tag{14}$$

$$\text{and} \quad \phi(g^2) \equiv \int^g dg' \frac{I(g')}{\beta(g')}, \tag{15}$$

F being an arbitrary function.

In perturbation theory

$$\beta(g) \approx -g^3(b_1 + b_2 g^2 + \cdots) \tag{16}$$

leading to

$$K(g) \approx \frac{1}{2 b_1 g^2} \left[1 + \frac{b_2}{b_1} g^2 \ln \left(\frac{1}{g^2} + \frac{b_2}{b_1} \right) + O(g^4) \right] \tag{17}$$

Now, both R and $D \equiv (q^2 \partial/\partial q^2) \Pi$ satisfy the homogeneous equation and so depend only on the single variable $z \equiv (q^2/\mu^2) e^{2K(g)}$. Thus, for such quantities, $q^2 \to \infty$ is equivalent to $g^2 \to 0^+$ This, of course, is just the asymptotic freedom connection, namely that the asymptotic q^2 behavior of R is deriveable from its small g^2 behavior and so, can presumably be systematically calculated via perturbation theory.

Conversely, it is clear that the small q^2 behavior is equivalent to $g^2 \to 0^-$. Thus, if perturbation theory [i.e. eq. (9)] were convergent, so that D or R were analytic in g^2 at

$g^2 = 0$, then one would have proven a remarkable theorem [15], namely, that their infra-red and ultra-violet behaviors had to be identical! Put slightly differently one could state this as saying that the difference between the IR and UV behaviors reflects the lack of analyticity at $g^2 = 0$. Indeed, if one knew the precise nature of the singularity at $g^2 = 0$ then one would know the IR behavior of the theory! Thus the problem of the large coefficients in QCD is presumably linked to the problem of its IR behavior.

It is well-known that causality implies that Π be an analytic function of q^2 for complex q^2 except for possible singularities along the positive real axis. This is normally expressed via a dispersion relation. Now, asymptotic freedom dictates that for $q^2 \to \infty$, $\Pi \sim ln\ q^2$ thereby requiring (at least) one subtraction [14]. This subtraction in the dispersion relation is, in fact, intimately related to the extra subtraction needed to renormalize Π and, consequently, to the inhomogeneity in the renormalization group equation. Using this it is easy to write a dispersion representation for D:

$$D\left(\frac{q^2}{\mu^2}, g^2\right) = \frac{q^2}{12\ \pi^2} \int_o^\infty \frac{dq'^2}{(q'^2 - q^2)^2}\ R\left(\frac{q'^2}{\mu^2}, g^2\right) \tag{18}$$

$$= \frac{q^2}{\mu^2} e^{2K(g)} \int_o^\infty \frac{dz}{\pi} \frac{f(z)}{\left[z - \frac{q^2}{\mu^2} e^{2K(g)}\right]^2} \tag{19}$$

In writing the second line, the RG constraint that both D and R be functions of z only [i.e. $R = 12\ \pi^2 f(z)$] have been explicitly incorporated.

Notice that no assumption about a mass gap has been made here. For this amplitude, the appearance of a mass gap (beginning at $4\ m_\pi^2$) is related to the introduction of quark masses. For the glueball channel, however, where a similar representation holds, it is generally expected that there exists a mass gap even in the massless quark limit [16]. In such a case the corresponding D will have a Taylor series expansion in q^2:

$$D(q^2/\mu^2, g^2) = \sum_{n=0}^\infty d_n(g^2)(q^2)^n \tag{20}$$

This expansion is the complement to the perturbation series where the expansion is in powers of g^2 with q^2-dependent coefficients. However, in contrast to that expansion which is asymptotic at best, this expansion necessarily has a finite radius of convergence (given by the square of the glueball mass, or, in the case we have been considering, $4\ m_\pi^2$ if quark masses are introduced).

Requiring that D be a function of z only determines the full g^2-dependence of the d_n:

$$D(q^2/\mu^2, g^2) = \sum_{n=0}^\infty \tilde{d}_n \left[q^2/\mu^2 e^{2K(g)}\right]^n \tag{21}$$

$$\approx \sum_{n=0}^\infty \tilde{d}_n\ e^{n/b_1 g^2} \left(\frac{1}{g^2} - \frac{b_2}{b_1}\right)^{nb_2/b_1^2} \left(\frac{q^2}{\mu^2}\right)^n \tag{22}$$

where $\tilde{d}_n$ are numbers independent of q^2, μ^2 or g^2 and, in the second line, we have used eq. (17). The question that must now be faced is how can this (exact) expression, which contains explicit non-analytic pieces in g^2 - indeed, essential singularities - ever be cast into the form of a perturbative expansion in g^2, as in eq. (9), which naïvely treats D as if it were analytic in g^2 at $g^2 = 0$? In order to begin to answer this question we need to digress a little into some properties of series expansions.

IV   DIGRESSION ON SERIES EXPANSIONS

Consider the following series expansion

$$f(x) = \sum_{n=0}^{\infty} \frac{(-1)^n}{n!} a_n x^n \qquad (23)$$

I first want to invert this to obtain a formula for the a_n in terms of $f(x)$. The trick is to make use of properties of the gamma-function[17] $\Gamma(s)$, namely, that it has a string of simple poles at $s = -n$ with residue $(-1)^n/n!$. Thus, (23) can be expressed as a contour integral

$$f(x) = \int_c \frac{ds}{2\pi i} \Gamma(s) a(s) x^{-s} \qquad (24)$$

where $a(s)$ is the analytic continuation of a_n such that $a(-n) \equiv a_n$ and C is the contour shown in the fig. 4. Now, if $a(s)$ has no singularities in the left-hand plane and the integrand is sufficiently convergent, C can be replaced by a line L parallel to the imaginary axis. The resulting representation will be recognized as an inverse Mellin transform from which one can read off the desired result:

$$a(s) = \frac{1}{\Gamma(s)} \int_o^{\infty} dx \, x^{s-1} f(x) \qquad (25)$$

An alternative form for $a(s)$ that is sometimes useful can be obtained by analytically continuing the integrand into the complex x-plane:

$$a(s) = \Gamma(1 - s) \int_c \frac{dx}{2\pi i} (-x)^{s-1} f(x) \qquad (26)$$

The contour C wraps around the cut defined along the positive real axis necessary to define x^{s-1} as a single-valued function. It can be opened up to pick up the singularities of $f(x)$. For example, if these only occur on the negative real axis then (26) reads

$$a(s) = \Gamma(1 - s) \int_o^{\infty} \frac{dx}{\pi} x^{s-1} Im \, f(-x) \qquad (27)$$

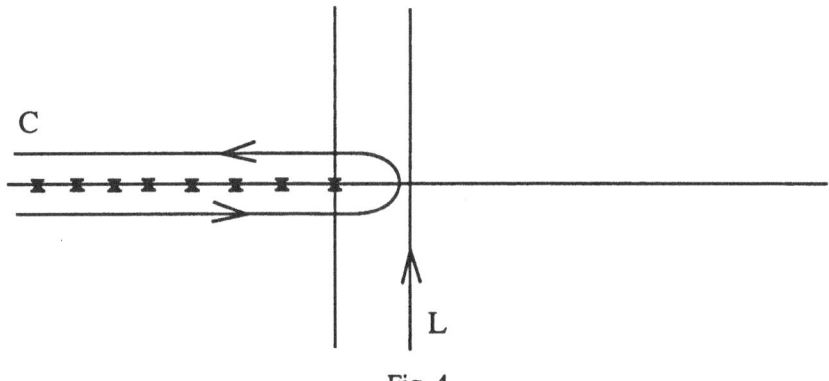

Fig. 4

Complex s-plane showing singularities of Γ (s) and
the contour C used in eq. (24). C can be deformed
to L and a Mellin transforms profound to obtain eq.
(25)

Suppose the series (23) were divergent; define a new series $g(x)$ from it by introducing coefficients b_n such that

$$g(x) = \sum_{n=0}^{\infty} \frac{(-1)^n}{n!} a_n b_n x^n \qquad (28)$$

This new series can always be made convergent by choosing the b_n to fall sufficiently fast for large n. Using the Mellin transform technique described in (A) above together with its convolution theorem leads to the following formula:

$$f(x) = \int_0^{\infty} \frac{du}{u} g(u) B(\frac{x}{u}) \qquad (29)$$

$$\text{where } B(x) \equiv \int_L \frac{ds}{2\pi i} \frac{x^{-s}}{b(s)} \qquad (30)$$

i.e. $1/b(s)$ is the Mellin transform of $B(x)$. As before $b(s)$ is the analytic continuation of b_n such that $b(-n) \equiv b_n$.

The idea of summability is then the following: choose the b_n to make the series for $g(x)$ convergent; insert its sum into (29) which, if it exists, gives a well-defined representation for $f(x)$. The best-known version of this is due to Borel[17]: choose $b_n = 1/n!$ then $B(s) = 1/\Gamma(1-s)$ leading to $B(x) = (1/x)e^{-1/x}$. This defines the Borel sum of $f(x)$:

$$f(x) = \frac{1}{x} \int_o^{\infty} du \, g(u) e^{-u/x} \qquad (31)$$

Other choices for the b_n are, of course, possible, however the Borel technique is the one that has received most attention. As an example of the Borel method consider the series generated by $a_n = (n!)^2$; this is clearly divergent. However, $g(x) = \sum(-x)^n$ can be summed to give $(1+x)^{-1}$ and so

$$f(x) = \int_o^{\infty} \frac{dv \, e^{-v}}{1+xv} \qquad (32)$$

This is supposed to be the true unique representation of $f(x)$. From this point of view the original divergent series simply arose from our "illegal" expansion of the integral as a power series in x.

The question arises as to when this technique does, in fact, give a unique and consistent representation of the function. There are, naturally, many important theorems and treatises dealing with such questions; however, this is not the place to review them. Roughly speaking, the method works when all the integrals converge uniformly. Of particular importance is the absence of singularities on the positive real u-axis. For example, if $a_n = (-1)^n (n!)^2$ then $g(u) = (1-x)^{-1}$ and the series is no longer Borel summable. Typically this means that undetermined essential singularities such as $e^{-a/x}$ cannot be excluded from $f(x)$.

Suppose that $f(x)$ has a power series expansion in some wedge of analyticity in the complex plane $\theta < \pi/2$. Consider

$$\left| f(x) - \sum_{n=0}^{N} \frac{(-1)^n}{n!} a_n x^n \right| \equiv R_N(x) \tag{33}$$

For the sorts of series that we are interested in $R_N(x) \overset{<}{\sim} C_N x^{N+1}$ so that $R_N(x)/x^N \to \infty$ when $N \to \infty$ for x fixed indicating zero radius of convergence. On the other hand $R_N(x)/x^N \to 0$ for N fixed and $x \to 0$. This is Poincaré's definition of an asymptotic series. As shall be demonstrated below, a general feature of quantum field theories is that $C_N \sim cb^{N+1}\Gamma(N+a)$ where a, b, and c are constants. It is easy to confirm that $R_N(x)$ minimizes when

$$N = N_o \approx 1/bx - a \tag{34}$$

and that

$$R_{N_o}(x) \approx c(2\pi bx)^{\frac{1}{2}} e^{-1/bx} \tag{35}$$

This is a remarkable result which demonstrates the character of asymptotic series for it shows that $R_N(x) \to 0$ for x sufficiently small. **Thus when $N = N_o$, the series exponentially approaches the correct value of the function $f(x)$ for sufficiently small x even though it diverges! If further terms are added to the partial sum one is driven further from the correct result.** Thus, if one believes that in QED the appropriate expansion parameter is α/π and that the series is asymptotic in Poincaré's sense then it is not until π/α terms that one need be concerned! Furthermore one can approach within $e^{-\pi/\alpha}$ of the exact result!

V   LARGE n-BEHAVIOR

As explained in the Introduction the main thrust of this paper is to gain some possible insight into the occurrence of large coefficients in the perturbation expansion. The question of summability discussed above, though intimately related to this problem, will be discussed elsewhere. The rest of the paper is therefore devoted to the question of the large n behaviour of the coefficients $a_n(q^2/\mu^2)$. We shall first review how this can be attacked using the "bare" path integral representation. Unfortunately this leaves several questions unanswered, especially for gauge theories[18]. We therefore turn to the representation (19) which incorporates q^2-analyticity and renormalizability and therefore, implicitly, the complete g^2-dependence. Furthermore, in contrast to the path integral it is a representation for the truly physical amplitude.

With a generalization to the path integral in mind, consider functions $f(x)$ which have the following representation

$$f(x) = \frac{1}{\sqrt{x}} \int_{-\infty}^{\infty} du\, e^{-A(u)/x} \tag{36}$$

For example, if $A(u) = u^2 + u^4$ then

$$f(x) = \int_{-\infty}^{\infty} du\, e^{-(u^2 + xu^4)} \tag{37}$$

which is the "zero-dimensional" limit of a Euclidean ϕ^4 field theory with coupling strength x. Notice that if this is naïvely expanded in powers of x, one obtains

$$f(x) \approx \sum_{n=0}^{\infty} (-1)^n \frac{\Gamma(2n + \frac{1}{2})}{\Gamma(n+1)} x^n \tag{38}$$

which is clearly divergent. Indeed it can be Borel-summed to reconstruct the original representation (37). Inserting (36) into the coefficient generating formula, eq. (25), gives

$$a(s) = \frac{\Gamma(\frac{1}{2} - s)}{\Gamma(s)} \int_{-\infty}^{\infty} du\, e^{(s - \frac{1}{2})\ln A(u)} \tag{39}$$

Thus $\ln A(u)$ generates the coefficients. Let us suppose, again with our eye on field theory, that $A(u) \sim u^2$ when $u \to 0$, then by continuing u into the complex plane, it is possible to re-express (39) in the form

$$a(s) = \frac{\Gamma(\frac{1}{2} - s)\Gamma(1 - s)}{\cos \pi s} e^{2\pi i s} \int_c \frac{du}{2\pi i} e^{(s - \frac{1}{2})\ln A(u)} \tag{40}$$

where the contour C wraps around the cut on the positive real axis necessary to define $\ln u$. This expression is ripe for exploitation by the method of steepest descents. Saddle points occur when $[\ln A(u)]' = 0$ *i.e.* $A'(u) = 0$. Notice that, although $A'(u) = 0$ when $u = 0$, this is not so for $[\ln A(u)]'$; thus, even though $u = 0$ is a saddle point of the original representation of $f(x)$ and is the point about which perturbation theory is developed, it is **not** a saddle point of the coefficient generating function, eq. (40). Typically $[\ln A(u)]'' > 0$ at the saddle point (u_0) and we find that for $s = -n \to -\infty$

$$a(-n) = a_n \approx \frac{[\Gamma(1 + n)]^2}{(n+1)} \frac{[-A(u_o)]^{-n}}{[2\pi A''(u_o)]^{\frac{1}{2}}} \tag{41}$$

provided $A(u_o) < 0$ which is generally valid for polynomial $A(u)$. It is straightforward to check that for the example (37), this formula agrees with eq. (38). **Eq. (41) shows that the effective expansion parameter is actually not x, but rather $x/A(u_o)$.** Furthermore, note that if $A(u_o) > 0$ then, naïvely, a factor $(-1)^n$ is induced in (41) which would imply that the series is no longer Borel summable. We shall return to this situation below.

The extension of the above analysis is straightforwardly generalizeable to a path integral representation. The vacuum-vacuum amplitude for a scalar field theory with action functional $A[\phi]$ is given by

$$W(g) = \frac{1}{\sqrt{g}} \int \mathcal{D}\phi e^{-A[\phi]/g} \tag{42}$$

where g is the coupling constant. This can be expanded in the usual Feynman graph perturbation series as a power expansion in g. The coefficients can be determined, as above, using eq. (25):

$$a(s) = \frac{\Gamma(1-s)\Gamma(\frac{1}{2}-s)}{2\pi i \cos \pi s} e^{2\pi i s} \int \mathcal{D}\phi e^{(s-\frac{1}{2})\ln A[\phi]} \tag{43}$$

Thus *ln* $A[\phi]$ **acts as the effective action for determining the** a_n. As before the trivial local minimum of A at $\phi = 0$, though being the starting point for perturbation theory, does not contribute to $a(s)$. The saddle points (at $\phi = \phi_o$, say), satisfy the classical equations of motion and have an action given by

$$A[\phi_o] = -\int d^4 x \phi_0^4 = -8\pi^2/3 < 0 \tag{44}$$

The functional integral can be evaluated at the saddle point and an answer analogous to (41) derived. Care must be taken in properly accounting for zero-modes etc. with the result that

$$a_n \approx n[\Gamma(n+1)]^2 \{-A_o[\phi_o]\}^{-n-\frac{5}{2}} \tag{45}$$

Thus the expansion parameter is not g but rather $(3g/8\pi^2)$.

In the literature this formula was originally derived using eq. (27) for $a(s)$[18]. This requires an immediate analytic continuation in g. Now for $Re\ g\ < 0$, eq. (42) diverges indicating that singularities occur only in the left-hand plane. To determine the nature of these singularities, the path integral itself needs to be analytically continued in ϕ. One finds a cut beginning at $g = 0$ extending along the negative $Re\ g$ axis. An evaluation of the discontinuity across this cut gives a result in agreement with eq. (45). Deriving the result this way makes a connection with Dyson's original argument since an imaginary part only develops if there are other vacua that are not stable.

In attempting to extend this technique to non-abelian gauge theories such as QCD serious problems arise. First, there is the classic problem of maintaining gauge invariance for physical quantities. Secondly, these theories lead to non-trivial saddle-points with **positive** action. As already emphasized this precludes a straightforward application of summability. On the other hand these additional minima of the action (typically referred to as instantons and the like) have a topological characterisitc associated with them. In that sense they give rise to a more general expansion beyond ordinary perturbation theory in terms of topological sectors, as represented in eq. (11). The problem is then to determine how much, if any, of the instanton-like contributions feed back to what is usually thought of as ordinary perturbabion theory. Because of problems such as these it has been difficult to apply these techniques directly to the path integral representation.

The representation, eq. (19), incorporates both causality and renormalizability and, as such, explicitly contains information that is not directly encoded in the path integral. In this sense it is potentially more useful for our purposes since it contains an essential feature of perturbation theory, absent in the path integral, namely renormalizability. In this section I shall therefore attempt to exploit (19) to determine the large n behavior of the coefficients.

Let us first express the perturbative expansion of D in the form

$$D\left(\frac{q^2}{\mu^2},g^2\right) \approx \sum_{n=0}^{\infty} \frac{(-1)^n}{n!} A_n\left(\frac{q^2}{\mu^2}\right)(g^2)^n \tag{46}$$

The coefficients that we are actually interested in are those occurring in the expansion of R as in eq. (9). These can be derived from the A_n via the formula:

$$\bar{a}_n = \left(-4\pi^2\right)^{n+1} \frac{3}{\pi n b_1} \left[\frac{\mathrm{Im}\, A_{n+1}}{(n+1)!} + \frac{b_2}{b_1}\frac{\mathrm{Im}\, A_n}{n!} + \cdots\cdots\right] \tag{47}$$

For large n only the first term need be kept since we anticipate that $A_n \sim (n!)^2$. In order to avoid apparent essential singularities at the origin, arising from the renormalization group, it is convenient to transform to the variable $k \equiv 1/g^2$. Using eqs. (26) and (19) the coefficients in (46) can be obtained from

$$A\left(s,\frac{q^2}{\mu^2}\right) = \Gamma(1-s)\frac{q^2}{\mu^2}\int_0^\infty \frac{dz}{\pi} f(z) \int_c \frac{dk}{2\pi i} \frac{(-k)^{-(1+s)}e^{2K}}{[z-q^2/\mu^2 e^{2K}]^2} \tag{48}$$

with K given by eq. (17). We are interested in the behavior of this expression when $s \to -\infty$. As before, this can be estimated using a steepest descents technique. The structure of the complex k-plane is evidently quite complicated as can be seen from fig. 5. There are three distinct types of singularity: (i) the familiar cut on the positive real axis necessary to define $(-k)^{-s-1}$; (ii) an infinite sequence of poles, (at $k = k_N$, say) arising from the vanishing of the denominator: $k_N/b_1 + \cdots\cdots \approx \ln(q^2z/\mu^2) \pm 2\pi Ni(N = 0,1\cdots)$ and finally (iii) cuts necessary to define potential logarithms in e^{2K}; (e.g. keeping only the first two terms in (16) or (17), there is a cut at $k = -b_2/b_1$ as shown in fig. 5).

As one might guess, this complex structure gives rise to a plethora of saddle points in the k-plane making an accurate estimate of (47) quite subtle. A more detailed discussion of this will be given in a later paper; however, roughly speaking, these saddle points fall into three categories that correspond to the three categories of singularity mentioned above. These occur at (a) $k \approx b_1(s+1) + b_2/b_1 + O(e^{-s})$ (b) $2K \approx \ln(q^2z/\mu^2) \pm 2\pi iN + O(1/s)$; (c) $\beta(g)/g^3 \approx 0$. Let us discuss each of these briefly. The first is the saddle point that we are most interested in, for it dominates the large s behavior. Notice that it corresponds to $g^2 \to 0^-$, as anticipated earlier. The second reflect the poles at $k = k_N$ in (ii) above. They generate the typical $(\ln q^2/\mu^2)^n + \cdots$ dependence of the coefficients familiar from asymptotic freedom. Since we want to compare with ref. 1 where $q^2 = \mu^2$ these saddle points are presumably not of interest here. Finally, there can be saddle points arising from other possible fixed points of $\beta(g)$. Notice that the usual "trivial" fixed point at $g^2 = 0$ (about which perturbation theory is actually generated) is excluded from this. This is analogous to

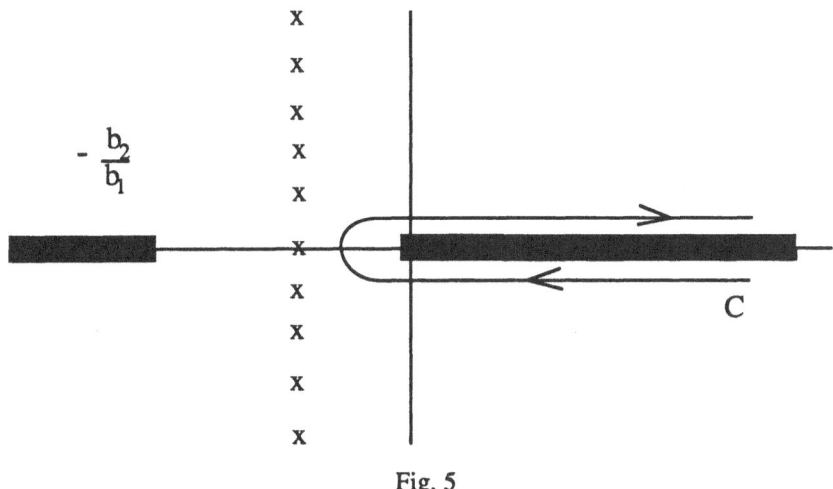

Fig. 5

Complex k-plane showing the contour C used in eq.
(48) together with the singularity structure

the situation we encountered above when dealing with the path integral formulation where the trivial saddle point at $A = 0$ does not explicitly contribute to the estimate. On the other hand, the "pseudo" fixed point at $k = -b_2/b_1$ generated by keeping only the first two terms, must be included. In QCD the signs of b_2 and b_1 are such that this occurs for Re $k < 0$ (or Re $g^2 < 0$). Had the sign of b_2/b_1 been different as in ϕ^4 theory (but not in QED!) then this cut migrates to the positive real axis overlapping the $(-k)^{-1-s}$ cut. In such a case it is not obvious that the coefficients can be determined and it is reasonable to speculate that this is related to the well-known claim that ϕ^4 is in fact a trivial theory.

Returning to eqs. (47) and (48) we can estimate the large n-behavior of $\tilde{a}_n$ by keeping only the single saddle- point at $k \approx b_1(s + 1)$. In that case we find, assuming $|s| \gg b_2/b_1$, that

$$\tilde{a}_n \sim -\frac{e}{\pi}(4\pi^2 eb_1)^{n-1}\frac{\Gamma(n)}{n^2} \tag{49}$$

Thus

$$\frac{\tilde{a}_{n+1}}{\tilde{a}_n} \approx -4\pi^2 eb_1 n \tag{50}$$

$$\approx -\frac{e}{4}(11 - 2/3N_f)n \tag{51}$$

indicating rapid growth of the coefficients with n. (In these expressions e is the base of natural logarithms, not to be confused with the electronic charge!) Perhaps the most striking aspect of this result is that **the effective expansion parameter is not α/π but rather** $\alpha_{eff} \equiv 4\pi^2 eb_1\alpha/\pi$. Repeating the analysis of Section IV C for the case here we find that the remainder R_n minimizes when $n \sim 1 + \alpha_{eff}^{-1} + (2 + \alpha\ eff.)^{-1} \sim 4.5$ [taking α itself to be ~ 0.15]. This is remarkable for it says that **an accurate result can be obtained by keeping only these first few (4-5) terms**. Indeed, an estimate of the error introduced by this process gives a contribution of only $\sim 10^{-3}$! This is all very encouraging; however, how accurately can we trust these estimates if n is so small? To get some idea, if we put n=2 in eq. (50) then $\tilde{a}_3/\tilde{a}_2 \sim 11$. This is indeed a relatively large number, although not big enough to account for the result of ref. 1. Furthermore, our asymptotic formula requires that succeeding terms alternate in sign, a characteristic which does not show up in 1. On the other hand the sign of $\tilde{a}_3$ agrees with our prediction, so the "problem" resides in $\tilde{a}_2$. One certainly would not expect our analysis to be valid for this coefficient so there is no serious contradiction. The problem is, of course, that even though corrections to (49) coming from expanding around the saddle point can be expected to be quite small, there are many other sub-asymptotic saddle points whose contribution may well be comparable to the leading contribution expressed in eq. (49). A more accurate analysis is therefore required to actually establish a firm estimate of $\tilde{a}_3$, for example, and to confirm the calculated result. Such an enterprise is currently being undertaken. It is worth noting that eq. (49) gives $\tilde{a}_3 \sim 12$, a factor 2π smaller than the calculated number of ref. 1.

It should also be pointed out that these leading estimates are both gauge and scheme-invariant, as one might expect. Ultimately one would like to be able to confidently extract α_s from the data (if it is sufficiently accurate!) which means that we need to know either where to stop the series or how to resum it. Our analysis indicates that stopping at $n \sim 4$ is sufficient. In that case one could simply add the estimate to the already calculated numbers.

Similarly one can resum the series beyond these terms using a variant of the Borel technique discussed in the previous Section. In any case it is clear that some consistent procedure or algorithm must eventually be invoked to control the divergence problem and the consequent large coefficients. In this talk I have attempted to show how this problem can be solved in principle and suggested some practical possibilities. A later paper will present details and pursue the solution.

References

[1] S. G. Gorishny, A. L. Kataev and S. A. Larin, Phys. Lett. **212B**, 238 (1988)

[2] F. Dyson, Phys. Rev. **85**, 631 (1952)

[3] F. Dyson, Phys. Rev. **83**, 608 (1951)

[4] R. P. Feynman, Solvay Conference 1959

[5] P. M. Stevenson, Phys. Rev. **D23**, 2916 (1981); C. J. Maxwell, Phys. Rev. **D28**, 2037 (1983); S. J. Brodsky, G. P. Lepage and P. B. Mackenzie, Phys. Rev. **D28**, 228 (1983)

[6] W. Celmaster and D. Sievers, Phys Rev. **D23**, 227 (1981)

[7] A. Dhar, Phys. Lett. **128B**, 407 (1983)

[8] R. Barbieri et al. Nuc. Phys. **B154**, 535 (1979)

[9] See, e.g., S. Raby, G. B. West and C. Hoffman, Phys. Rev. **390**, 828 (1989)

[10] F. Wilczek, Phys. Rev. Lett. **40**, 279 (1978)

[11] M. I. Visotsky, Phys. Lett. **97B**, 159 (1980); P. Nason, ibid. **175B**, 233 (1986)

[12] J. Lee-Franzini, Proc. XXIV Int. Conf. on High Energy Physics (Springer-Verlag, Berlin, 1989) p. 1432

[13] J. Fleischer et al., Univ. of Bielefeld preprint BI-TP 05/89

[14] See, e.g., G. B. West, Nuc. Phys. **B288**, 444 (1987)

[15] G. B. West, Phys. Lett. **145B**, 103 (1984)

[16] G. B. West, Nuc. Phys. B (Proc. Suppl.) **1A**, 57 (1987

[17] E.T. Whittaker and G. N. Watson, "A Course in Modern Analysis", Cambridge Univ. Press, 1950)

[18] For a review see J. Zinn-Justin, Phys. Rep. **70**, 109 (1981)

SUMMARY OF WORKING GROUP C: INTERPLAY OF QCD AND ELECTROWEAK CORRECTIONS

W. J. Stirling

Physics Department
Durham University
Durham, England

ABSTRACT

There were four papers presented in the discussion session on the interplay of QCD and electroweak corrections, covering physics at both LEP/SLC and HERA. In what follows, a brief review and critical discussion of the most important results is presented. More complete details can of course be found in the individual contributions.

QCD AT LEP/SLC

Perturbative QCD is an important component of the physics programme at high energy e^+e^- colliders such as LEP and SLC [1]. The two most important aspects of this are (i) *direct* QCD tests, involving for example the measurement of the strong coupling $\alpha_s(M_Z)$ and the study of multijet final states, and (ii) the role of QCD corrections in precision electroweak measurements. As an illustration of the latter, recall that part of the error on the number of light neutrinos, as measured via the total Z width, comes from the uncertainty on the value of Λ_{QCD} [1].

Before discussing these topics in more detail, it is worth emphasising the current status of precision α_s measurements. The original and arguably most precise determination of the strong coupling is from scaling violations in deep inelastic scattering. A conservative world average is [2] $\Lambda^{(4)}_{\overline{MS}} = 200^{+150}_{-100} \, MeV$, which is equivalent to $\alpha_s(34 \, GeV) = 0.126 \pm 0.014$. Another precision measurement comes from the ratio R of hadronic to $\mu^+\mu^-$ cross sections in e^+e^- annihilation. Several global analyses of data from DORIS, PETRA, PEP and TRISTAN have been performed recently [3,4]. The most up-to-date analysis of D'Agostini et al. [3] gives $\alpha_s(34 \, GeV) = 0.143 \pm 0.015$. Because of *asymptotic freedom*, the strong coupling decreases with increasing renormalisation scale. As a rough guide, the coupling changes by $\delta\alpha_s = -0.02$ in going from $\mu = 34 \, GeV$ to $\mu = M_Z$. An important conclusion from this is that QCD effects are *not* significantly smaller at LEP/SLC than say at PETRA/PEP, and therefore precision measurements of the strong coupling are in principal equally feasible. In fact, as will be discussed below, there is reason to believe that the measurements *will* be cleaner, since non-perturbative contributions tend to decrease as $1/\sqrt{s}$.

(a) Direct QCD Tests

A potentially important measurement of α_s comes from the ratio of hadronic to $\mu^+\mu^-$ cross sections on the Z peak:

$$R_Z = \frac{\Gamma(Z \to \text{hadrons})}{\Gamma(Z \to \mu^+\mu^-)} \tag{1}$$

First, recall the result for R_γ defined as the ratio of hadronic to $\mu^+\mu^-$ total cross sections in e^+e^- annihilation via a virtual photon i.e. far below the Z resonance:

$$R_\gamma(\sqrt{s}) = 3 \sum_q e_q^2 \left[1 + C_1 \frac{\alpha_s}{\pi} + C_2 \left(\frac{\alpha_s}{\pi}\right)^2 + C_3 \left(\frac{\alpha_s}{\pi}\right)^3 + ... \right] , \tag{2}$$

where $\alpha_s \equiv \alpha_s(\sqrt{s})$ and the $\overline{MS}$ scheme is assumed. Numerically, for five flavours of massless quarks, $C_1 = 1$, $C_2 \simeq 1.41$ and $C_3 \simeq 64.9$. The fact that the coefficient C_3, calculated by Gorishny et al. [5], is found to be "anomalously large" (i) casts doubt on the usefulness of special choices of renormalisation schemes and scales (FAC, PMS, BLM, ...) [6], and (ii) affects the determination of α_s in a non-trivial way: without C_3 the fitted value of α_s quoted above becomes $\alpha_s(34~GeV) = 0.158 \pm 0.020$ [3]. Because of this question mark over the convergence of the QCD perturbation series for R, it seems reasonable to assign a *theoretical uncertainty* on α_s of $\pm 10\%$ from this method.

In the standard model the leading order expression for R_Z depends on the weak rather than on the electric charge of the fermions:

$$R_Z = \frac{3 \sum_q (v_q^2 + a_q^2)}{v_\mu^2 + a_\mu^2} \tag{3}$$

with $a_\mu = a_d = -a_u = -1$ and $v_\mu = -1 + 4\sin^2\theta_W$, $v_d = -1 + \frac{4}{3}\sin^2\theta_W$, $v_u = 1 - \frac{8}{3}\sin^2\theta_W$. Thus $R_Z(u) = 3.43$, $R_Z(d) = 4.41$ and $R_Z \simeq 20.1$, in contrast to $R_\gamma = 11/3$ in lowest order, with five massless quark flavours.

The QCD corrections to R_γ and R_Z are *not* identical. In fact in general one has different corrections for the vector and axial contributions to the hadronic width [7,8,9]:

$$\begin{aligned} R_Z = R_V^{(0)} &\left[1 + C_1^V \frac{\alpha_s}{\pi} + C_2^V \left(\frac{\alpha_s}{\pi}\right)^2 + C_3^V \left(\frac{\alpha_s}{\pi}\right)^3 + ... \right] \\ + R_A^{(0)} &\left[1 + C_1^A \frac{\alpha_s}{\pi} + C_2^A \left(\frac{\alpha_s}{\pi}\right)^2 + C_3^A \left(\frac{\alpha_s}{\pi}\right)^3 + ... \right] \end{aligned} \tag{4}$$

The corrections differ significantly already at $O(\alpha_s)$ for massive quarks:

$$\begin{aligned} C_1^V &= 1 + 3\frac{m_q^2}{M_Z^2} + ... \\ C_1^A &= 1 + 3\frac{m_q^2}{M_Z^2} \log \frac{M_Z^2}{m_q^2} + ... \end{aligned} \tag{5}$$

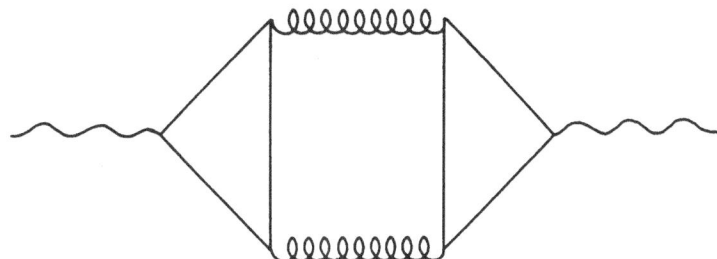

Fig.1. Diagram responsible for differences in the axial and vector current contributions to the hadronic width of the Z.

A complete analysis of the $O(\alpha_s^2)$ corrections has been performed by Kühn and Kniehl [9]. The coefficient C_2^A is different from C_2^V even for massless quarks. This arises from the axial current part of the triangle diagrams containing b and t quarks, Fig.1. Unlike the corresponding vector current contribution, a heavy top quark does not decouple from the loop correction:

$$
\begin{aligned}
C_2^V &= 1.41 \\
C_2^A &= 1.41 + \Delta C_2^A(M_Z/m_t)
\end{aligned}
\tag{6}
$$

with $\Delta C_2^A(x) \simeq 3\log x$ for $x \ll 1$. The net effect is illustrated in Fig.2 which shows the contribution of the various QCD corrections to R_Z. C_1, C_2 and C_3 refer to the standard (vector-current) corrections to $O(\alpha_s^3)$ for massless quarks. ΔC_2^A is the additional axial contribution defined above, and is shown for two values of the top quark mass. The effect of a non-zero b quark mass through $O(\alpha_s)$ is also shown. Evidently the latter two corrections are comparable in size to the C_2 correction, with the opposite sign. The C_3 contribution is still numerically larger than these.

What then are the prospects for extracting a precise value for α_s from data on the hadronic width of the Z? We note in passing that it is not possible to make an accurate measurement from the actual value of the hadronic width alone, since a change in α_s of $\pm 10\%$ corresponds to a change in Γ_h of only $\pm 6~MeV$, well below the expected experimental uncertainty. A much more accurate measurement comes from the *ratio* of the hadronic to leptonic cross-sections on resonance, i.e. R_Z. There is some uncertainty in the literature about the likely achievable experimental error on R_Z. In reference [10], an error of $\pm 1.5\%$ is quoted, but it may be possible to reduce this to below 1% [8]. We can turn the argument around and ask what precision on R_Z is necessary to achieve a competitive measurement of α_s. Using the above results we have, approximately,

$$
\frac{\delta R_Z}{R_Z} \simeq \frac{\alpha_s}{\pi} \frac{\delta \alpha_s}{\alpha_s}
\tag{7}
$$

and so the ratio has to be measured to better than about $0.3 - 0.4\%$ to reduce the experimental error on the strong coupling to below 10%. It is not clear at present whether this can in fact be achieved.

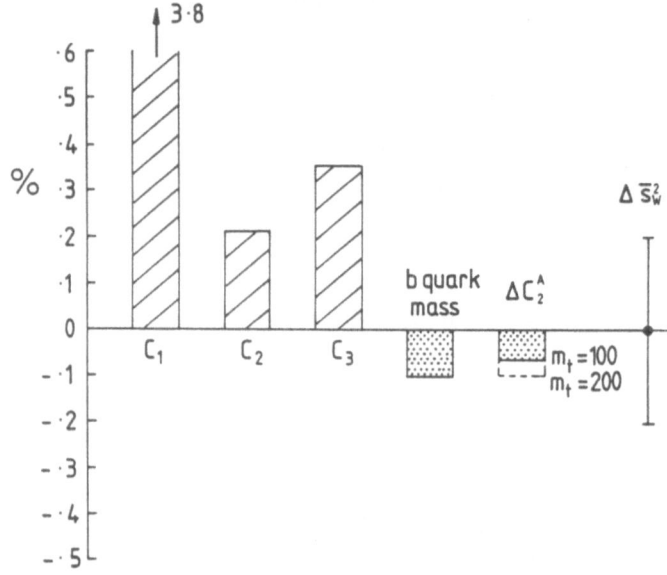

Fig.2. Corrections to the ratio of hadronic to leptonic Z decay widths, as discussed in the text.

A final comment on R_Z concerns the additional *theoretical* error on R_Z from the overall *weak* couplings. Unlike the pure QED result, in leading order the ratio R_Z depends on the weak mixing angle $\sin^2 \theta_W$ through the vector couplings v_f. One can either regard this as a parameter determined to a particular experimental precision or, better, as a parameter which can be calculated precisely in next-to-leading order electroweak perturbation theory given precise values of M_Z, α and G_F, and an estimate of the allowed range for m_t and M_H. A complete discussion of how best to estimate this uncertainty can be found in reference [11]. In summary, one uses an "effective Born approximation" for the partial widths, replacing the usual definition of $\sin^2 \theta_W$ by $\bar{s}_W^2$, thus taking into account the bulk of the light fermion and heavy top quark next-to-leading order contributions. As m_t varies from 50 to 230 GeV/c^2, $\bar{s}_W^2$ changes by approximately 0.005. The effect of this ± 0.0025 variation on the weak couplings and hence on R_Z is shown in Fig.2. It is evidently a non-negligible effect, corresponding to an additional $O(\pm 10\%)$ error on α_s. Of course this error will be substantially reduced once the top quark mass is pinned down more precisely either from measurements at LEP/SLC of the various asymmetries, A_{FB}, A_{LR}, etc., or from the discovery of the top quark itself.

The "traditional" method of determining α_s in e^+e^- annihilation is from 3-jet cross sections. The idea is that the strong coupling can be measured using a variety of *shape variables*:

$$\frac{d\sigma}{dX} = \frac{1}{F} \int d\Phi_3 |\mathcal{M}|^2 (e^+e^- \to q\bar{q}g)\delta(X - f(p_i))$$
$$= \frac{\alpha_s}{\pi}g(X) + ... \tag{8}$$

where X is a generic shape variable such as thrust (T), oblateness (O), aplanarity (A), heavy jet mass (M_H^2), etc. Equivalently, one can write

$$< X > = \frac{\alpha_s}{\pi} A_X + ... \qquad (9)$$

with A_X a number which can be calculated from the appropriate differential cross section.

The strong coupling constant can therefore be determined from fitting a variety of distributions or averages to the experimental data. In practice, the perturbation series for all the most commonly used quantities have been calculated to next-to-leading order. The generic result is

$$< X > = \frac{\alpha_s(\mu)}{\pi} A + \left(\frac{\alpha_s(\mu)}{\pi} \right)^2 \left[B + A b_0 \log \frac{\mu^2}{s} \right] + ... \qquad (10)$$

in, say, the $\overline{MS}$ scheme with $b_0 = (33 - 2n_f)/12$. The coefficient B comes from the real 4-parton final states combined with the 3-parton + virtual gluon final state.

Assuming that the above quantities can be measured with high statistical and systematic precision, there are two major theoretical uncertainties in extracting a precise value for α_s. The first concerns the unknown higher order perturbative corrections. It turns out that the relative size of the next-to-leading corrections varies from quantity to quantity. A very complete study of this question has been performed by Kunszt et al. [7]. The following table, extracted from reference [8], shows the quantity $r = B/A$ for various 3-jet variables X.

Table 1.

quantity	r
$< 1 - T >$	9.6
$< M_H^2 - M_L^2 >$	-0.1
$< O >$	-13.5
$\frac{1}{\sigma} \frac{d\Sigma}{d\chi}(60°)$	11.6
$f_3(0.08)$	11.1
$f_3(0.04)$	9.2

If one believes that the size of the first non-trivial term in the perturbation series gives a good indication of the size of the unknown higher order terms, then it is clear from the Table that some quantities are likely to be more reliable than others for extracting α_s. It is also true that the bigger the higher order coefficients (i.e. r) the stronger the scale (μ) dependence, an effect which must also be taken into account when assigning a theoretical error to a measured value of α_s. This is

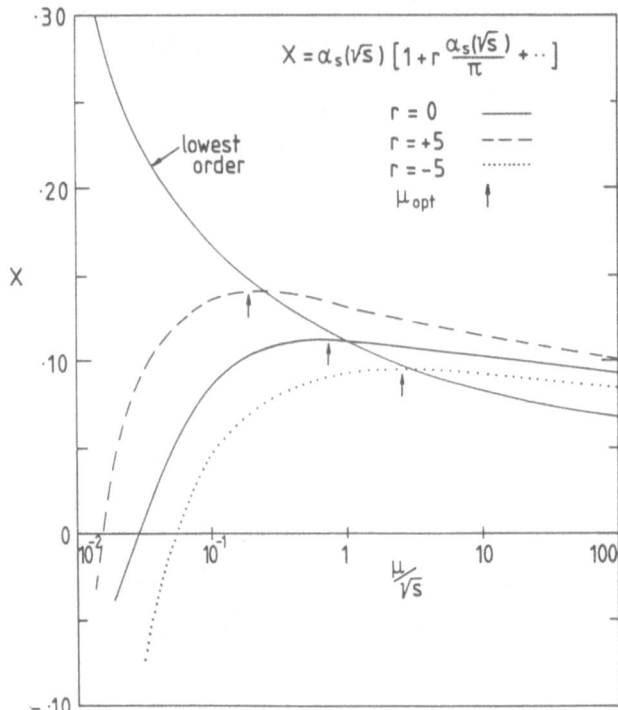

Fig.3. Dependence of a hypothetical shape variable X at LEP/SLC on the renormalisation scale μ, for various values of the next-to-leading order correction: $r = 0, \pm 5$.

illustrated in Fig.3, which shows the dependence of a hypothetical quantity X on the renormalisation scale μ, for various values of the next-to-leading order coefficient r.

The other uncertainty which must be taken into account arises from the hadronisation of the partons. It can be shown rather generally that the effects of fragmentation decrease at least as fast as $1/\sqrt{s}$. The size of the effect can be gauged by calculating distributions such as $d\sigma/dX$ using a parton-shower Monte Carlo and switching off the final hadronisation process. A detailed study is reported in reference [8]. The conclusion is that different quantities have quantitatively different hadronisation corrections, and that these corrections are significantly smaller than those at PETRA/PEP energies. Examples of quantities which have respectively large, small hadronisation corrections are the aplanarity and the three jet fraction defined according to the invariant mass clustering algorithm [12].

The most effective strategy appears to be to chose a quantity with demonstrably small hadronisation corrections, and to fit the next-to-leading order perturbative cross section to the corrected data to determine α_s. Theoretical uncertainties from scale dependence and hadronisation model dependence can then be added to give an overall error. In this way, it should be possible to obtain a measurement of α_s good to better than 10% at LEP and SLC.

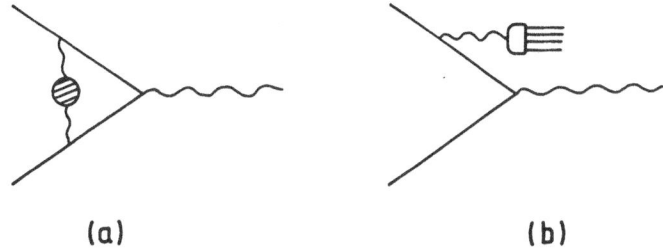

Fig.4. Feynman diagrams showing (a) the hadronic contribution to the $O(\alpha^2)$ correction to the electron form factor, and (b) real hadron emission.

(b) Hadronic Contribution to $O(\alpha^2)$ QED Corrections

QED radiative corrections are a vital component of precision measurements at LEP and SLC. $O(\alpha^2)$ corrections must be considered and among these are contributions involving quarks. For example, Fig.4 shows two types of 'hadronic correction', the first involving a form factor type correction to the e^+e^-Z vertex and the second describing the production of final state hadrons from initial state photon bremsstrahlung. Now in both cases the invariant mass of the hadronic system can be small, and a purely perturbative QCD treatment, i.e. replacement of the hadronic system by a $q\bar{q}$ state, is not applicable.

A detailed quantitative study of these type of corrections has been performed by Kühn et al., and is described in an accompanying article [13]. The essential idea is to relate the hadronic insertion in the photon propagator – via dispersion relations – to the measured hadronic cross section $R(\sqrt{s})$ in e^+e^- annihilation. Thus the contribution of Fig.4(a) to the form factor is [13]

$$\left(\frac{\alpha}{\pi}\right)^2 F^{(4)}(Q^2) = \frac{\alpha}{3\pi} \int \frac{dq'^2}{q'^2} R(q'^2) \frac{\alpha}{\pi} \rho\left(\frac{q^2}{q'^2 - i\epsilon}\right) \tag{11}$$

where ρ is obtained from the lowest order triangle graph with a massive vector propagator replacing the photon. Fig.5 shows the magnitude of the correction obtained in this way.

The *real* hadronic radiation can be treated in a similar way. It is useful to consider separately the cases of *soft* hadronic radiation, i.e. $\sigma_{\text{soft}}(s, \Delta)$ with $E_{\text{had}} < \Delta \ll \sqrt{s}$, and *hard* hadronic radiation, i.e. $\frac{d\sigma_{\text{hard}}}{dz}(s, z)$ with $M^2_{\text{had}} = zs$. In each case the cross section can be evaluated from an appropriate integral of the lowest order cross section for e^+e^- annihilation into a hadronic state of invariant mass Q^2. Detailed numerical results can be found in reference [13]. As an illustration, Fig.6 shows the differential cross section $d\sigma/dz$ for hard radiation of hadrons, compared with the corresponding cross section for lepton pair production.

(c) <u>QCD Corrections to Higgs Production and Decay</u>

Higgs production in Z decay, via the processes $Z \to H\mu^+\mu^-$ and $Z \to H\gamma$, should be detectable at LEP (first phase) and SLC for Higgs masses up to about 60 GeV/c^2 [14]. Because the Higgs couples to fermions with a strength proportional to the fermion mass, the dominant decays are expected to be $H \to b\bar{b}$, $c\bar{c}$, $\tau^+\tau^-$, etc. There are, in addition, small but non-negligible decay rates into $\gamma\gamma$ and gg final states. The QCD corrections to these partial decay rates have been discussed by Djouadi [15].

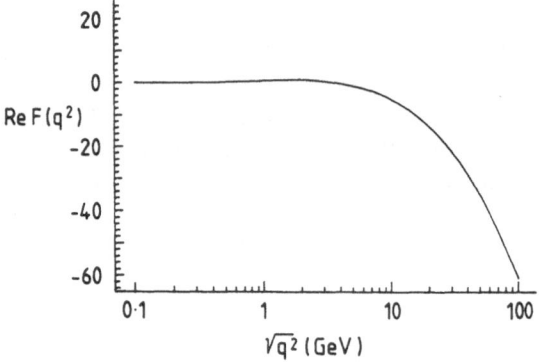

Fig.5. $O(\alpha^2)$ corrections to the electron form factor from the hadronic contribution shown in Fig.4(a), from reference [13].

The partial decay width into a pair of quarks of mass m_q is given by [14]

$$\Gamma(H \to q\bar{q}) = \Gamma_0[1 + \frac{\alpha_s}{\pi}\Delta_H(v) + ...]$$
$$\Gamma_0 = \frac{3G_F M_H}{4\sqrt{2}\pi} m_q^2 \, v \qquad (12)$$

with $v = \sqrt{1 - 4m_q^2/M_H^2}$. The $O(\alpha_s)$ QCD corrections were first calculated by Braaten and Leveille [16]. The coefficient $\Delta_H(v)$ is singular for both $v \to 0$ and $v \to 1$:

$$\Delta_H(v \to 0) \sim \frac{2\pi^2}{3}\frac{1}{v} - \frac{4}{3}$$
$$\Delta_H(v \to 1) \sim 4\log\frac{m_q}{M_H} + 3 \qquad (13)$$

In each case the leading singular behaviour can be resummed to all orders in perturbation theory. It is presumably the second of these that will be more relevant in

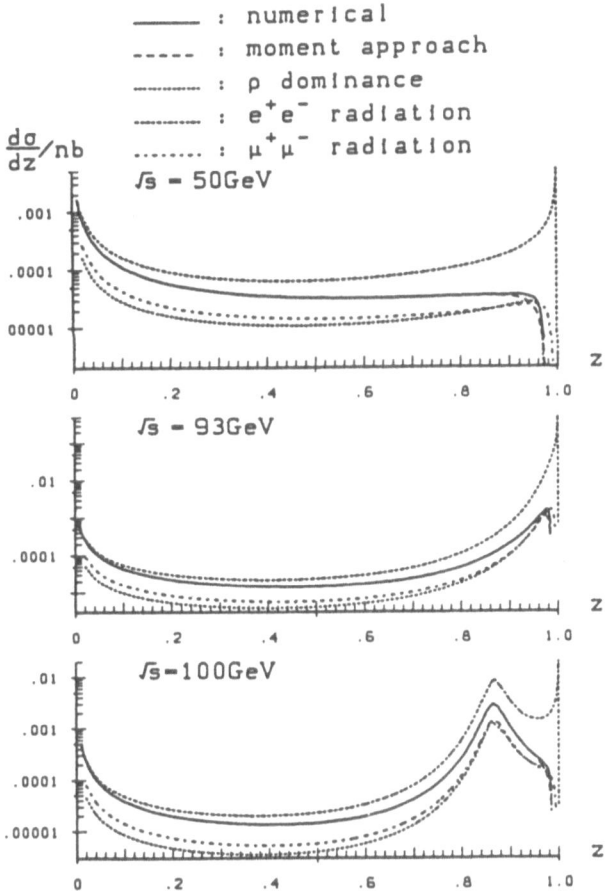

Fig.6. Differential cross section for hard radiation of hadrons, electrons and muons as a function of $z = M_{\text{had}}^2/s$, from reference [13].

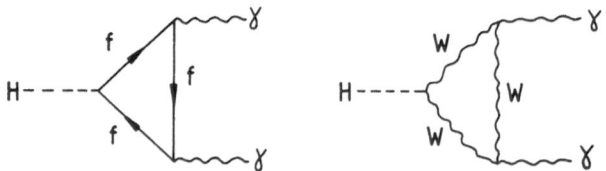

Fig.7. Feynman diagrams for Higgs decay to two photons.

practice. Thus, summing the leading mass singularities at each order gives [16]

$$\Gamma \sim \Gamma_0 \left[\frac{\alpha_s(M_H)}{\alpha_s(m_q)} \right]^{\frac{23}{24}} [1 + \frac{\alpha_s}{\pi} \overline{\Delta}_H + ...] \tag{14}$$

with $\overline{\Delta}_H$ now finite as $m_q \to 0$. Numerically, the effect of the QCD corrections is to *decrease* the leading order partial decay width by about 30% for $M_H \sim 40-50 \, GeV/c^2$ [15].

The Higgs partial decay width to two photons proceeds via a virtual fermion or W loop, Fig.7. To leading order [17]:

$$\Gamma_0 = \frac{\alpha^2 G_F M_H^3}{8\sqrt{2}\pi^3} \left| \sum_f N_c e_f^2 I_f + I_W \right|^2 \tag{15}$$

where I_f and I_W are the contributions of the corresponding loop integrals, depending respectively on m_f/M_H and M_W/M_H. Djouadi and Oualitsen have calculated the $O(\alpha_s)$ corrections to I_q [15]. Working in the limit where $\mu = M_H/m_q \ll 1$ they find

$$I_q \simeq \frac{1}{3}\left(1 + \frac{\mu^2}{20}\right)\left[1 + \frac{\alpha_s}{\pi}(1 - \frac{\mu^2}{4})\right] \tag{16}$$

Evidently the QCD correction is numerically small, and does not therefore enhance this particular decay channel relative to the $H \to f\bar{f}$ channels.

It is interesting that the same $(1 + \frac{\alpha_s}{\pi})$ correction also arises in the QCD correction to the Higgs production mechanism $Z \to H\gamma$, which is larger than $Z \to H\mu^+\mu^-$ for $55 \, GeV/c^2 \lesssim M_H < M_Z$ [14]. Again, the decay proceeds either by an internal fermion or W loop. However, even for a heavy top mass the W contribution dominates (by a factor of about 15) [15]. It turns out that the QCD correction to the quark loop, in the limit $m_q \gg M_H$, M_Z, is again $(1 + \frac{\alpha_s}{\pi})$ [15].

QCD AND ELECTROWEAK CORRECTIONS AT HERA

A detailed review of the interplay of electroweak and QCD effects at HERA has been discussed at this meeting by Hollik [18]. The primary physics aim at HERA is a detailed investigation of the short distance parton structure of the proton. The basic formula for the deep inelastic electron-proton scattering cross section is (photon exchange only)

$$\frac{d\sigma^{ep \to eX}}{dx dQ^2} = \frac{2\pi\alpha^2}{xQ^4}[1 + (1 - y)^2] \sum_f e_f^2 x (q_f(x, Q^2) + \bar{q}_f(x, Q^2)) \tag{17}$$

with $Q^2 = xys$. Similar formula exist for Z and W exchange, which become relevant at the typical HERA Q^2 values of order 10^4 GeV^2. The physics interest is in the determination of the quark distributions $q_f(x, Q^2)$, the precision measurement of scaling violations $\partial q_f(x, Q^2)/\partial Q^2$ to extract $\Lambda_{\overline{MS}}$, and the investigation of possible quark substructure signalled by anomalous Q^2-dependence of the cross section in eqn.17.

The basic $eq \to eq$ t-channel photon (and Z) exchange diagram – from which eqn.17 follows – receives electroweak radiative corrections. The standard procedure is to separate these into two classes. The first consists of the purely photonic (QED) corrections, i.e. the emission of real and virtual photons from the electron and quark lines. The second consists of the infra-red finite weak corrections involving Z-propagator insertions, Z-fermion vertex corrections, double Z exchange etc. The latter corrections have exact analogues in the LEP/SLC $e^+e^- \to f\bar{f}$ processes and contain, for example, next-to-leading order dependence on $m_{\rm top}$, M_H etc. It is relatively straightforward, therefore, to take them into account [19].

There *is* however a slight conceptual problem with the pure QED corrections. The 'leptonic' QED corrections, i.e. those involving photon emission from the electron line only, can give rise to large soft and collinear logarithms: $\log Q^2/m_e^2 \gg 1$. However the effect of these is simply to rescale the kinematic variables 'seen' by the quark, i.e. $x, y, Q^2 \to x', y', Q'^2$, with the largest corrections occuring near the edges of phase space. This is illustrated in Fig.8, taken from reference [20], showing the effect of the one-loop corrections $(1 + \delta_{NC}^1)$ on the neutral-current cross section for $x = 0.2$ and with different energy cuts on the bremsstrahlung photon. Again, it is in principle straightforward to take account of these corrections – for example, to extract *QED-corrected* structure functions $F_i(x, Q^2)$ – in a similar way as is done at LEP/SLC for initial state photon bremsstrahlung. What about the corresponding corrections to the *quark* line? Since $m_q \gg m_e$ one might expect that the corrections are intrinsically smaller. However the same formalism cannot be applied since the quarks are *not* on-mass-shell. Recall that when a quark emits a gluon, before or after being struck by a virtual photon, the effect can be incorporated in the Altarelli-Parisi equations and gives rise to QCD scaling violations. The emission of a collinear photon – rather than a gluon – can be incorporated in exactly the same way. This gives rise to a modified Altarelli-Parisi equation [21]:

$$Q^2 \frac{\partial}{\partial Q^2} q(x, Q^2) = \frac{\alpha_s(Q^2)}{2\pi} \left[\int_x^1 \frac{dy}{y} \left(P^{q\to qg}(y) q(\frac{x}{y}, Q^2) + P^{g\to q\bar{q}}(y) g(\frac{x}{y}, Q^2) \right) \right]$$
$$+ O(\alpha_s^2) + \frac{\alpha}{2\pi} \int_x^1 \frac{dy}{y} P^{q\to q\gamma}(y) q(\frac{x}{y}, Q^2) \tag{18}$$

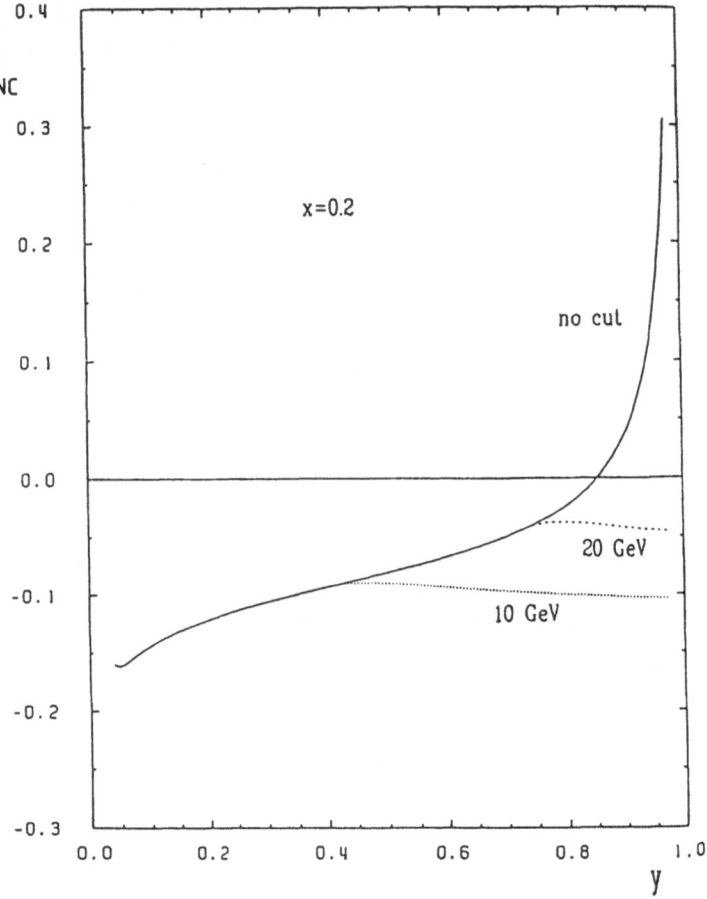

Fig.8. The one-loop QED correction to the neutral current cross section with different cuts on the bremsstrahlung photon, from reference [20].

with $P^{q \to q\gamma}(x) = e_q^2[(1 + x^2)/(1 - x)]_+$. As might be expected, the leading $O(\alpha)$ QED corrections turn out to be comparable in size to the next-to-leading $O(\alpha_s^2)$ QCD corrections, and therefore have only a small effect on the Q^2 evolution [21]. The correct procedure is therefore to take these QED corrections fully into account when analysing Q^2-dependent parton distributions. It is meaningless to try to correct the parton distributions themselves for this effect, since at *any* scale Q^2 the quark is surrounded by a virtual cloud of gluons *and* photons, and a 'primordial' quark distribution – corrected for the presence of gluons and photons – has no meaning.

At HERA, therefore, the general structure of the fully-corrected electron-proton deep inelastic scattering cross section is

$$\sigma^{ep} = e(x, Q^2) * q(x, Q^2) * \hat{\sigma}^{eq} \tag{19}$$

where $e(x, Q^2)$ is the QED electron structure function, $q(x, Q^2)$ is the (QCD + QED) corrected quark structure function in the proton, and $\hat{\sigma}$ incorporates the finite, nonfactorising (QCD + electroweak) higher order corrections. A similar structure obtains for both neutral- and charged-current processes.

CONCLUSIONS

Many interesting tests of QCD are possible at LEP/SLC from the process $e^+e^- \rightarrow Z^0 \rightarrow$ hadrons. There are interesting theoretical effects in the higher order QCD corrections to the ratio of hadronic to leptonic widths, but a competitive *precision* measurement of the strong coupling from this quantity looks difficult. More promising in this respect are the jet measures (shape parameters) which appear to suffer less from non-perturbative corrections at high energies. The errors on α_s will be dominated by the theoretical uncertainty in unknown higher order perturbative corrections, and the model-dependent systematic uncertainties from the parton hadronisation.

QCD corrections are also important in precision electroweak measurements, but appear to be under control in most cases. The typical QCD correction is $1 + \frac{\alpha_s(M_Z)}{\pi}$ which is only a 5% effect with a $\pm0.5\%$ uncertainty.

The situation is complementary at HERA, where the principle physics interest is testing the strong interaction at very short distances. Relative to the QCD effects (e.g. in scaling violations of structure functions) the electroweak radiative corrections are not significantly large and appear well under control.

ACKNOWLEDGEMENTS

It is a pleasure to thank Professor N. Dombey and Dr. F. Boudjema for organising such an enjoyable and stimulating meeting.

REFERENCES

1. See for example: *Z Physics at LEP1*, Eds. G. Altarelli, R. Kleiss and C. Verzegnassi, CERN Yellow Report 89-08 (1989), Vols. 1-3.

2. M. Barnett, I. Hinchliffe and W.J. Stirling in *Review of Particle Properties*, Phys. Lett. **204B** (1988).

3. G. D'Agostini, W. de Boer and G. Grindhammer, preprint DESY 89/057 (1989).

4. R. Marshall, Z. Phys. **C43** (1989) 595.

5. S.G. Gorishny, A.L. Kataev and S.A. Larin, Phys. Lett. **212B**, 238 (1988).

6. See for example W.J. Stirling, Proceedings of the Europhysics Conference on High Energy Physics, Madrid, 1989.

7. Z. Kunszt, these proceedings.

8. Z. Kunszt et al. in *Z Physics at LEP1*, Eds. G. Altarelli, R. Kleiss and C.

Verzegnassi, Vol.1: Standard Physics, CERN Yellow Report 89-08 (1989), page 373.

9. B.A. Kniehl and J.H. Kühn, Phys. Lett. **224B** (1989) 229.

10. G. Altarelli et al., *Physics at LEP*, eds. J. Ellis and R.D. Peccei, CERN Yellow Report 86-02 (1986), page 1.

11. M. Consoli et al. in *Z Physics at LEP1*, Eds. G. Altarelli, R. Kleiss and C. Verzegnassi, Vol.1: Standard Physics, CERN Yellow Report 89-08 (1989), page 7.

12. JADE collaboration: W. Bartel et al., Z. Phys. **C33** (1986) 23.

13. J.H. Kühn, these proceedings; B.A. Kniehl, M. Krawczyk, J.H. Kühn and R.G. Stuart, Phys. Lett. **209B** (1988) 337.

14. See for example P.J. Franzini et al. in *Z Physics at LEP1*, Eds. G. Altarelli, R. Kleiss and C. Verzegnassi, Vol.2: Higgs Search and New Physics, CERN Yellow Report 89-08 (1989), page 58.

15. A. Djouadi, these proceedings; A. Djouadi and F. Oualitsen, Aachen preprint PITHA-89/15 (1989).

16. E. Braaten and J.P. Leveille, Phys. Rev. **D22** (1980) 715.

17. J. Ellis, M.K. Gaillard and D.V. Nanopolous, Nucl. Phys. **B106** (1976) 292.

18. W. Hollik, these proceedings.

19. For a complete discussion see the report by Study Group 5, D.Yu. Bardin et al., on "Radiative Corrections" in the Proceedings of the HERA Workshop, ed. R.D. Peccei, DESY 1988, Vol.2, p. 577.

20. J. Kripfganz and H.J. Möhring, Z. Phys. **C38** (1988) 653.

21. J. Kripfganz and H. Perlt, in reference 16.

RADIATIVE CORRECTIONS TO
W^+W^-/ZZ PRODUCTION IN e^+e^-

Ansgar Denner [1]

Max Plank Institut für Physik und Astrophysik
Föhringer Ring 6, 8000 München, FRG

1 Introduction

SLC and LEP will soon provide important and accurate tests of the standard model (SM). The large cross sections at the Z-resonance will allow very precise measurements of the mass and the total and partial widths of the Z-boson, various asymmetries and other observables. Comparision of those measurements with thoroughly elaborated predictions of the SM will elucidate its validity and put limits on new physics. However, not all aspects of the SM can be investigated at the Z-peak like the nonabelian structure of the theory and the Higgs sector. Boson pair production, especially the reaction $e^+e^- \rightarrow W^+W^-$, which will become accessible at LEP200, is another important field for probing the SM. It will allow the direct investigation of the triple nonabelian gauge coupling and a precise determination of the W-mass. Confronting the measured value of the W-mass with the prediction from α, G_F and M_Z will be a stringent test of the SM at its quantum level and will provide bounds on the unknown top quark mass and on new physics.

The cross section for W-pair production at 200 GeV is about 20 pb which is ten times the muon pair cross section. This cross section together with the expected luminosity at LEP200 will result in an experimental accuracy at the percent level. To reach a comparable accuracy in theoretical predictions one loop radiative corrections (RC's) must be included. Compared to W-pair production the Z-pair production is of minor importance. The corresponding cross section is relatively suppressed by a factor ten. Nevertheless also this reaction should be checked for agreement with the SM.

In this talk I present the results for the cross sections for $e^+e^- \rightarrow W^+W^-, ZZ$ to lowest order and including one loop RC's. I also discuss the inclusion of the finite width of the produced bosons, which is necessary for a realistic description of the measurements via their decay products.

---

[1] On leave of absence from Physikalisches Institut, Universität Würzburg, 8700 Würzburg, FRG

2 Amplitudes, Born Cross Sections

The differential cross section for the process $e^+e^- \to B^+B^- (= W^+W^-, ZZ)$ is given by

$$\frac{d\sigma}{d\Omega} = \frac{\alpha^2}{4s}\sqrt{1 - \frac{4M_B^2}{s}} |\mathcal{M}|^2 , \tag{1}$$

where $s = E_{CMS}^2$ is the center of mass energy squared and M_B the mass of the produced bosons which are assumed to be charge conjugate to each other. The scattering angle θ_{CMS} is defined as the angle between the electron and the boson $B^- (=W^-)$. Denoting the polarization vectors of the outgoing bosons by $\varepsilon^\pm(k_\pm, \lambda_\pm)$ and the spinors of the incoming electron and positron by $u(p_-, \sigma_-)$ and $\bar{v}(p_+, \sigma_+)$, where $k_\pm, \lambda_\pm$ and $p_\pm, \sigma_\pm$ are the momenta and helicities of the bosons respectively leptons, the matrix element $\mathcal{M}$ is given by

$$\mathcal{M}(p_+, \sigma_+, p_-, \sigma_-, k_+, \lambda_+, k_-, \lambda_-)$$
$$= \langle k_+, \lambda_+, k_-, \lambda_- \mid T \mid p_-, \sigma_-, p_+, \sigma_+ \rangle \tag{2}$$
$$= \bar{v}(p_+, \sigma_+) T_{\mu\nu} u(p_-, \sigma_-) \varepsilon_+^\mu(k_+, \lambda_+) \varepsilon_-^\nu(k_-, \lambda_-) .$$

Since for $s \gg m_e^2$ chiral symmetry is valid the matrix element is only nonvanishing for opposite lepton helicities ($\sigma = \sigma_- = -\sigma_+$). CP-invariance, which is obeyed at the one loop level for the process $e^+e^- \to B^+B^-$ within the SM, implies

$$\mathcal{M}(\sigma, \lambda_+, \lambda_-) = \mathcal{M}(\sigma, -\lambda_-, -\lambda_+) . \tag{3}$$

Consequently, there are 2×6 different helicity matrix elements left equivalent to 12 different polarized cross sections.

Corresponding to the 6 different combinations of boson helicities the general matrix element can be decomposed into a sum of CP-invariant standard matrix elements $\mathcal{M}_i^\sigma$ multiplied by formfactors $F_i^\sigma(s,t)$ depending on the Mandelstam variables $s = (p_+ + p_-)^2$ and $t = (p_+ - k_+)^2$. For convenience I use seven (linearly dependent) matrix elements

$$\mathcal{M}(\sigma, \lambda_+, \lambda_-, s, t) = \sum_{i=1}^{7} \mathcal{M}_i^\sigma F_i^\sigma(s,t) \tag{4}$$

defined by

$$\mathcal{M}_1^\sigma = \bar{v}(p_+) \, \slashed{\varepsilon}_+ (\slashed{k}_+ - \slashed{p}_+) \, \slashed{\varepsilon}_- \omega_\sigma u(p_-) ,$$

$$\mathcal{M}_2^\sigma = \bar{v}(p_+) \, \slashed{k}_+ \, \varepsilon_+ \cdot \varepsilon_- \omega_\sigma u(p_-) ,$$

$$\mathcal{M}_3^\sigma = \bar{v}(p_+) [\slashed{\varepsilon}_+ \, \varepsilon_- \cdot k_+ - \slashed{\varepsilon}_- \, \varepsilon_+ \cdot k_-] \omega_\sigma u(p_-) ,$$

$$\mathcal{M}_4^\sigma = \bar{v}(p_+) [\slashed{\varepsilon}_+ \, \varepsilon_- \cdot p_- - \slashed{\varepsilon}_- \, \varepsilon_+ \cdot p_+] \omega_\sigma u(p_-) , \tag{5}$$

$$\mathcal{M}_5^\sigma = \bar{v}(p_+) \, \slashed{k}_+ \omega_\sigma u(p_-) \, \varepsilon_+ \cdot k_- \, \varepsilon_- \cdot k_+ ,$$

$$\mathcal{M}_6^\sigma = \bar{v}(p_+) \, \slashed{k}_+ \omega_\sigma u(p_-) \, \varepsilon_+ \cdot p_+ \, \varepsilon_- \cdot p_- ,$$

$$\mathcal{M}_7^\sigma = \bar{v}(p_+) \, \slashed{k}_+ \omega_\sigma u(p_-) [\varepsilon_+ \cdot k_- \, \varepsilon_- \cdot p_- + \varepsilon_+ \cdot p_+ \, \varepsilon_- \cdot k_+] .$$

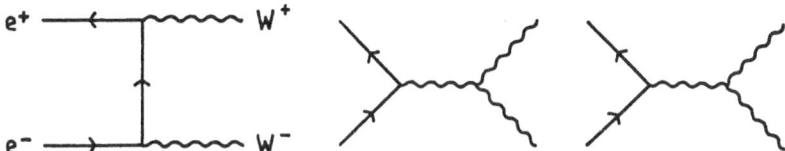

Figure 1. Lowest order diagrams for the process $e^+e^- \to W^+W^-$.

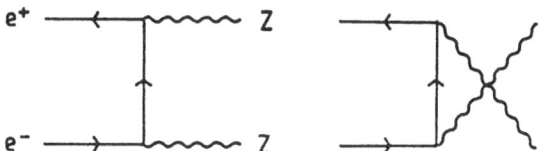

Figure 2. Lowest order diagrams for the process $e^+e^- \to ZZ$.

where $\omega_\sigma = (1 + 2\sigma\gamma_5)/2$ with $\sigma = \pm\frac{1}{2}$ projects on the lepton helicities. The relation between the seven standard matrix elements is

$$\mathcal{M}_7 = \mathcal{M}_5 + \frac{M_B^2 - t}{2}\mathcal{M}_3 + \frac{s}{2}(\mathcal{M}_4 - \mathcal{M}_1 - \mathcal{M}_2) . \tag{6}$$

In lowest order three Feynman diagrams are contributing to W-pair production (fig. 1), the t-channel ν-exchange diagram and the s-channel γ- and Z-annihilation diagrams containing the nonabelian triple gauge coupling. The analytic expressions read

$$\mathcal{M}_o(-, \lambda_+, \lambda_-, s, t) = \mathcal{M}_1^- \frac{1}{2s_w^2} \frac{1}{t} - \left[\mathcal{M}_2^- - \mathcal{M}_3^-\right] \left[\frac{2}{s} - \frac{c_w}{s_w} g_e^- \frac{2}{s - M_Z^2}\right] , \tag{7}$$

$$\mathcal{M}_o(+, \lambda_+, \lambda_-, s, t) = -\left[\mathcal{M}_2^+ - \mathcal{M}_3^+\right] \left[\frac{2}{s} - \frac{2}{s - M_Z^2}\right] . \tag{8}$$

In the case of Z-pair production the Born amplitude is built up by one t-channel and one u-channel matrix element (fig. 2) yielding

$$\mathcal{M}_o(\sigma, \lambda_+, \lambda_-, s, t) = g_+^\sigma g_-^\sigma (\frac{1}{t}\mathcal{M}_1^\sigma + \frac{1}{u}\mathcal{M}_{1,u}^\sigma) \tag{9}$$

with

$$\mathcal{M}_{1,u}^\sigma = \mathcal{M}_1^\sigma + 2\mathcal{M}_2^\sigma - 2\mathcal{M}_3^\sigma . \tag{10}$$

In both cases the Born amplitude involves only $\mathcal{M}_1$ and $\mathcal{M}_2 - \mathcal{M}_3$. The couplings between the gauge bosons and the leptons are given by

$$g_\gamma^\pm = 1 ,$$

$$g_Z^- = \frac{2s_w^2 - 1}{s_w c_w} , \qquad g_Z^+ = \frac{s_w}{c_w} , \tag{11}$$

$$g_W^- = \frac{1}{\sqrt{2}s_w} , \qquad g_W^+ = 0 .$$

with

$$c_w = \cos\theta_w = \frac{M_W}{M_Z} , \qquad s_w = \sqrt{1 - c_w^2} . \tag{12}$$

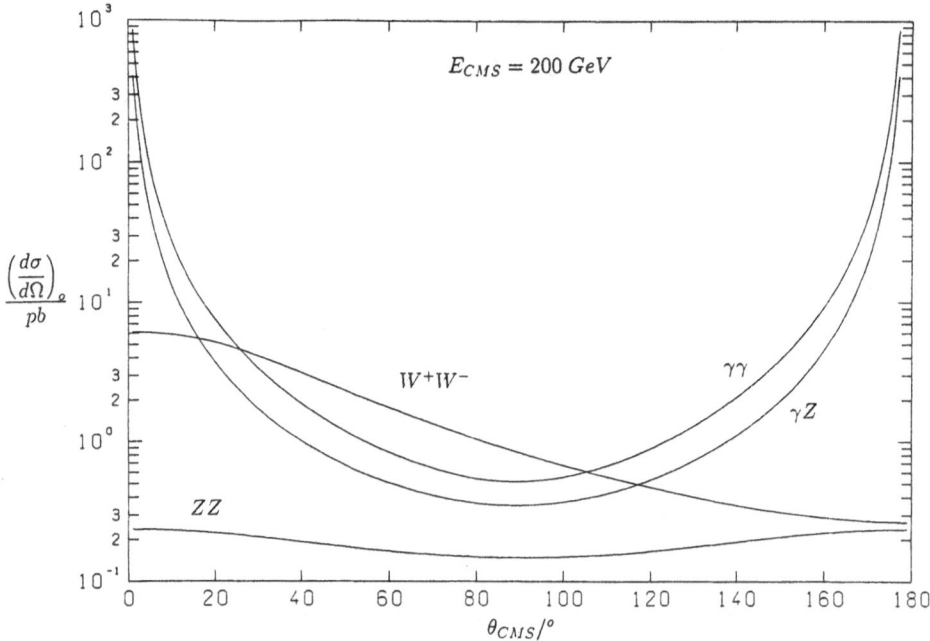

Figure 3. Lowest order differential cross section for e^+e^- annihilation into $\gamma\gamma$, γZ, ZZ and W^+W^- at a center of mass energy of 200 GeV.

The resulting differential cross sections are shown in fig. 3 at 200 GeV center of mass energy together with the cross sections for $\gamma\gamma$- and γZ- production. Bose symmetry respectively CP-invariance results in a symmetric cross section for the production of neutral gauge bosons, whereas the WW cross section is peaked in forward direction due to the t-channel contribution. As can be seen from fig. 6 the peaking gets stronger with increasing energy.

Fig. 4 shows the separate contributions to the total W-pair production cross section namely the t-channel exchange and the s-channel annihilation as a function of energy. Due to the longitudinally polarized W-bosons the separarte contributions of t-channel and s-channel diagrams rise proportional to s thus violating unitarity at high energies. However there is a strong destructive interference between the different Born diagrams resulting in a high energy behavior of the W-pair production cross section within the SM like $\frac{1}{s}\log s$ respecting unitarity. The cancellation reaches one order of magnitude at 400 GeV and two orders at 1 TeV. Since it depends crucially on the values of the couplings, small deviations from the SM will yield sizeable effects.

3   Radiative Corrections

3.1   Organization

The electroweak radiative corrections ($EWRC$'s) are calculated in a multiplicative on shell renormalization scheme using the finestructure constant α and the masses of the gauge bosons M_W, M_Z, of the fermions m_f and of the Higgs scalar M_H as physical (renormalized) parameters [1,2]. In the numerical evaluation the values $M_Z = 93\ GeV$, $M_H = 100\ GeV$ and $m_t = 50\ GeV$ were used. The mass of the

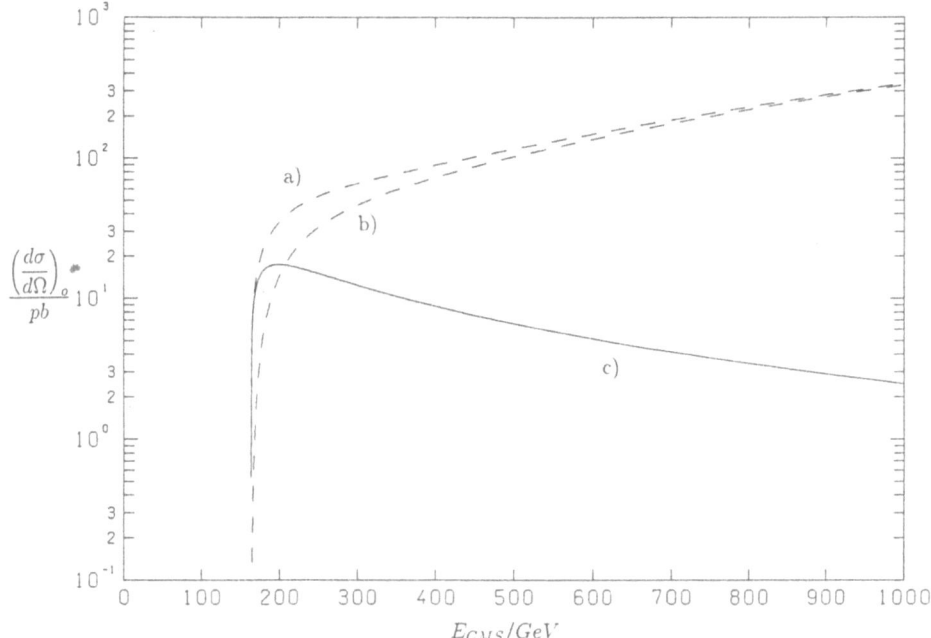

E_{CMS}/GeV

Figure 4. Lowest order contributions to the unpolarized W-pair production cross section: t-channel (a), s-channel (b) and SM cross section (c).

W-boson M_W is calculated from the experimentally well known Fermi constant G_F by solving the equation

$$M_W^2(1 - \frac{M_W^2}{M_Z^2}) = \frac{\pi\alpha}{\sqrt{2}G_F} \frac{1}{1 - \Delta r(\alpha, M_Z, M_W, M_H, m_f)} \ . \tag{13}$$

Δr contains the weak RC's to muon decay. The calculations were performed in a 't Hooft Feynman gauge. In order to compensate infrared divergencies soft photon bremsstrahlung with a cutoff on the photon energy $E_\gamma \leq \Delta E = 0.05\sqrt{s}$ was included.

The large number of Feynman diagrams (more than 200 for $e^+e^- \to W^+W^-$) enforces a systematic evaluation procedure. A convenient method is the one of Passarino and Veltman [3] which allows to decompose every one loop integral in terms of invariant integrals and Lorentz tensors. The invariant integrals can be reduced to the scalar integrals A_o, B_o, C_o and D_o which were calculated in [4]. Thus all the one loop corrections to the matrix elements can be expressed by the standard matrix elements $\mathcal{M}_i^\sigma$ times formfactors which are linear combinations of the scalar integrals

$$\delta\mathcal{M} = \sum_{i=1}^{7} \mathcal{M}_i^\sigma \delta F_i^\sigma(s,t) \ . \tag{14}$$

Finally the contribution of $\delta\mathcal{M}$ to the cross section is given by

$$\delta\left(\frac{d\sigma}{d\Omega}\right) = 2Re\{\mathcal{M}_o^*\delta\mathcal{M}\}\frac{\alpha^2}{4s}\sqrt{1 - \frac{4M_B^2}{s}} \ . \tag{15}$$

A detailed description of the evaluation procedure can be found in [7].

The cancellations already present at the Born level of W-pair production are even stronger at the level of RC's (partly due to the evaluation procedure outlined

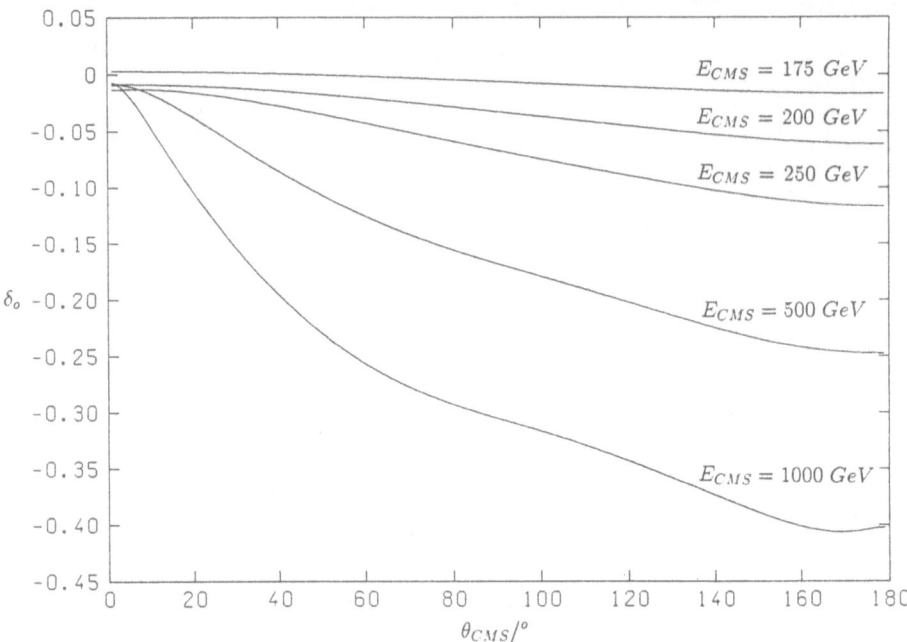

Figure 5. Angular dependence of the RC's relative to the Born cross section for $e^+e^- \to W^+W^-$.

above). Therefore the numerical stability of the computer routines had to be tested very carefully. The reliability of the results is founded on independent calculations agreeing with each other.

3.2 Results for $e^+e^- \to W^+W^-$

The $EWRC$'s to the reaction $e^+e^- \to W^+W^-$ were calculated by [5,6,7,8]. The numerical results of [7] and [8] agree within 0.3% of the cross section. The agreement between [5] and [7] is better than 0.5% where the cross section is sizeable. The calculation of [6] disagrees with the others.

Fig. 5 shows the angular dependence of the RC δ_o defined by

$$\left(\frac{d\sigma}{d\Omega}\right) = \left(\frac{d\sigma}{d\Omega}\right)_o (1 + \delta_o) \tag{16}$$

for different center of mass energies. The corresponding (unpolarized) Born cross section is plotted in fig. 6. For $LEP200$ energies the correction is typically 10 %. It rises up to -40 % with increasing energy in the backward direction. However, in the region where the relative corrections become big the cross section is small (see fig. 6) and the relative corrections to the integrated cross section stay below 15 % up to 1 TeV (fig. 7).

Until now the masses of the top quark and the Higgs boson are not known. The dependence of the W-pair cross section on these parameters is illustrated in figs. 8 and 9 where δ_1 is normalized to the corrected cross section for $m_t = 50$ GeV and $M_H = 100$ GeV

$$\left(\frac{d\sigma}{d\Omega}\right)(m_t, M_H) = \left(\frac{d\sigma}{d\Omega}\right)(m_t = 50, M_H = 100)(1 + \delta_1) . \tag{17}$$

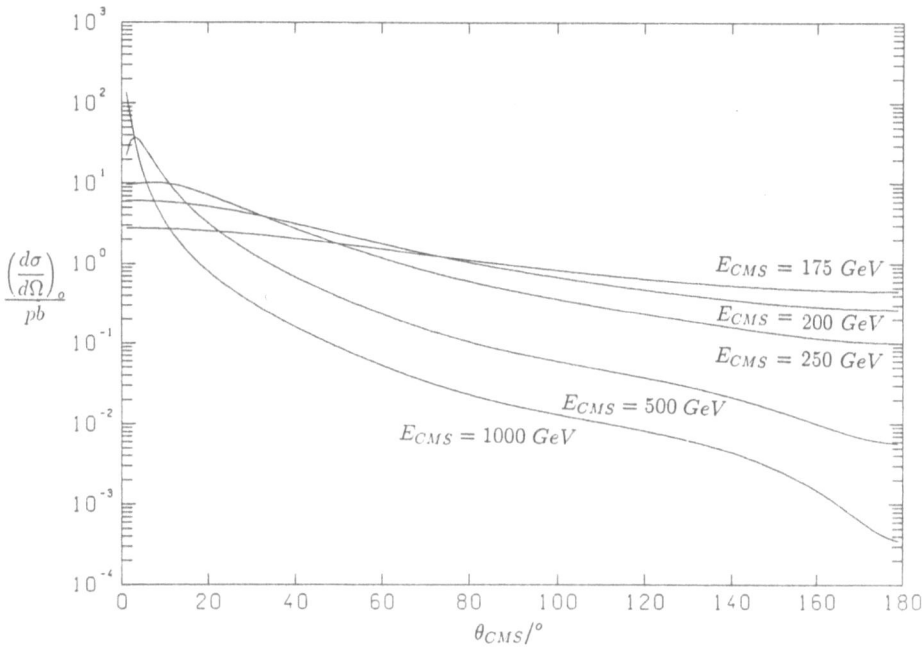

Figure 6. Angular dependence of the differential Born cross section for $e^+e^- \to W^+W^-$.

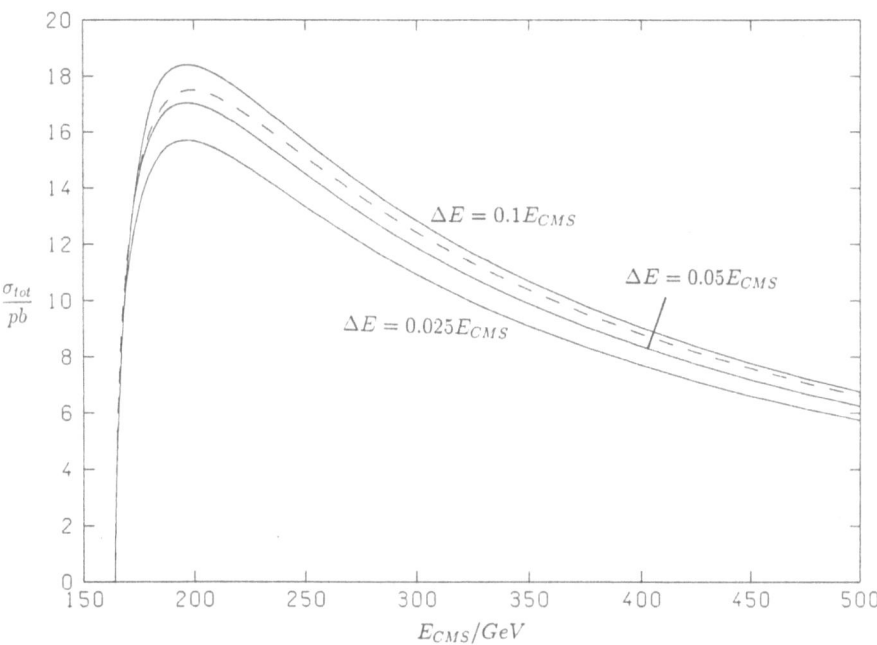

Figure 7. Integrated Born cross section for $e^+e^- \to W^+W^-$ to lowest order (dashed) and including one loop corrections for different soft photon energy cutoffs.

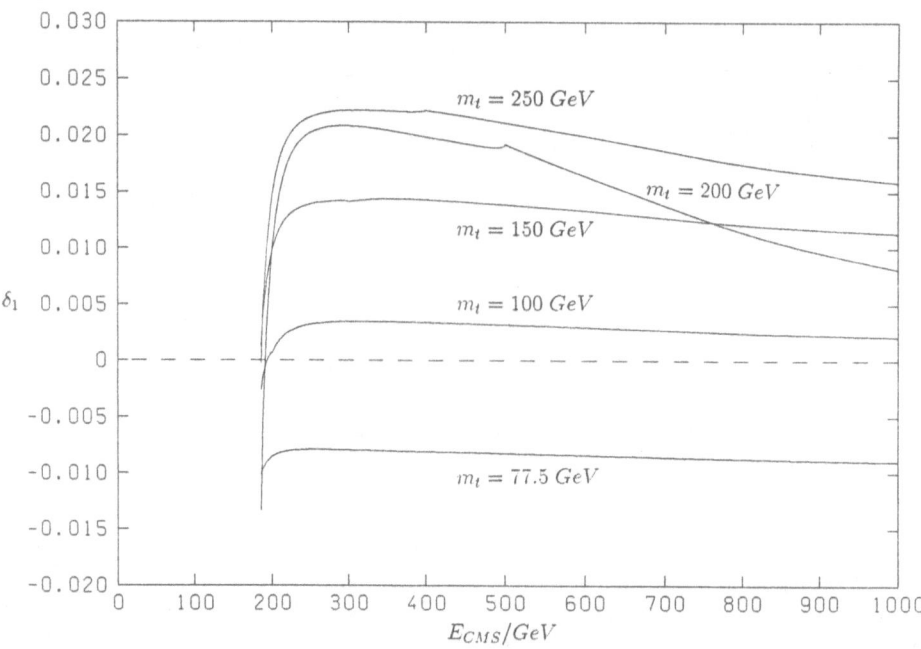

Figure 8. Dependence of the W-pair production cross section on the top quark mass.

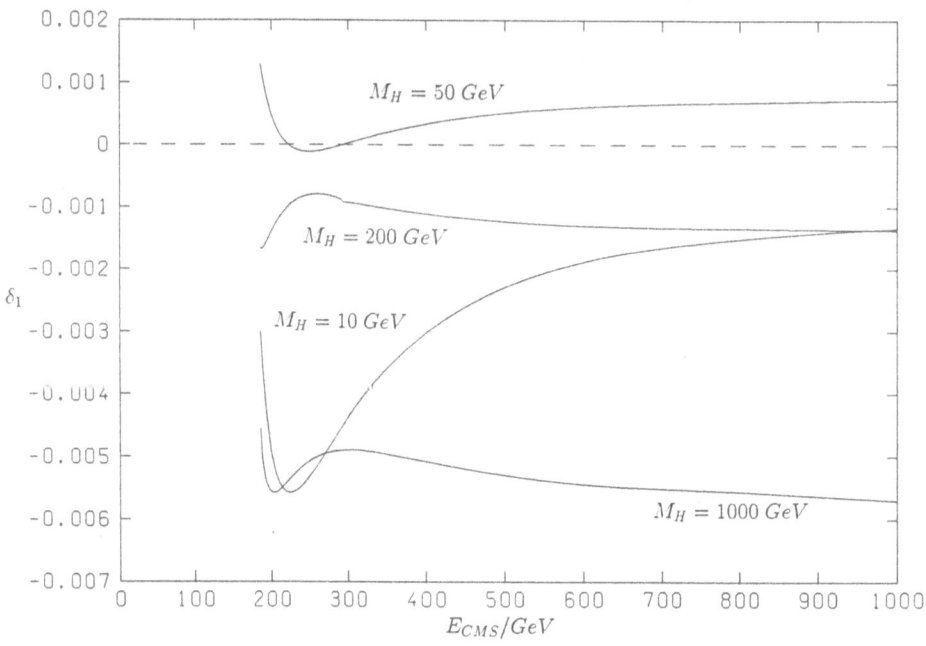

Figure 9. Dependence of the W-pair production cross section on the Higgs boson mass.

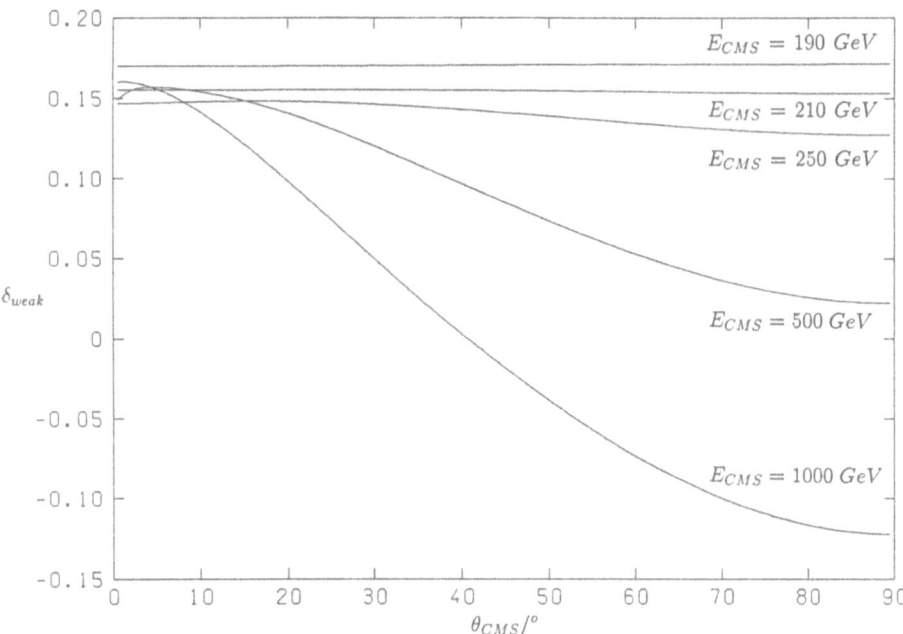

Figure 10. Angular dependence of the weak RC's relative to the Born cross section for $e^+e^- \to ZZ$.

Changing m_t from 50 to 250 GeV affects the cross section by 3.5 %, a variation of M_H between 10 and 1000 GeV by less than 1 %.

3.3 Results for $e^+e^- \to ZZ$

The $EWRC$'s to the reaction $e^+e^- \to ZZ$ were calculated by [9]. They can be separated in a gauge invariant way into electromagnetic δ_{em} and weak corrections δ_{weak}

$$\left(\frac{d\sigma}{d\Omega}\right) = \left(\frac{d\sigma}{d\Omega}\right)_o (1 + \delta_{em} + \delta_{weak}) . \tag{18}$$

The electromagnetic correction is about -17 % for a center of mass energy of 200 GeV and a soft photon energy cutoff $\Delta E = 0.05\sqrt{s}$. The weak corrections are shown in fig. 10 for different center of mass energies as a function of the scattering angle. At $LEP200$ energies they are dominated by fermion loop contributions which could be absorbed into a running α.

4 Finite Width Effects

W- and Z-bosons are unstable and decay before they reach the detector. The observed final states consist of the decay products of these bosons. Experimentally the unstable bosons are reconstructed by requiring the invariant mass M of the decay products to be close to the boson mass M_B

$$M_B - \Delta < M < M_B + \Delta \tag{19}$$

with Δ of the order of a few GeV.

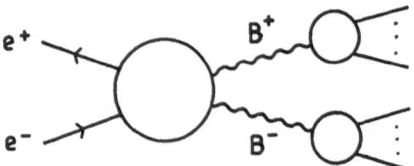

Figure 11. Structure of diagrams contributing to $e^+e^- \to B^+B^- \to final\ states$.

The diagrams describing the reaction $e^+e^- \to B^+B^- \to final\ states$ have the structure shown in fig. 11. They are enhanced by the resonance poles of the boson propagators. Other diagrams contributing to the same final state are relatively suppressed by a factor $\frac{\Delta\Gamma}{M_B^2} \approx 0.003$. The resonant diagrams lead to the following cross section [10,11]

$$\sigma(s) = \int ds_1 ds_2 \sigma(s, s_1, s_2)$$

$$\frac{1}{\pi} \frac{\sqrt{s_1}\,\Gamma(s_1)}{[s_1 - M_B]^2 + s_1[\Gamma(s_1)]^2} \frac{1}{\pi} \frac{\sqrt{s_2}\,\Gamma(s_2)}{[s_2 - M_B]^2 + s_2[\Gamma(s_2)]^2}\,, \tag{20}$$

where $\Gamma(s_i)$ is the 'off shell decay width' and $\sigma(s, s_1, s_2)$ the 'cross section' for the production of two off shell bosons with masses $\sqrt{s_1} = \sqrt{k_+^2}$ and $\sqrt{s_2} = \sqrt{k_-^2}$. s is the center of mass energy squared. In the limit $\Gamma(s_i) \to 0$

$$\frac{1}{\pi} \frac{\sqrt{s_i}\,\Gamma(s_i)}{[s_i - M_B]^2 + s_i[\Gamma(s_i)]^2} \to \delta(s_i - M_B^2)\,, \tag{21}$$

and $\sigma(s)$ approaches the cross section for the production of two stable on shell bosons

$$\sigma(s) \to \sigma(s, M_B^2, M_B^2)\,. \tag{22}$$

Since the width $\Gamma(s)$ is proportional to α in lowest order the effects of the finite width are expected to be comparable to the radiative corrections.

$\sigma(s, s_1, s_2)$ and $\Gamma(s_i)$ containing radiative corrections are in general complicated functions impeding the numerical evaluation of (20). Due to the Breit Wigner like factors in the integrand the main contribution to the integral comes from the region $s_1 \approx s_2 \approx M_B^2$. Consequently, it should be a good approximation to expand all functions with a smooth dependence on s_1 and s_2 around that point, keeping only the leading terms

$$\Gamma(s) \approx \Gamma(M_B^2)\frac{\sqrt{s}}{M_B}\,, \tag{23}$$

$$\sigma(s, s_1, s_2) = \sigma_o(s, s_1, s_2)(1 + \delta(s, s_1, s_2))$$

$$\approx \sigma_o(s, s_1, s_2)(1 + \delta(s, M_B^2, M_B^2))\,. \tag{24}$$

Inserting (23) and (24) in (20) yields

$$\sigma(s) = \left[\int\!\!\int ds_1 ds_2\, \sigma_o(s, s_1, s_2)\right.$$

$$\left.\frac{1}{\pi} \frac{s_1\Gamma(M_B^2)}{[s_1 - M_B]^2 + s_1^2[\Gamma(M_B^2)]^2} \frac{1}{\pi} \frac{s_2\Gamma(M_B^2)}{[s_2 - M_B]^2 + s_2^2[\Gamma(M_B^2)]^2}\right]$$

$$\frac{\sigma(s, M_B^2, M_B^2)}{\sigma_o(s, M_B^2, M_B^2)}\,. \tag{25}$$

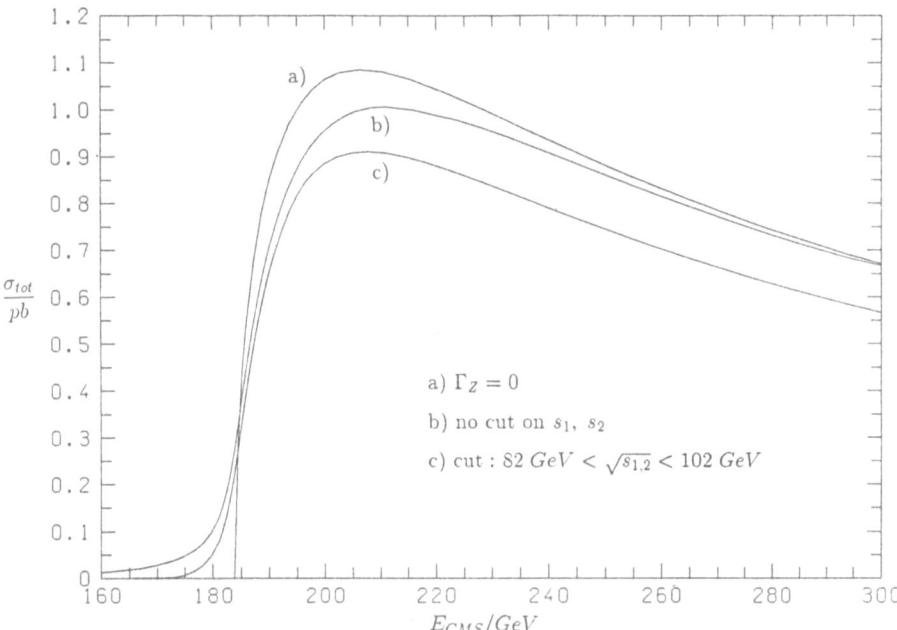

Figure 12. Total cross section for $e^+e^- \rightarrow ZZ$ including all one loop corrections for stable Z-bosons (a), including the finite width (b) and including the finite width and a cut on the invariant masses of the decay products (c).

It has been shown [10] that the error due to these approximations is below 0.1 % of the peak cross section for $e^+e^- \rightarrow ZZ$. The numerical evaluation, however, is simplified considerably. The effects of the Z-boson width on the total cross section including one loop corrections are illustrated in fig. 12 for a Z-boson mass $M_Z = 92\,GeV$. The plot contains the result for stable bosons and including the finite width with and without a cut on the invariant masses of the decay products. It turns out that the finite width effects are very important at threshold but become negligible at high energies.

In the case of W-pair production the finite width effects have been discussed for the Born cross section [11,12] and including large logarithmic and heavy quark corrections arising from fermion loops [13]. A simultaneous treatment of finite width effects and one loop radiative corrections is up to now not available. It is made difficult by the photon radiation from the off shell W-boson. It is also not clear whether the approximations (23,24) are applicable in this case. However, the inclusion of both the RC's as well as the finite width effects is inevitable for a proper measurement of the W-mass from the W-pair production threshold [14].

5   Conclusion

The reactions $e^+e^- \rightarrow W^+W^-/ZZ$ provide crucial tests of the SM. Especially W-pair production will allow a direct investigation of the triple gauge coupling and a direct determination of the W-boson mass. Because its cross section in the SM results from strong cancellations it is a good place to look for deviations from the SM.

The cross sections are large enough to allow an experimental accuracy of typically 1%. Therefore an adequate comparision of the experimental results with the *SM* predictions requires the inclusion of *RC*'s. Meanwhile reliable formulas and independent computer programs exist for the *EWRC*'s including soft photon Bremsstrahlung.

The dependence on the top quark mass is of the order of the expected experimental accuracy whereas the dependence on the Higgs mass is negligible.

Realistic calculations must also include hard photon bremsstrahlung and the effects arising from the finite width of the produced bosons. The inclusion of finite width effects is up to now only discussed for *Z*-pair production. It still has to be elaborated for *W*-pair production at the one loop level. Furthermore Monte Carlo programs for the hard photon bremsstrahlung are missing.

Acknowledgement

I would like to thank the organizers for the invitation to this workshop.

References

[1] A. Sirlin, *Phys. Rev.* **D22** (1980), 971; W. J. Marciano, A. Sirlin, *Phys. Rev.* **D22** (1980), 2695; A. Sirlin, W. J. Marciano, *Nucl. Phys.* **B189** (1981), 442.

[2] M. Böhm, W. Hollik and H. Spiesberger, *Fortschr. Phys.***34** (1986) 687.

[3] G. Passarino and M. Veltman, *Nucl. Phys.***B160**(1979) 151.

[4] G. 't Hooft and M. Veltman, *Nucl. Phys.* **B153** (1979) 365.

[5] M. Lemoine and M. Veltman, *Nucl. Phys.* **B164** (1980) 445.

[6] R. Philippe, *Phys. Rev.***D26** (1982) 1588.

[7] M. Böhm, A. Denner, T. Sack, W. Beenakker, F. Berends and W. Kuijff, *Nucl. Phys.***B304** (1988) 463.

[8] J. Fleischer, F. Jegerlehner and M. Zralek, *Z. Phys.***C42** (1989) 409.

[9] A. Denner and T. Sack, *Nucl. Phys.***B306** (1988) 221.

[10] A. Denner and T. Sack, Würzburg preprint (1988).

[11] T. Muta, R. Najima and S. Wakaizumi, *Mod. Phys. Lett.* **A1** (1986) 203.

[12] G. Barbiellini et al. in: Physics at LEP, ed J. Ellis and R. Peccei (CERN 86-02).

[13] B. Grzadkowski and Z. Hioki, *Phys. Lett.* **197B** (1987) 213.

[14] Z. Hioki, *Nucl. Phys.* **B316** (1989) 1 and references therein.

RADIATIVE CORRECTIONS TO WW SCATTERING IN THE STANDARD MODEL

R. Bouamrane

Randall Laboratory of Physics
University of Michigan
Ann Arbor, MI 48109, USA

1   Introduction

The standard model of electroweak interactions has been consistent with all known experimental data to this date. However, there remain some important components of this model as yet unverified experimentally, namely the vector boson self–interactions and the Higgs sector. Both are needed to insure renormalizability without spoiling gauge invariance.

If the Higgs exists and is heavy but still light enough for perturbation theory to remain valid ($m_{Higgs} < 1\ TeV$), this calculation might provide a basic guideline for measuring deviations from the minimal Standard Model in future high energy WW scattering experiments. On the other hand, if the Higgs mass is much higher than $1\ TeV$, this calculation allows us to *refine* the unitarity bounds by taking into account the on–loop radiative corrections to the J=0 partial–wave amplitude. (Of course, we do not know what higher order corrections will do.)

2   The model

The model we use is the minimal $SU(2) \times U(1)$ standard model. The Higgs sector is restricted to a single complex doublet. The free parameters of the model are:

g	weak coupling constant
s_θ, c_θ	the sine and cosine of the weak mixing angle
M	mass of the charged vector boson
m	mass of the Higgs boson
$m_t^\alpha,\ m_b^\alpha,\ m_e^\alpha$	masses of quarks and leptons (α is the generation index).

We assume massless neutrinos.

All our calculations are done in the limit $m^2 \gg s, t, u \gg M^2, m_t^2$ where s, t, and u are the usual Mandelstam variables. Only longitudinally polarized vector bosons are

Radiative Corrections, Edited by N. Dombey and
F. Boudjema, Plenum Press, New York, 1990

considered. Furthermore, we consider the following three independent amplitudes:

$$
\begin{aligned}
\mathcal{A}_a &= \mathcal{A}(W_L^+ W_L^- \to W_L^+ W_L^-) \\
\mathcal{A}_b &= \mathcal{A}(W_L^0 W_L^0 \to W_L^+ W_L^-) \\
\mathcal{A}_c &= \mathcal{A}(W_L^0 W_L^0 \to W_L^0 W_L^0)
\end{aligned}
\tag{1}
$$

3 Tree level calculations

The tree level amplitudes for $W_L W_L$ scattering have been extensively studied [1,2]. At energies very high with respect to the W mass, we have the following:

$$
\begin{aligned}
\mathcal{A}_a &= \frac{g^2 m^2}{4 M^2} \left(\frac{s}{-s + m^2} + \frac{t}{-t + m^2} \right) \\
\mathcal{A}_b &= \frac{g^2 m^2}{4 M^2} \left(\frac{s}{-s + m^2} \right) \\
\mathcal{A}_c &= \frac{g^2 m^2}{4 M^2} \left(\frac{s}{-s + m^2} + \frac{t}{-t + m^2} + \frac{u}{-u + m^2} \right)
\end{aligned}
$$

Taking $m^2 \gg s, t, u$ we have for the J=0 partial–wave amplitude:

$$
\begin{aligned}
a_a^0 &= \frac{g^2 s}{128 \pi M^2} \\
a_b^0 &= \frac{g^2 s}{64 \pi M^2} \\
a_c^0 &= 0
\end{aligned}
\tag{2}
$$

The most important feature of the J=0 partial–wave tree level amplitude for WW scattering is the cancellation at high energy ($s \gg m^2, M^2$) of the s^2 and s terms from the W exchange diagrams and similar terms from the four–W vertex and the Higgs exchange diagrams. This cancellation makes sure that tree level unitarity is respected at very high energies. This is true provided that the Higgs mass m is not too large (less than 1 TeV [1]). In the case $m^2 \gg s \gg M^2$, the J=0 partial–amplitude stays linear as a function of s. (It is constant for $W^0 W^0 \to W^0 W^0$). Tree level unitarity is violated at a critical energy $\sqrt{s^*} \simeq 1.7\ TeV$ [2]. This bound is obtained by considering the requirements of partial–wave unitarity on the two–channel system consisting of $W_L^+ W_L^-$, and $\frac{1}{\sqrt{2}} W_L^0 W_L^0$, with amplitudes given by (2). This is achieved by calculating the largest eigenvalue of the following 2×2 matrix:

$$
\begin{pmatrix}
a_a^0 & \frac{1}{\sqrt{2}} a_b^0 \\[2mm]
\frac{1}{\sqrt{2}} a_b^0 & \frac{1}{2} a_c^0
\end{pmatrix}
\tag{3}
$$

4 Renormalization

All renormalization schemes involve some sort of redefinition of the parameters and the fields in the Lagrangian. For example [3]:

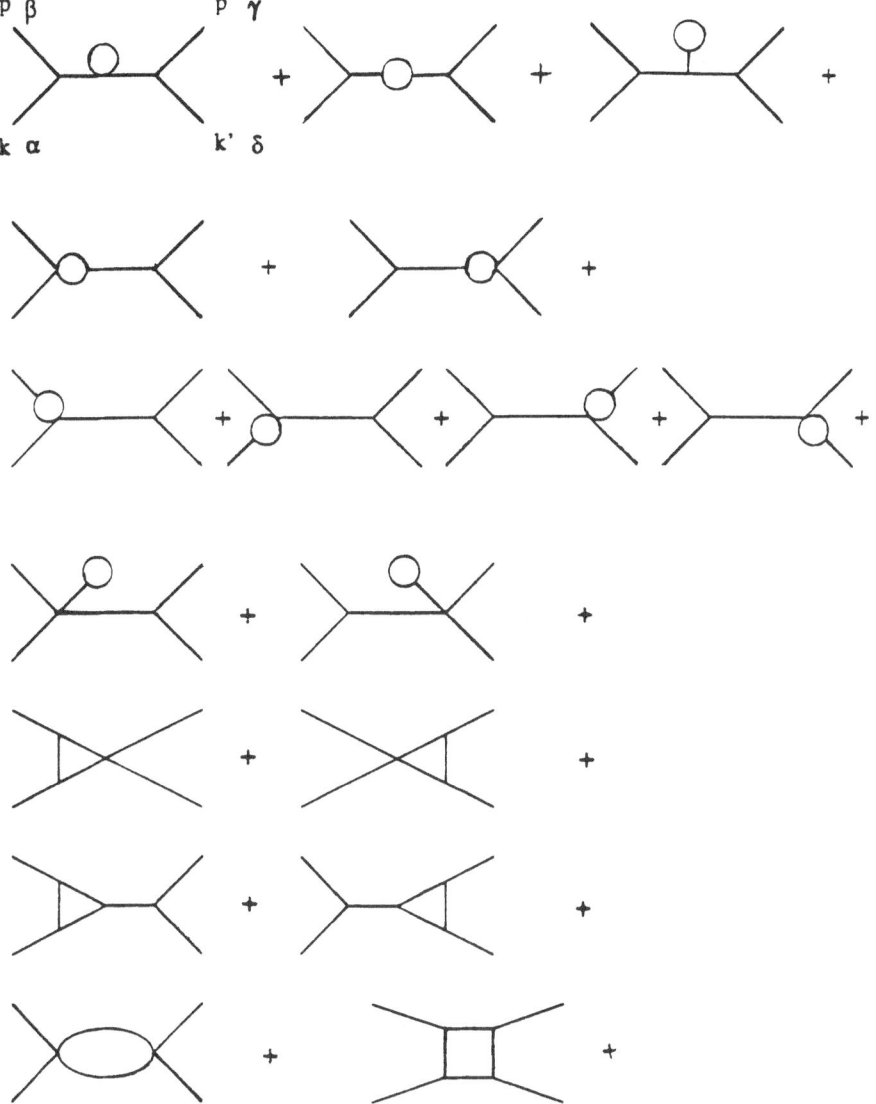

Figure 1. Topologies contributing to the one–loop corrections for WW scattering

$$
\begin{aligned}
g &\rightarrow g(1 + \delta_g) \\
M &\rightarrow M(1 + \delta_M) \\
c_\theta &\rightarrow c_\theta(1 + \delta_{c_\theta}) \\
m &\rightarrow m(1 + \delta_{\tilde{m}}) \\
W_\mu^\pm &\rightarrow W_\mu^\pm(1 + \delta_c) \\
W_\mu^0 &\rightarrow W_\mu^0(1 + \delta_0) + \delta_{0A} A_\mu \\
A_\mu &\rightarrow A_\mu(1 + \delta_A) + \delta_{A0} W_\mu^0 \\
\phi^\pm &\rightarrow \phi^\pm(1 + \delta_H) \\
\phi^0 &\rightarrow \phi^0(1 + \delta_H) \\
H &\rightarrow H(1 + \delta_H) + \frac{M}{g}\delta_t
\end{aligned}
\tag{4}
$$

Depending on the scheme prefered, the quantities δ_g, δ_M, etc. are chosen to compensate part of the one–loop corrections of the processes used as data input to fix g, M, etc. So, in principle, the one–loop corrections to these processes must be computed before any prediction on the four–W amplitude is made. This of course would be the case when doing an exact calculation. In the limit in which we are interested, namely $m^2 \gg s, t, u \gg M^2$, the quantities δ_g, δ_M, etc. can be chosen as:

$$
\delta = a_3 m^2 \log(m^2) + a_2 m^2 + a_1 \log(m^2) + a_0 \, ,
\tag{5}
$$

where the a_i's are constants to be determined.

The radiative corrections to the four–W amplitude are obtained by adding the contributions from all the diagrams shown in figs. 1 and 2. Diagrams with crosses involve the quantities δ_g, etc.. Their associated Feynman rules were derived from the extra terms generated in the Lagrangian after the redefinitions (4) were made. (For some examples, see fig. 3.)

For the three processes considered in (1), the counter-terms' contributions as a function of the δ's are:

$$
\mathcal{A}_a^{counter} = \frac{g^2}{32\pi^4 i M^2 m^2}(\delta_M - \delta_m + \delta_g + 2\delta_c)(t^2 + s^2)
\tag{6}
$$

$$
\mathcal{A}_b^{counter} = \frac{g^2}{64\pi^4 i M^2 m^2}(\delta_M - \delta_m + \delta_g + \delta_c + \delta_0 - \delta_{c_\theta})s^2
\tag{7}
$$

$$
\mathcal{A}_c^{counter} = \frac{g^2}{32\pi^4 i M^2 m^2}(\delta_M - \delta_m + \delta_g + 2\delta_0 - 2\delta_{c_\theta})(s^2 + t^2 + u^2)
\tag{8}
$$

We notice that only the m^2 terms (if any) contained in the δ's affect the one–loop order four–W amplitude. These terms were previously calculated by van der Bij and Veltman [3] and were shown to be unobservable at the one–loop level for processes not involving external Higgs lines. Thus, in the limit $m^2 \gg s, t, u \gg M^2$, the one–loop order amplitude does not depend at all on the physical processes chosen to fix the free parameters g, M, etc. The quantities δ_g, δ_c, δ_0, δ_{c_θ} are free from m^2 dependence. The only relevant quantity is:

$$
\delta_M - \delta_m = -2\pi^2 i \frac{m^2}{M^2}\left[\frac{3}{4}\frac{1}{n-4} + \frac{3}{8}\log(m^2) - \frac{25}{32} + \frac{9\pi}{32\sqrt{3}}\right]
\tag{9}
$$

The heart of the calculations is, of course, the evaluation of approximately 1000 loop diagrams using the algebraic manipulation program Schoonship [6]. We have

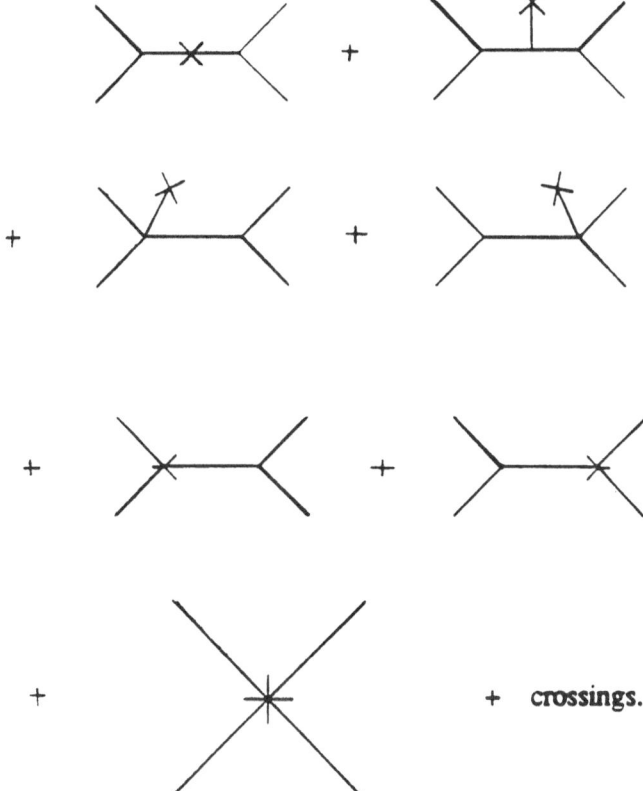

Figure 2. Counter–terms contribution to WW scattering

$$H \quad \text{-------} \times \qquad -m^2 M \delta_t$$

$$W_\mu^+ \quad \text{———} \times \text{———} \quad W_\nu^- \qquad - M^2(2\delta_c + 2\delta_M + \delta_t) + 2\delta_c \, (p.q \, \delta_{\mu\nu} - p_\mu q_\nu)$$

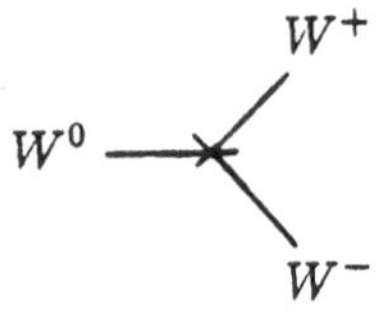

$$= \left[\delta_g + 2\delta_c + \tfrac{s_\theta}{c_\theta}\delta_{A0} + \delta_{c_\theta} \right] \times$$

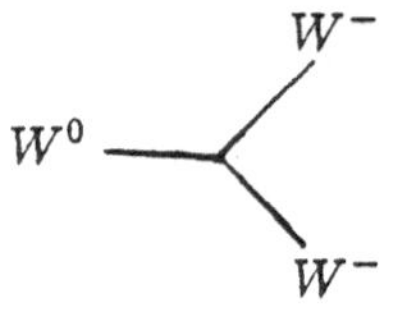

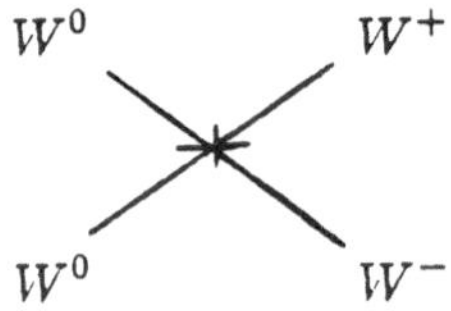

$$= \left[2\delta_g + 2\delta_c + 2\delta_0 + 2\tfrac{s_\theta}{c_\theta}\delta_{A0} + 2\delta_{c_\theta} \right] \times$$

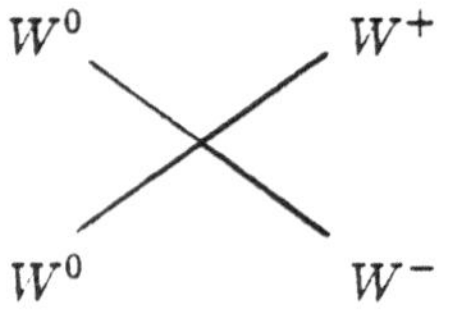

Figure 3. Examples of Feynman rules

kept the leading terms which are quadratic in s, t, u with or without logarithms such as $\log(s/m^2)$, etc..

These diagrams can essentially be grouped into two sets. The first set consists of diagrams with no Higgs propagators inside the loop. We have found that the contributions from these diagrams essentially reduce to two–point functions and completely cancel in all the three processes at which we have looked. The second set consists of diagrams with Higgs propagators occurring in the loop. After appropriately expanding all the integrands , again, only two–point functions are left. (For calculational details, see (refs. [4].)

5 Results

The radiative corrections for the processes considered are:

$$
\begin{aligned}
\mathcal{A}_a^1 &= -\frac{g^4}{16\pi^2 M^4}\Bigg\{\frac{1}{18}ts + \Big(\frac{5}{9} - \frac{9\pi}{32\sqrt{3}}\Big)(t^2 + s^2) \\
&\quad + \Big(\frac{1}{96}ts + \frac{5}{96}s^2\Big)\log(s/m^2) \\
&\quad + \Big(\frac{1}{96}ts + \frac{5}{96}t^2\Big)\log(t/m^2) \\
&\quad + \frac{1}{32}u^2\log(u/m^2)\Bigg\}
\end{aligned}
$$

$$
\begin{aligned}
\mathcal{A}_b^1 &= -\frac{g^4}{16\pi^2 M^4}\Bigg\{\Big(\frac{37}{72} - \frac{9\pi}{32\sqrt{3}}\Big)s^2 + \frac{1}{72}t^2 + \frac{1}{72}u^2 \\
&\quad + \frac{c_\theta^4}{4}(2s^2 - t^2 - u^2)\phi(s_\theta^2) \\
&\quad + \Big(\frac{1}{64}t^2 + \frac{1}{192}u^2 - \frac{1}{192}s^2\Big)\log(t/m^2) \\
&\quad + \Big(\frac{1}{64}u^2 + \frac{1}{192}t^2 - \frac{1}{192}s^2\Big)\log(u/m^2)\Bigg\}
\end{aligned}
$$

$$
\begin{aligned}
\mathcal{A}_c^1 &= -\frac{g^4}{16\pi^2 M^4}\Bigg\{\Big(\frac{13}{24} - \frac{9\pi}{32\sqrt{3}}\Big)(s^2 + t^2 + u^2) \\
&\quad + \frac{1}{16}s^2\log(s/m^2) + \frac{1}{16}t^2\log(t^2/m^2) + \frac{1}{16}u^2\log(u^2/m^2)\Bigg\} ,
\end{aligned}
$$

with

$$
\begin{aligned}
\phi(x) &= -\frac{x\sqrt{3 - 4x}}{2(1 - x)}\arctan(\sqrt{3 - 4x}) \\
&\quad + \frac{3x - 2}{4(1 - x)}\log(1 - x)
\end{aligned}
$$

The corresponding J=0 partial–wave amplitudes are:

$$
\begin{aligned}
a_a &= \frac{g^2 s}{128\pi^2 M^2}\Bigg[1 + \frac{g^2 s}{64\pi^2 M^2}\Big(-\frac{20}{9}\log(s/m^2) - \frac{2441}{108} + \frac{12\pi}{\sqrt{3}}\Big)\Bigg] \\
a_b &= \frac{g^2 s}{64\pi^2 M^2}\Bigg[1 + \frac{g^2 s}{64\pi^2 M^2}\Big(-\frac{5}{9}\log(s/m^2) - \frac{905}{108} + \frac{9\pi}{2\sqrt{3}} + \frac{16}{3}c_\theta^4\phi(s_\theta^2)\Big)\Bigg] \\
a_c &= \frac{g^2 s}{128\pi^2 M^2}\Bigg[0. + \frac{g^2 s}{64\pi^2 M^2}\Big(-\frac{10}{3}\log(s/m^2) - \frac{256}{9} + \frac{15\pi}{\sqrt{3}}\Big)\Bigg]
\end{aligned}
\tag{10}
$$

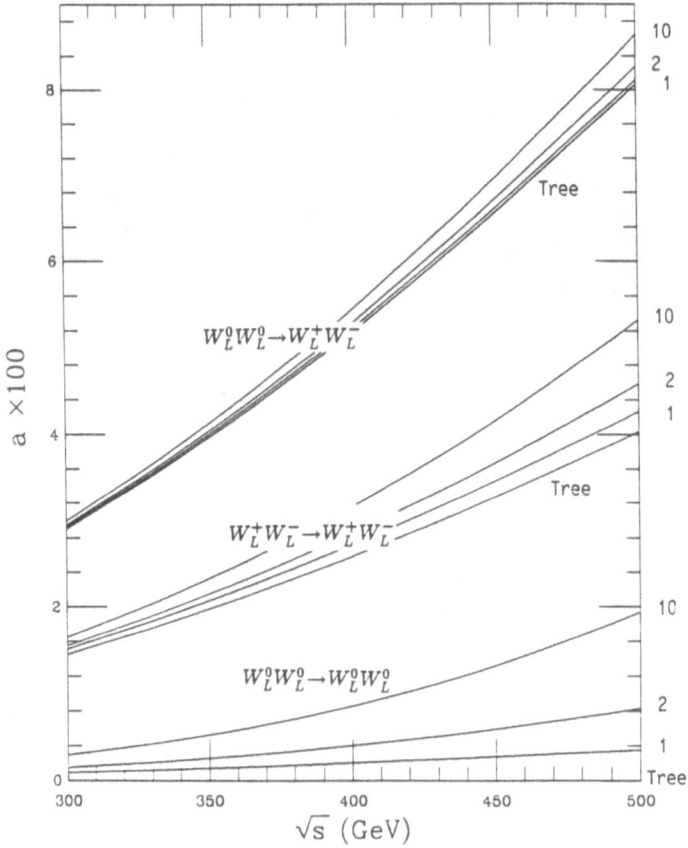

Higgs mass = 1,2 and 10 TeV

Figure 4. J=0 partial-wave amplitude in the energy window 300 to 500 GeV.

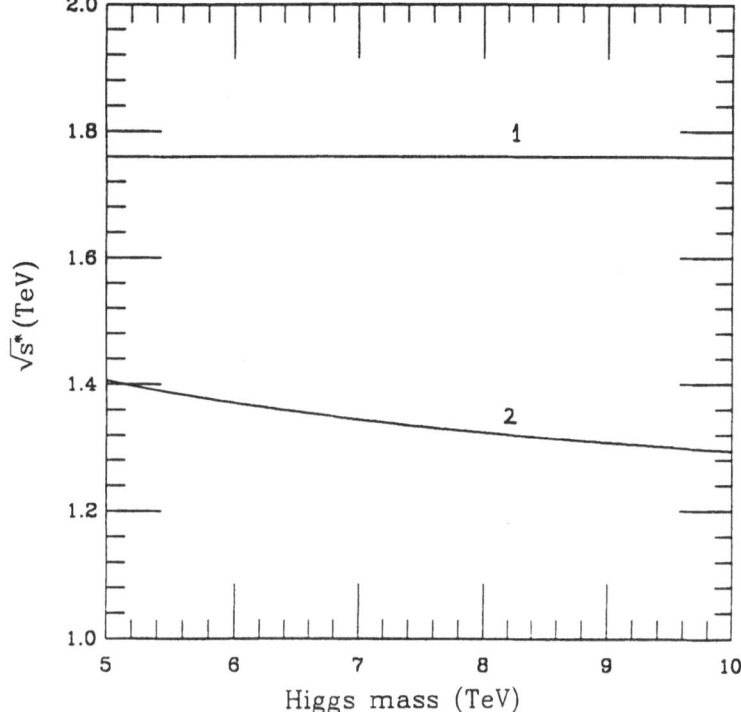

Figure 5. Critical energy as a function of the Higgs mass. Tree level (1). Tree level + one–loop corrections (2).

These results are plotted in fig. 3. In fig. 4, the critical energy $\sqrt{s^*}$ obtained by coupling the channels involving $W_L^+ W_L^-$ and $(\frac{1}{\sqrt{2}}) W_L^0 W_L^0$, with amplitudes given by (10), is plotted as a function of the Higgs mass. These results agree with [5].

6   Conclusion

The one–loop corrections are positive throughout the energy range of interest. Numerically speaking, for a Higgs mass of $1\ TeV$, the radiative corrections to the tree amplitude ratio remain within 5% for energies below $500 GeV$. For a heavier Higgs mass, the results we have presented should be taken only as an indication because perturbation theory is no longer valid. Nevertheless, for a very heavy Higgs (5–10 TeV), the critical energy at which the J=0 partial–wave reaches unity is reduced by (20–30%) when including the one–loop corrections.

Acknowledgements

The author thanks R. Akhoury and M. Veltman for discussions and for reading the manuscript; R. Chalabi for his help with the computer at the University of Sussex. This research was supported in part by the U.S. Department of Energy.

References

[1] B. Lee, C. Quigg, and H. Thacker, Phys. Rev. D16, 1519 (1977)

[2] M. Chanowitz and M. Gaillard, Nucl. Phys. B261, 379 (1985)

[3] J. Van der Bij and M. Veltman, Nucl. Phys. B231, 205 (1984)

[4] M. Veltman and F.J. Yndurain, *preprint* UM-TH-89-04
    R. Bouamrane, PhD thesis (1989), *unpublished*

[5] S. Dawson and S. Willenbrock, Phys. Rev. Let. 62, 1232 (1989)

[6] SCHOONSHIP, 68000 version Jan 1 1989.

NEW PHYSICS AS A BACKGROUND TO PRECISION MEASUREMENTS OF STANDARD MODEL RADIATIVE CORRECTIONS

C. Verzegnassi *)

Theoretical Physics Division
CERN, 1211 Geneva 23

1 Introduction

With the official opening of the operational stage at LEP 1 on 15 July 1989, a period of intensive searches will begin, whose outcome might well be some exciting breakthrough in the key mechanisms that govern the elementary interactions. Thanks to a remarkable joint experimental-theoretical effort, that will be the content of a next-to-appear CERN Yellow report [1], it will be possible to perform a systematic test of a large number of predictions of the Glashow-Salam-Weinberg Standard Model (SM) to the one-loop level of accuracy. If any substantial deviation between experimental and theory were discovered, a first strong support for some kind of New Physics beyond the SM would be thus set by an *indirect* effect.

Manifestations of New Physics at LEP 1 might, though, be of more spectacular nature if new particles not predicted by the SM were *directly* produced. Alternatively, observation of processes of production of conventional particles that would be *forbidden* in the SM would provide strong support for the existence of new interactions. A series of possibilities for a "minimal" number of theoretical proposals that include supersymmetric models, models with an extra gauge boson and composite models, has been thoroughly examined in Ref. [1]. The conclusion that emerges is that the chances of revealing new particles and/or interactions at LEP 1 are substantial in the range of values of the parameters (masses, mixing angles, scales...) that is schematically represented in Table 1. There one sees the new limits that would be set by LEP 1 compared to the before-LEP ones (several of which are in fact not existing at all), and one can draw conclusions that depend on how optimistic, or pessimistic, one feels.

A major problem that has to be faced whenever New Physics is looked for is the presence of a competitor SM background, that might mask, or simulate, the genuine effect (this applies both to *direct* production and to *indirect* searches). A successful search can be performed to the extent that

*)On leave of absence from: Department of Theoretical Physics, Trieste, and INFN, Sezione di Trieste, Italy.

TABLE 1

(Qualitative) limits on the parameters of some models of New Physics (Minimal Supersymmetric Standard Model, Composite models, models with an extra U(1)) derivable from negative searches at LEP 1, first phase. The results are taken from the work of R. Barbieri et al., F. Boudjema, F.M. Renard et al., Ref. [1].

	BEFORE LEP 1	AFTER LEP 1
$M_{\tilde{l}^\pm,\chi^\pm}$	25 GeV	40 GeV
$M_{\tilde{\nu}}$		40 GeV
M_χ		40 GeV
$M_{\nu'}$		$M_Z, \frac{M_Z}{2}$
$M_{f^{*+}}$	30 GeV	$\frac{M_Z}{2}$
$M_{S^\pm}$	30 GeV	$\frac{M_Z}{2}$
$M_{Y,Y_L,Z'}$	250-300 GeV	500-700 GeV
$|\theta_M|$	0.250	0.025

one is able to get rid of the unwanted SM background. In this way, realistic limits like those of Table 1 can be finally set for the parameters of the various models of New Physics.

An interesting example of the previous statement is represented by the searches of a new Z', e.g., of extended gauge origin via its "virtual" (in fact, of purely mixing kind) effects on Z. In this case, the SM background is represented by some still poorly known contributions to radiative corrections. A particularly dangerous one is the correction to the gauge boson propagators [2].

$$\Delta\rho(0) = \frac{\Pi_{ZZ}(0)}{M_Z^2} - \frac{\Pi_{WW}(0)}{M_W^2} \tag{1}$$

that, for large top mass, behaves as:

$$\Delta\rho(0) \simeq \frac{\alpha}{\pi}\frac{m_t^2}{M_Z^2} \Rightarrow 0.01 \quad (m_t = 2M_Z) \tag{2}$$

If the observation of some final hadronic states is involved, another potentially dangerous SM background comes from the still large error affecting the QCD coupling $\frac{\alpha_s}{\pi}$:

$$\delta\frac{\alpha_s}{\pi}(M_Z) \simeq \pm 0.01 \tag{3}$$

Since the size of the virtual Z' effects is typically at the level of a few per cent, the presence of unknown SM contributions potentially of $0(0.01)$ is clearly annoying. Luckily enough, properties of radiative corrections on Z resonance [3] make it possible to develop a strategy that eliminates systematically the "$\Delta\rho$ background" by introducing suitable combinations of "twiddled" observables (this technique can also, actually, include another kind of observables *not* on Z resonance) [4]. On the other hand, the uncertainty coming from α_s on various observables can also be either completely cancelled [5], or suitably decreased or controlled [6]. In conclusion, by properly treating the $\Delta\rho(0)$ *and* $\frac{\alpha_s}{\pi}$ background, one is led to the LEP 1 limits for a Z' of the most general E_6 [7] nature that are represented in the Figure, where also for comparison the existing limits derived by Amaldi et al. [8] (whose notations and conventions I follow) are shown. From the Figure, one can conclude that such a particle should not escape detection unless the values of the mixing angle θ_M meet the request:

$$| \theta_M | \leq 0.02 - 0.03 \tag{4}$$

(the precise value depends on the parameter $cos\beta$ that fixes the model). On the other hand, if a new Z' were actually seen via mixing effects, the intrinsic error on θ_M would be, intuitively, of the same size. Thus, whether such a new Z' can actually be "seen" at LEP 1 or not, there will always be an intrinsic uncertainty on θ_M given by Eq. (4), which means, in the most optimistic case:

$$\delta\theta_M = \pm 0.02 \tag{5}$$

The problem that I would like to consider here is whether this ultimate uncertainty can represent a trouble for some possible determinations of $\frac{\alpha_s}{\pi}$ and $\Delta\rho(0)$ from precision measurements (which remains after all a rather important goal). If this were the case (and I will show that it could actually

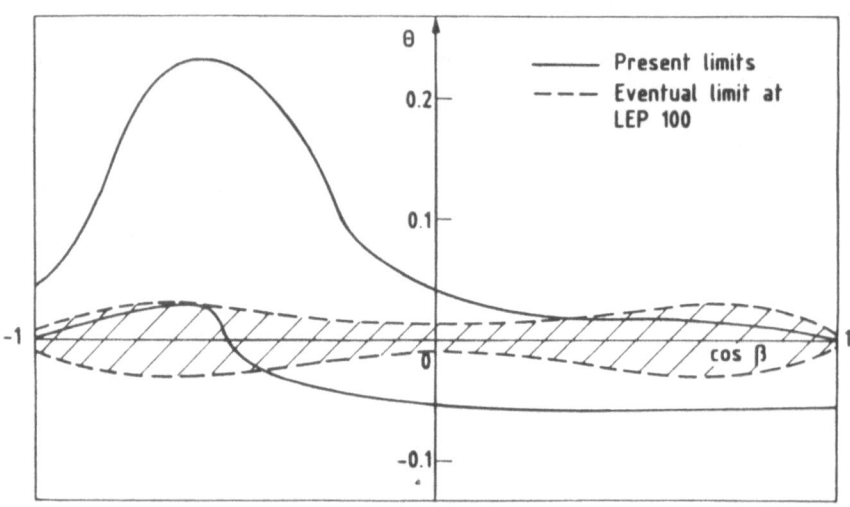

Figure 1

Limits on the mixing angle at variable $cos\beta$ (in the notation of Ref. (8)) from negative searches at LEP 1, taken from: F.M. Renard and C. Verzegnassi, Phys.Lett. B221 (1989) 197.

be), New Physics (NP) would then represent a background (!) to the determination of SM radiative corrections (completely overthrowing the previous picture). The question would then arise of whether some "NP background elimination" strategy can be worked out even in this case, and I will show a specific (and positive) example. In complete analogy to the more usual situation, one will thus be led to an "unbiased" determination of the SM parameter $\frac{\alpha_s}{\pi}$ that is free of extra gauge bosons possible backgrounds. This will only be possible if to the set of observables on Z resonance one will be able to add a very precise measurement of the M_W/M_Z ratio, as discussed in the next Section.

2 An unbiased sum rule for $\frac{\alpha_s}{\pi}$

In the SM, a possible way of deriving $\frac{\alpha_s}{\pi}$ (another one is discussed in great detail by Z. Kunszt in these Proceedings) would be to measure the hadronic/muonic ratio:

$$R'_5 \simeq \frac{\Gamma_{Z \to udscb}}{\Gamma_{Z \to \mu}} \tag{6}$$

which can be written as:

$$R'_5 \simeq R_5'^{(0)} \left[1 + \frac{1}{2} \Delta\rho(0) + \frac{\alpha_s}{\pi} \right] \tag{7}$$

(where by α_s I actually mean the sum of all known [9] contributions of $0(\alpha_s), 0(\alpha_s^2), 0(\alpha_s^3)$). The term $R_5'^{(0)}$ can be considered as known to better than one per cent in the SM [10]; to fix the scale, we can suppose that it has been evaluated for $m_t = 60$ GeV. The possible uncertainty coming from the top mass can thus be taken into account by writing, e.g.:

$$\Delta\rho(0) = (0.5 \pm 0.5)10^{-2} \tag{8}$$

that would correspond to assuming

$$m_t \simeq 130 \pm 70 GeV \tag{9}$$

The corresponding uncertainty on $\frac{\alpha_s}{\pi}$ would be:

$$\delta\frac{\alpha_s}{\pi}^{(\Delta\rho)} \simeq \pm 0.25 \times 10^{-2} \simeq 6.10^{-2} \cdot \frac{\alpha_s}{\pi} \tag{10}$$

The extra uncertainty on $\frac{\alpha_s}{\pi}$ would be coming from the experimental accuracy. Assuming a measurement of R'_s to a relative one per cent would give:

$$\delta\frac{\alpha_s}{\pi}^{(exp)} = 0.01 \tag{11}$$

so that, to achieve a precision of about a relative ten per cent, a measurement of R'_s as good as 0.5 per cent (relative) would be requested. Note that, if this were the case, a powerful test of the overall SM scheme would be provided by the comparison with the alternative derivation illustrated by Kunszt [9].

The situation would change, though, if the possible presence of New Physics, in particular of a new gauge boson, were taken into account either for theoretical reasons or because already strongly supported by some positive signals. This presence would mainly modify Eq. (7) as follows:

1. The one-loop radiative correction $\Delta\rho(0)$ should be replaced by a more complicated quantity that was called $\nabla\rho$ in Ref. [4]. This might contain positive virtual one-loop contributions from new heavy particles or sparticles, possible contributions coming at tree level from unconventional Higgses (which might be negative) and a positive shift due to the presence of the new Z'. In other words,

$$\nabla\rho = \Delta\rho(0) + \delta\rho_L^{(Higgses)} + \delta\rho_L^{(Z')} + \dots \tag{12}$$

where

$$\delta\rho_L^{(Z')} \equiv sin^2\theta_M \left[\frac{1}{\epsilon} - 1\right] > 0 \tag{13}$$

ϵ is the ratio of physical masses

$$\epsilon = \frac{M_Z^2}{m_{Z'}^2} \tag{14}$$

and θ_M is the mixing angle.

2. New contributions due to the modification of the Z couplings to fermions caused by the $Z - Z'$ mixing arise. In the most general case of a Z' generated by a previous gauge symmetry ($E_6, SU(2)_L \times SU(2)_R \times U(1)_{L-B}$) these expressions have been computed. Without entering the details of the computation, that can be found elsewhere [11], I will rather give the final expression of Eq. (7):

$$R_5' = R_5'^{(0)} \left[1 + \frac{1}{2}\nabla\rho + a\theta_M + \frac{\alpha_s}{\pi}\right] \tag{15}$$

where the model-dependent quantity a is, in fact, nearly constant and of order one. It is in fact, in the E_6 case,

$$a^{(E_6)} \simeq \frac{5}{6}[cos\beta + sin\beta] \tag{16}$$

and in the left-right symmetric case [12] (I use the same notations and convention of Durkin and Langacker [13])

$$a^{LR} \simeq \left[\frac{2}{15}\alpha_{LR} + \frac{2}{3}\alpha_{LR}\right] \tag{17}$$

with $\alpha_{LR} \simeq 1$.

As a consequence of the ultimate uncertainty on θ_M, Eq. (5), we see therefore that the determination of $\frac{\alpha_s}{\pi}$ via Eq. (15) would now be affected **at least** by the "NP uncertainty"

$$\delta\frac{\alpha_s}{\pi}^{NP} \simeq \pm0.02 \tag{18}$$

that would be catastrophically large. Note that I have assumed that the extra uncertainty from $\nabla\rho$ can be still neglected, which is not necessarily obvious now.

Since the merits of the variable R'_5, both from a theoretical and from an experimental point of view, are not irrelevant, one would be tempted to try to get rid of this "NP background". One possibility, that has already been suggested in the literature [11], is to consider the combination:

$$\hat{R} = R' + \frac{50}{59}\gamma \tag{19}$$

where

$$\gamma = \frac{9}{\alpha(M_Z)}\frac{\Gamma_{Z\to\mu}}{M_Z} \tag{20}$$

For the general class of Z' considered here, this combination is actually free of the pure $Z - Z'$ mixing effect (but not of the $\delta\rho_L^{(Z')}$ term!), this being a consequence of the assumed $SU(5)$ transformation properties of the various currents [11]. For this quantity, it would thus be possible to write:

$$\hat{R} = \hat{R}^{(0)}\left[1 + \frac{4}{3}\nabla\rho + \frac{\alpha_s}{\pi}\right] \tag{21}$$

where $\hat{R}^{(0)}$ can be considered as known to our requested level of accuracy. Note, though, that the error from $\nabla\rho$ would be now larger (at least $\sim 0.8 \times 10^{-2}$).

An interesting possibility would be provided by a measurement of the M_W/M_Z ratio to high accuracy. For the general class of considered Z', this ratio has in fact the remarkable property of providing the most direct measurement of the obscure block $\nabla\rho_L$ since, to a very good approximation, one has [4]:

$$\xi \equiv \frac{M_W^2}{M_Z^2\hat{C}_W^2} \simeq 1 + \frac{3}{2}\nabla\rho;\ \hat{C}_W^2 = \frac{1}{2}\left[1 + \sqrt{1 - \frac{4\mu^2}{M_Z^2}}\right];\quad \mu^2 = (38.7 GeV)^2 \tag{22}$$

Thus, we can write

$$\nabla\rho \simeq \frac{2}{3}[\xi - 1] \tag{23}$$

and replace it in Eq. (21), thus obtaining the following "sum rule" for $\frac{\alpha_s}{\pi}$:

$$\frac{\alpha_s}{\pi} = \frac{\hat{R}}{\hat{R}^{(0)}} - 1 - \frac{8}{9}[\xi - 1] \tag{24}$$

Equation (24) only contains either experimentally measurable quantities or known constants. It gives an expression of $\frac{\alpha_s}{\pi}$ that is true *both in the SM and in any generalization of the SM that produces one extra Z' of the considered nature*. In this sense, it could be used to obtain information on α_s completely free of a potentially dangerous New Physics background.

The price to pay for the elimination of the potentially dangerous background is represented by the addition of extra experimental error. Assuming that ξ can be measured to $\sim \pm 3 \times 10^{-3}$ at ACOL [14], we see that the only effective additional error, compared to the more conventional equation (7) (only valid in the SM) is that coming from $\delta(\frac{5}{6}\gamma)$. A reasonable estimate ($\sim$ one per cent) leads to conclude that this method might lead relatively simply to an evaluation of α_s with an error:

$$\delta\left(\frac{\alpha_s}{\pi}\right) \simeq 1 \times 10^{(-2)} \tag{25}$$

and, although this result is by a factor two worse than that expected in the alternative approach [9], its highly unbiased nature would suggest, to say the least, to compare these two different determinations of α_s.

A final comment is appropriate concerning the other interesting radiative correction $\Delta\rho(0)$, Eqs. (1), (2). In the SM, any information on $\Delta\rho(0)$ provides a corresponding hint on m_t (assuming $\rho_{tree} = 1$). In the more general case, $\Delta\rho(0)$ is replaced by $\nabla\rho$, Eq. (12) and the picture is more complicated. If one assumes $\rho_{tree} = 1$ (no Higgses that do not belong to $SU(2)_L$ singlets or doublets), a measurement of ξ provides only the sum of $\Delta\rho(0)$ and $\delta\rho_L^{Z'}$, Eq. (13). However, this time, the bounds on $\mid \theta_M \mid$, Eq. (4) would not limit $\delta\rho_L^{Z'}$. This can be seen better by rewriting $\delta\rho_L^{Z'}$ in an equivalent way:

$$\delta\rho_L^{Z'} = \mid Ptg\theta_M \mid \simeq \mid P\theta_M \mid \tag{26}$$

where $\mid P \mid$ is a model-dependent parameter which can be of order one in the considered models [15]. This introduces an intrinsic error from $\delta\rho_L^{Z'}$ of order $\sim 1.10^{-2}$ that is already beyond the relevant range of $\Delta\rho(0)$ and that cannot be eliminated by simple subtractions. For the determination of $\Delta\rho(0)$, thus, the NP background appears more subtle than that for the determination of α_s. Note, however, that a measurement of the full block $\nabla\rho$ could already be extremely interesting if extra independent *unbiased* limits on the top mass were available. From a combination of the two informations, one might then be led to some real prediction, or surprise!

3 Conclusion

I have shown that, for the specific problem of the determination of two of the most interesting parameters that appear in radiative corrections in the Standard Model, i.e., $\frac{\alpha_s}{\pi}$ and $\Delta\rho(0)$, a potentially dangerous background from New Physics exists. This can be completely eliminated in the case of $\frac{\alpha_s}{\pi}$, leading to the unbiased sum rule, Eq. (24), but the price is that two extra measurements (besides that of the ratio R_5'), i.e., that of the muonic Z width and that of the ratio M_W/M_Z, have to be introduced. For the quantity $\Delta\rho(0)$, the Z' background affects its derivation from M_W/M_Z to a large extent. At least another information, e.g., on m_t would be thus requested. The combination with the value M_W/M_Z might then lead to remarkable conclusions.

One general feature, that was already stressed in Ref. [4] and that it might be worth to stress again, is the need, and the remarkable usefulness, of performing several *combined* measurements. This appears to be the only way not to draw biased, unjustified conclusions. Luckily, the number of experimental results that should be available in the very near future seems to be sufficiently large to motivate, least to say, a moderate optimism.

References

[1] "Z Physics at LEP", to appear as CERN Yellow Report, G. Altarelli, R. Kleiss and C. Verzegnassi editors (1989).

[2] I use several definitions (but not the metric) of B.W. Lynn, M.E. Peskin and R.G. Stuart, CERN Yellow Report "Physics at LEP" CERN 86-02 (1986), J. Ellis and R. Peccei editors, p. 90.

[3] M. Cvetic and B.W. Lynn, Phys.Rev. **D35** (1987) 51.

[4] F. Boudjema, F.M. Renard and C. Verzegnassi, Nucl.Phys. **B314** (1989) 301.

[5] B.W. Lynn and C. Verzegnassi, Phys.Rev. **D35** (1987) 3326.

[6] A. Djouadi, Z.Phys. **C39** (1988) 561.

[7] D. London and J.L. Rosner, Phys.Rev. **D34** (1986) 1530.

[8] U. Amaldi et al., Phys.Rev. **D36** (1987) 1385.

[9] A complete review of the situation is given by Z. Kunszt in these Proceedings.

[10] See, e.g., F. Berends, These Proceedings.

[11] F.M. Renard and C. Verzegnassi, Phys.Lett. **B225** (1989) 431.

[12] See R.N. Mohapatra and G. Senjanovic, Phys.Rev. **D23** (1981) 165 for a list of references following the original J. Pati and A. Salam proposal.

[13] L.S. Durkin and P. Langacker, Phys.Lett. **166B** (1986) 436.

[14] I use the number quoted by G. Altarelli in the Yellow Report, [2].

[15] For a detailed discussion, see R. Casalbuoni, D. Dominici, F. Feruglio and R. Gatto in [1].

RADIATIVE CORRECTIONS AND COMPOSITE WEAK VECTOR BOSONS AT LEP100/SLC ENERGIES

Fawzi Boudjema

Physics Department, Sussex University
BN1 9QH Brighton, England

ABSTRACT

We discuss to which extent the idea that the usual weak vector bosons are composite can survive the stringent tests of precision measurements at LEP100 and SLC. It will be argued that, once the the universality of the weak coupling has been implemented within the composite picture for the observables at the Z peak, one recovers the bulk of the radiative corrections which one usually associates with the the standard model. The discrimination between the $SU(2) \times U(1)$ theory and composite W's can be achieved if one can probe the Higgs sector of the Standard Model or alternatively the additional vector bosons predicted within the composite scheme. This is within the spirit of the B-R-V (Boudjema-Renard-Verzegnassi) strategy that unambiguously identifies models with extra vector bosons once the uncertainties due to the yet unknown parameters of the $SU(2) \times U(1)$ theory (heavy top mass, ρ parameter, etc...) have been eliminated.

1   Introduction

The idea of compositeness may seem very incongruous to an audience of specialists in radiative corrections within a renormalisable gauge theory such as $SU(2) \times U(1)$ which, moreover, has been very successful in its low-energy predictions. However, it is fair to say that the Standard Model (SM)[1] will be promoted to the status of a fully-fledged theory only when its inner structure has been completely checked. This consists in probing the nature of the weak bosons self-interactions (whether the truly gauge non-abelian feature is present) and discovering the Higgs boson (thus hopefully revealing the nature of spontaneous symmetry breaking). In addition it is imperative to precisely quantify the effects of radiative corrections to have a test at the quantum level. At the same time, there is a widespread consensus among physicists to go beyond our present understanding or rather formulation of the Standard Model. This unease together with

Radiative Corrections, Edited by N. Dombey and
F. Boudjema, Plenum Press, New York, 1990

the present loose ends in the S.M justifies any alternative approach which up to now has managed to incorporate all the existing experimental data and whose goal is the resolution of some deficiencies inherent to $SU(2) \times (1)$.

Any of the contenders to the SM has in its manifesto a solution to the fundamental Higgs particle problem (naturalness and fine tuning) either by having a better formulation of the scalar sector (for instance supersymmetry very nicely evades the fine tuning problem) or by "making it do" without the Higgs. The latter proposition is quite specific to composite models whose other main motivation is an understanding of the proliferation of fermions and their neat embedding in repetitive generations. The last situation is reminiscent of the classification in the Mendeleev periodic table and of the regularity of the hadron spectrum both of which are a consequence of a construction with an economical set of fundamental objects. Another observation [2] is that weak interactions as described by the SM are the ONLY known FUNDAMENTAL short range forces. This has to be contrasted with the molecular and van der Waals forces and hadron interactions which are in fact residual effects of a fundamental long-range force ($U(1)_{em}$ and $SU(3)_{QCD}$). These compelling hints have led many to suspect that weak interactions and the mechanism of symmetry breaking are just a good parametrization of a more fundamental theory where the actual particles emerge as bound states.

2 The underlying theory

If it is indeed an underlying theory which has given birth to the so far successful description of weak interactions as a residual effective interaction, then one should suspect this fundamental theory to have some highly sophisticated features. One would think that this theory should be:

- renormalisable or even (ideally) finite.

- Higgs-free unless it is supersymmetric.

- confining at low energies, ie, energies below the Fermi scale.

In order to recover the symmetries of the effective description, the fundamental theory must have some GLOBAL (flavour) symmetry. Then in principle, starting from the underlying theory with the fundamental particles and interactions described by a $\mathcal{L}$agrangian $\mathcal{L}_F$ one should be able to derive a NEW effective $\mathcal{L}$agrangian, $\mathcal{L}_{eff}$, which contains a part that reproduces the S.M, albeit without the Higgs field, and additional pieces represented by high-dimension interactions. These are a manifestation of the New Physics and are suppresed by powers of the compositeness scale Λ

$$\mathcal{L}_F \Longrightarrow \mathcal{L}_{eff} \;=\; \underbrace{"\mathcal{L}_{SM}\,(no\ Higgs)"}_{\mathcal{L}_0} + \frac{\mathcal{L}_1}{\Lambda} + \frac{\mathcal{L}_2}{\Lambda^2} + \ldots \qquad (1)$$

One should not expect the derivation of the effective interaction from first principle to be at all obvious. For instance, knowing all the ingredients of QCD, one is still not able to derive the effective low-energy hadronic interaction. Nevertheless, the consequences of this approach is that one should have a drastic reduction in the number of free parameters since the parameters which appear in $\mathcal{L}_0$, $\mathcal{L}_1$, $\mathcal{L}_2$ are contained in $\mathcal{L}_F$ and therefore are related to each other. Consequently one should expect some interplay between different kinds of NEW operators especially those of dimension higher than 4 (non-renormalisable type) so that when considering a specific process some interferences might occur with the consequence that with a-priori anomalous couplings one still ends up with a cross section which has a good high energy behaviour. This is probably just a reflection of the unitarity present at the more fundamental level. Such an approach has been developed by our Director (Norman Dombey) and myself [3] , whereby starting with a primary interaction one is able to derive new Z^0 boson electromagnetic and self interactions whose parameters are directly related to the INDUCED weak mixing angle. This dynamical approach does give high dimension operators which do not break unitarity or do so at only very high energies where the full underlying theory should be operative anyway. This is in sharp contrast to the usual pedestrian way of writing effective operators.

3 Composite Weak Bosons Mimicking Gauge Bosons

The first major problem one has to resolve if one takes the view that the usual weak bosons are composite is : how come the weak coupling constant g_W is universal? One has got used to the idea that the universality of a coupling constant is a direct consequence of a LOCAL gauge symmetry with the ensuing elementarity of the gauge vector bosons. The second important question is that of the custodial GLOBAL SU(2) weak symmetry [4] responsible for the fact that the ρ parameter is equal to one (or at least very consistent with that value). In the SM this extra symmetry is implemented by having an ad-hoc (though, it has to be said, the most simple) Higgs representation and does not seem to be linked to the local SU(2) responsible for the universality of the weak coupling. Now, if instead of talking about W's I had mentioned the ρ mesons, replacing g_W by g_ρ, you would not have associated the notion of a local gauge symmetry with what we, now, know are composite mesons. And yet g_ρ is to a very good approximation a universal coupling. What fakes the gauge symmetry is Vector Meson Dominance (VMD for short) and the custodial SU(2) is manifest by having a triplet of spin one vector bosons. Take for instance the $\pi^+\pi^-\gamma$ or $\rho^+\rho^-\gamma$ vertex:

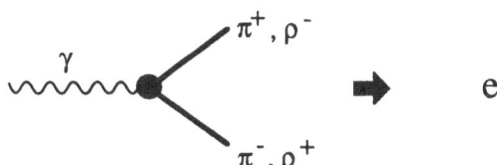

At zero momemtum transfer this gives the electric charge, i.e., the universal $U(1)_{QED}$ coupling, e. But we know that pions and ρ's are not point-like so that if the previous diagram was viewed under some microscope one would see the photon in fact coupling to (at this level) point-like constituents of the previous system.

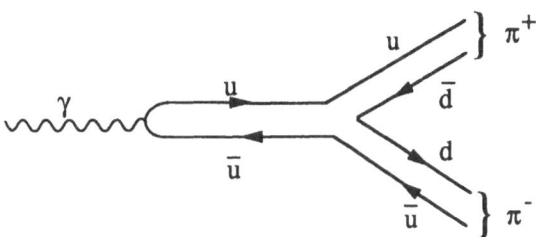

The diagram with the $u\bar{u}$ formation can also be viewed as

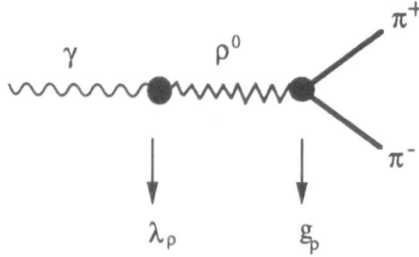

Having parametrized the $u\bar{u}$ formation by a component of the ρ^0, the interpretation is that the photonic interaction is dominated by the neutral vector meson. Then if the ρ^0 junction has strength λ_ρ and since this diagram is equivalent to the first diagram then one has:

$$e = \lambda_\rho \, g_\rho \tag{2}$$

This insures the universality of the coupling g_ρ . The global SU(2) is achieved by having the ρ 's in a triplet insuring that charged and neutral vector mesons couple with an identical strength. The same arguments can be translated to the W system. A consistent formal approach has been written down which has reproduced all the available low-energy data [5] . The important point to make is that VMD together with the global SU(2) and the local U(1) of electromagnetism has almost reproduced a local $SU(2) \times U(1)$ plus a global SU(2)!

In terms of the constituents of the W's one of the simplest constructions is in terms of a doublet of spin $1/2$ fermions in complete analogy with the ρ system [6] . Then this doublet would implement the needed SU(2) global. Also this construction unavoidably leads to the prediction of an isoscalar weak boson (analog of the ω^0) as well as excited W's (W^*'s) analogs of the ρ^* s. These constituents are assumed to be tightly bound by a new strong interaction (hypercolour) . They may or may not carry colour. In the previous VMD picture the fermions also are to be considered composite, sharing one of the W constituents. The simplest and most likely construction of the fermions is in terms of one spin-1/2 (constituent of the W's) and which carries the weak isospin quantum number and a spin-0 scalar which acts as a spectator as far as the W are concerned.

In the case where the W constituents are not coloured the spin-0 constituent of the quark carries the colour quantum number. In this case there are no gluonic corrections emerging from the W

and consequently there are no additional QCD corrections in β decay to be added to the ones we have in the conventional picture. Therefore" Veltman's diagram" [7]

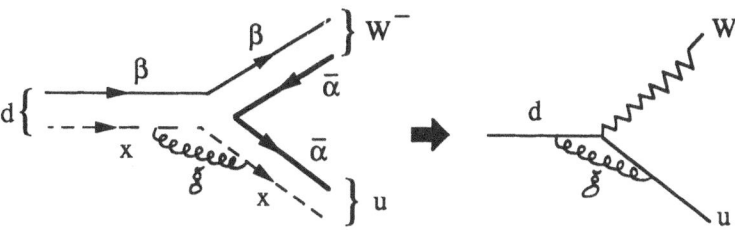

is not present.   The QCD corrections only involve the scalar spectator which solely belongs to the quark

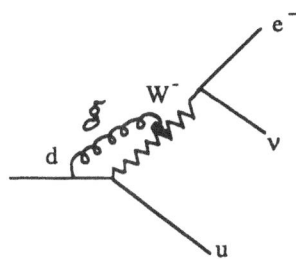

To abide by the principle of naturalness these constituent scalars should either be dynamical scalars (Goldstone bosons) or be supersymmetric partners of an elementary fermion. In the former case the scalar is not a fundamental scalar but is itself a composite state of yet more fundamental fermions. In this scenario one should postulate two scales: the lowest corresponding to the substructure fermion-scalar, the second associated to the mechanism responsible for the birth of the scalar [8] . This would be where the analogy with the hadron system stops. In QCD the symmetry is spontaneously broken and one ends up with massive composite fermions and massless pseudo-scalars. In fact recent investigations [9] based on general principles tend to suggest that in a vector-like theory which is not supersymmetric, like QCD, one can not expect bound-states fermions to form. This is the reason I have chosen to construct the fermions out of a spin-0 and a spin-1/2. One should add that some kind of symmetry (The most obvious bet is chiral symmetry) must be added to the above implementation to keep the fermions light. I also believe that it is the same mechanism (leading to massless fermions) which prevents (large) Flavour Changing Neutral Currents operators. Before moving to the issue of the inclusion of radiative corrections in the composite W hypothesis, let me remind you that in order to recover some basic well tested features of weak interactions two symmetry principles were used:

- The local $U(1)$ gauge symmetry of electromagnetism.

- A global SU(2) to which one should add the VMD parametrization.

4   Radiative Corrections at LEP/SLC energies with composite W's

Processes at LEP100 and SLC are dominated by four- fermion, Z- mediated (i.e., neutral current) interactions. The bulk of the radiative corrections are in fact pure Q.E.D corrections (initial and final states radiation). These specific corrections receive the same treatment in the previous realization of composite W's as in the gauge $SU(2) \times U(1)$ model, since the composite picture leaves the (vital) $U(1)_{QED}$ sector untouched. The other substantial electro-weak corrections are due to the running of the electromagnetic fine structure constant. Again because of the $U(1)_{em}$ and the VMD implementation they are readily included in the composite approach. Moreover it can be shown, that the quadratic large top mass dependence is also recovered [10] . In fact for four(light) fermions interactions one recovers almost all the standard corrections. I say almost because of course there are or there may be some small differences which probe the bosonic sector of the models. These include the effect of W loops in the vacuum polarization function.

They appear if the W electromagnetic and self-couplings do not have the standard gauge values. If this were the case the effect can be parametrized by the breaking of the above mentioned custodial global SU(2) weak symmetry leading to $\rho \neq 1$. Note that this need not be the case if the VMD approach is also implemented for the W couplings. And of course, there is the Higgs dependence. But thanks to the screening theorem [11] this effect is only logarithmic and moreover can be faked by the cut-off dependence of the effective composite description. In other words loop calculations made with the effective lagrangian $\mathcal{L}_0$ in equation(1) (no Higgs) are regularized by cutting the divergent integrals at the scale Λ which effectively replaces the Higgs mass dependence present in the standard description. Therefore as the previous arguments show, the above composite W picture not only has, so far , successfully reproduced the low energy data but, allowing for the experimental errors, may still be consistent with the SM description even at 100 GeV. The moral would be that the principles which allow the reproduction of features apparently specific to a gauge non-abelian theory not only pass the tests at the tree-level but might even pass the test at the quantum level and at much higher energies. This should make many of us frustrated in the event that no (clear-cut)discrepancy is found between the high-precision data at the Z peak and the predictions of the SM.

Nevertheless there are two important points I should make. First, if all the parameters of the SM were known then we would have an unambiguous check on the model. But this is certainly not the case. We do not know what the mass of the top is and if it is large it might induce large radiative corrections. This is also true if a new family existed with large mass splitting between the members of a doublet. This means that at present we can not give a definite prediction for an observable at the Z peak, say a partial width or the forward-bacward asymmetry. We can only give a number assuming a certain value for the ρ parameter, the top mass etc....
On the other hand there may well be some new structures beyond the standard model which only show up through indirect small effects. These effects can fake those of the yet undetermined virtual effects within the S.M. For some observables one can have the most unlucky situation where the contributions due to the new structure counter-

balance that due to the virtual radiative corrections present in both the new structure and the S.M. These kinds of observables would then hide to us , at the same time , not only the first manifestation of extensions to the S.M. but also some of the virtual radiative corrections effects. Therefore it is crucial to find a strategy to unambiguously uncover the genuine new physics effects and disentangle them from those which can be accomodated within the minimal $SU(2) \times U(1)$ model.

5   Selection Rules for New Physics

Evidence for New Physics from the ongoing LEP/SLC experiments will either be nakedly displayed in "broad day-light" by the direct production of a new species of particles or will present itself in a more subtle veiled form through an indirect anomalous contribution to standard processes, in much the same way that heavy particles belonging to representations of $SU(2) \times U(1)$ may reveal their presence through virtual loop effects. In the first manifestation not only will we have proof for the incompleteness of the SM but we could certainly attribute the discovery to a particular theory. In the latter case, the unveiling is more delicate. Since the effect is "virtual", one first have to wash away the possible virtual effects of the SM. Next, with the clean quantities one should be able to pinpoint to a particular model or theory, all this in the dark ...

5.1   The $\Delta\rho$ block

The hidden dark sector of the SM includes the top, the existence of extra doublets, the value of the ρ parameter and connected to this, the mass of the Higgs. For the virtual corrections as I stressed earlier, the latter corrections are small considering the expected accuracy on the measurement at LEP/SLC. But, luckily, the ambiguities caused by the remaining parameters can all be grouped into ONE SINGLE BLOCK which contributes to a generalised ρ parameter: $\Delta\rho$, which quantum mechanically breaks the custodial global SU(2). A substantial part of the radiative Higgs corrections is included in $\Delta\rho$. For instance with the usual three generations we have:

$$\Delta\rho = \frac{3G_\mu}{8\sqrt{2}\pi^2} \left[m_t^2 - m_W^2 \tan\theta_W \left(\ln\frac{M_H^2}{m_W^2} - \frac{5}{6} \right) \right] + \delta\rho_L \tag{3}$$

where G_μ is the muon decay constant and $\delta\rho_L$ is the deviation from one of the tree level ρ parameter. The top not only does not decouple from the theory but its effect increases quadratically. This effect reflects the isospin splitting within a doublet. Whereas the mild logarithmic Higgs contribution is a reflection of the mass splitting between the triplet of the spin-one vector bosons $W^\pm$ one the hand and Z^0 on the other, a consequence of the mixing between the primordial neutral vector bosons of the model. The contribution does indeed vanish when the mixing angle θ_W goes to zero. Each new doublet would have the effect of replacing

$$3\,m_t^2 \longrightarrow N_C \left(m_u^2 + m_d^2 - 2\frac{m_u^2 m_d^2}{m_u^2 - m_d^2} \ln(\frac{m_u^2}{m_d^2}) \right) \tag{4}$$

where N_C is the colour factor and u and d refer to the up and down members of a doublet.

I should however point at one notable exception concerning the observables at the Z peak. This is the $Zb\bar{b}$ vertex which receives a large top-mass corrections which can not be included in the $\Delta\rho$ bit. This effect signals the breakdown of universality at the quantum level. Hence one should be careful when using the $b\bar{b}$ width. But the effect is negligible when considering the forward-backward $b\bar{b}$ polarised asymmetry. Now, when this observation is made, it is clear how one should proceed to give unambiguous predictions. Within the SM we take a scheme with the following parameters:

- M_Z: the physical measured Z mass.

- $\alpha(m_W)$ the running fine structure constant at the scale m_W which has the effect of summing up the leading-log contributions.

- and G_μ the muon deacay constant.

Then each physical observable X_i may be written as

$$X_i = X_i^{(0)} + \alpha_i \Delta\rho \tag{5}$$

The direct corrections coming from vertices and boxes are negligible considering the experimental accuracy. Here $X_i^{(0)}$ represents the bulk of the radiative corrections and is only expressed in terms of the physical set $(M_Z, G_\mu, \alpha(m_W))$, hence it is compltely known once the Z mass has been measured. Although the $\Delta\rho$ contribution is not predicted the coefficient α_i is calculable and therefore known. Now consider the independent observable X_j which I write as:

$$X_j = X_j^{(0)} + \alpha_j \Delta\rho \tag{6}$$

Combining the two measurements I can form the "twiddled" quantity:

$$\bar{X}_j = X_j - \frac{\alpha_j}{\alpha_i} X_i = X_j^{(0)} - \frac{\alpha_j}{\alpha_i} X_i^{(0)} \tag{7}$$

This new quantity is a genuine unambiguous prediction of the SM. Therefore if after allowing for the experimental errors and the small theoretical QCD uncertainty in the value of α_S (when appropriate) a discrepancy is found then this would be a clear-cut manifestation of a new structure beyond the SM. The next question would be which new structure? Of course one should take all the high-precision observable data which are going to be available soon and form a set of "orthogonal" twiddled quantities. A deviation in just one of these is a signal of New Physics. The underlying structure has, apart from the "universal" common set of inputs $(M_Z, G_\mu, \alpha(m_W))$, some additional parameters say $\delta\sigma$ and $\delta\tau$ for the sake of clarity. For the strategy to be consistent it is imperative that one uses the same subset $(M_Z, G_\mu, \alpha(m_W))$ as in the SM. What is crucial is that the <u>physical</u> Z mass must be used. Then the observable X_i, X_j should now be written as

$$X_i = X_i^{(0)} + \alpha_i \Delta\rho' + \beta_i \Delta\sigma + \gamma_i \Delta\tau \tag{8}$$

$$X_j = X_j^{(0)} + \alpha_j \Delta\rho' + \beta_j \Delta\sigma + \gamma_j \Delta\tau \tag{9}$$

where the β's and γ's are known numbers. Note that the new theory might have a new, probably complicated formula for $\Delta\rho'$ but the twiddling will eliminate the block anyway. Now we see that we can "supertwiddle" the twiddled quantity to form a defining equation for the specific model by eliminating all the unknown parameters of the model. This equation constitutes a consistency check of the model and can be written as a sum rule

$$\mathcal{F}(\tilde{X}_j) = 0 \tag{10}$$

In other words in the multidimensional space spanned by the twidded set $\tilde{X}$, each model is represented by a specific surface. Each model will have a "personal" identifying track in the "theoretical detector" formed by the $\Delta\rho$ free quantities. It is only when the consistency check of the model is passed that one can start the job of extracting its parameters. This is the B-R-V [22] strategy developed by our Chairman this afternoon (Claudio Verzegnassi) Fernand Renard and myself generalising an original proposal by Cvetič and Brian Lynn [12] for the polarised asymmetries on the Z resonance.

5.2 Choosing Clean Observables

Clearly the strategy would work best, i.e be sensitive to New Physics, with high-precision observables. The other requirement is that these observables should be clean theoretically in the sense that the potentially large purely electromagnetic and strong interaction effects (initial, final state bremmstrahlung..) must be well under control in order not to blur the more interesting electroweak signals which are indicators of a new structure. In this respect the forward-backward asymmetries on Z resonance are not good candidates compared to the left-right and the polarised forward-bacward asymmetries. However in the first stages of LEP and SLC one has to content ourselves with unpolarised measurements. In addition to a very precise Z mass, the observables which can be measured with a high accuracy at Z peak and are suitable for vector boson searches are the following [13] :

1. The $Z \to e^+ e^-$ (or $\mu^+ \mu^-$) partial width (expected accuracy 2 %)

$$\gamma \equiv \frac{9}{\alpha} \cdot \frac{\Gamma(Z \to l^+ l^-)}{M_Z} \tag{11}$$

2. The ratio R'_Z of hadronic cross–section to $\mu^+ \mu^-$ cross–section (expected accuracy 0.5 %)

$$R'_Z \equiv \frac{3}{59} \cdot \frac{\Gamma(Z \to u, d, s, c, b)}{\Gamma(Z \to l^+ l^-)} \tag{12}$$

3. The τ polarization asymmetry can also be added in a subsequent analysis (it plays the same role as the left–right polarization asymmetry but with an accuracy limited to about ± 0.02)

$$A_\tau \equiv \frac{N(\tau_L) - N(\tau_R)}{N(\tau_L) + N(\tau_R)} \tag{13}$$

4. In the future one should also be able to add to these 3 observables the W/Z mass ratio with the accuracy expected from the $p\bar{p}$ colliders (0.5 %)

$$\xi \equiv \frac{2m_W^2}{M_Z^2 \left(1 + \sqrt{1 - 4\mu^2/M_Z^2}\right)} \tag{14}$$

where μ is given by

$$\mu \equiv \frac{\pi\alpha(m_W)}{\sqrt{2}G_\mu}$$

Following the B–R–V strategy, we can use the above observables to form three quantities which are free of large $(\Delta\rho)$ radiative corrections uncertainties. Two of them are accessible at LEP100

$$n \equiv \left[R_Z' - \frac{1}{2}\gamma\right] - \left[R_Z' - \frac{1}{2}\gamma\right]^{(0)} \tag{15}$$

$$q \equiv \left[R_Z' - \frac{1}{6}A_\tau\right] - \left[R_Z' - \frac{1}{6}A_\tau\right]^{(0)} \tag{16}$$

or equivalently

$$z \equiv \left[\gamma - \frac{1}{3}A_\tau\right] - \left[\gamma - \frac{1}{3}A_\tau\right]^{(0)} \tag{17}$$

the third one needs the W/Z mass ratio

$$p \equiv \left[R_Z' - \frac{1}{3}\xi\right] - \left[R_Z' - \frac{1}{3}\xi\right]^{(0)} \tag{18}$$

in which $[\ ]^{(0)}$ represents standard expectations (with $\Delta\rho = 0$).

In the event that polarised beams will become available the ideal observables are:

$$A_{LR} = \frac{\sigma\left(e^+e_L^- \to had.\right) - \sigma\left(e^+e_R^- \to had.\right)}{\sigma\left(e^+e_L^- \to had.\right) + \sigma\left(e^+e_R^- \to had.\right)} \tag{19}$$

$$A_{FB(pol)}^{(b,c)} = \frac{1}{P}\frac{\left(N_{F,P}^{b,c} - N_{F,-P}^{b,c}\right) - \left(N_{B,P}^{b,c} - N_{B,-P}^{b,c}\right)}{\left(N_{F,P}^{b,c} + N_{F,-P}^{b,c}\right) + \left(N_{B,P}^{b,c} + N_{B,-P}^{b,c}\right)} \tag{20}$$

$(b; c \equiv b\bar{b}; c\bar{c}, \ F; B \ 0 < \cos\theta_{CM} < 1; -1 < \cos\theta_{CM} < 0)$

P is the degree of polarization of the initial beam.

We would then form the "twiddled quantities"

$$x = A_{FB(pol)}^{(b)} - \frac{1}{15}A_{LR}^{(h)} - \frac{9}{13} \tag{21}$$

$$y = A_{FB(pol)}^{(c)} - \frac{9}{25}A_{LR}^{(h)} - \frac{9}{20} \tag{22}$$

5.3 Special twiddled quantities

Some of the above observables (x, y, q) are in fact just ratios of fermionic coupling constants. If one forms a set of twiddled quantities with these variables only, the twiddling will in effect not only eliminate $\Delta \rho$ but all of the oblique corrections (hence no $\Delta \alpha$) and the twiddled quntitities will be independent of the mass of the Z. We just get a pure number which is a combination of the isospin and hypercharge quantum numbers of the fermions. The reason for this is that the effect of radiative corrections in these measurables is universal (flavour independent) and amounts to a redifinition of the weak mixing angle. In a sense these quantitities are scheme independent and rightly so since they are tree-level quantities. Another such quantity is

$$\frac{\Gamma\left(Z \to \nu\bar\nu\right)}{\Gamma\left(Z \to \mu^+\mu-\right)} \equiv \frac{\Gamma_{\nu\bar\nu}}{\Gamma_{\mu\mu}} \tag{23}$$

Note that this observable is automatically twiddled, i.e, free of radiative corrections. Exploiting this observation, one can have an unambiguous way of counting neutrinos at the Z peak. In effect subtracting from the total width Γ_t the width into hadrons Γ_{had} and leptons we get the ratio

$$\Omega = \frac{\Gamma_{inv}}{\Gamma_{\mu\mu}} = \left(\frac{\Gamma_{tot}}{\Gamma_{\mu\mu}} - \frac{\Gamma_{had}}{\Gamma_{\mu\mu}} - 3\right) = N_\nu \frac{\Gamma_{\nu\bar\nu}}{\Gamma_{\mu\mu}} \tag{24}$$

where N_ν is the number of neutrinos. This a variation on Feldman's [15] proposal who first noted that because the hadronic width is about 70% of the total width, the experimental error on the quantity Γ_{inv} is greatly reduced. Abdelhak Djouadi and myself [25] have pointed out the theoretical merit of the method showing that this ratio (instead of Γ_{inv}) is free of radiative corrections and effectively measures the number of neutrinos if no New Physics were around. If new physics were around we presented a strategy combining this measurement with another twiddled quantity , namely n, which should be measured at the same time as Ω to achieve an unbiased way of counting neutrinos at the Z peak

6   Selecting between Models with New Z'

The prediction of at least one new Z' is a rather common feature to all models beyond the SM. Bounds on the mass of the least massive of these extra neutral bosons from existing data are as low as $120 GeV$ for a Z' belonging to a Grand Unification scheme [14] and about $250 GeV$ for genuine composite spin-ones [19] . High statistics experiments at LEP will either find evidence for these bosons or else set higher limits on the masses. Although by sitting at the Z peak one eliminates the physical Z' propagator effect one nonetheless rather amplifies the effects of this Z' on the usual Z properties due to the mixing of the Z' with the Z.

This mixing has two consequences. First there is mass mixing which leads to a shift between the value of the Z-mass as predicted by the SM and the value of the physical Z-mass in the presence of a structure beyond the SM. The latter is the physical mass as infered by the position of the peak. This shift contributes to $\Delta\rho'$. Second one has current mixing. The additional Z' is generally accompanied by the introduction of a new current different from the usual $SU(2)_L$ and the $U(1)_Y$ (hypercharge). Through mixing a component (at least) of this current couples to the physical Z with the consequence that at the Z peak there is a new current which also means that this extra piece unlike the previous effect is flavour dependent. Note that in the process of twiddling the first effect is eliminated so that one can concentrate on the new current.

As far as the composite picture is concerned, I hope I convinced you that (when the extra Z' are very heavy and decouple) the raditive corrections in the SM and the VMD implementation of compositeness are practically the same as in the SM. For the test one takes the effect of the extra Z' at zeroth order. To show you how the B-R-V strategy can make an identification between different models possible, I will compare composite Z' and gauge Z'.

6.1 Composite Z'

6.1.1 Isoscalar Weak Bosons

As I said earlier these can be viewed as the analog of ω^0 of hadron interactions. If one takes the compositeness of the W seriously as implemented by VMD then these isosclar must exist [18]. You see, VMD applied to W only gives you the isospin part of the electromagnetic current. To get the full electromagnetic current one has to add the hypercharge part. This implies the existence of the Y. There are two versions of the latter. The first assumes that the Y couples to the full hypercharge current. Nevertheless, this implementation begs the following question: why should the W which are apparently constructed as the Y only couple left-handed fermions? To remedy to this, one can assume that the Y also couples to only the left-handed part of the hypercharge current. But then one necessarily has to add a right-handed part either by hand (not in the spirit of the approach) or justify it it by requiring yet another extremely heavy right-handed isoscalar. This presupposes that in a way the compositeness scale in the right-handed sector is far higher than in the left-handed one. The model is reminiscent the Left-Right models. Again the coupling g_Y of these bosons to fermions is universal, due to VMD. Parametrising the $Y - \gamma$ junction by λ_Y we get

$$e = \lambda_Y g_Y \tag{25}$$

These models contain two extra parameters λ_Y and the mass of the Y boson.

6.1.2 Excited W

Again these are the analogs of the ρ' mesons [20]. One should expect their masses to set the scale of compositeness in the bosonic sector. Their presence would mean that the low lying W (the familiar ones) do not completely saturate the weak isospin current, but that the W^* contribute to that as well. Hence with obvious notations one would get

$$e = \lambda_\rho \, g_\rho + \lambda'_\rho \, g'_\rho \tag{26}$$

Low energy data resticts $x_W = \lambda'_\rho \, g'_\rho$ to be extremely small ($x_W/e << 1$) This model introduces three extra parameters (the mass of the W^* g'_ρ and λ'_ρ which not independent as in the case of Y's.

6.2 A Composite-Gauge Hybrid Model

Imagine that the Higgs particle is extremely heavy, say above $1 TeV$. "Weak" interactions would then enter the strong regime and one would loose the ability to use perturbation theory. In this strong regime the usual Higgs description is at best only effective and it would be more satisfying (and in a sense economical) to use an alternative picture devoid of any Higgs. Also in this strongly interacting system new bound states are "bound" to form. This should again remind you of the situation in hadronic interactions. There the Goldstone bosons (the familiar pseudoscalars) may be described by the non-linear σ model and you know that massive vector resonances $(\rho, \rho'....)$ behaving almost like Yang-Mills do form. This analogy has been exploited by Gatto and his co-workers [21]. For the electroweak model this suggests a non-linear realization of symmetry breaking and extending the initial $SU(2)_L \times SU(2)_R$ to $SU(2)_L \times SU(2)_R \times SU(2)_V$ making sure that the breaking leaves a diagonal $SU(2)_D$ which implements the custodial global symmetry I mentioned above. This approach leads to no physical scalar but a new triplet of massive vector bosons the Z_V appears in the spectrum. The model introduces two extra parameter and in many respect resembles the previous one.

6.3 Superstring Inspired Vector Bosons

Superstring theories have revived the interest of many in Grand Unified Theories. The models based on the "promising" E_6 predict at least one (possibly) low mass extra vector boson. This comes about from considering the most general breaking [16] of E_6, $E_6 \rightarrow SO(10) \times U(1)_\psi \rightarrow SU(5) \times U(1)_\psi \times U(1)_\chi$. The ensuing lowest mass extra Z' is a combination of the two $U(1)$'s such that

$$Z' = \cos \beta Z'_\chi + \sin \beta Z'_\psi \tag{27}$$

In the examples I have also considered a Z' from a Left-Right symmetric model.[17]

6.4 Theoretical Detectors

For all these models the selection rules have been written down.

First in the first stages of LEP when no polarization is available, Fig.1 shows that some disentangling between the various extra Z' can be performed albeit not complete. If polarization were made possible the variables defined in section 5.2 can be used. As one can see from Fig.2, this analysis could complement the previous one.

These particular observables really act as a discriminator between genuine gauge models and composite ones. For the latter the only deviation is along the z-axis. Hence if the combined experimental point falls outside this axis this would be a clear and unambiguous signal for models containing a Higgs explicitly. Therefore this three dimensional detector acts a Higgs detector and could in a very indirect way test our ideas about the mechanisms of symmetry breaking! [23]

7 Conclusions

I would like to conclude by saying that if you try to construct a model of composite weak bosons, you must implement it by a mechanism which systematically leads to the

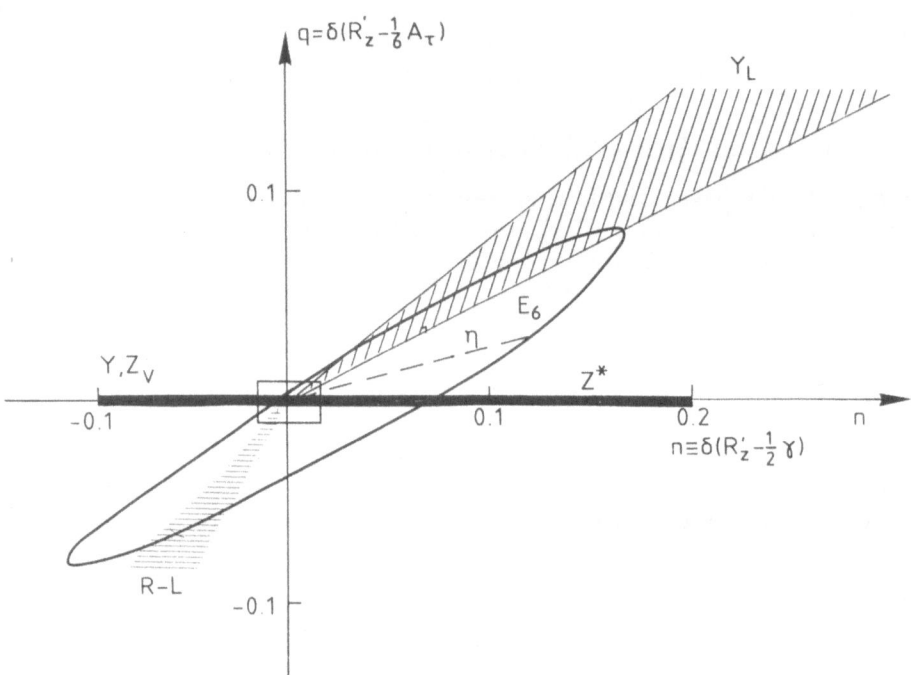

Figure 1. Allowed domains for the extra Z' in the q and n variables.

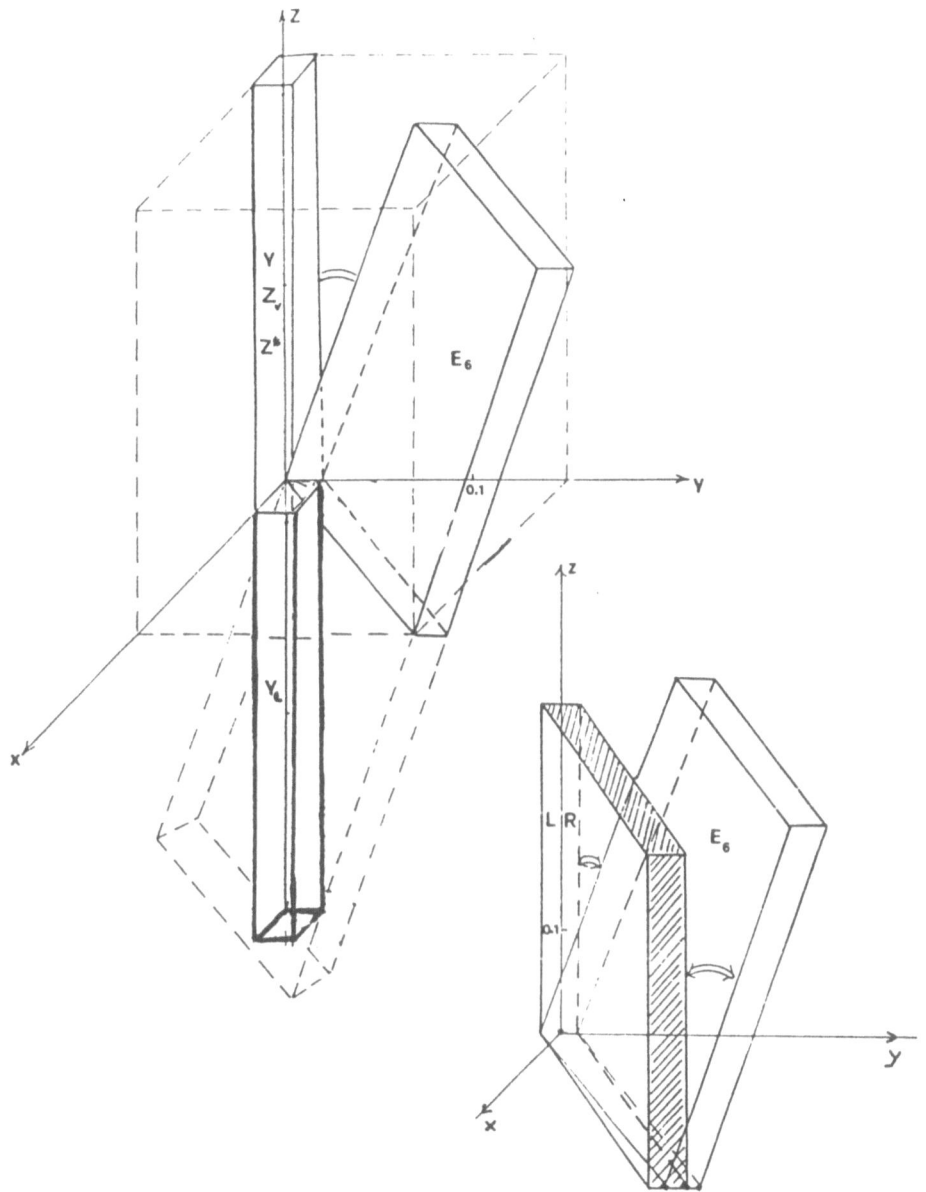

Figure 2. The characterising "wafers" for the models with extra Z' using polarized variables.

universality of the weak constant. The VMD approach does just that. In fact it does more, it reproduces for you most of the radiative corrections you normally take to be a feature of the $SU(2) \times U(1)$. If deviations from the SM are found in the forthcoming observables at LEP one should be very cautious when trying to associate them with a particular theory. It is imperative to combine high-precision measurements and write selection rules for the specific New Physics.

This said this New Physics might not *first* reveal itself in such a subtle way. We might be very lucky by observing a fantastic direct effect by the direct production of a new particle. As far as compositeness is concerned I would expect the production of excited neutrinos since these should be the Lightest Excited Particles. Abdelhak Djouadi and myself are investigating this possibilty and studying the signatures of these particles [24] . The properties of excited neutrinos bear some similarities to the much advertised LSP of supersymmetry and in some configurations to sequential neutrinos with mixing with ordianry neutrinos, but the prospect of LEP's at LEP seems so exciting.

References

[1] S.L. Glashow, Nucl. Phys. **22** 579 (1961).
S. Weinberg, Phys. Rev. Lett. **19** (1967) 1264.
A. Salam, Proc. 8th. Nobel Symposium, ed N. Svartholm, Almquist and Wiksells, Stockholm 1968,p.367.

[2] H. Harari, Proc. of the XVIIIth Solvay Conference On Physics, Austin, Texas, USA, November 1982, Phys. Rep. **104** (1984) 159, ed. by L. Van Hove.

[3] F. Boudjema and N. Dombey, Z. Phys. **C35** (1987) 499.
F. Boudjema, Ph.D. Thesis, Sussex University, September 1987, unpublished.

[4] M.A. Shifman, L. Vainshtein and V. Zakharov, Nucl. Phys. **179** (1981) 390.

[5] D. Shildknecht, Proc. of the 8th Int. Workshop on weak Interactions and Neutrinos, ed. A. Morales (World Scientific, Singapore 1983) p. 407.

[6] H. Fritzsch and G. Mandelbaum, Phys. Lett. **109B** (1982) 224.

[7] M. Veltman, Proc. of XXIIIrd Rencontre de Moriond, p. 133, ed. by J. Tran Thanh Van, édition Frontières 1988.

[8] F. Schrempp, Habilitationsschrift, June 1986. Preprint MPI-PAE/Pth 32/86.

[9] C. Vafa and E. Witten, Nucl. Phys. **B234** (1984) 173.

[10] G. Gounaris and D. Schildknecht, Z. Phys. **C42** (1080) 107.
The treatment of radiative corrections in this paper can easily be applied to the VMD approach.

[11] M. Veltman, Acta Phys. Pol. **B8** (1977) 475; Phys. Lett. **70B** (1977) 253.

[12] M. Cvetič and B.W. Lynn, **D35** (1987) 51.

[13] F. Boudjema and F.M. Renard, Conveners of "Compositeness at LEP" Working Group, CERN Yellow Book, edited By G. Altarelli, R. Kleiss and C. Verzegnassi, In Press.

[14] U. Amaldi et al., Phys. Rev. **D36** (1987) 1385.

[15] G. Feldman, Mark II/ SLC – PWG Note#2–24 (1987).

[16] See for instance G. Belanger and S. Godfrey, Phys. Rev. **D35** (1987) 378.

[17] V. Barger, E. Ma and K. Whisnant, Phys. Rev. **D28** (1983) 1618.
L. S. Durkin and P. Langacker, Phys. Lett. **166B** (1986) 436.

[18] M. Kuroda, D. Schildknecht and K. H. Schwarzer, Nucl. Phys. **B261** (1985) 432.

[19] U. Baur, M. Linder and K. H. Schwarzer, Nucl. Phys. **B291** (1987) 1.

[20] U. Baur, M. Linder and K. H. Schwarzer, Phys. Lett. **193B** (1987) 110.

[21] R. Casalbuoni et al., Nucl. Phys. **B282** (1987) 235.

[22] F. Boudjema, F.M. Renard and C. Verzegnassi, Nucl. Phys. **B314** (1989) 301;
Phys. Lett. **202B** (1988) 411.

[23] F. Boudjema and F.M. Renard, in Polarization at LEP, CERN 88-06 (1988) p.250,
ed. by G. Alexander et al..

[24] F. Boudjema and A. Djouadi, in preparation.

[25] F. Boudjema and A. Djouadi, RWTH-Aachen Preprint Pitha 89/13.

SUMMARY AND CONCLUDING REMARKS

A. Sirlin

New York University
New York
NY 10003, USA

It is clear that we are at the threshold of an exciting era of great precision at high energies.

It is good to remember, however, that high precision has been achieved for a long time at low energies. For example, the test of unitarity of the K–M matrix, which is the modern version of the universality of the weak interactions, currently reads

$$|V_{ud}|^2 + |V_{us}|^2 + |V_{ub}|^2 = 0.9971 \pm 0.0023$$

The first term, V_{ud}, is obtained from the comparison between μ and β decays and has required the work of many distinguished nuclear physicists such as D. Wilkinson, J.C. Hardy, I.S. Towner, R.J. Blin–Stoyle and, more recently, W.E. Ormand and B.A. Brown, as well as particle physicists. From our point of view what is particularly interesting is that, without radiative corrections, the r.h.s. would be $\simeq 1.037$ and the Standard Model (SM) would not be tenable. Indeed, a large radiative correction dominated by $1 + (3\alpha/2\pi)\ln(m_Z/2E_m) \simeq 1.035$ (E_m is the end–point energy of the positron in β decay) literally rescues the theory from contradiction. The detailed analysis of CVC and universality has recently required the careful study of two loop corrections of $O(Z\alpha^2)$ (Z is the charge of the daughter nucleus). Such accurate agreement at low energies can be used to put constraints on new physics: lower bounds for masses of additional Z's in some models, masses of super–symmetric partners, scale of compositeness, etc.. In his talk, Veltman mentioned an interesting observation of his that can be used to argue qualitatively against the idea of compositeness of vector bosons. His argument runs roughly as follows: if the W is composite then at the interaction vertex with quarks there is a black box that reflects the composite structure. Because of colour charge conservation, there must be a flow of colour charge inside the black box. But this implies that a gluon, for instance, can be emitted from the black box and absorbed by the external quarks. This diagram affects β but not μ decay. If we assume that the effect is of $O(\alpha_3(m_w)/2\pi) \simeq 0.02$ it would destroy the very good agreement with the three generation unitarity of the KM matrix. Radiative corrections have also played an important role in μ decay. For example, experimentally $\rho_{Michel} = 0.7518 \pm 0.0026$. But it has been estimated long ago that, without radiative corrections, the experimentalists would read $\simeq 0.71$, about 15σ away from $\rho_{Michel} = 0.75$ predicted by the two component theory of the neutrino. A similar observation holds for the δ parameter $\delta = 0.755 \pm 0.009$.

The early studies of μ and β decays at the one-loop level had also an interesting byproduct: the cancellation of mass singularities in total decay rates was found heuristically in these corrections. This, in turn, motivated the KLN theorem.

More recently, radiative corrections have played an interesting role in the analysis of neutral currents, m_w and m_z.[1,2]. Table 1 shows the effect of the radiative corrections in extracting $\sin^2\theta_w$ from the most precise observables.[1] It is clear that the corrections help considerably in bringing the three determinations into good agreement.

As Rankin has pointed out, several weeks ago when the central value at SLC appeared to be $m_z \simeq 90.4$ GeV, it seemed that we were ready for some excitement. A recent analysis[3] that employed existing neutral current and m_w data and assumed a further measurement $m_z = m_z^{exp} \pm 100$ MeV (m_z^{exp} is the central value of the hypothetical measurement) gave the 90% CL range allowed for m_t, as a function of m_z^{exp}. At $m_z^{exp} = 90.4$GeV, a large lower bound $m_t \geq 100$ GeV was implied by the analysis. At $m_z^{exp} \simeq 91$GeV, the situation is more conventional. The lower bound will be given by Fermilab and the upper bound of $m_t \leq 200$ GeV implied by the analysis of the radiative corrections is roughly the same as before. It is worth noting that a 91GeV value would be within 1 σ of the prediction $m_z = 91.8 \pm 0.9$GeV given in Ref. 3 on the basis of all the data existing at that time.

As the important issue of gauge invariance has been brought up in the discussion, I would like to make some further comments. The theory tells us what the physical observables are. In particular, they must be ξ-independent where ξ is the parameter that specifies the gauge within the class of renormalizable gauges. Recall, for example, that the free W propagator is of the form

$$D_{\mu\nu}(k) = -i\left[g_{\mu\nu} - \frac{k_\mu k_\nu (1-1/\xi)}{(k^2 - m_w^2/\xi)}\right] \frac{1}{(k^2 - m_w^2)}$$

A good illustration of the subtlety involved in the cancellation of gauge-dependent terms in physical amplitudes can be found in a recent analysis of the ξ-dependent parts in $\kappa^{(\nu_\mu e)}(q^2)$, the cofactor of $\sin^2\theta_w$ in $\nu_\mu e$ scattering.[4] One finds that part of the ξ dependent contributions to the $\gamma - Z$ mixing self energy $A_{\gamma Z}$ cancels against those of a subset of vertex diagrams while the remainder of $A_{\gamma Z}$ cancels against the other vertex parts and the WW box diagrams. In particular, as pointed out by Kleiss, me and others at the discussions, the WW box diagrams are not gauge invariant. It is also worth noting that $A_{\gamma Z}$ contains non-gauge invariant contributions proportional to q^2 as well as m_w^2.

An interesting illustration of the constraints imposed by the theory on the physical observables is provided by the "neutrino electromagnetic form factor" or its derivative at $q^2 = 0$, which is proportional to the "neutrino charge radius". In the early 60's J. Bernstein and T.D. Lee attempted to calculate the two $\gamma\nu\bar{\nu}$ vertex diagrams involving virtual W's in a non-perturbative way (ξ-limiting procedure), in a nonrenormalizable theory.[5] In the SM, however, it has been pointed out long ago by Bardeen, Gastmans and Lautrup that the ν charge radius is ξ dependent (and infinite in the unitary gauge) and therefore not a physical quantity.[6] Recently it has been shown how one can separate a gauge invariant, finite and target - independent part of $\kappa^{(\nu_\mu e)}(q^2)$ that can be interpreted as an electromagnetic form factor of the neutrino in the effective low energy theory derived from the SM at $|q^2| << m_w^2$.[4]

In my opinion, even in constructing improved or effective Born amplitudes, one should employ gauge invariant quantities. For example, if one considers the $O(\alpha)$ corrections proportional to $\sin^2\theta_w$ in neutral current amplitudes near the Z^0 resonance, the use of (cf. Eq(25a) of Ref. 7).

TABLE 1. Effect of the radiative corrections in the determination of $\sin^2\theta_W$ = $1 - m_W^2/m_Z^2$ from the most precise observables, according to Ref. 1. The central values assume m_t = 45 GeV and m_H = 100 GeV, while the errors allow the range $m_t \leqslant 100$ GeV, $m_H \leqslant 1$ TeV.

	$\sin^2\theta_W$ (no rad. corr.)	$\sin^2\theta_W$ (with rad. corr.)
R_ν	0.242 ± 0.006	0.233 ± 0.006
m_W	0.212 ± 0.007	0.229 ± 0.007
m_Z	0.208 ± 0.012	0.230 ± 0.012

$$\sin^2\theta_{eff}(m_z^2) = \sin^2\theta_W \left\{ 1 - \frac{c}{s} \, \mathrm{Re} \, \frac{A_{\gamma z}^{(f)}(m_z^2)}{m_z^2} \right.$$

$$\left. - \frac{c^2}{s^2} \, \mathrm{Re} \left[\frac{A_{zz}^{(f)}(m_z^2)}{m_z^2} - \frac{A_{ww}^{(f)}(m_w^2)}{m_w^2} \right] \right\}$$

where the superscript f means "fermionic contribution", would be all right because $\sin^2\theta_W$ is gauge invariant and we have absorbed potentially large corrections which are also finite and gauge invariant. To extend this to the bosonic sector, one must be careful: one obtains a finite expression by changing f→b (b means bosonic) and adding $-(c/s) \, A_{\gamma z}^{(b)}(0)$ to the expression between curly brackets, but it would not be gauge invariant. I would like to emphasize that a natural gauge invariant way of absorbing the dominant m_t dependence as well as the m_H dependence of $\kappa(q^2)$ is to employ $\sin^2\hat\theta_W(m_z)$, the MS definition at mass scale m_z. This has been explained long ago in the case of ν−lepton scattering.[8] Another gauge invariant way, which removes essentially the complete m_H and m_t dependence (except in e+ē→b+b̄), involves a combination of self energies and vertex parts evaluated exactly on resonance. As the latter are not universal, it is then necessary to choose a particular case, such as the Zeē amplitude on resonance, to define the effective angle. It is possible that either or both of these gauge invariant approaches will play an important role in physics at the Z^0 resonance. I, for one, would like to urge their application.

In his talk Veltman emphasized that the assumptions (a) all masses derive from the Higgs system (b) the Higgs system is the simplest one (linear σ model) lead to four important consequences:

A) $m_w = m_z\cos\theta_W$;   B) parity is violated in weak interactions;   C) $m\gamma=0$; D) parity is conserved in e.m. interactions. He also pointed out that this is not necessarily the case in many generalizations of these assumptions.

Altarelli mentioned the recent work of Consoli, Hollik, and Jegerlehner[9] and van der Bij and Hoogeveen[10] who have studied the large m_t effects in Δr to higher orders. Their conclusion is

$$\frac{1}{1-\Delta r} = \frac{1}{\left[1 + \mathrm{Re} \, \hat\pi_{\gamma\gamma}(m_z^2) \right]} \frac{1}{\left[1 + \frac{c^2}{s^2} \delta\rho \right]}$$

where $\delta\rho$ is the sum of irreducible one and higher loop contributions: $\delta\rho = 3x_t + 3x_t^2(19-2\pi^2)+...$ with $x_t = (G_\mu/\sqrt{2})(m_t^2/8\pi^2)$. One should remember that Δr contains other contributions not present in $\mathrm{Re}\,\hat\pi_{\gamma\gamma}(m_z^2)$ and $(c^2/s^2)\delta\rho$. This inclusion has been discussed in Ref. 9.

Altarelli also mentioned Blondel's observation that $R_\nu \equiv \sigma(\nu_\mu+N\rightarrow\nu_\mu+X)/\sigma(\nu_\mu+N\rightarrow\mu^-+X)$ essentially measures m_w^4/m_z^4, independently of m_t and ρ_0.[11] One has

$$R_\nu \simeq \frac{1}{2} \frac{(\rho_{NC})^2}{(\rho_{cc})^2} \left[1 - 2\kappa s^2 + \frac{10}{9} (1+r) (\kappa s^2)^2 \right]$$

At the tree level, this reduces to

$$R_\nu \simeq \frac{1}{2} \left[\frac{m_w}{m_z} \right]^4 \frac{1}{c^4} \left[1 - 2s^2 + \frac{10}{9} (1.39)s^4 \right]$$

Blondel points out that the explicit dependence on s^2 is very mild for $s^2 \simeq 0.23$. On the other hand, the independence from m_t, which is related and has been known for sometime,[7,12] can be understood from the relation $\delta s^2 = 0.964 \delta\rho - 0.230 \delta\kappa$ where $\delta\rho$ and $\delta\kappa$ describe the variation of the radiative corrections with m_t and δs^2 the corresponding change in s^2 extracted from R_ν.[13]

The determination of m_Z at Tristan, discussed by Altarelli and de Boer, is on the low side. On the basis of all data $m_Z = (88.6 \pm^{2.0}_{1.8})$ GeV (Tristan). According to Altarelli, in part this may be a problem of radiative corrections. The same data analysed by Haidt with a different program of radiative corrections leads to $m_Z = (89.2 \pm 1.6)$ GeV.

There has been much beautiful work on the line shape, as reported by Berends. The objective is to obtain the line shape from the theory such that the position is accurate within 15 MeV and the size is correct within 0.5%. There is an interesting electroweak effect in the propagator. Neglecting scaling violations proportional to m_b^2/m_z^2 (m_b is the mass of the bottom quark) the imaginary part of the Z self energy is proportional to s. This leads [14,15] to an s-dependent Breit-Wigner propagator proportional to $[s - m_z^2 + i(s/m_z^2)m_z\Gamma_z]^{-1}$. The factor (s/m_z^2) in the imaginary part shifts the position of the maximum by $\simeq -35$MeV.

The main effect on the line shape is due to initial state radiation. The corrected cross section is given by an integral of the form

$$\sigma(s) = \int_{z_0}^1 dz\, \sigma^0(sz)\, G(z) \quad (z_0 \geqslant 4\, m_f^2/s)$$

where $\sigma^0(s)$ is the total cross section with non-photonic corrections. In the calculation of $G(z)$ all the terms of $O(\alpha)$ are well established and so are the terms of $O(\alpha^2)$ containing L^2 and L ($L = \ln(s/m_e^2)$).[16] The corrections shift the peak position by $+112$MeV and decrease the peak height by a factor 0.74.

For a constant width Breit-Wigner cross section, in the absence of QED corrections, the position of the peak is $(\sqrt{s})_{peak} \simeq M_Z + 18$MeV. Including the QED corrections and the effect of the s-dependent width leads to $(\sqrt{s})$ peak $\simeq m_Z + 18 + 112 - 35 = m_Z + 95$MeV.

The structure function approach is another beautiful formulation to study higher order QED corrections and has been discussed and applied extensively by Nicrosini and Trentadue [17]. In his talk, Mario Greco compared this method with the coherent approach for multiple soft photon radiation.

There has been much work reported on Wednesday on Monte Carlo's with talks by Campagne, Bonvicini, Jadach, Martinez and Van der Marck. This is of course very important for the implementation of radiative corrections in the real world.

Tests of QCD in e^- e^+ collisions and the interplay of QCD and electroweak corrections were discussed by Kunszt, Djouadi and Kühn, while Hollik gave us a comprehensive review of electroweak rad. corr. in ep collisions. Kühn analysed hadronic initial state radiation and found it small, with a few funny events well under control; on the other hand, the QCD corrections to the tb doublet contributions to Δr are not negligible, being of the order of 2 to 5. 10^{-3} for m_t between 60 and 250GeV;[18] he also discussed QCD corrections to the $b\bar{b}$

asymmetry, which is a sensitive measure of $\sin^2\theta_W$. Burkhardt analysed in detail the contributions to $\mathrm{Re}\,\hat{\pi}_{\gamma\gamma}(s)$ associated with light quarks [19]. In particular, he concluded that there is no satisfactory set of fixed quark masses to describe $\mathrm{Re}\,\hat{\pi}_{\gamma\gamma}(s)$ over a broad range of energies.

Earlier in the meeting, Verzegnassi discussed the search for Higgs particles and additional Z's at LEP[20] and Ward and Hollik analysed the general situation concerning renormalization schemes.

Radiative corrections to W^+W^-/ZZ production are important for LEPII and were discussed by Denner. Barroso analysed the renormalization of flavour–changing neutral currents, giving simple arguments for some apparently mysterious cancellations of divergences, while Boudjema studied the search for signals of compositeness at LEP energies.

In the last talk, West analysed the breakdown of perturbation theory in QED and QCD using renormalization group and analyticity arguments. This seems to be an urgent problem in view of the huge coefficients encountered in recent higher order QCD calculations.

To conclude, I would like to say that it is wonderful to see many young people working so intensively and, in many cases, with considerable beauty and elegance, in the field of radiative corrections to weak interactions. When this was started, around 1955 and until the emergence of the SM, you could count us with the fingers of one (perhaps two) hands!

According to Rankin's transparency with the beautiful resonance curve emerging, we seem to be at close range of Z^0. Let us try to go closer and find its basic properties!

And, very importantly, many thanks to F. Boudjema, Norman Dombey and the University of Sussex for their wonderful hospitality!

REFERENCES

1. U. Amaldi et al., Phys. Rev. D36 (1987) 1385.
2. G. Costa et al., Nucl. Phys. B297 (1988) 244.
3. P. Langacker, W.J. Marciano and A. Sirlin, Phys. Rev. D36, (1987) 2191.
4. G. Degrassi, W.J. Marciano and A. Sirlin, Phys. Rev. D39, (1989) 287.
5. J. Bernstein and T.D. Lee, Phys. Rev. Letters 11, (1963) 512.
6. W.A. Bardeen, R. Gastmans and B. Lautrup, Nucl. Phys. B46 (1972) 319.
7. W.J. Marciano and A. Sirlin, Phys. Rev. D22 (1980) 2695.
8. S. Sarantakos, W.J. Marciano and A. Sirlin, Nucl. Physics B217 (1983) 84.
9. M. Consoli, W. Hollik and F. Jegerlehner, CERN-TH.5395/89.
10. J.J. Van der Bij and F. Hoogeveen, Nucl. Phys. B283 (1987) 477.
11. A. Blondel, CERN-EP/89-84.
12. R.G. Stuart, Z. Phys. C34 (1987) 445.
13. A. Sirlin, Comments on Nucl. and Part. Physics 17, (1987) 279.
14. F.A. Berends, G. Burgers, W. Hollik and W.L. van Neerven, Phys. Lett. 203B (1988) 177.
15. Yu Bardin, A. Leike, T. Riemann and M. Sachwitz, Phys. Lett. B206 (1988) 539.
16. F.A. Berends, G. Burgers, W.L. van Neerven, Phys. Lett. 185B (1987) 395; Nucl. Phys. B297 (1988) 429, E: Nucl. Phys. B 304 (1988) 921.
17. O. Nicrosini, L. Trentadue, Phys. Lett. 196B (1987) 551 ;Z. Phys. C39 (1988) 479.
18. B.A. Kniehl, J.H. Kühn and R.G. Stuart, MPI-PAE/PTh 36/88.
19. H. Burckhardt et al., in "Polarization at LEP", eds G. Alexander et al., CERN 88-06 (1988).
20. C. Verzegnassi, CERN-TH.5438/89.

INDEX